College Mathematics and Statistics

Custom Edition for University of Delaware

Larson | Hodgkins | Brase | Brase

CENGAGE Learning

Australia • Brazil • Japan • Korea • Mexico • Singapore • Spain • United Kingdom • United States

CENGAGE Learning

College Mathematics and Statistics: Custom Edition for University of Delaware

College Algebra with Applications for Business and the Life Sciences, Second Edition
Ron Larson | Anne V. Hodgkins

Understanding Basic Statistics, Seventh Edition
Charles Henry Brase | Corrine Pellillo Brase

This book contains select works from existing Cengage Learning resources and was produced by Cengage Learning Custom Solutions for collegiate use. As such, those adopting and/or contributing to this work are responsible for editorial content accuracy, continuity and completeness.

ISBN: 978-1-337-03872-0

Cengage Learning
20 Channel Center Street
Boston, MA 02210
USA

Cengage Learning is a leading provider of customized learning solutions with office locations around the globe, including Singapore, the United Kingdom, Australia, Mexico, Brazil, and Japan. Locate your local office at: **www.international.cengage.com/region.**

Cengage Learning products are represented in Canada by Nelson Education, Ltd.

For your lifelong learning solutions, visit **www.cengage.com/custom.**

Visit our corporate website at **www.cengage.com.**

Printed at CLDPC, USA, 01-19

Brief Contents

From:
College Algebra with Applications for Business and the Life Sciences
Larson | Hodgkins

From:
Understanding Basic Statistics
Brase | Brase

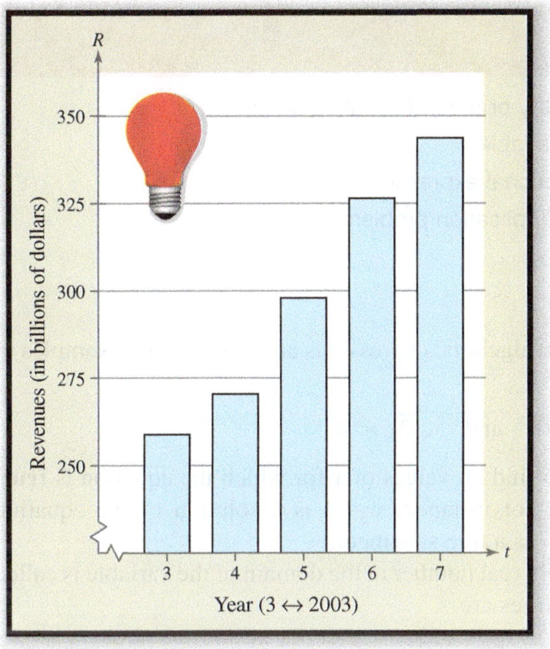

Example 9 on page 99 shows how a quadratic equation can be used to approximate revenues for the electric power industry in the United States for given years.

1 Equations and Inequalities

1.1 Linear Equations

- Classify an equation as an identity or a conditional equation.
- Solve a linear equation in one variable.
- Solve an equation involving fractional expressions.
- Use a linear model to solve an application problem.

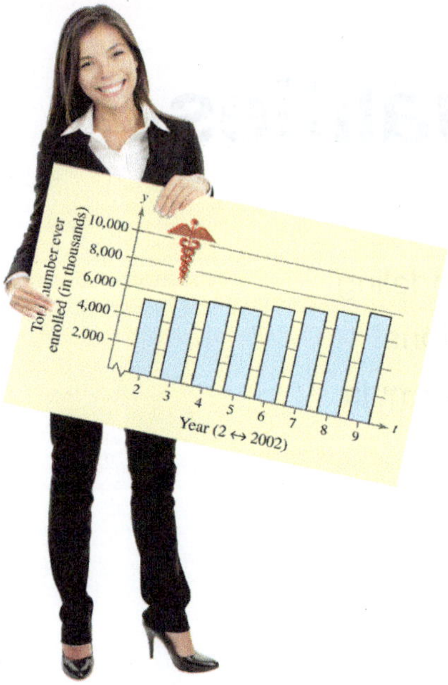

In Exercise 79 on page 79, you will use a linear model to determine the number of enrollees in a health insurance program.

Equations and Solutions

An **equation** is a statement that two algebraic expressions are equal. Some examples of equations in x are

$$3x - 5 = 7, \quad x^2 - x - 6 = 0, \quad \text{and} \quad \sqrt{2x} = 4.$$

To **solve** an equation in x means to find all values of x for which the equation is **true.** Such values are called **solutions.** For instance, $x = 4$ is a solution of the equation $3x - 5 = 7$, because $3(4) - 5 = 7$ is a true statement.

An equation that is true for *every* real number in the domain of the variable is called an **identity.** Two examples of identities are

$$x^2 - 9 = (x + 3)(x - 3) \quad \text{and} \quad \frac{x}{3x^2} = \frac{1}{3x}, \quad x \neq 0.$$

The first equation is an identity because it is a true statement for all real values of x. The second equation is an identity because it is true for all nonzero real values of x.

An equation that is true for just *some* (or even none) of the real numbers in the domain of the variable is called a **conditional equation.** For example, the equation $x^2 - 9 = 0$ is conditional because $x = 3$ and $x = -3$ are the only values in the domain that satisfy the equation.

Example 1 Classifying Equations

Determine whether each equation is an identity or a conditional equation.

a. $2(x + 3) = 2x + 6$

b. $2(x + 3) = x + 6$

c. $2(x + 3) = 2x + 3$

SOLUTION

a. This equation is an identity because it is true for every real value of x.

b. This equation is a conditional equation because $x = 0$ is the only value in the domain for which the equation is true.

c. This equation is a conditional equation because there are no real number values of x for which the equation is true.

✓ Checkpoint 1

Determine whether the equation $4(x + 1) = 4x + 4$ is an identity or a conditional equation.

Equations are used in algebra for two distinct purposes: (1) *identities* are usually used to state mathematical properties and (2) *conditional equations* are usually used to model and solve problems that occur in real life.

Linear Equations in One Variable

The most common type of conditional equation is a **linear equation.**

Definition of a Linear Equation

A **linear equation** in one variable x is an equation that can be written in the standard form

$$ax + b = 0$$

where a and b are real numbers with $a \neq 0$.

A linear equation in x has exactly one solution. To see this, consider the following steps. (Remember that $a \neq 0$.)

$ax + b = 0$ Original equation

$ax = -b$ Subtract b from each side.

$x = -\dfrac{b}{a}$ Divide each side by a.

So, the equation $ax + b = 0$ has exactly one solution,

$$x = -\frac{b}{a}.$$

To solve a linear equation in x, isolate x by forming a sequence of **equivalent** (and usually simpler) equations, each having the same solution as the original equation. The operations that yield equivalent equations come from the basic rules of algebra reviewed in Section 0.2.

Forming Equivalent Equations

A given equation can be transformed into an equivalent equation by one or more of the following steps.

	Given Equation	*Equivalent Equation*
1. Remove symbols of grouping, combine like terms, or simplify one or both sides of the equation.	$2x - x = 4$ $3(x - 2) = 5$	$x = 4$ $3x - 6 = 5$
2. Add (or subtract) the same quantity to (from) *each* side of the equation.	$x + 1 = 6$ $x - 4 = 3$	$x = 5$ $x = 7$
3. Multiply (or divide) *each* side of the equation by the same *nonzero* quantity.	$2x = 6$ $\dfrac{x}{8} = 4$	$x = 3$ $x = 32$
4. Interchange the two sides of the equation.	$2 = x$	$x = 2$

The steps for solving a linear equation in x written in standard form are shown in Example 2.

ALGEBRA TUTOR

For help in solving equations similar to the one in Example 2, see the review of solving equations on page 148.

Example 2 Solving a Linear Equation

Solve $3x - 6 = 0$.

SOLUTION

$$3x - 6 = 0 \quad \text{Write original equation.}$$
$$3x = 6 \quad \text{Add 6 to each side.}$$
$$x = 2 \quad \text{Divide each side by 3.}$$

✓ Checkpoint 2

Solve each equation.

a. $5 + 5x = 15$ **b.** $10 - x = 11$

After solving an equation, you should **check each solution** in the *original* equation. For instance, in Example 2, you can check that 2 is the solution by substituting 2 for x in the original equation

$$3x - 6 = 0.$$

CHECK

$$3x - 6 = 0 \quad \text{Write original equation.}$$
$$3(2) - 6 \stackrel{?}{=} 0 \quad \text{Substitute 2 for } x.$$
$$6 - 6 = 0 \quad \text{Solution checks. ✓}$$

Example 3 Solving a Linear Equation

Solve $6(x - 1) + 4 = 3(7x + 1)$.

SOLUTION

$$6(x - 1) + 4 = 3(7x + 1) \quad \text{Write original equation.}$$
$$6x - 6 + 4 = 21x + 3 \quad \text{Distributive Property}$$
$$6x - 2 = 21x + 3 \quad \text{Simplify.}$$
$$-15x = 5 \quad \text{Add 2 to, and subtract } 21x \text{ from, each side.}$$
$$x = -\tfrac{1}{3} \quad \text{Divide each side by } -15.$$

The solution is $x = -\frac{1}{3}$.

CHECK

$$6(x - 1) + 4 = 3(7x + 1) \quad \text{Write original equation.}$$
$$6\left(-\tfrac{1}{3} - 1\right) + 4 \stackrel{?}{=} 3\left[7\left(-\tfrac{1}{3}\right) + 1\right] \quad \text{Substitute } -\tfrac{1}{3} \text{ for } x.$$
$$6\left(-\tfrac{4}{3}\right) + 4 \stackrel{?}{=} 3\left(-\tfrac{7}{3} + 1\right) \quad \text{Simplify.}$$
$$-8 + 4 \stackrel{?}{=} -7 + 3 \quad \text{Simplify.}$$
$$-4 = -4 \quad \text{Solution checks. ✓}$$

STUDY TIP

You may think a solution to a problem looks easy when it is worked out in class, but you may not know where to begin when solving the problem on your own. Keep in mind that many problems involve some trial and error before a solution is found.

✓ Checkpoint 3

Solve each equation.

a. $3(x + 5) = x - 7$ **b.** $2(x + 2) + 6 = 4(2x - 3)$

Some equations in one variable have *infinitely many solutions*. To recognize an equation of this type, perform the regular steps for solving the equation. If, when writing equivalent equations, you reach a statement that is true for all values in the domain of the variable, then the equation is an identity and has infinitely many solutions.

Example 4 An Equation with Infinitely Many Solutions

Solve $x + 4(x - 2) = 3x + 2(x - 4)$.

SOLUTION

$x + 4(x - 2) = 3x + 2(x - 4)$ Write original equation.

$x + 4x - 8 = 3x + 2x - 8$ Distributive Property

$5x - 8 = 5x - 8$ Simplify.

$-8 = -8$ Subtract $5x$ from each side.

Because the last equation is true for every real value of x, the original equation is an identity and you can conclude that it has infinitely many solutions.

Checkpoint 4

Solve $x + 5 + 3(2x + 1) = 7x + 8$.

It is also possible for an equation in one variable to have *no solution*. When solving an equation of this type, you will reach a statement that is not true for any value of the variable.

Example 5 An Equation with No Solution

Solve $4x - 9 + 2(x + 8) = 1 + 6(x + 4)$.

SOLUTION

$4x - 9 + 2(x + 8) = 1 + 6(x + 4)$ Write original equation.

$4x - 9 + 2x + 16 = 1 + 6x + 24$ Distributive Property

$6x + 7 = 6x + 25$ Simplify.

$7 \neq 25$ Subtract $6x$ from each side.

Because the statement

$$7 = 25$$

is not true, you can conclude that the original equation has no solution.

Checkpoint 5

Solve $1 + 4(x + 1) = 4(2 + x)$.

Equations in one variable with infinitely many solutions or no solution are not linear because they cannot be written in the standard form

$$ax + b = 0.$$

Note that a linear equation in x has exactly one solution.

Equations Involving Fractional Expressions

To solve an equation involving fractional expressions, you can multiply every term in the equation by the least common denominator (LCD) of the terms.

Example 6 An Equation Involving Fractional Expressions

$$\frac{x}{3} + \frac{3x}{4} = 2 \qquad \text{Original equation}$$

$$(12)\frac{x}{3} + (12)\frac{3x}{4} = (12)2 \qquad \text{Multiply each term by the least common denominator.}$$

$$4x + 9x = 24 \qquad \text{Simplify.}$$

$$13x = 24 \qquad \text{Combine like terms.}$$

$$x = \frac{24}{13} \qquad \text{Divide each side by 13.}$$

✓Checkpoint 6

Solve $\frac{4x}{5} - \frac{x}{2} = 9$.

When multiplying or dividing an equation by a *variable expression*, it is possible to introduce an **extraneous** solution—one that does not satisfy the original equation. In such cases, a check is especially important.

Example 7 An Equation with an Extraneous Solution

Solve $\frac{1}{x-2} = \frac{3}{x+2} - \frac{6x}{x^2-4}$.

SOLUTION The least common denominator is $x^2 - 4 = (x+2)(x-2)$. Multiply each term by this LCD and simplify.

$$\frac{1}{x-2} = \frac{3}{x+2} - \frac{6x}{x^2-4} \qquad \text{Write original equation.}$$

$$\frac{1}{x-2}(x+2)(x-2) = \frac{3}{x+2}(x+2)(x-2) - \frac{6x}{x^2-4}(x+2)(x-2)$$

$$x + 2 = 3(x-2) - 6x, \quad x \neq \pm 2 \qquad \text{Simplify.}$$

$$x + 2 = 3x - 6 - 6x \qquad \text{Distributive Property}$$

$$4x = -8 \qquad \text{Simplify.}$$

$$x = -2 \qquad \text{Extraneous solution}$$

By checking $x = -2$, you can see that it yields a denominator of zero for the fraction

$$\frac{3}{x+2}.$$

So, $x = -2$ is extraneous, and the equation has *no solution*.

✓Checkpoint 7

Solve $\frac{1}{x-4} + \frac{1}{x} = \frac{4}{x(x-4)}$.

TECH TUTOR

Use the *table* feature of your graphing utility to check the solution in Example 3. In the equation editor, enter the expression to the left of the equal sign in y_1 and enter the expression to the right of the equal sign in y_2, as follows.

$$y_1 = 6(x-1) + 4$$

$$y_2 = 3(7x+1)$$

Set the *table* feature to ASK mode. When you enter the solution $-\frac{1}{3}$ for x, both y_1 and y_2 are -4, as shown.

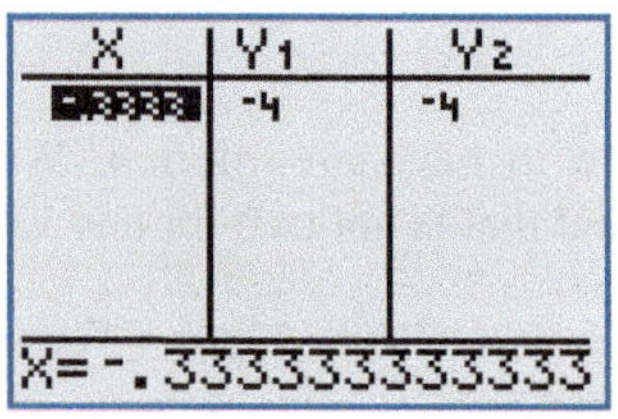

Similarly, a graphing utility can help you determine if a solution is extraneous. For instance, enter the equation from Example 7 into the graphing utility's equation editor. Then, use the *table* feature in ASK mode to enter -2 for x. You will see that the graphing utility displays ERROR in the y_2 column. So, the solution $x = -2$ is extraneous.

An equation with a *single fraction* on each side can be cleared of denominators by **cross-multiplying,** which is equivalent to multiplying each side of the equation by the least common denominator and then simplifying.

TECH TUTOR

When using the *equation editor* of a graphing utility, you must enter equations in terms of x. So, to enter an equation like the one shown in Example 8, replace y with x, as shown.

$$y_1 = (3x - 2)/(2x + 1)$$

$$y_2 = (6x - 9)/(4x + 3)$$

Example 8 Cross-Multiplying to Solve an Equation

Solve $\dfrac{3y - 2}{2y + 1} = \dfrac{6y - 9}{4y + 3}$.

SOLUTION

$$\frac{3y - 2}{2y + 1} = \frac{6y - 9}{4y + 3} \qquad \text{Write original equation.}$$

$$(3y - 2)(4y + 3) = (6y - 9)(2y + 1) \qquad \text{Cross-multiply.}$$

$$12y^2 + y - 6 = 12y^2 - 12y - 9 \qquad \text{Multiply.}$$

$$13y = -3 \qquad \text{Isolate } y\text{-term on left.}$$

$$y = -\frac{3}{13} \qquad \text{Divide each side by 13.}$$

The solution is $y = -\frac{3}{13}$. Check this in the original equation.

✓ Checkpoint 8

Solve $\dfrac{3x - 6}{x + 10} = \dfrac{3}{4}$.

Example 9 Using a Calculator to Solve an Equation

Solve $\dfrac{1}{9.38} - \dfrac{3}{x} = \dfrac{5}{0.3714}$.

SOLUTION To minimize roundoff error, solve for x in the equation before performing any calculations. The least common denominator is $(9.38)(0.3714)(x)$.

$$\frac{1}{9.38} - \frac{3}{x} = \frac{5}{0.3714}$$

$$(9.38)(0.3714)(x)\left(\frac{1}{9.38} - \frac{3}{x}\right) = (9.38)(0.3714)(x)\left(\frac{5}{0.3714}\right)$$

$$0.3714x - 3(9.38)(0.3714) = 5(9.38)(x), \quad x \neq 0$$

$$[0.3714 - 5(9.38)]x = 3(9.38)(0.3714)$$

$$x = \frac{3(9.38)(0.3714)}{0.3714 - 5(9.38)}$$

$$x \approx -0.225 \qquad \text{Round to three decimal places.}$$

The solution is $x \approx -0.225$. Check this in the original equation.

STUDY TIP

Because of roundoff error, a check of a decimal solution may not yield exactly the same values for each side of the original equation. The difference, however, should be quite small.

✓ Checkpoint 9

Solve $\dfrac{5}{x} + \dfrac{1}{2.7} = \dfrac{4}{0.6}$.

Application

Example 10 Hourly Earnings

The mean hourly earnings y (in dollars) of employees in the manufacturing industry in the United States from 2003 to 2008 can be modeled by the linear equation

$$y = 0.388t + 14.58, \quad 3 \leq t \leq 8$$

where t represents the year, with $t = 3$ corresponding to 2003. Use the model to estimate the year in which the mean hourly earnings were \$17.30. *(Source: U.S. Bureau of Labor Statistics)*

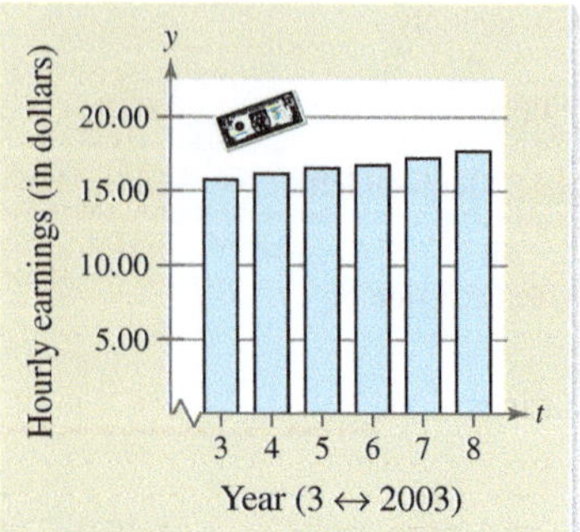

FIGURE 1.1

SOLUTION To determine when the mean hourly earnings were \$17.30, solve the model for t when $y = 17.3$.

$$y = 0.388t + 14.58 \qquad \text{Write original model.}$$

$$17.3 = 0.388t + 14.58 \qquad \text{Substitute 17.3 for } y.$$

$$2.72 = 0.388t \qquad \text{Subtract 14.58 from each side.}$$

$$t = \frac{2.72}{0.388} \qquad \text{Divide each side by 0.388.}$$

$$t \approx 7 \qquad \text{Use a calculator.}$$

Because $t = 3$ corresponds to 2003, it follows that $t = 7$ corresponds to 2007. See Figure 1.1. So, the mean hourly earnings were \$17.30 in 2007.

✓ Checkpoint 10

The mean hourly earnings y (in dollars) of employees in the trade, transportation, and utilities industry in the United States from 2000 to 2008 can be modeled by the linear equation

$$y = 0.350t + 13.29, \quad 0 \leq t \leq 8$$

where t represents the year, with $t = 0$ corresponding to 2000. Use the model to estimate the year in which the mean hourly earnings were \$14.35. *(Source: U.S. Bureau of Labor Statistics)*

SUMMARIZE (Section 1.1)

1. State the definitions of an identity and a conditional equation *(page 70)*. For an example of identifying identities and conditional equations, see Example 1.
2. State the definition of a linear equation and list the four steps that can be used to form an equivalent equation *(page 71)*. For examples of solving linear equations, see Examples 2 and 3.
3. Describe two methods for solving equations involving fractional expressions *(pages 74 and 75)*. For examples of solving equations involving fractional expressions, see Examples 6, 7, 8, and 9.
4. Describe a real-life example of how a linear equation can be used to analyze the mean hourly earnings of employees in a particular industry *(page 76, Example 10)*.

Johanna Goodyear/www.shutterstock.com

SKILLS WARM UP 1.1 The following warm-up exercises involve skills that were covered in earlier sections. You will use these skills in the exercise set for this section. For additional help, review Sections 0.5 and 0.7.

In Exercises 1–10, perform the indicated operations and simplify your answer.

1. $(2x - 4) - (5x + 6)$

2. $(3x - 5) + (2x - 7)$

3. $2(x + 1) - (x + 2)$

4. $-3(2x - 4) + 7(x + 2)$

5. $\frac{x}{2} + \frac{x}{7}$

6. $x - \frac{x}{4}$

7. $\frac{1}{x + 1} - \frac{1}{x}$

8. $\frac{5}{x} + \frac{3}{x}$

9. $\frac{4}{x} + \frac{3}{x - 2}$

10. $\frac{1}{x + 1} - \frac{1}{x - 1}$

Exercises 1.1

See www.CalcChat.com for worked-out solutions to odd-numbered exercises.

Classifying Equations In Exercises 1–6, determine whether the equation is an identity or a conditional equation. *See Example 1.*

1. $2(x - 1) = 2x - 2$

2. $3(x + 2) = 3x + 6$

3. $2(x - 1) = 3x + 4$

4. $3(x + 2) = 2x + 4$

5. $2(x + 1) = 2x + 1$

6. $3(x + 4) = 3x + 4$

Checking Solutions In Exercises 7–16, determine whether each value of x is a solution of the equation.

	Equation	*Values*	
7.	$5x - 3 = 3x + 5$	(a) $x = 0$	(b) $x = -5$
		(c) $x = 4$	(d) $x = 10$
8.	$7 - 3x = 5x - 17$	(a) $x = -3$	(b) $x = 0$
		(c) $x = 8$	(d) $x = 3$
9.	$3x^2 + 2x - 5 = 2x^2 - 2$	(a) $x = -3$	(b) $x = 1$
		(c) $x = 4$	(d) $x = -5$
10.	$5x^3 + 2x - 3 = 4x^3 + 2x - 11$	(a) $x = 2$	(b) $x = -2$
		(c) $x = 0$	(d) $x = 10$
11.	$\frac{5}{2x} - \frac{4}{x} = 3$	(a) $x = -\frac{1}{2}$	(b) $x = 4$
		(c) $x = 0$	(d) $x = \frac{1}{4}$
12.	$3 + \frac{1}{x + 2} = 4$	(a) $x = -1$	(b) $x = -2$
		(c) $x = 0$	(d) $x = 5$
13.	$(x + 5)(x - 3) = 20$	(a) $x = 3$	(b) $x = -2$
		(c) $x = 0$	(d) $x = -7$
14.	$(3x + 5)(2x - 7) = 0$	(a) $x = -\frac{5}{3}$	(b) $x = -\frac{2}{7}$
		(c) $x = \frac{2}{3}$	(d) $x = \frac{3}{2}$
15.	$\sqrt{2x - 3} = 3$	(a) $x = 6$	(b) $x = -3$
		(c) $x = -\frac{1}{3}$	(d) $x = -2$
16.	$\sqrt[3]{x - 8} = 3$	(a) $x = 2$	(b) $x = -5$
		(c) $x = 35$	(d) $x = 8$

Solving an Equation In Exercises 17–32, solve the equation and check your solution. (Some equations have no solution.) *See Examples 2, 3, 4, and 5.*

17. $x + 10 = 15$

18. $9 - x = 13$

19. $7 - 2x = 15$

20. $7x + 2 = 16$

21. $8x - 5 = 3x + 10$

22. $7x + 3 = 3x - 13$

23. $5x - x = 4x - 4$

24. $3 - x = 4x - 5x$

25. $4x - 8 = -4(2 - x)$

26. $3(3 - x) = 9 - 3x$

27. $2(x + 5) - 7 = 3(x - 2)$

28. $2(13t - 15) + 3(t - 19) = 0$

29. $6[x - (2x + 3)] = 8 - 5x$

30. $3[2x - (x + 7)] = 5(x - 3)$

31. $x - 3(2x + 3) = 8 - 5x$

32. $9x - 10 = 5x + 2(2x - 5)$

An Equation Involving Fractional Expressions In Exercises 33–48, solve the equation and check your solution. (Some equations have no solution.) *See Examples 6, 7, and 8.*

33. $\frac{5x}{4} + \frac{1}{2} = x - \frac{1}{2}$

34. $\frac{x}{5} - \frac{x}{2} = 3$

35. $\frac{3}{2}(z + 5) = \frac{1}{4}(z + 24)$

36. $\frac{3x}{2} + \frac{1}{4}(x - 2) = 10$

37. $\frac{100 - 4u}{3} = \frac{5u + 6}{4} + 6$

38. $\frac{17 + y}{y} + \frac{32 + y}{y} = 100$

39. $\frac{5x - 4}{5x + 4} = \frac{2}{3}$

40. $\frac{10x + 3}{5x + 6} = \frac{1}{2}$

41. $10 - \frac{13}{x} = 4 + \frac{5}{x}$

42. $\frac{15}{x} - 4 = \frac{6}{x} + 3$

43. $\frac{1}{x - 3} + \frac{1}{x + 3} = \frac{10}{x^2 - 9}$

44. $\frac{1}{x - 2} + \frac{3}{x + 3} = \frac{4}{x^2 + x - 6}$

45. $\frac{6}{(x - 3)(x - 1)} = \frac{3}{x - 3} + \frac{4}{x - 1}$

46. $\frac{2}{(x - 4)(x - 2)} = \frac{1}{x - 4} + \frac{2}{x - 2}$

47. $\frac{x - 3}{2x + 1} = \frac{x - 5}{2x + 4}$

48. $\frac{2x - 3}{6x - 5} = \frac{x + 2}{3x + 4}$

Solving an Equation In Exercises 49–56, solve the equation and check your solution. (Some equations have no solution.)

49. $0.25x + 0.75(10 - x) = 3$

50. $0.60x + 0.40(100 - x) = 50$

51. $(x + 2)^2 + 5 = (x + 3)^2$

52. $(x + 1)^2 + 2(x - 2) = (x + 1)(x - 2)$

53. $(x + 2)^2 - x^2 = 4(x + 1)$

54. $4(x + 1) - 3x = x + 5$

55. $(2x + 1)^2 = 4(x^2 + x + 1)$

56. $(2x - 1)^2 = 4(x^2 - x + 6)$

The symbol indicates an exercise in which you are instructed to use graphing technology or a symbolic computer algebra system. The solutions of other exercises may also be facilitated by use of appropriate technology.

57. Error Analysis A student states that the solution of the equation

$$\frac{2}{x(x - 2)} + \frac{5}{x} = \frac{1}{x - 2}$$

is $x = 2$. Describe and correct the student's error.

58. Error Analysis A student states that the equation

$$-3(x + 2) = -3x + 6$$

is an identity. Describe and correct the student's error.

59. Writing Explain why a solution of an equation involving fractional expressions may be extraneous.

60. Writing Describe two methods you can use to check a solution of an equation involving fractional expressions.

61. Writing What is meant by "equivalent equations?" Give an example of two equivalent equations.

62. Think About It For what value(s) of b does the equation $7x + 3 = 7x + b$ have infinitely many solutions? no solution?

Using a Calculator to Solve an Equation In Exercises 63–68, use a calculator to solve the equation. (Round your solution to three decimal places.) *See Example 9.*

63. $0.275x + 0.725(500 - x) = 300$

64. $2.763 - 4.5(2.1x - 5.1432) = 6.32x + 5$

65. $\frac{x}{0.6321} + \frac{x}{0.0692} = 1000$

66. $(x + 5.62)^2 + 10.83 = (x + 7)^2$

67. $\frac{2}{7.398} - \frac{4.405}{x} = \frac{1}{x}$

68. $\frac{x}{2.625} + \frac{x}{4.875} = 1$

69. Think About It What method or methods would you recommend for checking the solutions to Exercises 63–68 using your graphing utility?

70. Think About It In Exercises 63–68, your answers are rounded to three decimal places. What effect does rounding have as you check a solution?

Roundoff Error In Exercises 71–74, evaluate the expression in two ways. (a) Calculate entirely on your calculator using appropriate parentheses, and then round the answer to two decimal places. (b) Round both the numerator and the denominator to two decimal places before dividing, and then round the final answer to two decimal places. (c) Then describe why roundoff error is introduced by the second method.

71. $\frac{1 + 0.73205}{1 - 0.73205}$

72. $\frac{1 + 0.86603}{1 - 0.86603}$

73. $\dfrac{333 + \dfrac{1.98}{0.74}}{4 + \dfrac{6.25}{3.15}}$

74. $\dfrac{1.73205 - 1.19195}{3 - (1.73205)(1.19195)}$

75. Personal Income The per capita personal income y (in dollars) in West Virginia from 2000 through 2009 can be approximated by the linear equation

$$y = 1114.4t + 21{,}952, \quad 0 \le t \le 9$$

where t represents the year, with $t = 0$ corresponding to 2000. Use the model to estimate the year in which the per capita personal income was about \$32,000. *(Source: Bureau of Economic Analysis)*

76. Annual Sales The annual sales S (in billions of dollars) of Oracle Corporation from 2004 through 2009 can be approximated by the linear equation

$$S = 2.903t - 1.98, \quad 4 \le t \le 9$$

where t represents the year, with $t = 4$ corresponding to 2004. Use the model to estimate the year in which Oracle's annual sales were about \$18 billion. *(Source: Oracle Corporation)*

Forensics In Exercises 77 and 78, use the following information. The relationship between the length of an adult's femur (thigh bone) and the height of the adult can be approximated by the linear equations

$y = 0.432x - 10.44$ Female

$y = 0.449x - 12.15$ Male

where y is the length of the femur in inches and x is the height of the adult in inches (see figure).

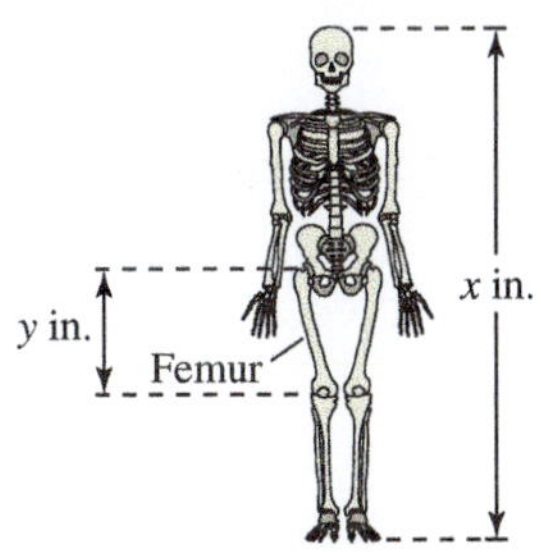

77. A crime scene investigator discovers a femur belonging to an adult human female. The bone is 18 inches long. Estimate the height of the female.

78. Officials search a forest for a missing man who is 6 feet 2 inches tall. They find an adult male femur that is 21 inches long. Is it possible that the femur belongs to the missing man?

79. Health Insurance The Children's Health Insurance Program (CHIP) tracks the total number of children ever enrolled in the program. For the years 2002 through 2009, the total numbers ever enrolled y (in thousands) are shown in the bar graph and can be approximated by

$$y = 322.5t + 4788, \quad 2 \le t \le 9$$

where t is the year, with $t = 2$ corresponding to 2002. *(Source: Centers for Medicare and Medicaid Services)*

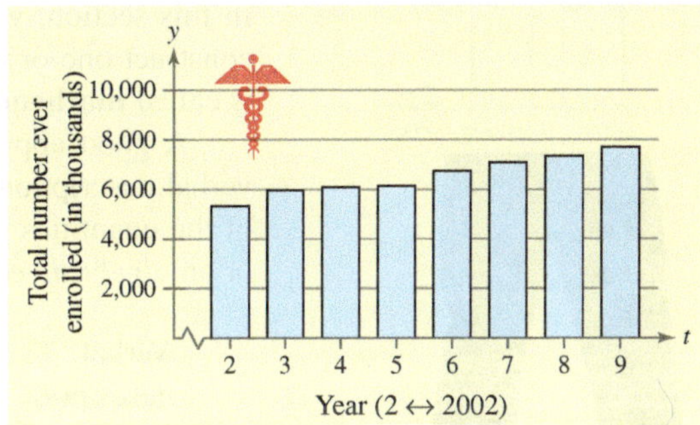

(a) In what year did the total number of CHIP enrollees exceed 7 million?

(b) Use the model to predict the year in which the total number of CHIP enrollees will reach 8.3 million.

80. HOW DO YOU SEE IT? From 1997 through 2006, the federal minimum wage was \$5.15 per hour. Adjusting for inflation, the federal minimum wage's value y (in 1996 dollars) during each of these years is shown in the bar graph. *(Source: U.S. Department of Labor)*

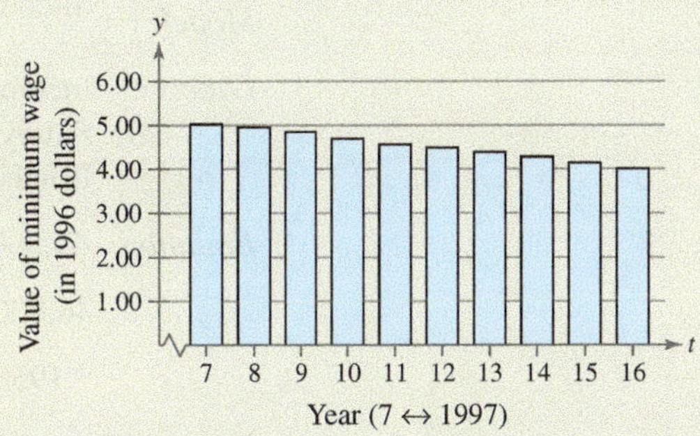

(a) In which year was the value of the minimum wage equal to \$4.85 in 1996 dollars?

(b) By 2009, the minimum wage increased to \$7.25, a value of \$5.30 in 1996 dollars. Suppose it had not increased. Use the linear relationship shown in the bar graph to predict the value (in 1996 dollars) of a 2009 minimum wage of \$5.15 per hour.

1.2 Mathematical Modeling

- Construct a mathematical model from a verbal model.
- Model and solve percent and distance problems.
- Model and solve mixture problems.
- Use common formulas to solve geometry problems.

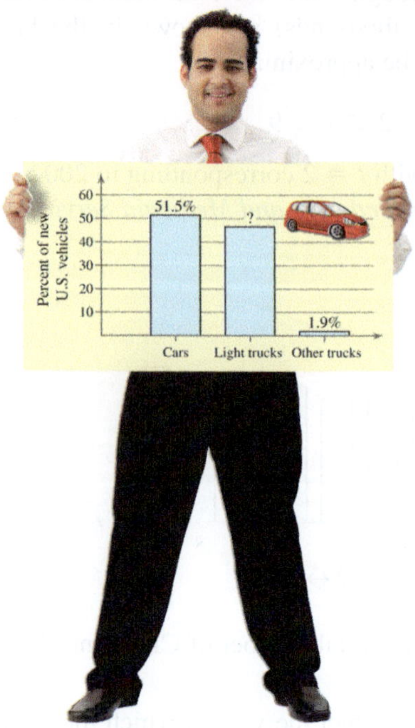

In Exercise 37 on page 89, you will use a mathematical model to determine the number of new cars that were sold or leased in 2009.

Introduction to Problem Solving

In this section, you will use algebra to solve real-life problems. To do this, you will construct one or more equations that represent each real-life problem. This procedure is called **mathematical modeling.**

A good approach to mathematical modeling is to use two stages. First, use the verbal description of the problem to form a *verbal model.* Then, assign labels to each of the quantities in the verbal model and use the labels to form a *mathematical model* or an *algebraic equation.*

Verbal description → Verbal model → Algebraic equation

When you are trying to construct a verbal model, it is helpful to look for a *hidden equality*—a statement that two algebraic expressions are equal. These two expressions might be explicitly stated as being equal, or they might be known to be equal (based on prior knowledge or experience).

Example 1 Using a Verbal Model

You accept a job with an annual income of \$47,800. This includes your salary and a \$1000 year-end bonus. You are paid twice a month. What is your salary per pay period?

SOLUTION Because there are 12 months in a year and you are paid twice a month, it follows that there are 24 pay periods during the year.

Verbal Model: Income for year = 24 pay periods · Salary per pay period + Bonus

Labels: Income for year = 47,800 (dollars)
Salary per pay period = x (dollars)
Bonus = 1000 (dollars)

Equation: $47{,}800 = 24x + 1000$

$46{,}800 = 24x$

$1950 = x$

Your salary per pay period is \$1950.

CHECK

$47{,}800 = 24x + 1000$ Write original equation.

$47{,}800 \stackrel{?}{=} 24(1950) + 1000$ Substitute 1950 for x.

$47{,}800 = 47{,}800$ Solution checks. ✓

✓ Checkpoint 1

You accept a job with an annual income of \$51,000. This includes your salary and a \$1600 year-end bonus. You are paid weekly. What is your salary per pay period?

Andresr/Shutterstock.com

Translating Key Words and Phrases

Key Words and Phrases	*Verbal Description*	*Algebraic Statement*
Consecutive		
Next, subsequent	Consecutive integers	$n, n + 1$
Addition		
Sum, plus, greater, increased by, more than, exceeds, total of	The sum of 5 and x Seven more than y	$5 + x$ $y + 7$
Subtraction		
Difference, minus, less than, decreased by, subtracted from, reduced by, the remainder	Four decreased by b Three less than z Five subtracted from w	$4 - b$ $z - 3$ $w - 5$
Multiplication		
Product, multiplied by, twice, times, percent of	Two times x	$2x$
Division		
Quotient, divided by, per	The quotient of x and 8	$\frac{x}{8}$

Example 2 **Constructing Mathematical Models**

a. A salary of \$50,700 is increased by 6%. Write an equation that represents the new salary.

Verbal Model: New salary $=$ 6%(Original salary) $+$ Original salary

Labels: Original salary $= 50{,}700$ (dollars)
New salary $= S$ (dollars)
Percent $= 0.06$ (percent in decimal form)

Equation: $S = 0.06(50{,}700) + 50{,}700$

b. An e-reader is marked down 10% to \$225. Write an equation you can use to find the original price.

Verbal Model: Original price $-$ 10%(Original price) $=$ Sale price

Labels: Original price $= p$ (dollars)
Sale price $= 225$ (dollars)
Percent $= 0.1$ (percent in decimal form)

Equation: $p - 0.1p = 225$

Checkpoint 2

A bid price of \$6800 for an auction item is increased by 5%. Write an equation that you can use to find the new bid price.

In Example 2, notice that part of the labeling process is to list the unit of measure for each labeled quantity. Developing this habit helps in checking the validity of a verbal model.

Using Mathematical Models

Example 3 Finding the Percent of a Raise

You accept a job that pays \$12.50 an hour. You are told that after a two-month probationary period, your hourly wage will be increased to \$13 an hour. What percent raise will you receive after the two-month period?

SOLUTION

Verbal Model: Raise = Percent · Old wage

Labels: Old wage = 12.5 (dollars)
Raise = 13 − 12.5 = 0.5 (dollars)
Percent = r (percent in decimal form)

Equation: $0.5 = r \cdot 12.5$

By solving this equation, you can find that you will receive a raise of

$$r = \frac{0.5}{12.5} = 0.04$$

or 4%.

Checkpoint 3

You buy stock at \$25 per share. You sell the stock at \$30 per share. What is the percent increase of the stock's value?

Example 4 Finding the Percent of Employees

The number of people employed by the U.S. federal government for a recent year is 2,730,040. The table shows the number of people employed in the executive branch of the federal government. What percent of employees in the U.S. federal government are in the executive branch? *(Source: U.S. Office of Personnel Management)*

	Number of Employees
Executive Office of the President	1717
Executive Departments	1,740,979
Independent Agencies	923,744

The President presides over the executive branch of the federal government, which employs more than 2.5 million people.

SOLUTION

Verbal Model: Executive branch employees = Percent · Federal government employees

Labels: Federal government employees = 2,730,040 (people)
Executive branch employees = 2,666,440 (people)
Percent = r (percent in decimal form)

Equation: $2{,}666{,}440 = r \cdot 2{,}730{,}040$

By solving this equation, you can find that the percent of federal government employees in the executive branch is $r = 2{,}666{,}440/2{,}730{,}040$, or about 97.7%.

Checkpoint 4

Your income last year was \$42,000. By the end of the year, you paid a total of \$648 for parking fees. What percent of your income was paid for parking fees?

Vladislav Gurfinkel/Shutterstock.Com

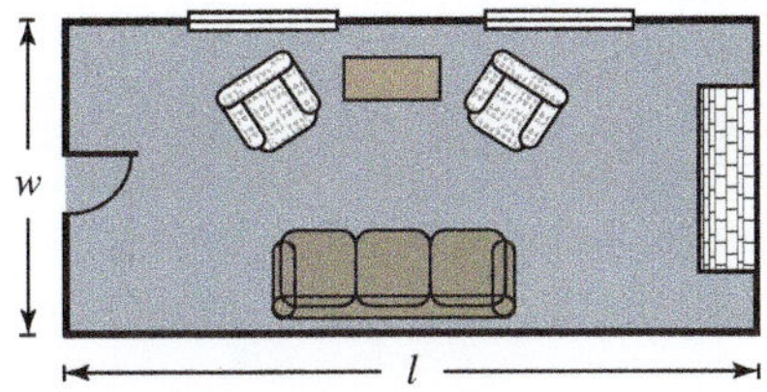

FIGURE 1.2

Example 5 Finding the Dimensions of a Room

A rectangular family room is twice as long as it is wide, and its perimeter is 84 feet. Find the dimensions of the family room.

SOLUTION For this problem, it helps to sketch a diagram, as shown in Figure 1.2.

Verbal Model: $2 \cdot$ Length $+ 2 \cdot$ Width $=$ Perimeter

Labels: Perimeter $= 84$ (feet)
Width $= w$ (feet)
Length $= l = 2w$ (feet)

Equation: $2(2w) + 2w = 84$

$$4w + 2w = 84$$
$$6w = 84$$
$$w = 14 \text{ feet}$$
$$l = 2w = 2(14) = 28 \text{ feet}$$

The dimensions of the room are 14 feet by 28 feet.

✓ Checkpoint 5

A rectangular driveway is three times as long as it is wide, and its perimeter is 120 feet. Find the dimensions of the driveway.

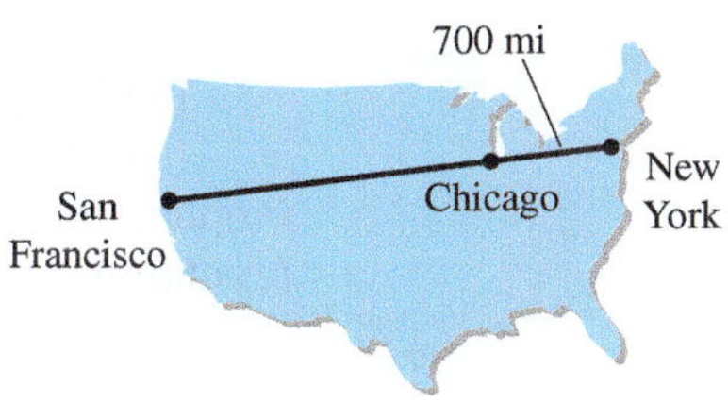

FIGURE 1.3

Example 6 A Distance Problem

A plane travels nonstop from New York to San Francisco, a distance of about 2600 miles. It takes 1.5 hours to fly from New York to Chicago, a distance of about 700 miles (see Figure 1.3). Assuming the plane flies at a constant speed, how long does the entire trip take? What time (EST) should the plane leave New York to arrive in San Francisco by 5 P.M. PST (8 P.M. EST)?

SOLUTION To solve this problem, use the formula that relates distance, rate, and time. That is, (distance) = (rate)(time). Because it took the plane 1.5 hours to travel a distance of about 700 miles, you can conclude that its rate (or speed) is

$$\text{Rate} = \frac{\text{distance}}{\text{time}}$$
$$= \frac{700 \text{ miles}}{1.5 \text{ hours}} \approx 466.67 \text{ miles per hour.}$$

Because the entire trip is about 2600 miles, the time for the entire trip is

$$\text{Time} = \frac{\text{distance}}{\text{rate}}$$
$$= \frac{2600 \text{ miles}}{466.67 \text{ miles per hour}} \approx 5.57 \text{ hours.}$$

Because 0.57 hour represents about 34 minutes, you can conclude that the entire trip takes about 5 hours and 34 minutes. So, the plane must leave New York by 2:26 P.M. EST in order to arrive in San Francisco by 8 P.M. EST.

✓ Checkpoint 6

A boat travels at full speed to an island 11 miles away. It takes 0.3 hour to travel the first 3 miles. How long does the entire trip take?

Another way to solve the distance problem in Example 6 is to use the concept of **ratio and proportion.** To do this, let x represent the time required to fly from New York to San Francisco, set up the following proportion, and solve for x.

$$\frac{\text{Time to San Francisco}}{\text{Time to Chicago}} = \frac{\text{Distance to San Francisco}}{\text{Distance to Chicago}}$$

$$\frac{x}{1.5} = \frac{2600}{700}$$

$$1.5 \cdot \frac{x}{1.5} = 1.5 \cdot \frac{2600}{700}$$

$$x \approx 5.57$$

Notice how ratio and proportion are used with a property from geometry to solve the problem in the following example.

Example 7 An Application Involving Similar Triangles

To determine the height of Petronas Tower 1 (in Kuala Lumpur, Malaysia), you measure the shadow cast by the building to be 113 meters long, as shown in Figure 1.4. Then you measure the shadow cast by a 100-centimeter post and find that its shadow is 25 centimeters long. Use this information to determine the height of Petronas Tower 1.

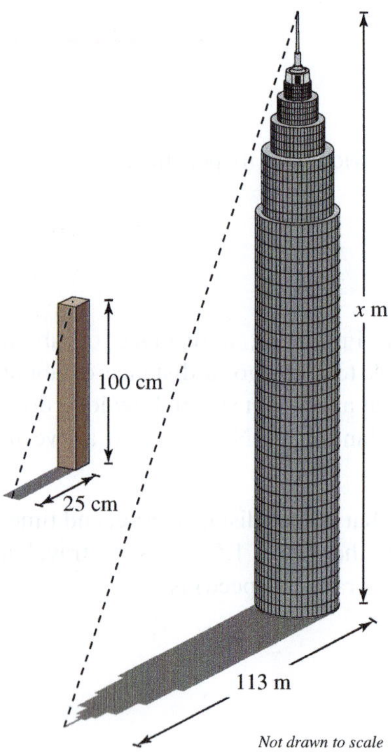

FIGURE 1.4

SOLUTION To find the height of the tower, you can use a property from geometry that states that the ratios of corresponding sides of similar triangles are equal.

Verbal Model: $\frac{\text{Height of tower}}{\text{Length of tower's shadow}} = \frac{\text{Height of post}}{\text{Length of post's shadow}}$

Labels:
Height of tower $= x$ (meters)
Length of tower's shadow $= 113$ (meters)
Height of post $= 100$ (centimeters)
Length of post's shadow $= 25$ (centimeters)

Equation:

$$\frac{x}{113} = \frac{100}{25}$$

$$113 \cdot \frac{x}{113} = 113 \cdot \frac{100}{25}$$

$$x = 113 \cdot 4$$

$$x = 452 \text{ meters}$$

The Petronas Tower 1 is 452 meters high.

CHECK

$$\frac{x}{113} = \frac{100}{25}$$ Write original equation.

$$\frac{452}{113} = \frac{100}{25}$$ Substitute 452 for x.

$$4 = 4$$ Solution checks. ✓

✓ Checkpoint 7

A tree casts a shadow that is 24 feet long. At the same time, a four-foot tall mailbox casts a shadow that is 3 feet long. How tall is the tree?

Mixture Problems

The next example is called a **mixture problem** because it involves two different unknown quantities that are *mixed* in a specific way. Watch for a *hidden product* in the verbal model.

TECH TUTOR

You can write a program for a programmable calculator to solve simple interest problems. Use a program with Example 8 to find how much interest was earned on just the portion of the money invested at $2\frac{1}{2}\%$.

Example 8 A Simple Interest Problem

You invested a total of $10,000 in accounts that earned $1\frac{1}{2}\%$ and $2\frac{1}{2}\%$ simple interest. In 1 year, the two accounts earned $207.50 in interest. How much did you invest in each account?

SOLUTION The formula for simple interest is

$$I = Prt$$

where I is the interest, P is the principal, r is the annual interest rate (in decimal form), and t is the time in years.

Verbal Model: Interest from $1\frac{1}{2}\%$ + Interest from $2\frac{1}{2}\%$ = Total interest

You can let x represent the amount invested at $1\frac{1}{2}\%$. Because the total amount invested at $1\frac{1}{2}\%$ and $2\frac{1}{2}\%$ is $10,000, you can let $10{,}000 - x$ represent the amount invested at $2\frac{1}{2}\%$.

Labels: Amount invested at $1\frac{1}{2}\% = x$ (dollars)

Amount invested at $2\frac{1}{2}\% = 10{,}000 - x$ (dollars)

Interest from $1\frac{1}{2}\% = Prt = (x)(0.015)(1)$ (dollars)

Interest from $2\frac{1}{2}\% = Prt = (10{,}000 - x)(0.025)(1)$ (dollars)

Total interest $= 207.50$ (dollars)

Equation:
$$\begin{aligned} 0.015x + 0.025(10{,}000 - x) &= 207.50 \\ 0.015x + 250 - 0.025x &= 207.50 \\ -0.01x &= -42.50 \\ x &= 4250 \end{aligned}$$

STUDY TIP

In Example 8, notice the hidden products in the two terms on the left side of the equation. Both hidden products come from the common formula $I = Prt$.

So, the amount invested at $1\frac{1}{2}\%$ is $4250 and the amount invested at $2\frac{1}{2}\%$ is

$$\begin{aligned} 10{,}000 - x &= 10{,}000 - 4250 \\ &= \$5750. \end{aligned}$$

Check these results in the original statement of the problem.

CHECK

Interest from $1\frac{1}{2}\%$ | Interest from $2\frac{1}{2}\%$ | Total interest

$$\begin{aligned} 0.015(4250) + 0.025(10{,}000 - 4250) &\stackrel{?}{=} 207.50 \\ 63.75 + 143.75 &\stackrel{?}{=} 207.50 \\ 207.50 &= 207.50 \end{aligned}$$

Solution checks. ✓

✓ Checkpoint 8

You invested a total of $2500 in accounts that earned 1% and 2% simple interest. In 1 year, the two accounts earned a total of $38 in interest. How much did you invest in each account?

Common Formulas

Many common types of geometric, scientific, and investment problems use ready-made equations, called **formulas.** Knowing formulas such as those in the following lists will help you translate and solve a wide variety of real-life problems involving perimeter, area, volume, temperature, interest, and distance.

Common Formulas for Area, Perimeter, and Volume

Square

$A = s^2$

$P = 4s$

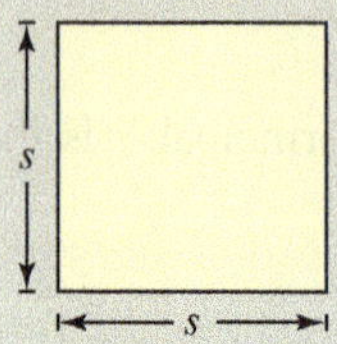

Rectangle

$A = lw$

$P = 2l + 2w$

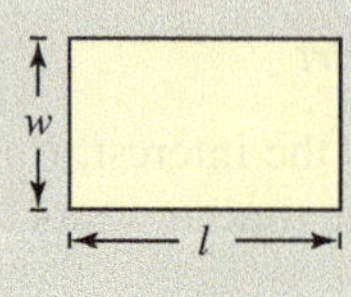

Circle

$A = \pi r^2$

$C = 2\pi r$

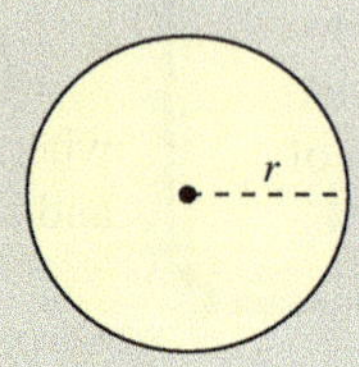

Triangle

$A = \frac{1}{2}bh$

$P = a + b + c$

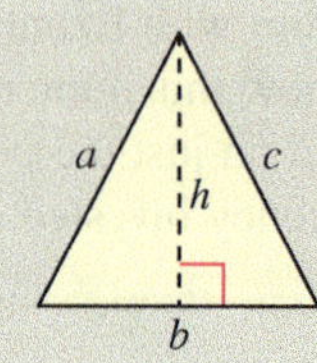

Cube

$V = s^3$

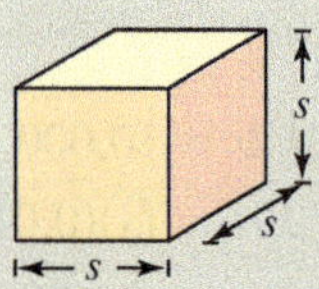

Rectangular Solid

$V = lwh$

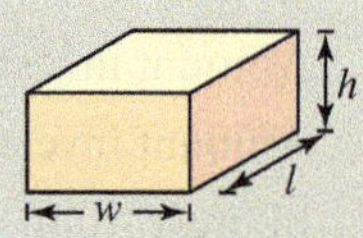

Circular Cylinder

$V = \pi r^2 h$

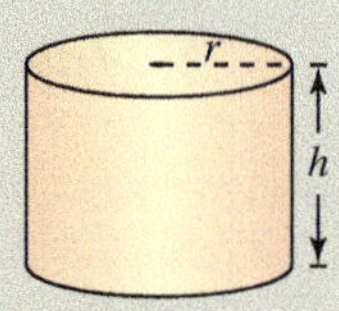

Sphere

$V = \frac{4}{3}\pi r^3$

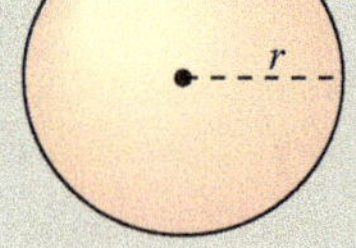

Miscellaneous Common Formulas

Temperature: $F = \frac{9}{5}C + 32$

$C = \frac{5}{9}(F - 32)$

$F =$ degrees Fahrenheit

$C =$ degrees Celsius

Simple Interest: $I = Prt$

$I =$ interest

$P =$ principal

$r =$ interest rate

$t =$ time

Distance: $d = rt$

$d =$ distance traveled

$r =$ rate

$t =$ time

When working with applied problems, you often need to rewrite common formulas. For instance, the formula

$$P = 2l + 2w$$

for the perimeter of a rectangle can be rewritten or solved for w to produce

$$w = \frac{1}{2}(P - 2l).$$

Example 9 Using a Formula

A cylindrical shipping container has a volume of 15,607 cubic inches and a radius of 12 inches, as shown in Figure 1.5. Find the height of the container.

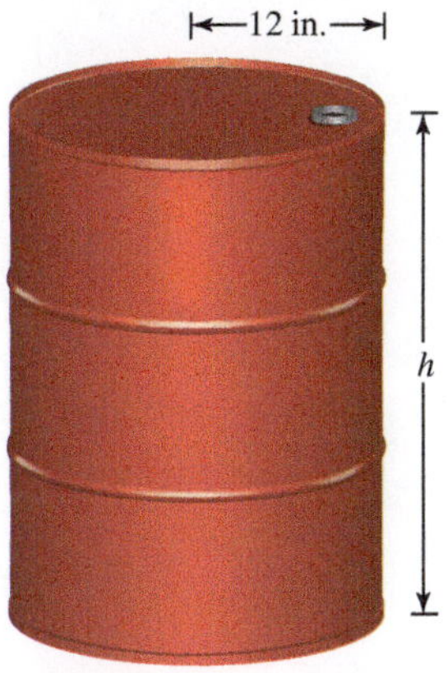

FIGURE 1.5

SOLUTION The formula for the *volume of a cylinder* is $V = \pi r^2 h$. To find the height of the container, solve for h.

$$h = \frac{V}{\pi r^2}$$

Then, using $V = 15{,}607$ and $r = 12$, find the height.

$$h = \frac{15{,}607}{\pi(12)^2} \qquad \text{Substitute 15,607 for } V \text{ and 12 for } r.$$

$$= \frac{15{,}607}{144\pi} \qquad \text{Simplify denominator.}$$

$$\approx 34.5 \qquad \text{Use a calculator.}$$

So, the height of the container is about 34.5 inches. You can use unit analysis to check that your answer is reasonable.

$$\frac{15{,}607 \text{ in.}^{\not{3}}}{144\pi \not{\text{in.}^2}} \approx 34.5 \text{ in.}$$

✓ Checkpoint 9

One cubic foot of water fills a cylindrical pipe with a radius of 0.5 foot. What is the height of the pipe?

Strategy for Solving Word Problems

1. *Search* for the hidden equality—two expressions said to be equal or known to be equal. A sketch may be helpful.
2. *Write* a verbal model that equates these two expressions. Identify any *hidden* products.
3. *Assign* numbers to the known quantities and letters (or algebraic expressions) to the unknown quantities.
4. *Rewrite* the verbal model as an algebraic equation using the assigned labels.
5. *Solve* the resulting algebraic equation.
6. *Check* that the answer satisfies the word problem as stated. (Remember that "solving for x" or another variable may not completely answer the question.)

SUMMARIZE (Section 1.2)

1. Describe the procedure of mathematical modeling *(page 80)*. For examples of constructing and using mathematical models, see Examples 1, 2, 3, 4, 5, 6, and 7.
2. Describe a mixture problem *(page 85)*. For an example of a mixture problem, see Example 8.
3. State the definition of a formula *(page 86)*. For an example that uses a volume formula, see Example 9.

SKILLS WARM UP 1.2 The following warm-up exercises involve skills that were covered in earlier sections. You will use these skills in the exercise set for this section. For additional help, review Section 1.1.

In Exercises 1–10, solve the equation (if possible) and check your solution.

1. $3x - 42 = 0$

2. $64 - 16x = 0$

3. $2 - 3x = 14 + x$

4. $7 + 5x = 7x - 1$

5. $5[1 + 2(x + 3)] = 6 - 3(x - 1)$

6. $2 - 5(x - 1) = 2[x + 10(x - 1)]$

7. $\frac{x}{3} + \frac{x}{2} = \frac{1}{3}$

8. $\frac{2}{x} + \frac{2}{5} = 1$

9. $1 - \frac{2}{z} = \frac{z}{z + 3}$

10. $\frac{x}{x + 1} - \frac{1}{2} = \frac{4}{3}$

Exercises 1.2

See www.CalcChat.com for worked-out solutions to odd-numbered exercises.

Writing an Algebraic Expression In Exercises 1–6, (a) use the verbal description to write a verbal model, (b) assign labels to the quantities in the verbal model, and (c) use the labels to write an algebraic expression.

1. The sum of two consecutive natural numbers

2. The product of two natural numbers whose sum is 25

3. Distance Traveled The distance traveled in t hours by a car traveling at 50 miles per hour

4. Travel Time The travel time for a plane that is traveling at a rate of r miles per hour for 200 miles

5. Geometry The perimeter of a rectangle whose width is x and whose length is twice the width

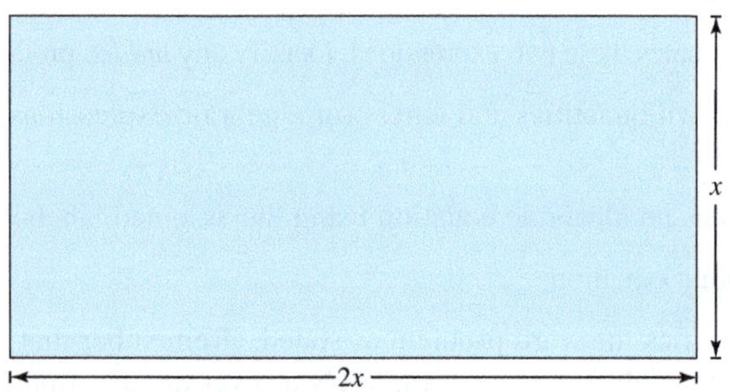

6. Geometry The area of a triangle whose base is 20 inches and whose height is h inches

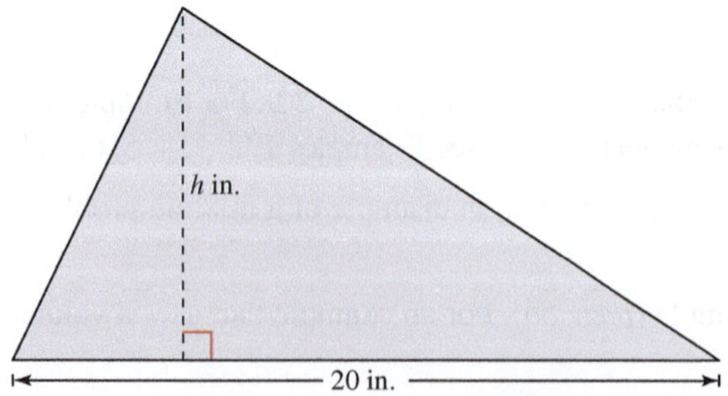

Writing an Algebraic Equation In Exercises 7–12, write an equation that represents the verbal description. *See Example 2.*

7. The sum of 5 and x equals 8.

8. The difference of n and 7 is 4.

9. The quotient of r and 2 is 9.

10. The product of x and 6 equals -9.

11. A salary of \$60,000 is increased by 8%.

12. A food processor is marked down 30% to \$105.

Constructing a Mathematical Model In Exercises 13–22, (a) use the verbal description to write a verbal model, (b) assign labels to the quantities in the verbal model, (c) use the labels to write a mathematical model, and (d) solve the problem. *See Examples 3 and 4.*

13. Find two consecutive numbers whose sum is 525.

14. Find three consecutive natural numbers whose sum is 804.

15. One positive number is five times another positive number. The difference of the two numbers is 148. Find the numbers.

16. One positive number is one-fifth of another number. The difference of the two numbers is 76. Find the numbers.

17. Salary Your weekly salary increases from \$440 to \$506. What percent is your raise?

18. Salary Your monthly salary increases from \$2200 to \$2376. What percent is your raise?

19. Discount Rate A satellite radio system for your car has been discounted by \$30. The sale price is \$119. What percent of the original list price is the discount?

20. Discount Rate The price of a shirt has been discounted by \$20. The sale price is \$29.95. What percent of the original list price is the discount?

21. Weekly Paycheck Your weekly paycheck is 12% *more* than your coworker's. Your two paychecks total \$848. Find the amount of each paycheck.

22. Weekly Paycheck Your weekly paycheck is 12% *less* than your coworker's. Your two paychecks total \$848. Find the amount of each paycheck.

Movie Series In Exercises 23–28, use the table, which shows Harry Potter movies in the order of their releases. Find the percent increase or decrease in the U.S. gross revenue of the two indicated movies. *(Source: Nash Information Services)*

Movie Title: *Harry Potter and the . . .*	U.S. gross revenue
Sorcerer's Stone (2001)	\$317,557,891
Chamber of Secrets (2002)	\$261,987,880
Prisoner of Azkaban (2004)	\$249,538,952
Goblet of Fire (2005)	\$290,013,036
Order of the Phoenix (2007)	\$292,004,738
Half-Blood Prince (2009)	\$301,959,197

23. *Sorcerer's Stone* (2001) to *Chamber of Secrets* (2002)

24. *Chamber of Secrets* (2002) to *Prisoner of Azkaban* (2004)

25. *Prisoner of Azkaban* (2004) to *Goblet of Fire* (2005)

26. *Goblet of Fire* (2005) to *Order of the Phoenix* (2007)

27. *Order of the Phoenix* (2007) to *Half-Blood Prince* (2009)

28. *Sorcerer's Stone* (2001) to *Half-Blood Prince* (2009)

Size Inflation In Exercises 29–32, use the following information. Restaurants tend to serve food in larger portions now than they have in the past. Several examples are shown in the table. Find the percent increase in size from the past to 2011 for the indicated food item.

Food or drink item	Past size	2011 size
Small soft drink	7 fl oz	16 fl oz
Small French fries	2.4 oz	2.5 oz
Large French fries	3.5 oz	5.4 oz
Pizza	10 in.	12 in.

29. Small soft drink

30. Small French fries

31. Large French fries

32. Pizza (Compare the areas.)

33. Comparing Calories A lunch consisting of a large deluxe cheeseburger, large fries, and a large soft drink contains 1560 calories. A lunch consisting of a small cheeseburger, small fries, and a small soft drink contains 680 calories. Find the percent change in calories from the larger to the smaller lunch.

34. Comparing Calories One slice (or one-eighth) of a 14-inch cheese pizza has 250 calories. One slice of a pepperoni pizza has 280 calories. Find the percent change in calories from a slice of the cheese pizza to a slice of the pepperoni pizza.

35. Salary You accept a new job with a starting salary of \$35,000. You receive an 8% raise at the start of your second year, a 7.8% raise at the start of your third year, and a 9.4% raise at the start of your fourth year.

(a) Find your salary for the second year.

(b) Find your salary for the third year.

(c) Find your salary for the fourth year.

36. Salary You accept a new job with a starting salary of \$48,000. You receive a 4% raise at the start of your second year, a 5.5% raise at the start of your third year, and an 11.4% raise at the start of your fourth year.

(a) Find your salary for the second year.

(b) Find your salary for the third year.

(c) Find your salary for the fourth year.

37. New Vehicle Sales In 2009, about 10,601,000 new motor vehicles (cars, light trucks, and other types of trucks) were sold or leased in the United States. The bar graph shows the percent of vehicles sold or leased by type. *(Source: U.S. Bureau of Economic Analysis)*

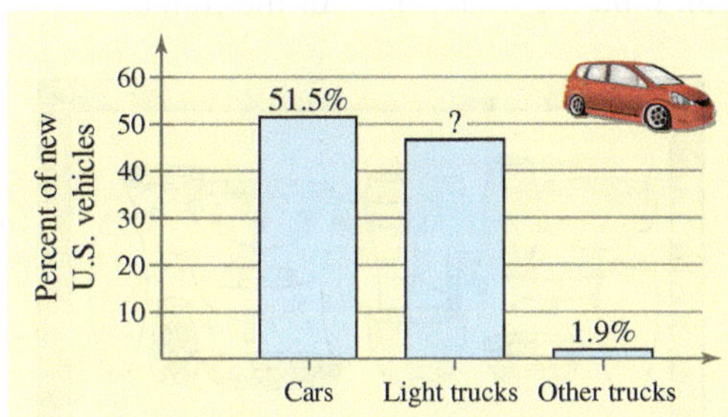

(a) Determine the percent of new motor vehicles that were light trucks.

(b) Determine the numbers of new cars, light trucks, and other types of trucks that were sold or leased in 2009.

(c) Determine the percent of new trucks that were light trucks.

38. Media Usage It is projected that the average person will spend 3515 hours using various media in 2012. The bar graph shows these hours by media type (a person can use more than one type of media at a time). *(Source: Veronis Suhler Stevenson)*

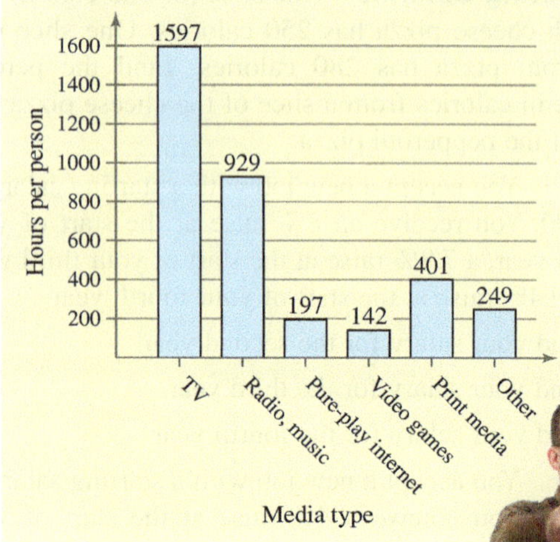

(a) Find the percent of the total media time spent reading print media.

(b) Find the percent of the total media time spent on something other than reading print media.

(c) Find the percent of the total media time spent playing video games and watching TV.

39. Geometry A picture frame (see figure) has a total perimeter of 3 feet. The width of the frame is 0.62 times its length. Find the dimensions of the frame.

40. Geometry A room is 1.5 times as long as it is wide, and its perimeter is 75 feet. Sketch a diagram of the room. Then find the dimensions of the room.

41. Simple Interest You invest \$2500 at 4.5% simple interest. How many years will it take for the investment to earn \$1000 in interest?

42. Simple Interest An investment earns \$1075.20 interest over a seven-year period. What is the rate of simple interest on a \$4800 principal investment?

43. Course Grade To get an A in a course, you need an average of 90% or better on four tests that are worth 100 points each. Your scores on the first three tests were 87, 92, and 84. What must you score on the fourth test to get an A for the course?

44. Course Grade To get an A in a course, you need an average of 90% or better on four tests. The first three tests are worth 100 points each and the fourth is worth 200 points. Your scores on the first three tests are 87, 92, and 84. What must you score on the fourth test to get an A for the course?

45. List Price The price of a swimming pool has been discounted 15%. The sale price is \$1200. Find the original list price of the swimming pool.

46. List Price The price of a home theater system has been discounted 10%. The sale price is \$499. Find the original list price of the system.

47. Wholesale Price A store marks up an electric power drill 60% from its wholesale price. In a clearance sale, the price is discounted by 25%. The sale price is \$21.60. What was the wholesale price of the power drill?

48. Wholesale Price A store marks up a picture frame 80% from its wholesale price. In a clearance sale, the price is discounted by 40%. The sale price is \$28.08. What was the wholesale price of the picture frame?

49. Weekly Salary Due to economic factors, your employer has reduced your weekly wage by 15%. Before the reduction, your weekly salary was \$425.

(a) What is your reduced salary?

(b) What percent raise must you receive to bring your weekly salary back up to \$425? Explain why the percent raise is different from the percent reduction.

50. HOW DO YOU SEE IT? The number of Internet users in the world increased by approximately the same amount from 2004 to 2007 as from 2007 to 2009, as shown in the bar graph. Was the *percent* increase from 2004 to 2007 approximately the same as from 2007 to 2009? Explain. *(Source: ITU World Telecommunication)*

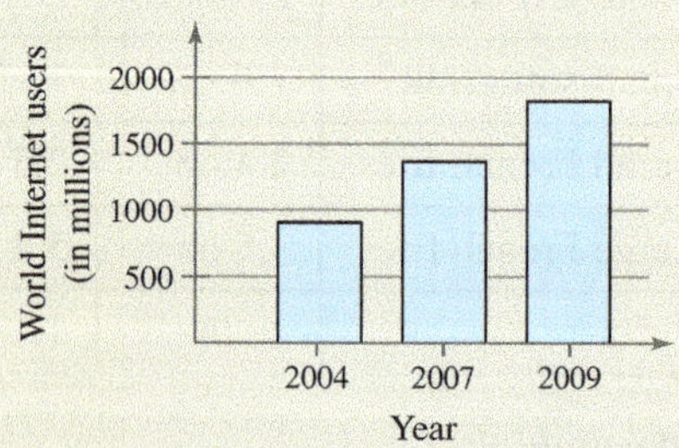

dragon_fang/Shutterstock.com

51. **Travel Time** You are driving to a college 150 miles from home. It takes 28 minutes to travel the first 30 miles. At this rate, how long is your entire trip?

52. **Catch-Up Time** Students are traveling in two cars to a football game 135 miles away. One car travels at an average speed of 45 miles per hour. The second car starts $\frac{1}{2}$ hour later and travels at an average speed of 55 miles per hour. How long will it take the second car to catch up to the first car?

53. **Height of a Tree** To determine the height of a tree, you measure its shadow and the shadow of a five-foot lamppost, as shown in the figure. How tall is the tree?

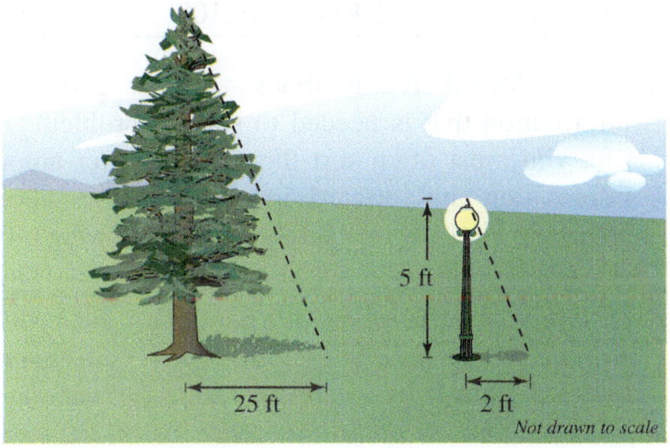

54. **Height of a Building** To determine the height of a building, you measure the building's shadow and the shadow of a four-foot stake, as shown in the figure. How tall is the building?

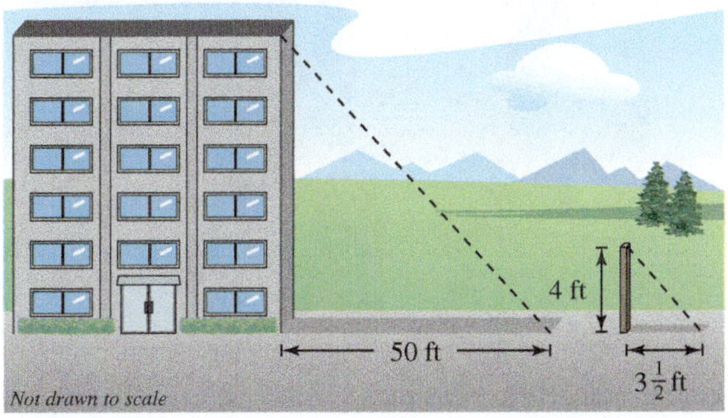

55. **Radio Waves** Radio waves travel at the same speed as light, 3.0×10^8 meters per second. Find the time required for a radio wave to travel from mission control in Houston to NASA astronauts on the surface of the moon 3.84×10^8 meters away.

56. **Distance to a Star** Find the distance (in miles) to a star that is 50 light years (distance traveled by light in 1 year) away. (Light travels at 186,000 miles per second.)

57. **Investment Mix** You invest $15,000 in two funds paying 6.5% and 7.5% simple interest. The total annual interest is $1020. How much do you invest in each fund?

58. **Investment Mix** You invest $30,000 in two funds paying 3% and $4\frac{1}{2}$% simple interest. The total annual interest is $1230. How much do you invest in each fund?

59. **Stock Mix** You invest $5000 in two stocks. In one year, the value of stock A increases by 9.8% and the value of stock B increases by 6.2%. The total value of the stocks is now $5389.20. How much did you originally invest in each stock?

60. **Stock Mix** You invest $4000 in two stocks. In one year, the value of stock A increases by 5.4% and the value of stock B increases by 12.8%. The total value of the stocks is now $4401. How much did you originally invest in each stock?

Production Limit In Exercises 61 and 62, use the following information. *Variable costs* depend on the number of units produced. *Fixed costs* are the same regardless of how many units are produced. Find the greatest number of units the company can produce each month.

61. The company has fixed monthly costs of $15,000 and variable monthly costs of $8.75 per unit. The company has $90,000 available each month to cover costs.

62. The company has fixed monthly costs of $10,000 and variable monthly costs of $9.30 per unit. The company has $85,000 available each month to cover costs.

Rewriting Formulas In Exercises 63–70, solve for the indicated variable.

63. **Area of a Triangle**

Solve for h in $A = \frac{1}{2}bh$.

64. **Volume of a Rectangular Prism**

Solve for l in $V = lwh$

65. **Perimeter of a Rectangle**

Solve for l in $P = 2l + 2w$.

66. **Markup**

Solve for C in $S = C + RC$.

67. **Discount**

Solve for L in $S = L - RL$.

68. **Investment at Simple Interest**

Solve for r in $A = P + Prt$.

69. **Investment at Compound Interest**

Solve for P in $A = P\left(1 + \frac{r}{n}\right)^{nt}$.

70. **Area of a Trapezoid**

Solve for b in $A = \frac{1}{2}(a + b)h$.

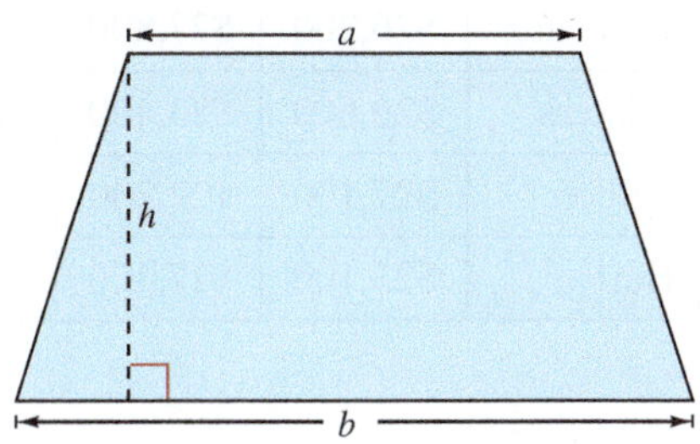

71. Length of a Tank The diameter of a cylindrical propane gas tank is 4 feet (see figure). The total volume of the tank is 603.2 cubic feet. Find the length of the tank.

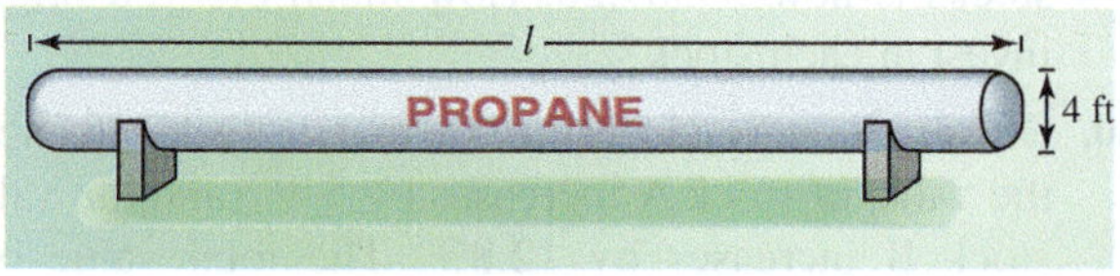

72. Water Depth A trough is 12 feet long, 3 feet deep, and 3 feet wide (see figure). Find the depth of the water when the trough contains 70 gallons of water. (1 gallon ≈ 0.13368 cubic foot.)

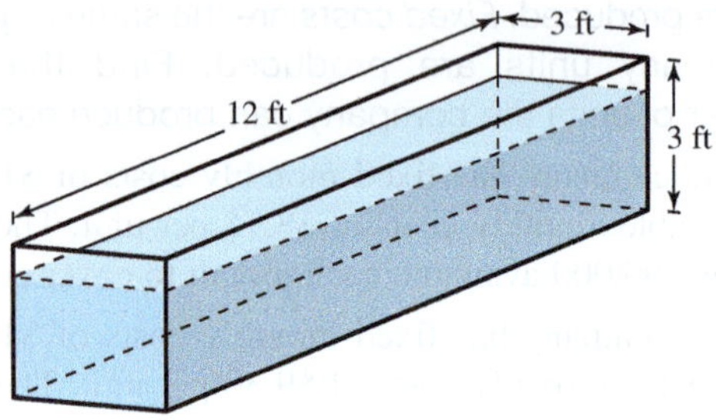

73. Mixture A 55-gallon barrel contains a mixture with a concentration of 40%. How much of this mixture must be withdrawn and replaced by 100% concentrate to bring the mixture up to 75% concentration? (See figure.)

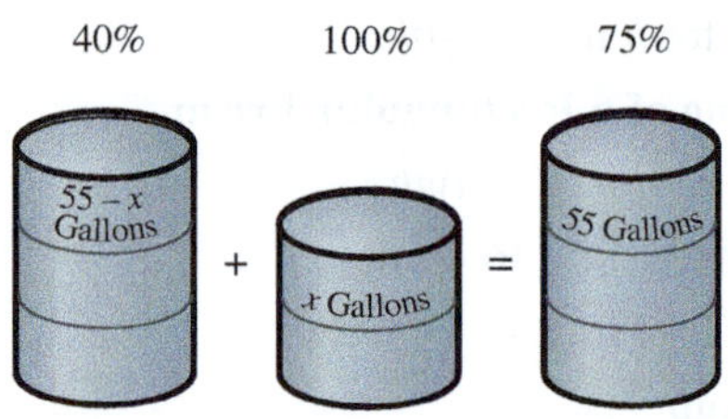

74. Fuel Mixture You have two gallons of a fuel mixture that is 32 parts gasoline and 1 part two-cycle engine oil. How much gasoline do you need to add to form a new mixture that is 40 parts gasoline and 1 part oil?

75. Monthly Sales The table shows the monthly sales of a sales team for the second quarter of the calendar year. Find the average monthly sales for each salesperson. Then find the team's average sales for each month.

Name	April	May	June
Williams	\$25,000	\$28,800	\$21,000
Gonzalez	\$26,200	\$27,800	\$29,500
Walters	\$26,600	\$23,400	\$26,900
Gilbert	\$27,100	\$22,200	\$29,000
Hart	\$23,100	\$27,400	\$22,800

76. Monthly Sales The table shows the monthly sales of a sales team for the fourth quarter of the calendar year. Use a spreadsheet software program to find the average monthly sales for each salesperson. Then find the team's average sales for each month.

Name	October	November	December
Williams	\$20,000	\$25,100	\$23,900
Gonzalez	\$24,200	\$23,600	\$18,500
Walters	\$31,900	\$23,800	\$18,400
Gilbert	\$24,600	\$23,100	\$30,700
Hart	\$32,400	\$19,100	\$28,600

77. Think About It Textbooks usually give just the information that is needed to solve a problem. In real life, you must often sort through a greater amount of information and discard irrelevant facts, called **red herrings.** Find any red herrings in the following problem.

Beneath the ocean surface, pressure changes at a rate of approximately 4.4 pounds per square inch for every 10-foot change in depth. A diver takes 30 minutes to ascend 25 feet from a depth of 150 feet. What change in pressure does the diver experience?

Courtesy of CitiKitty, Inc.

Business Capsule

CitiKitty, Inc. was founded in 2005 by 26-year-old Rebecca Rescate after she moved into a small apartment in New York City with no place to hide her cat's litter box. Finding no easy-to-use cat toilet training kit, she created one, and CitiKitty was born with an initial investment of \$20,000. Today the company flourishes with an expanded product line. Revenues in 2010 reached \$350,000.

78. Research Project Use your school's library, the Internet, or some other reference source to find information about the start-up costs of beginning a business, such as the example above. Write a short paper about the company.

1.3 Quadratic Equations

- Solve a quadratic equation by factoring.
- Solve a quadratic equation by extracting square roots.
- Construct and use a quadratic model to solve an application problem.

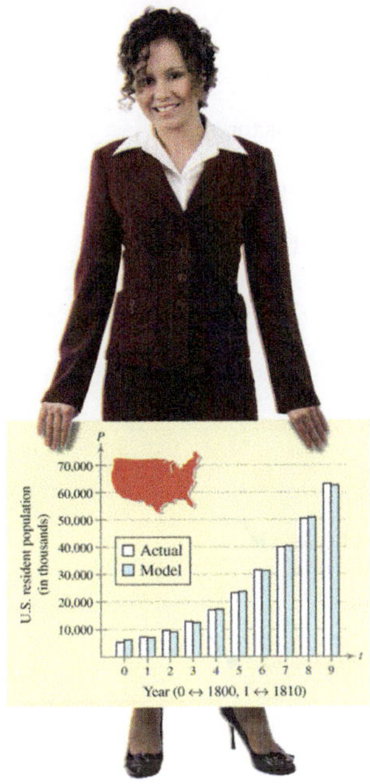

In Exercise 87 on page 103, you will use a quadratic model to determine the population of the United States.

Solving Quadratic Equations by Factoring

In the first two sections of this chapter, you studied linear equations in one variable. In this and the next section, you will study quadratic equations.

Definition of a Quadratic Equation

A **quadratic equation** in x is an equation that can be written in the general form

$$ax^2 + bx + c = 0$$

where a, b, and c are real numbers with $a \neq 0$. Another name for a quadratic equation in x is a **second-degree polynomial equation in x.**

There are three basic techniques for solving quadratic equations: factoring, extracting square roots, and the *Quadratic Formula*. (The Quadratic Formula is discussed in the next section.) The factoring technique is based on the following property.

Zero-Factor Property

If $ab = 0$, then $a = 0$ or $b = 0$.

To use this property, rewrite the left side of the general form of a quadratic equation as the product of two linear factors. Then find the solutions of the quadratic equation by setting each linear factor equal to zero.

STUDY TIP

The Zero-Factor Property applies *only* to equations written in general form (in which one side of the equation is zero). So, be sure that all terms are collected on one side *before* factoring. For instance, in the equation

$$(x - 5)(x + 2) = 8$$

it is *incorrect* to set each factor equal to 8. Can you solve this equation correctly?

Example 1 Solving a Quadratic Equation by Factoring

Solve $x^2 - 3x - 10 = 0$.

SOLUTION

$x^2 - 3x - 10 = 0$ Write original equation.

$(x - 5)(x + 2) = 0$ Factor.

$x - 5 = 0 \Rightarrow x = 5$ Set 1st factor equal to 0.

$x + 2 = 0 \Rightarrow x = -2$ Set 2nd factor equal to 0.

The solutions are $x = 5$ and $x = -2$. Check these in the original equation.

✓ Checkpoint 1

Solve each equation.

a. $x^2 - x - 12 = 0$

b. $x^2 + 10x + 24 = 0$

Karlova Irina/www.shutterstock.com

Example 2 Solving a Quadratic Equation by Factoring

Solve $6x^2 - 3x = 0$.

SOLUTION

$$
\begin{aligned}
6x^2 - 3x &= 0 && \text{Write original equation.}\\
3x(2x - 1) &= 0 && \text{Factor out common factor.}\\
3x &= 0 \quad \Rightarrow \quad x = 0 && \text{Set 1st factor equal to 0.}\\
2x - 1 &= 0 \quad \Rightarrow \quad x = \tfrac{1}{2} && \text{Set 2nd factor equal to 0.}
\end{aligned}
$$

The solutions are $x = 0$ and $x = \frac{1}{2}$.

CHECK

$$
\begin{aligned}
6x^2 - 3x &= 0 && \text{Write original equation.}\\
6(0)^2 - 3(0) &\stackrel{?}{=} 0 && \text{Substitute 0 for } x.\\
0 - 0 &= 0 && \text{First solution checks. ✓}\\
6\left(\tfrac{1}{2}\right)^2 - 3\left(\tfrac{1}{2}\right) &\stackrel{?}{=} 0 && \text{Substitute } \tfrac{1}{2} \text{ for } x.\\
\tfrac{6}{4} - \tfrac{3}{2} &= 0 && \text{Second solution checks. ✓}
\end{aligned}
$$

✓ Checkpoint 2

Solve $4x^2 - 8x = 0$.

If the two factors of a quadratic expression are the same, the corresponding solution of the quadratic equation is a **double** or **repeated** solution.

TECH TUTOR

To check the solution in Example 3 with your graphing utility, you should first write the equation in general form.

$$9x^2 - 6x + 1 = 0$$

Then enter the expression $9x^2 - 6x + 1$ into y_1 of the equation editor. Now you can use the ASK mode of the *table* feature of your graphing utility to check the solution. For instructions on how to use the *table* feature, see Appendix A.

Example 3 A Quadratic Equation with a Repeated Solution

Solve $9x^2 - 6x = -1$.

SOLUTION

$$
\begin{aligned}
9x^2 - 6x &= -1 && \text{Write original equation.}\\
9x^2 - 6x + 1 &= 0 && \text{Write in general form.}\\
(3x - 1)^2 &= 0 && \text{Factor.}\\
3x - 1 &= 0 && \text{Set repeated factor equal to 0.}\\
x &= \tfrac{1}{3} && \text{Solution}
\end{aligned}
$$

The only solution is $x = \frac{1}{3}$.

CHECK

$$
\begin{aligned}
9x^2 - 6x &= -1 && \text{Write original equation.}\\
9\left(\tfrac{1}{3}\right)^2 - 6\left(\tfrac{1}{3}\right) &\stackrel{?}{=} -1 && \text{Substitute } \tfrac{1}{3} \text{ for } x.\\
1 - 2 &\stackrel{?}{=} -1 && \text{Simplify.}\\
-1 &= -1 && \text{Solution checks. ✓}
\end{aligned}
$$

✓ Checkpoint 3

Solve $x^2 + 4x = -4$.

Extracting Square Roots

There is a shortcut for solving equations of the form $u^2 = d$, where $d > 0$. By factoring, you can see that this equation has two solutions.

$$u^2 = d \qquad \text{Write original equation.}$$

$$u^2 - d = 0 \qquad \text{Write in general form.}$$

$$(u + \sqrt{d})(u - \sqrt{d}) = 0 \qquad \text{Factor.}$$

$$u + \sqrt{d} = 0 \Rightarrow u = -\sqrt{d} \qquad \text{Set 1st factor equal to 0.}$$

$$u - \sqrt{d} = 0 \Rightarrow u = \sqrt{d} \qquad \text{Set 2nd factor equal to 0.}$$

Solving an equation of the form $u^2 = d$ without going through the steps of factoring is called **extracting square roots.**

Extracting Square Roots

The equation $u^2 = d$, where $d > 0$, has exactly two solutions:

$$u = \sqrt{d} \quad \text{and} \quad u = -\sqrt{d}.$$

These solutions can also be written as $u = \pm\sqrt{d}$.

Example 4 Extracting Square Roots

Solve $4x^2 = 12$.

SOLUTION

$$4x^2 = 12 \qquad \text{Write original equation.}$$

$$x^2 = 3 \qquad \text{Divide each side by 4.}$$

$$x = \pm\sqrt{3} \qquad \text{Extract square roots.}$$

The solutions are $x = \sqrt{3}$ and $x = -\sqrt{3}$. Check these in the original equation.

✓ Checkpoint 4

Solve $2x^2 = 8$.

Example 5 Extracting Square Roots

Solve $(x - 3)^2 = 7$.

SOLUTION

$$(x - 3)^2 = 7 \qquad \text{Write original equation.}$$

$$x - 3 = \pm\sqrt{7} \qquad \text{Extract square roots.}$$

$$x = 3 \pm \sqrt{7} \qquad \text{Add 3 to each side.}$$

The solutions are $x = 3 \pm \sqrt{7}$. Check these in the original equation.

✓ Checkpoint 5

Solve $(x - 1)^2 = 16$.

Lucian Coman/Shutterstock.com

Applications

Quadratic equations often occur in problems dealing with area. Here is a simple example.

A square room has an area of 144 square feet. Find the dimensions of the room.

To solve this problem, let x represent the length of each side of the room. Then use the formula for the area of a square to write and solve the equation

$$x^2 = 144$$

to conclude that each side of the room is 12 feet long. Note that although the equation $x^2 = 144$ has two solutions, $x = -12$ and $x = 12$, the negative solution makes no sense (for this problem), so you should choose the positive solution.

Example 6 Finding the Dimensions of a Room

A rectangular sunroom is 3 feet longer than it is wide (see Figure 1.6) and has an area of 154 square feet. Find the dimensions of the room.

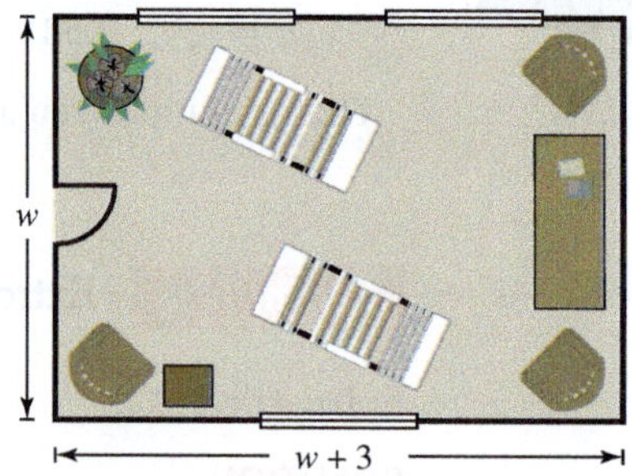

FIGURE 1.6

SOLUTION Use the same type of problem-solving strategy that was presented in Section 1.2.

Verbal Model: Width of room · Length of room = Area of room

Labels: Area of room = 154 (square feet)
Width of room = w (feet)
Length of room = $w + 3$ (feet)

Equation:

$$w(w + 3) = 154$$

$$w^2 + 3w - 154 = 0$$

$$(w - 11)(w + 14) = 0$$

$$w - 11 = 0 \Rightarrow w = 11$$

$$w + 14 = 0 \Rightarrow w = -14$$

Choosing the positive value, you can conclude that the width is 11 feet and the length is $w + 3 = 14$ feet. Check this in the original statement of the problem, as shown.

CHECK

The length of 14 feet is 3 feet more than the width of 11 feet. ✓

The area of the sunroom is $(11)(14) = 154$ square feet. ✓

✓ Checkpoint 6

A rectangular kitchen is 8 feet longer than it is wide and has an area of 84 square feet. Find the dimensions of the kitchen.

STUDY TIP

The position equation described here ignores air resistance. This implies that it is appropriate to use the position equation only to model falling objects that have little air resistance and that fall over short distances.

Another application of quadratic equations involves an object that is falling (or is vertically projected into the air). The equation that gives the height of such an object is called a **position equation,** and on Earth's surface it has the form

$$s = -16t^2 + v_0t + s_0.$$

In this equation, s represents the height of the object (in feet), v_0 represents the initial velocity of the object (in feet per second), s_0 represents the initial height of the object (in feet), and t represents the time (in seconds).

Example 7 Falling Object

A construction worker accidentally drops a wrench from a height of 235 feet and yells "Look out below!" (see Figure 1.7). Could a person at ground level hear this warning in time to get out of the way of the falling wrench?

SOLUTION Because sound travels at about 1100 feet per second, it follows that a person at ground level hears the warning within 1 second of the time the wrench is dropped. To set up a mathematical model for the height of the wrench, use the position equation

$$s = -16t^2 + v_0t + s_0.$$

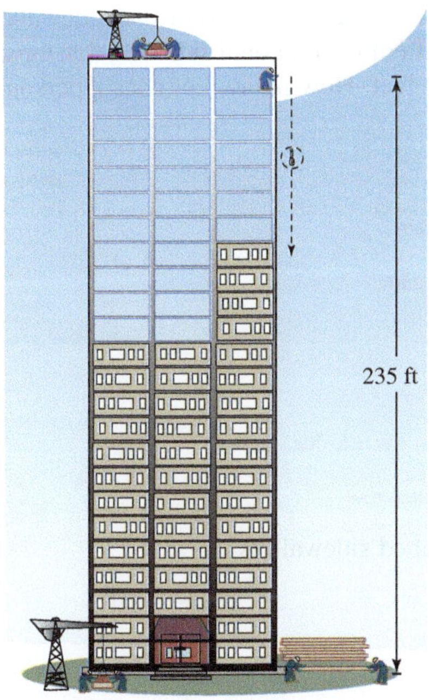

FIGURE 1.7

Because the object is dropped rather than thrown, the initial velocity is $v_0 = 0$ feet per second. Moreover, because the initial height is $s_0 = 235$ feet, you have the following model.

$$\begin{aligned} s &= -16t^2 + (0)t + 235 \\ &= -16t^2 + 235 \end{aligned}$$

After falling for 1 second, the height of the wrench is

$$-16(1)^2 + 235 = 219 \text{ feet.}$$

After falling for 2 seconds, the height of the wrench is

$$-16(2)^2 + 235 = 171 \text{ feet.}$$

To find the number of seconds it takes the wrench to hit the ground, let the height s be zero and solve the equation for t.

$s = -16t^2 + 235$ Write position equation.

$0 = -16t^2 + 235$ Substitute 0 for s.

$16t^2 = 235$ Add $16t^2$ to each side.

$t^2 = \dfrac{235}{16}$ Divide each side by 16.

$t = \dfrac{\sqrt{235}}{4}$ Extract positive square root.

$t \approx 3.83$ Use a calculator.

The wrench will take about 3.83 seconds to hit the ground. If the person hears the warning within 1 second after the wrench is dropped, the person still has almost 3 seconds to get out of the way.

Checkpoint 7

You drop a rock from a height of 144 feet. How long does it take the rock to hit the ground?

A third type of application using a quadratic equation involves the hypotenuse of a right triangle. Recall from geometry that the sides of a right triangle are related by a formula called the **Pythagorean Theorem.** This theorem states that if a and b are the lengths of the legs of the triangle and c is the length of the hypotenuse (see Figure 1.8),

$$a^2 + b^2 = c^2. \quad \text{Pythagorean Theorem}$$

Notice how this formula is used in the next example.

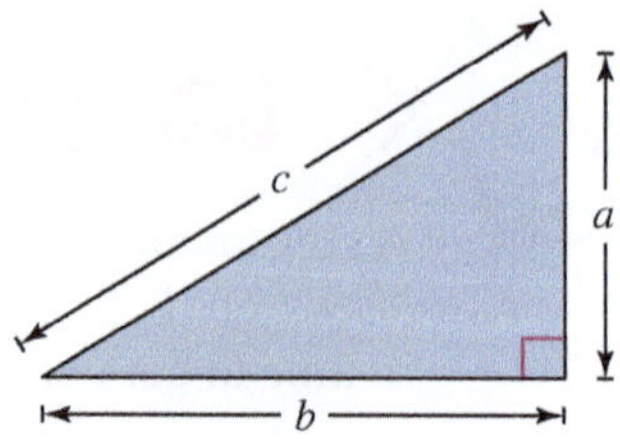

FIGURE 1.8

Example 8 Cutting Across the Lawn

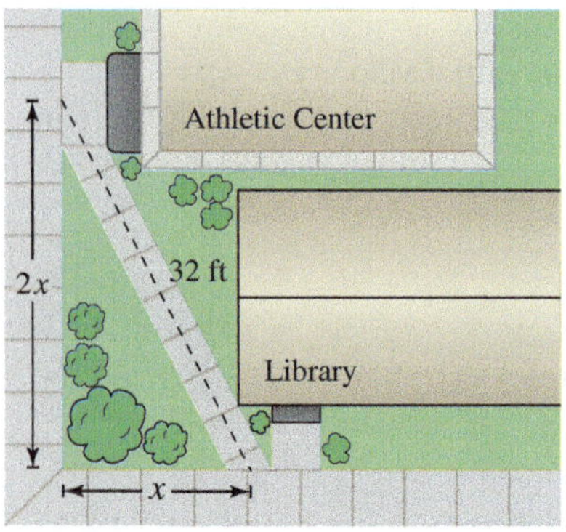

FIGURE 1.9

An L-shaped sidewalk from the athletic center to the library on a college campus is shown in Figure 1.9. The sidewalk was constructed so that the length of one sidewalk forming the L is twice as long as the other. The length of the diagonal sidewalk that cuts across the grounds between the two buildings is 32 feet. How many feet does a person save by walking on the diagonal sidewalk?

SOLUTION Use the Pythagorean Theorem.

$a^2 + b^2 = c^2$ — Pythagorean Theorem

$x^2 + (2x)^2 = 32^2$ — Substitute for a, b, and c.

$5x^2 = 1024$ — Combine like terms.

$x^2 = 204.8$ — Divide each side by 5.

$x = \pm\sqrt{204.8}$ — Take the square root of each side.

$x = \sqrt{204.8}$ — Extract positive square root.

The total distance covered by walking on the L-shaped sidewalk is

$$\begin{aligned} x + 2x &= 3x \\ &= 3\sqrt{204.8} \\ &\approx 42.9 \text{ feet.} \end{aligned}$$

Walking on the diagonal sidewalk saves a person about

$$42.9 - 32 = 10.9 \text{ feet.}$$

Checkpoint 8

In Example 8, suppose the length of one sidewalk forming the L is three times as long as the other. How many feet does a person save by walking on the 32-foot diagonal sidewalk?

A fourth type of application of a quadratic equation is one in which a quantity y is changing over time t according to a quadratic model. In the next example, y is exchanged with R because R is a better descriptor in the model.

Example 9 Electric Power Revenues

From 2003 to 2007, the yearly revenues R (in billions of dollars) for the electric power industry in the United States can be modeled by

$$R = 2.237t^2 + 239.12, \quad 3 \le t \le 7$$

where t represents the year, with $t = 3$ corresponding to 2003 (see Figure 1.10). Use the model to approximate the year that revenues were about \$325,000,000,000. *(Source: Edison Electric Institute)*

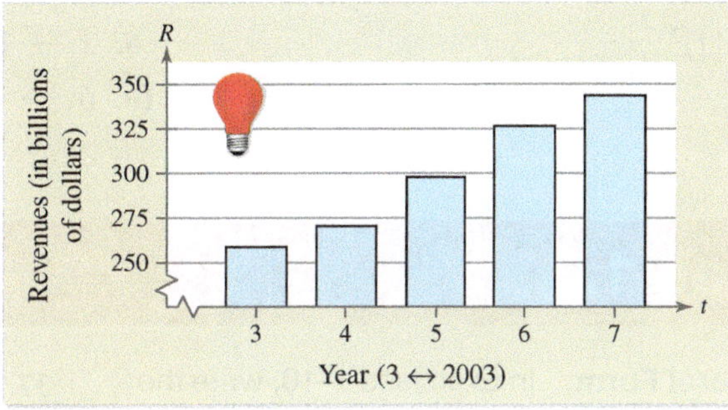

FIGURE 1.10

SOLUTION To solve this problem, let the yearly revenues R be 325 billion dollars and solve the equation for t.

$$2.237t^2 + 239.12 = 325 \quad \text{Substitute 325 for } R.$$

$$2.237t^2 = 85.88 \quad \text{Subtract 239.12 from each side.}$$

$$t^2 \approx 38.391 \quad \text{Divide each side by 2.237.}$$

$$t \approx \sqrt{38.391} \quad \text{Extract positive square root.}$$

$$t \approx 6.196 \quad \text{Simplify.}$$

The solution is $t \approx 6$. Because $t = 3$ represents 2003, you can conclude that, according to the model, revenues were about \$325 billion in the year 2006.

Checkpoint 9

Use the model in Example 9 to predict the year that revenues will be about \$500 billion.

SUMMARIZE (Section 1.3)

1. Describe how to solve a quadratic equation by factoring *(page 93)*. For examples of solving a quadratic equation by factoring, see Examples 1, 2, and 3.
2. Describe how to solve a quadratic equation by extracting square roots *(page 95)*. For examples of solving a quadratic equation by extracting square roots, see Examples 4 and 5.
3. Describe how a quadratic model can be used to represent data, such as the revenues of a particular industry *(page 99, Example 9)*.

SKILLS WARM UP 1.3 The following warm-up exercises involve skills that were covered in earlier sections. You will use these skills in the exercise set for this section. For additional help, review Sections 0.4 and 0.6.

In Exercises 1–4, simplify the expression.

1. $\sqrt{\frac{7}{50}}$ **2.** $\sqrt{32}$

3. $\sqrt{7^2 + 3 \cdot 7^2}$ **4.** $\sqrt{\frac{1}{4} + \frac{3}{8}}$

In Exercises 5–10, factor the expression.

5. $3x^2 + 5x$ **6.** $4x^2 - 25$

7. $16 - (x - 11)^2$ **8.** $x^2 + 7x - 18$

9. $10x^2 + 13x - 3$ **10.** $6x^2 - 73x + 12$

Exercises 1.3

See www.CalcChat.com for worked-out solutions to odd-numbered exercises.

Writing the General Form In Exercises 1–10, write the quadratic equation in general form.

1. $2x^2 = 3 - 5x$ **2.** $4x^2 - 2x = 9$

3. $x^2 = 19x$ **4.** $10x^2 = 70$

5. $(x - 3)^2 = 5$ **6.** $12 - 3(x + 7)^2 = 0$

7. $x(x + 2) = 3x^2 + 1$ **8.** $x(x + 5) = 2(x + 5)$

9. $\dfrac{3x^2 - 10}{5} = 12x$ **10.** $\dfrac{x^2 - 7}{3} = 2x$

Solving a Quadratic Equation by Factoring In Exercises 11–22, solve the quadratic equation by factoring. *See Examples 1, 2, and 3.*

11. $x^2 - 2x - 8 = 0$ **12.** $x^2 - 10x + 9 = 0$

13. $6x^2 + 3x = 0$ **14.** $9x^2 - 1 = 0$

15. $x^2 + 10x + 25 = 0$

16. $16x^2 + 56x + 49 = 0$

17. $3 + 5x - 2x^2 = 0$ **18.** $2x^2 = 19x + 33$

19. $x^2 + 4x = 12$ **20.** $x^2 + 4x = 21$

21. $-x^2 - 7x = 10$ **22.** $-x^2 + 8x = 12$

Extracting Square Roots In Exercises 23–40, solve the quadratic equation by extracting square roots. When a solution is irrational, list both the exact solution *and* its approximation rounded to two decimal places. *See Examples 4 and 5.*

23. $x^2 = 16$ **24.** $x^2 = 144$

25. $x^2 = 7$ **26.** $x^2 = 27$

27. $3x^2 = 36$ **28.** $9x^2 = 25$

29. $(x - 12)^2 = 144$ **30.** $(x + 13)^2 = 225$

31. $(x + 2)^2 = 12$ **32.** $(x + 5)^2 = 20$

33. $12x^2 = 300$ **34.** $6x^2 = 250$

35. $5x^2 = 190$ **36.** $15x^2 = 620$

37. $3x^2 + 2(x^2 - 4) = 15$

38. $x^2 + 3(x^2 - 5) = 10$

39. $6x^2 - 3(x^2 + 1) = 23$

40. $2x^2 + 5(x^2 - 2) = 29$

Choosing a Method In Exercises 41–62, solve the quadratic equation using any convenient method. *See Examples 1, 2, 3, 4, and 5.*

41. $x^2 = 64$ **42.** $7x^2 = 16$

43. $x^2 - 2x + 1 = 0$ **44.** $x^2 - 6x + 5 = 0$

45. $16x^2 - 9 = 0$ **46.** $11x^2 + 33x = 0$

47. $4x^2 - 12x + 9 = 0$

48. $x^2 - 14x + 49 = 0$

49. $(x + 4)^2 = 49$ **50.** $(x - 3)^2 = 36$

51. $4x = 4x^2 - 3$ **52.** $80 + 6x = 9x^2$

53. $50 + 5x = 3x^2$ **54.** $144 - 73x + 4x^2 = 0$

55. $12x = x^2 + 27$ **56.** $26x = 8x^2 + 15$

57. $50x^2 - 60x + 10 = 0$

58. $9x^2 + 12x + 3 = 0$

59. $(x + 3)^2 - 4 = 0$ **60.** $(x - 2)^2 - 9 = 0$

61. $(x + 1)^2 = x^2$ **62.** $(x + 1)^2 = 4x^2$

63. Writing Consider the expression $(x + 2)^2$. How would you convince someone in your class that $(x + 2)^2 \neq x^2 + 4$? Give an argument based on the rules of algebra. Give an argument using your graphing utility.

64. Writing Consider the expression $\sqrt{a^2 + b^2}$. How would you convince someone in your class that $\sqrt{a^2 + b^2} \neq a + b$? Give an argument based on the rules of algebra or geometry. Give an argument using your graphing utility.

65. Geometry A one-story building is 14 feet longer than it is wide (see figure). The building has 1632 square feet of floor space. What are the dimensions of the building?

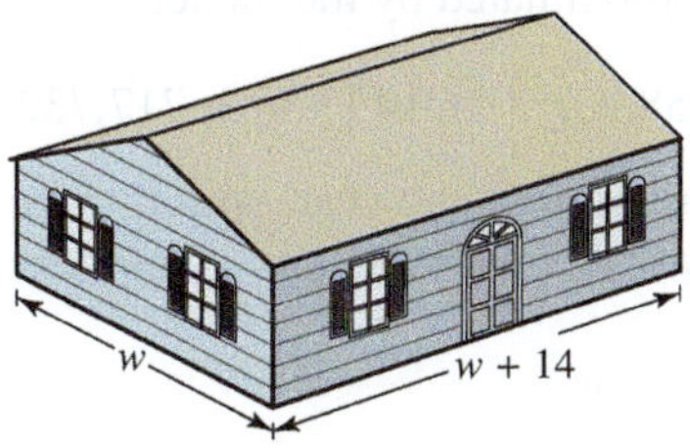

66. Geometry A billboard is 10 feet longer than it is high (see figure). The billboard has 336 square feet of advertising space. What are the dimensions of the billboard?

67. Geometry A triangular sign has a height that is equal to its base. The area of the sign is 4 square feet. Find the base and height of the sign.

68. Geometry The building lot shown in the figure has an area of 8000 square feet. What are the dimensions of the lot?

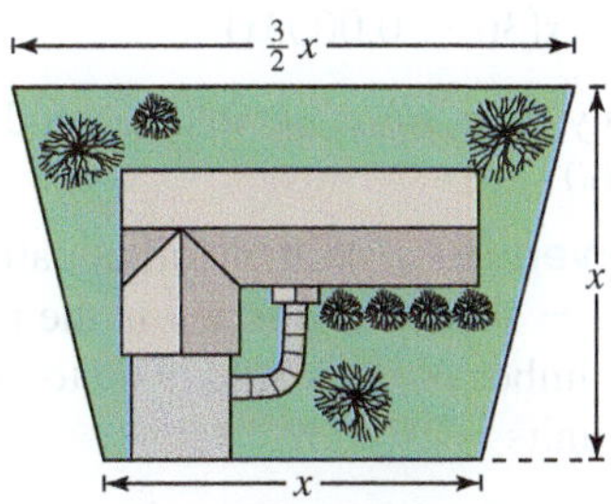

69. Geometry A rectangular garden that is 30 feet long and 20 feet wide is surrounded on all four sides by a rock path that is x feet wide. The total area of the garden and the rock path is 1200 square feet. What is the width of the path?

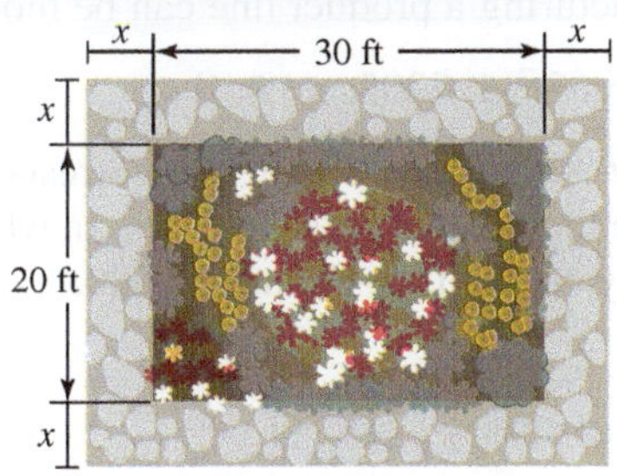

70. Geometry A rectangular pool is 30 feet wide and 40 feet long. It is surrounded on all four sides by a wooden deck that is x feet wide. The total area enclosed within the perimeter of the deck is 3000 square feet. What is the width of the deck?

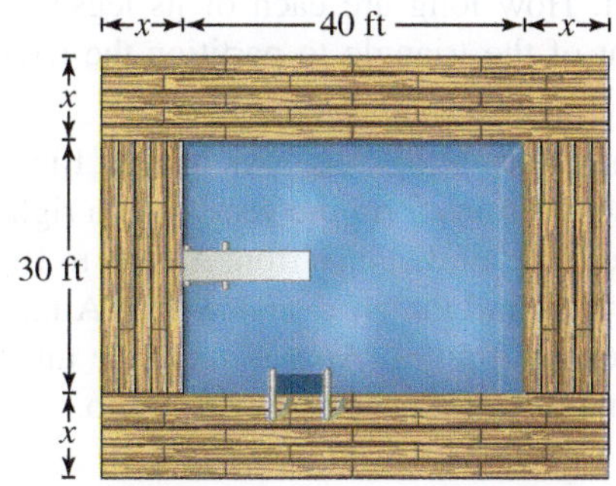

In Exercises 71–76, assume that air resistance is negligible, which implies that the position equation $s = -16t^2 + v_0t + s_0$ is a reasonable model.

71. Falling Object A rock is dropped from the top of a 200-foot cliff that overlooks the ocean. How long will it take for the rock to hit the water?

72. Royal Gorge Bridge The Royal Gorge Bridge near Canon City, Colorado is one of the highest suspension bridges in the world. The bridge is 1053 feet above the Arkansas river. A rock is dropped from the bridge. How long does it take the rock to hit the water?

73. Olympic Diver The high-dive platform in the Olympics is 10 meters above the water. A diver wants to perform an armstand dive, which means she will drop to the water from a handstand position. How long will the diver be in the air? (*Hint:* 1 meter $\approx$ 3.2808 feet)

74. The Owl and the Mouse An owl is flying at a height of 100 feet and spots a mouse directly below. The owl folds up its wings and drops into a dive at the mouse. How long does the mouse have to escape?

75. Wind Resistance At the same time a skydiver jumps from an airplane 13,000 feet above the ground, a steel ball is dropped from the plane. Because of air resistance, it takes the skydiver 67 seconds to freefall to a height of 3000 feet, where the parachute opens. The steel ball has relatively no air resistance, so its height can be modeled by the position equation. How much faster does the ball reach a height of 3000 feet than the skydiver?

76. Wind Resistance At the same time a skydiver jumps from an airplane 8900 feet above the ground, a steel ball is dropped from the plane. Because of air resistance, it takes the skydiver 44 seconds to freefall to a height of 2500 feet, where the parachute opens. The steel ball has relatively no air resistance, so its height can be modeled by the position equation. How much faster does the ball reach a height of 2500 feet than the skydiver?

77. Geometry The hypotenuse of an isosceles right triangle is 6 centimeters long. How long are the legs? (An isosceles right triangle is one whose two legs are of equal length.)

78. Geometry An equilateral triangle has a height of 3 feet. How long are each of its legs? (*Hint:* Use the height of the triangle to partition the triangle into two right triangles of the same size.)

79. Flying Distance A commercial jet flies to three cities whose locations form the vertices of a right triangle (see figure). The air distance from Atlanta to Buffalo is about 703 miles and the air distance from Atlanta to Chicago is about 583 miles. Approximate the air distance from Atlanta to Buffalo *by way of* Chicago.

In Exercises 80 and 81, use the following information. The sum of the angles of a triangle is 180°. Also, if two angles of a triangle are equal, then the lengths of the sides opposite those angles are equal.

80. Depth of a Whale The sonar on a research ship detects a whale that is 3000 feet from the ship. The angle formed by the ocean surface and a line from the ship to the whale is 45° (see figure). How deep is the whale?

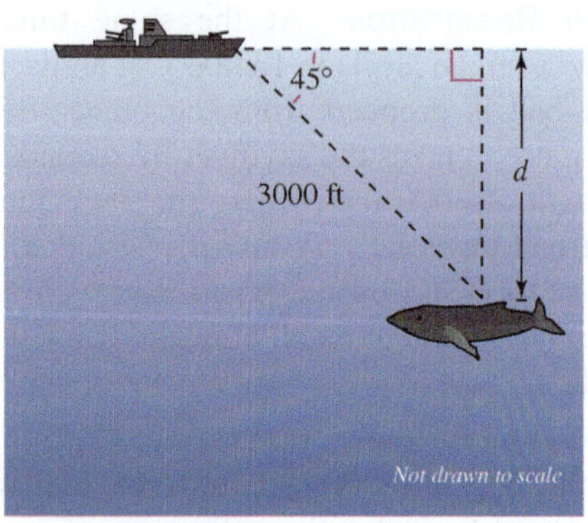

81. Depth of a Whale Shark A research ship is tracking the movements of a whale shark that is 700 meters from the ship. The angle formed by the ocean surface and a line from the ship to the whale shark is 45°. How deep is the whale shark?

82. Tennis The winnings W (in dollars) of the men's singles champions of the Wimbledon Tennis Tournament from 2004 through 2009 can be approximated by the model

$$W = 8258.93t^2 - 60{,}437.5t + 717{,}732, \quad 4 \le t \le 9$$

where t represents the year, with $t = 4$ corresponding to 2004. *(Source: Wimbledon Championships)*

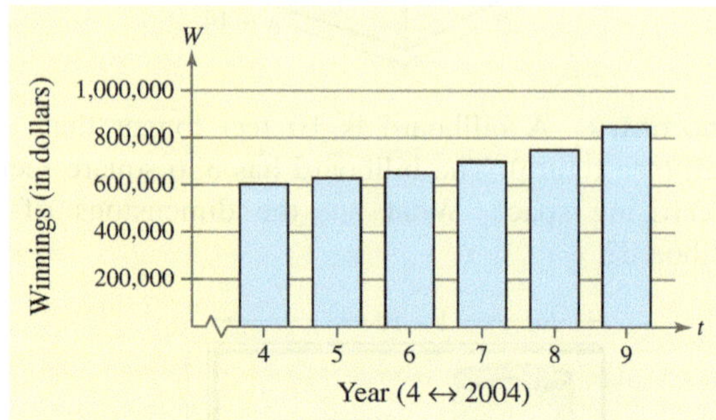

(a) According to the model, what is the first year the men's singles champion won $1 million or more?

(b) The Wimbledon men's singles champion won $1 million in 2010. Based on your answer in part (a), do you think this is a good model for making predictions after 2009? Explain.

83. Total Revenue The demand equation for a product is $p = 36 - 0.0003x$, where p is the price per unit and x is the number of units sold. The total revenue R from selling x units is given by

$$R = xp = x(36 - 0.0003x).$$

How many units must be sold to produce a revenue of $1,080,000?

84. Total Revenue The demand equation for a product is $p = 40 - 0.0005x$, where p is the price per unit and x is the number of units sold. The total revenue R from selling x units is given by

$$R = xp = x(40 - 0.0005x).$$

How many units must be sold to produce a revenue of $800,000?

85. Production Cost A company determines that the average monthly cost C (in dollars) of raw materials for manufacturing a product line can be modeled by

$$C = 35.65t^2 + 7205, \quad t \ge 0$$

where t is the year, with $t = 0$ corresponding to 2010. Use the model to estimate the year in which the average monthly cost will be about $10,000.

86. Blue Oak The blue oak tree, native to California, is known for its slow rate of growth. Fencing enclosures protect seedlings from herbivore damage and promote faster growth. The height H (in inches) of an enclosed blue oak tree is expected to change according to the model

$$H = 0.74t^2 + 25, \quad 0 \le t \le 5$$

where t represents the year, with $t = 0$ corresponding to 2010. Approximate the height of the tree in 2012. Then, use the model to approximate the year in which the height of the tree will be about 37 inches.

87. U.S. Population The resident population P (in thousands) of the United States from 1800 to 1890 can be approximated by the model

$$P = 694.59t^2 + 6179, \quad 0 \le t \le 9$$

where t represents the year, with $t = 0$ corresponding to 1800, $t = 1$ corresponding to 1810, and so on (see figure). *(Source: U.S. Census Bureau)*

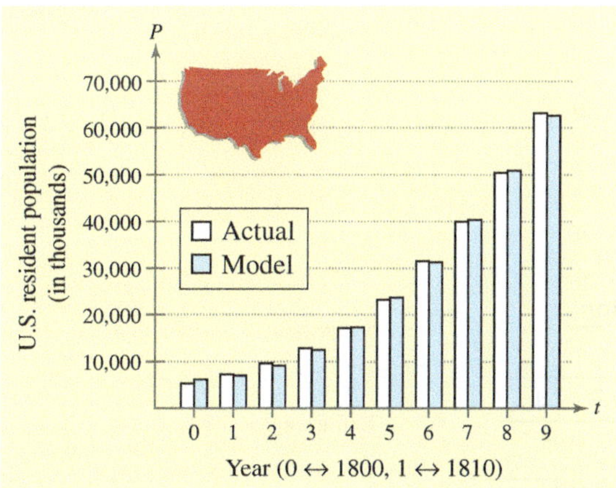

(a) Assume this model had continued to be valid up through the present time. In what year would the resident population of the United States have reached 250,000,000?

(b) Judging from the figure, would you say that this model is a good representation of the resident population through 1890?

(c) Use the model to predict the resident population in 2010. How does your answer compare with the actual 2010 resident population of approximately 310,000,000?

(d) The U.S. Census Bureau predicts that the resident population in 2050 will be 439,010,253. Would you say that the model supports this prediction? Explain.

88. HOW DO YOU SEE IT? Each figure shows a person or a ball about to go into motion through the air. Match each figure with the position equation that represents the situation. Without calculating, compare the terms of the equations to determine which situation involves the least amount of time in the air. Explain your reasoning.

(i)  (ii)  (iii) 

(a) $s = -16t^2 + 33t$

(b) $s = -16t^2 + 10$

(c) $s = -16t^2 - 20t + 10$

89. Monthly Cost A company determines that the average monthly cost C (in dollars) of staffing temporary positions can be modeled by

$$C = 135.47t^2 + 13{,}702, \quad t \ge 0$$

where t represents the year, with $t = 0$ corresponding to 2010. Use the model to predict the year in which the average monthly cost will be about \$25,000.

90. Education The enrollment E in an early childhood development program for a school district from 2000 to 2012 can be approximated by the model

$$E = 1.678t^2 + 1025, \quad 0 \le t \le 12$$

where t represents the year, with $t = 0$ corresponding to 2000. Use the model to approximate the year in which the early childhood enrollment was about 1160 children. Can you use the model to estimate early childhood enrollment for the year 1980? Explain.

91. Meteorology The temperature T (in degrees Fahrenheit) during a certain day can be approximated by

$$T = 0.31t^2 + 32.9, \quad 7 \le t \le 15$$

where t represents the hour of the day, with $t = 7$ corresponding to 7 A.M. Use the model to approximate the time when the temperature was 85° F. Can you use this model to predict the temperature at 7 P.M.? Explain.

92. Electromedical Exports From 2004 through 2009, the yearly exports E (in billions of dollars) of electromedical equipment from the United States can be approximated by

$$E = 0.119t^2 + 10.7, \quad 4 \le t \le 9$$

where t represents the year, with $t = 4$ corresponding to 2004. Use the model to estimate the year when the value of these exports was about \$15 billion. *(Source: AeA)*

1.4 The Quadratic Formula

- Develop the Quadratic Formula by completing the square.
- Use the discriminant to determine the number of real solutions of a quadratic equation.
- Solve a quadratic equation using the Quadratic Formula.
- Use the Quadratic Formula to solve an application problem.

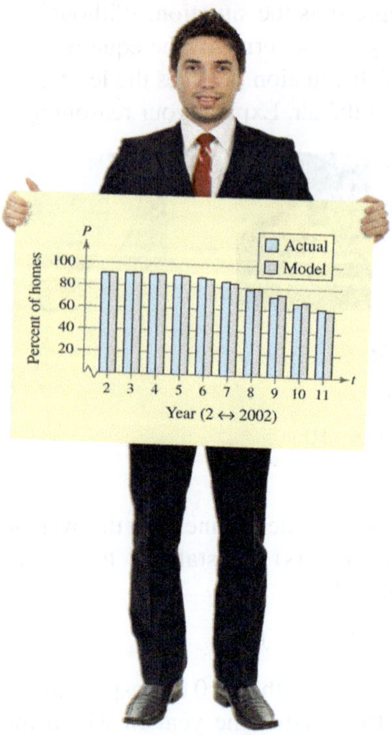

In Exercise 71 on page 112, you will use the Quadratic Formula to determine the year when only 60% of households still had VCRs.

Development of the Quadratic Formula

In Section 1.3, you studied two methods for solving quadratic equations. These two methods are efficient for special quadratic equations that are factorable or that can be solved by extracting square roots. There are, however, many quadratic equations that cannot be solved efficiently by either of these two techniques. Fortunately, there is a general formula that can be used to solve *any* quadratic equation. It is called the **Quadratic Formula.** This formula is derived using a process called **completing the square.**

$$ax^2 + bx + c = 0 \qquad \text{General form, } a \neq 0$$

$$ax^2 + bx = -c \qquad \text{Subtract } c \text{ from each side.}$$

$$x^2 + \frac{b}{a}x = -\frac{c}{a} \qquad \text{Divide each side by } a.$$

$$x^2 + \frac{b}{a}x + \left(\frac{b}{2a}\right)^2 = -\frac{c}{a} + \left(\frac{b}{2a}\right)^2 \qquad \text{Complete the square.}$$

$$\left(\text{half of } \frac{b}{a}\right)^2$$

$$\left(x + \frac{b}{2a}\right)^2 = \frac{b^2 - 4ac}{4a^2} \qquad \text{Simplify.}$$

$$x + \frac{b}{2a} = \pm\sqrt{\frac{b^2 - 4ac}{4a^2}} \qquad \text{Extract square roots.}$$

$$x = \frac{-b \pm \sqrt{b^2 - 4ac}}{2a} \qquad \text{Solutions}$$

The Quadratic Formula

The solutions of

$$ax^2 + bx + c = 0, \quad a \neq 0$$

are given by the **Quadratic Formula,**

$$x = \frac{-b \pm \sqrt{b^2 - 4ac}}{2a}.$$

To help you remember the Quadratic Formula, try to memorize a verbal statement of the rule, such as:

> "The opposite of *b*, plus or minus the square root of *b* squared minus 4*ac*, all divided by 2*a*."

Blaj Gabriel/ww.shutterstock.com

The Discriminant

In the Quadratic Formula, the quantity under the radical sign, $b^2 - 4ac$, is called the **discriminant** of the quadratic expression $ax^2 + bx + c$.

$$b^2 - 4ac \qquad \text{Discriminant}$$

It can be used to determine the number of real solutions of a quadratic equation.

Solutions of a Quadratic Equation

The solutions of a quadratic equation

$$ax^2 + bx + c = 0, \quad a \neq 0$$

can be classified by the discriminant, $b^2 - 4ac$, as follows.

1. If $b^2 - 4ac > 0$, then the equation has *two* distinct real solutions.

2. If $b^2 - 4ac = 0$, then the equation has *one* repeated real solution.

3. If $b^2 - 4ac < 0$, then the equation has *no* real solutions.

If the discriminant of a quadratic equation is negative, as in case 3 above, then its square root is imaginary (not a real number) and the Quadratic Formula yields two complex solutions. You will study complex solutions in Section 3.5.

Example 1 Using the Discriminant

Use the discriminant to determine the number of real solutions to each of the following quadratic equations.

a. $4x^2 - 20x + 25 = 0$ **b.** $13x^2 + 7x + 1 = 0$ **c.** $5x^2 = 8x$

SOLUTION

a. Using $a = 4$, $b = -20$, and $c = 25$, the discriminant is

$$b^2 - 4ac = (-20)^2 - 4(4)(25)$$
$$= 400 - 400 = 0.$$

Because $b^2 - 4ac = 0$, there is *one* repeated real solution.

b. Using $a = 13$, $b = 7$, and $c = 1$, the discriminant is

$$b^2 - 4ac = (7)^2 - 4(13)(1)$$
$$= 49 - 52 = -3.$$

Because $b^2 - 4ac < 0$, there are *no* real solutions.

c. In general form, this equation is $5x^2 - 8x = 0$, with $a = 5$, $b = -8$, and $c = 0$, which implies that the discriminant is

$$b^2 - 4ac = (-8)^2 - 4(5)(0)$$
$$= 64.$$

Because $b^2 - 4ac > 0$, there are *two* distinct real solutions.

✓ Checkpoint 1

Use the discriminant to determine the number of real solutions of $x^2 + 6x + 9 = 0$.

Using the Quadratic Formula

When using the Quadratic Formula, remember that *before* the formula can be applied, you must first write the quadratic equation in general form.

TECH TUTOR

You can write a program for a graphing utility to solve equations using the Quadratic Formula. Use a program to solve the equation in Example 2.

Example 2 Two Distinct Solutions

Solve $x^2 + 3x = 9$.

SOLUTION

$$x^2 + 3x = 9 \qquad \text{Write original equation.}$$

$$x^2 + 3x - 9 = 0 \qquad \text{Write in general form.}$$

$$x = \frac{-3 \pm \sqrt{(3)^2 - 4(1)(-9)}}{2(1)} \qquad \text{Quadratic Formula}$$

$$x = \frac{-3 \pm \sqrt{45}}{2} \qquad \text{Simplify.}$$

$$x = \frac{-3 \pm 3\sqrt{5}}{2} \qquad \text{Simplify.}$$

The solutions are

$$x = \frac{-3 + 3\sqrt{5}}{2} \quad \text{and} \quad x = \frac{-3 - 3\sqrt{5}}{2}.$$

Check these in the original equation.

Checkpoint 2

Solve $x^2 + 2x = 2$.

Example 3 One Repeated Solution

Solve $8x^2 - 24x + 18 = 0$.

SOLUTION Begin by dividing each side by the common factor 2.

$$8x^2 - 24x + 18 = 0 \qquad \text{Write original equation.}$$

$$4x^2 - 12x + 9 = 0 \qquad \text{Divide each side by 2.}$$

$$x = \frac{-(-12) \pm \sqrt{(-12)^2 - 4(4)(9)}}{2(4)} \qquad \text{Quadratic Formula}$$

$$x = \frac{12 \pm \sqrt{0}}{8} \qquad \text{Simplify.}$$

$$x = \frac{3}{2} \qquad \text{Repeated solution}$$

The only solution is

$$x = \tfrac{3}{2}.$$

Check this in the original equation.

Checkpoint 3

Solve $27x^2 - 18x = -3$.

The discriminant in Example 3 is a perfect square (zero in this case), and you could have factored the quadratic as

$$4x^2 - 12x + 9 = 0$$
$$(2x - 3)^2 = 0$$

and concluded that the solution is $x = \frac{3}{2}$. Because factoring is easier than applying the Quadratic Formula, try factoring first when solving a quadratic equation. If factors cannot easily be found, then use the Quadratic Formula. For instance, try solving the quadratic equation

$$x^2 - x - 12 = 0$$

in two ways—by factoring and by the Quadratic Formula—to see that you get the same solutions either way.

When using a calculator with the Quadratic Formula, you should get in the habit of using the memory key to store intermediate steps. This will save steps and minimize roundoff error.

Example 4 Using a Calculator with the Quadratic Formula

Solve $16.3x^2 - 197.6x + 7.042 = 0$.

SOLUTION In this case, $a = 16.3$, $b = -197.6$, $c = 7.042$, and you have

$$x = \frac{-(-197.6) \pm \sqrt{(-197.6)^2 - 4(16.3)(7.042)}}{2(16.3)}.$$

To evaluate these solutions, begin by calculating the square root of the discriminant, as follows.

Scientific Calculator Keystrokes

197.6 [+/−] [x^2] [−] 4 [×] 16.3 [×] 7.042 [=] [√]

Graphing Calculator Keystrokes

[√] [(] [(] [(−)] 197.6 [)] [x^2] [−] 4 [×] 16.3 [×] 7.042 [)] [ENTER]

In either case, the result is 196.434777. Storing this result and using the recall key, you can find the following two solutions.

$$x \approx \frac{197.6 + 196.434777}{2(16.3)} \approx 12.087 \qquad \text{Add stored value.}$$

$$x \approx \frac{197.6 - 196.434777}{2(16.3)} \approx 0.036 \qquad \text{Subtract stored value.}$$

✓ Checkpoint 4

Solve $4.7x^2 - 3.2x - 5.9 = 0$.

TECH TUTOR

Try to calculate the value of x in Example 4 by using additional parentheses instead of storing the intermediate result, 196.434777, in your calculator.

Applications

In Section 1.3, you studied four basic types of applications involving quadratic equations: area, falling bodies, the Pythagorean Theorem, and quadratic models. The solution to each of these types of problems can involve the Quadratic Formula. For instance, Example 5 shows how the Quadratic Formula can be used to analyze a quadratic model for a patient's blood oxygen level.

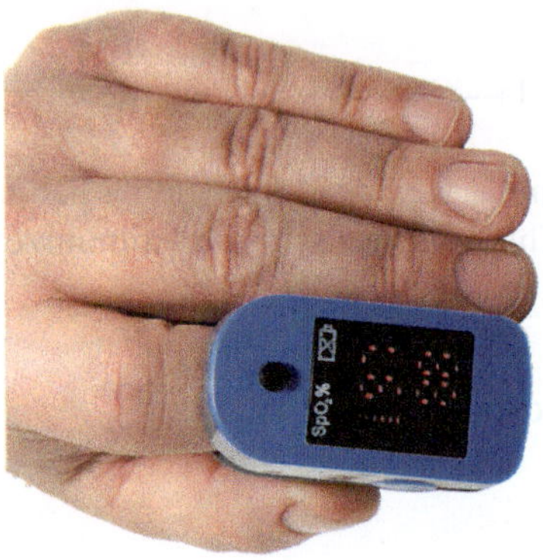

A pulse oximeter helps monitor the oxygen saturation of an individual's blood, as well as his or her heart rate.

Example 5 Blood Oxygen Level

Doctors treated a patient at an emergency room from 1:00 p.m. to 5:00 p.m. The patient's blood oxygen level L (in percent) during this time period can be modeled by

$$L = -0.25t^2 + 3.0t + 87, \quad 1 \le t \le 5$$

where t represents the time of day, with $t = 1$ corresponding to 1:00 p.m. Use the model to estimate the time when the patient's blood oxygen level was 95%.

SOLUTION To find the hour when the patient's blood oxygen level was 95%, solve the equation

$$95 = -0.25t^2 + 3.0t + 87.$$

To begin, write the equation in general form.

$$-0.25t^2 + 3.0t - 8 = 0$$

Then apply the Quadratic Formula with $a = -0.25$, $b = 3.0$, and $c = -8$.

$$\begin{aligned} t &= \frac{-3 \pm \sqrt{3^2 - 4(-0.25)(-8)}}{2(-0.25)} \\ &= \frac{-3 \pm \sqrt{9 - 8}}{-0.5} \\ &= \frac{-3 \pm \sqrt{1}}{-0.5} \\ &= 4 \text{ or } 8 \end{aligned}$$

Of the two possible solutions, only $t = 4$ makes sense in the context of the problem because $t = 8$ is not in the domain of L. Because $t = 1$ corresponds to 1:00 P.M., it follows that $t = 4$ corresponds to 4:00 P.M. So, from the model you can conclude that the patient's blood oxygen level was 95% at about 4:00 P.M. Figure 1.11 shows the patient's oxygen level recorded every 30 minutes.

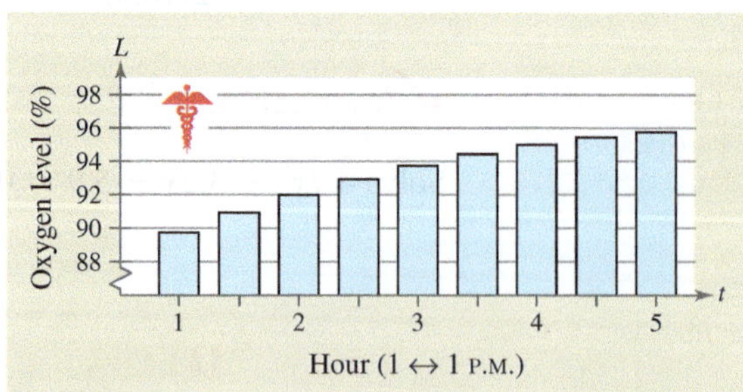

FIGURE 1.11

✓ Checkpoint 5

Use the model in Example 5 to estimate the time when the patient's blood oxygen level was 92%.

STUDY TIP

Note in the position equation

$$s = -16t^2 + v_0t + s_0$$

that the initial velocity v_0 is positive when an object is rising and negative when an object is falling.

TECH TUTOR

Use the *last entry* feature of your graphing calculator to find the time in the air on Earth for the rock in Example 6. Simply replace -2.7 with -16 in the position equation.

In Section 1.3, you learned that the position equation for a falling object is of the form

$$s = -16t^2 + v_0t + s_0$$

where s is the height (in feet) of the object, v_0 is the initial velocity (in feet per second), t is the time (in seconds), and s_0 is the initial height (in feet). This equation is valid only for free-falling objects near Earth's surface. Because of differences in gravitational force, position equations are different on other planets or moons. The next example looks at a position equation for a falling object on our moon.

Example 6 Throwing an Object on the Moon

An astronaut standing on the surface of the moon throws a rock straight up at 27 feet per second from a height of 6 feet. The height s (in feet) of the rock is given by

$$s = -2.7t^2 + 27t + 6$$

where t is the time (in seconds). How much time elapses before the rock strikes the lunar surface?

SOLUTION Because s gives the height of the rock at any time t, you can find the time that the rock hits the surface of the moon by setting s equal to zero and solving for t.

$$-2.7t^2 + 27t + 6 = 0 \qquad \text{Substitute 0 for } s.$$

$$t = \frac{-27 \pm \sqrt{(27)^2 - 4(-2.7)(6)}}{2(-2.7)} \qquad \text{Quadratic Formula}$$

$$= \frac{-27 \pm \sqrt{793.8}}{-5.4} \qquad \text{Simplify.}$$

$$\approx 10.2 \text{ seconds} \qquad \text{Choose positive solution.}$$

So, about 10.2 seconds elapse before the rock hits the lunar surface.

✓ Checkpoint 6

In Example 6, suppose the rock is thrown straight up at 13 feet per second from a height of 4 feet. The height s (in feet) of the rock is given by

$$s = -2.7t^2 + 13t + 4.$$

How much time (in seconds) elapses before the rock strikes the lunar surface? ■

SUMMARIZE (Section 1.4)

1. State the definition of the discriminant of a quadratic expression and describe how it is used to determine the number of real solutions of a quadratic equation *(page 105)*. For an example of using the discriminant, see Example 1.
2. Describe how to solve a quadratic equation using the Quadratic Formula *(pages 106 and 107)*. For examples that use the Quadratic Formula, see Examples 2, 3, and 4.
3. Describe a real-life example that uses the Quadratic Formula *(pages 108 and 109, Examples 5 and 6)*.

Kurhan/Shutterstock.com

SKILLS WARM UP 1.4 The following warm-up exercises involve skills that were covered in earlier sections. You will use these skills in the exercise set for this section. For additional help, review Sections 0.4 and 1.3.

In Exercises 1–4, simplify the expression.

1. $\sqrt{25 - 4(3)(-12)}$

2. $\sqrt{49 - 4(2)(3)}$

3. $\sqrt{12^2 - 4(3)(4)}$

4. $\sqrt{15^2 + 4(9)(12)}$

In Exercises 5–10, solve the quadratic equation by factoring.

5. $x^2 - x - 2 = 0$

6. $2x^2 + 3x - 9 = 0$

7. $x^2 - 4x = 5$

8. $2x^2 + 13x = 7$

9. $x(x - 7) = -10$

10. $x(x - 3) = 4$

Exercises 1.4

See www.CalcChat.com for worked-out solutions to odd-numbered exercises.

Using the Discriminant In Exercises 1–8, use the discriminant to determine the number of real solutions of the quadratic equation. *See Example 1.*

1. $x^2 - 4x + 4 = 0$

2. $2x^2 - x - 1 = 0$

3. $3x^2 + 4x + 1 = 0$

4. $x^2 + 2x + 4 = 0$

5. $2x^2 - 5x = -5$

6. $3 - 6x = -3x^2$

7. $\frac{1}{5}x^2 + \frac{6}{5}x - 8 = 0$

8. $\frac{1}{3}x^2 - 5x + 25 = 0$

Using the Quadratic Formula In Exercises 9–30, use the Quadratic Formula to solve the quadratic equation. *See Examples 2 and 3.*

9. $2x^2 + x - 1 = 0$

10. $2x^2 - x - 1 = 0$

11. $16x^2 + 8x - 3 = 0$

12. $25x^2 - 20x + 3 = 0$

13. $6x = 4 - x^2$

14. $2 + 2x = x^2$

15. $x^2 - 10x + 22 = 0$

16. $x^2 + 14x + 44 = 0$

17. $x^2 + 8x - 4 = 0$

18. $4x^2 - 4x - 4 = 0$

19. $12x - 9x^2 = -3$

20. $16x^2 + 22 = 40x$

21. $36x^2 + 24x = 7$

22. $3x + x^2 = 1$

23. $4x^2 + 4x = 7$

24. $16x^2 - 40x = -5$

25. $28x - 49x^2 = 4$

26. $9x^2 + 24x = -16$

27. $8t = 5 + 2t^2$

28. $25h^2 + 61 = -80h$

29. $(y - 5)^2 = 2y$

30. $(x + 6)^2 = -2x$

Using a Calculator In Exercises 31–36, use a calculator to solve the quadratic equation. (Round your answer to three decimal places.) *See Example 4.*

31. $5.1x^2 - 1.7x - 3.2 = 0$

32. $-0.067x^2 - 0.852x + 1.277 = 0$

33. $12.67x^2 + 31.55x + 8.09 = 0$

34. $422x^2 - 506x - 347 = 0$

35. $-0.003x^2 + 0.025x - 0.98 = 0$

36. $-0.005x^2 + 0.101x - 0.193 = 0$

Choosing a Method In Exercises 37–46, solve the quadratic equation using any convenient method. *See Examples 2, 3, and 4.*

37. $2x^2 + 7 = 2x^2 - x - 4$

38. $x^2 - 2x + 5 = x^2 - 5$

39. $4x^2 - 15 = 25$

40. $3x^2 - 16 = 38$

41. $x^2 - 2x - 1 = 0$

42. $x^2 + 3x - 4 = 0$

43. $(x - 1)^2 = 9$

44. $(x - 5)^2 = 16$

45. $3x^2 + 5x - 11 = 4(x - 2)$

46. $2x^2 + 4x - 9 = 2(x - 1)$

Writing Real-Life Problems In Exercises 47–50, solve the number problem *and* write a real-life problem that could be represented by this verbal model. For instance, an applied problem that could be represented by Exercise 47 is shown below.

The sum of the length and width of a one-story house is 100 feet. The house has 2500 square feet of floor space. What are the length and width of the house?

47. Find two numbers whose sum is 100 and whose product is 2500.

48. One number is 1 more than another number. The product of the two numbers is 72. Find the numbers.

49. One number is 1 more than another number. The sum of their squares is 113. Find the numbers.

50. One number is 2 more than another number. The product of the two numbers is 440. Find the numbers.

Cost Equation In Exercises 51–54, use the cost equation to find the number of units x that a manufacturer can produce for cost C. (Round your answer to the nearest positive integer.)

51. $C = 0.125x^2 + 20x + 5000,\quad C = \$14{,}000$

52. $C = 0.5x^2 + 15x + 5000,\quad C = \$11{,}500$

53. $C = 800 + 0.04x + 0.002x^2, \quad C = \$1680$

54. $C = 312.5 - 10x + 0.4x^2, \quad C = \$900$

55. Seating A lecture hall has 108 chairs arranged in rows with the same number of chairs in each row. You eliminate three rows by adding six chairs to each of the other rows. How many rows are there now? How many chairs are in each row?

56. Child Care A child care center has 200 feet of fencing to enclose two adjacent rectangular safe play areas (see figure).

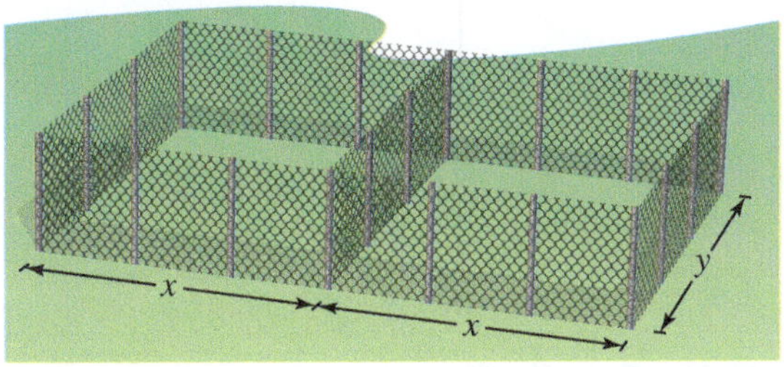

(a) Write an expression for the total area A of the play areas in terms of x.

(b) Find the dimensions for which the enclosed area is 1600 square feet.

57. Geometry An open box is made from a square piece of material by cutting two-inch squares from the corners and turning up the sides (see figure). The volume of the finished box is 200 cubic inches. Find the size of the original piece of material.

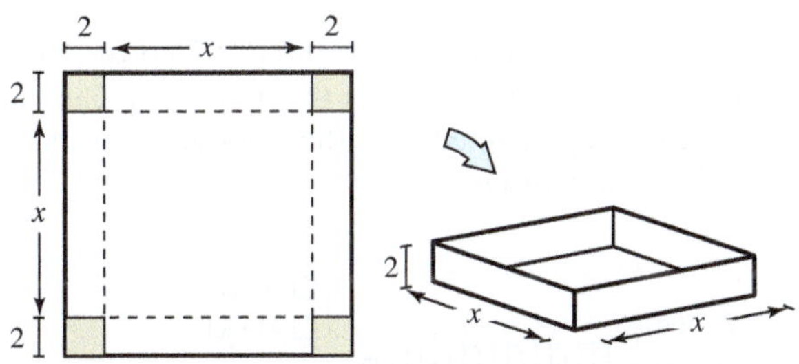

58. Geometry The swimming pool shown is 4 feet deep and has a volume of 1024 cubic feet. Find the dimensions of the base of the pool.

59. Stratosphere You are at the Stratosphere Tower in Las Vegas. While riding the X-Scream tower ride, you accidently drop a coin from a height of 866 feet with an initial velocity of -5 feet per second. How long does it take the coin to hit the ground?

60. Sports You throw a baseball straight up into the air at a velocity of 80 feet per second. You release the baseball at a height of 5.5 feet and catch it when it falls back to a height of 5.5 feet.

(a) Use the position equation to write a mathematical model for the height of the baseball.

(b) Find the height of the baseball after 1 second, 2 seconds, and 3 seconds. What must have occurred sometime in the interval $2 \le t \le 3$? Explain.

(c) For how many seconds is the baseball in the air?

61. On the Moon An astronaut on the moon throws a rock straight upward into space. The height s (in feet) of the rock is given by

$$s = -2.7t^2 + 40t + 5$$

where the initial velocity is 40 feet per second, the initial height is 5 feet, and t is the time (in seconds). How long will it take the rock to hit the surface of the moon? Suppose a rock is thrown with the same initial velocity and height on Earth. How long will it take the rock to hit the ground?

62. Hot Air Balloon Two people are floating in a hot air balloon 200 feet above a lake. One person tosses out a coin with an initial velocity of 20 feet per second. One second later, the balloon is 20 feet higher and the other person drops another coin (see figure). The position equation for the first coin is $s = -16t^2 + 20t + 200$, and the position equation for the second coin is $s = -16t^2 + 220$. Which coin will hit the water first? (*Hint:* Remember that the first coin was tossed one second before the second coin was dropped.)

63. Comparing Gravity Does a rock thrown upward from an initial height of 6 feet with an initial velocity of 27 feet per second take longer to reach the ground on Earth or on the moon? (See Example 6.)

64. Comparing Gravity Does a rock thrown downward from an initial height of 6 feet with an initial velocity of -14 feet per second take longer to reach the ground on Earth or on the moon? (See Example 6.)

65. Aviation A commercial jet flies to three cities whose locations form the vertices of a right triangle (see figure). The total flight distance (from Oklahoma City to Austin to New Orleans and back to Oklahoma City) is approximately 1348 miles. It is 560 miles between Oklahoma City and New Orleans. Approximate the other two distances.

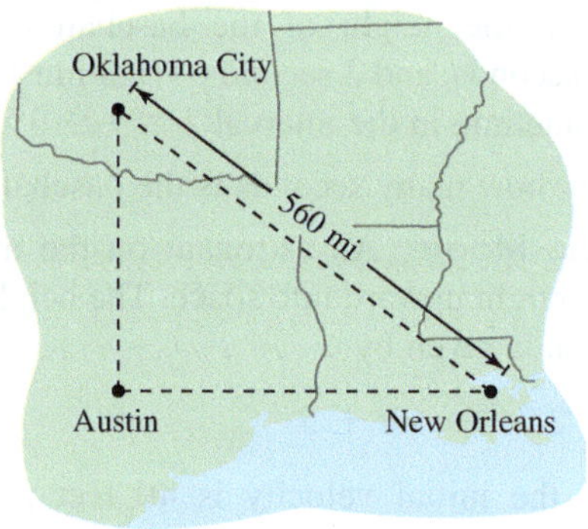

66. Distance from a Dock A windlass is used to tow a boat to a dock. The figure shows the boat when it is 75 feet from the windlass. How far is the boat from the dock?

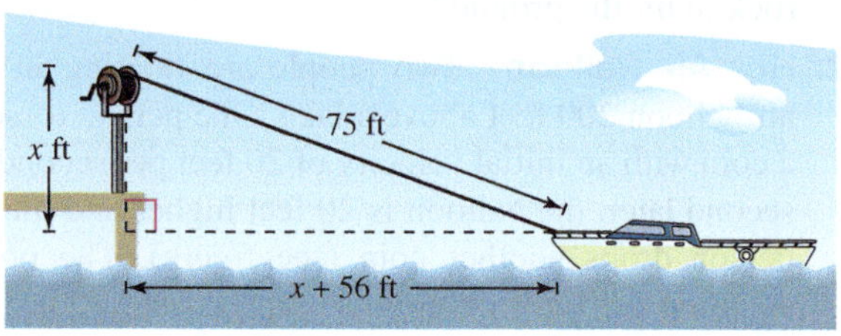

67. Blood Oxygen Level Doctors treated a patient at an emergency room from 2:00 P.M. to 7:00 P.M. The patient's blood oxygen level L (in percent) during this time period can be modeled by

$$L = -0.270t^2 + 3.59t + 83.1, \quad 2 \le t \le 7$$

where t represents the time of day, with $t = 2$ corresponding to 2:00 P.M. Use the model to estimate the time (rounded to the nearest hour) when the patient's blood oxygen level was 93%.

68. Green Mountain Coffee Roasters The sales S (in millions of dollars) for Green Mountain Coffee Roasters from 2004 through 2009 can be approximated by

$$S = 31.61t^2 - 283.4t + 776, \quad 4 \le t \le 9$$

where t represents the year, with $t = 4$ corresponding to 2004. Use the model to estimate the year when sales were about \$340 million. *(Source: Green Mountain Coffee Roasters, Inc.)*

69. Flying Speed Two planes leave simultaneously from the same airport, one flying due east and the other due south (see figure). The eastbound plane is flying 50 miles per hour faster than the southbound plane. After 3 hours the planes are 2440 miles apart. Find the speed of each plane.

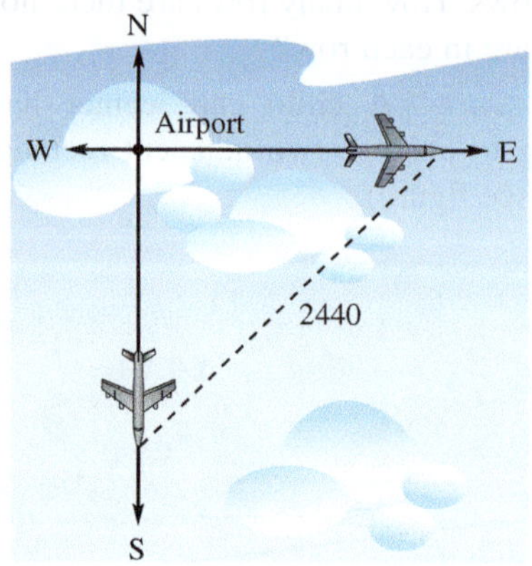

70. Flying Speed Two planes leave simultaneously from the same airport, one flying due east and the other due south. The eastbound plane is flying 100 miles per hour faster than the southbound plane. After 2 hours the planes are 1500 miles apart. Find the speed of each plane.

71. VCRs The percent P of homes in the United States with VCRs from 2002 through 2011 can be approximated by the model

$$P = -0.496t^2 + 3.01t + 87.2, \quad 2 \le t \le 11$$

where t represents the year, with $t = 2$ corresponding to 2002. The bar graph shows the actual percents and the percents represented by the model. *(Source: The Nielsen Company)*

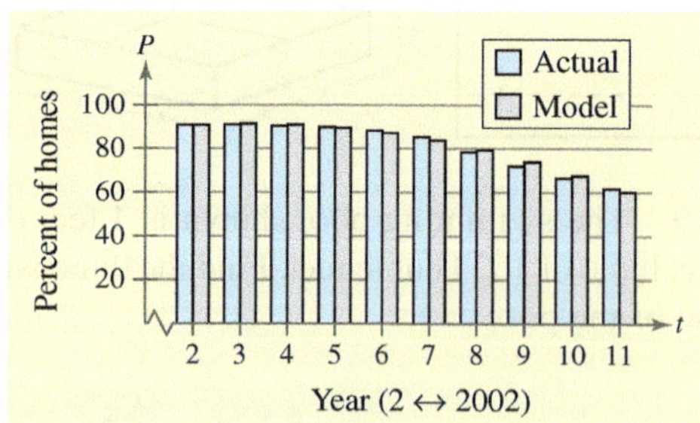

(a) Use the model to estimate the year when the percent of homes with VCRs was about 60%.

(b) Use the model to estimate the year when the percent of homes with VCRs was about 80%.

(c) Use the model to predict the last year that VCRs will exist in U.S. households. Would you say that the model is valid to represent data up to that year?

72. HOW DO YOU SEE IT? The bar graph shows the growth of a baby girl. A model that approximates her body length l (in centimeters) at m months of age is

$$l = -0.0813m^2 + 3.028m + 50.0, \quad 0 \le m \le 15.$$

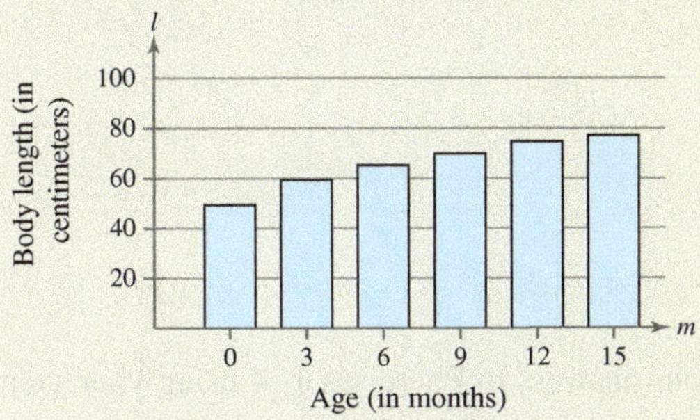

(a) During which 3-month period shown in the bar graph did the baby girl grow the most?

(b) For $l = 50$ centimeters, the model simplifies to

$$0 = -0.0813m^2 + 3.028m.$$

Is the discriminant positive, negative, or zero?

(c) Based on your answer in part (b) and the bar graph, will the original model work well for predicting the girl's height as she gets older? Explain.

73. Biology The metabolic rate of an ectothermic organism increases with increasing temperature within a certain range. The graph shows experimental data for the oxygen consumption C (in microliters per gram per hour) of a beetle at certain temperatures. The data can be approximated by the model

$$C = 0.45x^2 - 1.65x + 50.75, \quad 10 \le x \le 25$$

where x is the temperature in degrees Celsius.

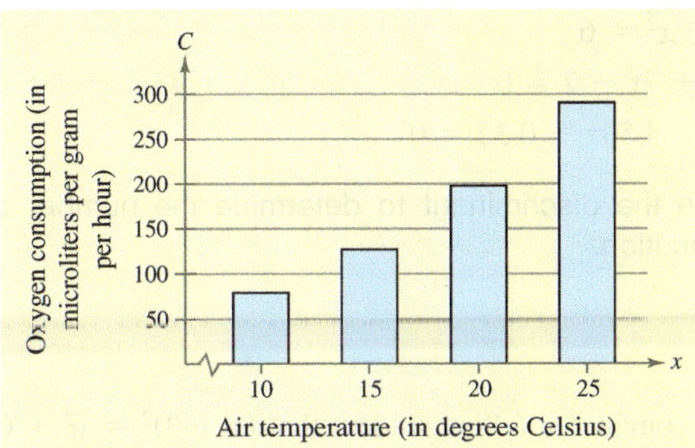

(a) The oxygen consumption is 150 microliters per gram per hour. What is the air temperature?

(b) The temperature is increased from 10°C to 20°C. The oxygen consumption is increased by approximately what factor?

74. Total Revenue The demand equation for a product is

$$p = 60 - 0.0004x$$

where p is the price per unit and x is the number of units sold. The total revenue R from selling x units is given by

$$R = xp.$$

How many units must be sold to produce a revenue of \$220,000?

75. Total Revenue The demand equation for a product is

$$p = 50 - 0.0005x$$

where p is the price per unit and x is the number of units sold. The total revenue R from selling x units is given by

$$R = xp.$$

How many units must be sold to produce a revenue of \$250,000?

76. Think About It When the Quadratic Formula is used to solve certain problems, such as the problem in Example 5 on page 108, why is only one solution used?

77. Think About It The discriminants of two quadratic equations are 5 and -10. Can the equations have the same solutions? Explain.

78. Writing List the methods you have learned for solving quadratic equations. Identify the type of quadratic equation that is most easily solved by each method. Give an example showing the solutions for each type.

79. Project: Revenue For a project involving the revenues for Buffalo Wild Wings from 2002 to 2010, visit this text's website at *www.cengagebrain.com*. *(Source: Buffalo Wild Wings, Inc.)*

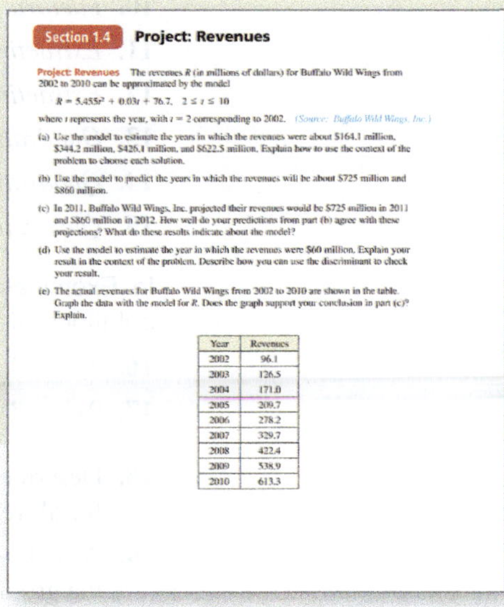

Section 1.4 **Project: Revenues**

Project: Revenues The revenues R (in millions of dollars) for Buffalo Wild Wings from 2002 to 2010 can be approximated by the model

$$R = 5.455t^2 + 0.03t + 76.7, \quad 2 \le t \le 10$$

where t represents the year, with $t = 2$ corresponding to 2002. *(Source: Buffalo Wild Wings, Inc.)*

(a) Use the model to estimate the years in which the revenues were about \$164.1 million, \$344.2 million, \$426.1 million, and \$622.5 million. Explain how to use the context of the problem to choose each solution.

(b) Use the model to predict the years in which the revenues will be about \$725 million and \$860 million.

(c) In 2011, Buffalo Wild Wings, Inc. projected their revenues would be \$725 million in 2011 and \$860 million in 2012. How well do your predictions from part (b) agree with these projections? What do these results indicate about the model?

(d) Use the model to estimate the year in which the revenues were \$60 million. Explain your result in the context of the problem. Describe how you can use the discriminant to check your result.

(e) The actual revenues for Buffalo Wild Wings from 2002 to 2010 are shown in the table. Graph the data with the model for R. Does the graph support your conclusion in part (c)? Explain.

Year	Revenues
2002	96.1
2003	126.5
2004	171.0
2005	209.7
2006	278.2
2007	329.7
2008	422.4
2009	538.9
2010	613.3

QUIZ YOURSELF

See www.CalcChat.com for worked-out solutions to odd-numbered exercises.

Take this quiz as you would take a quiz in class. When you are done, check your work against the answers given in the back of the book.

In Exercises 1–4, solve the equation and check your solution.

1. $3(x - 2) - 4(2x + 5) = 4$

2. $\frac{3x + 3}{5x - 2} = \frac{3}{4}$

3. $\frac{2}{x(x - 1)} + \frac{1}{x} = \frac{1}{x - 4}$

4. $(x + 3)^2 - x^2 = 6(x + 2)$

5. Describe how you can check your answers to Exercises 1–4 using your graphing utility.

In Exercises 6 and 7, use a calculator to solve the equation. (Round your solution to three decimal places.)

6. $\frac{x}{2.004} - \frac{x}{5.128} = 100$

7. $0.378x + 0.757(500 - x) = 215$

In Exercises 8 and 9, write an algebraic equation for the verbal description. Find the solution if possible and check.

8. A company has fixed costs of \$30,000 per month and variable costs of \$8.50 per unit manufactured. The company has \$200,000 available each month to cover monthly costs. How many units can the company manufacture?

9. The demand equation for a product is $p = 75 - 0.0002x$, where p is the price per unit and x is the number of units sold. The total revenue R from selling x units is given by $R = xp$. How many units must be sold to produce a revenue of \$300,000?

In Exercises 10–15, solve the quadratic equation by the indicated method.

10. *Factoring:* $3x^2 + 13x = 10$

11. *Extracting square roots:* $3x^2 = 15$

12. *Extracting square roots:* $(x + 3)^2 = 17$

13. *Quadratic Formula:* $2x + x^2 = 6$

14. *Quadratic Formula:* $3x^2 + 7x - 2 = 0$

15. *Quadratic Formula:* $3x^2 - 4.50x - 0.32 = 0$

In Exercises 16 and 17, use the discriminant to determine the number of real solutions of the quadratic equation.

16. $2x^2 - 4x + 9 = 0$

17. $9x^2 - 3x + \frac{1}{4} = 0$

18. Describe how you would convince a fellow student that $(x + 3)^2 = x^2 + 6x + 9$ for all real values of x.

19. A rock is dropped from a height of 324 feet. How long will it take the rock to hit the ground?

20. An open box has a square base and a height of 6 inches. The volume of the box is 486 cubic inches. Find the dimensions of the box.

1.6 Linear Inequalities

In Exercise 85 on page 136, you will use an inequality to predict when the average National Football League player's salary will exceed \$2.2 million.

- Write bounded and unbounded intervals using inequalities or interval notation.
- Develop an understanding of the properties of inequalities.
- Solve and graph a linear inequality.
- Solve an inequality involving absolute value.
- Construct and use a linear inequality to solve an application problem.

Introduction

Simple inequalities are used to *order* real numbers. The inequality symbols $<$, $\leq$, $>$, and $\geq$ are used to compare two numbers and to denote subsets of real numbers. For instance, the simple inequality

$$x \geq 3$$

denotes all real numbers x that are greater than or equal to 3.

In this section, you will expand your work with inequalities to include more involved statements such as

$$5x - 7 > 3x + 9$$

and

$$-3 \leq 6x - 1 < 3.$$

As with an equation, you **solve an inequality** in the variable x by finding all values of x for which the inequality is true. Such values are **solutions** and are said to **satisfy** the inequality. The set of all real numbers that are solutions of an inequality is the **solution set** of the inequality. For instance, the solution set of

$$x + 3 > 4$$

is all real numbers that are greater than 1.

The set of all points on the real number line that represent the solution set of an inequality is the **graph** of the inequality. Graphs of many types of inequalities consist of intervals on the real number line. The four different types of **bounded** intervals are summarized below.

Bounded Intervals on the Real Number Line

Let a and b be real numbers such that $a < b$. The following intervals on the real number line are **bounded.** The numbers a and b are the **endpoints** of each interval.

Notation	*Interval Type*	*Inequality*	*Graph*
$[a, b]$	Closed	$a \leq x \leq b$	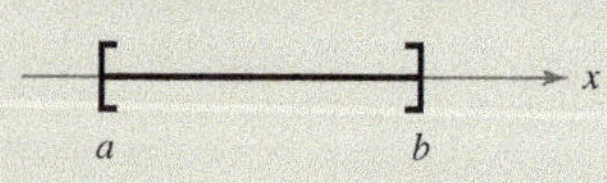
(a, b)	Open	$a < x < b$	
$[a, b)$		$a \leq x < b$	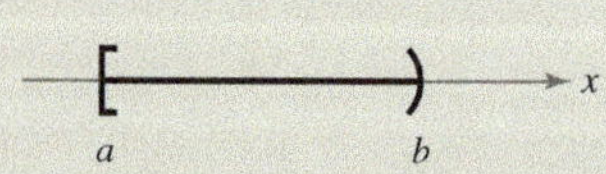
$(a, b]$		$a < x \leq b$	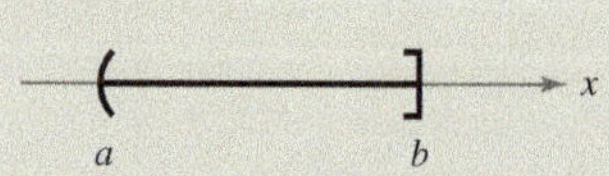

Benjamin Thorn/Shutterstock.com

Note that a closed interval contains both of its endpoints and an open interval does not contain either of its endpoints. Often, the solution of an inequality is an interval on the real line that is **unbounded.** For instance, the interval consisting of all positive numbers is unbounded. The symbols ∞, **positive infinity,** and $-\infty$, **negative infinity,** do not represent real numbers. They are simply convenient symbols used to describe the unboundedness of an interval such as $(1, \infty)$.

Unbounded Intervals on the Real Number Line

Let a and b be real numbers. The following intervals on the real number line are **unbounded.**

Notation	*Interval Type*	*Inequality*	*Graph*
$[a, \infty)$		$x \geq a$	
(a, ∞)	Open	$x > a$	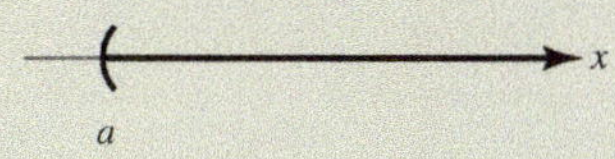
$(-\infty, b]$		$x \leq b$	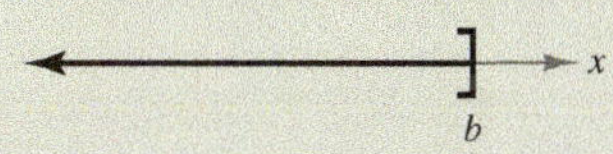
$(-\infty, b)$	Open	$x < b$	
$(-\infty, \infty)$	Entire real line	$-\infty < x < \infty$	

Example 1 **Intervals and Inequalities**

Write an inequality to represent each interval. Then state whether the interval is bounded or unbounded.

a. $(-3, 5]$

b. $(-4, \infty)$

c. $[0, 2]$

d. $(-\infty, 0)$

SOLUTION

a. $(-3, 5]$ corresponds to $-3 < x \leq 5$. Bounded

b. $(-4, \infty)$ corresponds to $x > -4$. Unbounded

c. $[0, 2]$ corresponds to $0 \leq x \leq 2$. Bounded

d. $(-\infty, 0)$ corresponds to $x < 0$. Unbounded

✓ Checkpoint 1

Write an inequality to represent each interval. Then state whether the interval is bounded or unbounded.

a. $[2, 7)$

b. $(-\infty, 6)$

Properties of Inequalities

The procedures for solving linear inequalities in one variable are much like those for solving linear equations. To isolate the variable, you can make use of the **properties of inequalities.** These properties are similar to the properties of equality, but there are two important exceptions. When each side of an inequality is multiplied or divided by a negative number, the direction of the inequality symbol must be reversed. Here is an example.

$-2 < 5$ Original inequality

$(-3)(-2) > (-3)(5)$ Multiply each side by -3 and reverse the inequality symbol.

$6 > -15$ Simplify.

Notice in the example above that when the inequality symbol is not reversed, you obtain the false statement

$6 < -15.$ False statement

Two inequalities that have the same solution set are **equivalent.** For instance, the inequalities

$x + 2 < 5$ and $x < 3$

are equivalent. To obtain the second inequality from the first, you can subtract 2 from each side of the inequality. The following list describes operations that can be used to create equivalent inequalities.

Properties of Inequalities

Let a, b, c, and d be real numbers.

1. *Transitive Property*

$a < b$ and $b < c \Rightarrow a < c$

2. *Addition of Inequalities*

$a < b$ and $c < d \Rightarrow a + c < b + d$

3. *Addition of a Constant*

$a < b \Rightarrow a + c < b + c$

4. *Multiplication by a Constant*

For $c > 0$, $a < b \Rightarrow ac < bc$

For $c < 0$, $a < b \Rightarrow ac > bc$ Reverse direction of inequality symbol.

Each of the properties above is true if the symbol $<$ is replaced by $\leq$ and the symbol $>$ is replaced by $\geq$. For instance, another form of the multiplication property would be as shown.

For $c > 0$, $a \leq b \Rightarrow ac \leq bc$

For $c < 0$, $a \leq b \Rightarrow ac \geq bc$

On your own, try to verify each of the properties of inequalities by using several examples with real numbers.

Solving a Linear Inequality

The simplest type of inequality to solve is a **linear inequality** in a single variable. For instance,

$$2x + 3 > 4$$

is a linear inequality in x.

As you read through the following examples, pay special attention to the steps in which the inequality symbol is reversed. Remember that when you multiply or divide by a negative number, you must reverse the inequality symbol.

ALGEBRA TUTOR *xy*

For help in solving inequalities similar to the one in Example 2, see the review of solving inequalities on page 149.

Example 2 Solving a Linear Inequality

Solve $5x - 7 > 3x + 9$.

SOLUTION

$5x - 7 > 3x + 9$ Write original inequality.

$2x - 7 > 9$ Subtract $3x$ from each side.

$x > 8$ Add 7 to each side and then divide each side by 2.

The solution set is all real numbers that are greater than 8, which is denoted by $(8, \infty)$. The graph is shown in Figure 1.13.

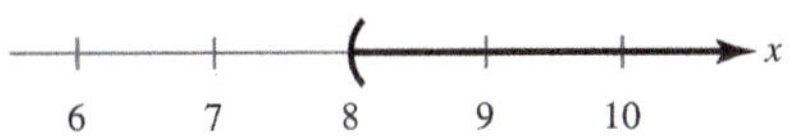

Solution Interval: $(8, \infty)$
FIGURE 1.13

✓ Checkpoint 2

Solve $3x < 2x + 1$.

Checking the solution set of an inequality is not as simple as checking the solutions of an equation. You can, however, get an indication of the validity of a solution set by substituting a few convenient values of x to see whether the original inequality is satisfied. For instance, substitute several values of x that are greater than 8 into the original inequality in Example 2 and check to see if the inequality is satisfied.

Example 3 Solving a Linear Inequality

Solve $1 - \frac{3x}{2} \geq x - 4$.

SOLUTION

$1 - \frac{3x}{2} \geq x - 4$ Write original inequality.

$2 - 3x \geq 2x - 8$ Multiply each side by 2.

$2 - 5x \geq -8$ Subtract $2x$ from each side.

$-5x \geq -10$ Subtract 2 from each side.

$x \leq 2$ Divide each side by -5 and reverse inequality symbol.

The solution set is all real numbers that are less than or equal to 2, which is denoted by $(-\infty, 2]$. The graph is shown in Figure 1.14.

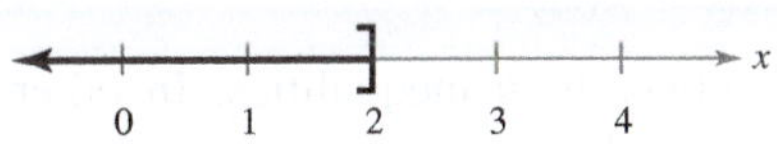

Solution Interval: $(-\infty, 2]$
FIGURE 1.14

✓ Checkpoint 3

Solve $-\frac{4x}{3} \leq 2 - x$. Then graph the solution set on the real number line.

Sometimes it is convenient to write two inequalities as a **double inequality.** For instance, you can write the two inequalities

$$-4 \le 5x - 2 \quad \text{and} \quad 5x - 2 < 7$$

more simply as

$$-4 \le 5x - 2 < 7.$$

This enables you to solve the two inequalities together, as demonstrated in Example 4.

Example 4 Solving a Double Inequality

Solve $-3 \le 6x - 1 < 3$.

SOLUTION To solve a double inequality, you can isolate x as the middle term.

$$-3 \le 6x - 1 < 3 \qquad \text{Write original inequality.}$$

$$-3 + 1 \le 6x - 1 + 1 < 3 + 1 \qquad \text{Add 1 to each part.}$$

$$-2 \le 6x < 4 \qquad \text{Simplify.}$$

$$\frac{-2}{6} \le \frac{6x}{6} < \frac{4}{6} \qquad \text{Divide each part by 6.}$$

$$-\frac{1}{3} \le x < \frac{2}{3} \qquad \text{Simplify.}$$

The solution set is all real numbers that are greater than or equal to $-\frac{1}{3}$ and less than $\frac{2}{3}$. The interval notation for this solution set is $\left[-\frac{1}{3}, \frac{2}{3}\right)$. The graph of this solution set is shown in Figure 1.15.

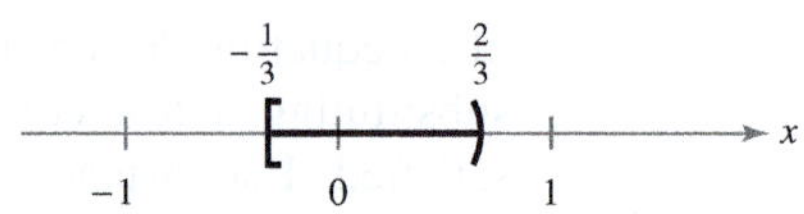

Solution Interval: $\left[-\frac{1}{3}, \frac{2}{3}\right)$

FIGURE 1.15

✓Checkpoint 4

Solve $-1 < 3 - 2x \le 5$. Then graph the solution set on the real number line.

The double inequality in Example 4 could have been solved in two parts as follows.

$$-3 \le 6x - 1 \quad \text{and} \quad 6x - 1 < 3$$

$$-2 \le 6x \qquad\qquad 6x < 4$$

$$-\frac{1}{3} \le x \qquad\qquad x < \frac{2}{3}$$

The solution set consists of all real numbers that satisfy *both* inequalities. In other words, the solution set is the set of all values of x for which

$$-\tfrac{1}{3} \le x < \tfrac{2}{3}.$$

When combining two inequalities to form a double inequality, be sure that the inequalities satisfy the Transitive Property. For instance, it is *incorrect* to combine the inequalities $3 < x$ and $x \le -1$ as $3 < x \le -1$. This is obviously wrong because 3 is not less than -1.

Inequalities Involving Absolute Value

STUDY TIP

When an absolute value inequality is solved, the solution may consist of a bounded interval on the real number line (Example 5) or two unbounded intervals on the real number line (Example 6).

Solving an Absolute Value Inequality

Let x be a variable or an algebraic expression and let a be a real number such that $a \geq 0$.

1. The solutions of $|x| < a$ are all values of x that lie between $-a$ and a.

   $|x| < a$ if and only if $-a < x < a$.

2. The solutions of $|x| > a$ are all values of x that are less than $-a$ or greater than a.

   $|x| > a$ if and only if $x < -a$ or $x > a$.

These rules are also valid if $<$ is replaced by $\leq$ and $>$ is replaced by $\geq$.

Example 5 Solving an Absolute Value Inequality

Solve $|x - 5| < 2$.

SOLUTION

$|x - 5| < 2$ Write original inequality.

$-2 < x - 5 < 2$ Equivalent inequality

$-2 + 5 < x - 5 + 5 < 2 + 5$ Add 5 to each part.

$3 < x < 7$ Simplify.

The solution set consists of all real numbers that are greater than 3 and less than 7, which is denoted by $(3, 7)$. The graph is shown in Figure 1.16.

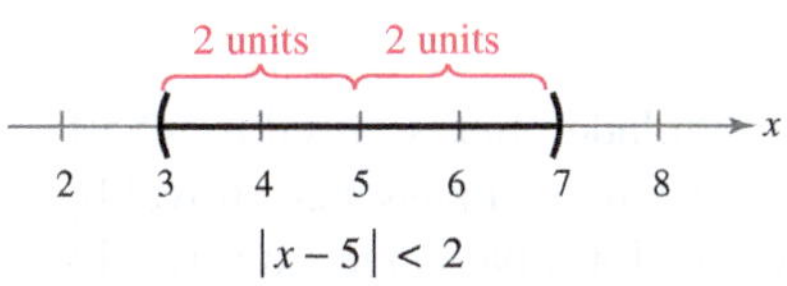

FIGURE 1.16

Checkpoint 5

Solve $|x + 2| \leq 7$. Then graph the solution set on the real number line.

Example 6 Solving an Absolute Value Inequality

Solve $|x + 3| \geq 7$.

SOLUTION

$|x + 3| \geq 7$ Write original inequality.

$x + 3 \leq -7$ or $x + 3 \geq 7$ Equivalent inequalities

$x + 3 - 3 \leq -7 - 3$ $\quad$ $x + 3 - 3 \geq 7 - 3$ Subtract 3 from each side.

$x \leq -10$ $\quad$ $x \geq 4$ Simplify.

The solution set is all real numbers that are less than or equal to -10 *or* greater than or equal to 4, which is denoted by

$$(-\infty, -10] \cup [4, \infty)$$

(see Figure 1.17). The symbol $\cup$ (union) means *or*.

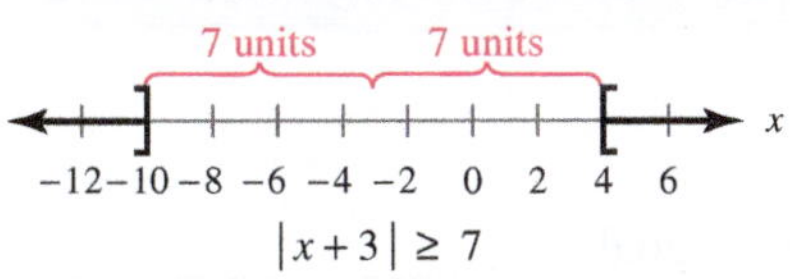

FIGURE 1.17

Checkpoint 6

Solve $|x + 1| > 3$. Then graph the solution set on the real number line.

Applications

Example 7 Comparing Taxi Fares

The cost of traveling by taxi in Las Vegas is \$3.20 for the initial charge and \$2 for each mile driven. The cost in Philadelphia is \$2.30 for the initial charge and \$2.10 for each mile driven. How many miles must a passenger travel by taxi in Philadelphia before the cost is more than the cost for a passenger to travel by taxi in Las Vegas?

SOLUTION

Verbal Model: [Cost for taxi in Philadelphia] > [Cost for taxi in Las Vegas]

Labels: Miles driven $= m$ (miles)
Cost for taxi in Las Vegas $= 3.2 + 2m$ (dollars)
Cost for taxi in Philadelphia $= 2.3 + 2.1m$ (dollars)

Inequality: $2.3 + 2.1m > 3.2 + 2m$

$$0.1m > 0.9$$

$$m > 9$$

When a passenger travels more than 9 miles, the cost for a taxi in Philadelphia is more than the cost for a taxi in Las Vegas.

✓ Checkpoint 7

In Example 7, suppose the initial charge for a taxi in Philadelphia is raised to \$2.50 and the cost per mile is raised to \$2.20. How many miles must a passenger travel by taxi in Philadelphia before the cost is more than the cost for a passenger to travel by taxi in Las Vegas?

Example 8 Exercise Program

A 225-pound man begins an exercise and diet program that is designed to reduce his weight by at least 2 pounds per week. Find the maximum number of weeks before the man's weight will reach his goal of 192 pounds.

SOLUTION

Verbal Model: [Desired weight] ≤ [Current weight] − [2 pounds per week] · [Number of weeks]

Labels: Desired weight $= 192$ (pounds)
Current weight $= 225$ (pounds)
Number of weeks $= x$ (weeks)

Inequality: $192 \le 225 - 2x$

$$-33 \le -2x$$

$$16.5 \ge x$$

It will take at most $16\frac{1}{2}$ weeks for the man to reach his goal.

✓ Checkpoint 8

In Example 8, find the maximum number of weeks before the man's weight will reach 200 pounds.

Rido/Shutterstock.com

Example 9 Accuracy of a Measurement

You go to a candy store to buy chocolates that cost \$9.89 per pound. The scale used in the store has a state seal of approval that indicates the scale is accurate to within one-half ounce. According to the scale, your purchase weighs one-half pound and costs \$4.95. How much might you have been undercharged or overcharged due to an error in the scale?

SOLUTION To solve this problem, let x represent the *true* weight of the candy. Because the scale is accurate to within one-half ounce (or $\frac{1}{32}$ pound), you can conclude that the absolute value of the difference of the exact weight (x) and the scale weight $\left(\frac{1}{2}\text{ pound}\right)$ is less than or equal to $\frac{1}{32}$ pound. That is,

$$\left|x - \frac{1}{2}\right| \le \frac{1}{32}.$$

You can solve this inequality as shown.

$$-\frac{1}{32} \le x - \frac{1}{2} \le \frac{1}{32}$$

$$\frac{15}{32} \le x \le \frac{17}{32}$$

$$0.46875 \le x \le 0.53125$$

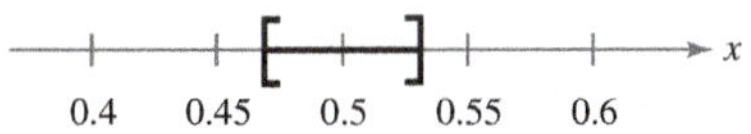

Solution Interval: [0.46875, 0.53125]
FIGURE 1.18

The graph of this solution set is shown in Figure 1.18. In other words, your "one-half" pound of candy could have weighed as little as 0.46875 pound (which would have cost $0.46875 \cdot \$9.89 \approx \$4.64$) or as much as 0.53125 pound (which would have cost $0.53125 \cdot \$9.89 \approx \$5.25$). So, you could have been undercharged by as much as \$0.30 or overcharged by as much as \$0.31.

✓ Checkpoint 9

You go to a grocery store to buy ground beef that costs \$3.96 per pound. The scale used in the store is accurate to within $\frac{1}{3}$ ounce $\left(\text{or } \frac{1}{48}\text{ pound}\right)$. According to the scale, your purchase weighs 7.5 pounds and costs \$29.70. How much might you have been undercharged or overcharged due to an error in the scale?

SUMMARIZE (Section 1.6)

1. Make a list of the different types of bounded and unbounded intervals on the real number line *(pages 126 and 127)*. For an example on bounded and unbounded intervals, see Example 1.
2. Make a list of the properties of inequalities *(page 128)*.
3. Describe how to solve a linear inequality *(pages 128 and 129)*. For examples of solving linear inequalities, see Examples 2 and 3.
4. State the definition of a double inequality *(page 130)*. For an example of solving a double inequality, see Example 4.
5. Describe how to solve an absolute value inequality *(page 131)*. For examples of solving absolute value inequalities, see Examples 5 and 6.
6. Describe a real-life example that uses an inequality *(pages 132 and 133, Examples 7, 8, and 9)*.

Iryna Rasko/Shutterstock.com

SKILLS WARM UP 1.6 The following warm-up exercises involve skills that were covered in earlier sections. You will use these skills in the exercise set for this section. For additional help, review Sections 0.1 and 0.2.

In Exercises 1–4, determine which of the two numbers is greater.

1. $-\frac{1}{2}, -7$ **2.** $-\frac{1}{3}, -\frac{1}{6}$

3. $-\pi, -3$ **4.** $-6, -\frac{13}{2}$

In Exercises 5–8, use inequality notation to describe the subset of real numbers.

5. x is nonnegative.

6. z is strictly between -3 and 10.

7. P is no more than 2.

8. W is at least 200.

In Exercises 9 and 10, evaluate the expression at each value of x.

9. $|x - 10|, x = 12, x = 3$ **10.** $|2x - 3|, x = \frac{3}{2}, x = 1$

Exercises 1.6

See www.CalcChat.com for worked-out solutions to odd-numbered exercises.

Intervals and Inequalities In Exercises 1–6, write an inequality that represents the interval. Then state whether the interval is bounded or unbounded. *See Example 1.*

1. $[-1, 5]$ **2.** $(2, 10]$

3. $(11, \infty)$

4. $[-5, \infty)$

5. $(-\infty, -2)$ **6.** $(-\infty, 7]$

Matching In Exercises 7–14, match the inequality with its graph. [The graphs are labeled (a), (b), (c), (d), (e), (f), (g), and (h).]

(a)
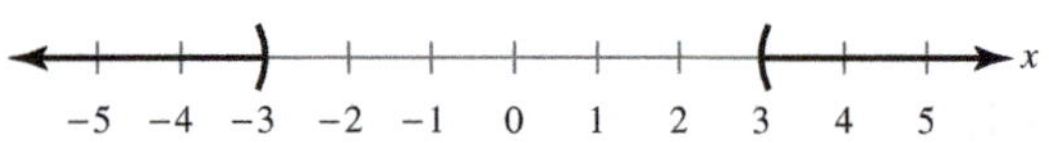

(b)
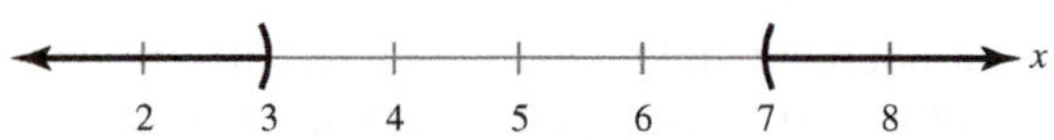

(c)
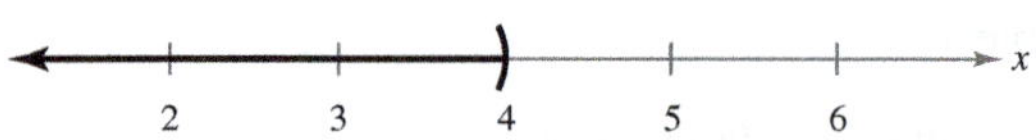

(d)
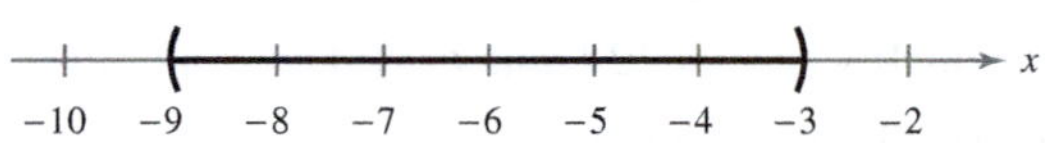

(e)
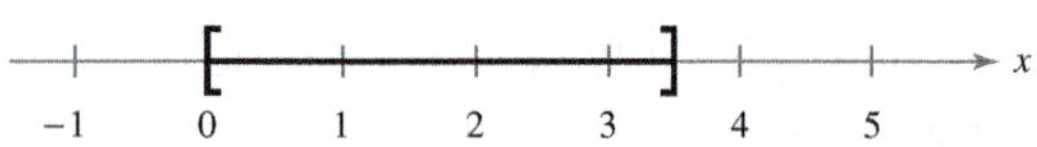

(f)
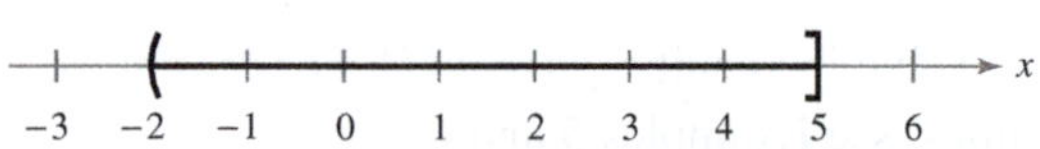

(g)
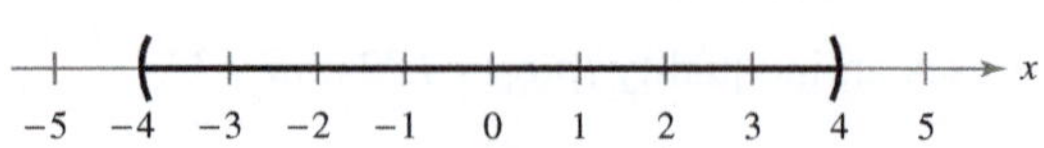

(h)
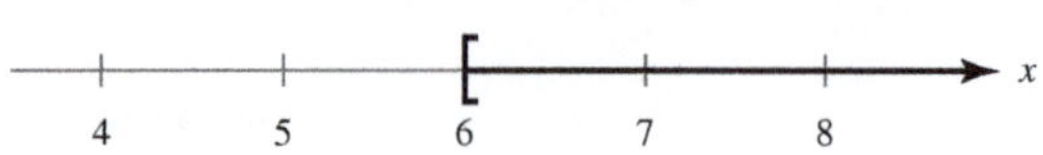

7. $x < 4$ **8.** $x \geq 6$

9. $-2 < x \leq 5$ **10.** $0 \leq x \leq \frac{7}{2}$

11. $|x| < 4$ **12.** $|x| > 3$

13. $|x - 5| > 2$ **14.** $|x + 6| < 3$

Checking Solutions In Exercises 15–22, determine whether each value of x is a solution of the inequality.

15. $5x - 12 > 0$

(a) $x = 3$ (b) $x = -3$ (c) $x = \frac{5}{2}$ (d) $x = \frac{3}{2}$

16. $4x - 5 < 3$

(a) $x = 0$ (b) $x = 2$ (c) $x = -2$ (d) $x = 4$

17. $0 < \dfrac{x - 2}{4} < 2$

(a) $x = 4$
(b) $x = 10$
(c) $x = 0$
(d) $x = \frac{7}{2}$

18. $-1 < \dfrac{3 - x}{2} \leq 1$

(a) $x = 0$
(b) $x = -5$
(c) $x = 1$
(d) $x = 5$

19. $|x + 2| \leq 10$

(a) $x = -15$
(b) $x = -4$
(c) $x = 1$
(d) $x = 8$

20. $|x - 10| \geq 3$

(a) $x = 13$
(b) $x = -1$
(c) $x = 14$
(d) $x = 9$

21. $|2x - 3| < 15$

(a) $x = -6$
(b) $x = 0$
(c) $x = 12$
(d) $x = 7$

22. $|3x + 5| > 7$

(a) $x = -5$
(b) $x = -2$
(c) $x = \frac{1}{3}$
(d) $x = 10$

Solving an Inequality In Exercises 23–30, copy and complete the statement using the correct inequality symbol.

23. If $2x > 6$, then x __?__ 3.

24. If $3x > 9$, then x __?__ 3.

25. If $2x \le -8$, then x __?__ -4.

26. If $3x \le -15$, then x __?__ -5.

27. If $2 - 4x > -10$, then x __?__ 3.

28. If $5 - 3x > -7$, then x __?__ 4.

29. If $-\frac{2}{3}x \ge -6$, then x __?__ 9.

30. If $-\frac{3}{4}x \ge -12$, then x __?__ 16.

Solving a Linear Inequality In Exercises 31–44, solve the inequality. Then graph the solution set on the real number line. *See Examples 2 and 3.*

31. $\frac{3}{2}x \ge 9$ **32.** $\frac{2}{5}x > 7$

33. $-10x < 40$ **34.** $-6x > 15$

35. $\frac{3}{5}x - 7 < 8$ **36.** $\frac{5}{4}x + 1 \le 11$

37. $2x + 7 < 3 + 4x$ **38.** $6x - 4 \le 2 + 8x$

39. $2x - 1 \ge 5x + 1$ **40.** $3x + 1 \ge 2 + x$

41. $3(x + 2) + 7 < 2x - 5$

42. $2(x + 7) - 4 \ge 5(x - 3)$

43. $-3(x - 1) + 7 < 2x + 8$

44. $5 - 3x > -5(x + 4) + 6$

Solving a Double Inequality In Exercises 45–52, solve the inequality. Then graph the solution set on the real number line. *See Example 4.*

45. $3 \le 2x - 1 < 7$ **46.** $3 > 1 - \frac{x}{2} > -3$

47. $-11 \le 1 + 4(x + 1) < 13$

48. $-8 \le 1 - 3(x - 2) < 13$

49. $-4 < \frac{x - 3}{3} < 4$ **50.** $0 \le \frac{x + 3}{2} < 5$

51. $\frac{3}{4} > x + 1 > \frac{1}{4}$ **52.** $-\frac{1}{2} < x - 1 < \frac{3}{4}$

Solving an Absolute Value Inequality In Exercises 53–70, solve the inequality. Then graph the solution set on the real number line. *See Examples 5 and 6.*

53. $|x| < 6$ **54.** $|x| > 8$

55. $\left|\frac{x}{2}\right| > 3$ **56.** $|5x| > 10$

57. $|x + 3| < 5$ **58.** $|x - 7| < 6$

59. $|x - 5| < 0$ **60.** $|x - 5| \ge 0$

61. $|2x - 5| > 6$ **62.** $|2x - 20| \le 4$

63. $2|x + 10| \ge 9$ **64.** $3|4 - 5x| \le 9$

65. $\left|\frac{x - 3}{2}\right| \ge 5$ **66.** $\left|\frac{2x + 1}{2}\right| < 6$

67. $|9 - x| - 2 < -1$ **68.** $|x + 14| + 3 > 17$

69. $2|2x - 5| - 4 > 6$ **70.** $2|5 - 3x| + 7 < 21$

Using Absolute Value In Exercises 71–78, use absolute value notation to define the solution set.

71.
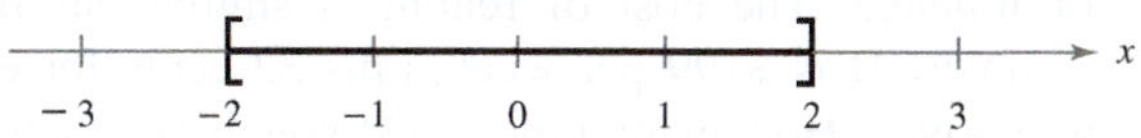

72.
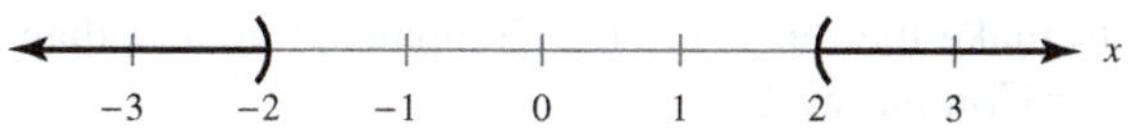

73.
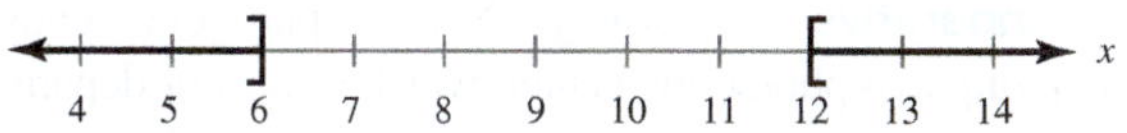

74.
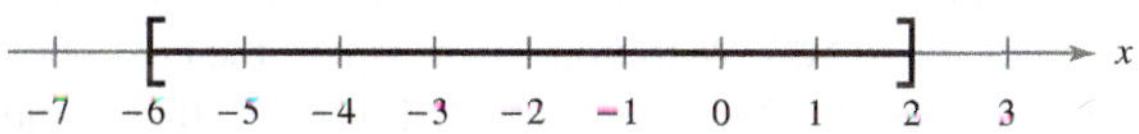

75. All real numbers at most 10 units from 12

76. All real numbers at least 5 units from 8

77. All real numbers whose distances from -3 are more than 5

78. All real numbers whose distances from -6 are no more than 7

79. Annual Operating Cost A utility company has a fleet of vans. The annual operating cost C per van is

$C = 0.52m + 2500$

where m is the number of miles traveled by a van in a year. What number of miles will yield an annual operating cost that is less than \$12,000?

80. IQ Scores The admissions office of a college wants to determine whether there is a relationship between IQ scores x and grade-point averages y after the first year of school. An equation that models the data obtained by the admissions office is

$y = 0.068x - 4.753.$

Estimate the values of x that predict a grade-point average of at least 3.0.

81. Brand Name Drugs The average prices B (in dollars) of brand name prescription drugs from 2005 through 2009 can be approximated by

$B = 14.516t + 24.65, \quad 5 \le t \le 9$

where t represents the year, with $t = 5$ corresponding to 2005. Use the model to estimate the first year in which the average price exceeded \$140. *(Source: National Association of Chain Drug Stores)*

82. Generic Drugs The average prices G (in dollars) of generic prescription drugs from 2005 through 2009 can be approximated by

$$G = 2.649t + 15.00, \quad 5 \leq t \leq 9$$

where t represents the year, with $t = 5$ corresponding to 2005. Use the model to estimate the first year in which the average price exceeded \$30. *(Source: National Association of Chain Drug Stores)*

83. Comparative Shopping The cost of renting a car from Company A is \$279 per week with no extra charge for mileage. The cost of renting a similar car from Company B is \$199 per week, plus 32 cents for each mile driven. How many miles must you drive in a week to make the rental fee for Company B greater than that for Company A?

84. Comparative Shopping Your department sends its copying to a photocopy center, which bills your department \$0.08 per page. You are considering buying a departmental copier for \$2500. With your own copier, the cost per page would be \$0.025. The expected life of the copier is 4 years. How many copies must you make in the four-year period to justify purchasing the copier?

85. Football Salaries The average National Football League player's salary S (in millions of dollars) each year from 1998 through 2009 can be approximated by

$$S = 0.1215t - 0.351, \quad 8 \leq t \leq 19$$

where t represents the year, with $t = 8$ corresponding to 1998 (see figure). *(Source: National Football League Players Association)*

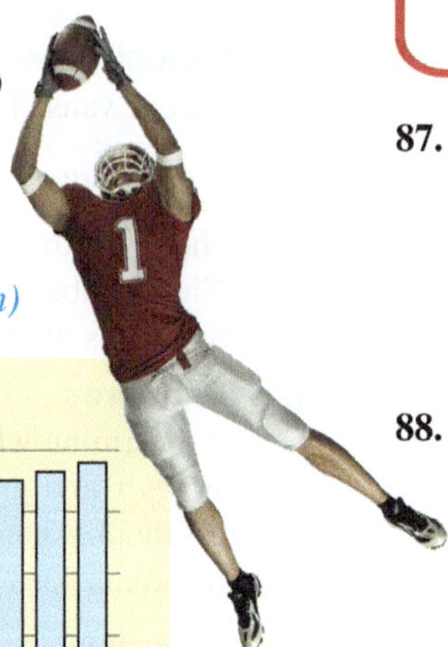

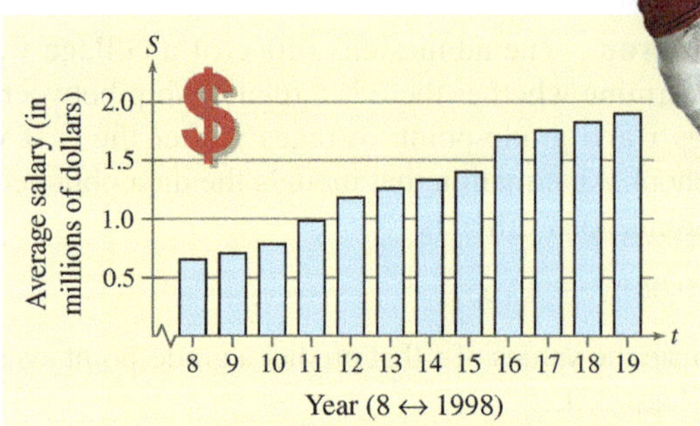

(a) Use the model to estimate the years when the average National Football League player's salary was less than \$1.5 million. Do your results agree with the bar graph? Explain.

(b) Use the model to predict the first year that the average National Football League player's salary will exceed \$2.2 million. Do your results seem reasonable? Explain.

86. Public College Enrollment The projected public college enrollment E (in thousands) in the United States from 2012 to 2019 can be approximated by

$$E = 265.3t + 14{,}157, \quad 2 \leq t \leq 9$$

where t represents the year, with $t = 2$ corresponding to 2012 (see figure). *(Source: U.S. National Center for Education Statistics)*

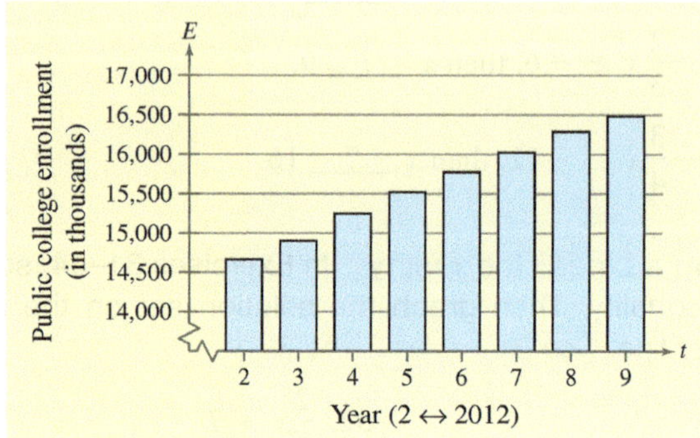

(a) Use the model to predict the first year that the public college enrollment will be greater than 16 million students. Do your results agree with the bar graph? Explain.

(b) Use the model to predict the first year that the public college enrollment will be greater than 17 million students. Do your results seem reasonable? Explain.

87. Weight Loss Program A person enrolls in a weight loss program that guarantees a loss of at least $1\frac{1}{2}$ pounds per week. The person's weight at the beginning of the program is 180 pounds. Find the maximum number of weeks it will take for the person to attain a weight of 130 pounds.

88. Salary Increase You accept a new job with a starting salary of \$34,800. You are told that you will receive an annual raise of at least \$1500. What is the maximum number of years you must work before your annual salary will be \$45,000?

89. Maximum Width An overnight delivery company will not accept any package whose combined length and girth (perimeter of a cross section) exceeds 132 inches. The package shown is 72 inches long and has square cross sections. Describe the lengths of x for which the company will deliver the package.

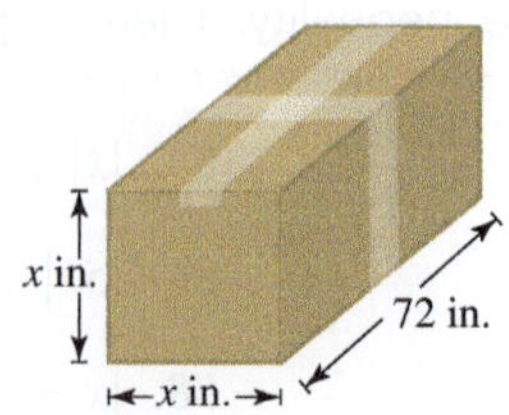

90. Break-Even Analysis The revenue R from selling x units of a product is

$$R = 139.95x.$$

The cost C of producing x units is

$$C = 97x + 850.$$

In order to obtain a profit, the revenue must be greater than the cost.

(a) Complete the table.

x	10	20	30	40	50	60
R						
C						

(b) For what values of x will this product return a profit?

91. Break-Even Analysis The revenue R from selling x units of a product is $R = 25.95x$. The cost C of producing x units is

$$C = 13.95x + 125{,}000.$$

In order to obtain a profit, the revenue must be greater than the cost. For what values of x will the product return a profit?

92. HOW DO YOU SEE IT? A company sells all the units of a product that it produces. The double bar graph shows the daily cost and revenue based on the numbers of units x produced.

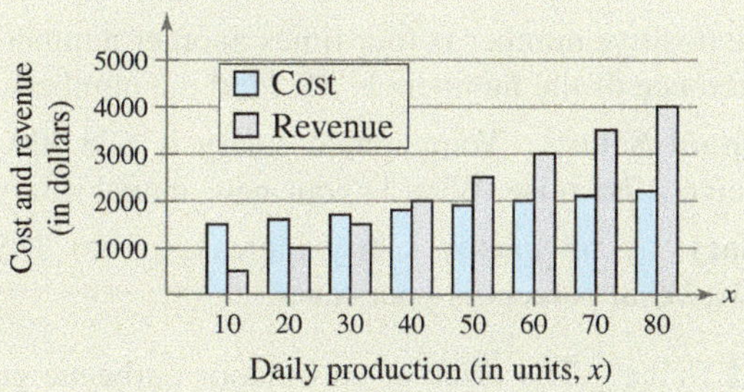

(a) Use the double bar graph to estimate the number of units the company must produce in a day to break even, with cost equal to revenue.

(b) Use your answer in part (a) to write an inequality that represents the daily numbers of units x for which the company receives a profit.

(c) Write an inequality that represents the daily numbers of units x for which the company receives a profit of at least \$1000.

93. Simple Interest For \$1500 to grow to more than \$1680 in 3 years, what must the simple interest rate be?

94. Simple Interest For \$2000 to grow to more than \$2180 in 2 years, what must the simple interest rate be?

95. Geometry You measure the side of a square as 10.4 inches with a possible error of $\frac{1}{16}$ inch. Using these measurements, determine the interval containing the possible areas of the square.

96. Geometry You measure the side of a square as 24.2 centimeters with a possible error of 0.25 centimeter. Using these measurements, determine the interval containing the possible areas of the square.

97. Human Height The heights h of two-thirds of a population satisfy the inequality

$$|h - 68.5| \leq 2.7$$

where h is measured in inches. Determine the interval on the real number line in which these heights lie.

98. Time Study A time study was conducted to determine the length of time required to perform a particular task in a manufacturing process. The times required by approximately two-thirds of the workers in the study satisfied the inequality

$$\left|\frac{t - 15.6}{1.9}\right| < 1$$

where t is time in minutes. Determine the interval on the real number line in which these times lie.

99. Humidity Control The specifications for an electronic device state that it is to be operated in a room with relative humidity h defined by

$$|h - 50| \leq 30.$$

What are the minimum and maximum relative humidities for the operation of this device?

100. Body Temperature Physicians consider an adult's body temperature x (in degrees Fahrenheit) to be normal when it satisfies the inequality

$$|x - 98.6| \leq 1.$$

Determine the range of temperatures that are considered to be normal.

101. Accuracy of Measurement You buy six T-bone steaks that cost \$8.99 per pound. The weight listed on the package is 5.72 pounds. The scale that weighed the package is accurate to within $\frac{1}{2}$ ounce. How much money might you have been undercharged or overcharged?

102. Daily Sales A doughnut shop sells a dozen doughnuts for \$7.95. Beyond the fixed costs (rent, utilities, and insurance) of \$165 per day, it costs \$1.45 for enough materials (flour, sugar, and so on) and labor to produce a dozen doughnuts. The daily profit from doughnut sales varies between \$400 and \$1200. Between what numbers of doughnuts (in dozens) do the daily sales vary?

Review Exercises

See www.CalcChat.com for worked-out solutions to odd-numbered exercises.

Classifying Equations In Exercises 1–4, determine whether the equation is an identity or a conditional equation.

1. $5(x - 3) = 2x + 9$ **2.** $3(x + 2) = 3x + 6$

3. $4(x + 2) = 4x + 8$ **4.** $2(x + 3) = 3x + 2$

Checking Solutions In Exercises 5–8, determine whether each value of x is a solution of the equation.

5. $5(x - 1) = 3x + 1$

(a) $x = 0$ (b) $x = -1$ (c) $x = 3$ (d) $x = 2$

6. $3x^2 + 7x + 5 = x^2 + 9$

(a) $x = 0$ (b) $x = \frac{1}{2}$ (c) $x = -4$ (d) $x = -1$

7. $6 + \dfrac{3}{x - 4} = 5$

(a) $x = 5$ (b) $x = 0$ (c) $x = -2$ (d) $x = 7$

8. $\sqrt{x + 4} = 3$

(a) $x = 0$ (b) $x = -1$ (c) $x = 5$ (d) $x = 12$

Solving an Equation In Exercises 9–16, solve the equation and check your solution. (Some equations have no solution.)

9. $x + 7 = 20$ **10.** $2x + 15 = 43$

11. $4(x + 3) - 3 = 2(4 - 3x) - 4$

12. $(x + 3) + 2(x - 4) = 5(x + 3)$

13. $5x + x = 6x + 3$ **14.** $4 - x = 2x - 3x + 1$

15. $8(x - 2) = 8x - 16$ **16.** $-(3 - 2x) = 2x - 3$

An Equation Involving Fractional Expressions In Exercises 17–20, solve the equation and check your solution.

17. $\dfrac{3x - 2}{5x - 1} = \dfrac{3}{4}$

18. $\dfrac{3}{x - 4} + \dfrac{8}{2x + 5} = \dfrac{11}{2x^2 - 3x - 20}$

19. $\dfrac{x}{x + 3} - \dfrac{4}{x + 3} + 2 = 0$

20. $7 - \dfrac{3}{x} = 8 + \dfrac{5}{x}$

Using a Calculator to Solve an Equation In Exercises 21–24, use a calculator to solve the equation. (Round your solution to three decimal places.)

21. $0.375x - 0.75(300 - x) = 200$

22. $0.235x + 2.6(-x - 4) = 30$

23. $\dfrac{x}{0.055} + \dfrac{x}{0.085} = 1$ **24.** $\dfrac{x}{0.0645} + \dfrac{x}{0.098} = 2$

25. Stores The numbers N of Safeway stores in operation from 2003 through 2009 can be approximated by the linear equation

$$N = -15.5t + 1859, \quad 3 \le t \le 9$$

where t represents the year, with $t = 3$ corresponding to 2003. Use the model to estimate the year in which the number of Safeway stores decreased to 1720. *(Source: Safeway, Inc.)*

26. Federal Programs The numbers of students S (in millions) participating in the school breakfast program from 2001 through 2009 can be approximated by the linear equation

$$S = 0.42t + 7.3, \quad 1 \le t \le 9$$

where t represents the year, with $t = 1$ corresponding to 2001. Use the model to estimate the year in which 9.4 million students participated in the school breakfast program. *(Source: U.S. Department of Agriculture, Food and Nutrition Service)*

Constructing a Mathematical Model In Exercises 27–30, use the verbal description to (a) write a verbal model, (b) assign labels to the quantities in the verbal model, (c) use the labels to write a mathematical model, and (d) solve the problem.

27. Three consecutive even integers have a sum of 42. Find the smallest of these integers.

28. One positive number is four times another number. The difference of the numbers is 51. Find the numbers.

29. Annual Salary Your annual salary is \$28,900. You receive a 7% raise. What is your new annual salary?

30. Salary Your weekly salary increases from \$550 to \$594. What percent is your raise?

31. List Price The price of an outdoor barbecue grill is discounted 15%. The sale price is \$139. Find the original list price of the grill.

32. Discount Rate The price of a home gym is discounted by \$300. The sale price is \$599.99. What percent of the original price is the discount?

33. Travel Time Two cars start at the same time at a given point and travel in the same direction at average speeds of 45 miles per hour and 50 miles per hour. After how long are the cars 10 miles apart?

34. Exercise Two bicyclists start at the same time at a given point and travel in the same direction at average speeds of 8 miles per hour and 10 miles per hour. After how long are the bicyclists 5 miles apart?

35. **Box Office** James Cameron's 1997 movie *Titanic* earned U.S. gross revenue of \$600,788,188. Cameron's 2009 movie *Avatar* earned U.S. gross revenue of \$760,507,625. Find the percent increase in the U.S. gross revenue from *Titanic* to *Avatar*. *(Source: Nash Information Services)*

36. **Oil Imports** The United States imported 1594 million barrels of crude oil from members of OPEC (Organization of Petroleum Exporting Countries) in 2009. Use the bar graph to determine the amount imported from each of the five top contributing countries. *(Source: U.S. Energy Information Administration)*

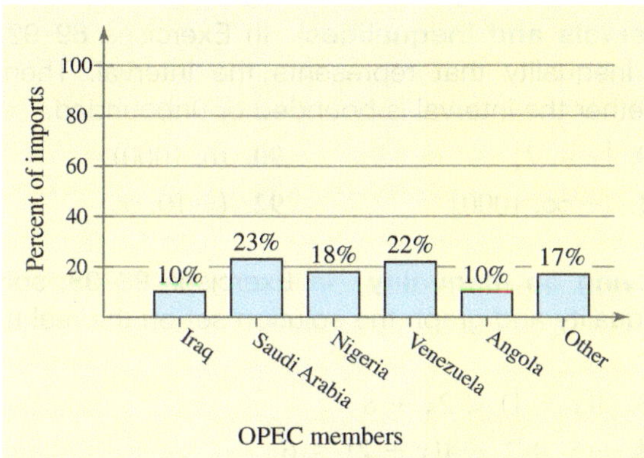

37. **Mixture** A car radiator contains 10 quarts of a 10% antifreeze solution. The car's owner wishes to create a 10-quart solution that is 30% antifreeze. How many quarts will have to be replaced with pure antifreeze?

38. **Mixture** A three-gallon acid solution contains 3% boric acid. How many gallons of 20% boric acid solution should be added to make a final solution that is 8% boric acid?

39. **Geometry** A volleyball court is twice as long as it is wide, and its perimeter is 177 feet. Find the dimensions of the volleyball court.

40. **Geometry** A room is 1.25 times as long as it is wide, and its perimeter is 90 feet. Find the dimensions of the room.

41. **Simple Interest** You deposit \$500 in a savings account earning 4% simple interest. How much interest will you earn in 1 year?

42. **Simple Interest** You deposit \$800 in a money market account. One year later the account balance is \$814. What was the simple interest rate?

Solving a Quadratic Equation by Factoring In Exercises 43–46, solve the quadratic equation by factoring. Check your solutions.

43. $6x^2 = 5x + 4$

44. $-x^2 = 15x + 36$

45. $x^2 - 11x + 24 = 0$

46. $4 - 4x + x^2 = 0$

Extracting Square Roots In Exercises 47–50, solve the quadratic equation by extracting square roots. When a solution is irrational, list both the exact solution and its approximation rounded to two decimal places.

47. $x^2 = 11$

48. $16x^2 = 25$

49. $(x + 4)^2 = 18$

50. $(x - 1)^2 = 5$

51. **Geometry** A billboard is 12 feet longer than it is high. The billboard has 405 square feet of advertising space. What are the dimensions of the billboard?

52. **Grand Canyon** The Grand Canyon is 6000 feet deep at its deepest part. A rock is dropped over the deepest part of the canyon. How long does the rock take to hit the water in the Colorado River below?

53. **Total Revenue** The demand equation for a product is

$$p = 60 - 0.0001x$$

where p is the price per unit and x is the number of units sold. The total revenue R from selling x units is given by

$$R = xp = x(60 - 0.0001x).$$

How many units must be sold to produce a revenue of \$8,000,000?

54. **Depth of an Underwater Cable** A ship's sonar locates a cable 2000 feet from the ship (see figure). The angle between the surface of the water and a line from the ship to the cable is 45°. How deep is the cable?

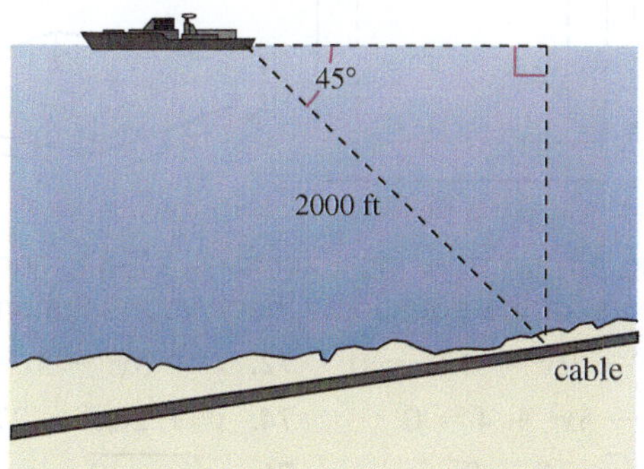

Using the Discriminant In Exercises 55–58, use the discriminant to determine the number of real solutions of the quadratic equation.

55. $x^2 + 11x + 24 = 0$

56. $x^2 + 5x + 12 = 0$

57. $3x^2 - 4x + 2 = 0$

58. $4x^2 + 12x + 9 = 0$

Using the Quadratic Formula In Exercises 59–64, use the Quadratic Formula to solve the quadratic equation. Check your solutions.

59. $x^2 - 12x + 30 = 0$

60. $5x^2 + 16x - 12 = 0$

61. $(y + 7)^2 = -5y$

62. $6x = 7 - 2x^2$

63. $x^2 + 6x - 3 = 0$

64. $10x^2 - 11x = 2$

Using a Calculator In Exercises 65–68, use a calculator to solve the quadratic equation. (Round your answers to three decimal places.)

65. $3.6x^2 - 5.7x - 1.9 = 0$

66. $2.3x^2 + 6.6x - 3.9 = 0$

67. $34x^2 - 296x + 47 = 0$

68. $39x^2 + 75x - 21 = 0$

69. On the Moon An astronaut standing on the edge of a cliff on the moon drops a rock over the cliff. The height s of the rock after t seconds is given by

$$s = -2.7t^2 + 200.$$

The rock's initial velocity is 0 feet per second and its initial height is 200 feet. Determine how long it will take the rock to hit the lunar surface. How long does it take a rock dropped from a height of 200 feet on Earth to hit the ground?

70. Geometry An open box is made from a square piece of material by cutting three-inch squares from the corners and turning up the sides (see figure). The volume of the finished box is 363 cubic inches. Find the size of the original piece of material.

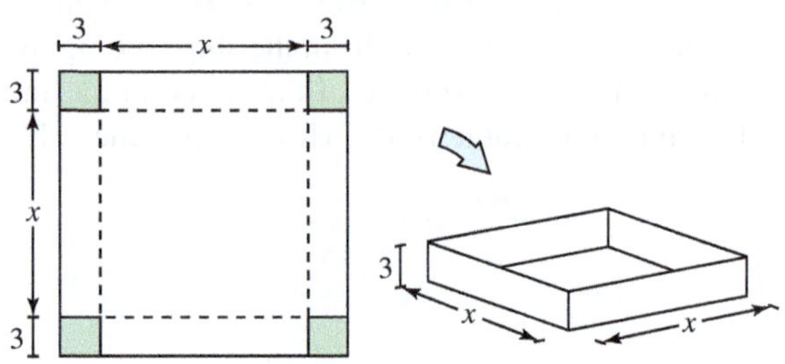

Solving Equations In Exercises 71–84, find the real solution(s) of the equation. Check your solution(s).

71. $3x^3 - 9x^2 - 12x = 0$

72. $x^4 + 3x^3 - 5x - 15 = 0$

73. $x^4 - 5x^2 + 4 = 0$

74. $x^6 + 26x^3 - 27 = 0$

75. $2\sqrt{x} - 5 = 0$

76. $\sqrt{3x - 2} + x = 4$

77. $2\sqrt{x - 3} - 4 = 3x$

78. $\sqrt[3]{3x + 5} = 5$

79. $(x^2 - 5)^{2/3} = 9$

80. $(x^2 - 5x - 6)^{4/3} = 16$

81. $|5x + 4| = 11$

82. $|x^2 + 4x| - 2x = 8$

83. $\dfrac{5}{x + 1} + \dfrac{3}{x + 3} = 1$

84. $x + \dfrac{3}{x + 2} = 2$

85. Sharing the Cost Three students are planning to share the expense of renting a condominium at a resort for 1 week. By adding a fourth person to the group, each person could save \$75 in rental fees. How much is the rent for the week?

86. Sharing the Cost A college charters a bus for \$1800 to take a group to a museum. When four more students join the trip, the cost per student decreases by \$5. How many students were in the original group?

87. Market Research The demand equation for a product is given by

$$p = 45 - \sqrt{0.002x + 1}$$

where x is the number of units demanded per day and p is the price per unit. Find the demand when the price is set at \$19.95.

88. Cash Advance You take out a cash advance of \$500 on a credit card. After 3 months, the amount you owe is \$525.29. What is the annual percentage rate for this cash advance? (Assume that the interest is compounded monthly and that you have made no payments.)

Intervals and Inequalities In Exercises 89–92, write an inequality that represents the interval. Then state whether the interval is bounded or unbounded.

89. $[-4, 7]$

90. $(6, 1000)$

91. $(-\infty, 1000]$

92. $(-10, \infty)$

Solving an Inequality In Exercises 93–98, solve the inequality and graph the solution set on the real number line.

93. $3(x - 1) < 2x + 8$

94. $-5 \le 2 - 4(x + 2) \le 6$

95. $-3 < \dfrac{2x + 1}{4} < 3$

96. $-1 \le -5 - 3x < 4$

97. $|x + 10| + 3 < 5$

98. $|2x - 3| - 4 > 2$

99. Break-Even Analysis The revenue R from selling x units of a product is

$$R = 89.95x.$$

The cost C of producing x units is

$$C = 35x + 2500.$$

For what values of x will this product return a profit?

100. Accuracy of Measurement You buy a 16-inch gold chain that costs \$9.95 per inch. The chain is measured accurately to within $\frac{1}{16}$ inch. How much money might you have been undercharged or overcharged?

Finding Test Intervals In Exercises 101–104, find the test intervals of the inequality.

101. $x^2 - 1 < 0$

102. $x^2 - 3x - 4 \ge 0$

103. $\dfrac{x + 3}{x - 1} \le 2$

104. $\dfrac{x + 5}{x + 7} > 3$

Solving an Inequality In Exercises 105–110, solve the inequality and graph the solution set on the real number line.

105. $5(x + 1)(x - 3) < 0$

106. $(x + 4)^2 \le 4$

107. $x^3 - 9x < 0$

108. $\dfrac{x + 5}{x + 8} \ge 2$

109. $\frac{2 + 3x}{4 - x} < 2$

110. $\frac{1}{x + 1} \geq \frac{1}{x + 5}$

Using a Calculator In Exercises 111–114, use a calculator to solve the inequality. (Round each number in your answer to two decimal places.)

111. $-1.2x^2 + 4.76 > 1.32$

112. $3.5x^2 + 4.9x - 6.1 < 2.4$

113. $\frac{1}{3.7x - 6.1} > 2.9$

114. $\frac{3}{5.4x - 2.7} < 8.9$

Finding the Domain of an Expression In Exercises 115–118, find the domain of the expression.

115. $\sqrt{x - 10}$

116. $\sqrt[4]{2x + 5}$

117. $\sqrt{x^2 - 15x + 54}$

118. $\sqrt{81 - 4x^2}$

119. Height of a Flare A flare is fired straight upward from ground level with an initial velocity of 100 feet per second. During what time period will its height exceed 150 feet?

120. Path of a Soccer Ball A soccer player kicks the ball toward the center of the goal. The path of the ball can be modeled by

$$y = -0.054x^2 + 1.43x$$

where x is the horizontal distance (in feet) from where the ball is kicked and y is the corresponding height (in feet).

(a) The soccer goal is 8 feet high. Write an inequality to determine the values of x for which the ball is low enough to go into the goal.

(b) Solve the inequality from part (a).

(c) The player kicks the ball 15 feet from the goal and the ball is not blocked. Does the player score a goal? Explain your reasoning.

121. Geometry A rectangular field with a perimeter of 80 meters is to have an area of at least 380 square meters (see figure). Describe the possible lengths l of the field.

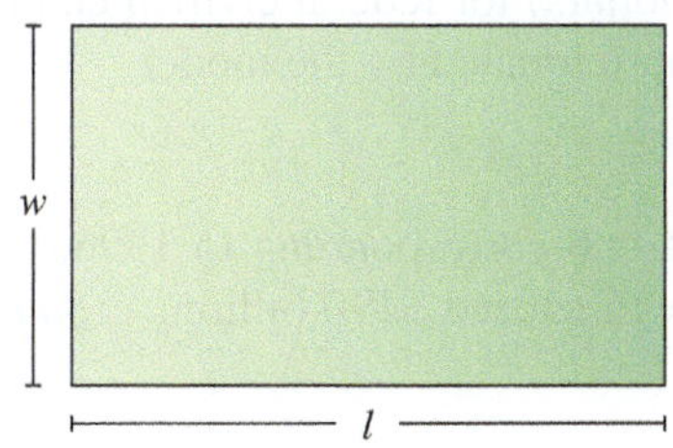

122. Compound Interest P dollars, invested at interest rate r compounded annually, increases to an amount A given by

$$A = P(1 + r)^5$$

in 5 years. An investment of \$1000 increases to an amount greater than \$1400 in 5 years. The interest rate must be greater than what percent?

123. Compound Interest P dollars, invested at an interest rate r compounded semiannually, increases to an amount

$$A = P(1 + r/2)^{2 \cdot 8}$$

in 8 years. An investment of \$2000 increases to an amount greater than \$2800 in 8 years. The interest rate must be greater than what percent?

124. Company Profits The revenue R and cost C for a product are given by

$$R = x(75 - 0.0005x)$$

and

$$C = 25x + 100{,}000$$

where R and C are measured in dollars and x represents the number of units sold. How many units must be sold to obtain a profit of at least \$500,000?

125. Price of a Product In Exercise 124, the revenue equation

$$R = x(75 - 0.0005x)$$

implies that the demand equation is

$$p = 75 - 0.0005x$$

where p is the price per unit. What prices per unit can the company set to obtain a profit of at least \$1,000,000?

126. Income The total investment income I (in millions of dollars) of Aflac each year from 2001 through 2009 can be approximated by the model

$$I = 6.07t^2 + 90.3t + 1448, \quad 1 \leq t \leq 9$$

where t represents the year, with $t = 1$ corresponding to 2001. *(Source: Aflac Incorporated)*

(a) Complete the table. Round each value of I to the nearest whole number.

t	1	3	5	9
I				

(b) Aflac expected to earn investment income of greater than \$3 billion in 2010. Does the model support this? Explain.

(c) Aflac predicts that the yearly investment income will be at least \$3.5 billion by 2015. Does the model support this prediction? Explain.

TEST YOURSELF

See www.CalcChat.com for worked-out solutions to odd-numbered exercises.

Take this test as you would take a test in class. When you are done, check your work against the answers given in the back of the book.

1. Solve the equation $3(x + 2) - 8 = 4(2 - 5x) + 7$.
2. Find the domain of (a) $\sqrt[3]{2x + 3}$ and (b) $\sqrt{9 - x^2}$.
3. In May, the total profit for a company was 8% less than it was in April. The total profit for the 2 months was \$625,509.12. Find the profit for each month.

In Exercises 4–13, solve the equation. Check your solution(s).

4. *Factoring:* $6x^2 + 7x = 5$
5. *Factoring:* $12 + 5x - 2x^2 = 0$
6. *Extracting square roots:* $x^2 - 5 = 10$
7. *Quadratic Formula:* $(x + 5)^2 = -3x$
8. *Quadratic Formula:* $3x^2 - 11x = 2$
9. *Quadratic Formula:* $5.4x^2 - 3.2x - 2.5 = 0$
10. $|2x - 3| = 10$
11. $\sqrt{x - 3} + x = 5$
12. $x^4 - 10x^2 + 9 = 0$
13. $(x^2 - 9)^{2/3} = 9$

14. The demand equation for a product is $p = 40 - 0.0001x$, where p is the price per unit and x is the number of units sold. The total revenue R from selling x units is given by $R = xp$. How many units must be sold to produce a revenue of \$2,000,000? Explain your reasoning.

In Exercises 15–18, solve the inequality and graph the solution set on the real number line.

15. $\dfrac{3x + 1}{5} < 2$
16. $|4 - 5x| \geq 24$
17. $\dfrac{x + 3}{x + 7} > 2$
18. $3x^3 - 12x \leq 0$

19. The revenue R and cost C for a product are given by

$$R = x(90 - 0.0004x)$$

and

$$C = 25x + 300{,}000$$

where R and C are measured in dollars and x represents the number of units sold. How many units must be sold to obtain a profit of at least \$800,000?

20. The total payroll P (in billions of dollars) for federal civilian employees each year from 1996 through 2009 can be approximated by the model

$$P = 0.145t^2 + 0.79t + 107.4, \quad 6 \leq t \leq 19$$

where t represents the year, with $t = 6$ corresponding to 1996. Use the model to predict the year when the payroll will exceed \$190 billion. *(Source: U.S. Office of Personnel Management)*

2 Functions and Graphs

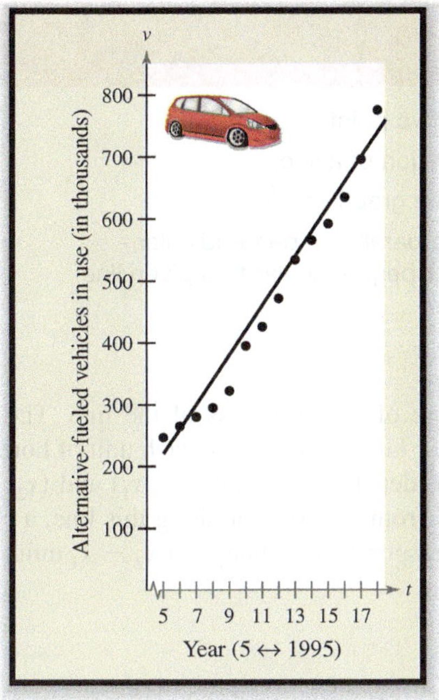

Example 8 on page 189 shows how a scatter plot can be used to find a linear model that approximates the number of alternative-fueled vehicles in use in the United States for a given year.

GG Pro Photo/www.shutterstock.com
Kurhan/Shutterstock.com

2.2 Lines in the Plane

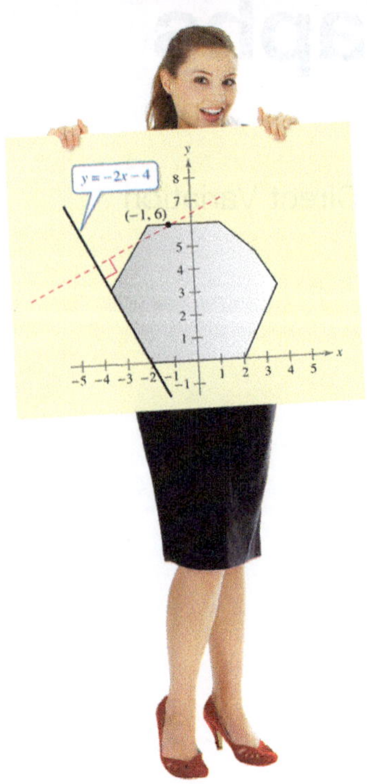

In Exercise 96 on page 182, you will use perpendicular lines to write an equation representing the line of a cut in making a memorial stone.

- Find the slope of a line passing through two points.
- Use the point-slope form to find the equation of a line.
- Use the slope-intercept form to sketch the graph of a line.
- Use slope to determine whether lines are parallel or perpendicular, and write the equation of a line parallel or perpendicular to a given line.

The Slope of a Line

The **slope** of a nonvertical line is a measure of the steepness of the line. The slope represents the number of units the line rises or falls vertically for each unit of horizontal change from left to right. For instance, consider the two points (x_1, y_1) and (x_2, y_2) on the line shown in Figure 2.21. As you move from left to right along this line, a change of $y_2 - y_1$ units in the vertical direction corresponds to a change of $x_2 - x_1$ units in the horizontal direction. That is,

$$y_2 - y_1 = \text{the change in } y$$

and

$$x_2 - x_1 = \text{the change in } x.$$

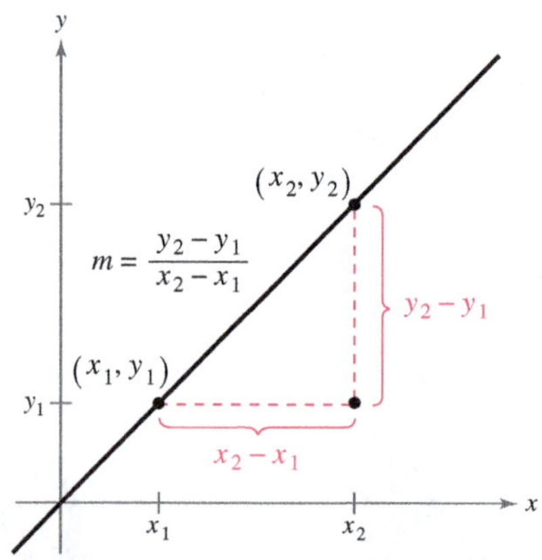

FIGURE 2.21

The slope of the line is defined as the quotient of these two changes.

STUDY TIP

When using the formula for slope, the *order of subtraction* is important. Given two points on a line, you can label either one of them as (x_1, y_1) and the other as (x_2, y_2). Once this is done, you must form the numerator and denominator using the same order of subtraction.

$$m = \frac{y_2 - y_1}{x_2 - x_1} \quad \text{Correct}$$

$$m = \frac{y_1 - y_2}{x_1 - x_2} \quad \text{Correct}$$

$$m = \frac{y_2 - y_1}{x_1 - x_2} \quad \text{Incorrect}$$

Definition of the Slope of a Line

The **slope** m of the nonvertical line passing through the points (x_1, y_1) and (x_2, y_2) is

$$m = \frac{y_2 - y_1}{x_2 - x_1} = \frac{\text{change in } y}{\text{change in } x}$$

where $x_1 \neq x_2$.

The change in x is sometimes called the *run* and the change in y is sometimes called the *rise*.

In real-life problems, such as finding the steepness of a ramp or the increase in the value of a product, the slope of a line can be interpreted as either a *ratio* or a *rate*. If the x-axis and the y-axis have the same units of measure, then the slope has no units and is a *ratio*. If the x-axis and the y-axis have different units of measure, then the slope is a *rate* or *rate of change*. You will learn more about rates of change in Section 2.3.

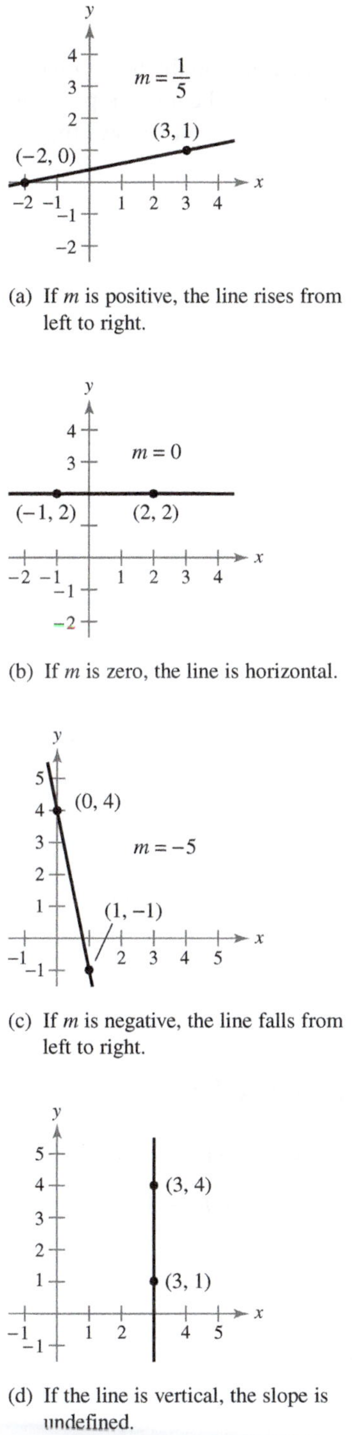

(a) If m is positive, the line rises from left to right.

(b) If m is zero, the line is horizontal.

(c) If m is negative, the line falls from left to right.

(d) If the line is vertical, the slope is undefined.

FIGURE 2.22

Example 1 Finding the Slope of a Line Through Two Points

Find the slope of the line passing through each pair of points.

a. $(-2, 0)$ and $(3, 1)$ **b.** $(-1, 2)$ and $(2, 2)$

c. $(0, 4)$ and $(1, -1)$ **d.** $(3, 4)$ and $(3, 1)$

SOLUTION

a. Letting $(x_1, y_1) = (-2, 0)$ and $(x_2, y_2) = (3, 1)$, you obtain a slope of

$$m = \frac{y_2 - y_1}{x_2 - x_1} \qquad \frac{\text{Difference in } y\text{-values}}{\text{Difference in } x\text{-values}}$$

$$= \frac{1 - 0}{3 - (-2)}$$

$$= \frac{1}{5}.$$

b. The slope of the line passing through $(-1, 2)$ and $(2, 2)$ is

$$m = \frac{2 - 2}{2 - (-1)}$$

$$= \frac{0}{3}$$

$$= 0.$$

c. The slope of the line passing through $(0, 4)$ and $(1, -1)$ is

$$m = \frac{-1 - 4}{1 - 0}$$

$$= \frac{-5}{1}$$

$$= -5.$$

d. The slope of the line passing through $(3, 4)$ and $(3, 1)$ is undefined. Applying the formula for slope, you have

$$m = \frac{1 - 4}{3 - 3} = \frac{-3}{0}. \qquad \text{Division by zero is undefined.}$$

Because division by zero is not defined, the slope of a vertical line is not defined.

The graphs of the four lines are shown in Figure 2.22.

✓Checkpoint 1

Find the slope of the line passing through the points $(-1, 2)$ and $(3, 4)$.

From Example 1, you can make the following generalizations about the slope of a line.

Slope of a Line

1. A line with positive slope $(m > 0)$ *rises* from left to right.
2. A line with negative slope $(m < 0)$ *falls* from left to right.
3. A line with zero slope $(m = 0)$ is *horizontal*.
4. A line with undefined slope is *vertical*.

The Point-Slope Form

When you know the slope of a line and the coordinates of one point on the line, you can find an equation of the line. For instance, in Figure 2.23, let (x_1, y_1) be a given point on the line whose slope is m. If (x, y) is *any other* point on the line, then it follows that

$$\frac{y - y_1}{x - x_1} = m.$$

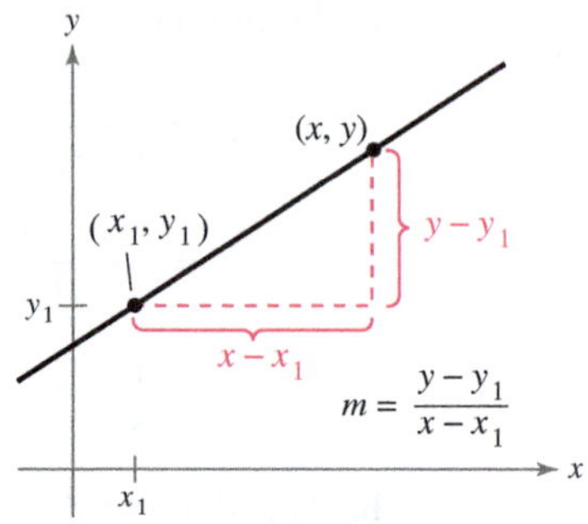

Any two points on a line can be used to determine the slope of the line.

FIGURE 2.23

ALGEBRA TUTOR *xy*

For help with identifying and plotting points in the Cartesian plane, see the *Chapter 2 Algebra Tutor* on pages 240 and 241.

This equation in the variables x and y can be rewritten to produce the following **point-slope form** of the equation of a line.

> **Point-Slope Form of the Equation of a Line**
>
> The **point-slope form** of the equation of the line that passes through the point (x_1, y_1) and has a slope of m is
>
> $$y - y_1 = m(x - x_1).$$

Example 2 The Point-Slope Form of the Equation of a Line

Find an equation of the line that passes through $(1, -2)$ and has a slope of 3.

SOLUTION Use the point-slope form with $(x_1, y_1) = (1, -2)$ and $m = 3$.

$$y - y_1 = m(x - x_1) \qquad \text{Point-slope form}$$

$$y - (-2) = 3(x - 1) \qquad \text{Substitute } y_1 = -2,\ x_1 = 1, \text{ and } m = 3.$$

$$y + 2 = 3x - 3 \qquad \text{Simplify.}$$

$$y = 3x - 5 \qquad \text{Equation of line}$$

The graph of this line is shown in Figure 2.24.

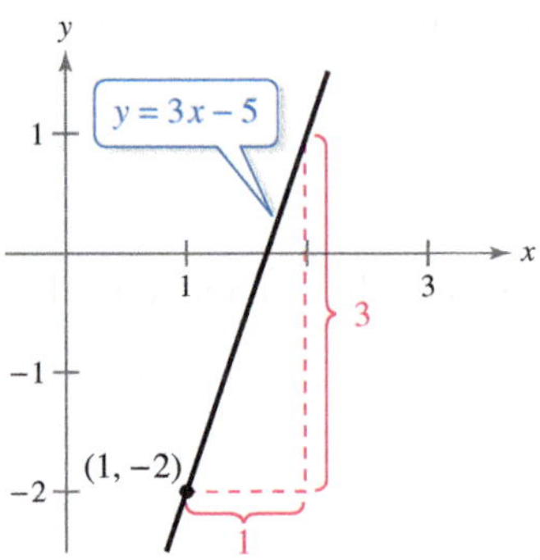

FIGURE 2.24

Checkpoint 2

Find an equation of the line that passes through the given point and has the given slope.

a. $(2, 4)$, $m = -2$

b. $(-8, -3)$, $m = \frac{3}{2}$

c. $(5, 1)$, $m = 0$

The point-slope form can be used to find the equation of a line passing through two points (x_1, y_1) and (x_2, y_2). First, use the formula for the slope of a line passing through two points. Then, use the point-slope form to obtain

$$y - y_1 = \frac{y_2 - y_1}{x_2 - x_1}(x - x_1).$$

This is sometimes called the **two-point form** of the equation of a line.

> **TECH TUTOR**
>
> Several models of graphing utilities have programs that use the two-point form to find an equation of a line. After you enter the coordinates of two points, the program outputs the slope and y-intercept of the line that passes through the points. Try using one of these programs to find the linear equation in Example 3.

Example 3 A Linear Model for Sales Prediction

During the first two quarters of the calendar year, Apple Inc. had sales of \$13.5 billion and \$15.7 billion, respectively.

a. Write a linear equation giving the sales y in terms of the quarter x.

b. Use the equation to predict the sales during the fourth quarter. Can you assume that sales will follow this linear pattern? *(Source: Apple Inc.)*

SOLUTION

a. Let $(x_1, y_1) = (1, 13.5)$ and $(x_2, y_2) = (2, 15.7)$ be two points on the line. Use the two-point form to find an equation of the line.

$$y - 13.5 = \frac{15.7 - 13.5}{2 - 1}(x - 1)$$ Substitute for x_1, y_1, x_2, and y_2 in two-point form.

$$y - 13.5 = 2.2(x - 1)$$ Simplify quotient.

$$y = 2.2x + 11.3$$ Equation of line

b. Using the equation from part (a), the fourth-quarter sales $(x = 4)$ should be

$$y = 2.2(4) + 11.3$$ Substitute 4 for x.

$$= \$20.1 \text{ billion.}$$ Simplify.

See Figure 2.25. Without more data, you cannot assume that the sales pattern will be linear. Many factors, such as seasonal demand and past sales history, help to determine the sales pattern.

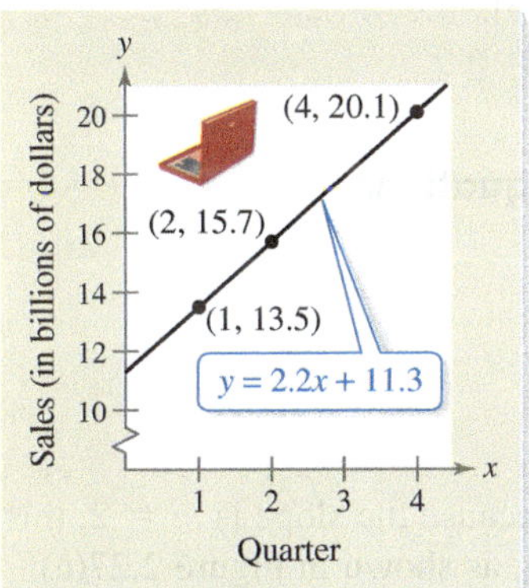

FIGURE 2.25

✓ Checkpoint 3

A company has sales of \$1.2 million and \$1.4 million in its first two years. Write a linear equation giving the sales y in terms of the year x.

The estimation method illustrated in Example 3 is called **linear extrapolation.** Note in Figure 2.26(a) that for linear extrapolation, the estimated point lies to the *right* of the given points. When the estimated point lies *between* two given points, the procedure is called **linear interpolation,** as shown in Figure 2.26(b).

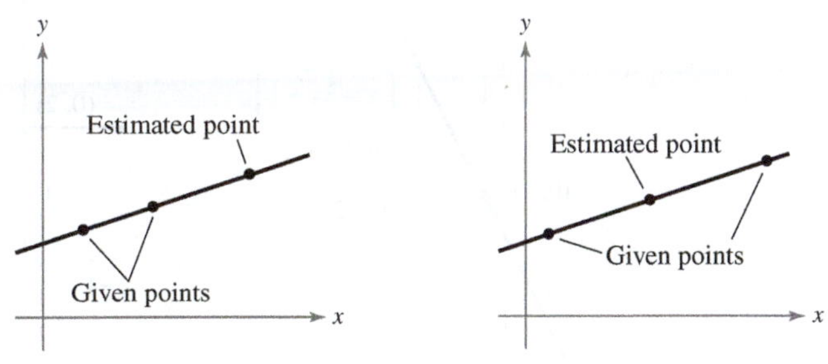

(a) Linear extrapolation (b) Linear interpolation

FIGURE 2.26

Edyta Pawlowska/www.shutterstock.com

Sketching Graphs of Lines

You have seen that for *finding the equation of a line* it is convenient to use the point-slope form. This formula, however, is not particularly useful for *sketching the graph of a line*. The form that is better suited to graphing linear equations is the **slope-intercept form** of the equation of a line. You can derive the slope-intercept form as shown.

$$y - y_1 = m(x - x_1)$$ Point-slope form

$$y = mx - mx_1 + y_1$$ Solve for y.

$$y = mx + (y_1 - mx_1)$$ Commutative Property of Addition

$$y = mx + b$$ Slope-intercept form $(b = y_1 - mx_1)$

Slope-Intercept Form of the Equation of a Line

The graph of the equation

$$y = mx + b$$

is a line whose slope is m and whose y-intercept is $(0, b)$.

Example 4 Sketching the Graphs of Linear Equations

Sketch the graph of each linear equation.

a. $y = 2x + 1$ **b.** $y = 2$ **c.** $x + y = 2$

SOLUTION

a. Because $b = 1$, the y-intercept is $(0, 1)$. Moreover, because the slope is $m = 2$, this line *rises* two units for each unit it moves to the right, as shown in Figure 2.27(a).

b. By writing the equation $y = 2$ in the form $y = (0)x + 2$ you can see that the y-intercept is $(0, 2)$ and the slope is zero. A zero slope implies that the line is horizontal, as shown in Figure 2.27(b).

c. By writing the equation $x + y = 2$ in slope-intercept form $y = -x + 2$ you can see that the y-intercept is $(0, 2)$. Moreover, because the slope is $m = -1$, this line *falls* one unit for each unit it moves to the right, as shown in Figure 2.27(c).

✓ Checkpoint 4

Sketch the graph of each linear equation.

a. $y = 4x + 1$ **b.** $y - 2x = -3$

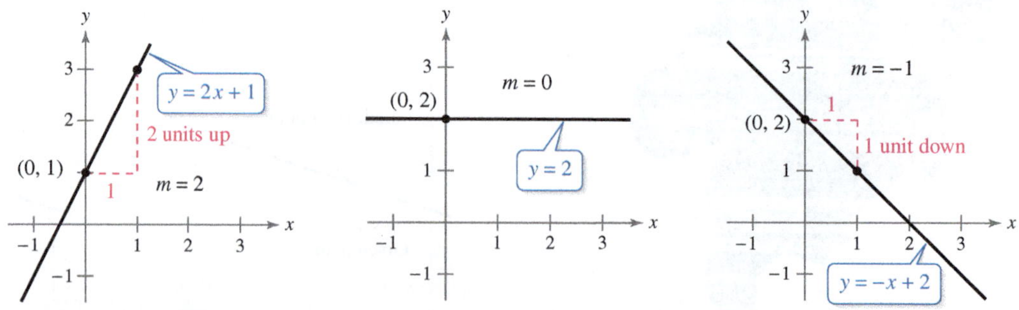

(a) When m is positive, the line rises from left to right.

(b) When m is zero, the line is horizontal.

(c) When m is negative, the line falls from left to right.

FIGURE 2.27

From the slope-intercept form of the equation of a line, you can see that a horizontal line ($m = 0$) has an equation of the form

$$y = (0)x + b \quad \text{or} \quad y = b. \qquad \text{Horizontal line}$$

This is consistent with the fact that each point on a horizontal line through $(0, b)$ has a y-coordinate of b, as shown in Figure 2.28.

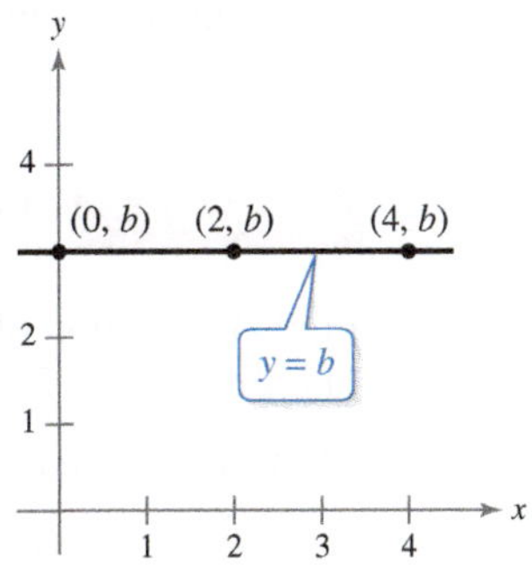

Horizontal Line

FIGURE 2.28

Similarly, each point on a vertical line through $(a, 0)$ has an x-coordinate of a, as shown in Figure 2.29. So, a vertical line has an equation of the form

$$x = a. \qquad \text{Vertical line}$$

This equation cannot be written in slope-intercept form because the slope of a vertical line is undefined. However, *every* line has an equation that can be written in the **general form**

$$Ax + By + C = 0 \qquad \text{General form}$$

where A and B are not *both* zero. If $A = 0$ (and $B \neq 0$), then the general equation can be reduced to the form $y = b$, which represents a horizontal line. If $B = 0$ (and $A \neq 0$), then the general equation can be reduced to the form $x = a$, which represents a vertical line.

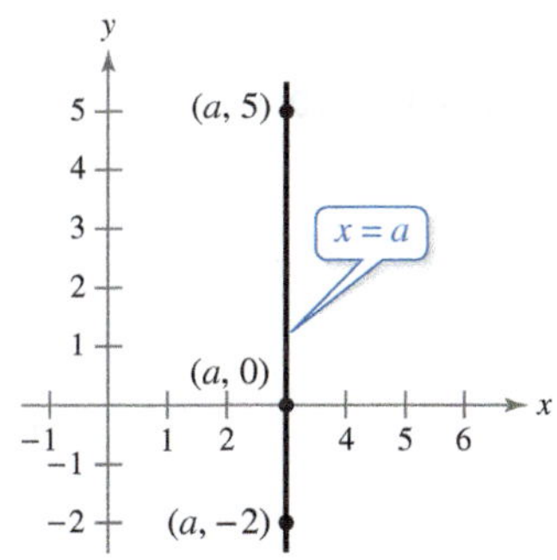

Vertical Line

FIGURE 2.29

Summary of Equations of Lines

1. General form:	$Ax + By + C = 0$
2. Vertical line:	$x = a$
3. Horizontal line:	$y = b$
4. Slope-intercept form:	$y = mx + b$
5. Point-slope form:	$y - y_1 = m(x - x_1)$

Make sure you understand that one form of the equation of a line may be more efficient than another form for different problem situations. For instance, the point-slope form (or two-point form) is more efficient for problems in which two points are given. Note that another form of an equation of a line is the *two-intercept* or *intercept form*. This form is introduced in Exercises 59–64 in this section.

Parallel and Perpendicular Lines

The slope of a line is a convenient tool for determining whether two lines are parallel, perpendicular, or neither.

Parallel Lines

Two distinct nonvertical lines are **parallel** if and only if their slopes are equal.

Example 5 Equations of Parallel Lines

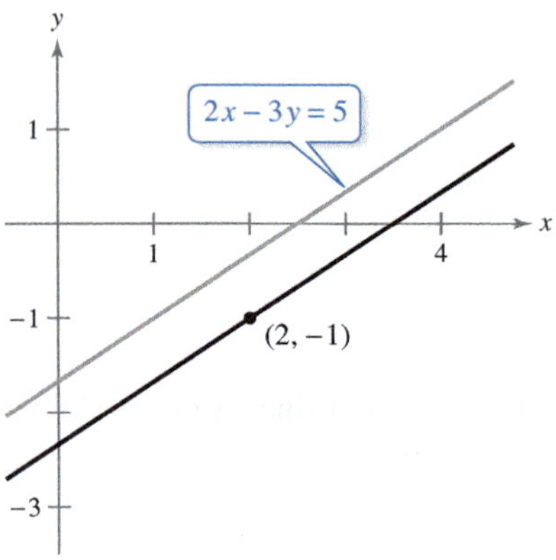

FIGURE 2.30

Find an equation of the line that passes through the point $(2, -1)$ and is parallel to the line $2x - 3y = 5$, as shown in Figure 2.30.

SOLUTION Start by rewriting the given equation in slope-intercept form.

$$2x - 3y = 5 \quad \text{Write original equation.}$$

$$-3y = -2x + 5 \quad \text{Subtract } 2x \text{ from each side.}$$

$$y = \frac{2}{3}x - \frac{5}{3} \quad \text{Write in slope-intercept form.}$$

So, the given line has a slope of

$$m = \tfrac{2}{3}.$$

Because any line parallel to the given line must also have a slope of $\frac{2}{3}$, the required line through $(2, -1)$ has the following equation.

$$y - y_1 = m(x - x_1) \quad \text{Point-slope form}$$

$$y - (-1) = \frac{2}{3}(x - 2) \quad \text{Substitute for } y_1, x_1, \text{ and } m.$$

$$y + 1 = \frac{2}{3}x - \frac{4}{3} \quad \text{Simplify.}$$

$$y = \frac{2}{3}x - \frac{4}{3} - 1 \quad \text{Solve for } y.$$

$$y = \frac{2}{3}x - \frac{7}{3} \quad \text{Write in slope-intercept form.}$$

Notice the similarity between the slope-intercept form of the original equation and the slope-intercept form of the parallel equation.

✓ Checkpoint 5

Find an equation of the line that passes through the given point and is parallel to the line $2y - 6x = 2$.

a. $(2, 4)$

b. $(-2, 0)$

You have seen that two nonvertical lines are parallel if and only if they have the same slope. Two nonvertical lines are *perpendicular* if and only if their slopes are negative reciprocals of each other. For instance, the lines

$$y = 2x \quad \text{and} \quad y = -\tfrac{1}{2}x$$

are perpendicular because one has a slope of $2 = \frac{2}{1}$ and the other has a slope of $-\frac{1}{2}$.

Perpendicular Lines

Two nonvertical lines are **perpendicular** if and only if their slopes are negative reciprocals of each other. That is,

$$m_1 = -\frac{1}{m_2}.$$

Example 6 Equations of Perpendicular Lines

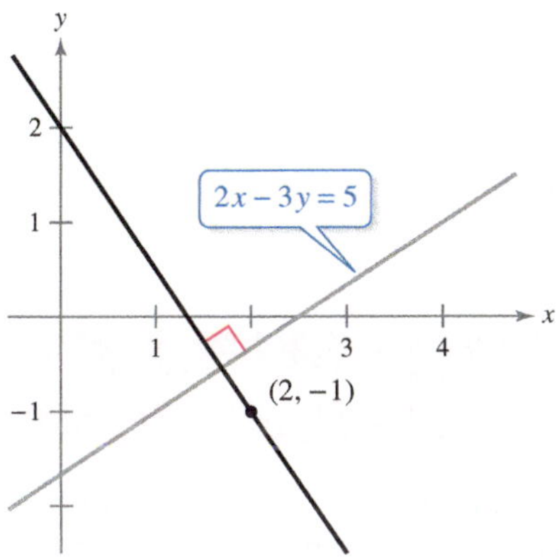

FIGURE 2.31

Find an equation of the line that passes through the point $(2, -1)$ and is perpendicular to the line $2x - 3y = 5$, as shown in Figure 2.31.

SOLUTION By rewriting the given equation in slope-intercept form

$$y = \frac{2}{3}x - \frac{5}{3}$$

you can see that the line has a slope of $\frac{2}{3}$. So, any line that is perpendicular to this line must have a slope of $-\frac{3}{2}$ because $-\frac{3}{2}$ is the negative reciprocal of $\frac{2}{3}$. The required line through the point $(2, -1)$ has the following equation.

$$y - y_1 = m(x - x_1) \quad \text{Point-slope form}$$

$$y - (-1) = -\frac{3}{2}(x - 2) \quad \text{Substitute for } y_1, x_1, \text{ and } m.$$

$$y + 1 = -\frac{3}{2}x + 3 \quad \text{Simplify.}$$

$$y = -\frac{3}{2}x + 3 - 1 \quad \text{Solve for } y.$$

$$y = -\frac{3}{2}x + 2 \quad \text{Write in slope-intercept form.}$$

✓ Checkpoint 6

Find an equation of the line that passes through the point $(-2, 12)$ and is perpendicular to the line $y = \frac{1}{4}x - 2$.

SUMMARIZE (Section 2.2)

1. State the definition of the slope of a line and describe how to find the slope of a line passing through two points *(page 172)*. For an example of finding the slope of a line through two points, see Example 1.
2. State the point-slope form of the equation of a line *(page 174)*. For examples of using the point-slope form of the equation of a line, see Examples 2 and 3.
3. State the slope-intercept form of the equation of a line and describe how it is used to sketch the graph of a line *(page 176)*. For an example of using the slope-intercept form to sketch the graphs of lines, see Example 4.
4. Describe the slopes of two nonvertical parallel lines and the slopes of two nonvertical perpendicular lines *(pages 178 and 179)*. For examples of finding equations of parallel and perpendicular lines, see Examples 5 and 6.

SKILLS WARM UP 2.2

The following warm-up exercises involve skills that were covered in earlier sections. You will use these skills in the exercise set for this section. For additional help, review Sections 0.2 and 1.2.

In Exercises 1 and 2, perform the indicated operations.

1. $\dfrac{4-(-4)}{-3-(-1)}$

2. $\dfrac{-5-8}{0-(-3)}$

3. Evaluate $-1/m$ when $m = 4/5$.

4. Evaluate $-1/m$ when $m = -3$.

In Exercises 5–10, solve for y in terms of x.

5. $2x - 3y = 6$

6. $4x + 2y = 0$

7. $y - (-4) = 3[x - (-1)]$

8. $y - 7 = \frac{2}{3}(x - 3)$

9. $y - (-1) = \dfrac{3-(-1)}{2-4}(x - 4)$

10. $y - 5 = \dfrac{3-5}{0-2}(x - 2)$

Exercises 2.2

See www.CalcChat.com for worked-out solutions to odd-numbered exercises.

Finding the Slope of a Line In Exercises 1–4, use the change in y and the change in x to find the slope of the line.

1.

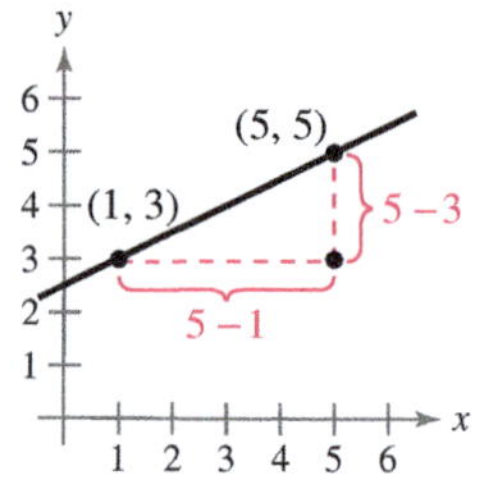

2.

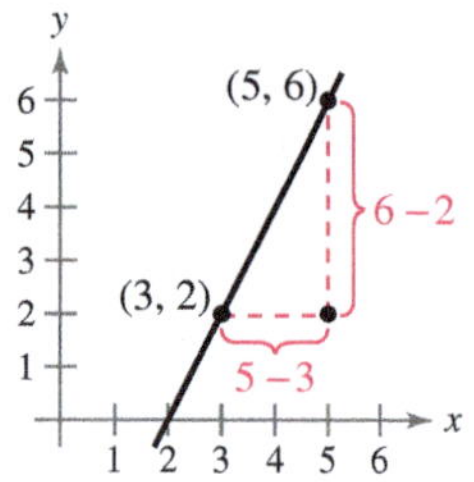

3.

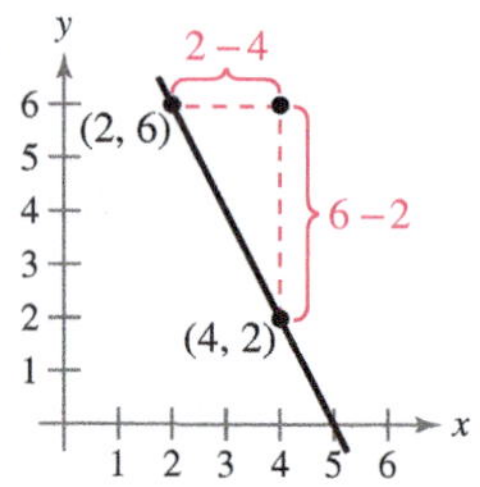

4.

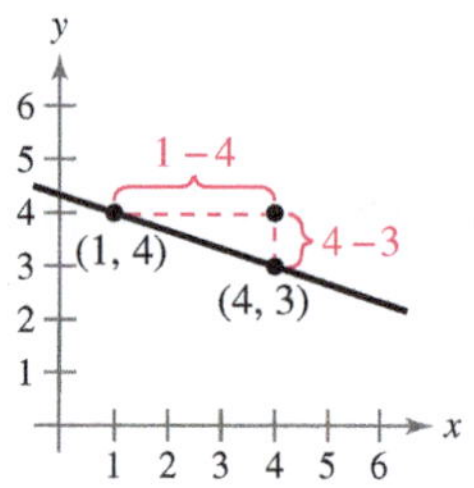

Using Slope In Exercises 5 and 6, identify the line that has the indicated slope.

5. (a) $m = \frac{1}{3}$
(b) m is undefined.
(c) $m = -2$

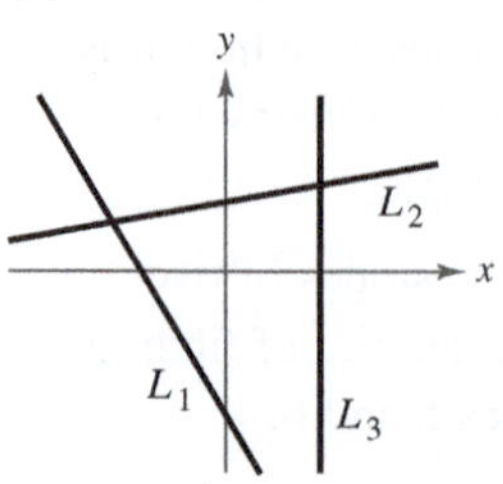

6. (a) $m = 0$
(b) $m = -\frac{3}{4}$
(c) $m = 1$

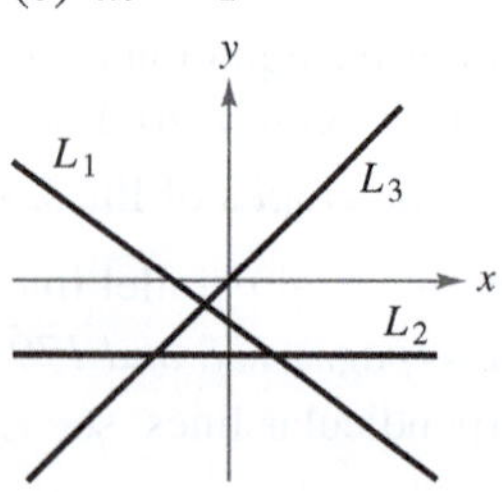

Sketching the Graphs of Lines In Exercises 7 and 8, sketch the graphs of the lines through the point with the indicated slopes in the same coordinate plane.

	Point	*Slopes*	
7.	$(-3, 4)$	(a) -2	(b) $\frac{2}{3}$
		(c) 0	(d) Undefined
8.	$(-2, -5)$	(a) -1	(b) $\frac{3}{4}$
		(c) 0	(d) Undefined

Finding the Slope of a Line Through Two Points In Exercises 9–14, plot the points and find the slope of the line passing through the points. *See Example 1.*

9. $(6, 9), (-4, -1)$

10. $(2, 4), (4, -4)$

11. $(-6, -1), (-6, 4)$

12. $(0, -10), (-4, 0)$

13. $\left(-\frac{1}{3}, 1\right), \left(-\frac{2}{3}, \frac{5}{6}\right)$

14. $\left(\frac{7}{8}, \frac{3}{4}\right), \left(\frac{5}{4}, -\frac{1}{4}\right)$

Using Slope In Exercises 15–22, use the point on the line and the slope of the line to find three additional points through which the line passes. (There are many correct answers.)

	Point	*Slope*
15.	$(5, -2)$	$m = 0$
16.	$(-3, 4)$	$m = 0$
17.	$(2, -5)$	m is undefined.
18.	$(-1, 3)$	m is undefined.
19.	$(5, -6)$	$m = 1$
20.	$(10, -6)$	$m = -1$
21.	$(-6, -1)$	$m = \frac{1}{2}$
22.	$(7, -5)$	$m = -\frac{2}{3}$

The Point-Slope Form of the Equation of a Line In Exercises 23–32, find an equation of the line that passes through the point and has the indicated slope. Then sketch the graph of the line. *See Example 2.*

	Point	*Slope*
23.	$(7, 0)$	$m = 1$
24.	$(0, -4)$	$m = -1$
25.	$(-2, 0)$	$m = -4$
26.	$(1, 3)$	$m = 3$
27.	$(4, 0)$	$m = -\frac{1}{3}$
28.	$(-8, 3)$	$m = -\frac{1}{2}$
29.	$(-2, -7)$	$m = 0$
30.	$(-10, 4)$	$m = 0$
31.	$\left(4, \frac{5}{2}\right)$	$m = \frac{4}{3}$
32.	$\left(-\frac{1}{2}, \frac{3}{2}\right)$	$m = -3$

Writing an Equation of a Line In Exercises 33–44, find an equation of the line passing through the points.

33. $(0, 0), (-1, -4)$
34. $(6, -1), (0, 0)$
35. $(7, -4), (-7, 3)$
36. $(4, 3), (-4, -4)$
37. $(-9, 11), (-9, 14)$
38. $(3, 5), (3, -2)$
39. $(-1, 7), (3, 7)$
40. $(3, -2), (-8, -2)$
41. $\left(2, \frac{1}{2}\right), \left(\frac{1}{2}, \frac{5}{4}\right)$
42. $(1, 1), \left(6, -\frac{2}{3}\right)$
43. $(1, 0.6), (-2, -0.6)$
44. $(-8, 0.6), (2, -2.4)$

45. Discovery Use a graphing utility to graph each equation in the same viewing window. What is the y-intercept of each graph?

$y_1 = x + 2 \qquad y_2 = \frac{1}{4}x + 2 \qquad y_3 = -x + 2$

46. Discovery In Exercise 45, what can you determine about each graph from the coefficient of x in its equation?

Sketching the Graph of a Linear Equation In Exercises 47–56, find the slope and y-intercept (if possible) of the line specified by the equation. Then sketch the graph of the line. *See Example 4.*

47. $y = 2x - 1$
48. $y = 3 - x$
49. $4x - y - 6 = 0$
50. $2x + 3y - 9 = 0$
51. $8 - 3x = 0$
52. $2x + 5 = 0$
53. $7x + 6y - 30 = 0$
54. $x - y - 10 = 0$
55. $2y - 7 = 0$
56. $8 - 5y = 0$

57. Writing A student is confused about why the slope of a vertical line is undefined. Describe how you would help this student understand the concept of undefined slope.

58. Writing A student makes the comment, "I do not understand why a horizontal line has zero slope and how that is different from undefined or no slope." Describe how you would explain the concepts of zero slope and undefined slope and how they are different from each other.

The Intercept Form of the Equation of a Line In Exercises 59–64, use the *intercept form* to find an equation of the line with the given intercepts. The intercept form of the equation of a line with intercepts $(a, 0)$ and $(0, b)$ is

$$\frac{x}{a} + \frac{y}{b} = 1, \quad a \neq 0, \quad b \neq 0.$$

59. x-intercept: $(1, 0)$
y-intercept: $(0, -4)$

60. x-intercept: $(-3, 0)$
y-intercept: $(0, 4)$

61. x-intercept: $(-2, 0)$
y-intercept: $(0, -2)$

62. x-intercept: $(5, 0)$
y-intercept: $(0, 1)$

63. x-intercept: $\left(-\frac{1}{6}, 0\right)$
y-intercept: $\left(0, -\frac{2}{3}\right)$

64. x-intercept: $\left(-\frac{2}{3}, 0\right)$
y-intercept: $\left(0, \frac{1}{2}\right)$

65. Discovery Use a graphing utility to graph the equation $y = x$ in a viewing window with a *square* setting, in which the horizontal and vertical tick marks have equal spacing. Then adjust the viewing window several times by increasing and decreasing the number of units on the y-axis. How does changing the viewing window affect the graph? Which setting best represents the slope of the line?

66. Discovery Use a graphing utility to graph each equation in the same viewing window with a *square* setting (see Exercise 65).

$y_1 = \frac{4}{7}x - 1 \qquad y_2 = \frac{4}{7}x + 2 \qquad y_3 = -\frac{7}{4}x + 8$

(a) What is the geometric relationship between the two lines that have the same slope?

(b) In each pair of perpendicular lines, what is the numerical relationship between their slopes?

Parallel and Perpendicular Lines In Exercises 67–74, the equations of two lines are given. Determine whether the lines L_1 and L_2 are parallel, perpendicular, or neither.

67. L_1: $y = 3x + 4$; L_2: $y = x - \frac{1}{4}$
68. L_1: $y = \frac{3}{4}x + 1$; L_2: $y = -\frac{4}{3}x + 3$
69. L_1: $2x - y = 1$; L_2: $x + 2y = -1$
70. L_1: $x - 5y = -2$; L_2: $-3x + 15y = 6$
71. L_1: $x - 3y = -3$; L_2: $2x - 6y = 6$
72. L_1: $4x - y = -2$; L_2: $8x - 2y = 6$
73. L_1: $2x - 3y - 15 = 0$; L_2: $3x + 2y + 8 = 0$
74. L_1: $x - 4y - 12 = 0$; L_2: $3x - 4y - 8 = 0$

Parallel and Perpendicular Lines In Exercises 75–82, determine whether the lines L_1 and L_2 passing through the indicated pairs of points are parallel, perpendicular, or neither.

75. L_1: $(-5, 0), (-2, 1)$; L_2: $(0, 1), (3, 2)$
76. L_1: $(-1, 6), (1, 4)$; L_2: $(3, -3), (6, -9)$
77. L_1: $(0, -1), (5, 9)$; L_2: $(0, 3), (4, 1)$
78. L_1: $(3, 6), (-6, 0)$; L_2: $(0, -1), \left(5, \frac{7}{3}\right)$
79. L_1: $(-2, -1), (1, 5)$; L_2: $(1, 3), (5, -5)$

80. L_1: $(4, 8), (-4, 2)$; L_2: $(3, -5), \left(-1, \frac{1}{3}\right)$
81. L_1: $(-1, 7), (-6, 4)$; L_2: $(0, 1), (5, 4)$
82. L_1: $(-1, 3), (2, -5)$; L_2: $(3, 0), (2, -7)$

Equations of Parallel and Perpendicular Lines In Exercises 83–88, write equations of the lines passing through the given point that are (a) parallel to the given line and (b) perpendicular to the given line. *See Examples 5 and 6.*

	Point	*Line*
83.	$(6, 2)$	$y = 2x - 1$
84.	$(-5, 4)$	$y = -x + 8$
85.	$\left(\frac{1}{4}, -\frac{2}{3}\right)$	$2x - 3y = 5$
86.	$\left(\frac{7}{8}, \frac{3}{4}\right)$	$5x + 3y = 0$
87.	$(-1, 0)$	$y = -3$
88.	$(2, 5)$	$x = 4$

89. Wheelchair Ramp The maximum recommended slope of a wheelchair ramp is $\frac{1}{12}$. A business is installing a wheelchair ramp that rises 34 inches over a horizontal length of 30 feet. Is the ramp steeper than recommended? *(Source: Americans with Disabilities Act Handbook)*

90. HOW DO YOU SEE IT? Match the description of the situation with its graph. Then write the equation of the line.

(a) You are paying \$10 per week to repay a \$100 loan.

(b) An employee is paid \$12.50 per hour plus \$1.50 for each unit produced per hour.

(c) A sales representative receives \$30 per day for food plus \$0.51 for each mile traveled.

(d) A computer that was purchased for \$800 depreciates \$100 per year.

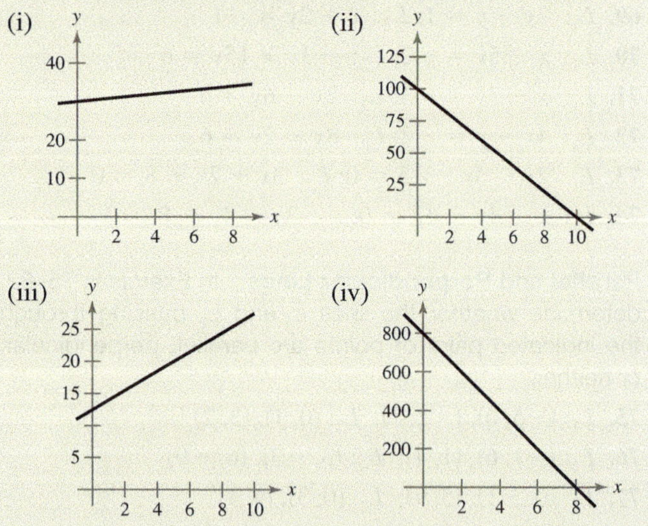

91. Temperature Find an equation of the line that gives the relationship between the temperature in degrees Celsius C and the temperature in degrees Fahrenheit F. Remember that water freezes at 0° Celsius (32° Fahrenheit) and boils at 100° Celsius (212° Fahrenheit).

92. Temperature Use the result of Exercise 91 to complete the table. Is there a temperature for which the Fahrenheit reading is the same as the Celsius reading? If so, what is it?

C		$-10°$	$10°$			$177°$
F	$0°$			$68°$	$90°$	

93. Sales During the first two quarters of the calendar year, a business had sales of \$158,000 and \$165,000, respectively. Write a linear equation giving the sales y in terms of the quarter x. Use the equation to predict the fourth quarter sales. Can you assume that sales will follow this linear pattern?

94. Retail Stores AutoZone operated 3483 stores in 2004. By 2007, AutoZone operated 4056 stores. Assuming the number of stores followed a linear pattern, how many stores would AutoZone have operated in 2010? The actual number of stores operated by AutoZone in 2010 was 4627. Was the actual increase in the number of stores approximately linear? *(Source: AutoZone)*

95. Scuba Diving The pressure (in atmospheres) exerted on a scuba diver's body has a linear relationship with the diver's depth. At sea level (or a depth of 0 feet), the pressure exerted on a diver is 1 atmosphere. At a depth of 99 feet, the pressure exerted on a diver is 4 atmospheres. Write a linear equation to describe the pressure p (in atmospheres) in terms of the depth d (in feet) below the surface of the sea. What is the rate of change in pressure with respect to depth? *(Source: PADI Open Water Diver Manual)*

96. Stone Cutting A stone cutter is making a 6-foot-tall memorial stone. The diagram shows coordinates labeled in feet. The stone cutter plans to cut through the point $(-1, 6)$ along a line that is perpendicular to one side of the stone, as indicated by the dashed line in the figure. Find an equation of the line of the cut.

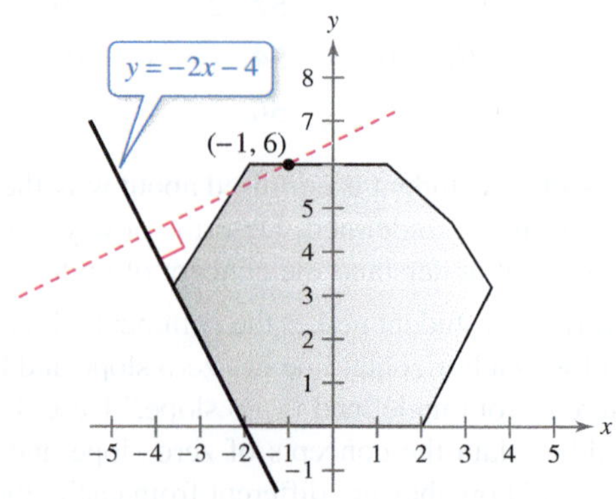

2.3 Linear Modeling and Direct Variation

In Exercise 41 on page 192, you will use data points from a scatter plot to create a linear model describing the number of autistic children receiving disability services.

- Use a mathematical model to approximate a set of data points.
- Construct a linear model to relate quantities that vary directly.
- Construct and use a linear model with slope as the rate of change.
- Use a scatter plot to find a linear model that fits a set of data.

Mathematical Models

The primary objective of applied mathematics is to find equations or **mathematical models** that describe real-world situations. In developing a mathematical model to represent actual data, you should strive for two (often conflicting) goals—accuracy and simplicity. That is, you want the model to be simple enough to be workable, yet accurate enough to produce meaningful results.

You have already studied some techniques for fitting models to data. For instance, in Section 2.2, you learned how to find the equation of a line that passes through two points. In this section, you will study other techniques for fitting models to data: *direct variation, rates of change,* and *linear regression.*

Example 1 A Mathematical Model

The weight of a puppy recorded every two months is shown in the table.

Age (in months)	2	4	6	8	10	12
Weight (in pounds)	24	45	67	93	117	130

A linear model that approximates the puppy's weight w (in pounds) in month t is

$$w = 11.03t + 2.1, \quad 2 \le t \le 12.$$

How closely does the model represent the data?

SOLUTION By graphing the data points with the linear model (see Figure 2.32), you can see that the model is a "good fit" for the actual data. The table shows how each actual weight w compares with the weight $w*$ given by the model.

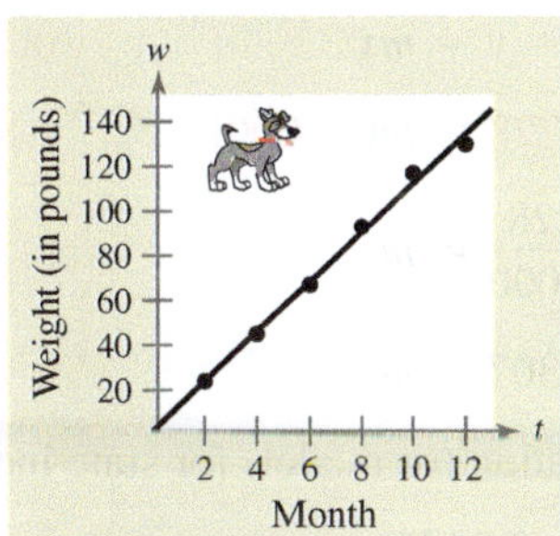

FIGURE 2.32

t	w	$w*$
2	24	24.16
4	45	46.22
6	67	68.28
8	93	90.34
10	117	112.4
12	130	134.46

✓ Checkpoint 1

In Example 1, what are the best and worst approximations given by the model? ■

Direct Variation

There are two basic types of linear models in x and y. The more general model has a y-intercept that is nonzero:

$$y = mx + b, \quad b \neq 0.$$

The simpler model,

$$y = mx$$

has a y-intercept that is zero. In the simpler model, y is said to **vary directly** as x, or to be **directly proportional** to x.

Direct Variation

The following statements are equivalent.

1. y **varies directly** as x.
2. y is **directly proportional** to x.
3. $y = mx$ for some nonzero constant m, where m is the **constant of variation** or the **constant of proportionality.**

Example 2 State Income Tax

In Pennsylvania, state income tax is directly proportional to *taxable income.* For a taxable income of \$40,000, the Pennsylvania state income tax is \$1228. Find a mathematical model that gives the Pennsylvania state income tax in terms of taxable income.

SOLUTION

Verbal Model: State income tax $= m \cdot$ Taxable income

Labels: State income tax $= y$ (dollars)
Taxable income $= x$ (dollars)
Income tax rate $= m$ (percent in decimal form)

Equation: $y = mx$

Find m by substituting the given information into the equation $y = mx$.

$y = mx$ Direct variation model

$1228 = m(40{,}000)$ Substitute $y = 1228$ and $x = 40{,}000$.

$\dfrac{1228}{40{,}000} = m$ Divide each side by 40,000.

$0.0307 = m$ Income tax rate

An equation (or model) for state income tax in Pennsylvania is

$$y = 0.0307x.$$

So, Pennsylvania has a state income tax rate of 3.07% of taxable income. The graph of this equation is shown in Figure 2.33.

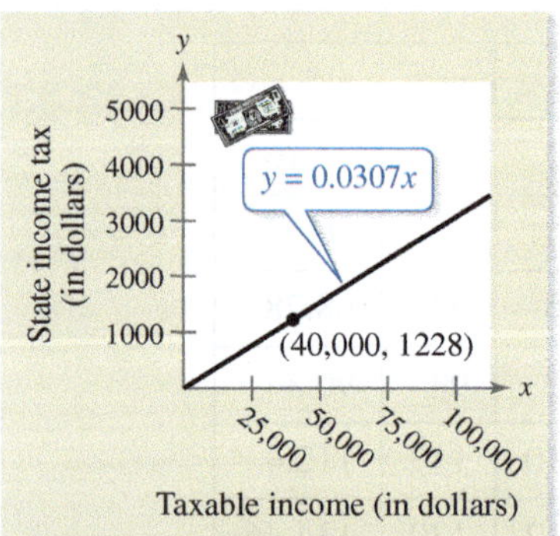

FIGURE 2.33

Checkpoint 2

You buy a flash drive for \$14.50 and pay sales tax of \$0.87. The sales tax is directly proportional to the price. Find a mathematical model that gives the sales tax in terms of the price.

Most measurements in the English system and the metric system are directly proportional. The next example shows how to use a direct proportion to convert between centimeters per day and inches per day.

Example 3 The English and Metric Systems

Bamboos are some of the fastest growing plants in the world. One type of bamboo has been measured to grow 47.6 inches, or 120.9 centimeters, per day. Use this information to find a mathematical model that relates inches per day to centimeters per day. *(Source: American Bamboo Society)*

SOLUTION Let y represent the growth rate in inches per day and let x represent the growth rate in centimeters per day. Then y and x are related by the equation

$$y = mx.$$

Use the fact that $y = 47.6$ when $x = 120.9$ to find the value of m.

$y = mx$ Direct variation model

$47.6 = m(120.9)$ Substitute $y = 47.6$ and $x = 120.9$.

$\frac{47.6}{120.9} = m$ Divide each side by 120.9.

$0.39371 \approx m$ Use a calculator.

So, the conversion factor from centimeters per day to inches per day is approximately 0.39371, and the model is

$$y = 0.39371x.$$

The graph of this equation is shown in Figure 2.34.

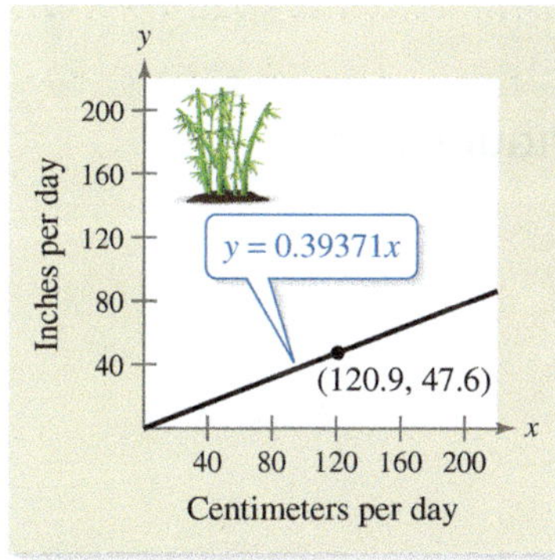

FIGURE 2.34

Checkpoint 3

You buy an ice bucket with a capacity of 44 ounces, or 1.3 liters. Write a mathematical model that relates ounces to liters.

You can use the model from Example 3 to convert any growth rate in centimeters per day to inches per day, as shown in the table.

Centimeters per day	Inches per day
40	15.7
80	31.5
120	47.2
160	63.0
200	78.7

Rates of Change

A second common type of linear model is one that involves a known rate of change. In the linear equation

$$y = mx + b$$

you know that m represents the slope of the line. In real-life problems, the slope can often be interpreted as the **rate of change** of y with respect to x. Rates of change should always be listed in appropriate units of measure.

Example 4 Mountain Climbing

A mountain climber is climbing up a 500-foot cliff. At 1 P.M., the climber is 115 feet up the cliff. By 4 P.M., the climber has reached a height of 280 feet, as shown in Figure 2.35.

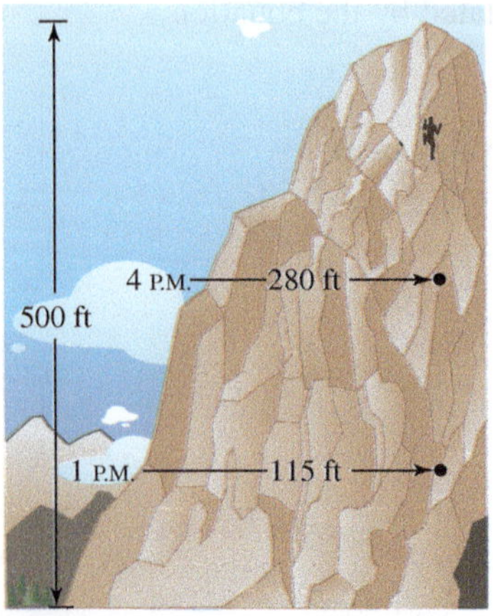

FIGURE 2.35

a. Find the average rate of change of the climber. Use this rate of change to find an equation that relates the height of the climber to the time.

b. Use the equation to estimate the time when the climber reaches the top of the cliff.

SOLUTION

a. Let y represent the climber's height on the cliff (in feet). Let $t = 1$ represent 1 P.M., and let t be measured in hours. Then the two points that represent the climber's two positions are

$$(t_1, y_1) = (1, 115) \quad \text{and} \quad (t_2, y_2) = (4, 280).$$

So, the average rate of change of the climber is

$$\begin{aligned} \text{Average rate of change} &= \frac{y_2 - y_1}{t_2 - t_1} \\ &= \frac{280 - 115}{4 - 1} \\ &= 55 \text{ feet per hour.} \end{aligned}$$

An equation that relates the height of the climber to the time is

$$y - y_1 = m(t - t_1) \qquad \text{Point-slope form}$$

$$y - 115 = 55(t - 1) \qquad \text{Substitute } y_1 = 115,\ t_1 = 1, \text{ and } m = 55.$$

$$y = 55t + 60. \qquad \text{Linear model}$$

If you had chosen to use the point (t_2, y_2) to determine the equation, then you would have obtained a different equation initially: $y - 280 = 55(t - 4)$. However, simplifying this equation yields the same linear model $y = 55t + 60$.

b. To estimate the time when the climber reaches the top of the cliff, let $y = 500$ and solve for t.

$$500 = 55t + 60 \qquad \text{Substitute 500 for } y.$$

$$440 = 55t \qquad \text{Subtract 60 from each side.}$$

$$8 = t \qquad \text{Divide each side by 55.}$$

Because $t = 8$ corresponds to 8 P.M., at the average rate of change, the climber will reach the top at 8 P.M.

✓ Checkpoint 4

At what time does the climber in Example 4 reach a height of 390 feet?

Example 5 Population of San Francisco, California

Between 2000 and 2009, the population of San Francisco, California increased at an average rate of approximately 4292 people per year. In 2000, the population was 776,733. Find a mathematical model that gives the population of San Francisco in terms of the year, and use the model to predict the population in 2012. *(Source: U.S. Census Bureau)*

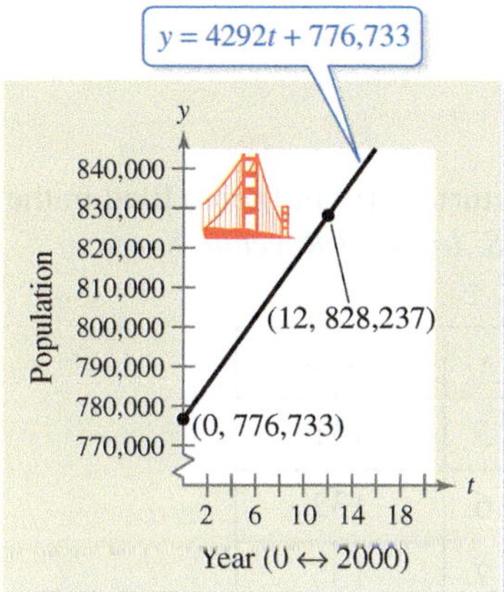

FIGURE 2.36

SOLUTION Let y represent the population of San Francisco and let t represent the calendar year, with $t = 0$ corresponding to 2000. It is convenient to let $t = 0$ correspond to 2000 because you are given the population in 2000. Now, using the rate of change of 4292 people per year, you have

Rate of change	2000 population

$$y = mt + b$$

$$y = 4292t + 776{,}733.$$

Using this model, you can predict the 2012 population to be

$$\begin{aligned} \text{2012 population} &= 4292(12) + 776{,}733 \\ &= 828{,}237. \end{aligned}$$

The graph is shown in Figure 2.36.

✓ Checkpoint 5

Use the model in Example 5 to predict the population of San Francisco in 2014.

Example 6 Straight-Line Depreciation

A racing team buys a \$4750 welder that has a useful life of 10 years. The salvage value of the welder at the end of the 10 years is \$400. Write a linear equation that describes the value of the welder throughout its useable life.

SOLUTION Let V represent the value of the welder (in dollars) at the end of year t. Because the value of the welder at the time of purchase is \$4750, you can represent the initial value of the welder by the ordered pair $(t_1, V_1) = (0, 4750)$. You can represent the salvage value by the ordered pair $(t_2, V_2) = (10, 400)$. The slope of the line is

$$\begin{aligned} m &= \frac{400 - 4750}{10 - 0} \\ &= \frac{-4350}{10} \\ &= -435 \end{aligned}$$

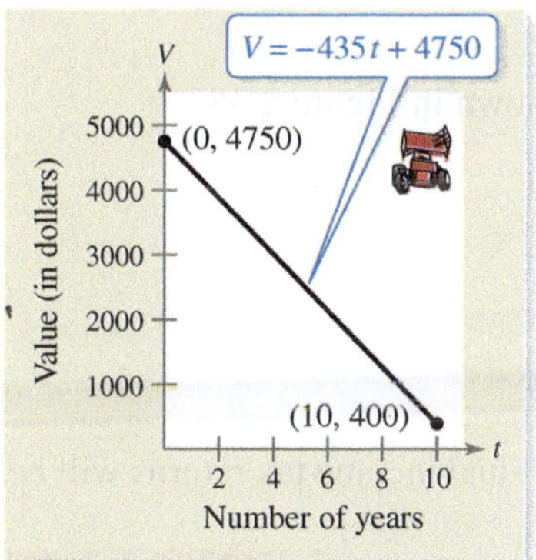

FIGURE 2.37

which represents the annual depreciation in *dollars per year.* Using the slope-intercept form, you can write the equation of the line.

$$V = -435t + 4750 \qquad \text{Slope-intercept form}$$

The graph of the equation is shown in Figure 2.37.

✓ Checkpoint 6

Write a linear equation to model the value of a new machine that costs \$2300 and is worth \$350 after 10 years.

TECH TUTOR

When you use the *regression* feature of your graphing utility, you may obtain an "*r*-value," which gives a measure of how well the model fits the data (see figure).

```
LinReg
 y=ax+b
 a=1.454545455
 b=125.0545455
 r²=.9465588641
 r=.9729125676
```

The closer the value of $|r|$ is to 1, the better the fit. For the data in Example 7, $r \approx 0.973$, which implies that the model is a good fit. For instructions on how to use the *regression* feature, see Appendix A.

Scatter Plots and Regression Analysis

Another type of linear modeling is a graphical approach that is commonly used in statistics. To find a mathematical model that approximates a set of actual data points, plot the points in a rectangular coordinate system. This collection of points is called a **scatter plot.** You can use the statistical features of a graphing utility to calculate the equation of the best-fitting line for the data in your scatter plot. The statistical method of fitting a line to a collection of points is called **linear regression.** A discussion of linear regression is beyond the scope of this text, but the program in most graphing utilities is easy to use and allows you to analyze linear data that may not be convenient to graph by hand.

Example 7 Individual Tax Returns

The table shows the number of individual income tax returns y (in millions) filed in the United States in the years 2000 to 2009. *(Source: U.S. Internal Revenue Service)*

Year	x	Returns, y
2000	0	125
2001	1	127
2002	2	129
2003	3	130
2004	4	130

Year	x	Returns, y
2005	5	131
2006	6	132
2007	7	135
2008	8	138
2009	9	139

a. Use the *regression* feature of a graphing utility to find a linear model for the data. Let $x = 0$ represent 2000.

b. Use the graphing utility to graph the linear model along with a scatter plot of the data.

c. Use the linear model to predict the number of individual tax returns in 2011.

SOLUTION

a. Enter the data into the graphing utility. Then, using the *regression* feature of the graphing utility, you should obtain a linear model for the data that can be rounded to the following:

$$y = 1.45x + 125.1, \quad 0 \le x \le 9.$$

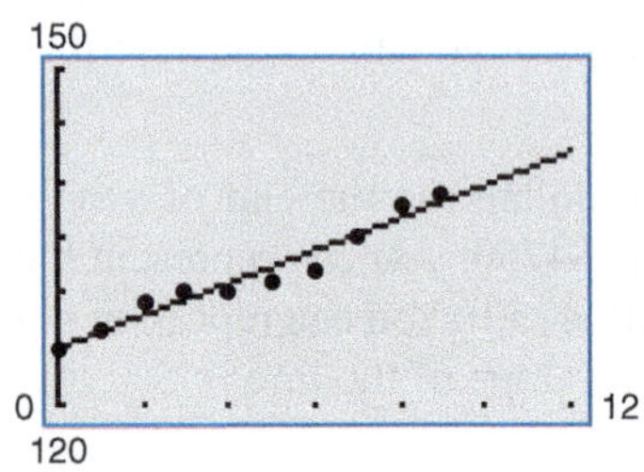

FIGURE 2.38

b. The graph of the equation and the scatter plot are shown in Figure 2.38.

c. Substitute $x = 11$ into the equation found in part (a).

$y = 1.45x + 125.1$ Write equation.

$y = 1.45(11) + 125.1$ Substitute 11 for x.

$y = 141.05$ Simplify.

So, according to the model, about 141,000,000 individual income tax returns will be filed in the United States in 2011.

✓ Checkpoint 7

Repeat Example 7 using only the data for the years 2004–2009.

Example 8 Alternative-Fueled Vehicles

The number of alternative-fueled vehicles v (in thousands) in use in the United States each year from 1995 through 2008 is shown in the table. Construct a scatter plot that represents the data and find a linear model that approximates the data. *(Source: U.S. Energy Information Administration)*

Year	1995	1996	1997	1998	1999	2000	2001
v	246.9	265.0	280.2	295.0	322.3	394.7	425.5

Year	2002	2003	2004	2005	2006	2007	2008
v	471.1	534.0	565.5	592.1	634.6	695.8	775.7

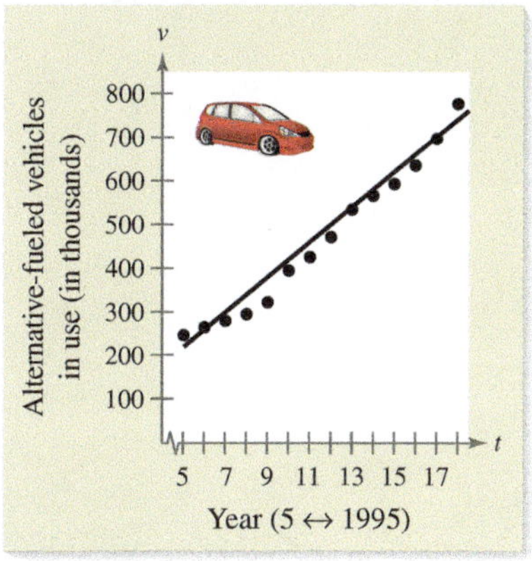

FIGURE 2.39

SOLUTION Let $t = 5$ represent 1995. The scatter plot of the data is shown in Figure 2.39. Draw a line on the scatter plot that approximates the data. To find an equation of the line, approximate two points on the line:

$$(t_1, v_1) = (7, 300)$$

and

$$(t_2, v_2) = (17, 700).$$

So, the slope of the line is

$$m \approx \frac{v_2 - v_1}{t_2 - t_1} = \frac{700 - 300}{17 - 7} = 40.$$

Using the point-slope form, you can determine that an equation of the line is

$$v - 300 = 40(t - 7) \quad \text{Point-slope form}$$

$$v = 40t + 20. \quad \text{Slope-intercept form}$$

To check this model, compare the actual v-values with the v-values given by the model.

✓ Checkpoint 8

Repeat Example 8 using only the data for 2003 through 2008.

SUMMARIZE (Section 2.3)

1. State the definition of a mathematical model *(page 183)*. For an example that uses a mathematical model, see Example 1.
2. Describe what it means when two quantities vary directly *(page 184)*. For examples of direct variation, see Examples 2 and 3.
3. Describe the relationship between slope and rate of change *(page 186)*. For examples that involve rates of change, see Examples 4, 5, and 6.
4. State the definition of linear regression *(page 188)*. For examples of linear regression, see Examples 7 and 8.

Edyta Pawlowska /Shutterstock.com

SKILLS WARM UP 2.3 The following warm-up exercises involve skills that were covered in earlier sections. You will use these skills in the exercise set for this section. For additional help, review Sections 2.1 and 2.2.

In Exercises 1–4, sketch the graph of the line.

1. $y = 2x$

2. $y = \frac{1}{2}x$

3. $y = 2x + 1$

4. $y = \frac{1}{2}x + 1$

In Exercises 5 and 6, find an equation of the line that has the given slope and *y*-intercept.

5. Slope: 1; y-intercept: $(0, 2)$

6. Slope: $\frac{3}{2}$; y-intercept: $(0, 3)$

In Exercises 7–10, find an equation of the line that passes through the two points.

7. $(1, 3)$ and $(6, 8)$

8. $(0, 4)$ and $(7, 10)$

9. $(1, 5.2)$ and $(5, 4.7)$

10. $(2, 6.5)$ and $(8, 3.6)$

Exercises 2.3

See www.CalcChat.com for worked-out solutions to odd-numbered exercises.

1. Dog Growth The weight of a puppy recorded every two months is shown in the table.

Age (in months)	2	4	6
Weight (in pounds)	21	44	63

Age (in months)	8	10	12
Weight (in pounds)	82	92	101

A linear model that approximates the puppy's weight w (in pounds) in month t is $w = 8.0t + 11$, $2 \le t \le 12$. Plot the actual data with the model. How closely does the model represent the data?

2. Non-Wage Earners The numbers of working-age civilians (in millions) in the United States who were not involved in the labor force from 2000 through 2009 are given by the ordered pairs.

(2000, 70.0) (2001, 71.4) (2002, 72.7)
(2003, 74.7) (2004, 76.0) (2005, 76.8)
(2006, 77.4) (2007, 78.7) (2008, 79.5)
(2009, 81.7)

A linear model that approximates the data is

$$y = 1.22t + 70.4, \; 0 \le t \le 9$$

where y is the number of civilians (in millions) and $t = 0$ represents 2000. Plot the actual data with the model. How closely does the model represent the data? *(Source: U.S. Bureau of Labor Statistics)*

3. Aeropostale Sales The yearly sales y (in millions of dollars) of Aeropostale from 2001 through 2009 are given by the ordered pairs.

(2001, 304.8) (2002, 550.9) (2003, 734.9)
(2004, 964.2) (2005, 1204.3) (2006, 1413.2)
(2007, 1590.9) (2008, 1885.5) (2009, 2230.1)

Use a graphing utility to create a scatter plot of the data. Let $x = 1$ represent 2001. Then use the *regression* feature of the graphing utility to find a best-fitting line for the data. Graph the model and the data together. How closely does the model represent the data? *(Source: Aeropostale, Inc.)*

4. Consumer Price Index For urban consumers of educational and communication materials, the Consumer Price Index y, giving the dollar amount equal to the buying power of \$100 in December 1997, is shown for each year from 1998 through 2009 by the ordered pairs.

(1998, 100.3) (1999, 101.2) (2000, 102.5)
(2001, 105.2) (2002, 107.9) (2003, 109.8)
(2004, 111.6) (2005, 113.7) (2006, 116.8)
(2007, 119.6) (2008, 123.6) (2009, 127.4)

Use a graphing utility to create a scatter plot of the data. Let $x = 8$ represent 1998. Then use the *regression* feature of the graphing utility to find a best-fitting line for the data. Graph the model and the data together. How closely does the model represent the data? *(Source: U.S. Bureau of Labor Statistics)*

Direct Variation In Exercises 5–10, y is proportional to x. Use the x- and y-values to find a linear model that relates y and x.

5. $x = 8,\ y = 3$ **6.** $x = 5,\ y = 9$

7. $x = 15,\ y = 300$ **8.** $x = 12,\ y = 204$

9. $x = 7,\ y = 3.2$ **10.** $x = 11,\ y = 1.5$

Direct Variation In Exercises 11–14, write a linear model that relates the variables.

11. H varies directly as p; $H = 27$ when $p = 9$

12. s is proportional to t; $s = 32$ when $t = 4$

13. c is proportional to d; $c = 12$ when $d = 20$

14. r varies directly as s; $r = 25$ when $s = 40$

15. Simple Interest The simple interest received from an investment is directly proportional to the amount of the investment. By investing \$2000 in a corporate bond, you obtain interest of \$120 at the end of 1 year. Find a mathematical model that gives the interest I at the end of 1 year in terms of the amount invested P.

16. Simple Interest The simple interest received from an investment is directly proportional to the amount of the investment. By investing \$4000 in a municipal bond, you obtain interest of \$280 at the end of 1 year. Find a mathematical model that gives the interest I at the end of 1 year in terms of the amount invested P.

17. Property Tax Your property tax is based on the assessed value of your property. (The assessed value is often lower than the actual value of the property.) A house that has an assessed value of \$165,000 has a property tax of \$2211.

(a) Find a mathematical model that gives the amount of property tax y in terms of the assessed value x of the property.

(b) Use the model to find the property tax on a house that has an assessed value of \$185,000.

18. State Sales Tax An item that sells for \$145.99 has a sales tax of \$10.22.

(a) Find a mathematical model that gives the amount of sales tax y in terms of the retail price x.

(b) Use the model to find the sales tax on a purchase that has a retail price of \$540.50.

19. Miles and Kilometers While driving, your speedometer indicates that your speed is 64 miles per hour, or 103 kilometers per hour.

(a) Use this information to find a mathematical model that relates miles per hour to kilometers per hour.

(b) Use the model to complete the table.

Kilometers per hour	40	60	80	100	120
Miles per hour					

20. Liters and Gallons You are buying gasoline and notice that 14 gallons of gasoline is the same as 53 liters.

(a) Use this information to find a mathematical model that relates gallons to liters.

(b) Use the model to complete the table.

Gallons	5	10	20	25	30
Liters					

Writing a Linear Equation In Exercises 21–26, you are given the 2012 value of a product *and* the rate at which the value is expected to change during the next 5 years. Use this information to write a linear equation that gives the dollar value V of the product in terms of the year. (Let $t = 12$ represent 2012.)

	2012 Value	*Rate*
21.	\$2540	\$140 increase per year
22.	\$156	\$4.50 increase per year
23.	\$20,400	\$2142 decrease per year
24.	\$45,000	\$2800 decrease per year
25.	\$154,000	\$10,780 increase per year
26.	\$245,000	\$5600 increase per year

27. Parachuting After opening the parachute, the descent of a parachutist follows a linear model. At 2:08 P.M., the height of the parachutist is 7000 feet. At 2:10 P.M., the height is 4600 feet.

(a) Write a linear equation that gives the height h of the parachutist in terms of the time t. (Let $t = 0$ represent 2:08 P.M., and let t be measured in seconds.)

(b) Use the equation in part (a) to find the time when the parachutist will reach the ground.

28. Distance Traveled by a Car You are driving at a constant speed. At 4:30 P.M., you drive by a sign that gives the distance to Montgomery, Alabama as 84 miles. At 4:59 P.M., you drive by another sign that gives the distance to Montgomery as 56 miles.

(a) Write a linear equation that gives your distance from Montgomery in terms of time t. (Let $t = 0$ represent 4:30 P.M. and let t be measured in minutes.)

(b) Use the equation in part (a) to find the time when you will reach Montgomery.

29. Straight-Line Depreciation A business purchases a piece of equipment for \$875. After 5 years, the equipment will have no value. Write a linear equation giving the value V of the equipment during the 5 years.

30. Straight-Line Depreciation A business purchases a piece of equipment for \$25,000. The equipment will be replaced in 10 years, at which time its salvage value is expected to be \$2000. Write a linear equation giving the value V of the equipment during the 10 years.

31. Hourly Wages A manufacturer pays its assembly line workers \$11.50 per hour. In addition, workers receive a piecework rate of \$0.75 per unit produced. Write a linear equation for the hourly wages W in terms of the number of units x produced per hour.

32. Sales Commission A salesperson receives a monthly salary of \$2500 plus a commission of 7% of sales. Write a linear equation for the salesperson's monthly wage W in terms of the person's monthly sales S.

33. Hair Growth Rate A beautician cuts a customer's hair to a length of 4 inches. During the next year, the customer's hair grows at a rate of 0.5 inch per month.

(a) Write a linear equation giving the hair length H (in inches) in terms of the number of months t.

(b) The hair keeps growing at this constant rate. Predict the hair length in 1.5 years.

34. Pest Management The cost of implementing an invasive species management system in a forest is related to the area of the forest. It costs \$630 to implement the system in a forest area of 10 acres. It costs \$1070 in a forest area of 18 acres.

(a) Write a linear equation giving the cost of the invasive species management system in terms of the number of acres x of forest.

(b) Use the equation in part (a) to find the cost of implementing the system in a forest area of 30 acres.

Finding an Equation of a Line In Exercises 35–40, can the data be approximated by a linear model? If so, sketch the line that best approximates the data. Then find an equation of the line.

35.

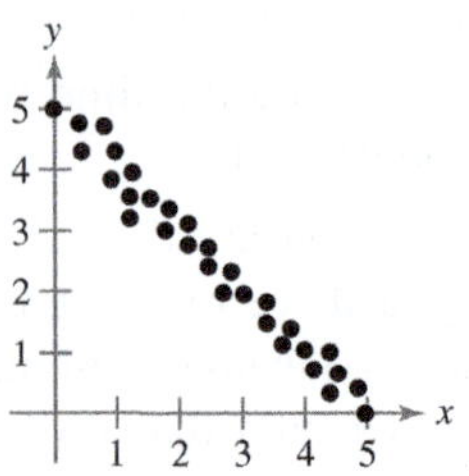

36.

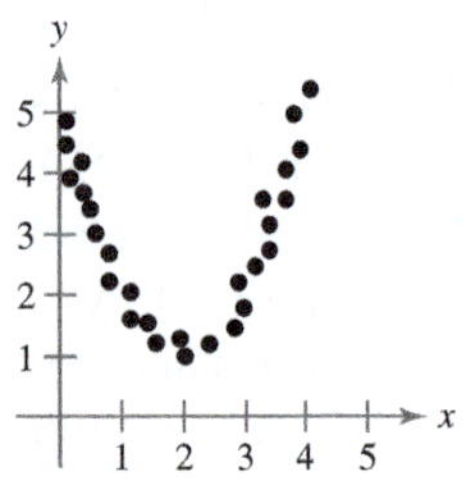

37.

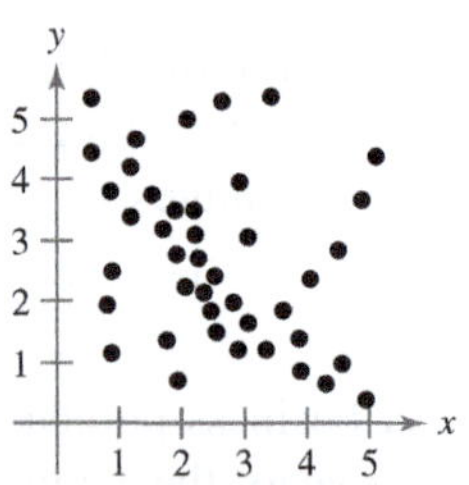

38.

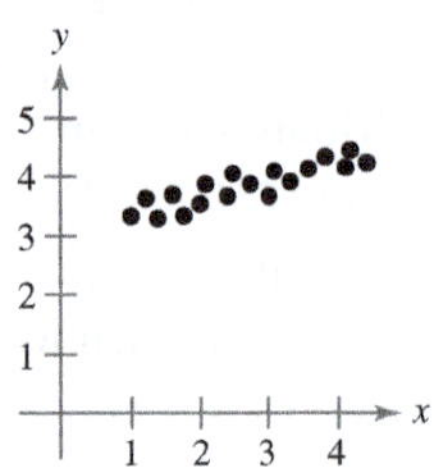

39.

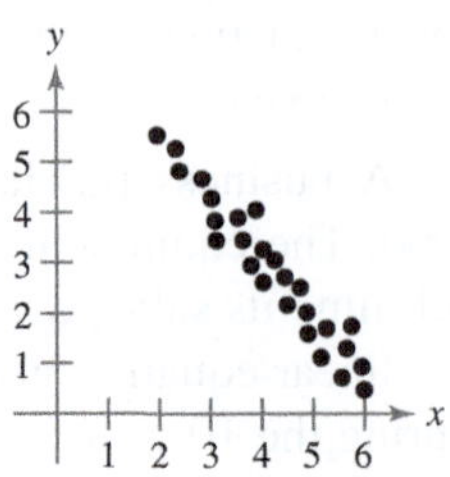

40.

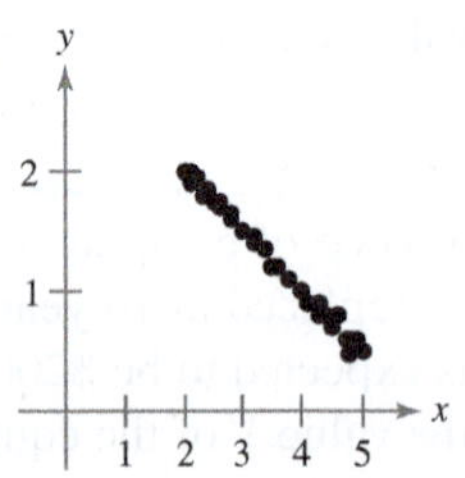

41. Autism The numbers of autistic children C (in thousands) receiving disability services for the years 2001 through 2009 are shown in the table. *(Source: U.S. Department of Education)*

Year	2001	2002	2003	2004	2005	2006	2007	2008	2009
Autistic children, C	98.6	118.8	141.1	166.5	193.8	224.6	258.0	292.8	333.0

(a) Use a graphing utility to create a scatter plot of the data. Let $t = 1$ represent 2001. Do the data appear linear?

(b) Use the *regression* feature of the graphing utility to find a linear model for the data.

(c) State the slope of the graph of the linear model from part (b) and interpret its meaning in the context of the problem.

(d) Use the linear model to predict the numbers of autistic children receiving disability services in 2011 and 2012. Are your predictions reasonable?

42. Japan The population of Japan is expected to drop by 30% over the next 50 years as the percent of its citizens that are elderly increases. Projections for Japan's population through 2050 are shown in the table. *(Source: National Institute of Population and Social Security Research)*

Year, t	2005	2010	2020	2030	2040	2050
Population, P (in millions)	127.7	127.5	124.1	117.6	109.3	100.6

(a) Use a graphing utility to create a scatter plot of the data. Let $t = 5$ represent 2005. Do the data appear linear?

(b) Use the *regression* feature of the graphing utility to find a linear model for the data.

(c) Identify the slope of the model from part (b) and interpret its meaning in the context of the problem.

(d) Use the linear model to predict the populations in 2025, 2035, and 2060. Are these predictions reasonable?

43. Yearly Revenue The yearly revenues y (in millions of dollars) for California Pizza Kitchen for the years 2000 through 2009 are given by the ordered pairs. *(Source: California Pizza Kitchen, Inc.)*

(2000, 210.8) (2001, 249.3)
(2002, 306.3) (2003, 359.9)
(2004, 422.5) (2005, 479.6)
(2006, 554.6) (2007, 632.9)
(2008, 677.1) (2009, 664.7)

(a) Use a graphing utility to create a scatter plot of the data. Let $t = 0$ represent 2000.

(b) Use two points on the scatter plot to find an equation of a line that approximates the data.

(c) Use the *regression* feature of the graphing utility to find a linear model for the data. Use this model and the model from part (b) to predict the revenues in 2010 and 2011.

(d) California Pizza Kitchen projected the revenues in 2010 and 2011 to be \$645 million and \$685 million, respectively. How close are these projections to the predictions from the models?

(e) California Pizza Kitchen also expects the yearly revenue to reach \$850 million in 2013, 2014, or 2015. Do the models from parts (b) and (c) support this? Explain your reasoning.

44. Revenue per Share The revenues per share of stock y (in dollars) for California Pizza Kitchen for the years 2000 through 2009 are given by the ordered pairs. *(Source: California Pizza Kitchen, Inc.)*

(2000, 7.86) (2001, 9.02)
(2002, 10.90) (2003, 12.66)
(2004, 14.64) (2005, 16.26)
(2006, 19.16) (2007, 22.32)
(2008, 28.37) (2009, 27.47)

(a) Use a graphing utility to create a scatter plot of the data. Let $t = 0$ represent 2000.

(b) Use two points on the scatter plot to find an equation of a line that approximates the data.

(c) Use the *regression* feature of the graphing utility to find a linear model for the data. Use this model and the model from part (b) to predict the revenues per share in 2010 and 2011.

(d) California Pizza Kitchen projected the revenues per share in 2010 and 2011 to be \$25.80 and \$27.40, respectively. How close are these projections to the predictions from the models?

(e) California Pizza Kitchen also expects the revenue per share to reach \$34.00 in 2013, 2014, or 2015. Do the models from parts (b) and (c) support this? Explain your reasoning.

45. Purchasing Power The value y (in 1982 dollars) of each dollar received by producers in each of the years from 1994 through 2008 in the United States is represented by the ordered pairs. *(Source: U.S. Bureau of Labor Statistics)*

(1994, 0.797) (1995, 0.782)
(1996, 0.762) (1997, 0.759)
(1998, 0.765) (1999, 0.752)
(2000, 0.725) (2001, 0.711)
(2002, 0.720) (2003, 0.698)
(2004, 0.673) (2005, 0.642)
(2006, 0.623) (2007, 0.600)
(2008, 0.565)

(a) Use a spreadsheet software program to generate a scatter plot of the data. Let $t = 4$ represent 1994. Do the data appear linear?

(b) Use the *regression* feature of the spreadsheet software program to find a linear model for the data.

(c) Use the model to predict the value (in 1982 dollars) of 1 dollar received by producers in 2010 and in 2011. Discuss the reliability of your predictions based on your scatter plot and the graph of your linear model for the data.

46. Purchasing Power The value y (in 1982–1984 dollars) of each dollar paid by consumers in each of the years from 1994 through 2008 in the United States is represented by the ordered pairs. *(Source: U.S. Bureau of Labor Statistics)*

(1994, 0.675) (1995, 0.656)
(1996, 0.638) (1997, 0.623)
(1998, 0.613) (1999, 0.600)
(2000, 0.581) (2001, 0.565)
(2002, 0.556) (2003, 0.543)
(2004, 0.529) (2005, 0.512)
(2006, 0.496) (2007, 0.482)
(2008, 0.464)

(a) Use a spreadsheet software program to generate a scatter plot of the data. Let $t = 4$ represent 1994. Do the data appear linear?

(b) Use the *regression* feature of the spreadsheet software program to find a linear model for the data.

(c) Use the model to predict the value (in 1982–1984 dollars) of 1 dollar paid by consumers in 2010 and in 2011. Discuss the reliability of your predictions based on your scatter plot and the graph of your linear model for the data.

47. Health Services The numbers of employees E (in thousands) in the health services industry for the years 2004 through 2009 are shown in the table. *(Source: U.S. Bureau of Labor Statistics)*

Year	2004	2005	2006
Employees, E	14,190	14,536	14,925

Year	2007	2008	2009
Employees, E	15,380	15,798	16,101

(a) Use a graphing utility to create a scatter plot of the data. Let $t = 4$ represent 2004. Do the data appear linear?

(b) Use the *regression* feature of the graphing utility to find a linear model for the data.

(c) Use the model to predict the numbers of employees in 2011 and 2013.

(d) Graph the linear model along with the scatter plot of the data. Comparing the data with the model, are the predictions in part (c) most likely high, low, or just about right? Explain your reasoning.

48. Health Care The total yearly health care expenditures E (in billions of dollars) in the United States for the years 2000 through 2009 are shown in the table. *(Source: U.S. Centers for Medicare and Medicaid Services)*

Year	2000	2001	2002	2003	2004
Expenditures, E	1353	1469	1602	1735	1855

Year	2005	2006	2007	2008	2009
Expenditures, E	1983	2113	2240	2339	2486

(a) Use a graphing utility to create a scatter plot of the data. Let $t = 0$ represent 2000. Do the data appear linear?

(b) Use the *regression* feature of the graphing utility to find a linear model for the data.

(c) Use the model to predict the health care expenditures in 2011 and 2012.

(d) Graph the linear model along with the scatter plot. The government's projections for 2011 and 2012 are \$2703 billion and \$2850 billion, respectively. Compare these projections with the predictions in part (c). Are the government's projections high, low, or just about right? Explain your reasoning.

49. Think About It Annual data from three years are used to create linear models for the population and the yearly snowfall of Reno, Nevada. Which model is more likely to give better predictions for future years? Discuss the appropriateness of using only three data points in each situation.

50. HOW DO YOU SEE IT? Match the description of the situation with its graph. Find the slope of the line and describe how it is interpreted in the real-life situation. [The graphs are labeled (i), (ii), (iii), and (iv).]

(a) You pay \$0.06 in sales tax for every dollar spent.

(b) A person pays \$40 to reserve a banquet hall, plus \$12 for each hour it is occupied.

(c) You pay \$800 for a new refrigerator that depreciates \$50 per year.

(d) You pay \$34.99 per day to rent a car.

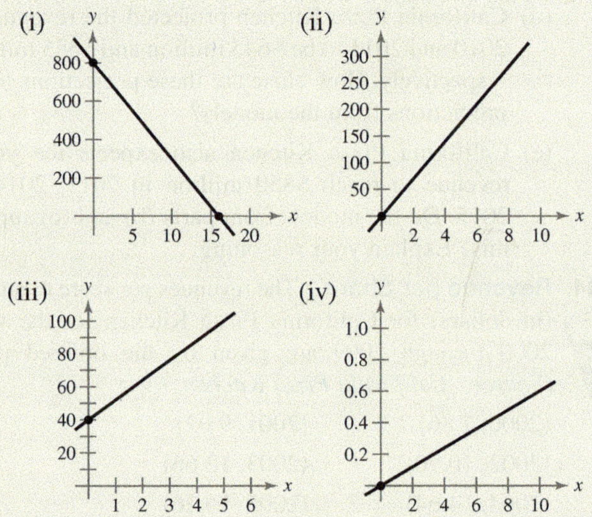

51. Think About It You begin a dance video game with 800 points. You lose 50 points each time you incorrectly complete a dance move. Does this description match graph (i) in Exercise 50? Explain.

52. Think About It You start with \$12 and save \$40 per week. Does this description match graph (iii) in Exercise 50? Explain.

53. Think About It How can you use the rate of change of y with respect to x to determine whether the data points $(1, 10)$, $(2, 20)$, $(3, 30)$, $(4, 40)$, and $(5, 50)$ are linear?

54. Writing Use your campus library, the Internet, or some other reference source to locate data that you think describes a linear relationship. Use a graphing utility to create a scatter plot of the data and find a linear model that represents the points. Interpret the slope and y-intercept in the context of the data. Write a summary of your findings.

2.4 Functions

In Exercise 86 on page 206, you will use a piecewise-defined function to approximate the numbers of Coast Guard interdictions in various years.

- Determine whether a correspondence is a function, and use function notation.
- Find the domain of a function.
- Write a function that relates quantities in an application problem.

Introduction to Functions and Function Notation

Many everyday phenomena involve two quantities that are related to each other by some rule of correspondence. Here are some examples.

1. The simple interest I earned on \$1000 for 1 year is related to the annual interest rate r by the formula $I = 1000r$.
2. The distance d traveled on a bicycle in 2 hours is related to the speed s of the bicycle by the formula $d = 2s$.
3. The area A of a circle is related to its radius r by the formula $A = \pi r^2$.

Not all correspondences between two quantities have simple mathematical formulas. For instance, people commonly match up athletes with jersey numbers and hours of the day with temperatures. In each of these cases, however, there is some rule of correspondence that matches each item from one set with exactly one item from a different set. Such a rule of correspondence is called a **function.**

Definition of a Function

A **function** f from a set A to a set B is a rule of correspondence that assigns to each element x in the set A exactly one element y in the set B. The set A is the **domain** (or set of inputs) of the function f, and the set B contains the **range** (or set of outputs).

To get a better idea of this definition, look at the function that relates the time of day to the temperature in Figure 2.40. This function can be represented by the following set of ordered pairs.

$$\{(1, 9°), (2, 13°), (3, 15°), (4, 15°), (5, 12°), (6, 4°)\}$$

In each ordered pair, the first coordinate (x-value) is the input and the second coordinate (y-value) is the output. In this example, note the following characteristics of a function.

1. Each element of A (the domain) must be matched with an element of B (the range).
2. Some elements of B may not be matched with any element of A.
3. Two or more elements of A may be matched with the same element of B.
4. An element of A cannot be matched with two different elements of B.

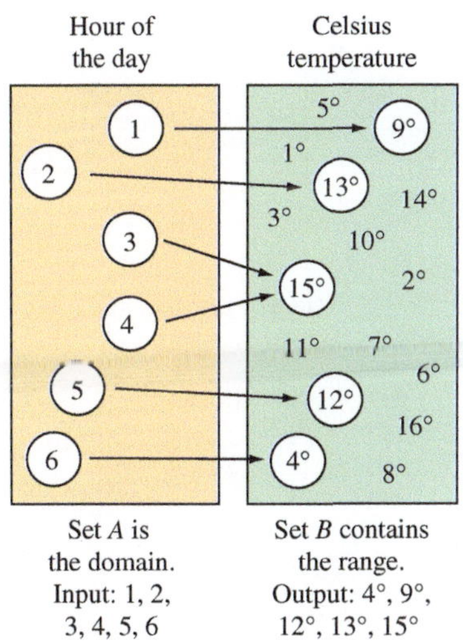

Function from Set A to Set B

FIGURE 2.40

To determine whether a correspondence is a function, you must decide whether each element of the domain A is matched with exactly one element of the range B. When any element of A is matched with two or more elements of B, the correspondence is not a function. For example, people are not a function of their birthday month because many people are born in any given month.

Example 1 Testing for Functions

Let $A = \{a, b, c\}$ and $B = \{1, 2, 3, 4, 5\}$. Which of the following sets of ordered pairs or figures represent functions from set A to set B?

a. $\{(a, 2), (b, 3), (c, 4)\}$

b. $\{(a, 4), (b, 5)\}$

c.

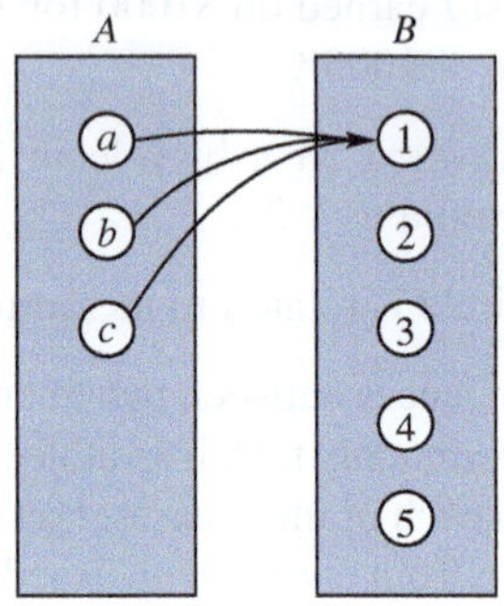

d.

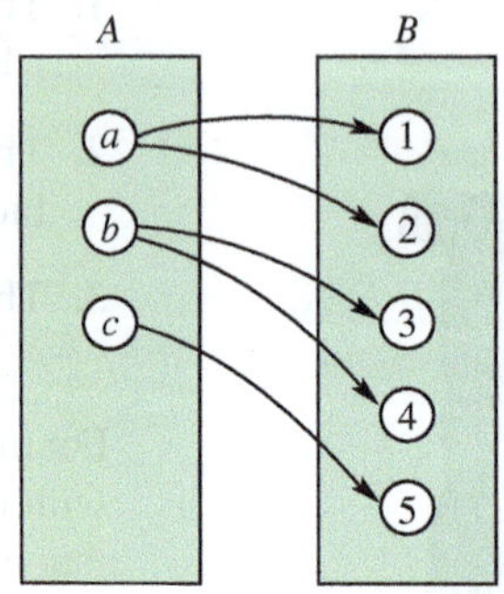

SOLUTION

a. This collection of ordered pairs *does* represent a function from A to B. Each element of A is matched with exactly one element of B.

b. This collection of ordered pairs *does not* represent a function from A to B. Not every element of A is matched with an element of B. The element c of A is not matched with an element of B.

c. This figure *does* represent a function from A to B. It does not matter that each element of A is matched with the same element of B.

d. This figure *does not* represent a function from A to B. The element a of A is matched with *two* elements of B. This is also true of the element b.

✓Checkpoint 1

Let $A = \{a, b, c, d\}$ and $B = \{1, 3, 5, 7\}$. Decide whether each set of ordered pairs represents a function from set A to set B.

a. $\{(a, 1), (c, 5), (d, 7)\}$

b. $\{(a, 3), (b, 7), (c, 1), (d, 3)\}$

c. $\{(b, 3), (b, 5)\}$

Representing functions by sets of ordered pairs is a common practice in *discrete mathematics*. In algebra, however, it is more common to represent functions by equations or formulas involving two variables. For instance, the equation

$$y = x^2 \qquad y \text{ is a function of } x.$$

represents the variable y as a function of the variable x. In this equation, x is the **independent variable** and y is the **dependent variable.** The domain of the function is the set of all values taken on by the independent variable x, and the range of the function is the set of all values taken on by the dependent variable y.

Example 2 Testing for Functions Represented by Equations

Which of the equations represent(s) y as a function of x?

a. $x^2 + y = 1$ **b.** $-x + y^2 = 1$

SOLUTION To determine whether y is a function of x, try to solve for y in terms of x.

a. Solving for y yields

$x^2 + y = 1$ Write original equation.

$y = 1 - x^2.$ Solve for y.

To each value of x there corresponds exactly one value of y. So, y *is* a function of x.

b. Solving for y yields

$-x + y^2 = 1$ Write original equation.

$y^2 = 1 + x$ Add x to each side.

$y = \pm\sqrt{1 + x}.$ Solve for y.

The $\pm$ indicates that to a given value of x there correspond two values of y. So, y *is not* a function of x.

✓ Checkpoint 2

Determine whether each equation represents y as a function of x.

a. $y - 2 = x^2$ **b.** $x^2 + y^2 = 25$

STUDY TIP

Make sure you understand how function notation is written and how it is read. Frequently, $f(x)$ is misinterpreted as "f times x" rather than "f of x."

When an equation is used to represent a function, it is convenient to name the function so that it can be referenced easily. For example, you know that the equation $y = 1 - x^2$ describes y as a function of x. Suppose you give this function the name "f." Then you can use the **function notation** shown below.

Input	*Output*	*Equation*
x	$f(x)$	$f(x) = 1 - x^2$

The symbol $f(x)$ is read as the **value of f at x** or simply **f of x.** The symbol $f(x)$ corresponds to the y-value for a given x. So, you can write $y = f(x)$. Keep in mind that f is the *name* of the function, whereas $f(x)$ is the *value* of the function at x. For instance, the function given by

$$f(x) = 3 - 2x$$

has *function values* denoted by $f(-1)$, $f(0)$, $f(2)$, and so on. To find these values, substitute the specified input values into the given equation.

For $x = -1$, $f(-1) = 3 - 2(-1)$

$= 3 + 2$

$= 5.$

For $x = 0$, $f(0) = 3 - 2(0)$

$= 3 - 0$

$= 3.$

For $x = 2$, $f(2) = 3 - 2(2)$

$= 3 - 4$

$= -1.$

Although f is often used as a convenient function name and x is often used as the independent variable, you can use other letters. For instance,

$$f(x) = x^2 - 4x + 7, \quad f(t) = t^2 - 4t + 7, \quad \text{and} \quad g(s) = s^2 - 4s + 7$$

all define the same function. In fact, the role of the independent variable in a function is simply that of a "placeholder." Consequently, the function above could be described by the form

$$f(\quad) = (\quad)^2 - 4(\quad) + 7.$$

Example 3 Evaluating a Function

Let $g(x) = -x^2 + 4x + 1$. Evaluate $g(x)$ at each specified value and simplify.

a. $g(2)$ **b.** $g(t)$ **c.** $g(x + 2)$

SOLUTION

a. Replace x with 2 in $g(x) = -x^2 + 4x + 1$.

$$g(2) = -(2)^2 + 4(2) + 1 = -4 + 8 + 1 = 5$$

b. Replace x with t.

$$g(t) = -(t)^2 + 4(t) + 1 = -t^2 + 4t + 1$$

c. Replace x with $x + 2$.

$$\begin{aligned} g(x+2) &= -(x+2)^2 + 4(x+2) + 1 \\ &= -(x^2 + 4x + 4) + 4x + 8 + 1 \\ &= -x^2 - 4x - 4 + 4x + 8 + 1 \\ &= -x^2 + 5 \end{aligned}$$

STUDY TIP

In Example 3(c), note that $g(x + 2)$ is not equal to $g(x) + g(2)$. In general, $g(u + v) \neq g(u) + g(v)$.

✓ Checkpoint 3

Let $h(x) = 2x^2 + x - 4$. Find $h(-1)$.

A function defined by two or more equations over a specified domain is called a **piecewise-defined function.**

Example 4 A Piecewise-Defined Function

Evaluate the function when $x = -1$, 0, and 1.

$$f(x) = \begin{cases} x^2 + 1, & x < 0 \\ x - 1, & x \geq 0 \end{cases}$$

SOLUTION For each x-value, only one part of the piecewise function is used. Because $x = -1$ is less than 0, use $f(x) = x^2 + 1$ to obtain

$$f(-1) = (-1)^2 + 1 = 2.$$

For $x = 0$, use $f(x) = x - 1$ to obtain $f(0) = (0) - 1 = -1$. For $x = 1$, use $f(x) = x - 1$ to obtain $f(1) = (1) - 1 = 0$. The graph of f is shown in Figure 2.41.

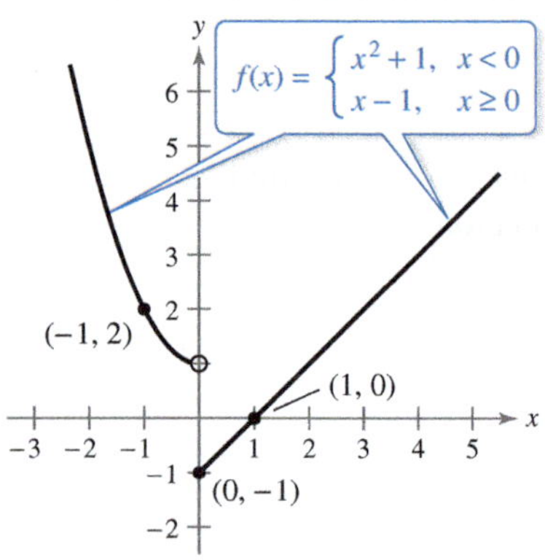

FIGURE 2.41

✓ Checkpoint 4

Evaluate the function in Example 4 when $x = -3$ and 3.

Finding the Domain of a Function

The domain of a function can be described explicitly or it can be *implied* by the expression used to define the function. The **implied domain** is the set of all real numbers for which the expression is defined. For instance, the function given by

$$f(x) = \frac{1}{x^2 - 4}$$ Domain excludes x-values that result in division by zero.

has an implied domain that consists of all real x other than $x = \pm 2$. These two values are excluded from the domain because division by zero is undefined. Another common type of implied domain results from the restrictions needed to avoid even roots of negative numbers. For example, the function given by

$$f(x) = \sqrt{x}$$ Domain excludes x-values that result in even roots of negative numbers.

is defined only for $x \geq 0$. So, its implied domain is the interval $[0, \infty)$. In general, the domain of a function *excludes* values that would cause division by zero *or* result in the even root of a negative number.

Example 5 Finding the Domain of a Function

Find the domain of each function.

a. f: $\{(-3, 0), (-1, 4), (0, 2), (2, 2), (4, -1)\}$ **b.** $g(x) = \dfrac{1}{x + 5}$

c. Volume of a sphere: $V = \dfrac{4}{3}\pi r^3$ **d.** $h(x) = \sqrt{4 - x^2}$

e. $r(x) = \sqrt[3]{x + 3}$

SOLUTION

a. The domain of f consists of all first coordinates in the set of ordered pairs.

$$\text{Domain} = \{-3, -1, 0, 2, 4\}$$

b. Excluding x-values that yield zero in the denominator, the domain of g is the set of all real numbers x such that $x \neq -5$.

c. Because this function represents the volume of a sphere, the values of the radius r must be positive. So, the domain is the set of all real numbers r such that $r > 0$.

d. This function is defined only for x-values for which

$$4 - x^2 \geq 0.$$

Using the methods described in Section 1.7, you can conclude that $-2 \leq x \leq 2$. So, the domain of h is the interval $[-2, 2]$.

e. Because the cube root of any real number is defined, the domain of r is the set of all real numbers, or $(-\infty, \infty)$.

✓ Checkpoint 5

Find the domain of the function $f(x) = 6 - x^3$.

In Example 5(c), note that the domain of a function may be implied by the physical context. For instance, from the equation $V = \frac{4}{3}\pi r^3$, you would have no reason to restrict r to positive values, but the physical context implies that a sphere cannot have a negative or zero radius.

Applications

Example 6 The Dimensions of a Container

You are working with a cylindrical beaker in a chemistry lab experiment. The height of the beaker is 4 times the radius, as shown in Figure 2.42.

a. Write the volume of the beaker as a function of the radius r.

b. Write the volume of the beaker as a function of the height h.

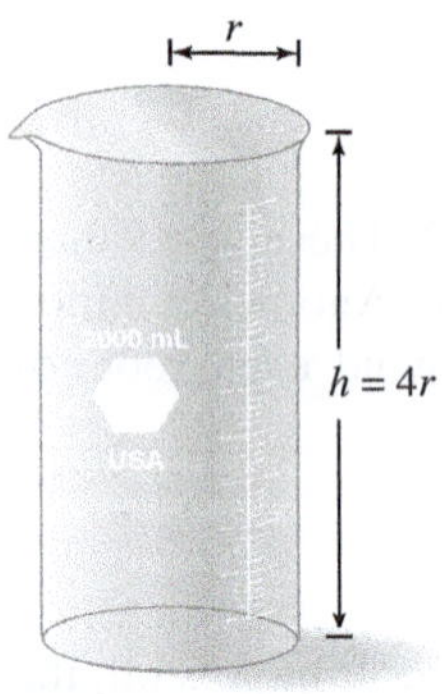

FIGURE 2.42

SOLUTION

a. $V = \pi r^2 h$ Write volume formula.

$= \pi r^2(4r)$ Substitute $4r$ for h.

$= 4\pi r^3$ V is a function of r.

b. $V = \pi\left(\frac{h}{4}\right)^2 h$ Substitute $\frac{h}{4}$ for r.

$= \frac{\pi h^3}{16}$ V is a function of h.

Checkpoint 6

In Example 6, write the volume of the beaker as a function of the height h when the radius is twice the height.

Example 7 The Path of a Baseball

A baseball is hit at a point 3 feet above home plate at a velocity of 100 feet per second and an angle of 45°. The path of the baseball is given by the function

$$y = -0.0032x^2 + x + 3$$

where y and x are measured in feet. Will the baseball clear a 10-foot fence located 300 feet from home plate?

SOLUTION When $x = 300$, the height of the baseball is given by

$$y = -0.0032(300)^2 + 300 + 3 = 15 \text{ feet.}$$

Yes, the ball will clear the fence, as shown in Figure 2.43.

STUDY TIP

Notice that in Figure 2.43, the baseball is not at the point (0, 0) when it is initially hit. This is because the baseball is hit at a point 3 feet above home plate. So, it is hit at the point (0, 3).

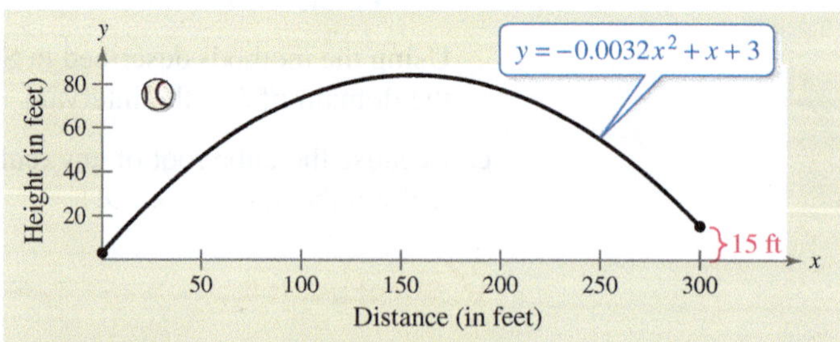

FIGURE 2.43

Checkpoint 7

In Example 7, will the baseball clear a 35-foot fence located 280 feet from home plate?

Example 8 Houses Sold

The number H (in thousands) of new privately owned one-family houses sold in the United States increased in a quadratic pattern from 1998 through 2004. Then, in 2005, the pattern changed from a quadratic to a linear pattern (see Figure 2.44). These two patterns can be approximated by the function

$$H = \begin{cases} 15.036t^2 - 278.68t + 2159.3, & 8 \le t \le 14 \\ -238.20t + 4843.4, & 15 \le t \le 19 \end{cases}$$

with $t = 8$ corresponding to 1998. Use this function to approximate the total number of new privately owned one-family houses sold between 1998 and 2009. *(Source: U.S. Census Bureau and U.S. Department of Housing and Urban Development)*

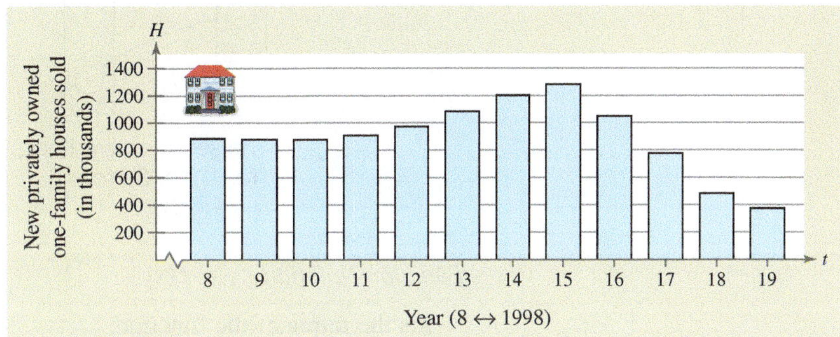

FIGURE 2.44

SOLUTION From 1998 through 2004, use the equation $H = 15.036t^2 - 278.68t + 2159.3$ to approximate the number of houses sold, as shown in the table.

t	8	9	10	11	12	13	14
H	892.2	869.1	876.1	913.2	980.3	1077.5	1204.8

From 2005 through 2009, use the equation $H = -238.20t + 4843.4$ to approximate the number of houses sold, as shown in the table.

t	15	16	17	18	19
H	1270.4	1032.2	794.0	555.8	317.6

To approximate the total number of houses sold from 1998 through 2009, you can add the amounts for the 12 years to obtain the sum 10,783.2. Because H is measured in thousands, you can conclude that the total number of new privately owned one-family houses sold between 1998 and 2009 was approximately 10,783,200.

✓ Checkpoint 8

The median sales price P (in thousands of dollars) of new privately owned one-family houses sold in the United States from 1999 through 2009 can be modeled by the function

$$P = \begin{cases} 8.66t + 82.3, & 9 \le t \le 13 \\ -4.680t^2 + 153.12t - 1004.5, & 14 \le t \le 19 \end{cases}$$

with $t = 9$ corresponding to 1999. Use the function to approximate the median sales price for each year from 1999 through 2009. *(Source: U.S. Census Bureau and U.S. Department of Housing and Urban Development)* ■

Stephen Coburn/www.shutterstock.com

Summary of Function Terminology

Function: A **function** is a relationship between two variables such that to each value of the independent variable there corresponds exactly one value of the dependent variable.

For instance, let $A = \{a, b, c\}$ and $B = \{1, 2, 3, 4\}$.

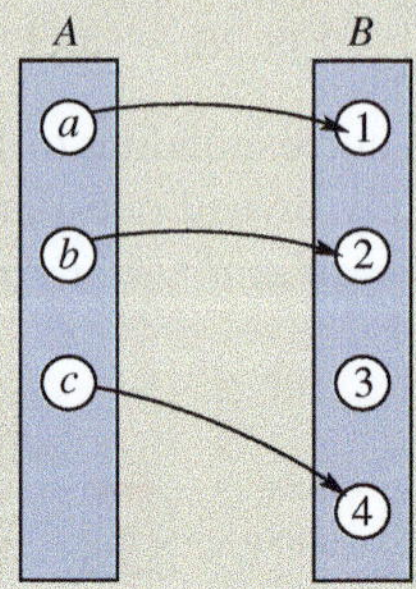

The set of ordered pairs $\{(a, 1), (b, 2), (c, 4)\}$ *is a function.*

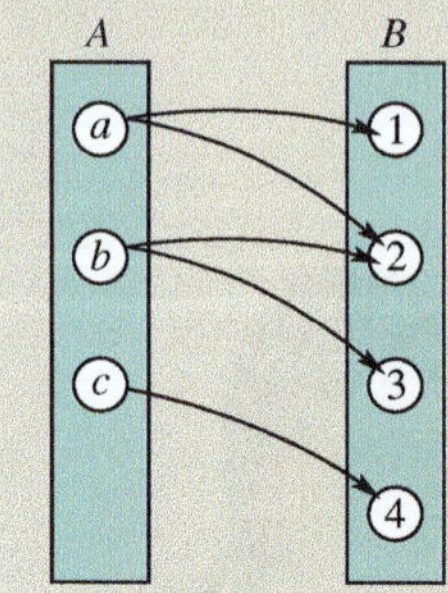

The set of ordered pairs $\{(a, 1), (a, 2), (b, 2), (b, 3), (c, 4)\}$ *is not a function.*

Function Notation: $y = f(x)$

f is the **name** of the function.

y is the **dependent variable.**

x is the **independent variable.**

$f(x)$ is the **value of the function at x.**

Domain: The **domain** of a function is the set of all values (inputs) of the independent variable for which the function is defined. If x is in the domain of f, then f is said to be **defined** at x. If x is not in the domain of f, then f is said to be **undefined** at x.

Range: The **range** of a function is the set of all values (outputs) assumed by the dependent variable (that is, the set of all function values).

Implied Domain: If f is defined by an algebraic expression and the domain is not specified, then the **implied domain** consists of all real numbers for which the expression is defined.

SUMMARIZE (Section 2.4)

1. State the definition of a function *(page 195)*. For examples of determining functions, see Examples 1 and 2.
2. Describe function notation *(pages 197 and 198)*. For examples of using function notation, see Examples 3 and 4.
3. State the definition of an implied domain of a function *(page 199)*. For an example of finding the domain of a function, see Example 5.
4. Describe a real-life example of how a function can be used to analyze the path of an object *(page 200, Example 7)*.

holbox/www.shutterstock.com

SKILLS WARM UP 2.4 The following warm-up exercises involve skills that were covered in earlier sections. You will use these skills in the exercise set for this section. For additional help, review Sections 0.2, 1.2, and 1.7.

In Exercises 1–4, simplify the expression.

1. $2(-3)^3 + 4(-3) - 7$

2. $4(-1)^2 - 5(-1) + 4$

3. $(x + 1)^2 + 3(x + 1) - 4 - (x^2 + 3x - 4)$

4. $(x - 2)^2 - 4(x - 2) - (x^2 - 4)$

In Exercises 5 and 6, solve for y in terms of x.

5. $2x + 5y - 7 = 0$

6. $y^2 = x^2$

In Exercises 7–10, solve the inequality.

7. $x^2 - 4 \geq 0$

8. $9 - x^2 \geq 0$

9. $x^2 + 2x + 1 \geq 0$

10. $x^2 - 3x + 2 \geq 0$

Exercises 2.4

See www.CalcChat.com for worked-out solutions to odd-numbered exercises.

Testing for Functions In Exercises 1–4, decide whether the set of figures represents a function from A to B.

$A = \{a, b, c\}$

$B = \{1, 2, 3, 4\}$

Give a reason for your answer. *See Example 1.*

1.

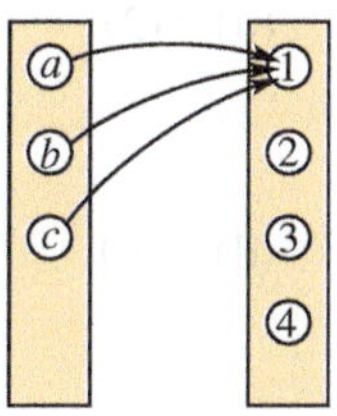

2.

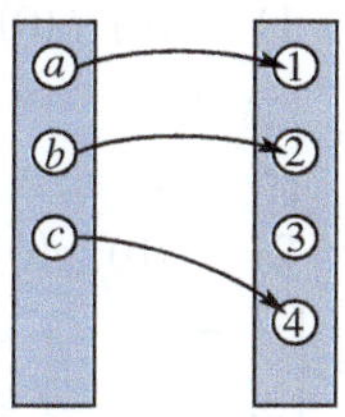

3.

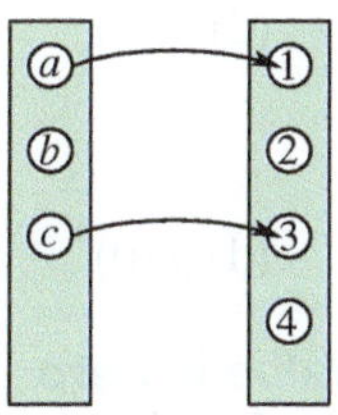

4.

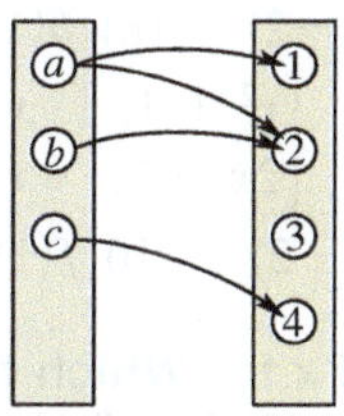

Testing for Functions In Exercises 5–8, decide whether the set of ordered pairs represents a function from A to B.

$A = \{0, 1, 2, 3\}$

$B = \{-2, -1, 0, 1, 2\}$

Give a reason for your answer. *See Example 1.*

5. $\{(0, 1), (1, -2), (2, 0), (3, 2)\}$

6. $\{(0, -1), (2, 2), (1, -2), (3, 0), (1, 1)\}$

7. $\{(0, 0), (1, 0), (2, 0), (3, 0)\}$

8. $\{(0, 2), (3, 0), (1, 1)\}$

Testing for Functions In Exercises 9–14, decide whether the set of ordered pairs represents a function from A to B.

$A = \{a, b, c\}$

$B = \{0, 1, 2, 3\}$

Give a reason for your answer. *See Example 1.*

9. $\{(a, 1), (c, 2), (c, 3), (b, 3)\}$

10. $\{(a, 3), (c, 1)\}$

11. $\{(0, a), (2, c), (3, b)\}$

12. $\{(a, 0), (c, 1), (b, 3)\}$

13. $\{(a, 1), (b, 2), (c, 3)\}$

14. $\{(c, 0), (b, 0), (a, 3)\}$

Testing for Functions Represented by Equations In Exercises 15–24, determine whether the equation represents y as a function of x. *See Example 2.*

15. $x^2 + y^2 = 4$

16. $x = y^2$

17. $x^2 + y = 4$

18. $y = \sqrt{x + 5}$

19. $2x + 3y = 4$

20. $5x - 7 = 3y$

21. $y^2 = x^2 - 1$

22. $x + y^2 = 4$

23. $x^2y - x^2 + 4y = 0$

24. $xy - y - x - 2 = 0$

Evaluating a Function In Exercises 25–28, fill in the blank and simplify. *See Example 3.*

25. $f(x) = 6 - 4x$

(a) $f(3) = 6 - 4(\quad)$

(b) $f(-7) = 6 - 4(\quad)$

(c) $f(t) = 6 - 4(\quad)$

(d) $f(c + 1) = 6 - 4(\quad)$

26. $f(s) = \dfrac{1}{s + 1}$

(a) $f(4) = \dfrac{1}{(\quad) + 1}$ (b) $f(0) = \dfrac{1}{(\quad) + 1}$

(c) $f(4x) = \dfrac{1}{(\quad) + 1}$

(d) $f(x + 1) = \dfrac{1}{(\quad) + 1}$

27. $g(x) = \dfrac{1}{x^2 - 2x}$

(a) $g(1) = \dfrac{1}{(\quad)^2 - 2(\quad)}$

(b) $g(-3) = \dfrac{1}{(\quad)^2 - 2(\quad)}$

(c) $g(t) = \dfrac{1}{(\quad)^2 - 2(\quad)}$

(d) $g(t + 1) = \dfrac{1}{(\quad)^2 - 2(\quad)}$

28. $f(t) = \sqrt{25 - t^2}$

(a) $f(3) = \sqrt{25 - (\quad)^2}$

(b) $f(5) = \sqrt{25 - (\quad)^2}$

(c) $f(x + 5) = \sqrt{25 - (\quad)^2}$

(d) $f(2x) = \sqrt{25 - (\quad)^2}$

Evaluating a Function In Exercises 29–40, evaluate the function at each specified value of the independent variable and simplify. *See Example 3.*

29. $f(x) = 2x - 3$
(a) $f(1)$
(b) $f(-3)$
(c) $f(x - 1)$
(d) $f\left(\frac{1}{4}\right)$

30. $g(y) = 7 - 3y$
(a) $g(0)$
(b) $g\left(\frac{7}{3}\right)$
(c) $g(s)$
(d) $g(s + 2)$

31. $h(t) = t^2 - 2t$
(a) $h(2)$
(b) $h(-1)$
(c) $h(x + 2)$
(d) $h(1.5)$

32. $k(b) = 2b^2 + 7b + 3$
(a) $k(0)$
(b) $k\left(-\frac{1}{2}\right)$
(c) $k(a)$
(d) $k(x + 2)$

33. $V(r) = \dfrac{4}{3}\pi r^3$
(a) $V(3)$
(b) $V(0)$
(c) $V\left(\frac{3}{2}\right)$
(d) $V(2r)$

34. $A(s) = \dfrac{\sqrt{3}s^2}{4}$
(a) $A(1)$
(b) $A(0)$
(c) $A(2x)$
(d) $A(3)$

35. $f(y) = 3 - \sqrt{y}$
(a) $f(4)$
(b) $f(100)$
(c) $f(4x^2)$
(d) $f(0.25)$

36. $f(x) = \sqrt{x + 3} - 2$
(a) $f(-3)$
(b) $f(1)$
(c) $f(x - 3)$
(d) $f(x + 4)$

37. $c(x) = \dfrac{1}{x^2 - 16}$
(a) $c(4)$
(b) $c(0)$
(c) $c(y + 2)$
(d) $c(y - 2)$

38. $q(t) = \dfrac{2t^2 + 3}{t^2}$
(a) $q(2)$
(b) $q(0)$
(c) $q(x)$
(d) $q(-x)$

39. $f(x) = \dfrac{|x|}{x}$
(a) $f(2)$
(b) $f(-2)$
(c) $f(x^2)$
(d) $f(x - 1)$

40. $f(x) = |x| + 4$
(a) $f(2)$
(b) $f(-2)$
(c) $f(x^2)$
(d) $f(x + 2)$

A Piecewise-Defined Function In Exercises 41–44, evaluate the function at each specified value of the independent variable and simplify. *See Example 4.*

41. $f(x) = \begin{cases} 3x - 1, & x < 0 \\ 2x + 3, & x \geq 0 \end{cases}$

(a) $f(-1)$ (b) $f(0)$ (c) $f(-2)$ (d) $f(2)$

42. $g(x) = \begin{cases} \frac{1}{4}x + 2, & x \leq 4 \\ 8 - x, & x > 4 \end{cases}$

(a) $g(-2)$ (b) $g(0)$ (c) $g(4)$ (d) $g(6)$

43. $h(t) = \begin{cases} t^2 - 2, & t < 2 \\ 3, & t \geq 2 \end{cases}$

(a) $h(-2)$ (b) $h(1)$ (c) $h(2)$ (d) $h(5)$

44. $f(x) = \begin{cases} x^2 + 1, & x \leq 1 \\ 2x - 3, & x > 1 \end{cases}$

(a) $f(-2)$ (b) $f(1)$ (c) $f\left(\frac{3}{2}\right)$ (d) $f(0)$

Values of x for Which the Function Value is Zero In Exercises 45–52, find all real values of x for which $f(x) = 0$.

45. $f(x) = 15 - 3x$

46. $f(x) = \dfrac{2x - 5}{3}$

47. $f(x) = x^2 - 9$

48. $f(x) = 2x^2 - 11x + 5$

49. $f(x) = x^3 - x$

50. $f(x) = x^3 - 3x^2 - 4x + 12$

51. $f(x) = \dfrac{3}{x - 1} + \dfrac{4}{x - 2}$

52. $f(x) = 3 + \dfrac{2}{x - 1}$

Finding the Domain of a Function In Exercises 53–70, find the domain of the function. *See Example 5.*

53. $f: \{(0, 1), (3, 4), (5, 7), (8, 11), (9, 12)\}$

54. $g: \{(10, -4), (20, 1), (30, 6), (40, 9), (50, 13)\}$

55. $h: \{(-2, 7.4), (-1, 6.5), (0, 4.8), (1, 4.1), (2, 3.3)\}$

56. $s: \{(-5.7, 9.1), (-3.9, 5.1), (-2.9, 3.2), (-1.5, 1.1)\}$

57. $g(x) = 1 - 2x^2$

58. $f(x) = 5x^2 + 2x - 1$

59. $h(t) = \dfrac{4}{t}$

60. $s(y) = \dfrac{3y}{y + 5}$

61. $g(y) = \sqrt[3]{y - 10}$

62. $f(t) = \sqrt[3]{t + 4}$

63. $f(x) = \sqrt[4]{1 - x^2}$

64. $g(x) = \sqrt{x + 1}$

65. $g(x) = \dfrac{1}{x} - \dfrac{3}{x + 2}$

66. $h(x) = \dfrac{10}{x^2 - 2x}$

67. $f(x) = \dfrac{\sqrt{x + 1}}{x - 2}$

68. $f(s) = \dfrac{\sqrt{s - 1}}{s - 4}$

69. $f(x) = \dfrac{x - 4}{\sqrt{x}}$

70. $f(x) = \dfrac{x - 5}{\sqrt{x^2 - 9}}$

Writing a Function as a Set of Ordered Pairs In Exercises 71–74, the domain of f is the set

$A = \{-2, -1, 0, 1, 2\}.$

Write the function as a set of ordered pairs.

71. $f(x) = x^2$

72. $f(x) = \dfrac{2x}{x^2 + 1}$

73. $f(x) = \sqrt{x + 2}$

74. $f(x) = |x + 1|$

75. Think About It Consider

$$f(x) = \sqrt{x - 2} \quad \text{and} \quad g(x) = \sqrt[3]{x - 2}.$$

Why are the domains of f and g different?

76. Think About It A student says that the domain of

$$f(x) = \frac{\sqrt{x + 1}}{x - 3}$$

is all real numbers except $x = 3$. Is the student correct? Explain.

77. Think About It Set A contains n elements and set B contains m elements. How many ordered pairs are there in a set that represents a function f from set A to set B?

78. Discovery Use a graphing utility to graph the functions

$$f(x) = \sqrt{4 - x^2} \quad \text{and} \quad g(x) = \sqrt{x^2 - 4}.$$

What is the domain of each function? Do the domains of these two functions overlap? If so, for what values?

79. Volume of a Box An open box is to be made from a square piece of material 18 inches on a side by cutting equal squares from the corners and turning up the sides (see figure).

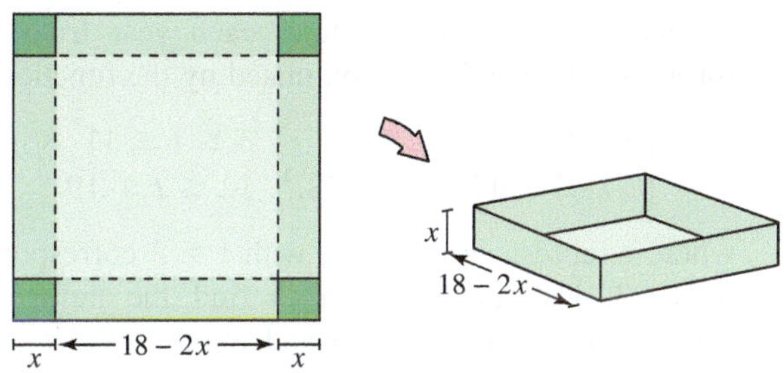

(a) Write the volume V of the box as a function of its height x.

(b) What is the domain of the function?

(c) Determine the volume of a box with a height of 4 inches.

80. Height of a Balloon A balloon carrying a transmitter ascends vertically from a point 2000 feet from the receiving station (see figure). Let d be the distance between the balloon and the receiving station. Write the height h of the balloon as a function of d. What is the domain of this function?

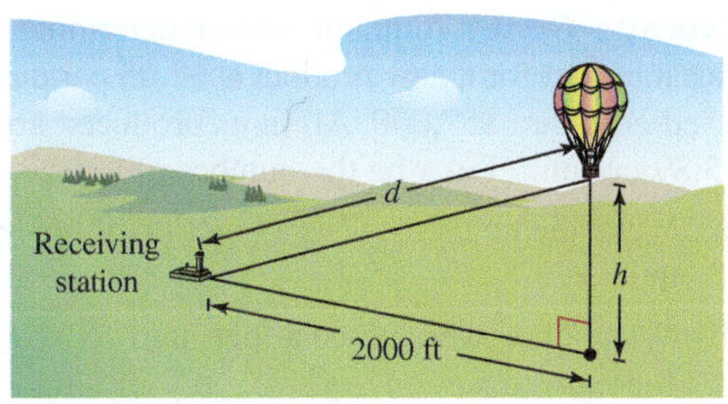

81. Path of a Ball A player throws a baseball toward a teammate. The height y (in feet) of the baseball is given by the function

$$y = -\frac{1}{10}x^2 + 3x + 6$$

where x is the horizontal distance (in feet) from where the ball is thrown. The teammate is 30 feet away and can reach to a maximum height of 7 feet. Can the teammate catch the ball without backing up? Explain.

82. Path of a Salmon Part of the life cycle of a salmon is migration for reproduction. Salmon are anadromous fish. This means that they swim from the ocean to fresh water streams to lay their eggs. During migration, salmon must jump waterfalls to reach their destination. A migrating salmon initiates a jump 4 feet from a waterfall that is 3 feet high. Its path through the air is given by the function

$$h = -0.42x^2 + 2.52x$$

where h is the height (in feet) and x is the horizontal distance (in feet) from where the salmon leaves the water. Will the salmon clear the waterfall?

83. Farmland The number of acres A (in millions) of farmland in the United States each year from 1993 through 2009 can be approximated by the function

$$A = \begin{cases} -3.4t + 979, & 3 \le t \le 11 \\ 0.28t^2 - 11.9t + 1043, & 12 \le t \le 19 \end{cases}$$

where t represents the year, with $t = 3$ corresponding to 1993. Use the function to find the numbers of acres of farmland in 2000 and 2009. *(Source: U.S. Department of Agriculture)*

84. Cost, Revenue, and Profit A company produces a product for which the variable cost is \$11.75 per unit and the fixed costs are \$112,000. The product sells for \$21.95 per unit. Let x be the number of units produced and sold.

(a) Add the variable cost and the fixed costs to write the total cost C as a function of the number of units x.

(b) Write the revenue R as a function of the number of units x.

(c) Use the formula $P = R - C$ to write the profit P as a function of the number of units x.

85. Average Cost A manufacturer determines that the variable cost for a new product is \$2.05 per unit and the fixed costs are \$57,000. All units produced are sold for \$5.89 per unit. Let x be the number of units.

(a) Write the total cost C as a function of the number of units x.

(b) Write the average cost per unit $\overline{C} = C/x$ as a function of x.

(c) Complete the table.

x	100	1000	10,000	100,000
$\overline{C}$				

(d) Write a paragraph analyzing the data in the table. What do you observe about the average cost per unit as x gets larger?

86. Migrant Interdictions The U.S. Coast Guard interdicts undocumented migrants at sea to prevent them from entering the United States, its territories, and its possessions. The number N of migrants interdicted each year from 2000 through 2010 can be approximated by the function

$$N = \begin{cases} 556.50t^2 - 1096.5t + 4280, & 0 \le t \le 3 \\ -1481.9t + 16{,}793, & 4 \le t \le 10 \end{cases}$$

where t represents the year, with $t = 0$ corresponding to 2000 (see figure). *(Source: U.S. Coast Guard)*

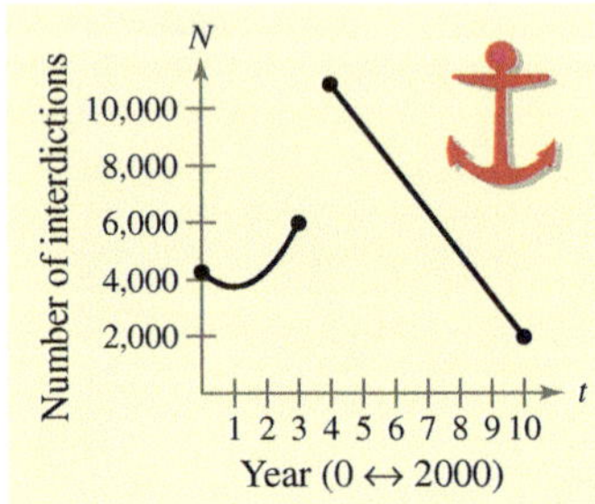

(a) Use the function to approximate the numbers of migrants interdicted in 2003, 2004, and 2010.

(b) Use the function to approximate the total number of migrants interdicted from 2000 through 2010.

87. Profits Johnson & Johnson manufactures and sells health care products. The net profits P (in billions of dollars) of Johnson & Johnson for the years 2000 through 2009 are shown in the table. *(Source: Johnson & Johnson)*

Year	2000	2001	2002	2003	2004
Profit, P	4.80	5.89	6.81	8.10	9.30

Year	2005	2006	2007	2008	2009
Profit, P	10.55	11.13	12.09	12.95	12.91

(a) Use a graphing utility to create a scatter plot of the data. Let t represent the year, with $t = 0$ corresponding to 2000.

(b) Use the *regression* feature of the graphing utility to find a linear model and a quadratic model for the data.

(c) Use each model to approximate the net profits for the years 2000 through 2009. Compare the values generated by each model with the actual values shown in the table. Which model is a better fit? Explain.

88. Dividends The dividends D (in dollars) per share declared by Coca-Cola for the years 1995 through 2010 are shown in the table. *(Source: Coca-Cola Company)*

Year	Dividend, D	Year	Dividend, D
1995	0.44	2003	0.88
1996	0.50	2004	1.00
1997	0.56	2005	1.12
1998	0.60	2006	1.24
1999	0.64	2007	1.36
2000	0.68	2008	1.52
2001	0.72	2009	1.64
2002	0.80	2010	1.76

(a) Use a graphing utility to create a scatter plot of the data. Let t represent the year, with $t = 5$ corresponding to 1995.

(b) Use the *regression* feature of the graphing utility to find a linear model and a quadratic model for the data.

(c) Use the graphing utility to graph each model from part (b) with the data.

(d) Which model do you think better fits the data? Explain your reasoning.

(e) Use the model you selected in part (d) to predict the dividends per share in 2011 and 2015. Coca-Cola predicts the dividends per share to be about \$1.88 in 2011 and to reach \$2.48 by one of the years from 2013 to 2015. Do your predictions support those of Coca-Cola? Explain.

89. Ripples in a Pond A stone is thrown into the middle of a calm pond, causing ripples to form in concentric circles. The radius r of the outermost ripple increases at the rate of 0.75 foot per second.

(a) Write a function for the radius r of the circle formed by the outermost ripple in terms of time t.

(b) Write a function for the area A enclosed by the outermost ripple. Complete the table.

Time, t	1	2	3	4	5
Radius, r (in feet)					
Area, A (in square feet)					

(c) Compare the ratios $A(2)/A(1)$ and $A(4)/A(2)$. What do you observe? Based on your observation, predict the area when $t = 8$. Verify by checking $t = 8$ in the area function.

90. HOW DO YOU SEE IT? You pop the cork from a bottle at an outdoor event. The graph represents the height h of the cork after t seconds.

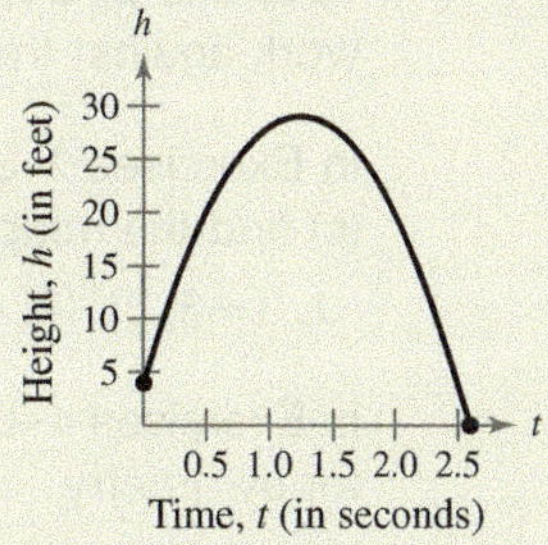

(a) Explain why h is a function of t.

(b) Approximate the height of the cork after 0.5 second and after 1.25 seconds.

(c) Approximate the domain of h.

(d) Is t a function of h? Explain.

Think About It In Exercises 91 and 92, determine whether the statements use the word *function* in ways that are *mathematically* correct. Explain your reasoning.

91. (a) The sales tax on a purchased item is a function of the selling price.

(b) Intelligence is a function of education.

92. (a) The market value of a home is a function of its size.

(b) The speed at which a free-falling baseball strikes the ground is a function of its initial height.

93. Project: Expenditures For a project involving the personal expenditures for Internet access in the United States from 2000 to 2009, visit this text's website at *www.cengagebrain.com*. *(Source: Bureau of Economic Analysis)*

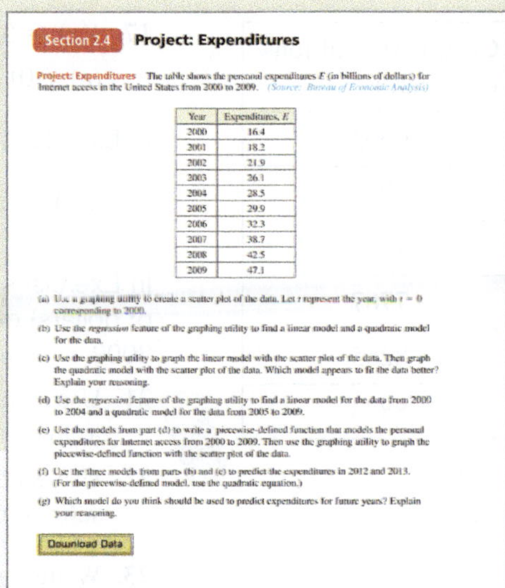

Section 2.4 **Project: Expenditures**

Project: Expenditures The table shows the personal expenditures E (in billions of dollars) for Internet access in the United States from 2000 to 2009. *(Source: Bureau of Economic Analysis)*

Year	Expenditures, E
2000	16.4
2001	18.2
2002	21.9
2003	26.1
2004	28.5
2005	29.9
2006	32.3
2007	38.7
2008	42.5
2009	47.1

(a) Use a graphing utility to create a scatter plot of the data. Let t represent the year, with $t = 0$ corresponding to 2000.

(b) Use the *regression* feature of the graphing utility to find a linear model and a quadratic model for the data.

(c) Use the graphing utility to graph the linear model with the scatter plot of the data. Then graph the quadratic model with the scatter plot of the data. Which model appears to fit the data better? Explain your reasoning.

(d) Use the *regression* feature of the graphing utility to find a linear model for the data from 2000 to 2004 and a quadratic model for the data from 2005 to 2009.

(e) Use the models from part (d) to write a piecewise-defined function that models the personal expenditures for Internet access from 2000 to 2009. Then use the graphing utility to graph the piecewise-defined function with the scatter plot of the data.

(f) Use the three models from parts (b) and (e) to predict the expenditures in 2012 and 2013. (For the piecewise-defined model, use the quadratic equation.)

(g) Which model do you think should be used to predict expenditures for future years? Explain your reasoning.

Download Data

QUIZ YOURSELF

See www.CalcChat.com for worked-out solutions to odd-numbered exercises.

Take this quiz as you would take a quiz in class. When you are done, check your work against the answers given in the back of the book.

In Exercises 1–3, (a) plot the points, (b) find the distance between the points, and (c) find the midpoint of the line segment joining the points.

1. $(-3, 2), (4, -5)$ **2.** $(1.3, -4.5), (-3.7, 0.7)$ **3.** $(4, -2), \left(-1, -\frac{5}{2}\right)$

In Exercises 4–6, sketch the graph of the equation. Identify any intercepts and test for symmetry.

4. $y = 9 - x^2$ **5.** $y = x\sqrt{x + 4}$ **6.** $y = |x - 3|$

7. The point $(0, 3)$ lies on a circle whose center is at $(4, 5)$. Write the standard form of the equation of this circle.

8. Write the equation of the circle $x^2 + y^2 - 2x + 4y - 4 = 0$ in standard form. Then sketch the circle.

In Exercises 9–12, find an equation of the line that passes through the point and has the indicated slope. Then sketch the graph of the line.

	Point	*Slope*
9.	$(3, 5)$	$m = \frac{2}{3}$
10.	$(-2, 4)$	$m = 0$
11.	$(2, -3)$	m is undefined.
12.	$(-2, -5)$	$m = -2$

13. A city had a population of 233,134 in 2010 and 244,288 in 2012. Assuming the population followed a linear pattern, predict the population of the city in 2015.

14. Determine whether the equation $x^2 + y = 3$ represents y as a function of x.

In Exercises 15 and 16, let $A = \{6, 8, 10\}$ and $B = \{1, 2, 4, 16\}$. Decide whether the set of ordered pairs represents a function from A to B.

15. $\{(6, 1), (8, 1), (10, 2)\}$ **16.** $\{(6, 2), (10, 4)\}$

In Exercises 17 and 18, evaluate the function at each specified value of the independent variable and simplify.

17. $f(x) = 3(x + 2) - 4$
(a) $f(0)$ (b) $f(-3)$

18. $g(t) = 2t^3 - t^2$
(a) $g(1)$ (b) $g(-2)$

In Exercises 19 and 20, find the domain of the function.

19. $h(x) = \sqrt{x - 4}$ **20.** $f(x) = \dfrac{x}{x + 2}$

In Exercises 21 and 22, use the table, which shows the federal cost C (in millions of dollars) of the National School Lunch Program for each year from 2004 through 2009. *(Source: U.S. Department of Agriculture)*

21. Use a graphing utility to create a scatter plot of the data and use the regression feature to find a linear model and a quadratic model for the data. Let $t = 4$ represent 2004.

22. Use each model you found in Exercise 21 to predict the federal costs of the national school lunch program in 2010 and 2011.

Year	Cost, C (in millions of dollars)
2004	6663
2005	7055
2006	7389
2007	7707
2008	8265
2009	8873

Table for 21 and 22

23. Write the area A of a circle as a function of its circumference C.

2.5 Graphs of Functions

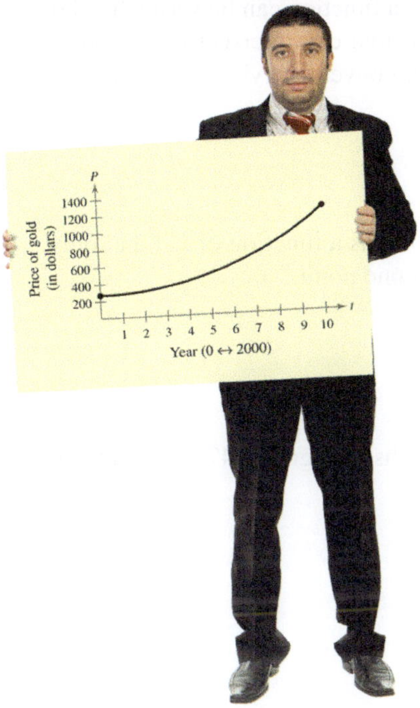

In Exercise 79 on page 218, you will use a graph to determine in which years the price of an ounce of gold was increasing or decreasing.

- Find the domain and range of a function using its graph.
- Identify the graph of a function using the Vertical Line Test.
- Describe the increasing, decreasing, and constant behavior of a function.
- Find the relative minima and relative maxima of the graph of a function.
- Sketch the graph of a step function.
- Classify a function as even or odd.
- Identify six common graphs and use them to sketch the graphs of functions.

The Graph of a Function

In Section 2.4, you studied functions from an algebraic point of view. In this section, you will study functions from a graphical perspective.

The **graph of a function f** is the collection of ordered pairs $(x, f(x))$ such that x is in the domain of f. As you study this section, remember that x is the directed distance from the y-axis and $f(x)$ is the directed distance from the x-axis, as shown in Figure 2.45. If the graph of a function has an x-intercept at $(a, 0)$, then a is a **zero** of the function. In other words, the zeros of a function are the values of x for which $f(x) = 0$. For instance, the function given by $f(x) = x^2 - 4$ has two zeros: -2 and 2.

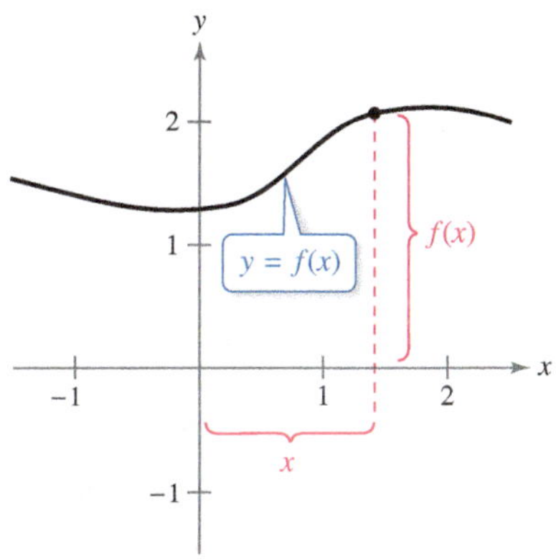

FIGURE 2.45

The **range** of a function (the set of values assumed by the dependent variable) is often easier to determine graphically than algebraically.

Example 1 Finding the Domain and Range of a Function

Use the graph of the function f, shown in Figure 2.46, to find (a) the domain of f, (b) the function values $f(-1)$ and $f(2)$, and (c) the range of f.

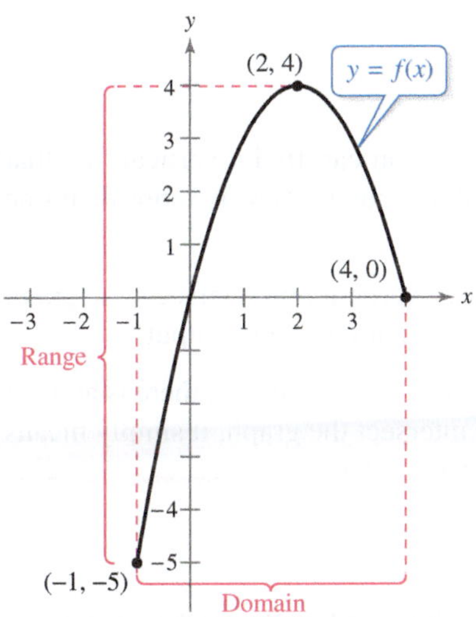

FIGURE 2.46

SOLUTION

a. Because the graph does not extend beyond $x = -1$ (on the left) and $x = 4$ (on the right), the domain of f is all x in the interval $[-1, 4]$.

b. Because $(-1, -5)$ is a point on the graph of f, it follows that

$$f(-1) = -5.$$

Similarly, because $(2, 4)$ is a point on the graph of f, it follows that

$$f(2) = 4.$$

c. Because the graph does not extend below $f(-1) = -5$ or above $f(2) = 4$, the range of f is the interval $[-5, 4]$.

Checkpoint 1

Use the graph of $f(x) = x^2 - 3$ to find the domain and range of f.

Vertical Line Test

By the definition of a function, at most one y-value corresponds to a given x-value. This means that the graph of a function cannot have two or more different points with the same x-coordinate, and no two points on the graph of a function can be vertically above or below each other. It follows, then, that a vertical line can intersect the graph of a function at most once. This observation provides a convenient visual test called the **Vertical Line Test** for functions.

> **Vertical Line Test for Functions**
>
> A set of points in a coordinate plane is the graph of y as a function of x if and only if no vertical line intersects the graph at more than one point.

Example 2 Using the Vertical Line Test

Use the Vertical Line Test to decide whether the graphs in Figure 2.47 represent y as a function of x.

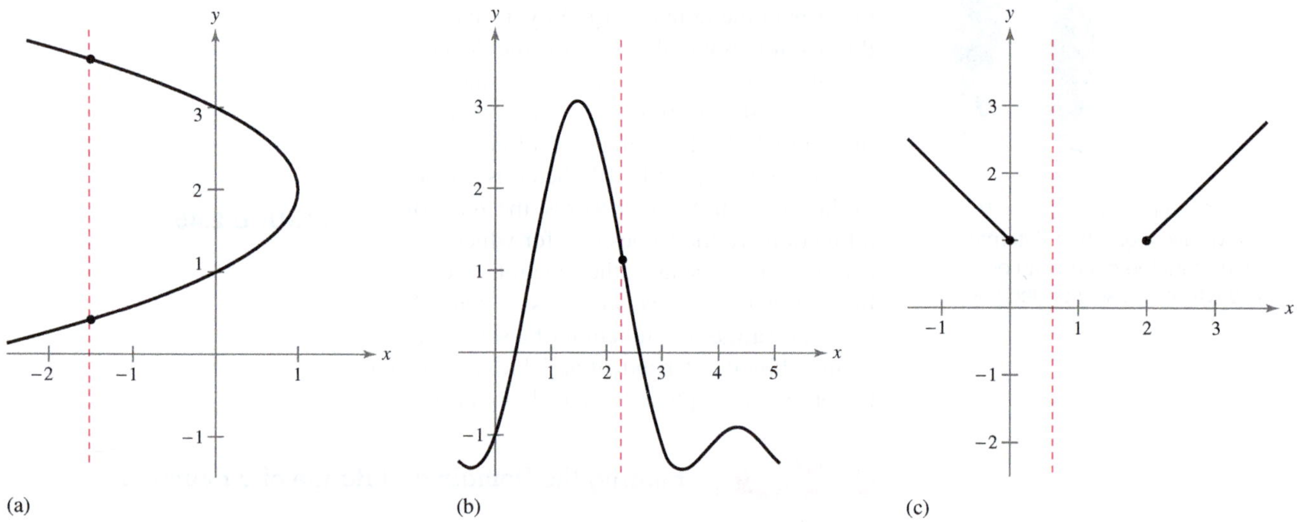

FIGURE 2.47

SOLUTION

a. This *is not* a graph of y as a function of x because you can find a vertical line that intersects the graph twice. That is, for a particular input x, there is more than one output y.

b. This *is* a graph of y as a function of x because every vertical line intersects the graph at most once. That is, for a particular input x, there is at most one output y.

c. This *is* a graph of y as a function of x. That is, for a particular input x, there is at most one output y. Note that if a vertical line does not intersect the graph, it simply means that the function is undefined for that particular value of x.

✓ Checkpoint 2

Use the Vertical Line Test to decide whether the graph of each equation represents y as a function of x.

a. $x^2 + y = 2$ **b.** $y = 2x^3$

Increasing and Decreasing Functions

The more you know about the graph of a function, the more you know about the function itself. For example, consider the graph that is shown in Figure 2.48. As you move from *left to right*, this graph decreases, then is constant, and then increases.

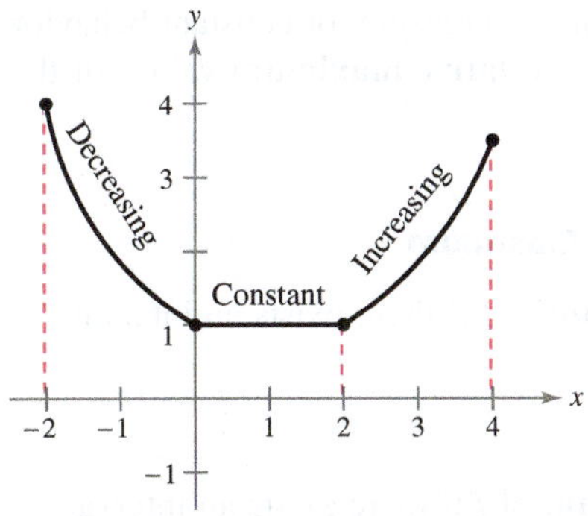

As x increases from left to right, notice how y decreases, then remains constant, and then increases.

FIGURE 2.48

Increasing, Decreasing, and Constant Functions

A function f is **increasing** on an interval if, for any x_1 and x_2 in the interval, $x_1 < x_2$ implies

$$f(x_1) < f(x_2).$$

A function f is **decreasing** on an interval if, for any x_1 and x_2 in the interval, $x_1 < x_2$ implies

$$f(x_1) > f(x_2).$$

A function f is **constant** on an interval if, for any x_1 and x_2 in the interval,

$$f(x_1) = f(x_2).$$

Example 3 Describing Function Behavior

Describe the increasing, decreasing, and constant behavior of each function shown in Figure 2.49.

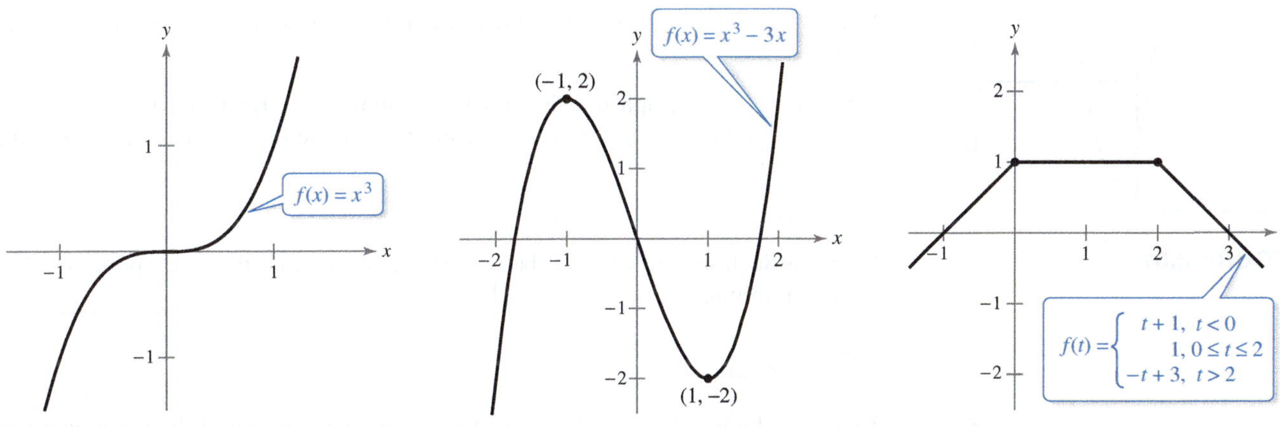

(a) (b) (c)

FIGURE 2.49

SOLUTION

a. This function is increasing over the entire real line.

b. This function is increasing on the interval $(-\infty, -1)$, decreasing on the interval $(-1, 1)$, and increasing on the interval $(1, \infty)$.

c. This function is increasing on the interval $(-\infty, 0)$, constant on the interval $(0, 2)$, and decreasing on the interval $(2, \infty)$.

✓ Checkpoint 3

Describe the increasing, decreasing, and constant behavior of the function $f(x) = x^2 + 3x$.

Make sure you understand that the behavior of functions (increasing, decreasing, or constant) is determined by the y-values, but that the intervals are written in terms of x.

Relative Minima and Relative Maxima

The points at which a function changes its increasing, decreasing, or constant behavior are helpful in determining the **relative minimum** or **relative maximum** values of the function.

Definition of Relative Minimum and Relative Maximum

A function value $f(a)$ is called a **relative minimum** of f if there exists an interval (x_1, x_2) that contains a such that

$$x_1 < x < x_2 \quad \text{implies} \quad f(a) \le f(x).$$

A function value $f(a)$ is called a **relative maximum** of f if there exists an interval (x_1, x_2) that contains a such that

$$x_1 < x < x_2 \quad \text{implies} \quad f(a) \ge f(x).$$

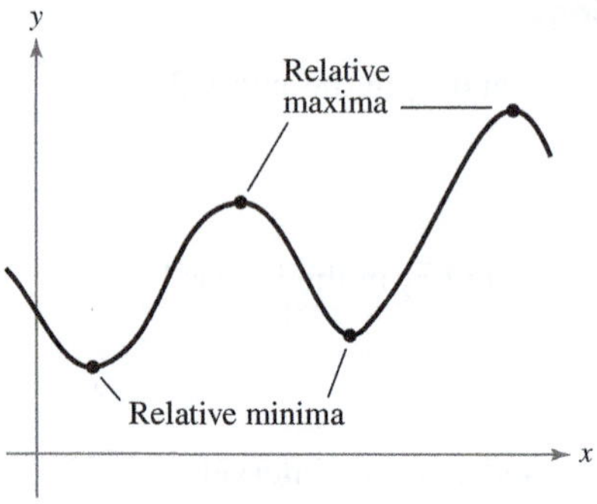

FIGURE 2.50

Figure 2.50 shows several examples of relative minima and relative maxima. In Section 3.1, you will study a technique for finding the *exact point* at which a second-degree polynomial function has a relative minimum or relative maximum. For the time being, however, you can use a graphing utility to find reasonable approximations of these points.

Example 4 Approximating a Relative Minimum

Use a graphing utility to approximate the relative minimum of the function given by $f(x) = 3x^2 - 4x - 2$.

SOLUTION The graph of f is shown in Figure 2.51. By using the *zoom* and *trace* features of a graphing utility, you can estimate that the function has a relative minimum at the point

$(0.67, -3.33)$. Relative minimum

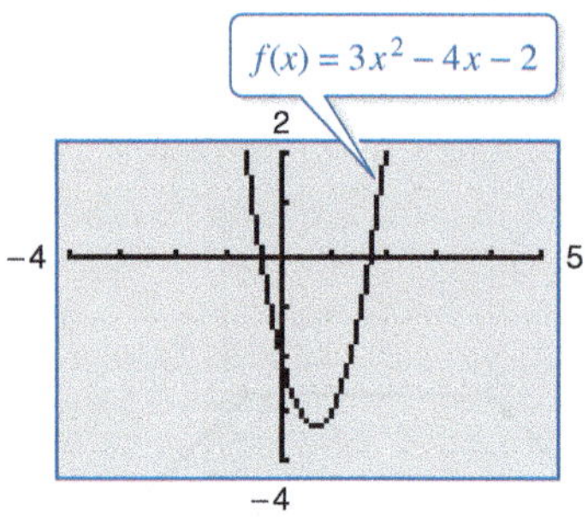

FIGURE 2.51

Later, in Section 3.1, you will be able to determine that the exact point at which the relative minimum occurs is $\left(\frac{2}{3}, -\frac{10}{3}\right)$.

✓Checkpoint 4

Use a graphing utility to approximate the relative maximum of the function given by $f(x) = -x^2 + 4x - 2$. ■

TECH TUTOR

For instructions on how to use the *table* feature and the *minimum* feature, see Appendix A.

You can also use the *table* feature of a graphing utility to approximate numerically the relative minimum of the function in Example 4. Using a table that begins at 0.6 and increments the value of x by 0.01, you can approximate the minimum of $f(x) = 3x^2 - 4x - 2$ to be -3.33, which occurs at $(0.67, -3.33)$. A third way to find the relative minimum is to use the *minimum* feature of a graphing utility.

TECH TUTOR

When using a graphing utility to estimate the x- and y-values of a relative minimum or relative maximum, the *zoom* feature will often produce graphs that are nearly flat. To overcome this problem, you can manually change the vertical setting of the viewing window. The graph will stretch vertically if the values of Ymin and Ymax are closer together.

Step Functions

The **greatest integer function** is denoted by $[\![x]\!]$ and is defined as

$$f(x) = [\![x]\!] = \text{the greatest integer less than or equal to } x.$$

The graph of this function is shown in Figure 2.52. Note that the graph of the greatest integer function jumps vertically one unit at each integer and is constant (a horizontal line segment) between each pair of consecutive integers. Because of the jumps in its graph, the greatest integer function is an example of a type of function called a **step function.** Some values of the greatest integer function are as follows.

$[\![-1]\!] = -1$ $\qquad$ $[\![-0.5]\!] = -1$

$[\![0]\!] = 0$ $\qquad$ $[\![0.5]\!] = 0$

$[\![1]\!] = 1$ $\qquad$ $[\![1.5]\!] = 1$

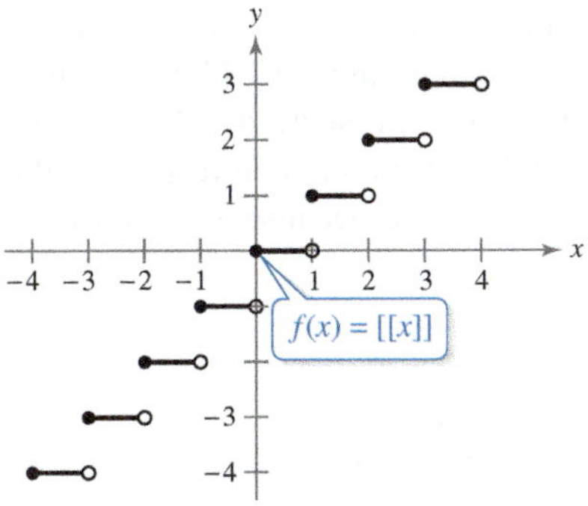

Greatest Integer Function
FIGURE 2.52

The range of the greatest integer function is the set of all integers.

When using a graphing utility to graph a step function, set the utility to *dot* mode rather than *connected* mode.

Example 5 The Cost of a Machinery Rental

The cost of renting a warehouse forklift is \$200 for up to, but not including, the first day and \$125 for each additional day (or portion of a day). The greatest integer function

$$C = 200 + 125[\![t]\!], \quad t > 0$$

can be used to model the cost of this rental, where C is the total cost of the rental (in dollars) and t is the length of the rental (in days).

a. Sketch the graph of this function.

b. How long can a company rent the forklift without spending more than \$1000?

SOLUTION

a. For rentals up to, but not including, 1 day, the cost is \$200. For rentals between 1 and 2 days, the cost is \$325, for rentals between 2 and 3 days, the cost is \$450, and so on, as shown in the table. Using these and other values, you can sketch the graph shown in Figure 2.53.

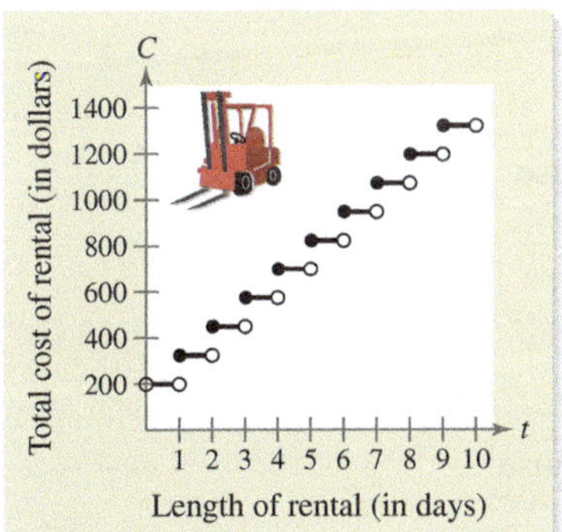

FIGURE 2.53

Length of rental, t	Cost of rental, C
$0 < t < 1$	\$200
$1 \le t < 2$	\$325
$2 \le t < 3$	\$450
⋮	⋮
$9 \le t < 10$	\$1325

b. From the graph, you can see that the company must limit the rental period to 7 days to avoid spending more than \$1000.

Checkpoint 5

Repeat Example 5 when the cost of each additional day (or portion of a day) is \$100.

Even and Odd Functions

In Section 2.1, you studied different types of symmetry of a graph. A function is said to be **even** if its graph is symmetric with respect to the y-axis and **odd** if its graph is symmetric with respect to the origin. The symmetry tests in Section 2.1 yield the following tests for even and odd functions. Even though symmetry with respect to the x-axis is introduced in Section 2.1, it will not be discussed here because a graph that is symmetric with respect to the x-axis is not a function.

Tests for Even and Odd Functions

A function given by $y = f(x)$ is even if, for each x in the domain of f,

$$f(-x) = f(x).$$

A function given by $y = f(x)$ is odd if, for each x in the domain of f,

$$f(-x) = -f(x).$$

Example 6 Identifying Even and Odd Functions

Decide whether each function is even, odd, or neither.

a. $g(x) = x^3 - x$ **b.** $h(x) = x^2 + 1$

SOLUTION

a. The function given by $g(x) = x^3 - x$ is odd because

$$\begin{aligned} g(-x) &= (-x)^3 - (-x) \\ &= -x^3 + x \\ &= -(x^3 - x) = -g(x). \end{aligned}$$

b. The function given by $h(x) = x^2 + 1$ is even because

$$\begin{aligned} h(-x) &= (-x)^2 + 1 \\ &= x^2 + 1 = h(x). \end{aligned}$$

The graphs of the two functions are shown in Figure 2.54.

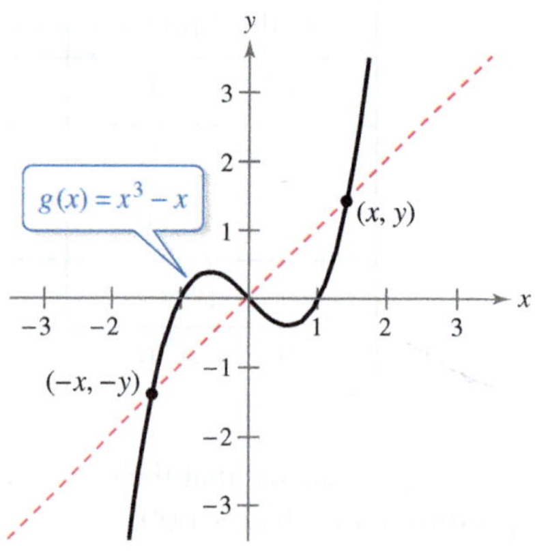

(a) Odd function (symmetric about origin)

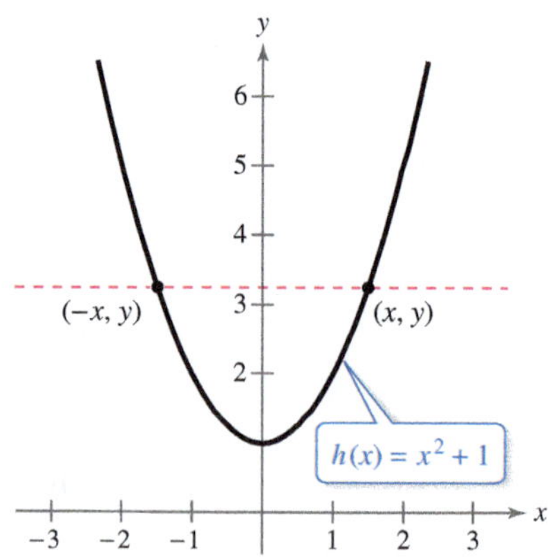

(b) Even function (symmetric about y-axis)

FIGURE 2.54

✓ Checkpoint 6

Decide whether the function $f(x) = -2x^2 + x - 1$ is even, odd, or neither. ■

Common Graphs

Figure 2.55 shows the graphs of six common functions. You need to be familiar with these graphs. They can be used as an aid when sketching other graphs. For instance, the graph of the absolute value function given by

$$f(x) = |x - 2|$$

is V-shaped.

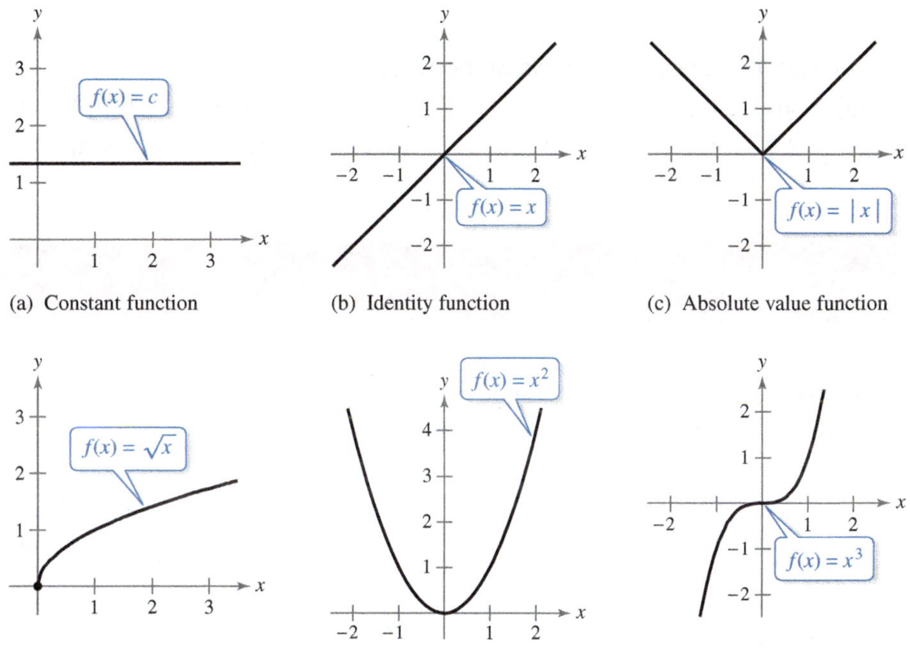

(a) Constant function (b) Identity function (c) Absolute value function

(d) Square root function (e) Squaring function (f) Cubing function

FIGURE 2.55

SUMMARIZE (Section 2.5)

1. State the definition of the graph of a function *(page 209)*. For an example of finding the domain and range of a function using its graph, see Example 1.
2. Describe the Vertical Line Test *(page 210)*. For an example that uses the Vertical Line Test, see Example 2.
3. Describe when a function is increasing, decreasing, or constant on an interval *(page 211)*. For an example of describing the increasing, decreasing, or constant behavior of a function, see Example 3.
4. State the definitions of relative minimum and relative maximum *(page 212)*. For an example of approximating a relative minimum using a graphing utility, see Example 4.
5. Describe the greatest integer function *(page 213)*. For an example that uses a greatest integer function, see Example 5.
6. State the definitions of an even function and an odd function *(page 214)*. For an example of identifying an even or odd function, see Example 6.

SKILLS WARM UP 2.5

The following warm-up exercises involve skills that were covered in earlier sections. You will use these skills in the exercise set for this section. For additional help, review Sections 1.4, 1.5, and 2.4.

1. Find $f(2)$ for $f(x) = -x^3 + 5x$.

2. Find $f(6)$ for $f(x) = x^2 - 6x$.

3. Find $f(-x)$ for $f(x) = 3/x$.

4. Find $f(-x)$ for $f(x) = x^2 + 3$.

In Exercises 5 and 6, solve the equation.

5. $x^3 - 16x = 0$

6. $2x^2 - 3x + 1 = 0$

In Exercises 7–10, find the domain of the function.

7. $g(x) = 4(x - 4)^{-1}$

8. $f(x) = 2x/(x^2 - 9x + 20)$

9. $h(t) = \sqrt[4]{5 - 3t}$

10. $f(t) = t^3 + 3t - 5$

Exercises 2.5

See www.CalcChat.com for worked-out solutions to odd-numbered exercises.

Finding the Domain and Range of a Function In Exercises 1–8, find the domain and range of the function. Then evaluate f at the given x-value. *See Example 1.*

1. $f(x) = \sqrt{x - 1}$, $x = 1$

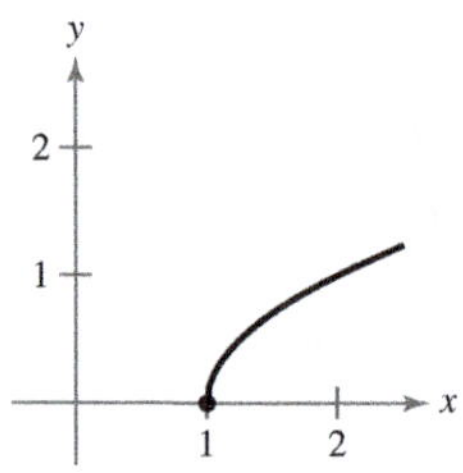

2. $f(x) = \sqrt{x^2 - 4}$, $x = -2$

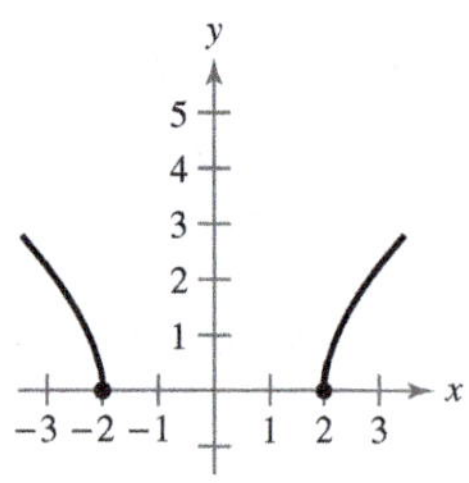

3. $f(x) = 4 - x^2$, $x = 0$

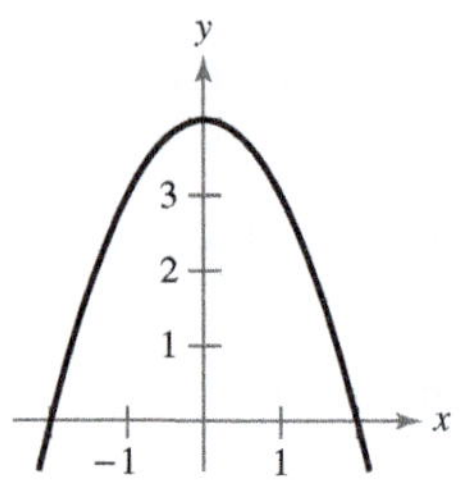

4. $f(x) = |x - 2|$, $x = 2$

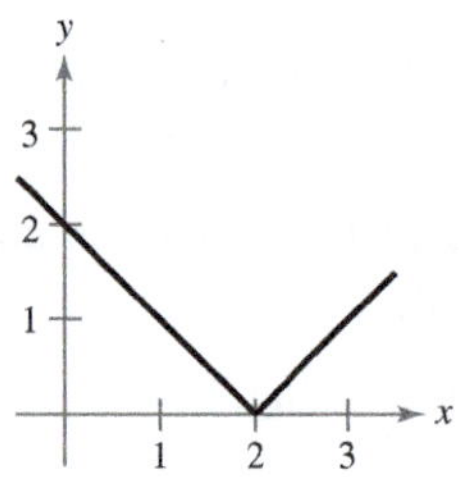

5. $f(x) = x^3 - 1$, $x = 0$

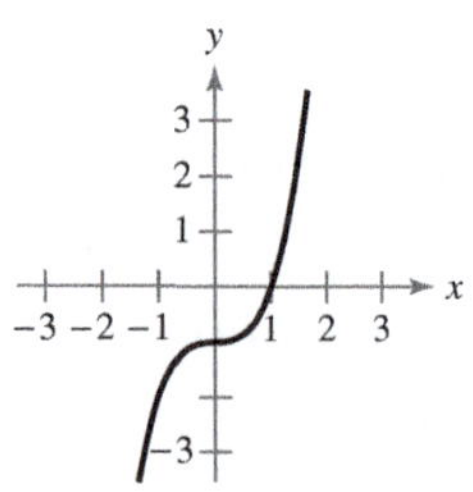

6. $f(x) = \dfrac{|x|}{x}$, $x = 5$

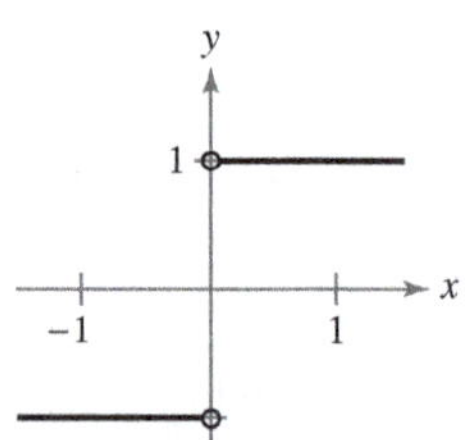

7. $f(x) = \sqrt{25 - x^2}$, $x = 0$

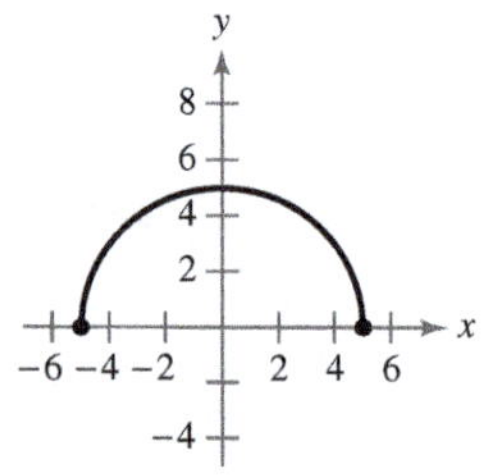

8. $f(x) = \sqrt{x^2 - 9}$, $x = 3$

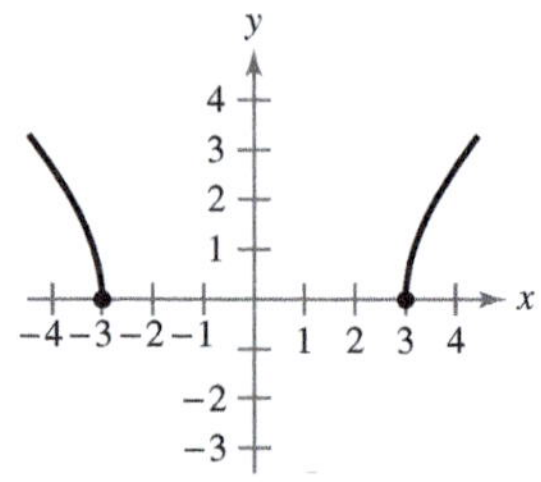

Using the Vertical Line Test In Exercises 9–16, use the Vertical Line Test to decide whether y is a function of x. *See Example 2.*

9. $y = x^2$

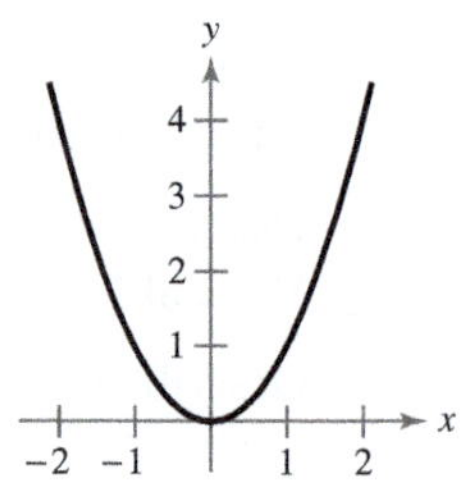

10. $x^2 = xy - 1$

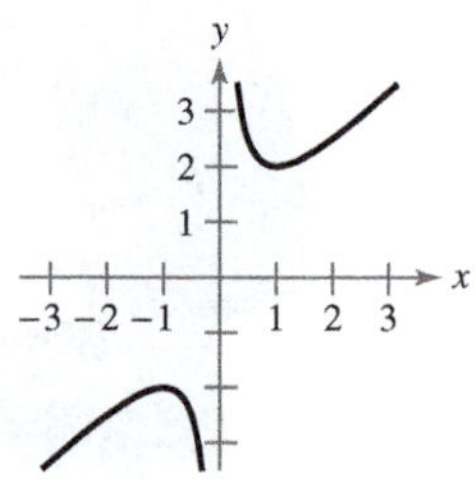

11. $x^2 + y^2 = 9$

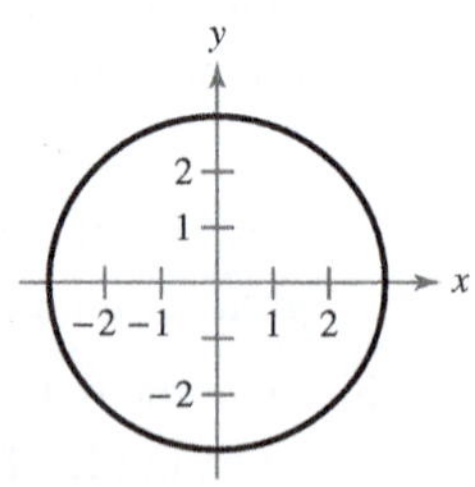

12. $x - y^2 = 0$

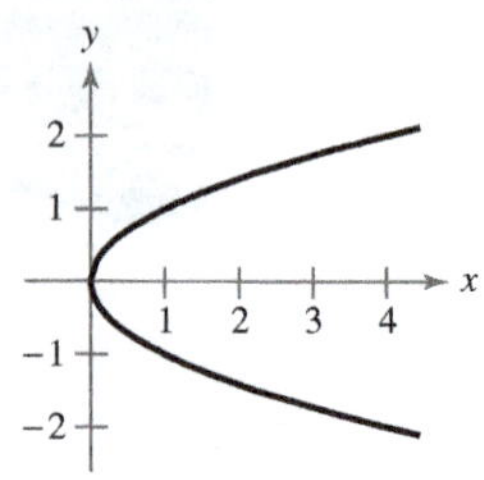

13. $y = \frac{1}{2}x^2$

14. $y = \frac{1}{4}x^3$

15. $x^2 + y^2 = 25$

16. $x = |y + 2|$

Describing Function Behavior In Exercises 17–28, describe the increasing, decreasing, and constant behavior of the function. Find the point or points where the behavior of the function changes. *See Example 3.*

17. $f(x) = 2x$

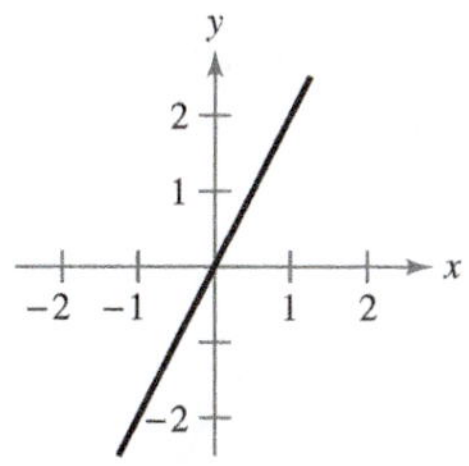

18. $f(x) = x^2 - 2x$

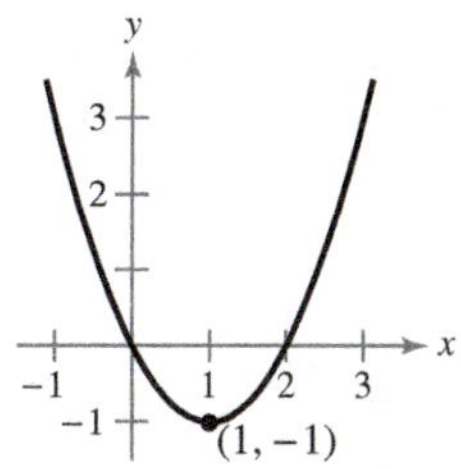

19. $f(x) = x^3 - 3x^2$

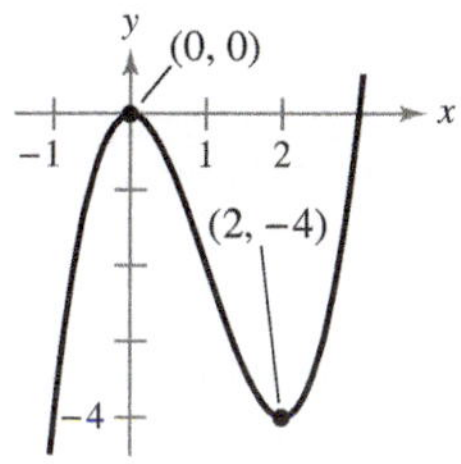

20. $f(x) = \sqrt{x^2 - 4}$

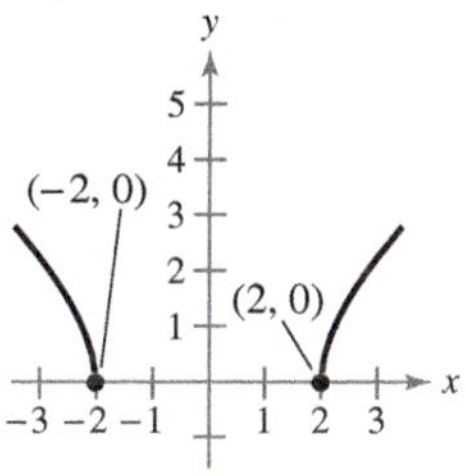

21. $f(x) = 3x^4 - 6x^2$

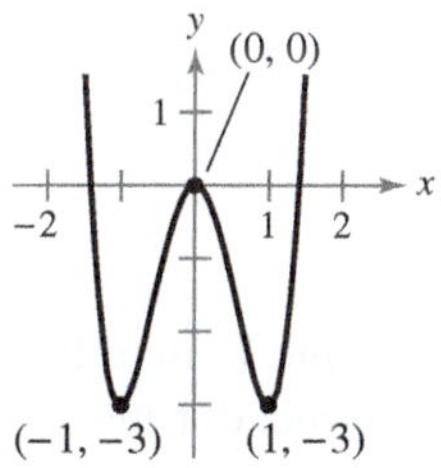

22. $f(x) = x^{2/3}$

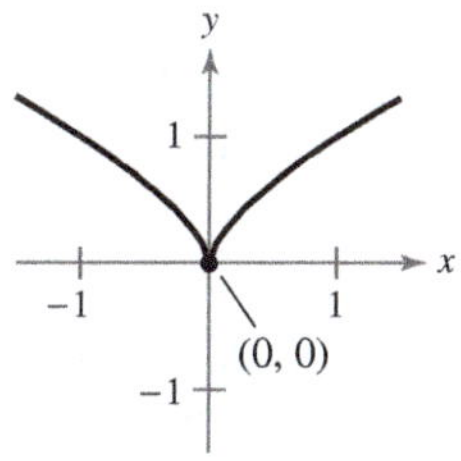

23. $y = x\sqrt{x + 3}$

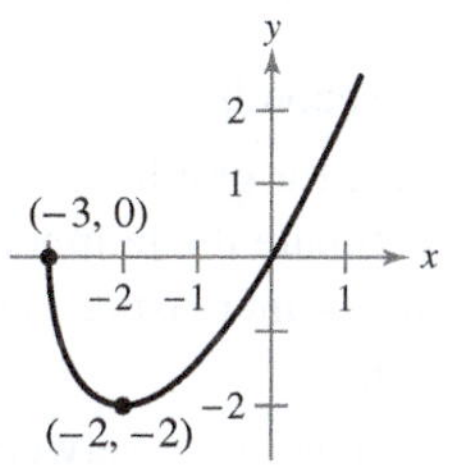

24. $y = |x + 1| + |x - 1|$

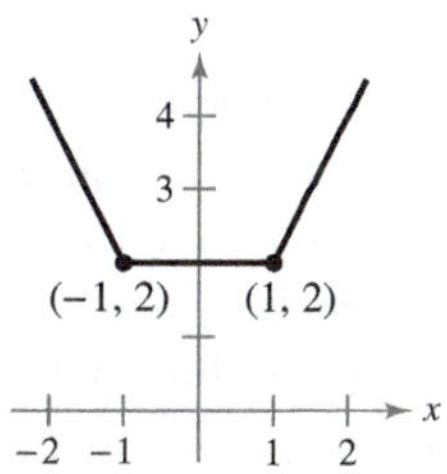

25. $f(x) = x^2 - 4x$

26. $f(x) = \frac{3}{2}x$

27. $f(x) = x^3 - 3x^2 + 2$

28. $f(x) = \sqrt{x^2 - 1}$

Approximating a Relative Minimum and Maximum In Exercises 29–34, use a graphing utility to graph the function, approximate the relative minimum or maximum of the function, and estimate the open intervals on which the function is increasing or decreasing. *See Example 4.*

29. $f(x) = x^2 - 4x + 1$

30. $f(x) = -x^2 + 6x + 3$

31. $f(x) = x^3 - 3x^2$

32. $f(x) = -x^3 + 3x + 1$

33. $f(x) = \frac{1}{4}(-4x^4 - 5x^3 + 10x^2 + 8x + 6)$

34. $f(x) = \frac{1}{4}(x^4 + x^3 - 10x^2 + 2x - 15)$

Using the Greatest Integer Function In Exercises 35–38, evaluate the function at each specified value of the independent variable.

35. $f(x) = [\![x]\!]$

(a) $f(2)$ (b) $f(2.5)$

(c) $f(-2.5)$ (d) $f(-4)$

36. $f(x) = [\![-x]\!]$

(a) $f(3)$ (b) $f(6.1)$

(c) $f(-5.9)$ (d) $f(-9)$

37. $f(x) = [\![x - 1.8]\!]$

(a) $f(4)$ (b) $f(3.7)$

(c) $f(-5.8)$ (d) $f(-6.3)$

38. $f(x) = [\![x + 0.3]\!]$

(a) $f(2.9)$ (b) $f(4.6)$

(c) $f(-2.3)$ (d) $f(-4.2)$

Identifying Even and Odd Functions In Exercises 39–44, decide whether the function is even, odd, or neither. *See Example 6.*

39. $f(x) = x^6 - 2x^2 + 3$

40. $f(t) = t^2 + 3t - 10$

41. $g(x) = x^3 - 5x$

42. $h(x) = x^3 + 3$

43. $f(x) = x\sqrt{4 - x^2}$

44. $g(s) = 4s^{2/3}$

Sketching Graphs of Functions In Exercises 45–58, sketch the graph of the function and determine whether the function is even, odd, or neither.

45. $f(x) = 3$

46. $g(x) = x$

47. $f(x) = 5 - 3x$

48. $h(x) = x^2 - 4$

49. $g(s) = \dfrac{s^3}{4}$

50. $f(t) = -t^4$

51. $f(x) = \sqrt{1 - x}$

52. $g(t) = \sqrt[3]{t - 1}$

53. $f(x) = x^{3/2}$

54. $f(x) = |x + 2|$

55. $f(x) = \begin{cases} x^2 + 1, & x \le 1 \\ 3x - 1, & x > 1 \end{cases}$

56. $f(x) = \begin{cases} 2x - 1, & x \le -1 \\ x^2 - 1, & x > -1 \end{cases}$

57. $f(x) = \begin{cases} x + 1, & x \le 0 \\ 4, & 0 < x \le 2 \\ 3x - 1, & x > 2 \end{cases}$

58. $f(x) = \begin{cases} 2x - 1, & x \le 1 \\ 3, & 1 < x \le 3 \\ 2x + 1, & x > 3 \end{cases}$

Graphing Functions In Exercises 59–64, use a graphing utility to graph the function and determine whether the function is even, odd, or neither.

59. $f(x) = x^2 - x^4$

60. $g(x) = 2x^3 + 1$

61. $h(x) = x^5 - 2x^3 + x$

62. $f(x) = x^5 - 2x^4 + x - 2$

63. $g(x) = 2 - x^6 - x^8$

64. $h(x) = x^9 + 3x^5 - x^3 + x$

Sketching Graphs of Functions In Exercises 65–78, sketch the graph of the function.

65. $f(x) = 4 - x$

66. $f(x) = 4x + 2$

67. $f(x) = x^2 - 9$

68. $f(x) = x^2 - 4x$

69. $f(x) = 1 - x^4$

70. $f(x) = x^4 - 4x^2$

71. $f(x) = \frac{1}{3}(3 + |x|)$

72. $f(x) = -1(1 + |x|)$

73. $f(x) = \sqrt{x + 3}$

74. $f(x) = \sqrt{x - 1}$

75. $f(x) = -[\![x]\!]$

76. $f(x) = 2[\![x]\!]$

77. $f(x) = [\![x - 1]\!]$

78. $f(x) = [\![x + 1]\!]$

79. Price of Gold The price P (in dollars) of an ounce of gold from 2000 through 2010 can be approximated by the model

$$P = 0.038214t^4 - 0.48703t^3 + 11.0120t^2 - 6.018t + 276.49, \quad 0 \le t \le 10$$

where t represents the year, with $t = 0$ corresponding to 2000. *(Source: World Gold Council)*

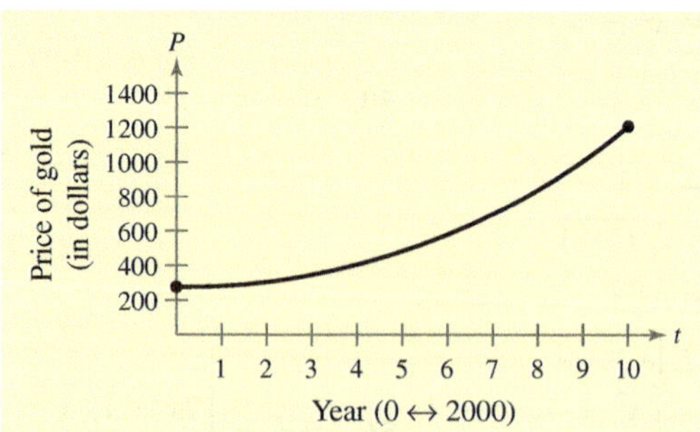

(a) Use the graph of P to find the maximum price of gold between 2000 and 2010.

(b) During which year(s) was the price decreasing? During which year(s) was the price increasing?

(c) Is it realistic to assume that the price of gold will continue to follow this model?

80. Sales The sales S (in billions of dollars) of petroleum and coal products from 1999 through 2009 can be approximated by the model

$$S = -1.28418t^4 + 67.9435t^3 - 1316.592t^2 + 11{,}169.74t - 34{,}707.3, \quad 9 \le t \le 19$$

where t represents the year, with $t = 9$ corresponding to 1999. *(Source: U.S. Census Bureau)*

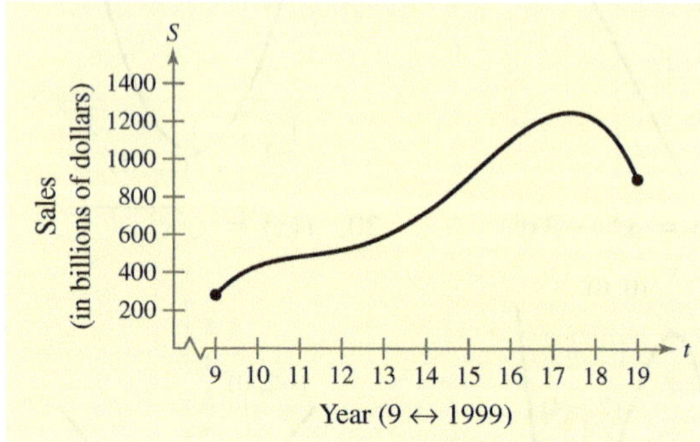

(a) Use the graph of S to find the maximum sales of these products between 1999 and 2009.

(b) During which year(s) were sales decreasing? During which year(s) were sales increasing?

(c) Is it realistic to assume that sales will continue to follow this model?

81. Lung Volume The change in volume V (in milliliters) of the lungs as they expand and contract during a breath can be approximated by the model

$$V = (-6.549s^2 + 26.20s - 3.8)^2, \quad 0 \le s \le 4$$

where s represents the number of seconds.

(a) Graph the volume function with a graphing utility and use the *trace* feature to estimate the numbers of seconds in which the volume is increasing and in which the volume is decreasing.

(b) Find the maximum change in volume between 0 and 4 seconds.

82. Maximum Profit The marketing department of a company estimates that the demand for a product is given by $p = 100 - 0.0001x$, where p is the price per unit and x is the number of units. The cost C of producing x units is given by

$$C = 350{,}000 + 30x$$

and the profit P for producing and selling x units is given by

$$P = R - C = xp - C.$$

Sketch the graph of the profit function and estimate the number of units that would produce a maximum profit. Verify your estimate using a graphing utility.

83. Maximum Profit The marketing department of a company estimates that the demand for a product is given by $p = 125 - 0.0002x$, where p is the price per unit and x is the number of units. The cost C of producing x units is given by

$$C = 225{,}000 + 80x$$

and the profit P for producing and selling x units is given by

$$P = R - C = xp - C.$$

Sketch the graph of the profit function and estimate the number of units that would produce a maximum profit. Verify your estimate using a graphing utility.

84. HOW DO YOU SEE IT? For the years 1995 through 2009, the book value B (in dollars) of a share of Wells Fargo stock can be approximated by the model

$$B = 0.0497t^2 - 0.163t + 3.53, \quad 5 \le t \le 19$$

where t represents the year, with $t = 5$ corresponding to 1995 (see figure). *(Source: Wells Fargo)*

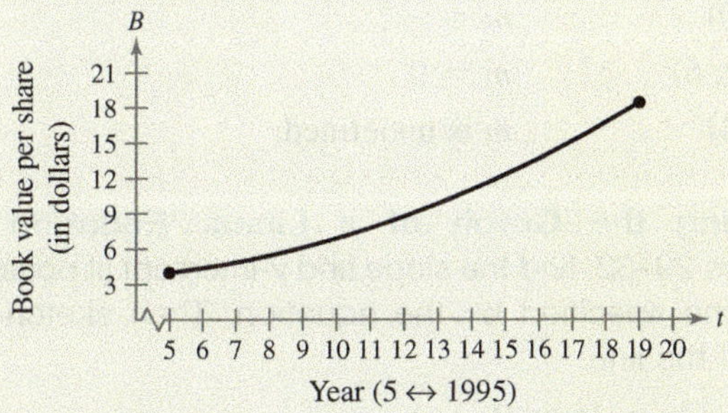

(a) Estimate the maximum book value per share from 1995 through 2009.

(b) Estimate the minimum book value per share from 1995 through 2009.

(c) Assume the book value B continues to follow the model through 2011. In which year is B at a maximum?

85. Cost of Overnight Delivery The cost of sending an overnight package from New York to Atlanta is \$26.10 for up to, but not including, the first pound and \$4.25 for each additional pound (or portion of a pound). A model for the total cost C of sending the package is $C = 26.1 + 4.25[\![x]\!]$, $x > 0$, where x is the weight of the package (in pounds). Sketch the graph of this function.

86. Cost of Overnight Delivery The cost of sending an overnight package from Los Angeles to Miami is \$31.60 for up to, but not including, the first pound and \$5.50 for each additional pound (or portion of a pound). A model for the total cost C of sending the package is $C = 31.6 + 5.5[\![x]\!]$, $x > 0$, where x is the weight of the package (in pounds). Sketch the graph of this function.

87. Strategic Reserve The total volume V (in millions of barrels) of the Strategic Oil Reserve in the United States from 1998 through 2009 can be approximated by

$$V = \begin{cases} -2.2292t^4 + 94.319t^3 - 1472.15t^2 + \\ \quad 10{,}050.1t - 24{,}773, \quad 8 \le t \le 13 \\ 0.796t^3 - 38.13t^2 + 611.8t - 2601, \\ \quad 14 \le t \le 19 \end{cases}$$

where t represents the year, with $t = 8$ corresponding to 1998. Sketch the graph of this function. *(Source: U.S. Energy Information Administration)*

88. Grade Level Salaries The 2010 salary S (in dollars) for federal employees at the Step 1 level can be approximated by the model

$$S = \begin{cases} 3109.2x + 13{,}012, & x = 1, 2, \ldots, 10 \\ 12{,}310.5x - 86{,}725, & x = 11, \ldots, 15 \end{cases}$$

where x represents the "GS" grade. Sketch a *bar graph* that represents this function. *(Source: U.S. Office of Personnel Management)*

89. Revenues The revenues of Symantec Corporation (in millions of dollars) from 2000 through 2009 are given by the ordered pairs. *(Source: Symantec Corporation)*

(2000, 853.6), (2001, 1071.4), (2002, 1406.9), (2003, 1870.1), (2004, 2582.8), (2005, 4143.4), (2006, 5199.4), (2007, 5874.4), (2008, 6149.9), (2009, 5985.0)

(a) Use the *regression* feature of a graphing utility to find a linear model for the data from 2000 through 2004. Let t represent the year, with $t = 0$ corresponding to 2000. Then determine the domain and range of the function.

(b) Use the *regression* feature of the graphing utility to find a quadratic model for the data from 2005 through 2009. Let t represent the year, with $t = 5$ corresponding to 2005. Then determine the domain and range of the function.

(c) Use your results from parts (a) and (b) to construct a piecewise model for all of the data. Use the graphing utility to graph the function.

(d) During which years did the revenues increase? During which year did the revenue decrease?

90. Think About It If f is an even function, determine whether g is even, odd, or neither. Explain.

(a) $g(x) = -f(x)$ (b) $g(x) = f(-x)$

(c) $g(x) = f(x) - 2$ (d) $g(x) = f(x - 2)$

Think About It In Exercises 91–94, find the coordinates of a second point on the graph of a function f when the given point is on the graph and the function is (a) even and (b) odd.

91. $\left(-\frac{3}{2}, 4\right)$ **92.** $\left(-\frac{5}{3}, -7\right)$

93. $(4, 9)$ **94.** $(5, -1)$

Review Exercises

See www.CalcChat.com for worked-out solutions to odd-numbered exercises.

Using the Distance and Midpoint Formulas In Exercises 1–4, (a) plot the points, (b) find the distance between the points, and (c) find the midpoint of the line segment joining the points.

1. $(3, 2), (-3, -5)$

2. $(-9, 3), (5, 7)$

3. $(3.45, 6.55), (-1.06, -3.87)$

4. $(-6.7, -3.9), (5.1, 8.2)$

Using the Distance Formula In Exercises 5 and 6, find the values of x for which the distance between the points is 25.

5. $(10, 10), (x, -5)$

6. $(x, -5), (-15, 10)$

Solution of an Equation In Exercises 7 and 8, determine whether each point is a solution of the equation.

7. $y = 2x^2 - 7x - 15$

(a) $(5, 0)$ (b) $(-2, 7)$

8. $y = \sqrt{16 - x^2}$

(a) $(1, 5)$ (b) $(4, 0)$

Sketching the Graph of an Equation In Exercises 9 and 10, complete the table. Use the resulting solution points to sketch the graph of the equation.

9. $y = -\frac{1}{2}x + 2$

x	-2	0	2	3	4
y					

10. $y = x^2 - 3x$

x	-1	0	1	2	3
y					

Sketching the Graph of an Equation In Exercises 11–16, sketch the graph of the equation. Identify any intercepts and test for symmetry.

11. $y = x^2 + 3$

12. $y^2 = x$

13. $y = 3x - 4$

14. $y = \sqrt{9 - x}$

15. $y = x^3 + 1$

16. $y = |x - 3|$

Finding the Equation of a Circle In Exercises 17 and 18, find the standard form of the equation of the circle.

17. Center: $(-1, 2)$; radius: 6

18. Endpoints of a diameter: $(-2, -3), (4, 5)$

Completing the Square to Sketch a Circle In Exercises 19 and 20, write the equation of the circle in standard form. Then sketch the circle.

19. $x^2 + y^2 - 4x + 6y - 12 = 0$

20. $4x^2 + 4y^2 - 4x - 8y - 11 = 0$

Finding the Slope of a Line Through Two Points In Exercises 21–24, plot the points and find the slope of the line passing through the points.

21. $(3, 7), (2, -1)$

22. $(3, -2), (-1, -2)$

23. $(3, 4), (3, -2)$

24. $(-1, 5), (2, -3)$

The Point-Slope Form of the Equation of a Line In Exercises 25–28, find an equation of the line that passes through the point and has the indicated slope. Then sketch the graph of the line.

	Point	*Slope*
25.	$(0, -5)$	$m = \frac{3}{2}$
26.	$(3, 0)$	$m = -\frac{1}{3}$
27.	$(-2, 6)$	$m = 0$
28.	$(5, 4)$	m is undefined.

Sketching the Graph of a Linear Equation In Exercises 29–32, find the slope and y-intercept (if possible) of the line specified by the equation. Then sketch the graph of the line.

29. $8x - 4y + 16 = 0$

30. $3y - 2 = 0$

31. $17 - 5x = 10$

32. $16x + 12y - 24 = 0$

Parallel and Perpendicular Lines In Exercises 33–36, determine whether the lines L_1 and L_2 passing through the pairs of points are parallel, perpendicular, or neither.

33. L_1: $(0, 3), (-2, 1)$; L_2: $(-8, -3), (4, 9)$

34. L_1: $(-1, 0), (5, 5)$; L_2: $(2, 1), (8, 6)$

35. L_1: $(3, 6), (-1, -5)$; L_2: $(-2, 3), (4, 7)$

36. L_1: $(-1, 2), (-1, 4)$; L_2: $(7, 3), (4, 7)$

Equations of Parallel and Perpendicular Lines In Exercises 37–42, write equations of the lines passing through the given point that are (a) parallel to the given line and (b) perpendicular to the given line.

	Point	*Line*
37.	$(4, 3)$	$y = \frac{1}{2}x - 1$
38.	$(1, -4)$	$y = -2x + 3$
39.	$(3, -2)$	$5x - 4y = 8$

	Point	*Line*
40.	$(-8, 3)$	$2x + 3y = 5$
41.	$(-1, -2)$	$y = 2$
42.	$(0, 5)$	$x = -3$

Direct Variation In Exercises 43–46, y is proportional to x. Use the x- and y-values to find a linear model that relates y and x.

43. $x = 3, y = 7$ **44.** $x = 5, y = 7.5$

45. $x = 10, y = 3480$

46. $x = 14, y = 1.95$

Direct Variation In Exercises 47–50, write a linear model that relates the variables.

47. A varies directly as r; $A = 30$ when $r = 6$.

48. y varies directly as z; $y = 7$ when $z = 14$.

49. a is proportional to b; $a = 15$ when $b = 20$.

50. m varies directly as n; $m = 12$ when $n = 36$.

51. Property Tax The property tax in a city is based on the assessed value of the property. A house that has an assessed value of \$150,000 has a property tax of \$1260. Find a mathematical model that gives the amount of property tax y in terms of the assessed value of the property x. Use the model to find the property tax on a house that has an assessed value of \$175,000.

52. Feet and Meters You are driving and you notice a billboard that indicates it is 1000 feet or 305 meters to the next restaurant of a national fast-food chain. Use this information to find a linear model that relates feet to meters. Use the model to complete the table.

Feet	20	50	100	120
Meters				

53. Fourth-Quarter Sales During the second and third quarters of the year, a business had sales of \$2,950,000 and \$3,150,000, respectively. Assume the growth of the sales follows a linear pattern. What will sales be during the fourth quarter?

54. Dollar Value The dollar value of a product in 2011 is \$75 and the item is expected to increase in value at a rate of \$4.75 per year. Write a linear equation that gives the dollar value of the product in terms of the year. Use this model to predict the dollar value of the product in 2013. (Let $t = 11$ represent 2011.)

55. Straight-Line Depreciation A small business purchases a piece of equipment for \$135,000. After 10 years, the equipment will have to be replaced. Its salvage value at that time is expected to be \$5500. Write a linear equation giving the value V of the equipment during the 10 years it will be used.

56. Sales The sales S (in millions of dollars) for Intuit Corporation for the years 2005 through 2010 are shown in the table. *(Source: Intuit Corporation)*

Year	Sales S (in millions of dollars)
2005	2079.9
2006	2342.3
2007	2672.9
2008	3071.0
2009	3182.5
2010	3455.0

(a) Use a graphing utility to create a scatter plot of the data. Let t represent the year, with $t = 5$ corresponding to 2005. Do the data appear linear?

(b) Use the *regression* feature of the graphing utility to find a linear model for the data.

(c) Use the linear model from part (b) to predict sales in 2011.

(d) Intuit Corporation predicts sales of \$3820 million for 2011. Does your prediction from part (c) agree with that of Intuit Corporation? Which prediction do you think is more reasonable? Explain.

Testing for Functions Represented by Equations In Exercises 57–60, decide whether the equation represents y as a function of x.

57. $3x + y = 12$ **58.** $y^2 = x^2 - 9$

59. $y^2 = x + 3$ **60.** $x^2 + y^2 - 6x + 8y = 0$

Testing for Functions In Exercises 61 and 62, decide whether the set of ordered pairs represents a function from A to B.

$A = \{1, 2, 3\}$ $B = \{-3, -4, -7\}$

Give a reason for your answer.

61. $\{(1, -3), (2, -7), (3, -3)\}$

62. $\{(1, -4), (2, -3), (3, -9)\}$

Evaluating a Function In Exercises 63 and 64, evaluate the function at each specified value of the independent variable and simplify.

63. $f(x) = \sqrt{x + 4} - 5$

(a) $f(5)$ (b) $f(0)$ (c) $f(-4)$ (d) $f(x + 3)$

64. $f(x) = \begin{cases} 2x - 1, & x \le 1 \\ x^2 + 2, & x > 1 \end{cases}$

(a) $f(0)$ (b) $f(1)$ (c) $f(3)$ (d) $f(-4)$

Finding the Domain of a Function In Exercises 65–70, find the domain of the function.

65. $f(x) = 2x^2 + 7x + 3$

66. $g(t) = 3t^2 + 1$

67. $h(x) = \sqrt{x + 5}$

68. $f(t) = \sqrt[3]{t - 3}$

69. $g(t) = \dfrac{\sqrt{t - 1}}{t - 4}$

70. $h(x) = \dfrac{2}{x^2 - 4}$

71. Reasoning A student has difficulty understanding why the domains of

$$h(x) = \frac{x^2 - 4}{x} \quad \text{and} \quad k(x) = \frac{x}{x^2 - 4}$$

are different. How would you explain their respective domains algebraically? How could you use a graphing utility to explain their domains?

72. Volume of a Box An open box is to be made from a square piece of material 20 inches on a side by cutting equal squares from the corners and turning up the sides (see figure).

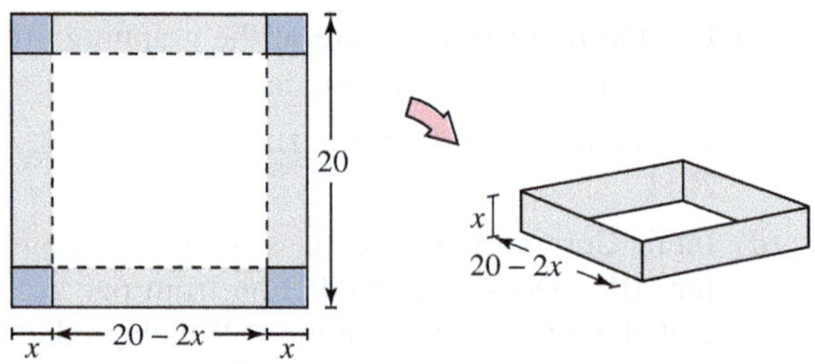

(a) Write the volume V of the box as a function of its height x.

(b) What is the domain of the function?

(c) Use a graphing utility to graph the function.

(d) Determine the volume of the box with a height of 3 inches.

73. Balance in an Account A person deposits \$6500 in an account that pays 5.75% interest compounded quarterly.

(a) Write the balance of the account in terms of the time t that the principal is left in the account.

(b) What is the domain of this function?

74. Vertical Motion The velocity v (in feet per second) of a ball thrown vertically upward from ground level is given by

$$v(t) = -32t + 80$$

where t is the time (in seconds).

(a) Find the velocity when $t = 1$.

(b) Find the time when the ball reaches its maximum height. [*Hint:* Find the time when $v(t) = 0$.]

(c) Find the velocity when $t = 3$.

Describing Function Behavior In Exercises 75–78, (a) determine the domain and range of the function, (b) determine the intervals over which the function is increasing, decreasing, or constant, (c) determine whether the function is even, odd, or neither, and (d) approximate any relative minimum or relative maximum values of the function.

75. $f(x) = x^2 + 1$

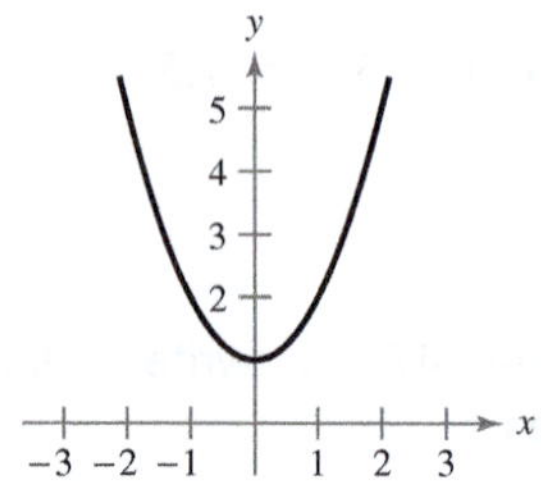

76. $f(x) = \sqrt{x^2 - 9}$

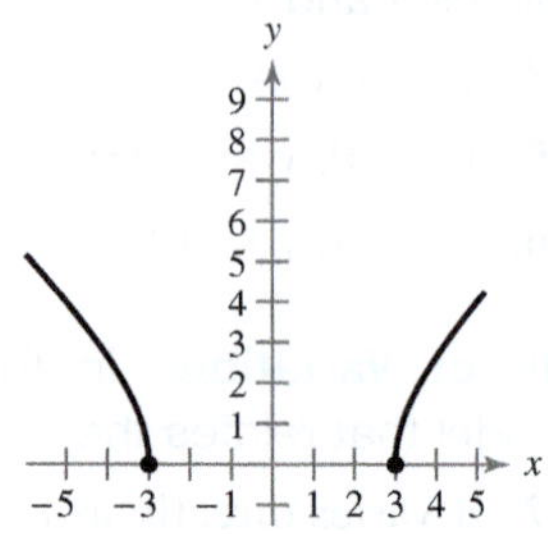

77. $f(x) = x^3 - 4x^2$

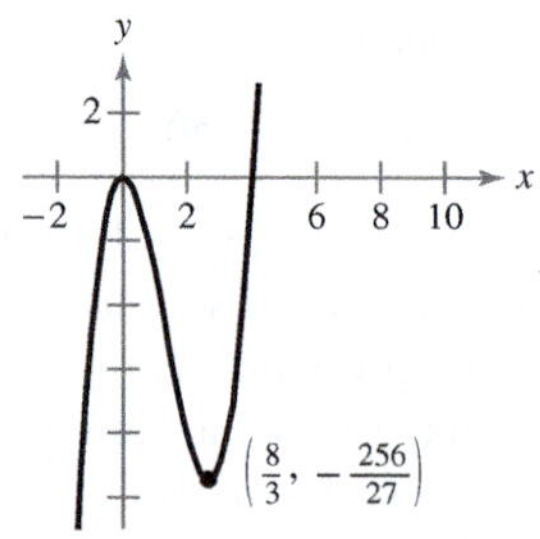

78. $f(x) = |x - 2|$

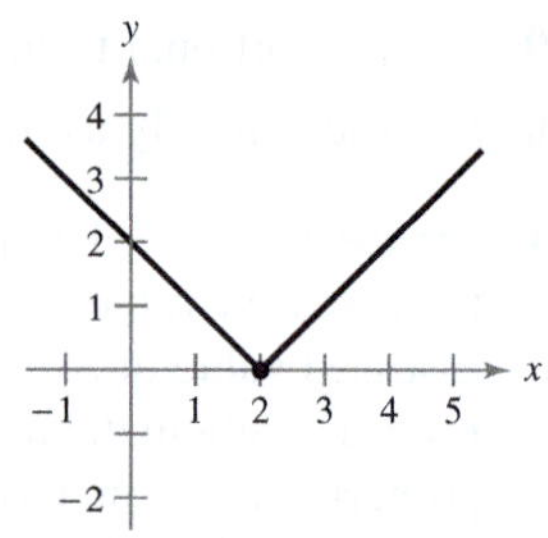

Using the Vertical Line Test In Exercises 79–82, use the Vertical Line Test to decide whether y is a function of x.

79. $y = \frac{1}{2}x^2$

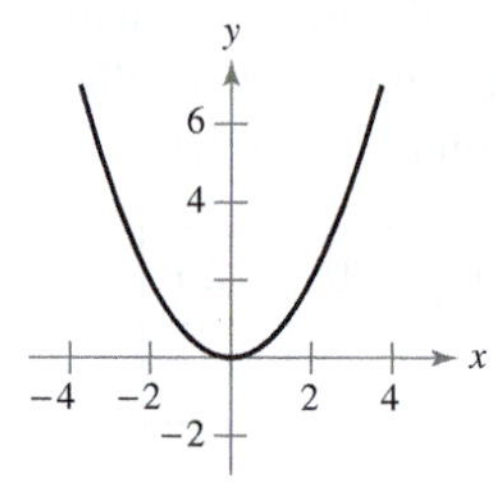

80. $y = \frac{1}{4}x^3$

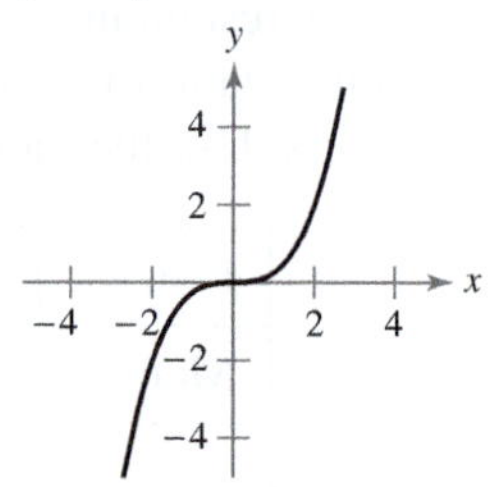

81. $x - y^2 = 1$

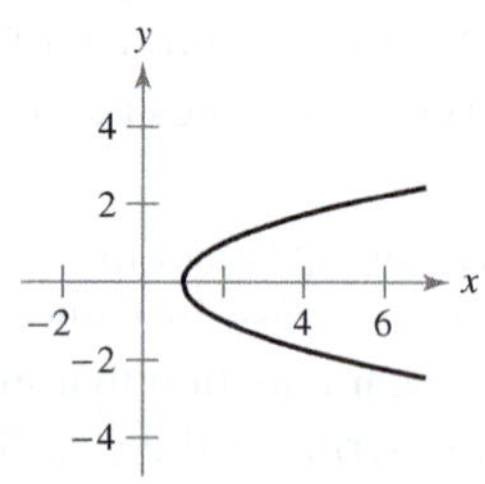

82. $x^2 + y^2 = 25$

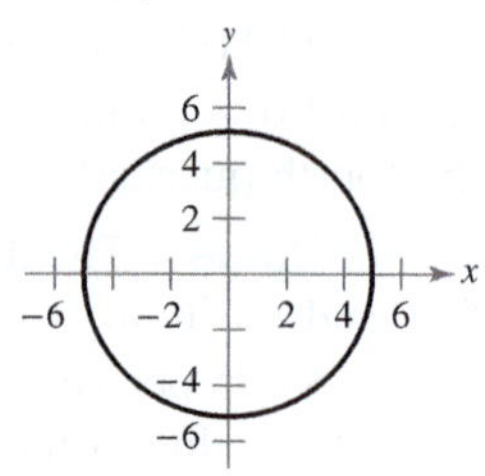

Sketching Graphs In Exercises 83–90, sketch the graph of the function.

83. $f(x) = |x + 3|$

84. $g(x) = \sqrt{x^2 - 16}$

85. $h(x) = -[\![x]\!] + 1$

86. $f(x) = 2[\![x]\!] + 2$

87. $g(x) = \begin{cases} x + 2, & x < 0 \\ 2, & x = 0 \\ x^2 + 2, & x > 0 \end{cases}$

88. $g(x) = \begin{cases} 3x + 1, & x < -1 \\ x^2 - 3, & x \geq -1 \end{cases}$

89. $h(x) = x^2 - 3x$

90. $f(x) = \sqrt{9 - x^2}$

Describing Transformations In Exercises 91 and 92, describe the sequence of transformations from $f(x) = x^2$ to g. Then sketch the graph of g by hand. Verify with a graphing utility.

91. $g(x) = -(x - 1)^2 - 2$

92. $g(x) = -x^2 + 3$

Describing Transformations In Exercises 93 and 94, describe the sequence of transformations from $f(x) = \sqrt{x}$ to g. Then sketch the graph of g by hand. Verify with a graphing utility.

93. $g(x) = \sqrt{x - 2}$ **94.** $g(x) = \sqrt{x} + 2$

Describing Transformations In Exercises 95 and 96, describe the sequence of transformations from $f(x) = \sqrt[3]{x}$ to g. Then sketch the graph of g by hand. Verify with a graphing utility.

95. $g(x) = \sqrt[3]{x + 2}$ **96.** $g(x) = 2\sqrt[3]{x} + 1$

Identifying Transformations In Exercises 97 and 98, identify the transformation shown in the graph and identify the associated common function. Write the equation of the graphed function.

97.

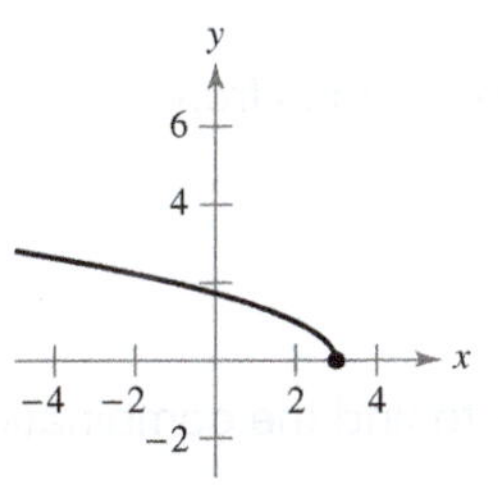

98.

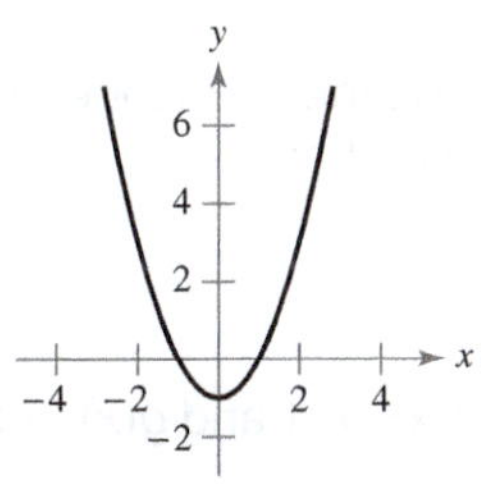

Finding Arithmetic Combinations of Functions In Exercises 99 and 100, find (a) $(f + g)(x)$, (b) $(f - g)(x)$, (c) $(fg)(x)$, and (d) $(f/g)(x)$. What is the domain of f/g?

99. $f(x) = 3x - 1, \quad g(x) = x^2 + 2x$

100. $f(x) = 3x, \quad g(x) = \sqrt{x^2 + 1}$

Evaluating Arithmetic Combinations of Functions In Exercises 101–104, evaluate the function for $f(x) = x^2 + 3x$ and $g(x) = 2x - 5$.

101. $(f + g)(2)$ **102.** $(f - g)(-1)$

103. $(fg)(3)$ **104.** $\left(\dfrac{f}{g}\right)(0)$

Composition of Functions In Exercises 105–108, find and determine the domains of (a) $f \circ g$ and (b) $g \circ f$.

105. $f(x) = x^2, \quad g(x) = x + 3$

106. $f(x) = 2x - 5, \quad g(x) = x^2 + 2$

107. $f(x) = \dfrac{1}{x}, \quad g(x) = 3x + x^2$

108. $f(x) = \dfrac{1}{x^2}, \quad g(x) = x^3$

Identifying a Composite Function In Exercises 109–112, find two functions f and g such that $(f \circ g)(x) = h(x)$. (There are many correct answers.)

109. $h(x) = (6x - 5)^2$

110. $h(x) = \sqrt[3]{x + 2}$

111. $h(x) = \dfrac{1}{(x - 1)^2}$

112. $h(x) = (x - 3)^3 + 2(x - 3)$

113. Area A square concrete foundation is prepared as a base for a large cylindrical aquatic tank that is to be used in ecology experiments (see figure).

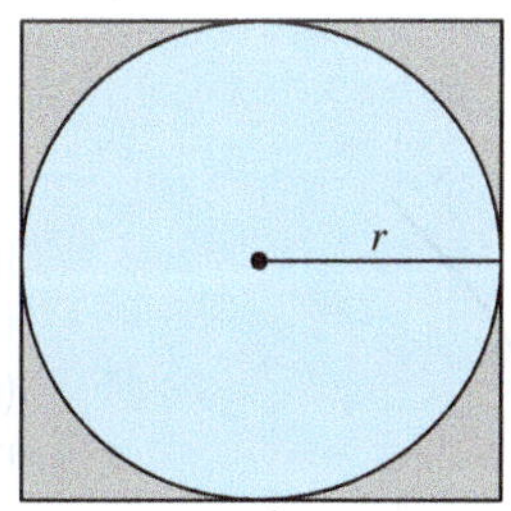

(a) Write the radius r of the tank as a function of the length x of the sides of the square.

(b) Write the area A of the circular base of the tank as a function of the radius r.

(c) Find and interpret $(A \circ r)(x)$.

114. Salary You are a sales representative for an automobile manufacturer. You are paid an annual salary plus a bonus of 4% of your sales over \$500,000. Consider the two functions given by

$f(x) = x - 500{,}000$

and

$g(x) = 0.04x.$

If x is greater than \$500,000, does $f(g(x))$ or $g(f(x))$ represent your bonus? Explain.

115. Bacteria The number N of bacteria in a petri dish is given by $N(T) = 8T^2 - 14T + 200, \ 2 \leq T \leq 32$, where T is the temperature (in degrees Fahrenheit). The temperature is $T(t) = 2t + 2, 0 \leq t \leq 15$, where t is the time in hours. Find and interpret $(N \circ T)(t)$.

TEST YOURSELF

See www.CalcChat.com for worked-out solutions to odd-numbered exercises.

Take this test as you would take a test in class. When you are done, check your work against the answers given in the back of the book.

In Exercises 1 and 2, find the distance between the points and the midpoint of the line segment joining the points.

1. $(-3, 2), (5, -2)$

2. $(3.25, 7.05), (-2.37, 1.62)$

3. Find the x- and y-intercepts of the graph of $y = (x + 5)(x - 3)$.

4. Describe the symmetry of the graph of

$$y = \frac{x}{x^2 - 4}.$$

5. Write the equation of the circle in standard form and sketch its graph.

$$x^2 + y^2 - 6x + 4y - 3 = 0$$

6. Find an equation of the line that passes through $(-3, 5)$ and has a slope of $\frac{2}{3}$.

In Exercises 7 and 8, decide whether the statement is true or false. Explain.

7. The equation $2x - 3y = 5$ represents y as a function of x.

8. If $A = \{3, 4, 5\}$ and $B = \{-1, -2, -3\}$, then the set $\{(3, -9), (4, -2), (5, -3)\}$ represents a function from A to B.

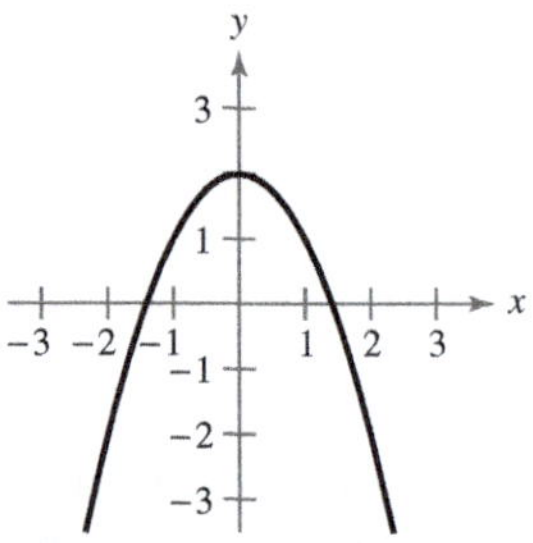

Figure for 9

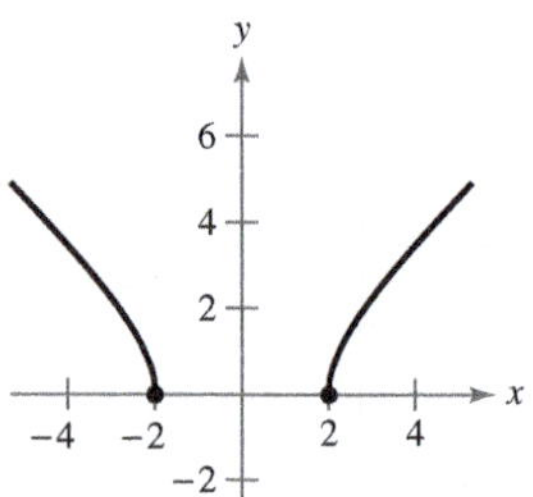

Figure for 10

In Exercises 9 and 10, (a) find the domain and range of the function, (b) determine the intervals over which the function is increasing, decreasing, or constant, (c) determine whether the function is even or odd, and (d) approximate any relative minimum or relative maximum values of the function.

9. $f(x) = 2 - x^2$ (See figure.)

10. $g(x) = \sqrt{x^2 - 4}$ (See figure.)

In Exercises 11 and 12, describe the sequence of transformations from $f(x) = x^2$ to g. Then sketch the graph of g by hand.

11. $g(x) = -(x + 2)^2$

12. $g(x) = (x - 3)^2 + 4$

In Exercises 13–16, use $f(x) = x^2 + 1$ and $g(x) = 2x - 2$ to find the combination or composition.

13. $(f - g)(x)$

14. $(fg)(x)$

15. $(f \circ g)(x)$

16. $(g \circ f)(x)$

17. A business purchases a piece of equipment for $30,000. After 5 years, the equipment will be worth only $3500. Write a linear equation that gives the value V of the equipment during the 5 years.

Year	Population, P (in millions)
2015	22.1
2020	22.8
2025	23.5
2030	24.2
2035	25.1
2040	26.1
2045	27.2
2050	28.1

Table for 18

18. The projected populations P (in millions) of children under the age of 5 in the United States for selected years from 2015 through 2050 are shown in the table. Use a graphing utility to create a scatter plot of the data and find a linear model for the data. Let t represent the year, with $t = 15$ corresponding to 2015. *(Source: U.S. Census Bureau)*

3 Polynomial and Rational Functions

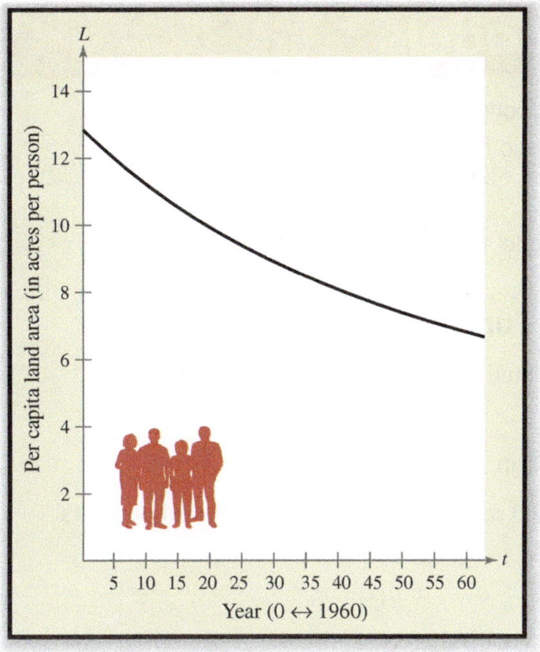

Example 8 on page 321 shows how a rational function can be used to model the per capita land area of the United States.

Morgan Lane Photography/www.shutterstock.com
Kurhan/Shutterstock.com

3.1 Quadratic Functions and Models

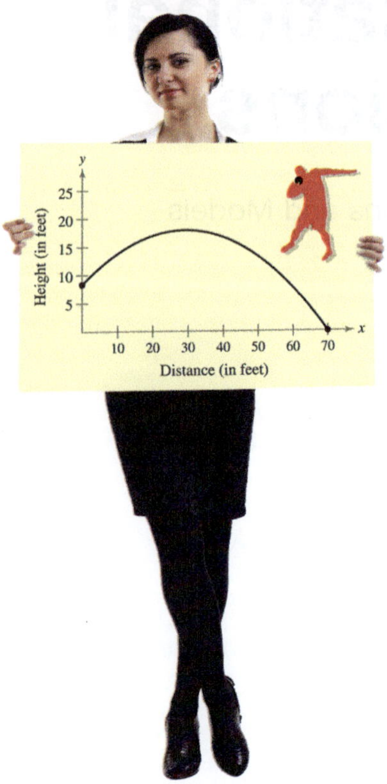

In Exercise 56 on page 260, you will use a quadratic function to determine the distance traveled by a shot in a shot put event.

- Sketch the graph of a quadratic function.
- Use the standard form of a quadratic function to sketch the graph of a parabola and find the equation of a parabola given its vertex and a point on its graph.
- Construct and use a quadratic model to solve an application problem.

The Graph of a Quadratic Function

In this and the next section, you will study the graphs of polynomial functions.

Definition of a Polynomial Function

Let n be a nonnegative integer and let $a_n, a_{n-1}, \ldots, a_2, a_1, a_0$ be real numbers with $a_n \neq 0$. The function given by

$$f(x) = a_n x^n + a_{n-1}x^{n-1} + \cdots + a_2x^2 + a_1x + a_0$$

is called a **polynomial function of x with degree n.**

For instance, each of the following functions is a polynomial function.

$$f(x) = 3x + 5 \qquad g(x) = \frac{1}{4}x^2 \qquad h(x) = 2x^3 + 7x^2 - 5x + 11$$

Polynomial functions are classified by degree. Recall that the degree of a polynomial is the highest degree of its terms. For instance, the polynomial function given by

$$f(x) = a, \quad a \neq 0 \qquad \text{Constant function}$$

has degree 0 and is called a **constant function.** In Chapter 2, you learned that the graph of this type of function is a horizontal line. The polynomial function given by

$$f(x) = ax + b, \quad a \neq 0 \qquad \text{Linear function}$$

has degree 1 and is called a **linear function.** In Chapter 2, you learned that the graph of the linear function given by $f(x) = ax + b$ is a line whose slope is a and whose y-intercept is $(0, b)$. In this section, you will study second-degree polynomial functions, which are called **quadratic functions.**

For instance, each of the following functions is a quadratic function.

$$f(x) = x^2 + 6x + 2 \qquad g(x) = 2(x + 1)^2 - 3 \qquad h(x) = (x - 2)(x + 1)$$

Definition of a Quadratic Function

Let a, b, and c be real numbers with $a \neq 0$. The function of x given by

$$f(x) = ax^2 + bx + c \qquad \text{Quadratic function}$$

is called a quadratic function.

The graph of a quadratic function is called a **parabola.** It is "∪"-shaped and can open upward or downward. Parabolas occur in many real-life applications, such as modeling the path of a baseball or the revenue and production costs of a company. You will learn about such applications later in this section.

All parabolas are symmetric with respect to a line called the **axis of symmetry,** or simply the **axis** of the parabola. The point at which the axis intersects the parabola is the **vertex** of the parabola, as shown in Figure 3.1. When the leading coefficient is positive, the graph of $f(x) = ax^2 + bx + c$ is a parabola that opens upward. When the leading coefficient is negative, the graph of $f(x) = ax^2 + bx + c$ is a parabola that opens downward. Later in this section, you will learn ways to find the coordinates of the vertex of a parabola.

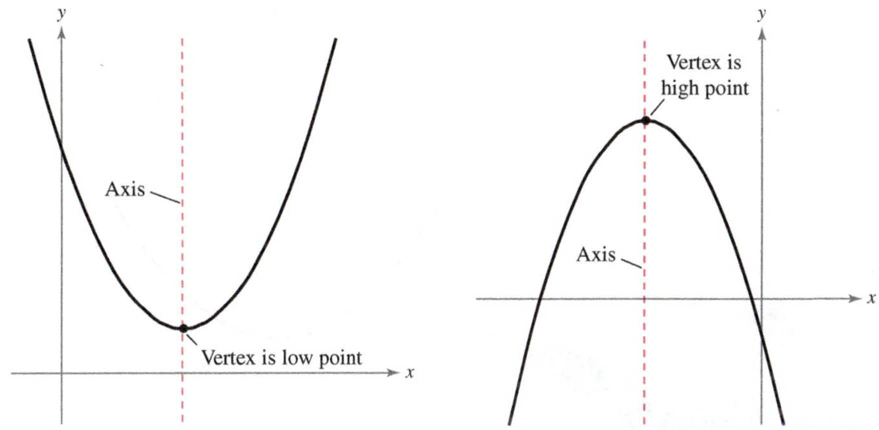

$a > 0$: Parabola opens upward. $a < 0$: Parabola opens downward.

FIGURE 3.1

The simplest type of quadratic function is

$$f(x) = ax^2.$$

Its graph is a parabola whose vertex is (0, 0). When $a > 0$, the vertex is the point with the *minimum* y-value on the graph, and when $a < 0$, the vertex is the point with the *maximum* y-value on the graph, as shown in Figure 3.2.

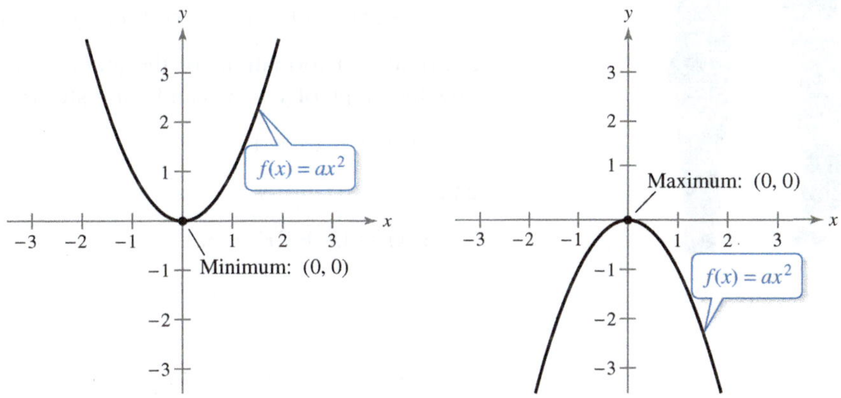

$a > 0$: Parabola opens upward. $a < 0$: Parabola opens downward.

FIGURE 3.2

When sketching the graph of

$$f(x) = ax^2$$

it is helpful to use the graph of $y = x^2$ as a reference, as discussed in Section 2.6. There you learned that when $a > 1$, the graph of $y = af(x)$ is a vertical stretch of the graph of $y = f(x)$. When $0 < a < 1$, the graph of $y = af(x)$ is a vertical shrink of the graph of $y = f(x)$. This is demonstrated again in Example 1.

Example 1 **Sketching the Graph of a Quadratic Function**

a. Compared with the graph of $y = x^2$, each output of $f(x) = \frac{1}{3}x^2$ vertically "shrinks" the graph by a factor of $\frac{1}{3}$, creating the wider parabola shown in Figure 3.3(a).

b. Compared with the graph of $y = x^2$, each output of $g(x) = 2x^2$ vertically "stretches" the graph by a factor of 2, creating the narrower parabola shown in Figure 3.3(b).

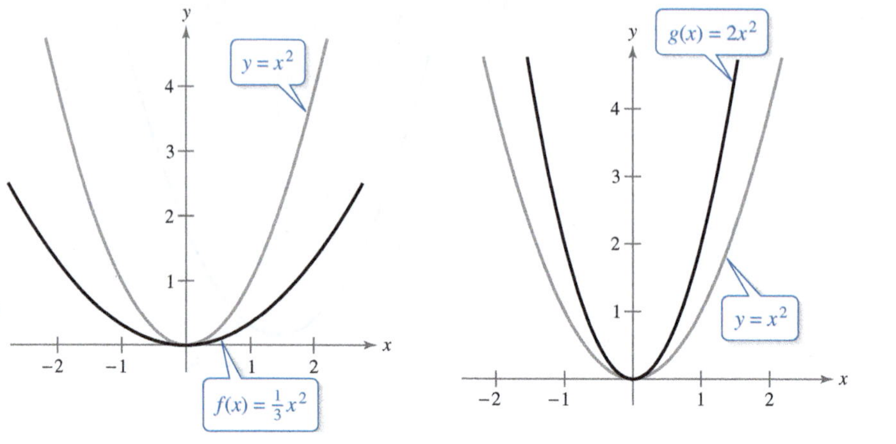

FIGURE 3.3

Checkpoint 1

Sketch the graph of $f(x) = 4x^2$. Then compare this graph with the graph of $y = x^2$.

Note that the coefficient a determines how widely the parabola given by $f(x) = ax^2$ opens. When $|a|$ is small, the parabola opens more widely than when $|a|$ is large.

Recall from Section 2.6 that the graphs of

$$y = f(x \pm c), \quad y = f(x) \pm c, \quad y = -f(x), \quad \text{and} \quad y = f(-x)$$

are rigid transformations of the graph of $y = f(x)$. For instance, in Figure 3.4, notice how the graph of $y = x^2$ can be transformed to produce the graphs of

$$f(x) = -x^2 + 1$$

and

$$g(x) = (x + 2)^2 - 3.$$

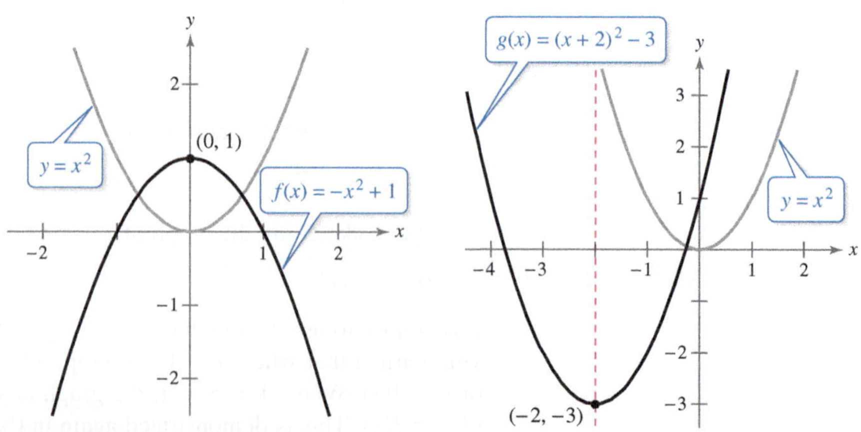

FIGURE 3.4

The Standard Form of a Quadratic Function

The **standard form** of a quadratic function is

$$f(x) = a(x - h)^2 + k.$$

This form is especially convenient for sketching a parabola because it identifies the vertex of the parabola.

ALGEBRA TUTOR *xy*

For help in factoring polynomial expressions, see the *Chapter 3 Algebra Tutor* on page 326.

Standard Form of a Quadratic Function

The quadratic function given by

$$f(x) = a(x - h)^2 + k, \quad a \neq 0$$

is said to be in **standard form.** The graph of f is a parabola whose axis is the vertical line $x = h$ and whose vertex is the point (h, k).

When $a > 0$, the parabola opens upward.

When $a < 0$, the parabola opens downward.

To write a quadratic function in standard form, you can use the process of *completing the square*, as illustrated in Example 2.

Example 2 Using Standard Form to Graph a Parabola

Sketch the graph of $f(x) = 2x^2 + 8x + 7$. Identify the vertex.

SOLUTION Begin by writing the quadratic function in standard form. The first step in completing the square is to factor out any coefficient of x^2 that is not 1.

$$\begin{aligned} f(x) &= 2x^2 + 8x + 7 && \text{Write original function.} \\ &= (2x^2 + 8x) + 7 && \text{Group } x\text{-terms.} \\ &= 2(x^2 + 4x) + 7 && \text{Factor 2 out of } x\text{-terms.} \\ &= 2(x^2 + 4x + 4 - 4) + 7 && \text{Add and subtract 4 within parentheses to complete the square.} \end{aligned}$$

$$\left(\tfrac{4}{2}\right)^2$$

After adding and subtracting 4 within the parentheses, regroup the terms to form a perfect square trinomial. The -4 can be removed from inside the parentheses. But, because of the 2 outside the parentheses, you must multiply -4 by 2 as shown below.

$$\begin{aligned} f(x) &= 2(x^2 + 4x + 4) - 2(4) + 7 && \text{Regroup terms.} \\ &= 2(x^2 + 4x + 4) - 8 + 7 && \text{Simplify.} \\ &= 2(x + 2)^2 - 1 && \text{Standard form} \end{aligned}$$

From this form, you can see that the graph of f is a parabola that opens upward with vertex $(-2, -1)$. This corresponds to a left shift of two units and a downward shift of one unit relative to the graph of $y = 2x^2$, as shown in Figure 3.5.

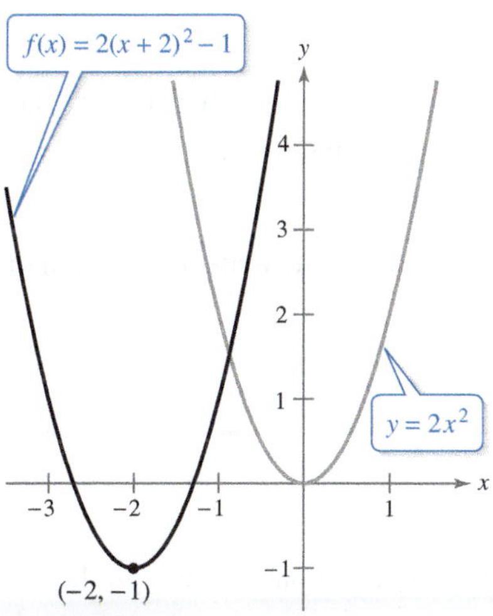

FIGURE 3.5

Checkpoint 2

Sketch the graph of each quadratic function. Identify the vertex.

a. $f(x) = x^2 + 2x + 4$

b. $f(x) = 2x^2 - 12x + 20$

Example 3 Using Standard Form to Graph a Parabola

Sketch the graph of $f(x) = -x^2 + 6x - 8$. Identify the vertex.

SOLUTION As in Example 2, begin by writing the quadratic function in standard form.

$$\begin{aligned} f(x) &= -x^2 + 6x - 8 && \text{Write original function.} \\ &= (-x^2 + 6x) - 8 && \text{Group } x\text{-terms.} \\ &= -(x^2 - 6x) - 8 && \text{Factor } -1 \text{ out of } x\text{-terms.} \\ &= -(x^2 - 6x + 9 - 9) - 8 && \text{Add and subtract 9 within parentheses to complete the square.} \\ & \qquad (-6/2)^2 && \\ &= -(x^2 - 6x + 9) - (-9) - 8 && \text{Regroup terms.} \\ &= -(x - 3)^2 + 1 && \text{Standard form} \end{aligned}$$

So, the graph of f is a parabola that opens downward with vertex at (3, 1), as shown in Figure 3.6.

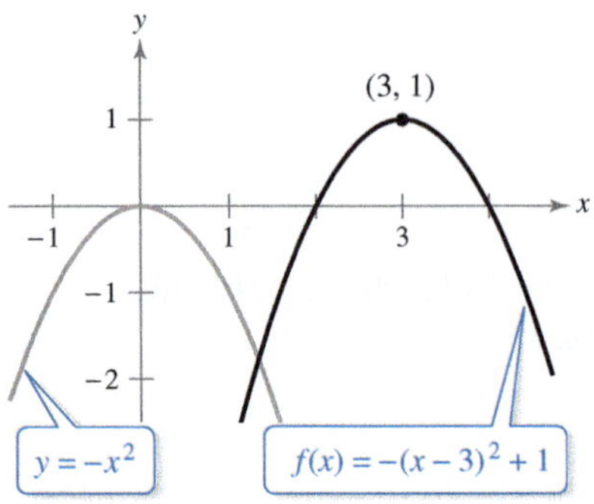

FIGURE 3.6

✓Checkpoint 3

Sketch the graph of $f(x) = -3x^2 + 12x + 1$. Identify the vertex.

Example 4 Finding an Equation of a Parabola

Find an equation of the parabola whose vertex is (1, 2) and that passes through the point (0, 0), as shown in Figure 3.7.

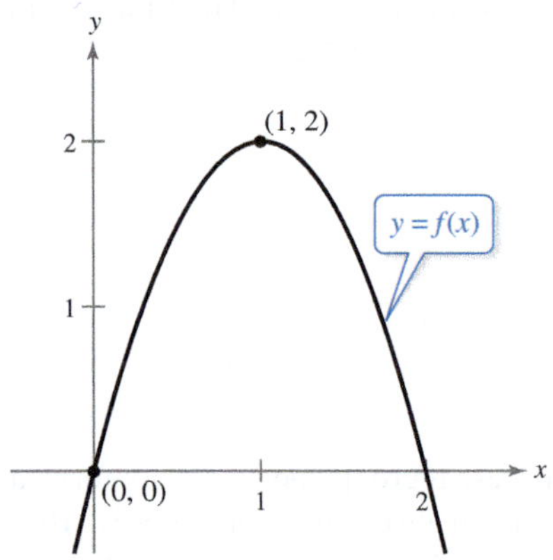

FIGURE 3.7

SOLUTION Because the parabola has a vertex at $(h, k) = (1, 2)$, the equation must have the form

$$f(x) = a(x - 1)^2 + 2. \qquad \text{Standard form}$$

Because the parabola passes through the point (0, 0), it follows that when $x = 0$, $f(x)$ must equal 0. Substitute 0 for x and 0 for $f(x)$ to obtain the equation

$$0 = a(0 - 1)^2 + 2.$$

By solving this equation for a, you find that $a = -2$. You can now write an equation of the parabola.

$$\begin{aligned} f(x) &= -2(x - 1)^2 + 2 && \text{Substitute for } a, h, \text{ and } k \text{ in standard form.} \\ &= -2x^2 + 4x && \text{Simplify.} \end{aligned}$$

✓Checkpoint 4

Find an equation of the parabola whose vertex is (3, 4) and that passes through the point (2, 5).

To find the x-intercepts of the graph of $f(x) = ax^2 + bx + c$, you must solve the equation

$$ax^2 + bx + c = 0.$$

If the expression $ax^2 + bx + c$ does not factor, then use the Quadratic Formula to determine the x-intercepts. Remember, however, that a parabola may have no x-intercepts.

Applications

Many applications involve finding the maximum or minimum value of a quadratic function. By writing $f(x) = ax^2 + bx + c$ in standard form, you can determine that the x-coordinate of the vertex is $-b/2a$. The y-coordinate of the vertex can then be found by evaluating the function at $x = -b/2a$.

TECH TUTOR

Your graphing utility may have *minimum* and *maximum* features that determine the minimum and maximum points of the graph of a function. You can use these features to find the vertex of a parabola. For instructions on how to use the *minimum* and *maximum* features, see Appendix A. Consult the user's guide for your graphing utility for more information.

Example 5 The Maximum Height of a Baseball

A baseball is hit 3 feet above the ground at a velocity of 100 feet per second and at an angle of 45° with respect to the ground. The path of the baseball is given by

$$f(x) = -0.0032x^2 + x + 3$$

where $f(x)$ is the height of the baseball (in feet) and x is the distance from home plate (in feet). What is the maximum height reached by the baseball?

SOLUTION For this quadratic function, you have

$$\begin{aligned} f(x) &= ax^2 + bx + c \\ &= -0.0032x^2 + x + 3. \end{aligned}$$

So, $a = -0.0032$ and $b = 1$. Because the function has a maximum when $x = -b/2a$, the baseball reaches its maximum height when it is

$$x = -\frac{b}{2a} = -\frac{1}{2(-0.0032)} = 156.25 \text{ feet}$$

from home plate. At this distance, the maximum height is

$$\begin{aligned} f(156.25) &= -0.0032(156.25)^2 + 156.25 + 3 \\ &= 81.125 \text{ feet.} \end{aligned}$$

The path of the baseball is shown in Figure 3.8.

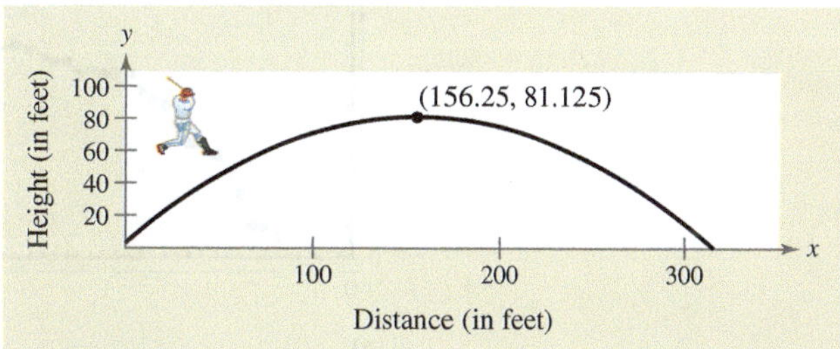

FIGURE 3.8

Checkpoint 5

In Example 5, the baseball is hit at a velocity of 70 feet per second. The path of the baseball is given by

$$f(x) = -0.007x^2 + x + 4$$

where $f(x)$ is the height of the baseball (in feet) and x is the distance from home plate (in feet). What is the maximum height reached by the baseball?

In Section 2.3, you plotted data points in the coordinate plane and estimated the best-fitting line. Fitting a quadratic model using this same process would be complicated. Most graphing utilities have a built-in statistical program that easily calculates the best-fitting quadratic model for a set of data points. Refer to the user's guide of your graphing utility for the required steps.

David Davis/www.shutterstock.com

Example 6 Fitting a Quadratic Function to Data

Each year, grizzly bears in Yellowstone National Park feed on moths along the mountain slopes of the park. The bears dig into the mountains and eat the moths they uncover. The table shows the number N of confirmed moth sites in the Greater Yellowstone Ecosystem each year from 1986 to 2008. Use a graphing utility to plot the data and find the quadratic model that best fits the data. Find the vertex of the graph of the quadratic model and interpret its meaning in the context of the problem. Let $x = 6$ represent the year 1986. *(Source: Yellowstone Grizzly Bear Investigations 2008)*

TECH TUTOR

For instructions on how to use the *regression* feature, see Appendix A. Consult the user's guide for your graphing utility for more information.

x	6	7	8	9	10	11	12	13	14	15	16	17
N	3	5	5	10	14	17	19	19	22	25	26	28

x	18	19	20	21	22	23	24	25	26	27	28
N	30	30	30	31	31	32	32	33	34	35	35

SOLUTION Begin by entering the data into your graphing utility and displaying the scatter plot. From the scatter plot that is shown in Figure 3.9(a), you can see that the points have a parabolic trend. Use the *quadratic regression* feature to find the quadratic function that best fits the data. The quadratic equation that best fits the data is

$$N = -0.076x^2 + 4.04x - 19.5, \quad 6 \le x \le 28.$$

Graph the data and the equation in the same viewing window, as shown in Figure 3.9(b). By using the *maximum* feature of your graphing utility, you can see that the vertex of the graph is approximately (26.6, 34.2), as shown in Figure 3.9(c). The vertex corresponds to the year in which the number of confirmed moth sites was the greatest. So, in 2006, the number of confirmed moth sites reached a maximum.

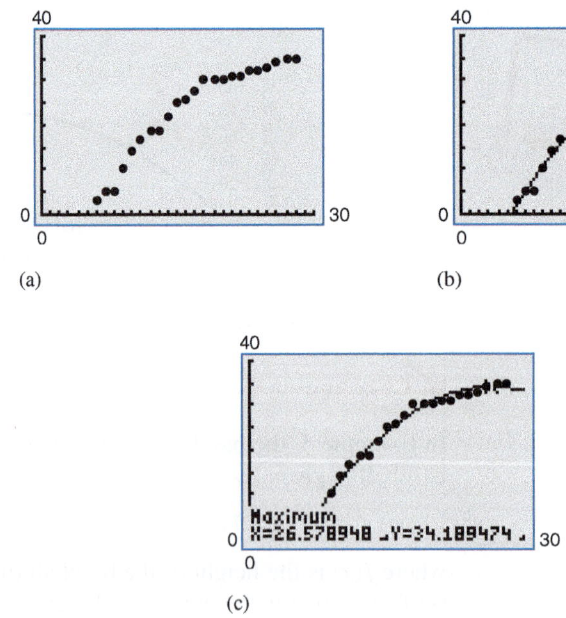

FIGURE 3.9

✓ Checkpoint 6

In Example 6, use the model to predict the number of confirmed moth sites in 2014. What do you notice?

Example 7 Charitable Contributions

The percent of a family's income that is given to charities is related to the income of the family. For families with annual incomes between \$5000 and \$100,000, the percent P can be modeled by

$$P(x) = 0.0014x^2 - 0.1529x + 5.855, \quad 5 \le x \le 100$$

where x is the annual income (in thousands of dollars). Use the model to estimate the income that corresponds to the minimum percent of income given to charities.

SOLUTION One way to answer the question is to sketch the graph of the quadratic function, as shown in Figure 3.10. From this graph, it appears that the minimum percent corresponds to an income of about \$55,000.

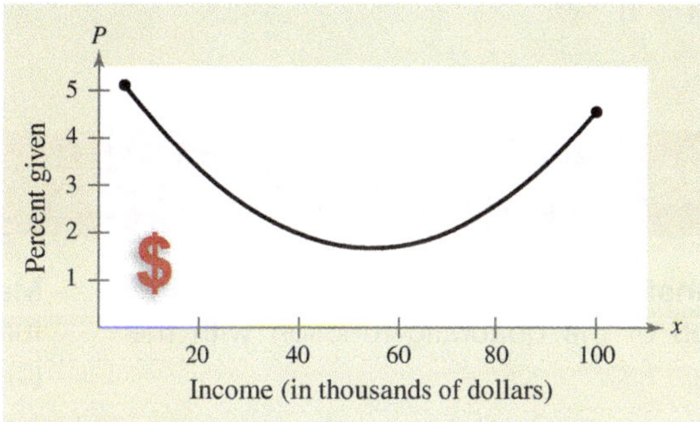

FIGURE 3.10

Another way to answer the question is to use the fact that the minimum point of the parabola occurs when $x = -b/2a$.

$$\begin{aligned} x &= -\frac{b}{2a} \\ &= -\frac{-0.1529}{2(0.0014)} \\ &\approx 54.6 \end{aligned}$$

From this x-value, you can conclude that the minimum percent corresponds to an income of about \$54,600.

✓ Checkpoint 7

A manufacturer has daily production costs C (in dollars per unit) of $C = 0.15x^2 - 9x + 700$, where x is the number of units produced. How many units should be produced each day to yield a minimum cost per unit? ■

SUMMARIZE (Section 3.1)

1. State the definition of a quadratic function and describe its graph *(pages 250–252)*. For an example of sketching the graph of a quadratic function, see Example 1.
2. State the standard form of a quadratic function *(page 253)*. For examples that use the standard form of a quadratic function, see Examples 2, 3, and 4.
3. Describe a real-life example of how a quadratic function can be used to analyze the path of a moving object *(page 255, Example 5)*.

SKILLS WARM UP 3.1 The following warm-up exercises involve skills that were covered in earlier sections. You will use these skills in the exercise set for this section. For additional help, review Sections 1.3 and 1.4.

In Exercises 1–4, solve the quadratic equation by factoring.

1. $2x^2 + 11x - 6 = 0$

2. $5x^2 - 12x - 9 = 0$

3. $3 + x - 2x^2 = 0$

4. $x^2 + 20x + 100 = 0$

In Exercises 5–10, use the Quadratic Formula to solve the quadratic equation.

5. $x^2 - 6x + 4 = 0$

6. $x^2 + 4x + 1 = 0$

7. $2x^2 - 16x + 25 = 0$

8. $3x^2 + 30x + 74 = 0$

9. $x^2 + 3x + 1 = 0$

10. $x^2 + 3x - 3 = 0$

Exercises 3.1

See www.CalcChat.com for worked-out solutions to odd-numbered exercises.

Describing a Transformation In Exercises 1 and 2, compare the graph of the quadratic function with the graph of $y = x^2$. *See Example 1.*

1. $f(x) = 5x^2$

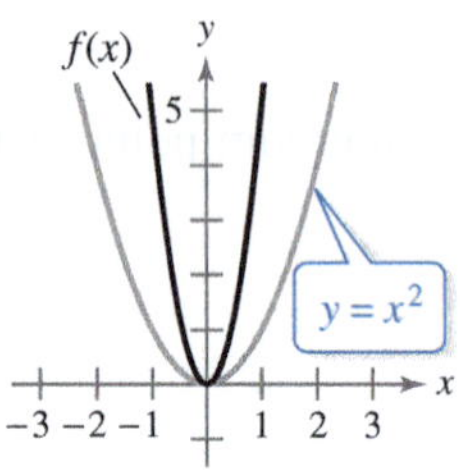

2. $f(x) = -\frac{1}{4}x^2$

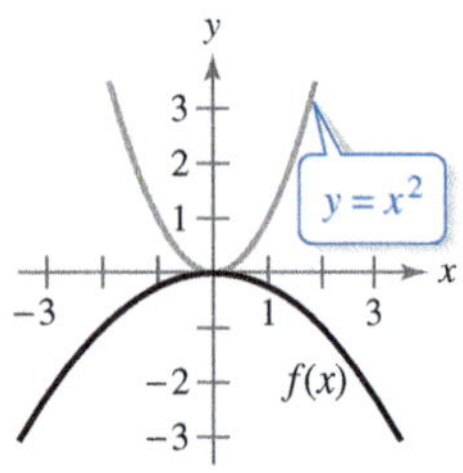

Sketching the Graph of a Quadratic Function In Exercises 3–6, sketch the graph of the quadratic function and compare it with the graph of $y = x^2$. *See Example 1.*

3. $f(x) = 3x^2$

4. $f(x) = -2x^2$

5. $g(x) = -\frac{1}{2}x^2$

6. $g(x) = \frac{3}{4}x^2$

Describing a Transformation In Exercises 7 and 8, compare the graph of the quadratic function with the graph of $y = x^2$.

7. $f(x) = -(x + 1)^2 + 1$

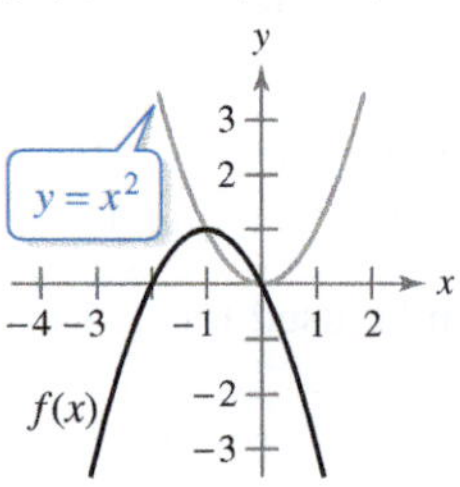

8. $f(x) = 3(x - 2)^2 - 1$

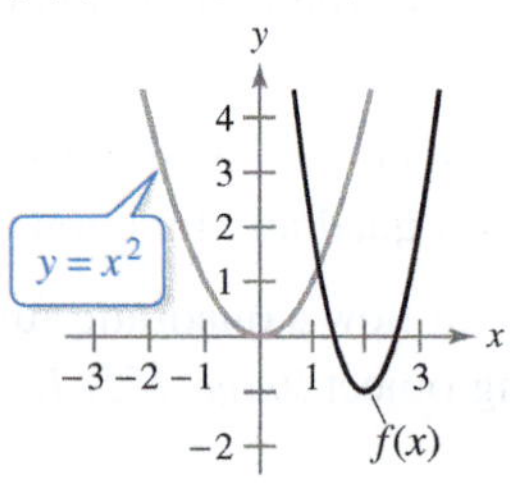

Matching In Exercises 9–14, match the quadratic function with its graph. [The graphs are labeled (a), (b), (c), (d), (e), and (f).]

(a)

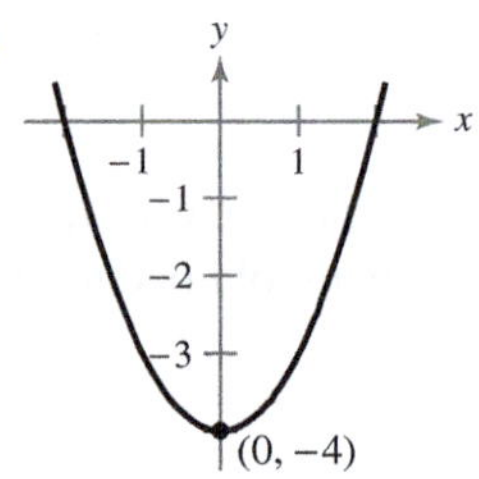

(b)

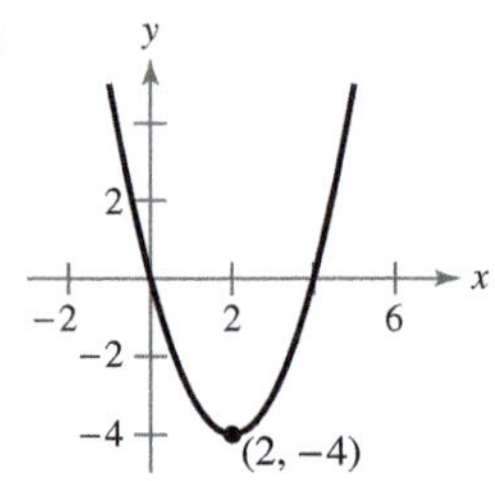

(c)

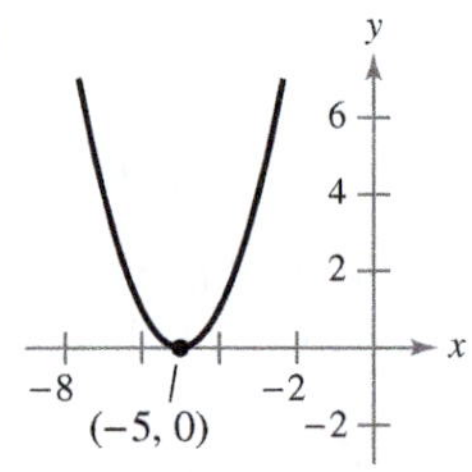

(d)

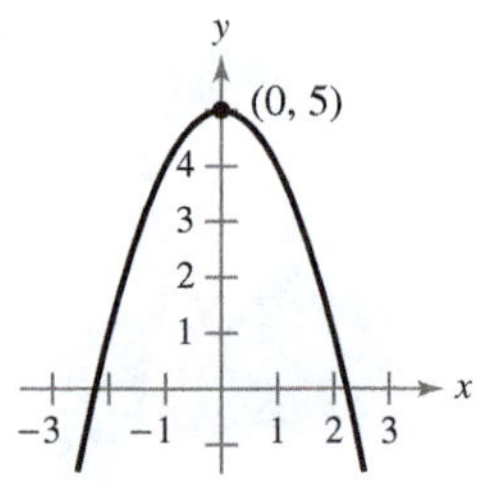

(e)

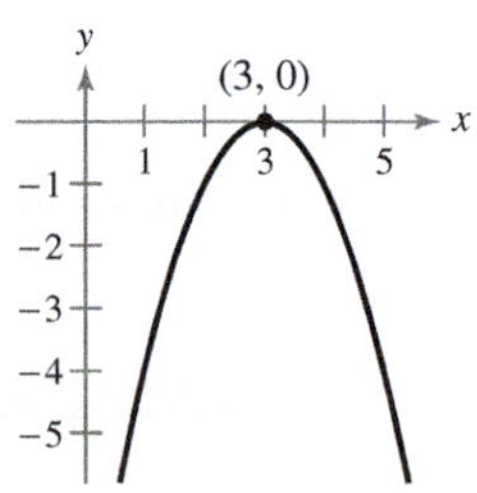

(f)

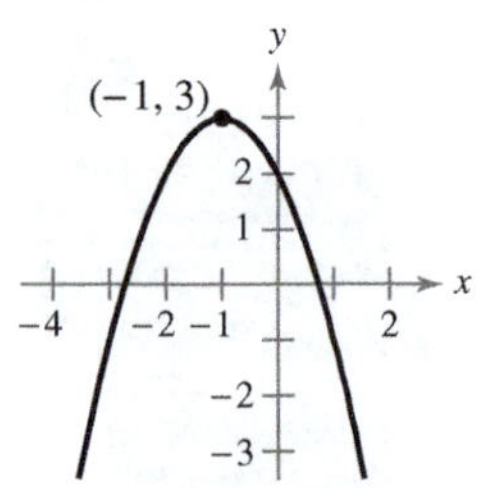

9. $f(x) = -(x - 3)^2$

10. $f(x) = (x + 5)^2$

11. $f(x) = x^2 - 4$

12. $f(x) = 5 - x^2$

13. $f(x) = -(x + 1)^2 + 3$

14. $f(x) = (x - 2)^2 - 4$

Using Standard Form to Graph a Parabola In Exercises 15–30, write the quadratic function in standard form (if necessary) and sketch its graph. Identify the vertex. *See Examples 2 and 3.*

15. $f(x) = 16 - x^2$

16. $h(x) = x^2 - 9$

17. $f(x) = (x + 5)^2 - 6$

18. $f(x) = (x - 6)^2 + 3$

19. $g(x) = x^2 + 2x + 1$

20. $h(x) = x^2 - 4x + 2$

21. $f(x) = -(x^2 + 2x - 3)$

22. $f(x) = -(x^2 + 6x - 3)$

23. $f(x) = x^2 - x + \frac{5}{4}$

24. $f(x) = x^2 + 3x + \frac{1}{4}$

25. $f(x) = -x^2 + 2x + 5$

26. $f(x) = -x^2 - 4x + 1$

27. $h(x) = 4x^2 - 4x + 21$

28. $f(x) = 2x^2 - x + 1$

29. $f(x) = \frac{1}{4}(x^2 - 16x + 32)$

30. $g(x) = \frac{1}{2}(x^2 + 4x - 2)$

Finding an Equation of a Parabola In Exercises 31–36, find an equation of the parabola. *See Example 4.*

31.

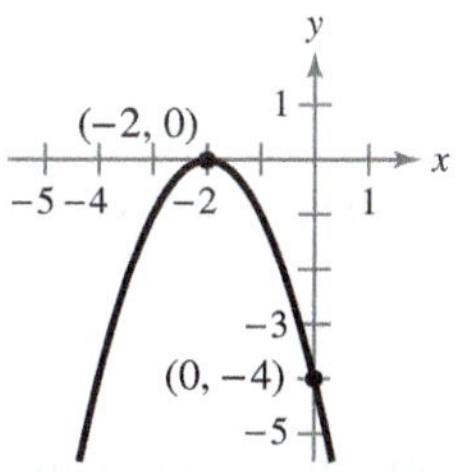

32.

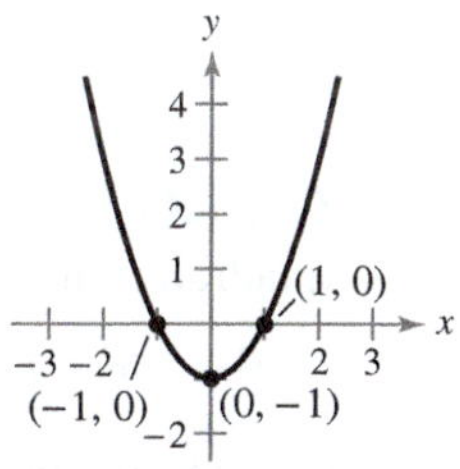

33.

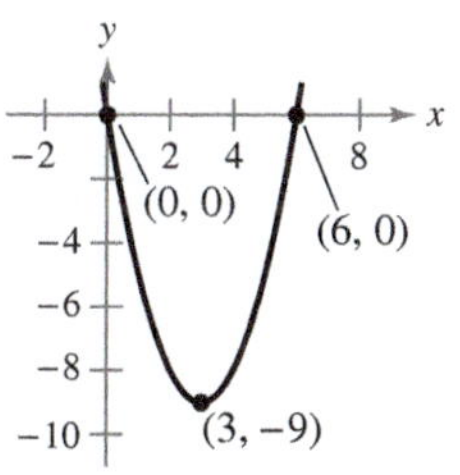

34.

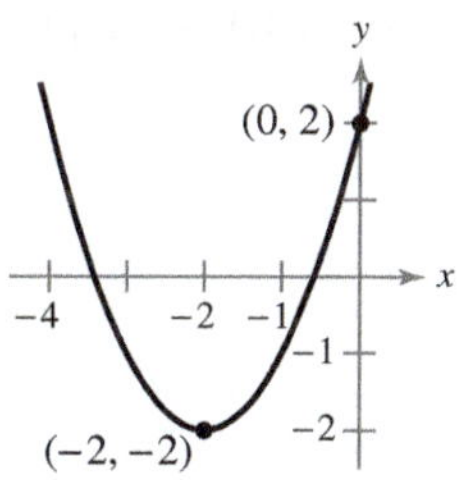

35.

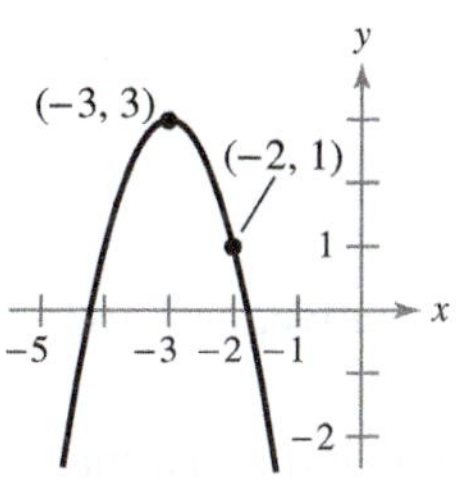

36.

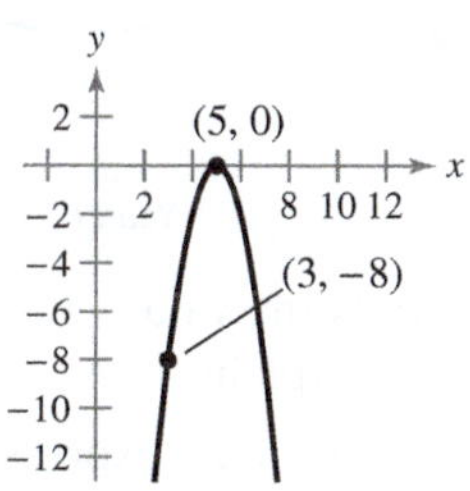

Finding an Equation of a Parabola In Exercises 37–40, find an equation of the parabola that has the indicated vertex and whose graph passes through the given point. *See Example 4.*

37. Vertex: $(2, -1)$; point: $(4, -3)$

38. Vertex: $(-3, 5)$; point: $(-6, -1)$

39. Vertex: $(5, 12)$; point: $(7, 15)$

40. Vertex: $(-2, -2)$; point: $(-1, 0)$

Finding the Slope of a Line In Exercises 41–46, find two quadratic functions whose graphs have the given x-intercepts. Find one function whose graph opens upward and another whose graph opens downward. (There are many correct answers.)

41. $(2, 0), (-1, 0)$

42. $(-4, 0), (0, 0)$

43. $(0, 0), (10, 0)$

44. $(4, 0), (8, 0)$

45. $(-3, 0), \left(-\frac{1}{2}, 0\right)$

46. $\left(-\frac{5}{2}, 0\right), (2, 0)$

47. Optimal Area The perimeter of a rectangle is 200 feet. Let x represent the width of the rectangle. Write a quadratic function for the area of the rectangle in terms of its width. Find the vertex of the graph of the quadratic function and interpret its meaning in the context of the problem.

48. Optimal Area The perimeter of a rectangle is 540 feet. Let x represent the width of the rectangle. Write a quadratic function for the area of the rectangle in terms of its width. Find the vertex of the graph of the quadratic function and interpret its meaning in the context of the problem.

49. Optimal Area A rancher has 1200 feet of fencing with which to enclose two adjacent rectangular corrals (see figure). What measurements will produce a maximum enclosed area?

50. Optimal Area An indoor physical-fitness room consists of a rectangular region with a semicircle on each end (see figure). The perimeter of the room is to be a 200-meter running track. What measurements will produce a maximum area of the rectangle?

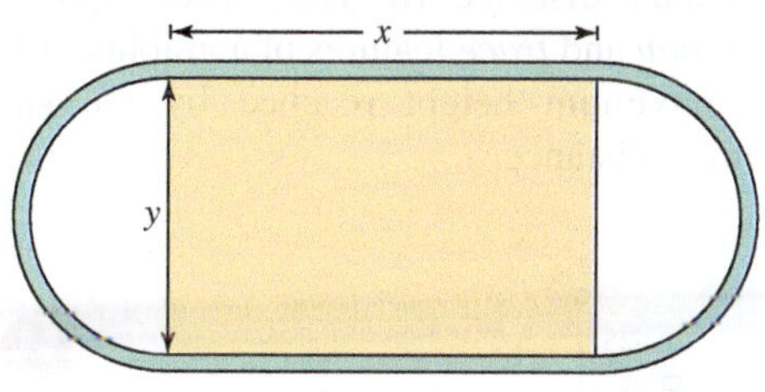

Optimal Revenue In Exercises 51 and 52, find the number of units that produces a maximum revenue. The revenue R is measured in dollars and x is the number of units produced.

51. $R = 1000x - 0.02x^2$

52. $R = 80x - 0.0001x^2$

53. Optimal Cost The daily production costs C (in dollars per unit) for a manufacturer of lighting fixtures are given by the quadratic function

$$C(x) = 800 - 10x + 0.25x^2$$

where x is the number of units produced. How many fixtures should be produced each day to yield a minimum cost per unit?

54. Optimal Profit The profit P (in dollars) for a manufacturer of sound systems is given by the quadratic function

$$P(x) = -0.0003x^2 + 150x - 375{,}000$$

where x is the number of units produced. What production level will yield a maximum profit?

55. Maximum Height of a Diver The path of a diver is given by

$$y = -\frac{4}{9}x^2 + \frac{24}{9}x + 10$$

where y is the height (in feet) and x is the horizontal distance from the end of the diving board (in feet). (See figure.) Use the *trace* or *maximum* feature of a graphing utility to find the maximum height of the diver.

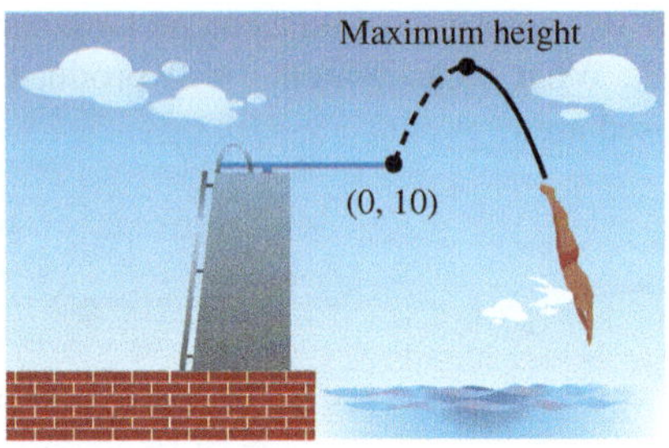

56. Maximum Height The path of the winning shot at a shot put event can be approximated by the quadratic function

$$y = -0.011x^2 + 0.65x + 8.3$$

where y is the height of the shot (in feet) and x is the horizontal distance (in feet). (See figure.) Use the *maximum* and *trace* features of a graphing utility to find the maximum height reached by the shot and its winning distance.

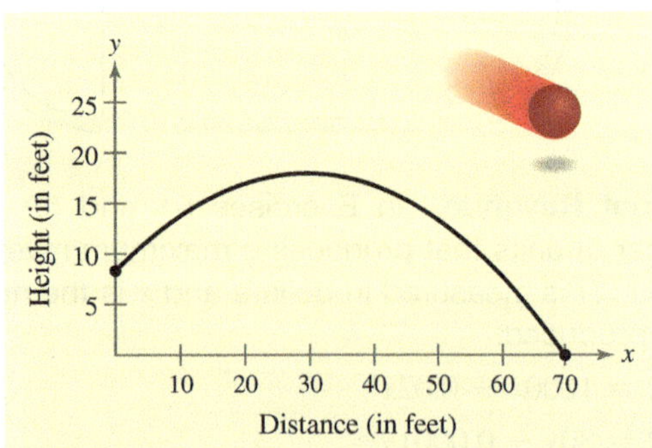

57. Revenue The table shows the total yearly revenues R (in millions of dollars) for golf courses and country clubs in the United States from 2003 through 2009. *(Source: U.S. Census Bureau)*

Year	2003	2004	2005	2006
Revenue, R	17,291	18,469	19,356	20,523

Year	2007	2008	2009
Revenue, R	21,195	21,044	20,326

(a) Use a graphing utility to create a scatter plot of the data. Let t represent the year, with $t = 3$ corresponding to 2003.

(b) Use the *regression* feature of the graphing utility to find a quadratic model for the data. Graph the model with the scatter plot from part (a).

(c) Use the graph of the model from part (b) to estimate when the yearly revenue was the greatest. Does this result agree with the actual data?

58. HOW DO YOU SEE IT? The graph shows a quadratic function of the form

$$P(t) = at^2 + bt + c$$

which represents the yearly profits for a company, where $P(t)$ is the profit in year t.

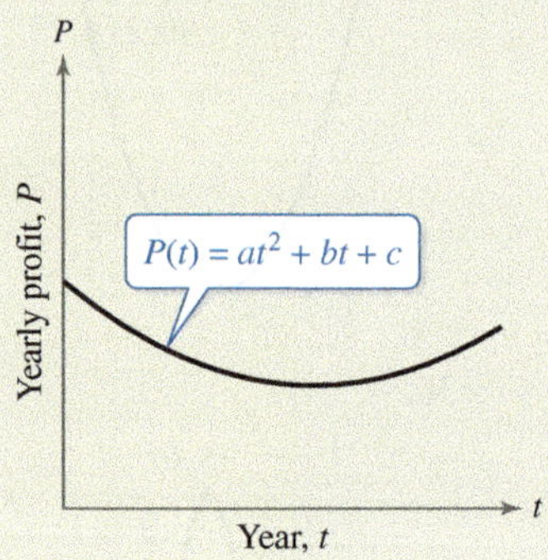

(a) Is the value of a positive, negative, or zero? Explain.

(b) Write an expression in terms of a and b that represents the year t when the company made the least profit.

(c) The company made the same yearly profits in 2004 and 2012. Estimate the year in which the company made the least profit.

(d) Assume that the model is still valid today. Are the yearly profits currently increasing, decreasing, or constant? Explain.

59. Osteopathic Physicians The table shows the numbers N (in thousands) of female osteopathic physicians from 2001 through 2010. *(Source: American Osteopathic Association)*

Year	2001	2002	2003	2004	2005
Physicians, N	10.9	11.8	12.9	14.0	15.1

Year	2006	2007	2008	2009	2010
Physicians, N	16.4	17.7	19.1	20.8	22.5

(a) Use a graphing utility to create a scatter plot of the data. Let t represent the year, with $t = 1$ corresponding to 2001.

(b) Use the *regression* feature of the graphing utility to find a quadratic model for the data.

(c) Use the graphing utility to graph the model from part (b) in the same viewing window as the scatter plot of the data.

(d) Use the graph of the model from part (c) to predict the number of female osteopathic physicians in 2012.

60. Kidney Transplants The table shows the numbers K of kidney transplants performed in the United States in the years 1999 through 2009. *(Source: U.S. Department of Health and Human Services)*

Year	1999	2000	2001	2002
Transplants, K	12,455	13,258	14,152	14,741

Year	2003	2004	2005	2006
Transplants, K	15,129	16,000	16,481	17,094

Year	2007	2008	2009
Transplants, K	16,624	16,517	16,829

(a) Use a graphing utility to create a scatter plot of the data. Let t represent the year, with $t = 9$ corresponding to 1999.

(b) Use the *regression* feature of the graphing utility to find a quadratic model for the data.

(c) Use the graphing utility to graph the model from part (b) in the same viewing window as the scatter plot of the data.

(d) Use the graph of the model from part (c) to predict the number of kidney transplants performed in 2010. Does your answer seem reasonable? Explain.

61. Analyzing Profits Let x be the number of units (in tens of thousands) that a computer company produces and let $p(x)$ be the profit (in hundreds of thousands of dollars). The table shows the profits for different levels of production.

Units, x	2	4	6	8	10
Profit, $p(x)$	270.5	307.8	320.1	329.2	325.0

Units, x	12	14	16	18	20
Profit, $p(x)$	311.2	287.8	254.8	212.2	160.0

(a) Use a graphing utility to create a scatter plot of the data.

(b) Use the *regression* feature of the graphing utility to find a quadratic model for $p(x)$.

(c) Use the graphing utility to graph your model for $p(x)$ with the scatter plot of the data.

(d) Find the vertex of the graph of the model from part (c). Interpret its meaning in the context of the problem.

(e) With these data and this model, the profit begins to decrease. Discuss how it is possible for production to increase and profit to decrease.

62. Baseball Distance Let x be the angle (in degrees) at which a baseball is hit with no spin at an initial speed of 40 meters per second, and let $d(x)$ be the distance (in meters) the ball travels. The table shows the distances for the different angles at which the ball is hit. *(Source: The Physics of Sports)*

Angle, x	10	15	30	36	42
Distance, $d(x)$	58.3	79.7	126.9	136.6	140.6

Angle, x	44	45	48	54	60
Distance, $d(x)$	140.9	140.9	139.3	132.5	120.5

(a) Use a graphing utility to create a scatter plot of the data.

(b) Use the *regression* feature of the graphing utility to find a quadratic model for $d(x)$.

(c) Use the graphing utility to graph your model for $d(x)$ with the scatter plot of the data.

(d) Find the vertex of the graph of the model from part (c). Interpret its meaning in the context of the problem.

63. Verifying a Formula Write the quadratic function $f(x) = ax^2 + bx + c$ in standard form to verify that the vertex occurs at $(-b/2a, f(-b/2a))$.

Review Exercises

See www.CalcChat.com for worked-out solutions to odd-numbered exercises.

Using Standard Form to Graph a Parabola In Exercises 1–4, write the quadratic function in standard form (if necessary) and sketch its graph. Identify the vertex and intercepts.

1. $f(x) = (x + 3)^2 - 5$
2. $g(x) = -(x - 1)^2 + 3$
3. $h(x) = 3x^2 - 12x + 11$
4. $f(x) = \frac{1}{3}(x^2 + 5x - 4)$

Finding an Equation of a Parabola In Exercises 5 and 6, find an equation of the parabola that has the indicated vertex and whose graph passes through the given point.

5. Vertex: $(-5, -1)$; point: $(-2, 6)$
6. Vertex: $(2, 5)$; point: $(4, 7)$

7. **Optimal Area** The perimeter of a rectangular archaeological dig site is 500 feet. Let x represent the width of the dig site. Write a quadratic function for the area of the rectangle in terms of its width. Of all possible dig sites with perimeters of 500 feet, what are the measurements of the one with the greatest area?

8. **Optimal Revenue** Find the number of units that produces a maximum revenue R (in dollars) for

$$R = 900x - 0.015x^2$$

where x is the number of units produced.

9. **Optimal Cost** A manufacturer of retinal imaging systems has daily production costs C (in dollars per unit) of

$$C = 25{,}000 - 50x + 0.065x^2$$

where x is the number of units produced.

(a) Use a graphing utility to graph the cost function.

(b) Graphically estimate the number of units that should be produced to yield a minimum cost per unit.

(c) Explain how to confirm the result of part (b) algebraically.

10. **Optimal Profit** The profit P (in dollars) for an electronics company is given by

$$P = -0.00015x^2 + 155x - 450{,}000$$

where x is the number of units produced.

(a) Use a graphing utility to graph the profit function.

(b) Graphically estimate the number of units that should be produced to yield a maximum profit.

(c) Explain how to confirm the result of part (b) algebraically.

11. **Batting Distance** In a study, baseballs were batted at angles of x degrees above the horizontal, each with an initial speed of 40 meters per second and a backspin of 30 revolutions per second. The distances $d(x)$ (in meters) traveled by the balls are shown in the table. *(Source: The Physics of Sports)*

x	10	15	30	36	42	43
$d(x)$	61.2	83.0	130.4	139.4	143.2	143.3

x	44	45	48	54	60
$d(x)$	142.8	142.7	140.7	132.8	119.7

(a) Use a graphing utility to create a scatter plot of the data.

(b) Use the *regression* feature of the graphing utility to find a quadratic model for the data.

(c) Use the graphing utility to graph the model from part (b) in the same viewing window as the scatter plot of the data.

(d) Find the vertex of the graph of the model from part (c). Interpret its meaning in the context of the problem.

12. **Dormitory Room Costs** The average costs C (in dollars) of a college dormitory room for the years 2003 through 2010 are shown in the table. *(Source: Digest of Education Statistics)*

Year	2003	2004	2005	2006
Cost, C	3179	3359	3569	3804

Year	2007	2008	2009	2010
Cost, C	4019	4214	4446	4657

(a) Use a graphing utility to create a scatter plot of the data. Let t represent the year, with $t = 3$ corresponding to 2003.

(b) Use the *regression* feature of the graphing utility to find a quadratic model for the data.

(c) Use the graphing utility to graph the model from part (b) in the same viewing window as the scatter plot of the data.

(d) Use the graph of the model from part (c) to predict the average cost of a dormitory room in 2011.

Sketching a Transformation of a Monomial Function In Exercises 13–16, use the graph of $y = x^5$ to sketch the graph of the function.

13. $f(x) = (x + 4)^5$ **14.** $f(x) = x^5 + 1$

15. $f(x) = 3 - \frac{1}{2}x^5$ **16.** $f(x) = 2(x + 3)^5$

Sketching a Transformation of a Monomial Function In Exercises 17–20, use the graph of $y = x^6$ to sketch the graph of the function.

17. $f(x) = x^6 - 2$ **18.** $f(x) = -\frac{1}{4}x^6$

19. $f(x) = -\frac{1}{2}x^6 - 5$ **20.** $f(x) = -(x + 7)^6 + 2$

Applying the Leading Coefficient Test In Exercises 21–24, describe the right-hand and left-hand behavior of the graph of the polynomial function.

21. $f(x) = \frac{1}{2}x^3 + 2x$ **22.** $f(x) = 5 + 4x^3 - x^5$

23. $f(x) = -x^6 + 3x^4 - x^2 + 6$

24. $f(x) = \frac{3}{4}(x^4 + 3x^2 + 2)$

Finding Zeros of a Polynomial Function In Exercises 25–28, find all real zeros of the function algebraically and use the results to sketch the graph *by hand*. Then use a graphing utility to confirm your sketch.

25. $f(x) = 16 - x^2$ **26.** $f(x) = x^4 - 6x^2 + 8$

27. $f(x) = x^3 - 7x^2 + 10x$

28. $f(x) = x^3 - 6x^2 - 3x + 18$

Long Division of Polynomials In Exercises 29 and 30, use long division to divide.

29. $(2x^3 - 5x^2 - x) \div (2x + 1)$

30. $(x^4 - 5x^3 + 10x^2 - 12) \div (x^2 - 2x + 4)$

Using Synthetic Division In Exercises 31 and 32, use synthetic division to divide.

31. $(x^3 - 6x + 9) \div (x + 3)$

32. $(x^5 - x^4 + x^3 - 13x^2 + x + 6) \div (x - 2)$

Using Synthetic Division In Exercises 33 and 34, use synthetic division to find each function value.

33. $f(x) = 6 + 2x^2 - 3x^3$ (a) $f(2)$ (b) $f(-1)$

34. $f(x) = 2x^4 + 3x^3 + 6$ (a) $f(\frac{1}{2})$ (b) $f(-1)$

Factoring a Polynomial In Exercises 35 and 36, (a) verify the given factors of $f(x)$, (b) find the remaining factors of $f(x)$, (c) use your results to write the complete factorization of $f(x)$, (d) list all real zeros of f, and (e) confirm your results by using a graphing utility to graph the function.

	Function	*Factors*
35.	$f(x) = x^3 - 4x^2 - 11x + 30$	$(x - 5), (x + 3)$
36.	$f(x) = 3x^3 + 23x^2 + 37x - 15$	$(3x - 1), (x + 5)$

Simplifying a Rational Expression In Exercises 37 and 38, simplify the rational expression.

37. $\dfrac{x^3 + 9x^2 + 2x - 48}{x - 2}$

38. $\dfrac{x^4 + 5x^3 - 20x - 16}{x^2 - 4}$

39. Profit The profit P (in dollars) from selling a motorcycle is given by

$$P = -42x^3 + 3000x^2 - 6000, \quad 0 \le x \le 65$$

where x is the advertising expense (in tens of thousands of dollars). For this motorcycle, the advertising expense was \$600,000 ($x = 60$) and the profit was \$1,722,000.

(a) Use a graphing utility to graph the function and use the result to find another advertising expense that would have produced the same profit.

(b) Use synthetic division to confirm the result of part (a) algebraically.

40. Profit The profit P (in dollars) from selling a novel is given by

$$P = -150x^3 + 7500x^2 - 450{,}000, \quad 0 \le x \le 45$$

where x is the advertising expense (in tens of thousands of dollars). For this novel, the advertising expense was \$400,000 ($x = 40$), and the profit was \$1,950,000.

(a) Use a graphing utility to graph the function and use the result to find another advertising expense that would have produced the same profit.

(b) Use synthetic division to confirm the result of part (a) algebraically.

Using the Rational Zero Test In Exercises 41 and 42, use the Rational Zero Test to list all possible rational zeros of f. Then use a graphing utility to graph the function. Use the graph to help determine which of the possible rational zeros are actual zeros of the function.

41. $f(x) = -4x^3 + 8x^2 - 3x + 15$

42. $f(x) = 3x^4 + 4x^3 - 5x^2 + 10x - 8$

Using the Rational Zero Test In Exercises 43–50, find all the real zeros of the function.

43. $f(x) = x^3 + 2x^2 - 5x - 6$

44. $g(x) = 2x^3 - 15x^2 + 24x + 16$

45. $h(x) = 3x^4 - 27x^2 + 60$

46. $f(x) = x^5 - 4x^3 + 3x$

47. $B(x) = 6x^3 - 19x^2 + 11x + 6$

48. $C(x) = 3x^4 + 3x^3 - 7x^2 - x + 2$

49. $p(x) = x^4 - x^3 - 2x - 4$

50. $q(x) = x^5 - 2x^4 + 2x^3 - 4x^2 - 3x + 6$

Using the Intermediate Value Theorem In Exercises 51 and 52, use the Intermediate Value Theorem to approximate the zero of f in the interval $[a, b]$. Give your approximation to the nearest tenth.

51. $f(x) = x^3 - 4x + 3, \quad [-3, -2]$

52. $f(x) = x^5 + 5x^2 + x - 1, \quad [0, 1]$

Approximating the Zeros of a Polynomial Function In Exercises 53 and 54, use the *zero*, *root*, or *zoom* and *trace* features of a graphing utility to approximate the real zeros of f. Give your approximations to the nearest thousandth.

53. $f(x) = 5x^3 - 11x - 3$

54. $f(x) = 2x^4 - 9x^3 - 5x^2 + 10x + 12$

55. Wholesale Revenue The revenues R (in millions of dollars) for Costco Wholesale for the years 2001 through 2010 are shown in the table. *(Source: Costco Wholesale)*

Year	Revenue, R
2001	34,797
2002	38,762
2003	42,546
2004	48,107
2005	52,935
2006	60,151
2007	64,909
2008	72,483
2009	71,449
2010	77,946

(a) Use a graphing utility to create a scatter plot of the data. Let t represent the year, with $t = 1$ corresponding to 2001.

(b) Use the *regression* feature of the graphing utility to find a linear model, a quadratic model, and a quartic model for the data.

(c) Use the graphing utility to graph each model with the data. How well does each model fit the data?

(d) Use each model to predict the year in which the revenue will be about \$87 billion. Explain any differences in the predictions.

56. Home Furnishing Sales The sales S (in millions of dollars) for Bed Bath & Beyond for the years 2000 through 2009 are shown in the table. *(Source: Bed Bath & Beyond)*

Year	Sales, S	Year	Sales, S
2000	2396.7	2005	5809.6
2001	2928.0	2006	6617.4
2002	3665.2	2007	7048.9
2003	4478.0	2008	7208.3
2004	5147.7	2009	7828.8

(a) Use a graphing utility to create a scatter plot of the data. Let t represent the year, with $t = 0$ corresponding to 2000.

(b) Use the *regression* feature of the graphing utility to find a linear model, a quadratic model, and a quartic model for the data.

(c) Use the graphing utility to graph each model with the data. How well does each model fit the data?

(d) Use each model to predict the year in which the sales will be about \$8.8 billion. Explain any differences in the predictions.

Writing Complex Numbers in Standard Form In Exercises 57–60, write the complex number in standard form and find its complex conjugate.

57. $\sqrt{-32}$ **58.** 12

59. $-3 + \sqrt{-16}$ **60.** $2 - \sqrt{-18}$

Adding, Subtracting, or Multiplying Complex Numbers In Exercises 61–70, perform the indicated operation and write the result in standard form.

61. $(7 - 3i) + (-2 + 5i)$ **62.** $(14 + 8i) - (-1 - 2i)$

63. $(5 + 8i)(5 - 8i)$ **64.** $\left(\frac{1}{2} + \frac{3}{4}i\right)\left(\frac{1}{2} - \frac{3}{4}i\right)$

65. $-2i(4 - 5i)$ **66.** $-3(-2 + 4i)$

67. $(3 + 4i)^2$ **68.** $(2 - 5i)^2$

69. $(3 + 2i)^2 + (3 - 2i)^2$ **70.** $(1 + i)^2 - (1 - i)^2$

Writing Quotients in Standard Form In Exercises 71–74, write the quotient in standard form.

71. $\dfrac{8 - i}{2 + i}$ **72.** $\dfrac{3 - 4i}{1 - 5i}$

73. $\dfrac{4 - 3i}{i}$ **74.** $\dfrac{2}{(1 + i)^2}$

Complex Solutions of a Quadratic Equation In Exercises 75–78, solve the quadratic equation.

75. $2x^2 - x + 3 = 0$ **76.** $3x^2 + 6x + 11 = 0$

77. $4x^2 + 11x + 3 = 0$

78. $9x^2 - 2x + 5 = 0$

Plotting Complex Numbers in the Complex Plane In Exercises 79 and 80, plot the complex number in the complex plane.

79. $-3 + 2i$

80. $-1 - 4i$

Finding the Zeros of a Polynomial Function In Exercises 81–86, find all the zeros of the function and write the polynomial as the product of linear factors.

81. $f(x) = x^4 - 81$

82. $h(x) = x^4 + 17x^2 + 16$

83. $f(t) = t^3 + 5t^2 + 3t + 15$

84. $h(x) = 2x^3 - 5x^2 + 4x - 10$

85. $g(x) = 4x^3 - 8x^2 + 9x - 18$

86. $f(x) = x^5 - 2x^4 + x^3 - x^2 + 2x - 1$

Finding a Polynomial Function with Given Zeros In Exercises 87 and 88, find a polynomial function with real coefficients that has the given zeros.

87. $1, 3i, -3i$

88. $1, -2, 1 - 3i, 1 + 3i$

Factoring a Polynomial In Exercises 89 and 90, write the polynomial (a) as the product of factors that are irreducible over the rationals, (b) as the product of linear and quadratic factors that are irreducible over the *reals*, and (c) in completely factored form.

89. $x^4 + 5x^2 - 24$

90. $x^4 - 2x^3 - 2x^2 - 14x - 63$
(*Hint:* One factor is $x^2 + 7$.)

Finding the Zeros of a Polynomial Function In Exercises 91–94, use the given zero of f to find all the zeros of f.

91. $f(x) = 4x^3 - x^2 + 64x - 16, \quad -4i$

92. $f(x) = 50 - 75x + 2x^2 - 3x^3, \quad 5i$

93. $f(x) = x^4 + 7x^3 + 24x^2 + 58x + 40, \quad -1 + 3i$

94. $f(x) = x^4 + 4x^3 + 8x^2 + 4x + 7, \quad -2 - \sqrt{3}i$

Finding the Domain and Asymptotes of a Rational Function In Exercises 95–98, find the domain of the function and identify any horizontal or vertical asymptotes.

95. $f(x) = \dfrac{-3}{x + 2}$

96. $f(x) = \dfrac{3x^2 + 7x - 5}{x^2 + 1}$

97. $f(x) = \dfrac{2x^2}{x^2 - 9}$

98. $f(x) = \dfrac{3x}{x^2 + x - 6}$

Sketching the Graph of a Rational Function In Exercises 99–102, sketch the graph of the rational function. To aid in sketching the graphs, check for intercepts, symmetry, vertical asymptotes, and horizontal asymptotes.

99. $P(x) = \dfrac{3 - x}{x + 2}$

100. $f(x) = \dfrac{4}{(x - 1)^2}$

101. $g(x) = \dfrac{1}{x^2 - 4} + 2$

102. $h(x) = \dfrac{-3x}{2x^2 + 3x - 5}$

Sketching the Graph of a Rational Function In Exercises 103 and 104, find all asymptotes (vertical, horizontal, and/or slant) of the given function. Sketch the graph of f.

103. $f(x) = \dfrac{x^2 - 16}{x - 4}$

104. $f(x) = \dfrac{x^3}{x^2 - 5}$

105. Population of Fish The Wildlife Commission introduces 60,000 game fish into a large lake. The population N (in thousands) of the fish is

$$N = \frac{20(3 + 5t)}{1 + 0.06t}, \quad t \geq 0$$

where t is time (in years).

(a) Find the populations when $t = 5$, 10, and 25.

(b) What is the limiting number of fish in the lake as time progresses?

106. Average Recycling Cost The cost C (in dollars) of recycling a waste product is $C = 325{,}000 + 8.5x$, for $x > 0$, where x is the number of pounds of waste. The average recycling cost $\overline{C}$ per pound is

$$\overline{C} = \frac{C}{x} = \frac{325{,}000 + 8.5x}{x}, \quad x > 0.$$

(a) Sketch the graph of $\overline{C}$.

(b) Find the average costs of recycling 1000, 10,000, 100,000, and 1,000,000 pounds of waste. What can you conclude?

107. Smokestack Emissions The cost C (in dollars) of removing p percent of the air pollutants in the stack emissions of a utility company that burns coal to generate electricity is

$$C = \frac{105{,}000p}{100 - p}, \quad 0 \leq p < 100.$$

(a) Find the cost of removing 25% of the pollutants.

(b) Find the cost of removing 60% of the pollutants.

108. Human Memory Model Consider the learning curve modeled by

$$P = \frac{0.7 + 0.65(n - 1)}{1 + 0.65(n - 1)}, \quad n \geq 0$$

where P is the percent of correct responses (in decimal form) after n trials.

(a) Complete the table.

n	1	2	3	4	5	6	7	8	9	10
P										

(b) According to this model, what is the limiting percent of correct responses as n increases?

TEST YOURSELF

See www.CalcChat.com for worked-out solutions to odd-numbered exercises.

Take this test as you would take a test in class. When you are done, check your work against the answers given in the back of the book.

1. Sketch the graph of the quadratic function given by

$$f(x) = -\tfrac{1}{2}(x - 1)^2 - 5.$$

Identify the vertex and intercept(s).

2. Describe the right-hand and left-hand behavior of the graph of f.

(a) $f(x) = 12x^3 - 5x^2 - 49x + 15$

(b) $f(x) = 5x^4 - 3x^3 + 2x^2 + 11x + 12$

3. Simplify $\dfrac{x^4 + 4x^3 - 19x^2 - 106x - 120}{x^2 - 3x - 10}$.

4. List all possible rational zeros of

$$f(x) = 4x^4 - 16x^3 + 3x^2 + 36x - 27.$$

Use synthetic division to show that $x = -\frac{3}{2}$ and $x = \frac{3}{2}$ are zeros of f. Using these results, completely factor the polynomial.

5. The sales per share S (in dollars) for Hormel Foods for the years 2001 through 2010 are shown in the table at the left. *(Source: Hormel Foods Corporation)*

(a) Use a graphing utility to create a scatter plot of the data. Let t represent the year, with $t = 1$ corresponding to 2001.

(b) Use the *regression* feature of the graphing utility to find a linear model, a quadratic model, and a quartic model for the data.

(c) Use the graphing utility to graph each model separately with the data in the same viewing window. How well does each model fit the data?

(d) Use each model to predict the year in which the sales per share will be about \$35. Then discuss the appropriateness of each model for predicting future values.

Year	Sales per share, S
2001	14.87
2002	14.13
2003	15.15
2004	17.33
2005	19.64
2006	20.92
2007	22.82
2008	25.11
2009	24.45
2010	27.15

Table for 5

In Exercises 6–9, perform the indicated operation and write the result in standard form.

6. $(12 + 3i) + (4 - 5i)$

7. $(10 - 2i) - (3 + 8i)$

8. $\left(5 + \sqrt{-12}\right)\left(3 - \sqrt{-12}\right)$

9. $(4 + 3i)(2 - 5i)$

10. Write the quotient in standard form: $\dfrac{1 + i}{1 - i}$.

In Exercises 11 and 12, solve the quadratic equation.

11. $x^2 + 5x + 7 = 0$

12. $2x^2 - 5x + 11 = 0$

13. Find a polynomial function with real coefficients that has 2, 5, $3i$, and $-3i$ as zeros.

14. Find all the zeros of $f(x) = x^3 + 2x^2 + 5x + 10$, given that $\sqrt{5}i$ is a zero.

15. Sketch the graph of

$$f(x) = \frac{3x}{x - 2}.$$

Label any intercepts and asymptotes. What is the domain of f?

4 Exponential and Logarithmic Functions

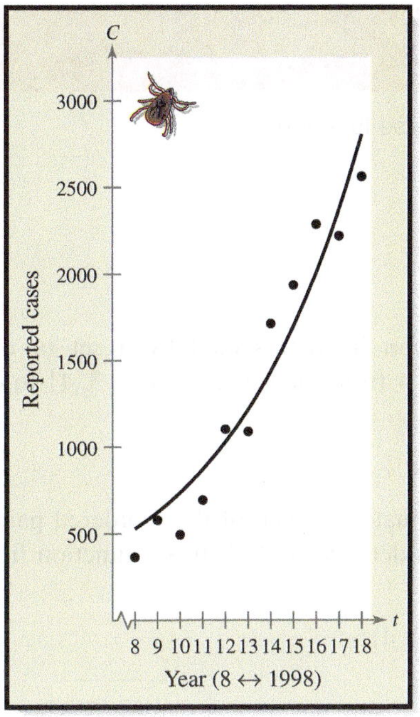

Example 1 on page 388 shows how an exponential growth function can be used to model the number of reported cases of Rocky Mountain spotted fever.

Kurhan/Shutterstock.com
wavebreakmedia ltd/www.shutterstock.com

4.1 Inverse Functions

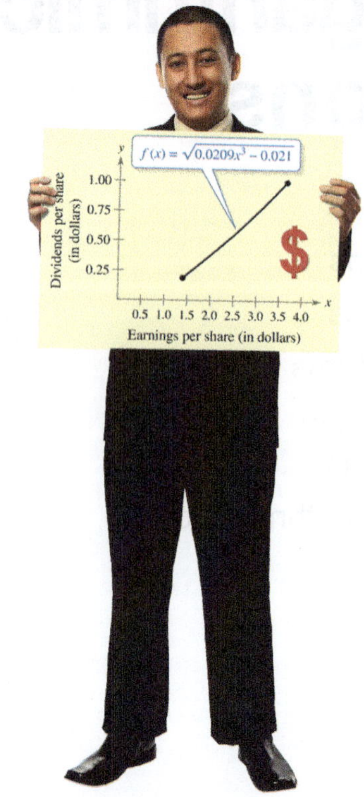

In Exercise 75 on page 345, you will use an inverse function to approximate the earnings per share in 2008 for Wal-Mart Stores.

- Determine whether a function has an inverse function.
- Find the inverse function of a function.
- Graph a function and its inverse function.

Inverse Functions

Recall from Section 2.4 that a function can be represented by a set of ordered pairs. For instance, the function $f(x) = x + 4$ from the set $A = \{1, 2, 3, 4\}$ to the set $B = \{5, 6, 7, 8\}$ can be written as

$$f(x) = x + 4:\ \{(1, 5), (2, 6), (3, 7), (4, 8)\}.$$

By interchanging the first and second coordinates of each of these ordered pairs, you can form the **inverse function** of f, which is denoted by f^{-1}. It is a function from the set B to the set A and can be written as

$$f^{-1}(x) = x - 4:\ \{(5, 1), (6, 2), (7, 3), (8, 4)\}.$$

Note that the domain of f is equal to the range of f^{-1} and vice versa, as shown in Figure 4.1. Also note that the functions f and f^{-1} have the effect of "undoing" each other. In other words, when you form the composition of f with f^{-1} or the composition of f^{-1} with f, you obtain the identity function, as shown.

$$f(f^{-1}(x)) = f(x - 4) = (x - 4) + 4 = x$$

$$f^{-1}(f(x)) = f^{-1}(x + 4) = (x + 4) - 4 = x$$

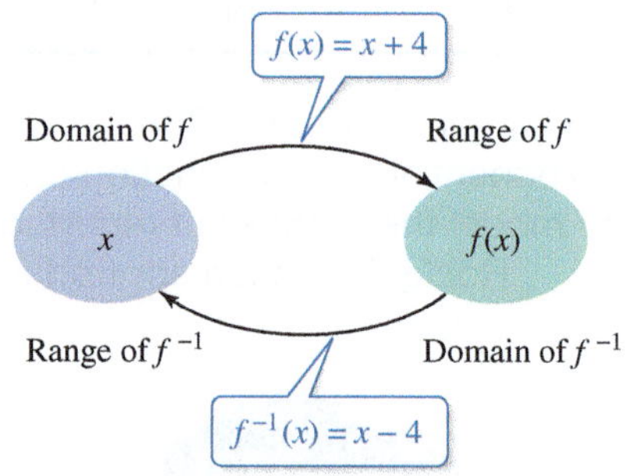

FIGURE 4.1

Example 1 Finding Inverse Functions Informally

Find the inverse function of $f(x) = 4x$. Then verify that both $f(f^{-1}(x))$ and $f^{-1}(f(x))$ are equal to the identity function.

SOLUTION The function $f(x) = 4x$ *multiplies* each input by 4. To "undo" this function, you need to *divide* each input by 4. So, the inverse function of $f(x) = 4x$ is

$$f^{-1}(x) = \frac{x}{4}.$$

You can verify that both $f(f^{-1}(x))$ and $f^{-1}(f(x))$ are equal to the identity function as shown.

$$f(f^{-1}(x)) = f\left(\frac{x}{4}\right) = 4\left(\frac{x}{4}\right) = x \qquad f^{-1}(f(x)) = f^{-1}(4x) = \frac{4x}{4} = x$$

✓ Checkpoint 1

Find the inverse function of $f(x) = \frac{x}{6}$. Then verify that both $f(f^{-1}(x))$ and $f^{-1}(f(x))$ are equal to the identity function.

leolintang/Shutterstock.com

Example 2 Finding Inverse Functions Informally

Find the inverse function of

$$f(x) = x - 6.$$

Then verify that both $f(f^{-1}(x))$ and $f^{-1}(f(x))$ are equal to the identity function.

SOLUTION The function $f(x) = x - 6$ *subtracts* 6 from each input. To "undo" this function, you need to *add* 6 to each input. So, the inverse function of $f(x) = x - 6$ is

$$f^{-1}(x) = x + 6.$$

You can verify that both $f(f^{-1}(x))$ and $f^{-1}(f(x))$ are equal to the identity function as shown.

$$f(f^{-1}(x)) = f(x + 6) \quad \text{Substitute } x + 6 \text{ for } f^{-1}(x).$$
$$= (x + 6) - 6 \quad \text{Substitute } x + 6 \text{ into } f(x).$$
$$= x \quad \text{Identity function}$$
$$f^{-1}(f(x)) = f^{-1}(x - 6) \quad \text{Substitute } x - 6 \text{ for } f(x).$$
$$= (x - 6) + 6 \quad \text{Substitute } x - 6 \text{ into } f^{-1}(x).$$
$$= x \quad \text{Identity function}$$

Checkpoint 2

Find the inverse function of $f(x) = x + 10$. Then verify that both $f(f^{-1}(x))$ and $f^{-1}(f(x))$ are equal to the identity function.

The formal definition of inverse function is shown below.

Definition of Inverse Function

Let f and g be two functions such that

$$f(g(x)) = x \quad \text{for every } x \text{ in the domain of } g$$

and

$$g(f(x)) = x \quad \text{for every } x \text{ in the domain of } f.$$

Under these conditions, the function g is the **inverse function** of the function f. The function g is denoted by f^{-1} (read "f-inverse"). So,

$$f(f^{-1}(x)) = x \quad \text{and} \quad f^{-1}(f(x)) = x.$$

The domain of f must be equal to the range of f^{-1}, and the range of f must be equal to the domain of f^{-1}.

Do not be confused by the use of -1 to denote the inverse function f^{-1}. In this text, f^{-1} *always* refers to the inverse function of the function f and *not* to the reciprocal of $f(x)$. That is,

$$f^{-1}(x) \neq \frac{1}{f(x)}.$$

If the function g is the inverse function of the function f, then it must also be true that the function f is the inverse function of the function g. For this reason, you can say that the functions f and g are *inverse functions of each other.*

Example 3 **Verifying Inverse Functions**

Show that the following functions are inverse functions.

$$f(x) = 2x^3 - 1 \quad \text{and} \quad g(x) = \sqrt[3]{\frac{x+1}{2}}$$

SOLUTION

$$f(g(x)) = f\left(\sqrt[3]{\frac{x+1}{2}}\right) = 2\left(\sqrt[3]{\frac{x+1}{2}}\right)^3 - 1$$

$$= 2\left(\frac{x+1}{2}\right) - 1$$

$$= x + 1 - 1$$

$$= x$$

$$g(f(x)) = g(2x^3 - 1) = \sqrt[3]{\frac{(2x^3 - 1) + 1}{2}}$$

$$= \sqrt[3]{\frac{2x^3}{2}}$$

$$= \sqrt[3]{x^3}$$

$$= x$$

Checkpoint 3

Show that the following functions are inverse functions.

$$f(x) = x^3 + 6 \quad \text{and} \quad g(x) = \sqrt[3]{x - 6}$$

Example 4 **Verifying Inverse Functions**

Which of the functions given by

$$g(x) = \frac{x-2}{5} \quad \text{and} \quad h(x) = \frac{5}{x} + 2$$

is the inverse function of $f(x) = \dfrac{5}{x-2}$?

SOLUTION By forming the composition of f with g, you can see that

$$f(g(x)) = f\left(\frac{x-2}{5}\right) = \frac{5}{[(x-2)/5] - 2} = \frac{25}{x - 12} \neq x.$$

Because this composition is not equal to the identity function x, it follows that g is *not* the inverse function of f. By forming the composition of f with h, you have

$$f(h(x)) = f\left(\frac{5}{x} + 2\right) = \frac{5}{[(5/x) + 2] - 2} = \frac{5}{5/x} = x.$$

So, it appears that h is the inverse function of f. You can confirm this result by showing that the composition of h with f is also equal to the identity function.

Checkpoint 4

Which of the functions given by $g(x) = \dfrac{x+4}{3}$ and $h(x) = \dfrac{x}{3} + 4$ is the inverse function of $f(x) = 3x - 4$?

Finding Inverse Functions

For simple functions (such as the ones in Examples 1 and 2), you can find inverse functions by inspection. For more complicated functions, it is best to use the following guidelines. The key step in these guidelines is switching the roles of x and y. This step corresponds to the fact that inverse functions have ordered pairs with the coordinates reversed.

STUDY TIP

Note in Step 3 of the guidelines for finding inverse functions that it is possible for a function to have no inverse function. For instance, the function given by $f(x) = x^2$ has no inverse function.

Guidelines for Finding Inverse Functions

1. In the equation for $f(x)$, replace $f(x)$ with y.
2. Interchange the roles of x and y.
3. Solve the new equation for y. If the new equation does not represent y as a function of x, then the function f does not have an inverse function. If the new equation does represent y as a function of x, then continue to Step 4.
4. Replace y with $f^{-1}(x)$ in the new equation.
5. Verify that f and f^{-1} are inverse functions of each other by showing that the domain of f is equal to the range of f^{-1}, the range of f is equal to the domain of f^{-1}, and

$$f(f^{-1}(x)) = x = f^{-1}(f(x)).$$

Example 5 **Finding Inverse Functions**

Find the inverse function of

$$f(x) = \frac{5 - 3x}{2}.$$

SOLUTION

$$f(x) = \frac{5 - 3x}{2}$$ Write original function.

$$y = \frac{5 - 3x}{2}$$ Replace $f(x)$ with y.

$$x = \frac{5 - 3y}{2}$$ Interchange x and y.

$$2x = 5 - 3y$$ Multiply each side by 2.

$$3y = 5 - 2x$$ Isolate the y-term.

$$y = \frac{5 - 2x}{3}$$ Solve for y.

$$f^{-1}(x) = \frac{5 - 2x}{3}$$ Replace y with $f^{-1}(x)$.

Note that both f and f^{-1} have domains and ranges that consist of the entire set of real numbers. Check that

$$f(f^{-1}(x)) = x \quad \text{and} \quad f^{-1}(f(x)) = x.$$

✓Checkpoint 5

Find the inverse function of $f(x) = 4x + 5$.

The Graph of an Inverse Function

The graphs of a function f and its inverse function f^{-1} are related to each other in the following way. If the point (a, b) lies on the graph of f, then the point (b, a) must lie on the graph of f^{-1}, and vice versa. This means that the graph of f^{-1} is a *reflection* of the graph of f in the line $y = x$, as shown in Figure 4.2.

TECH TUTOR

A graphing utility can help you check that the graphs of f and f^{-1} are reflections of each other in the line $y = x$. To do this, graph $y = f(x)$, $y = f^{-1}(x)$, and $y = x$ in the same viewing window, using a square setting.

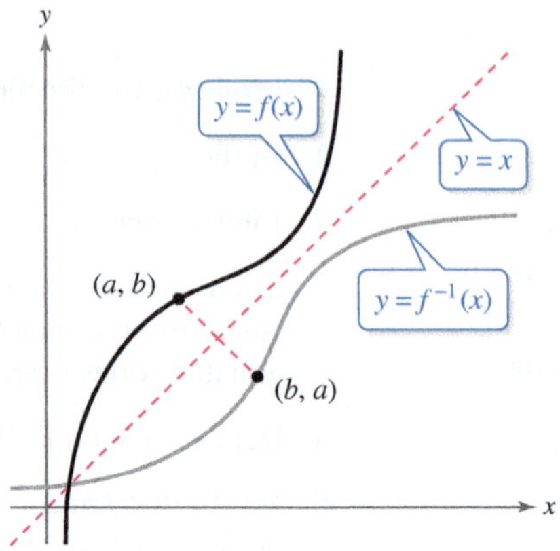

The graph of f^{-1} is a reflection of the graph of f in the line $y = x$.

FIGURE 4.2

Example 6 The Graphs of f and f^{-1}

Sketch the graphs of the inverse functions given by

$$f(x) = 2x - 3$$

and

$$f^{-1}(x) = \frac{1}{2}(x + 3)$$

in the same coordinate plane, and show that the graphs are reflections of each other in the line $y = x$.

SOLUTION The graphs of f and f^{-1} are shown in Figure 4.3. Visually, it appears that the graphs are reflections of each other in the line $y = x$. You can further verify this reflective property by testing a few points on each graph. Note in the following list that if the point (a, b) is on the graph of f, then the point (b, a) is on the graph of f^{-1}.

Graph of $f(x) = 2x - 3$	*Graph of* $f^{-1}(x) = \frac{1}{2}(x + 3)$
$(0, -3)$	$(-3, 0)$
$(1, -1)$	$(-1, 1)$
$(2, 1)$	$(1, 2)$
$(3, 3)$	$(3, 3)$

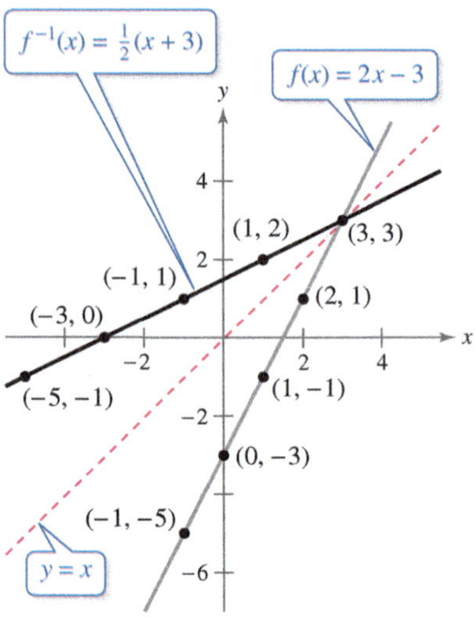

FIGURE 4.3

✓ Checkpoint 6

Sketch the graphs of the inverse functions given by

$$f(x) = \frac{2}{5}x + 2 \quad \text{and} \quad f^{-1}(x) = \frac{5}{2}x - 5$$

in the same coordinate plane, and show that the graphs are reflections of each other in the line $y = x$.

The Study Tip on page 339 mentioned that the function given by $f(x) = x^2$ has no inverse function. This means, *assuming the domain of f is the entire real line*, that the function given by $f(x) = x^2$ has no inverse function. If the domain of f is restricted to the nonnegative real numbers, however, then f does have an inverse function, as demonstrated in Example 7.

Example 7 The Graphs of f and f^{-1}

Sketch the graphs of the inverse functions given by

$$f(x) = x^2, \quad x \geq 0, \quad \text{and} \quad f^{-1}(x) = \sqrt{x}$$

in the same coordinate plane, and show that the graphs are reflections of each other in the line $y = x$.

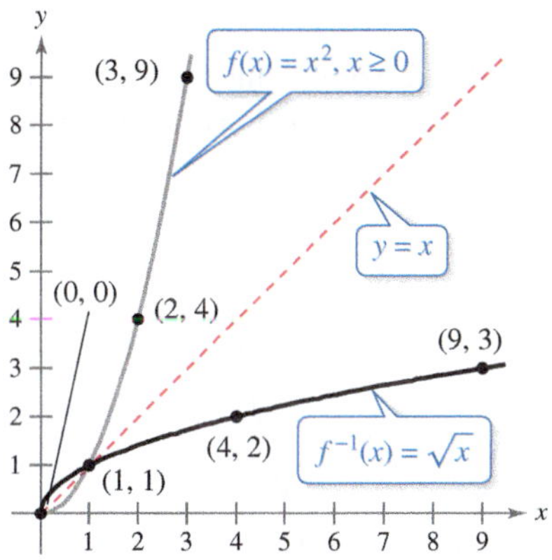

FIGURE 4.4

SOLUTION The graphs of f and f^{-1} are shown in Figure 4.4. Visually, it appears that the graphs are reflections of each other in the line $y = x$. You can further verify this reflective property by testing a few points on each graph. Note in the following list that if the point (a, b) is on the graph of f, then the point (b, a) is on the graph of f^{-1}.

Graph of $f(x) = x^2, \quad x \geq 0$	*Graph of* $f^{-1}(x) = \sqrt{x}$
(0, 0)	(0, 0)
(1, 1)	(1, 1)
(2, 4)	(4, 2)
(3, 9)	(9, 3)

You can verify algebraically that the functions are inverse functions of each other by showing that $f(f^{-1}(x)) = x$ and $f^{-1}(f(x)) = x$ as shown.

$$\begin{aligned} f(f^{-1}(x)) &= f(\sqrt{x}) \\ &= (\sqrt{x})^2 \\ &= x, \quad \text{if } x \geq 0 \end{aligned}$$

$$\begin{aligned} f^{-1}(f(x)) &= f^{-1}(x^2) \\ &= \sqrt{x^2} \\ &= x, \quad \text{if } x \geq 0 \end{aligned}$$

Checkpoint 7

Sketch the graphs of the inverse functions given by

$$f(x) = x^2 + 3, \quad x \geq 0, \quad \text{and} \quad f^{-1}(x) = \sqrt{x - 3}$$

in the same coordinate plane, and show that the graphs are reflections of each other in the line $y = x$.

The guidelines for finding the inverse function of a function include an *algebraic* test for determining whether a function has an inverse function. The reflective property of the graphs of inverse functions gives you a *geometric* test for determining whether a function has an inverse function. This test is called the **Horizontal Line Test** for inverse functions.

Horizontal Line Test for Inverse Functions

A function f has an inverse function if and only if no *horizontal* line intersects the graph of f at more than one point.

Example 8 Applying the Horizontal Line Test

Use the graph of f to determine whether the function has an inverse function.

a. $f(x) = x^3 - 1$ **b.** $f(x) = x^2 - 1$

SOLUTION

a. The graph of the function given by

$$f(x) = x^3 - 1$$

is shown in Figure 4.5(a). Because no horizontal line intersects the graph of f at more than one point, you can conclude that f does have an inverse function.

b. The graph of the function given by

$$f(x) = x^2 - 1$$

is shown in Figure 4.5(b). Because it is possible to find a horizontal line that intersects the graph of f at more than one point, you can conclude that f does not have an inverse function.

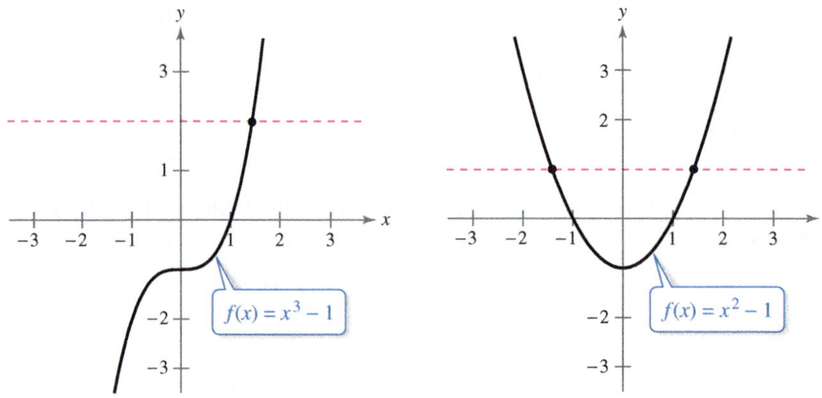

(a) (b)

FIGURE 4.5

Checkpoint 8

Use the graph of f to determine whether the function has an inverse function.

a. $f(x) = |x|$ **b.** $f(x) = \sqrt{x}$

SUMMARIZE (Section 4.1)

1. State the definition of an inverse function *(page 337)*. For examples of finding inverse functions informally and verifying inverse functions, see Examples 1, 2, 3, and 4.
2. Describe how to find an inverse function *(page 339)*. For an example of finding an inverse function, see Example 5.
3. Describe the relationship between the graph of a function and the graph of its inverse function *(page 340)*. For examples of sketching the graphs of functions and their inverse functions, see Examples 6 and 7.
4. Describe the Horizontal Line Test for inverse functions *(page 341)*. For an example of applying the Horizontal Line Test, see Example 8.

wavebreakmedia ltd/Shutterstock.com

SKILLS WARM UP 4.1 The following warm-up exercises involve skills that were covered in earlier sections. You will use these skills in the exercise set for this section. For additional help, review Sections 0.2, 0.4, 1.2, and 2.4.

In Exercises 1–4, find the domain of the function.

1. $f(x) = \sqrt[3]{x+1}$

2. $f(x) = \sqrt{x+1}$

3. $g(x) = \dfrac{2}{x^2 - 2x}$

4. $h(x) = \dfrac{x}{3x+5}$

In Exercises 5–8, simplify the expression.

5. $2\left(\dfrac{x+5}{2}\right) - 5$

6. $7 - 10\left(\dfrac{7-x}{10}\right)$

7. $\sqrt[3]{2\left(\dfrac{x^3}{2} - 2\right) + 4}$

8. $\sqrt[5]{(x+2)^5} - 2$

In Exercises 9 and 10, solve for x in terms of y.

9. $y = \dfrac{2x-6}{3}$

10. $y = \sqrt[3]{2x-4}$

Exercises 4.1

See www.CalcChat.com for worked-out solutions to odd-numbered exercises.

Finding Inverse Functions In Exercises 1–4, find the inverse function of the function f given by the set of ordered pairs.

1. $\{(1, 4), (2, 5), (3, 6), (4, 7)\}$

2. $\{(6, 2), (5, 3), (4, 4), (3, 5)\}$

3. $\{(-1, 1), (-2, 2), (-3, 3), (-4, 4)\}$

4. $\{(6, -2), (5, -3), (4, -4), (3, -5)\}$

Finding Inverse Functions Informally In Exercises 5–8, find the inverse function informally. Verify that $f(f^{-1}(x)) = x$ and $f^{-1}(f(x)) = x$. *See Examples 1 and 2.*

5. $f(x) = 2x$

6. $f(x) = -\dfrac{x}{4}$

7. $f(x) = x - 5$

8. $f(x) = x + 7$

Verifying Inverse Functions In Exercises 9–16, show that f and g are inverse functions by using the definition of inverse functions. *See Examples 3 and 4.*

9. $f(x) = 5x + 1, \quad g(x) = \dfrac{x-1}{5}$

10. $f(x) = 3 - 4x, \quad g(x) = \dfrac{3-x}{4}$

11. $f(x) = x^3, \quad g(x) = \sqrt[3]{x}$

12. $f(x) = \dfrac{1}{x}, \quad g(x) = \dfrac{1}{x}$

13. $f(x) = \sqrt{x-4}, \quad g(x) = x^2 + 4, \quad x \ge 0$

14. $f(x) = 9 - x^2, \quad x \ge 0, \quad g(x) = \sqrt{9-x}, \quad x \le 9$

15. $f(x) = 1 - x^3, \quad g(x) = \sqrt[3]{1-x}$

16. $f(x) = \dfrac{1}{1+x}, \quad x \ge 0, \quad g(x) = \dfrac{1-x}{x}, \quad 0 < x \le 1$

Sketching Inverse Functions In Exercises 17 and 18, use the graph of f to complete the table and to sketch the graph of f^{-1}.

17.

x	0	1	2	3	4
$f^{-1}(x)$					

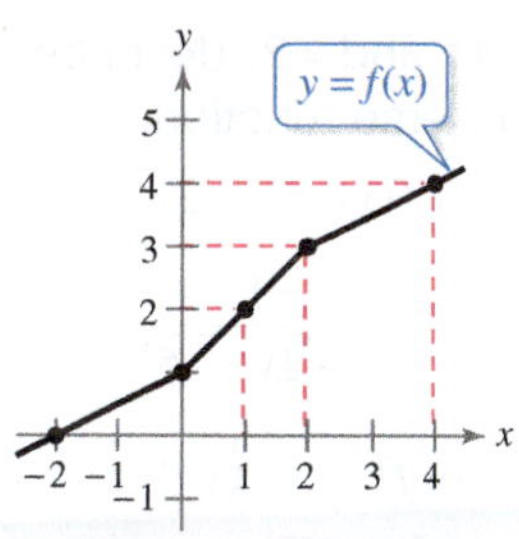

18.

x	0	2	4	6
$f^{-1}(x)$				

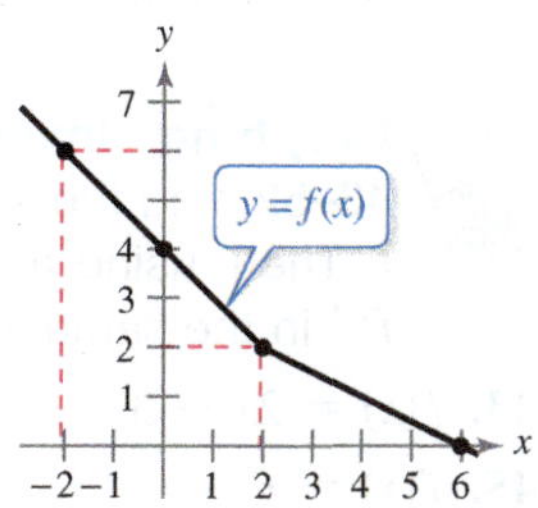

Finding Inverse Functions In Exercises 19–34, determine whether the function has an inverse function. If it does, find its inverse function. *See Example 5.*

19. $f(x) = x^4$ **20.** $f(x) = \dfrac{1}{x^2}$

21. $g(x) = \dfrac{x}{8}$ **22.** $f(x) = 3x + 5$

23. $p(x) = -4$ **24.** $f(x) = \dfrac{3x + 4}{5}$

25. $f(x) = (x + 3)^2,\ x \ge -3$

26. $q(x) = (x - 5)^2$

27. $h(x) = \dfrac{1}{x}$ **28.** $f(x) = |x - 2|,\ x \le 2$

29. $f(x) = \sqrt{2x + 3}$ **30.** $f(x) = \sqrt{x - 2}$

31. $g(x) = x^2 - x^4$ **32.** $f(x) = \dfrac{x^2}{x^2 + 1}$

33. $f(x) = 25 - x^2,\ \ x \le 0$

34. $f(x) = 36 + x^2,\ \ x \le 0$

The Graphs of f and f^{-1} In Exercises 35–40, sketch the graphs of the inverse functions in the same coordinate plane and show that the graphs are reflections of each other in the line $y = x$. *See Examples 6 and 7.*

35. $f(x) = \dfrac{x}{3},\quad f^{-1}(x) = 3x$

36. $f(x) = x + 8,\quad f^{-1}(x) = x - 8$

37. $f(x) = 7x + 1,\quad f^{-1}(x) = \dfrac{x - 1}{7}$

38. $f(x) = \dfrac{4}{x},\quad f^{-1}(x) = \dfrac{4}{x}$

39. $f(x) = x^5 - 2,\quad f^{-1}(x) = \sqrt[5]{x + 2}$

40. $f(x) = \sqrt{4 - x^2},\ \ 0 \le x \le 2,$
$f^{-1}(x) = \sqrt{4 - x^2},\ \ 0 \le x \le 2$

Error Analysis In Exercises 41 and 42, describe and correct the error in finding the inverse function.

41. Find the inverse function f^{-1} of $f(x) = \sqrt{2x - 5}$.

~~$f(x) = \sqrt{2x - 5}$, so $f^{-1}(x) = \dfrac{1}{\sqrt{2x - 5}}$.~~

42. Find the inverse function f^{-1} of $f(x) = \frac{3}{5}x + \frac{1}{3}$.

~~$f(x) = \frac{3}{5}x + \frac{1}{3}$, so $f^{-1}(x) = \frac{5}{3}x - 3$.~~

Graphing Inverse Functions In Exercises 43–52, find the inverse function f^{-1} of the function f. Then, using a graphing utility, graph both f and f^{-1} in the same viewing window.

43. $f(x) = 2x - 3$ **44.** $f(x) = 5x + 2$

45. $f(x) = x^5$ **46.** $f(x) = x^3 + 1$

47. $f(x) = \sqrt{x}$

48. $f(x) = x^2,\ x \ge 0$

49. $f(x) = \sqrt{16 - x^2},\quad 0 \le x \le 4$

50. $f(x) = \dfrac{3}{x + 1}$

51. $f(x) = \sqrt[3]{x + 2}$

52. $f(x) = x^{3/5} - 2$

Applying the Horizontal Line Test In Exercises 53–56, use the graph of f to determine whether the function has an inverse function. *See Example 8.*

53.

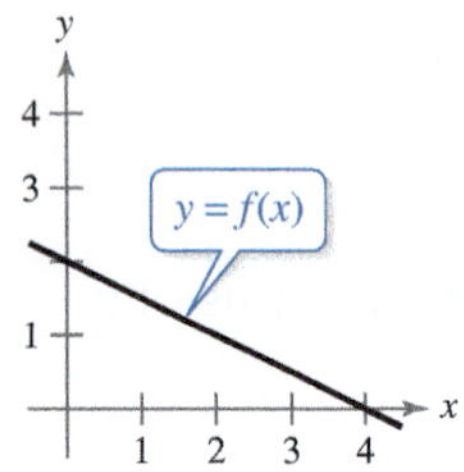

54.

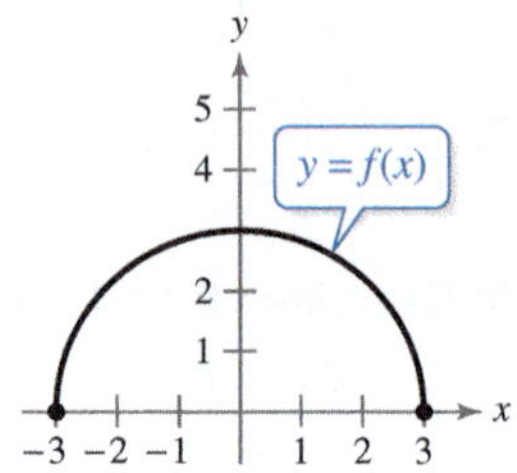

55.

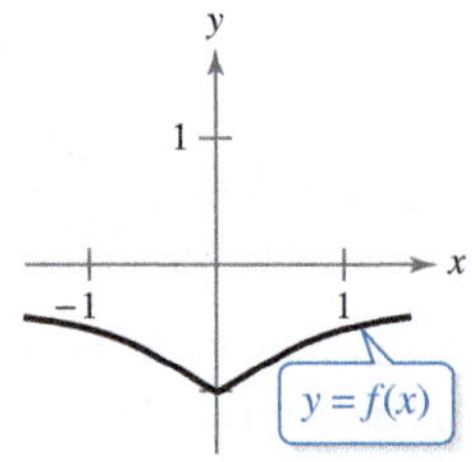

56.

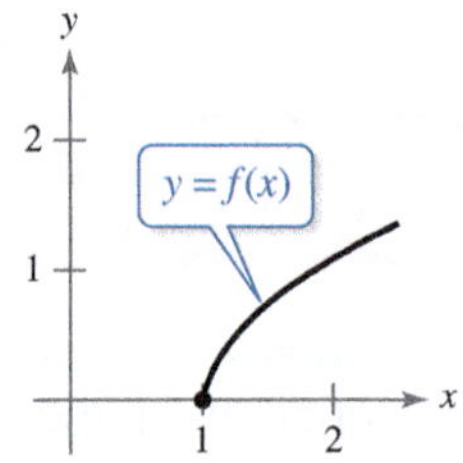

Applying the Horizontal Test In Exercises 57–62, graph the function and use the Horizontal Line Test to determine whether the function has an inverse function.

57. $g(x) = \dfrac{5 - 2x}{3}$ **58.** $f(x) = 10$

59. $h(x) = |x - 5|$ **60.** $g(x) = (x - 3)^2$

61. $f(x) = -\sqrt{9 - x^2}$ **62.** $f(x) = (x - 1)^3$

Finding the Composition of Functions In Exercises 63–66, use the functions given by

$f(x) = x + 4$ and $g(x) = 2x - 6$

to find the composition of functions.

63. $g^{-1} \circ f^{-1}$ **64.** $f^{-1} \circ g^{-1}$

65. $(f \circ g)^{-1}$ **66.** $(g \circ f)^{-1}$

Evaluating Functions In Exercises 67–70, use the functions given by

$f(x) = \dfrac{1}{8}x - 3$ and $g(x) = x^2$

to find the value.

67. $(f^{-1} \circ g^{-1})(1)$ **68.** $(g^{-1} \circ f^{-1})(-3)$

69. $(f^{-1} \circ f^{-1})(6)$ **70.** $(g^{-1} \circ g^{-1})(-4)$

71. Profit A company's profit P for producing x units is given by $P(x) = 47x - 5736$. Find the inverse function $P^{-1}(x)$ and explain what it represents. Describe the domains of $P(x)$ and $P^{-1}(x)$.

72. HOW DO YOU SEE IT? The cost C for a business to make personalized T-shirts is given by

$$C(x) = 7.50x + 1500$$

where x represents the number of T-shirts.

(a) The graphs of $C(x)$ and $C^{-1}(x)$ are shown below. Match each function with its graph.

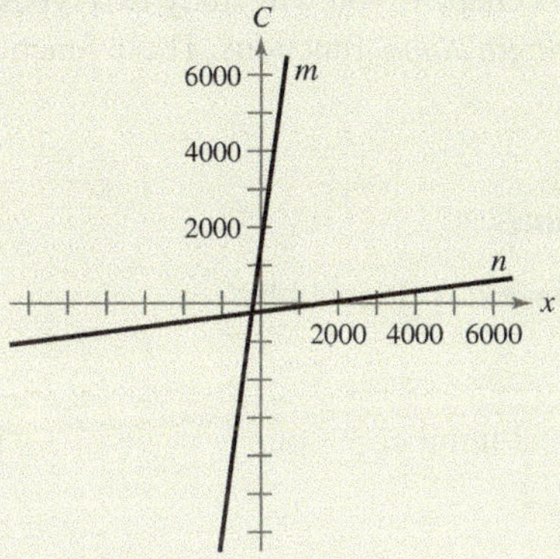

(b) Explain what $C(x)$ and $C^{-1}(x)$ represent in the context of the problem.

73. Movie Theaters The average prices of admission y (in dollars) to movie theaters for the years 2003 through 2010 are shown in the table. *(Source: National Association of Theater Owners)*

Year	2003	2004	2005	2006
Admission price, y	6.03	6.21	6.41	6.55

Year	2007	2008	2009	2010
Admission price, y	6.88	7.18	7.50	7.89

(a) Use a graphing utility to create a scatter plot of the data. Let t represent the year, with $t = 3$ corresponding to 2003.

(b) Use the *regression* feature of the graphing utility to find a linear model for the data.

(c) Algebraically find the inverse function of the model in part (b). Explain what this inverse function represents in a real-life context.

(d) Use the inverse function you found in part (c) to predict the year in which the average admission price to a movie theater will reach \$9.00.

74. Lead Exposure A project is conducted to study the amount of lead accumulated in the bones of humans. The concentration L (in micrograms per gram of bone mineral) of lead found in the tibia of a man is measured every five years. The results are shown in the table.

Age	15	20	25	30	35	40
Lead, L	3.2	5.4	9.2	12.2	13.8	16.0

(a) Use a graphing utility to create a scatter plot of the data. Let x represent the age (in years) of the man.

(b) Use the *regression* feature of the graphing utility to find a linear model for the data.

(c) Algebraically find the inverse function of the model in part (b). Explain what this inverse function represents in a real-life context.

(d) Use the inverse function you found in part (c) to estimate the age of the man when the concentration of lead in his tibia reaches 25 micrograms per gram of bone mineral.

75. Earnings-Dividend Ratio From 2000 through 2009, the earnings per share for Wal-Mart Stores were approximately related to the dividends per share by the function

$$f(x) = \sqrt{0.0209x^3 - 0.021}, \quad 1.40 \le x \le 3.66$$

where f represents the dividends per share (in dollars) and x represents the earnings per share (in dollars). (See figure.)

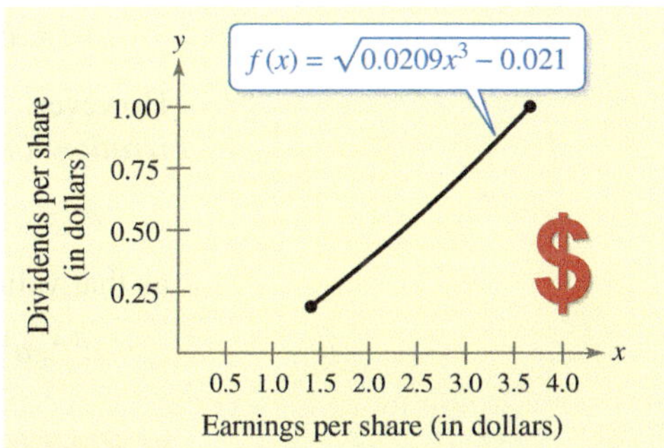

In 2008, Wal-Mart paid dividends of \$0.93 per share. Find the inverse function of f and use the inverse function to approximate the earnings per share in 2008. *(Source: Wal-Mart Stores, Inc.)*

76. Diesel Mechanics The function given by

$$y = 0.03x^2 + 245.5, \quad 0 < x < 100$$

approximates the exhaust temperature y for a diesel engine in degrees Fahrenheit, where x is the percent load for the diesel engine. Solve the equation for x in terms of y and use the result to find the percent load for a diesel engine when the exhaust temperature is 410°F.

4.2 Exponential Functions

- Evaluate an exponential expression.
- Sketch the graph of an exponential function.
- Evaluate and sketch the graph of the natural exponential function.
- Use the compound interest formulas.
- Use an exponential model to solve an application problem.

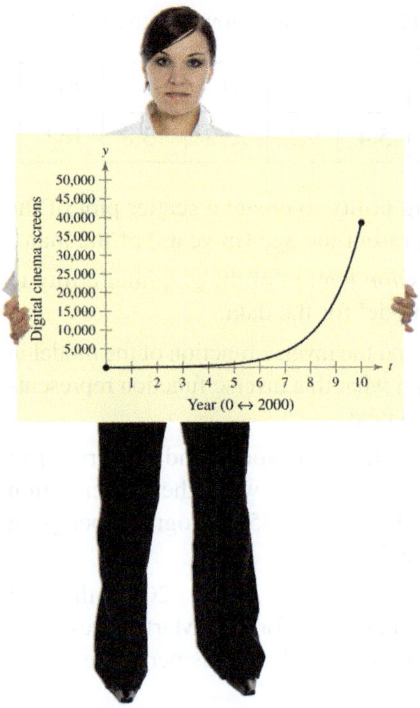

In Exercise 73 on page 356, you will use a natural exponential function to determine the number of digital cinema screens in the world.

Exponential Functions

So far, this text has dealt only with **algebraic functions,** which include polynomial functions and rational functions. In this chapter, you will study two types of nonalgebraic functions—*exponential* functions and *logarithmic* functions. These functions are examples of **transcendental functions.**

Definition of Exponential Function

The **exponential function f with base a** is denoted by

$$f(x) = a^x$$

where $a > 0$, $a \neq 1$, and x is any real number.

The base $a = 1$ is excluded because it yields

$$f(x) = 1^x = 1.$$

This is a constant function, not an exponential function.

You already know how to evaluate a^x for integer and rational values of x. For example, you know that

$$4^3 = 64 \quad \text{and} \quad 4^{1/2} = 2.$$

However, to evaluate 4^x for any real number x, you need to interpret forms with *irrational* exponents. For the purposes of this text, it is sufficient to think of

$$a^{\sqrt{2}} \qquad \left(\text{where } \sqrt{2} \approx 1.414214\right)$$

as that value having the successively closer approximations

$$a^{1.4}, a^{1.41}, a^{1.414}, a^{1.4142}, a^{1.41421}, a^{1.414214}, \ldots .$$

Example 1 **Evaluating an Exponential Expression**

Scientific Calculator

Number	*Keystrokes*	*Display*
$2^{-\pi}$	2 [y^x] π [+/−] [=]	0.113314732

Graphing Calculator

Number	*Keystrokes*	*Display*
$2^{-\pi}$	2 [∧] [(−)] π [ENTER]	.1133147323

✓ Checkpoint 1

Use a calculator to evaluate $2.2^{\sqrt{3}}$. Round your result to three decimal places.

Graphs of Exponential Functions

The graphs of all exponential functions have similar characteristics, as shown in Examples 2, 3, and 4.

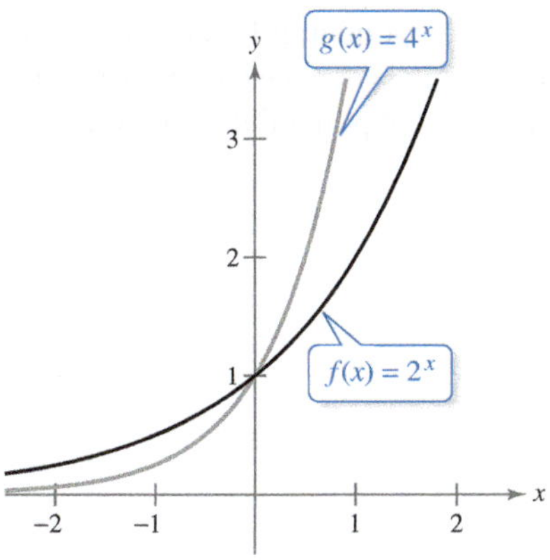

FIGURE 4.6

Example 2 Graphs of $y = a^x$

In the same coordinate plane, sketch the graph of each function.

a. $f(x) = 2^x$

b. $g(x) = 4^x$

SOLUTION The table below lists some values for each function, and Figure 4.6 shows their graphs. Note that both graphs are increasing. Moreover, the graph of $g(x) = 4^x$ is increasing more rapidly than the graph of $f(x) = 2^x$.

x	-2	-1	0	1	2	3
$f(x) = 2^x$	$2^{-2} = \frac{1}{2^2} = \frac{1}{4}$	$2^{-1} = \frac{1}{2^1} = \frac{1}{2}$	$2^0 = 1$	$2^1 = 2$	$2^2 = 4$	$2^3 = 8$
$g(x) = 4^x$	$4^{-2} = \frac{1}{4^2} = \frac{1}{16}$	$4^{-1} = \frac{1}{4^1} = \frac{1}{4}$	$4^0 = 1$	$4^1 = 4$	$4^2 = 16$	$4^3 = 64$

✓ Checkpoint 2

Sketch the graph of $f(x) = 5^x$.

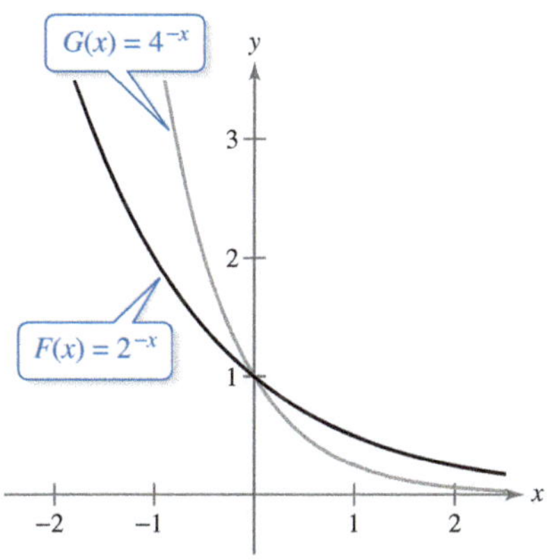

FIGURE 4.7

Example 3 Graphs of $y = a^{-x}$

In the same coordinate plane, sketch the graph of each function.

a. $F(x) = 2^{-x}$

b. $G(x) = 4^{-x}$

SOLUTION The table below lists some values for each function, and Figure 4.7 shows their graphs. Note that both graphs are decreasing. Moreover, the graph of $G(x) = 4^{-x}$ is decreasing more rapidly than the graph of $F(x) = 2^{-x}$.

x	-3	-2	-1	0	1	2
$F(x) = 2^{-x}$	8	4	2	1	$\frac{1}{2}$	$\frac{1}{4}$
$G(x) = 4^{-x}$	64	16	4	1	$\frac{1}{4}$	$\frac{1}{16}$

✓ Checkpoint 3

Sketch the graph of $F(x) = 5^{-x}$.

ALGEBRA TUTOR *xy*

For help with evaluating expressions involving exponents, see the *Chapter 4 Algebra Tutor* on page 400.

The tables in Examples 2 and 3 were evaluated by hand. You could use the *table* feature of a graphing utility to construct tables with more values.

In Example 3, note that the functions given by $F(x) = 2^{-x}$ and $G(x) = 4^{-x}$ can be rewritten with positive exponents.

$$F(x) = 2^{-x} = \left(\tfrac{2}{1}\right)^{-x} = \left(\tfrac{1}{2}\right)^x \quad \text{and} \quad G(x) = 4^{-x} = \left(\tfrac{4}{1}\right)^{-x} = \left(\tfrac{1}{4}\right)^x$$

Comparing the functions in Examples 2 and 3, observe that

$$F(x) = 2^{-x} = f(-x) \quad \text{and} \quad G(x) = 4^{-x} = g(-x).$$

Consequently, the graph of F is a reflection (in the y-axis) of the graph of f. The graphs of G and g have the same relationship.

The graphs in Figures 4.6 and 4.7 are typical of the exponential functions $y = a^x$ and $y = a^{-x}$. They have one y-intercept and one horizontal asymptote (the x-axis), and they are continuous. The basic characteristics of these exponential functions are summarized in Figures 4.8 and 4.9.

Characteristics of Exponential Functions

Graph of $y = a^x, a > 1$

- Domain: $(-\infty, \infty)$
- Range: $(0, \infty)$
- Intercept: $(0, 1)$
- Increasing
- x-axis is a horizontal asymptote ($a^x \to 0$ as $x \to -\infty$)
- Continuous

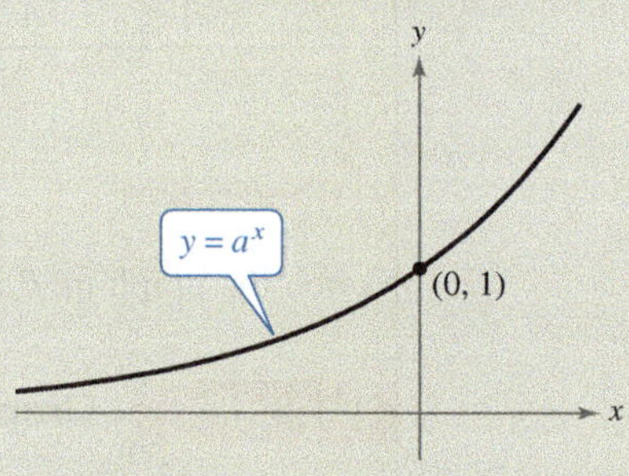

FIGURE 4.8

Graph of $y = a^{-x}, a > 1$

- Domain: $(-\infty, \infty)$
- Range: $(0, \infty)$
- Intercept: $(0, 1)$
- Decreasing
- x-axis is a horizontal asymptote ($a^{-x} \to 0$ as $x \to \infty$)
- Continuous
- Reflection of graph of $y = a^x$ in y-axis

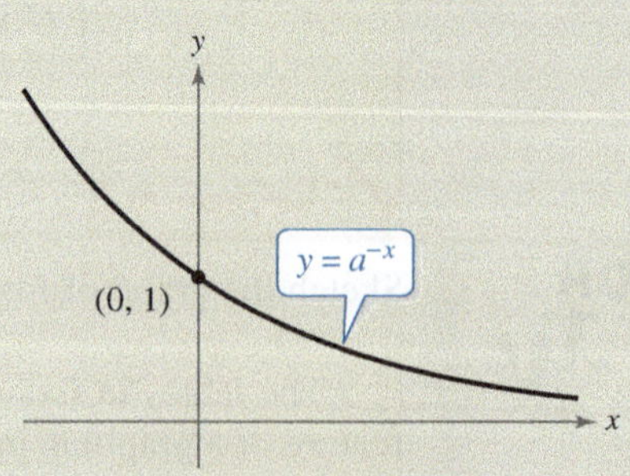

FIGURE 4.9

Dmitriy Pochitalin/Shutterstock.com

In the following example, notice how the graph of $y = a^x$ is used to sketch the graphs of functions of the form

$$f(x) = b \pm a^{x+c}.$$

Example 4 Transformations of Graphs of Exponential Functions

Each of the following graphs is a transformation of the graph of $f(x) = 3^x$, as shown in Figure 4.10.

a. Because $g(x) = 3^{x+1} = f(x + 1)$, the graph of g can be obtained by shifting the graph of f one unit to the *left*.

b. Because $h(x) = 3^x - 2 = f(x) - 2$, the graph of h can be obtained by shifting the graph of f *downward* two units.

c. Because $k(x) = -3^x = -f(x)$, the graph of k can be obtained by *reflecting* the graph of f in the x-axis.

d. Because $j(x) = 3^{-x} = f(-x)$, the graph of j can be obtained by *reflecting* the graph of f in the y-axis.

STUDY TIP

Notice in Example 4(b) that shifting the graph downward two units also shifts the horizontal asymptote of $f(x) = 3^x$ from the x-axis $(y = 0)$ to the line $y = -2$.

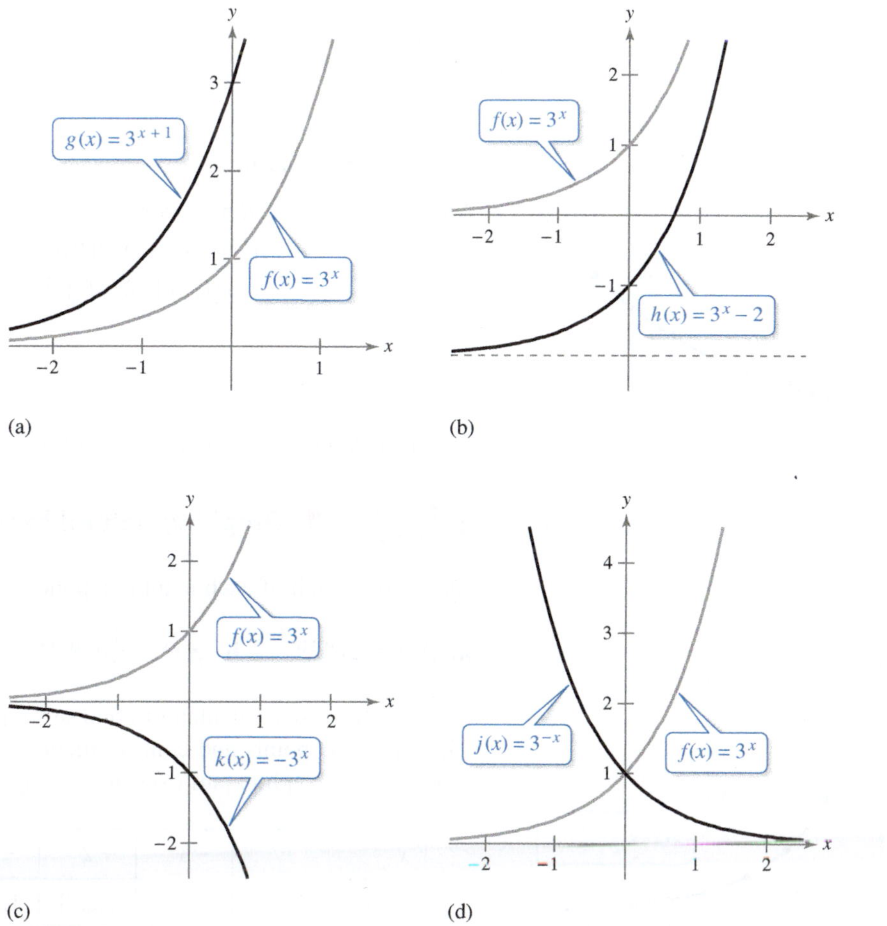

FIGURE 4.10

Checkpoint 4

Sketch the graph of

$$f(x) = 2^{x-1}.$$

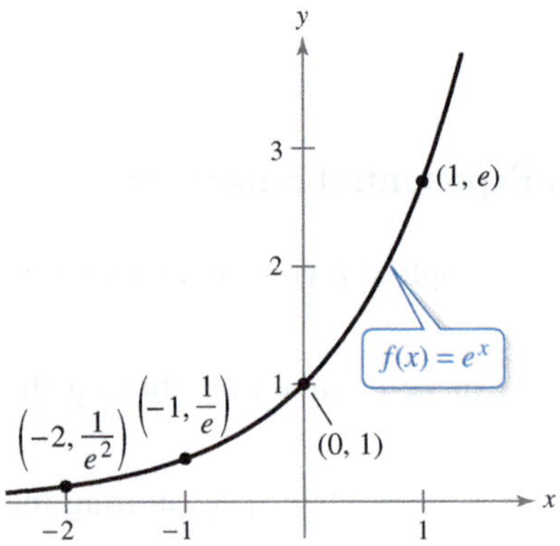

FIGURE 4.11

The Natural Base e

In many applications, the most convenient choice for a base is the irrational number

$$e = 2.718281828\ldots$$

called the **natural base.** The function given by

$$f(x) = e^x$$

is called the **natural exponential function.** Its graph is shown in Figure 4.11. The graph of the natural exponential function has the same basic characteristics as the graph of the exponential function given by $f(x) = a^x$ (see page 348). Be sure you see that for the exponential function given by $f(x) = e^x$, e is the constant $2.718281828\ldots$, whereas x is the variable.

Example 5 Evaluating the Natural Exponential Function

Use a calculator to evaluate the function given by $f(x) = e^x$ when $x = 2$ and $x = -1$.

SOLUTION

Scientific Calculator

Number	*Keystrokes*	*Display*
e^2	2 (2nd) [eˣ]	7.389056099
e^{-1}	1 (+/−) (2nd) [eˣ]	0.367879441

Graphing Calculator

Number	*Keystrokes*	*Display*
e^2	(2nd) [eˣ] 2 ( )) (ENTER)	7.389056099
e^{-1}	(2nd) [eˣ] (−) 1 ( )) (ENTER)	.3678794412

✓ Checkpoint 5

Use a calculator to evaluate $f(x) = e^x$ when $x = 6$ and $x = -3.5$.

Example 6 Graphing Natural Exponential Functions

Sketch the graph of each natural exponential function.

a. $f(x) = 2e^{0.24x}$ **b.** $g(x) = \frac{1}{2}e^{-0.58x}$

SOLUTION Use a calculator to find several points on each graph, as shown in the table. Then, plot the points and connect them with smooth curves, as shown in Figure 4.12. Note that the graph in part (a) is increasing, whereas the graph in part (b) is decreasing.

x	−3	−2	−1	0	1	2	3
$f(x) = 2e^{0.24x}$	0.974	1.238	1.573	2	2.542	3.232	4.109
$g(x) = \frac{1}{2}e^{-0.58x}$	2.849	1.595	0.893	0.5	0.280	0.157	0.088

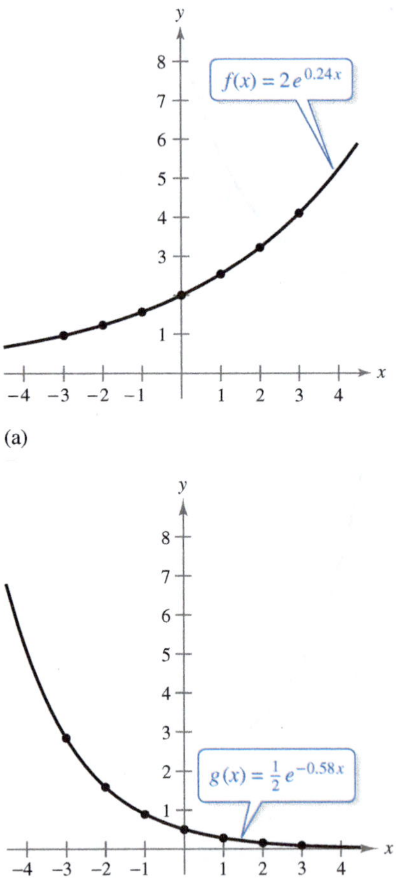

(a)

(b)

FIGURE 4.12

✓ Checkpoint 6

Sketch the graph of $f(x) = e^{0.5x}$.

Compound Interest

One of the most familiar examples of exponential growth is that of an investment earning **continuously compounded interest.** The formula for the balance in an account that is compounded n times per year is

$$A = P\left(1 + \frac{r}{n}\right)^{nt}$$

where A is the balance in the account, P is the initial deposit, r is the annual interest rate (in decimal form), and t is the number of years. Using exponential functions, you will *develop* this formula and show how it leads to continuous compounding.

Suppose a principal P is invested at an annual interest rate r, compounded once a year. The principal at the end of the first year, P_1, is equal to the initial deposit P plus the interest earned, Pr. So,

$$P_1 = P + Pr.$$

This can be rewritten by factoring out P from each term as shown.

$$\begin{aligned} P_1 &= P + Pr \\ &= P(1 + r) \end{aligned}$$

This pattern of multiplying the previous principal by $1 + r$ is then repeated each successive year, as shown below.

Year	*Balance After Each Compounding*
0	$P = P$
1	$P_1 = P(1 + r)$
2	$P_2 = P_1(1 + r) = P(1 + r)(1 + r) = P(1 + r)^2$
3	$P_3 = P_2(1 + r) = P(1 + r)^2(1 + r) = P(1 + r)^3$
	$\vdots$
t	$P_t = P(1 + r)^t$

To accommodate more frequent (quarterly, monthly, or daily) compounding of interest, let n be the number of compoundings per year and let t be the number of years. Then the rate per compounding is r/n and the account balance after t years is

$$A = P\left(1 + \frac{r}{n}\right)^{nt}.$$ Amount (balance) with n compoundings per year

When you let the number of compoundings n increase without bound, the process approaches what is called **continuous compounding.** In the formula for n compoundings per year, let $m = n/r$. This produces

$$A = P\left(1 + \frac{r}{n}\right)^{nt}$$ Amount with n compoundings per year

$$= P\left(1 + \frac{1}{m}\right)^{mrt}$$ Substitute mr for n and simplify.

$$= P\left[\left(1 + \frac{1}{m}\right)^{m}\right]^{rt}.$$ Property of exponents

As m increases without bound, it can be shown that

$$\left[1 + \left(\frac{1}{m}\right)\right]^{m}$$

approaches e. From this, you can conclude that the formula for continuous compounding is

$$A = Pe^{rt}.$$ Continuous compounding

Formulas for Compound Interest

After t years, the balance A in an account with principal P and annual interest rate r (in decimal form) is given by the following formulas.

1. For n compoundings per year: $A = P\left(1 + \frac{r}{n}\right)^{nt}$

2. For continuous compounding: $A = Pe^{rt}$

Be sure that the annual interest rate is written in decimal form. For instance, 6% should be written as 0.06 when using compound interest formulas.

Example 7 Compound Interest

You invest \$12,000 at an annual rate of 3%. Find the balance after 5 years when the interest is compounded (a) quarterly, (b) monthly, and (c) continuously. Which type of compounding earns the most money?

SOLUTION

a. For quarterly compounding, you have $n = 4$. So, in 5 years at 3%, the balance is

$$A = P\left(1 + \frac{r}{n}\right)^{nt} \qquad \text{Formula for compound interest}$$

$$= 12{,}000\left(1 + \frac{0.03}{4}\right)^{4(5)} \qquad \text{Substitute for } P, r, n, \text{ and } t.$$

$$\approx \$13{,}934.21. \qquad \text{Use a calculator.}$$

b. For monthly compounding, you have $n = 12$. So, in 5 years at 3%, the balance is

$$A = P\left(1 + \frac{r}{n}\right)^{nt} \qquad \text{Formula for compound interest}$$

$$= 12{,}000\left(1 + \frac{0.03}{12}\right)^{12(5)} \qquad \text{Substitute for } P, r, n, \text{ and } t.$$

$$\approx \$13{,}939.40. \qquad \text{Use a calculator.}$$

c. For continuous compounding, the balance is

$$A = Pe^{rt} \qquad \text{Formula for continuous compounding}$$

$$= 12{,}000e^{0.03(5)} \qquad \text{Substitute for } P, r, \text{ and } t.$$

$$\approx \$13{,}942.01. \qquad \text{Use a calculator.}$$

Continuous compounding yields more than quarterly and monthly compounding. This is typical of the two types of compounding. That is, for a given principal, interest rate, and time, continuous compounding will always yield a larger balance than compounding n times a year.

Checkpoint 7

You invest \$6000 at an annual rate of 4%. Find the balance after 7 years when the interest is compounded

a. quarterly.

b. monthly.

c. continuously.

Another Application

Example 8 Radioactive Decay

In 1986, a nuclear reactor accident occurred in Chernobyl in what was then the Soviet Union. The explosion spread highly toxic radioactive chemicals, such as plutonium, over hundreds of square miles, and the government evacuated the city and the surrounding area. Consider the model

$$P = 10\left(\frac{1}{2}\right)^{t/24,100}$$

which represents the amount of plutonium P that remains (from an initial amount of 10 pounds) after t years. Sketch the graph of this function over the interval from $t = 0$ to $t = 100,000$, where $t = 0$ represents 1986. How much of the 10 pounds of plutonium will remain in the year 2012? How much of the 10 pounds will remain after 100,000 years? Why is this city uninhabited?

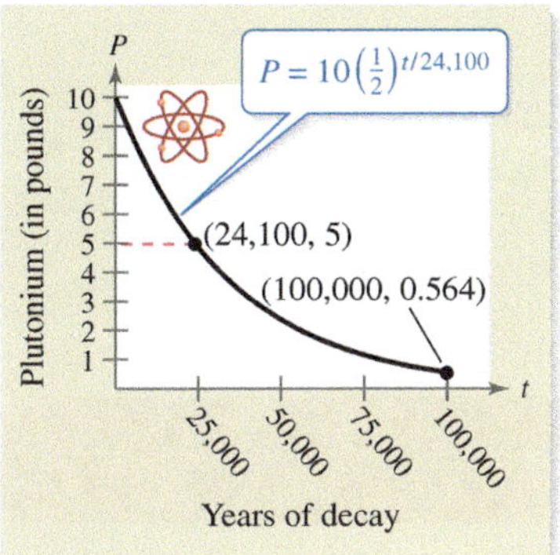

FIGURE 4.13

SOLUTION The graph of this function is shown in Figure 4.13. Plutonium has a *half-life* of about 24,100 years. That is, after 24,100 years, *half* of the original amount of plutonium will remain. After another 24,100 years, one-quarter of the original amount will remain, and so on. In the year 2012 ($t = 26$), there will still be

$$\begin{aligned} P &= 10\left(\frac{1}{2}\right)^{26/24,100} && \text{Substitute 26 for } t. \\ &\approx 10\left(\frac{1}{2}\right)^{0.0010788} && \text{Approximate exponent.} \\ &\approx 9.993 \text{ pounds} && \text{Use a calculator.} \end{aligned}$$

of plutonium remaining. After 100,000 years, there will still be

$$\begin{aligned} P &= 10\left(\frac{1}{2}\right)^{100,000/24,100} && \text{Substitute 100,000 for } t. \\ &\approx 10\left(\frac{1}{2}\right)^{4.149} && \text{Approximate exponent.} \\ &\approx 0.564 \text{ pound} && \text{Use a calculator.} \end{aligned}$$

of plutonium remaining. This city is uninhabited because much of the original amount of radioactive plutonium still remains.

✓ Checkpoint 8

In Example 8, how much of the initial 10 pounds of plutonium will remain in the year 2086?

SUMMARIZE (Section 4.2)

1. State the definition of an exponential function f with base a *(page 346)*. For an example of evaluating an exponential expression, see Example 1.
2. Describe the relationship between the graphs of the functions $f(x) = a^x$ and $g(x) = a^{-x}$, where $a > 1$ *(pages 347 and 348)*. For examples of graphing exponential functions, see Examples 2, 3, and 4.
3. State the definitions of the natural base and the natural exponential function *(page 350)*. For examples of evaluating and graphing natural exponential functions, see Examples 5 and 6.

SKILLS WARM UP 4.2 The following warm-up exercises involve skills that were covered in earlier sections. You will use these skills in the exercise set for this section. For additional help, review Sections 0.3 and 0.4.

In Exercises 1–12, use the properties of exponents to simplify the expression.

1. $5^{2x}(5^{-x})$ **2.** $3^{-x}(3^{3x})$ **3.** $\dfrac{4^{5x}}{4^{2x}}$

4. $\dfrac{10^{2x}}{10^{x}}$ **5.** $(4^x)^2$ **6.** $(4^{2x})^5$

7. $\left(\dfrac{2^x}{3^x}\right)^{-1}$ **8.** $(4^{6x})^{1/2}$ **9.** $(2^{3x})^{-1/3}$

10. $\left(\dfrac{3^{4x}}{5^{4x}}\right)^{1/4}$ **11.** $(16^x)^{1/4}$ **12.** $(27^x)^{1/3}$

Exercises 4.2

See www.CalcChat.com for worked-out solutions to odd-numbered exercises.

Evaluating an Exponential Expression In Exercises 1–6, use a calculator to evaluate the expression. Round your result to three decimal places. *See Example 1.*

1. $(2.6)^{1.3}$ **2.** $(1.12)^{20.5}$

3. $100(1.03)^{-9/5}$

4. $1500(2^{-5/2})$

5. $6^{-\sqrt{2}}$ **6.** $1.3^{-\sqrt{5}}$

Matching In Exercises 7–14, match the function with its graph. [The graphs are labeled (a), (b), (c), (d), (e), (f), (g), and (h).]

(a)
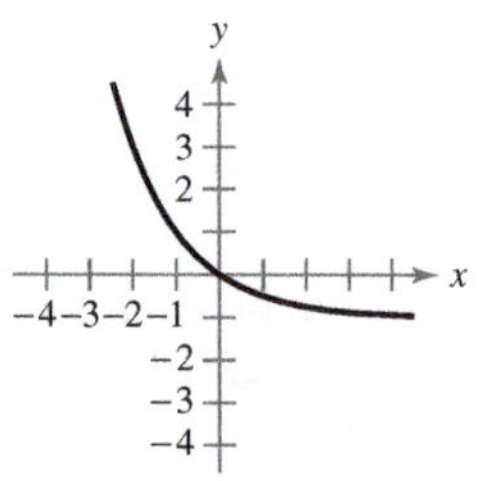

(b)
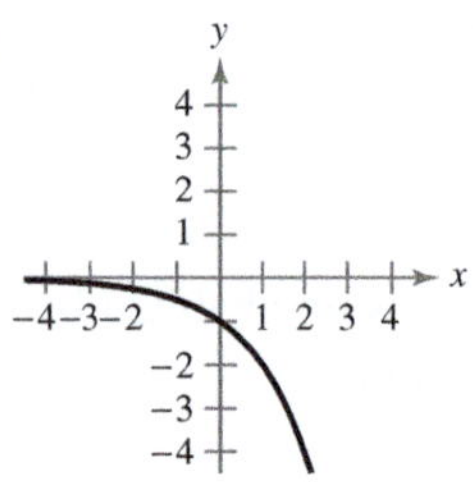

(c)
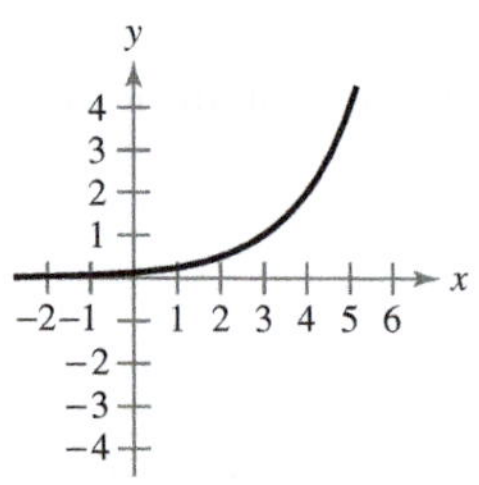

(d)
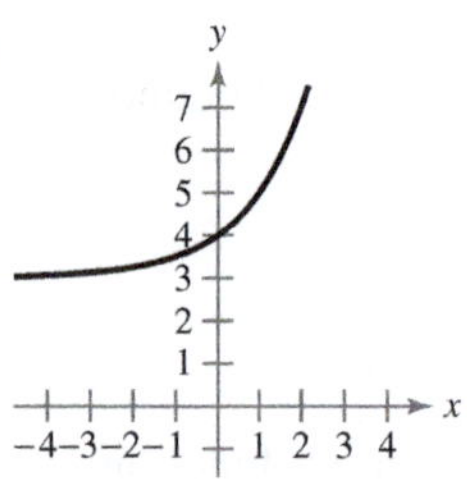

(e)
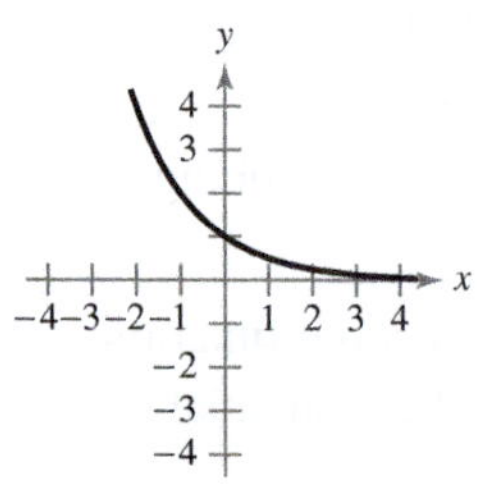

(f)
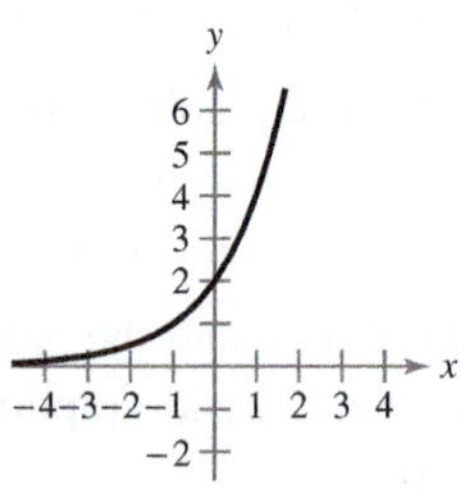

(g)
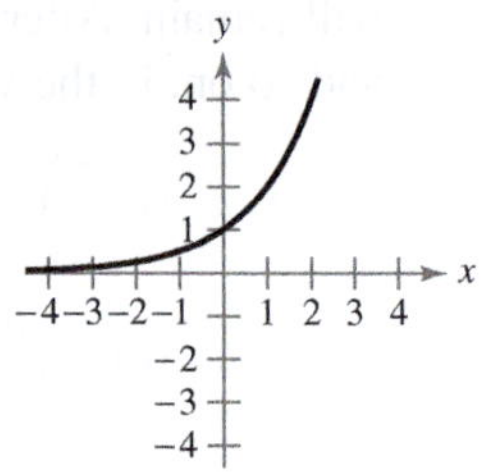

(h)
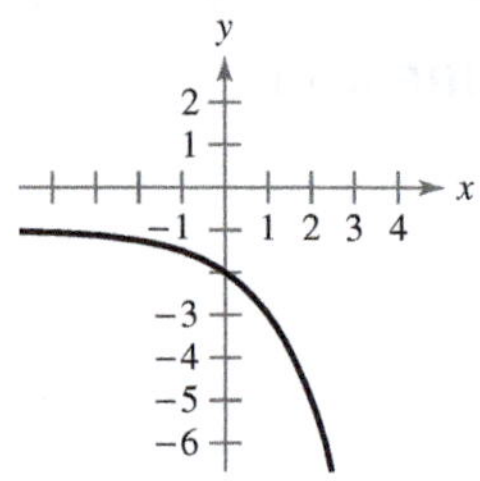

7. $f(x) = 2^x$ **8.** $f(x) = 2^{-x}$

9. $f(x) = -2^x$

10. $f(x) = -2^x - 1$

11. $f(x) = 2^x + 3$ **12.** $f(x) = 2^{-x} - 1$

13. $f(x) = 2^{x+1}$ **14.** $f(x) = 2^{x-3}$

Graphing an Exponential Function In Exercises 15–26, sketch the graph of the function. *See Examples 2, 3, and 4.*

15. $g(x) = 4^x$ **16.** $f(x) = \left(\frac{3}{2}\right)^x$

17. $f(x) = 4^{-x}$ **18.** $h(x) = \left(\frac{3}{2}\right)^{-x}$

19. $h(x) = 4^{x-3}$ **20.** $g(x) = \left(\frac{3}{2}\right)^{x+2}$

21. $g(x) = 4^{-x} - 2$

22. $f(x) = \left(\frac{3}{2}\right)^{-x} + 2$

23. $h(x) = -4^x$ **24.** $g(x) = -\left(\frac{3}{2}\right)^x$

25. $y = 2^{-x^2}$

26. $y = 3^{-x^2}$

Evaluating the Natural Exponential Function In Exercises 27–30, use a calculator to evaluate the function $f(x) = e^x$ for the given value of x. Round your result to three decimal places. *See Example 5.*

27. $x = 4$ **28.** $x = -5$

29. $x = -2.7$ **30.** $x = 0.3$

Graphing a Natural Exponential Function In Exercises 31–38, sketch the graph of the function. *See Example 6.*

31. $y = e^{-0.1x}$

32. $y = e^{0.2x}$

33. $f(x) = 2e^{0.12x}$

34. $f(x) = 3e^{-0.2x}$

35. $f(x) = e^{x+3}$

36. $h(x) = e^{x-2}$

37. $g(x) = 1 + e^{-x}$

38. $N(t) = 2 - e^{t}$

39. Think About It Is the natural exponential function also an exponential function with base a? Explain.

40. HOW DO YOU SEE IT? The figure shows the graphs of $y = 2^x$, $y = e^x$, $y = 10^x$, $y = 2^{-x}$, $y = e^{-x}$, and $y = 10^{-x}$. Match each function with its graph. (The graphs are labeled A through F.) Explain your reasoning.

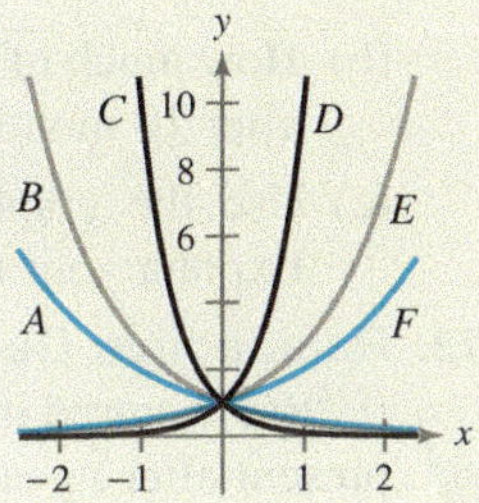

Compound Interest In Exercises 41–44, complete the table to find the balance A for P dollars invested at rate r for t years, compounded n times per year. *See Example 7.*

n	1	2	4	12	365	Continuous
A						

41. $P = \$5000$, $r = 8\%$, $t = 5$ years

42. $P = \$1000$, $r = 10\%$, $t = 10$ years

43. $P = \$2500$, $r = 5\%$, $t = 20$ years

44. $P = \$1000$, $r = 10\%$, $t = 40$ years

45. Compound Interest A bank offers two types of interest-bearing accounts. The first account pays 5% interest compounded quarterly. The second account pays 3% interest compounded continuously. Which account earns more money? Why?

46. Compound Interest A bank offers two types of interest-bearing accounts. The first account pays 6% interest compounded monthly. The second account pays 5% interest compounded continuously. Which account earns more money? Why?

Present Value The present value of money is the principal P you need to invest today so that it will grow to an amount A at the end of a specified time. The present value formula

$$P = A\left(1 + \frac{r}{n}\right)^{-nt}$$

is obtained by solving the compound interest formula

$$A = P\left(1 + \frac{r}{n}\right)^{nt}$$

for P. Recall that t is the number of years, r is the interest rate per year, and n is the number of compoundings per year. In Exercises 47–50, find the present value of amount A invested at rate r for t years, compounded n times per year.

47. $A = \$10{,}000$, $r = 6\%$, $t = 5$ years, $n = 4$

48. $A = \$50{,}000$, $r = 7\%$, $t = 10$ years, $n = 12$

49. $A = \$20{,}000$, $r = 8\%$, $t = 6$ years, $n = 4$

50. $A = \$1{,}000{,}000$, $r = 8\%$, $t = 20$ years, $n = 2$

Compound Interest In Exercises 51–54, complete the table to find the amount P that must be invested at rate r to obtain a balance of $A = \$100{,}000$ in t years.

t	1	10	20	30	40	50
P						

51. $r = 10\%$, compounded monthly

52. $r = 7\%$, compounded daily

53. $r = 9\%$, compounded continuously

54. $r = 5\%$, compounded continuously

Compound Interest In Exercises 55–58, you deposit a lump sum P in a trust fund on the day your child is born. The fund earns 6.5% interest compounded continuously. Find the amount P that will yield the given balance A on your child's 25th birthday.

55. $A = \$100{,}000$

56. $A = \$500{,}000$

57. $A = \$750{,}000$

58. $A = \$1{,}000{,}000$

Compound Interest In Exercises 59–62, you deposit a lump sum P in a trust fund on the day your grandchild is born. The fund earns 7.5% interest compounded continuously. Find the amount P that will yield the given balance A on your grandchild's 21st birthday.

59. $A = \$100{,}000$

60. $A = \$500{,}000$

61. $A = \$750{,}000$

62. $A = \$1{,}000{,}000$

63. Demand Function The demand function for a limited edition comic book is given by

$$p = 3000\left(1 - \frac{5}{5 + e^{-0.015x}}\right).$$

(a) Find the price p for a demand of $x = 75$ units.

(b) Find the price p for a demand of $x = 200$ units.

(c) Use a graphing utility to graph the demand function.

(d) Use the graph from part (c) to approximate the demand when the price is \$100.

64. Demand Function The demand function for a home theater system is given by

$$p = 7500\left(1 - \frac{7}{7 + e^{-0.003x}}\right).$$

(a) Find the price p for a demand of $x = 200$ units.

(b) Find the price p for a demand of $x = 900$ units.

(c) Use a graphing utility to graph the demand function.

(d) Use the graph from part (c) to approximate the demand when the price is \$400.

65. Bacteria Growth The number of a certain type of bacteria increases according to the model

$$P(t) = 100e^{0.01896t}$$

where t is time (in hours).

(a) Find $P(0)$. (b) Find $P(5)$.

(c) Find $P(10)$. (d) Find $P(24)$.

66. Bacteria Growth As a result of a medical treatment, the number of a certain type of bacteria decreases according to the model

$$P(t) = 100e^{-0.685t}$$

where t is time (in hours).

(a) Find $P(0)$. (b) Find $P(5)$.

(c) Find $P(10)$. (d) Find $P(24)$.

67. Population Growth The population P of a town increases according to the model

$$P(t) = 4500e^{0.0272t}$$

where t represents the year, with $t = 0$ corresponding to 2000. Use the model to predict the population in each year.

(a) 2010 (b) 2012

(c) 2015 (d) 2020

68. Population Growth The population P of a small city increases according to the model

$$P(t) = 36{,}000e^{0.0156t}$$

where t represents the year, with $t = 0$ corresponding to 2000. Use the model to predict the population in each year.

(a) 2009 (b) 2011

(c) 2015 (d) 2018

69. Radioactive Decay Strontium-90 has a half-life of 29.1 years. The amount S of 100 kilograms of strontium-90 present after t years is given by

$$S = 100e^{-0.0238t}.$$

How much of the 100 kilograms will remain after 50 years?

70. Radioactive Decay Neptunium-237 has a half-life of 2.1 million years. The amount N of 200 kilograms of neptunium-237 present after t years is given by

$$N = 200e^{-0.00000033007t}.$$

How much of the 200 kilograms will remain after 20,000 years?

71. Radioactive Decay Five pounds of the element polonium (^{210}Po) is released in a nuclear accident. The amount of polonium P that is present after t months is given by $P = 5e^{-0.1507t}$.

(a) Use a graphing utility to graph this function over the interval from $t = 0$ to $t = 10$.

(b) How much of the 5 pounds of polonium will remain after 10 months?

(c) Use the graph to estimate the half-life of ^{210}Po. Explain your reasoning.

72. Radioactive Decay One hundred grams of radium (^{226}Ra) is stored in a container. The amount of radium R present after t years is given by $R = 100e^{-0.0004335t}$.

(a) Use a graphing utility to graph this function over the interval from $t = 0$ to $t = 10{,}000$.

(b) How much of the 100 grams of radium will remain after 10,000 years?

(c) Use the graph to estimate the half-life of ^{226}Ra. Explain your reasoning.

73. Digital Cinema Screens The numbers y of digital cinema screens in the world from 2000 through 2010 can be modeled by

$$y = 25.87e^{0.7323t}, \quad 0 \le t \le 10$$

where t represents the year, with $t = 0$ corresponding to 2000 (see figure). *(Source: IHS Screen Digest)*

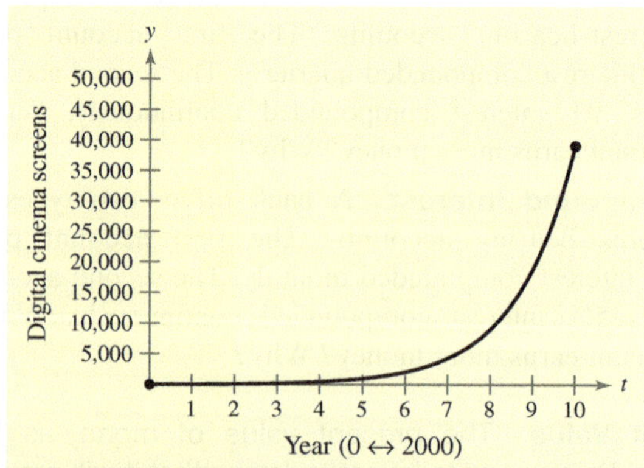

(a) Use the graph to estimate *graphically* the numbers of digital cinema screens in 2008 and 2010.

(b) Use the model to confirm *algebraically* the estimates obtained in part (a).

74. Revenue The revenue R (in millions of dollars) for Research in Motion each year from 2000 through 2009 can be modeled by

$$R = 162.53e^{0.5061t}, \quad 0 \le t \le 9$$

where t represents the year, with $t = 0$ corresponding to 2000 (see figure). *(Source: Research in Motion Limited)*

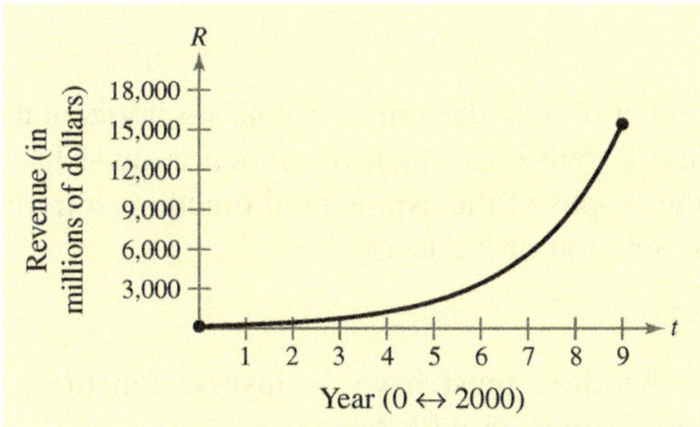

(a) Use the graph to estimate *graphically* the revenues for Research in Motion in 2003, 2007, and 2009.

(b) Use the model to confirm *algebraically* the estimates obtained in part (a).

75. Age at First Marriage For each year from 1980 through 2010, the median age A of an American woman at her first marriage can be modeled by

$$A = -45.10 + \frac{72.02}{1 + e^{-0.0523t - 2.6314}}, \quad 0 \le t \le 30$$

where t represents the year, with $t = 0$ corresponding to 1980 (see figure). *(Source: U.S. Census Bureau)*

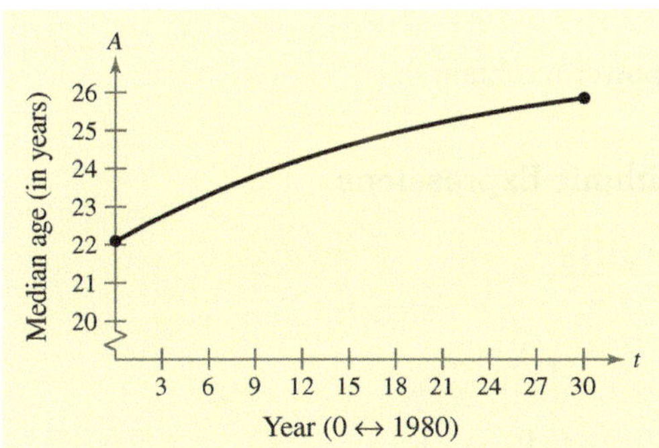

(a) Use the graph to estimate *graphically* the median age of an American woman at her first marriage in each of the years 1980, 1990, 2000, and 2010.

(b) Use the model to confirm *algebraically* the estimates obtained in part (a).

76. Prescriptions The numbers of mail-order prescriptions P (in millions) filled in the United States from 2002 through 2009 can be modeled by

$$P = 151 + \frac{89.24}{1 + e^{-0.0895t + 2.8739}}, \quad 2 \le t \le 9$$

where t represents the year, with $t = 2$ corresponding to 2002. *(Source: National Association of Chain Drug Stores)*

(a) Use a graphing utility to graph P for the years 2002 through 2009.

(b) Use the graph from part (a) to estimate the numbers of mail-order prescriptions filled in 2002, 2006, and 2009.

77. Gross Domestic Product The U.S. annual gross domestic product G (in billions of dollars) from 1985 through 2009 can be modeled by

$$G = 3355.4(1.0541)^t, \quad 5 \le t \le 29$$

where t represents the year, with $t = 5$ corresponding to 1985. *(Source: U.S. Bureau of Economic Analysis)*

(a) Use a graphing utility to graph G for the years 1985 through 2009.

(b) Use the graph from part (a) to estimate the gross domestic product in 1990, 2000, and 2009.

78. Writing Determine whether $e = \dfrac{271{,}801}{99{,}990}$. Justify your answer.

79. Project: Drug Concentration For a project involving the concentration of a drug in a patient's bloodstream, visit this text's website at *www.cengagebrain.com.*

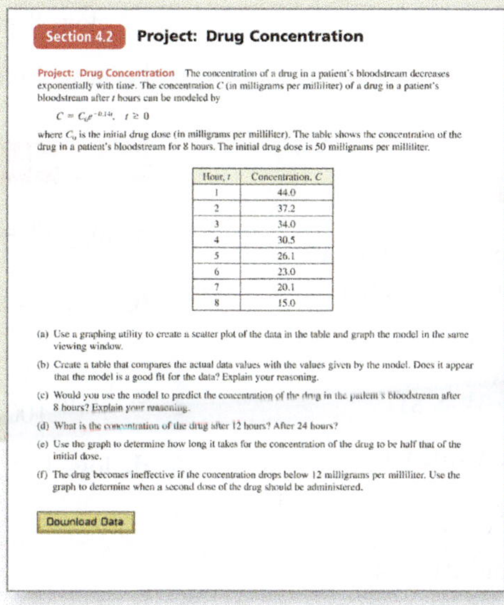

Section 4.2 **Project: Drug Concentration**

Project: Drug Concentration The concentration of a drug in a patient's bloodstream decreases exponentially with time. The concentration C (in milligrams per milliliter) of a drug in a patient's bloodstream after t hours can be modeled by

$$C = C_0e^{-0.14t}, \quad t \ge 0$$

where C_0 is the initial drug dose (in milligrams per milliliter). The table shows the concentration of the drug in a patient's bloodstream for 8 hours. The initial drug dose is 50 milligrams per milliliter.

Hour, t	Concentration, C
1	44.0
2	37.2
3	34.0
4	30.5
5	26.1
6	23.0
7	20.1
8	15.0

(a) Use a graphing utility to create a scatter plot of the data in the table and graph the model in the same viewing window.

(b) Create a table that compares the actual data values with the values given by the model. Does it appear that the model is a good fit for the data? Explain your reasoning.

(c) Would you use the model to predict the concentration of the drug in the patient's bloodstream after 8 hours? Explain your reasoning.

(d) What is the concentration of the drug after 12 hours? After 24 hours?

(e) Use the graph to determine how long it takes for the concentration of the drug to be half that of the initial dose.

(f) The drug becomes ineffective if the concentration drops below 12 milligrams per milliliter. Use the graph to determine when a second dose of the drug should be administered.

Download Data

4.3 Logarithmic Functions

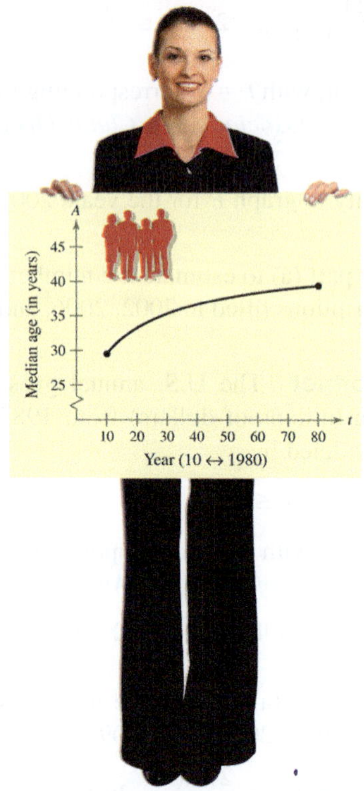

In Exercise 100 on page 367, you will use a natural logarithmic model to approximate the median age of the United States population.

- Recognize and evaluate a logarithmic function with base *a*.
- Sketch the graph of a logarithmic function.
- Recognize and evaluate the natural logarithmic function.
- Use a logarithmic model to solve an application problem.

Logarithmic Functions

In Section 4.1, you learned that if a function has the property that no horizontal line intersects the graph of the function more than once, then the function must have an inverse function. By looking back at the graphs of the exponential functions introduced in Section 4.2, you will see that every function of the form

$$f(x) = a^x \text{ (where } a > 0 \text{ and } a \neq 1)$$

passes the Horizontal Line Test and therefore must have an inverse function. This inverse function is called the **logarithmic function with base *a*.**

> **Definition of Logarithmic Function**
>
> For $x > 0$, $a > 0$, and $a \neq 1$,
>
> $$y = \log_a x \text{ if and only if } x = a^y.$$
>
> The function given by
>
> $$f(x) = \log_a x$$
>
> is called the **logarithmic function with base *a*.**

The equations $y = \log_a x$ and $x = a^y$ are equivalent. The first equation is in logarithmic form and the second is in exponential form.

When evaluating logarithms, remember that *a logarithm is an exponent.* This means that $\log_a x$ is the exponent to which a must be raised to obtain x. For instance,

$$\log_2 8 = 3$$

because 2 must be raised to the third power to obtain 8.

STUDY TIP

By the definition of a logarithmic function,

$$\overset{a}{3}^{\overset{y}{4}} = \overset{x}{81}$$

can be written as

$$\log_{\overset{a}{3}} \overset{x}{81} = \overset{y}{4}.$$

Example 1 Evaluating Logarithmic Expressions

a. $\log_2 32 = 5$ because $2^5 = 32$.

b. $\log_4 2 = \frac{1}{2}$ because $4^{1/2} = \sqrt{4} = 2$.

c. $\log_{10} \frac{1}{100} = -2$ because $10^{-2} = \frac{1}{10^2} = \frac{1}{100}$.

d. $\log_3 1 = 0$ because $3^0 = 1$.

✓ Checkpoint 1

Evaluate each expression.

a. $\log_9 81$ **b.** $\log_7 \frac{1}{49}$

EDHAR/Shutterstock.com

The logarithmic function with base 10 is called the **common logarithmic function.** On most calculators, this function is denoted by LOG.

STUDY TIP

Because $y = \log_a x$ is the inverse function of $y = a^x$, it follows that the domain of $y = \log_a x$ is the range of $y = a^x$, $(0, \infty)$. In other words, $y = \log_a x$ is defined only if x is positive.

Example 2 Evaluating Logarithmic Expressions on a Calculator

Scientific Calculator

Number	Keystrokes	Display
a. $\log_{10} 10$	10 LOG	1
b. $2 \log_{10} 2.5$	2.5 LOG × 2 =	0.795880017
c. $\log_{10}(-2)$	2 +/− LOG	ERROR

Graphing Calculator

Number	Keystrokes	Display
a. $\log_{10} 10$	LOG 10) ENTER	1
b. $2 \log_{10} 2.5$	2 LOG 2.5) ENTER	.7958800173
c. $\log_{10}(-2)$	LOG (−) 2) ENTER	ERROR

Many calculators display an error message (or a complex number) when you try to evaluate $\log_{10}(-2)$. This is because the domain of every logarithmic function is the set of *positive real numbers*. In other words, there is no real number power to which 10 can be raised to obtain -2.

✓ Checkpoint 2

Use a calculator to evaluate the expression $\log_{10} 200$. Round your result to three decimal places.

The following properties follow directly from the definition of the logarithmic function with base a.

Properties of Logarithms

1. $\log_a 1 = 0$ because $a^0 = 1$.
2. $\log_a a = 1$ because $a^1 = a$.
3. $\log_a a^x = x$ and $a^{\log_a x} = x$ — Inverse Properties
4. If $\log_a x = \log_a y$, then $x = y$. — One-to-One Property

Example 3 Using Properties of Logarithms

a. Solve the equation $\log_2 x = \log_2 3$ for x.

b. Solve the equation $\log_5 x = 1$ for x.

SOLUTION

a. Using the One-to-One Property (Property 4), you can conclude that $x = 3$.

b. Using Property 2, you can conclude that $x = 5$.

✓ Checkpoint 3

Solve the equation $\log_4 1 = x$ for x.

Graphs of Logarithmic Functions

To sketch the graph of $y = \log_a x$, you can use the fact that the graphs of inverse functions are reflections of each other in the line $y = x$.

Example 4 Graphs of Exponential and Logarithmic Functions

In the same coordinate plane, sketch the graph of each function.

a. $f(x) = 2^x$

b. $g(x) = \log_2 x$

SOLUTION

a. For $f(x) = 2^x$, construct a table of values.

x	-2	-1	0	1	2	3
$f(x) = 2^x$	$\frac{1}{4}$	$\frac{1}{2}$	1	2	4	8

By plotting these points and connecting them with a smooth curve, you obtain the graph of f shown in Figure 4.14.

b. Because $g(x) = \log_2 x$ is the inverse function of $f(x) = 2^x$, the graph of g is obtained by plotting the points $(f(x), x)$ and connecting them with a smooth curve. The graph of g is a reflection of the graph of f in the line $y = x$, as shown in Figure 4.14.

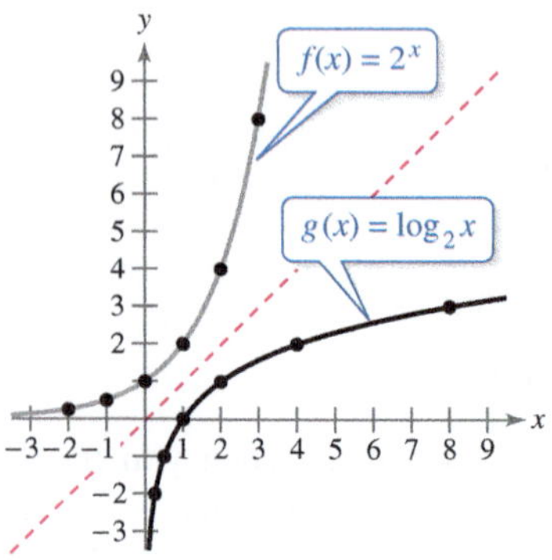

Inverse Functions
FIGURE 4.14

Checkpoint 4

In the same coordinate plane, sketch the graph of each function.

a. $f(x) = 4^x$

b. $g(x) = \log_4 x$

Before you can confirm the result of Example 4 with a graphing utility, you need to know how to enter $\log_2 x$. You will learn how to do this using the *change-of-base formula* discussed in Section 4.4.

Example 5 Sketching the Graph of a Logarithmic Function

Sketch the graph of the common logarithmic function given by $f(x) = \log_{10} x$.

SOLUTION Begin by constructing a table of values. Note that some of the values can be obtained without a calculator by using the properties of logarithms and the definition of a logarithmic function. Others require a calculator. Next, plot the points and connect them with a smooth curve, as shown in Figure 4.15.

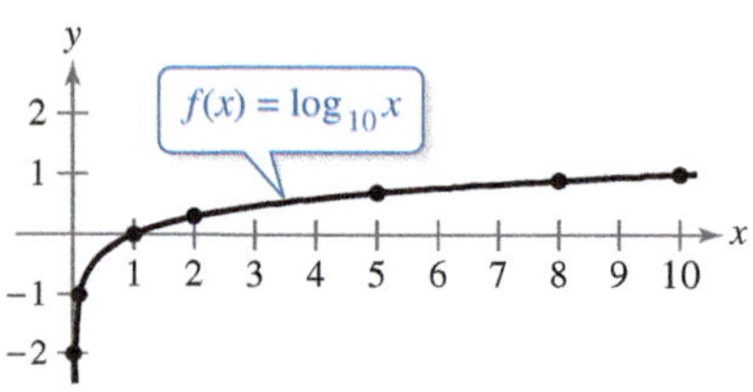

Common Logarithmic Function
FIGURE 4.15

	Without Calculator				With Calculator		
x	$\frac{1}{100}$	$\frac{1}{10}$	1	10	2	5	8
$f(x) = \log_{10} x$	-2	-1	0	1	0.301	0.699	0.903

Checkpoint 5

Sketch the graph of the function given by $f(x) = 2 \log_{10} x$.

The nature of the graph in Figure 4.15 is typical of functions of the form $f(x) = \log_a x$, $a > 1$. They have one x-intercept and one vertical asymptote. Notice how slowly the graph rises for $x > 1$. The basic characteristics of logarithmic graphs are summarized below and shown in Figure 4.16. Note that the vertical asymptote occurs at $x = 0$, where $\log_a x$ is *undefined.*

Characteristics of Logarithmic Functions

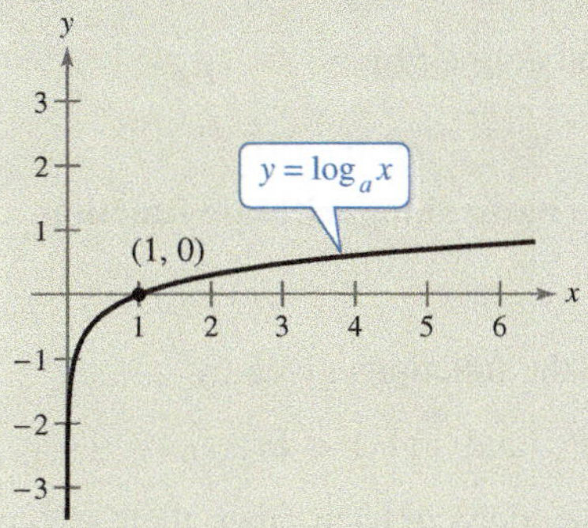

FIGURE 4.16

Graph of $y = \log_a x$, $a > 1$

- Domain: $(0, \infty)$
- Range: $(-\infty, \infty)$
- Intercept: $(1, 0)$
- Increasing
- Passes the Horizontal Line Test; therefore has an inverse function
- y-axis is a vertical asymptote ($\log_a x \rightarrow -\infty$ as $x \rightarrow 0^+$)
- Continuous
- Reflection of graph of $y = a^x$ about the line $y = x$

Remember that the domain of $y = \log_a x$ is the range of $y = a^x$ and the range of $y = \log_a x$ is the domain of $y = a^x$.

Example 6 Sketching the Graphs of Logarithmic Functions

The graph of each function below is similar to the graph of $f(x) = \log_{10} x$.

a. Because $g(x) = \log_{10}(x - 1) = f(x - 1)$, the graph of g can be obtained by shifting the graph of f one unit to the *right*. The domain of g is $(1, \infty)$ because $\log_{10}(x - 1)$ is defined only when $x - 1 > 0$. See Figure 4.17(a).

b. Because $h(x) = 2 + \log_{10} x = 2 + f(x)$, the graph of h can be obtained by shifting the graph of f two units *upward*. The domain of h is $(0, \infty)$ because $2 + \log_{10} x$ is defined only when $x > 0$. See Figure 4.17(b).

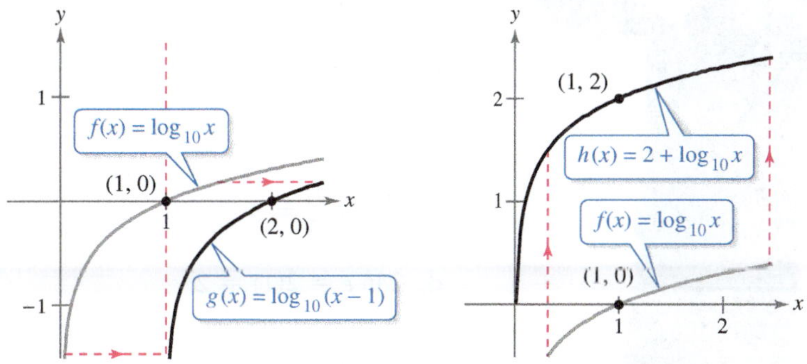

(a) Right shift of one unit
Domain of g: $(1, \infty)$

(b) Upward shift of two units
Domain of h: $(0, \infty)$

FIGURE 4.17

STUDY TIP

Notice in Example 6(a) that shifting the graph of $f(x)$ one unit to the right shifts the vertical asymptote from the y-axis ($x = 0$) to the line $x = 1$. It also shifts the x-intercept from $(1, 0)$ to $(2, 0)$.

✓ Checkpoint 6

Sketch the graph of $f(x) = \log_{10}(x + 3)$.

The Natural Logarithmic Function

By looking back at the graph of the natural exponential function introduced in Section 4.2, you will see that the graph of $f(x) = e^x$ passes the Horizontal Line Test. So, it has an inverse function. This inverse function is called the **natural logarithmic function** and is denoted by the special symbol ln x, read as "el en of x."

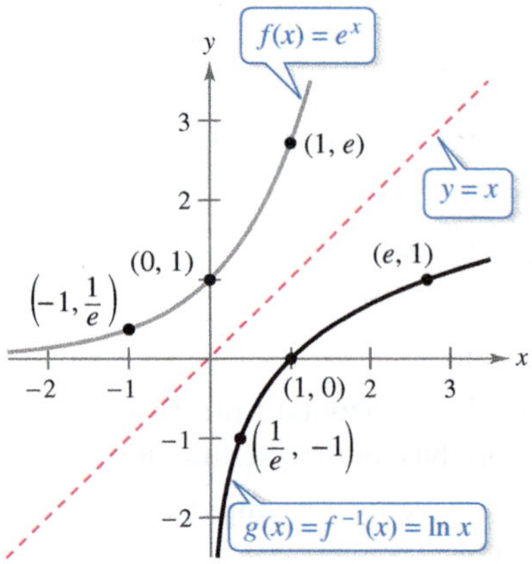

Inverse Functions

FIGURE 4.18

> **The Natural Logarithmic Function**
>
> The function defined by
>
> $$f(x) = \log_e x = \ln x, \quad x > 0$$
>
> is called the **natural logarithmic function.**

Because the functions given by

$$f(x) = e^x \quad \text{and} \quad g(x) = \ln x \qquad \text{Inverse functions}$$

are inverse functions of each other, their graphs are reflections of each other in the line $y = x$. This reflective property is illustrated in Figure 4.18. The four properties of logarithms listed on page 359 are also valid for natural logarithms.

> **Properties of Natural Logarithms**
>
> **1.** $\ln 1 = 0$ because $e^0 = 1$.
>
> **2.** $\ln e = 1$ because $e^1 = e$.
>
> **3.** $\ln e^x = x$ and $e^{\ln x} = x$ — Inverse Properties
>
> **4.** If $\ln x = \ln y$, then $x = y$. — One-to-One Property

Example 7 Using Properties of Natural Logarithms

Evaluate each logarithmic expression.

a. $\ln \frac{1}{e}$ **b.** $e^{\ln 5}$ **c.** $\frac{\ln 1}{3}$ **d.** $2 \ln e$

SOLUTION

a. $\ln \frac{1}{e} = \ln e^{-1} = -1$ — Inverse Property

b. $e^{\ln 5} = 5$ — Inverse Property

c. $\frac{\ln 1}{3} = \frac{0}{3} = 0$ — Property 1

d. $2 \ln e = 2(1) = 2$ — Property 2

✓ Checkpoint 7

Evaluate each logarithmic expression.

a. $\ln e^7$

b. $5 \ln 1$

c. $\frac{3}{4} \ln e$

Benjamin Thorn/www.shutterstock.com

On most calculators, the natural logarithm is denoted by [LN], as illustrated in Example 8.

Example 8 Evaluating the Natural Logarithmic Function

Use a calculator to evaluate each expression.

a. $\ln 2$ **b.** $\ln 0.3$ **c.** $\ln e^2$ **d.** $\ln(-1)$

SOLUTION

Scientific Calculator

Number	Keystrokes	Display
a. $\ln 2$	2 [LN]	0.69314718
b. $\ln 0.3$	.3 [LN]	−1.203972804
c. $\ln e^2$	2 [2nd] [e^x] [LN]	2
d. $\ln(-1)$	1 [+/−] [LN]	ERROR

Graphing Calculator

Number	Keystrokes	Display
a. $\ln 2$	[LN] 2 [)] [ENTER]	.6931471806
b. $\ln 0.3$	[LN] .3 [)] [ENTER]	−1.203972804
c. $\ln e^2$	[LN] [2nd] [e^x] 2 [)] [)] [ENTER]	2
d. $\ln(-1)$	[LN] [(−)] 1 [)] [ENTER]	ERROR

Checkpoint 8

Use a calculator to evaluate the expression $\ln 0.1$. Round your result to three decimal places.

In Example 8, note that $\ln(-1)$ gives an error message on most calculators. This occurs because the domain of $\ln x$ is the set of *positive real numbers* (see Figure 4.18). So, $\ln(-1)$ is undefined.

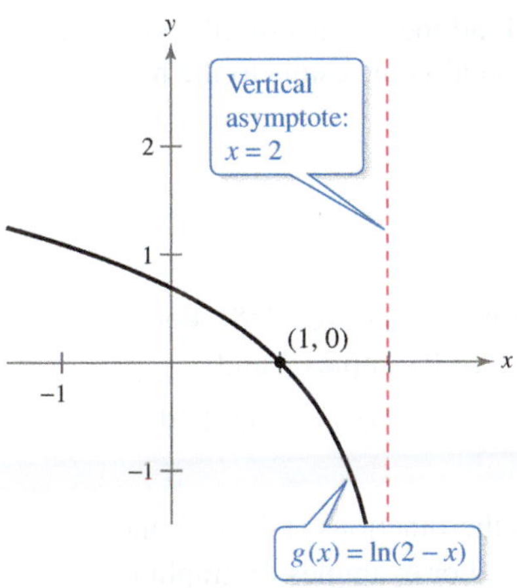

Domain of g: $(-\infty, 2)$
FIGURE 4.19

Example 9 Finding the Domains of Natural Logarithmic Functions

Find the domain of each function.

a. $f(x) = \ln(x - 2)$ **b.** $g(x) = \ln(2 - x)$ **c.** $h(x) = \ln x^2$

SOLUTION

a. Because $\ln(x - 2)$ is defined only when $x - 2 > 0$, it follows that the domain of f is $(2, \infty)$.

b. Because $\ln(2 - x)$ is defined only when $2 - x > 0$, it follows that the domain of g is $(-\infty, 2)$. The graph of g is shown in Figure 4.19.

c. Because $\ln x^2$ is defined only when $x^2 > 0$, it follows that the domain of h is all real numbers except $x = 0$.

Checkpoint 9

Find the domain of the function given by $f(x) = \ln(x + 5)$.

Application

Example 10 Human Memory Model

A group of students participating in a psychological experiment attended several lectures on a subject. Every month for a year after that, the students were tested to see how much of the material they remembered. The average scores $f(t)$ for the group are given by the *human memory model*

$$f(t) = 75 - 6\ln(t + 1), \quad 0 \le t \le 12$$

where t is the time (in months). Based on the results of the experiment, how many months can a student wait before retaking the exam and still expect to score 60 or better? (Do not count portions of months.)

SOLUTION To determine how many months a student can wait before retaking the exam and still expect to score 60 or better, use the model to create a table of values showing the scores for several months.

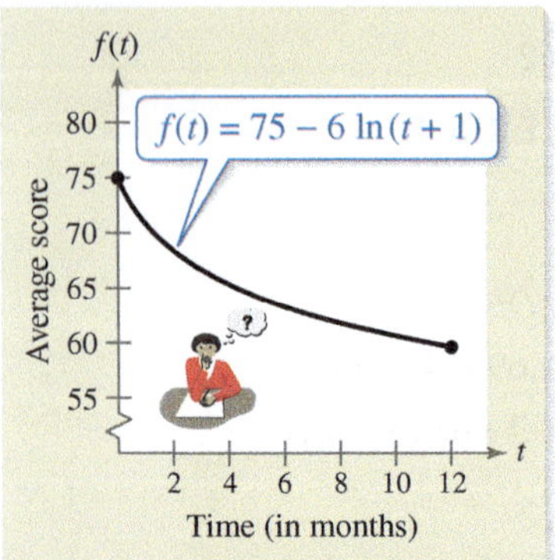

FIGURE 4.20

Month, t	0	1	2	3	4	5	6
Score, $f(t)$	75	70.84	68.41	66.68	65.34	64.25	63.32

Month, t	7	8	9	10	11	12
Score, $f(t)$	62.52	61.82	61.18	60.61	60.09	59.61

From the table, you can see that a student would need to retake the exam by the 11th month in order to score 60 or better. The graph of f is shown in Figure 4.20.

✓ Checkpoint 10

Biologists have found that an alligator's length l (in inches) can be approximated by the model

$$l = 27.1 \ln w - 32.8$$

where w is the weight (in pounds) of the alligator. Find the lengths of alligators for which $w = 150, 225, 380, 450,$ and 625 pounds. Round your results to the nearest tenth of an inch.

SUMMARIZE (Section 4.3)

1. State the definition of a logarithmic function with base a *(page 358)*. For examples of evaluating logarithmic expressions, see Examples 1 and 2.
2. Make a list of the properties of logarithms *(page 359)*. For an example that uses properties of logarithms, see Example 3.
3. Describe the relationship between the graphs of the functions $f(x) = a^x$ and $g(x) = \log_a x$, where $a > 1$ *(pages 360 and 361)*. For examples of graphing exponential and logarithmic functions, see Examples 4, 5, and 6.
4. State the definition of the natural logarithmic function and make a list of the properties of natural logarithms *(page 362)*. For examples of evaluating natural logarithms and using the properties of natural logarithms, see Examples 7 and 8.

SKILLS WARM UP 4.3 The following warm-up exercises involve skills that were covered in earlier sections. You will use these skills in the exercise set for this section. For additional help, review Sections 2.6 and 4.2.

In Exercises 1–4, determine the value of x.

1. $2^x = 8$ **2.** $4^x = 1$ **3.** $10^x = 0.1$ **4.** $e^x = e$

In Exercises 5 and 6, evaluate the expression. (Round the result to three decimal places.)

5. e^2 **6.** e^{-1}

In Exercises 7–10, describe how the graph of g is related to the graph of f.

7. $g(x) = f(x + 2)$ **8.** $g(x) = -f(x)$ **9.** $g(x) = -1 + f(x)$ **10.** $g(x) = f(-x)$

Exercises 4.3

See www.CalcChat.com for worked-out solutions to odd-numbered exercises.

Logarithmic Equations in Exponential Form In Exercises 1–6, match the logarithmic equation with its exponential form. [The exponential forms are labeled (a), (b), (c), (d), (e), and (f).]

1. $\log_4 16 = 2$ (a) $4^{1/2} = 2$

2. $\log_2 16 = 4$ (b) $2^{-4} = \frac{1}{16}$

3. $\log_2 \frac{1}{16} = -4$ (c) $4^2 = 16$

4. $\log_4 \frac{1}{16} = -2$ (d) $4^{-2} = \frac{1}{16}$

5. $\log_4 2 = \frac{1}{2}$ (e) $16^{1/2} = 4$

6. $\log_{16} 4 = \frac{1}{2}$ (f) $2^4 = 16$

Writing Logarithmic Equations In Exercises 7–12, write the exponential equation in logarithmic form. For example, the logarithmic form of $2^3 = 8$ is $\log_2 8 = 3$.

7. $4^4 = 256$ **8.** $7^3 = 343$

9. $81^{1/4} = 3$ **10.** $9^{3/2} = 27$

11. $6^{-2} = \frac{1}{36}$ **12.** $10^{-3} = 0.001$

Writing Exponential Equations In Exercises 13–20, write the logarithmic equation in exponential form. For example, the exponential form of $\log_5 125 = 3$ is $5^3 = 125$.

13. $\log_4 16 = 2$ **14.** $\log_{10} 1000 = 3$

15. $\log_2 \frac{1}{2} = -1$ **16.** $\log_3 \frac{1}{9} = -2$

17. $\log_5 0.2 = -1$ **18.** $\log_{10} 0.1 = -1$

19. $\log_{27} 3 = \frac{1}{3}$ **20.** $\log_8 2 = \frac{1}{3}$

Evaluating Logarithmic Expressions In Exercises 21–30, evaluate the expression without using a calculator. *See Example 1.*

21. $\log_3 9$ **22.** $\log_5 125$

23. $\log_2 \frac{1}{16}$ **24.** $\log_6 \frac{1}{36}$

25. $\log_8 2$ **26.** $\log_{64} 4$

27. $\log_7 7$ **28.** $\log_{12} 1$

29. $\log_{10} 0.0001$ **30.** $\log_{10} 100$

Evaluating Logarithmic Expressions on a Calculator In Exercises 31–36, use a calculator to evaluate the logarithm. Round your result to three decimal places. *See Example 2.*

31. $\log_{10} 345$ **32.** $\log_{10} 163$

33. $\log_{10} \frac{4}{5}$ **34.** $\log_{10} \frac{3}{4}$

35. $\log_{10} \sqrt{8}$ **36.** $\log_{10} \sqrt{3}$

Using Properties of Logarithms In Exercises 37–44, solve the equation for x. *See Example 3.*

37. $\log_5 5 = x$ **38.** $\log_6 6^2 = x$

39. $\log_2 2^{-1} = x$ **40.** $\log_4 4^3 = x$

41. $\log_2(x + 1) = \log_2 4$ **42.** $\log_2(x - 3) = \log_2 9$

43. $\log(2x + 1) = \log 15$ **44.** $\log(5x + 3) = \log 12$

Graphs of Exponential and Logarithmic Functions In Exercises 45–48, sketch the graphs of f and g in the same coordinate plane. *See Example 4.*

45. $f(x) = 7^x,\ g(x) = \log_7 x$

46. $f(x) = 5^x,\ g(x) = \log_5 x$

47. $f(x) = 6^x,\ g(x) = \log_6 x$

48. $f(x) = 10^x,\ g(x) = \log_{10} x$

Sketching the Graphs of Logarithmic Functions In Exercises 49–54, find the domain, vertical asymptote, and x-intercept of the logarithmic function. Then sketch its graph. *See Examples 5 and 6.*

49. $f(x) = \log_2 x$ **50.** $g(x) = \log_4 x$

51. $h(x) = \log_2(x + 4)$ **52.** $f(x) = \log_4(x - 3)$

53. $f(x) = -\log_2 x$ **54.** $h(x) = -\log_4(x - 1)$

Writing Logarithmic Equations In Exercises 55–58, write the exponential equation in logarithmic form.

55. $e^1 = e$ **56.** $e^4 = 54.5981\ldots$

57. $e^x = 4$ **58.** $e^{-x} = 2$

Writing Exponential Equations In Exercises 59–62, write the logarithmic equation in exponential form.

59. $\ln e = 1$ **60.** $\ln \frac{1}{e} = -1$

61. $\ln 1 = 0$ **62.** $\ln \frac{1}{2} = -0.693\ldots$

Using Properties of Natural Logarithms In Exercises 63–66, evaluate the expression without using a calculator. *See Example 7.*

63. $\ln e$ **64.** $\ln e^{10}$

65. $\ln e^{-4}$ **66.** $\ln \frac{1}{e^3}$

Evaluating the Natural Logarithmic Function In Exercises 67–72, use a calculator to evaluate the logarithm. Round your result to three decimal places. *See Example 8.*

67. $\ln 7$ **68.** $2 \ln 9$

69. $\ln 18.42$ **70.** $\ln 36.7$

71. $\ln \sqrt{6}$ **72.** $\ln \sqrt{10}$

Using Properties of Logarithms In Exercises 73–76, use the One-to-One Property to solve the equation for x.

73. $\ln(x + 2) = \ln 6$ **74.** $\ln(x - 4) = \ln 2$

75. $\ln(x^2 - 2) = \ln 23$ **76.** $\ln(x^2 - x) = \ln 6$

Graphs of Exponential and Logarithmic Functions In Exercises 77 and 78, sketch the graphs of f and g in the same coordinate plane.

77. $f(x) = e^x,\ g(x) = \ln x$ **78.** $f(x) = e^{2x},\ g(x) = \frac{1}{2} \ln x$

Finding the Domains of Natural Logarithmic Functions In Exercises 79–82, find the domain, vertical asymptote, and x-intercept of the logarithmic function. Then sketch its graph. *See Example 9.*

79. $g(x) = \ln(-x)$ **80.** $f(x) = \ln(3 - x)$

81. $h(x) = \ln(x + 1)$ **82.** $f(x) = 3 + \ln x$

Matching In Exercises 83–86, match the function with its graph. [The graphs are labeled (a), (b), (c), and (d).]

(a)

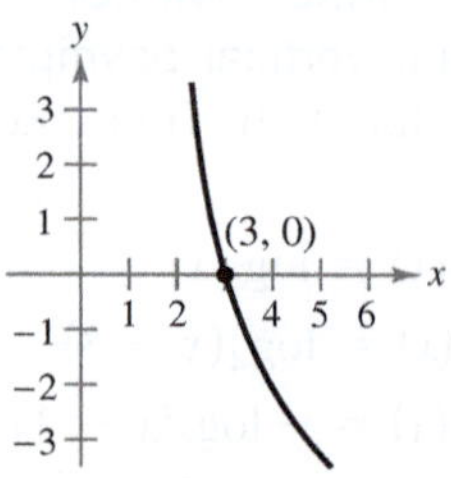

(b)

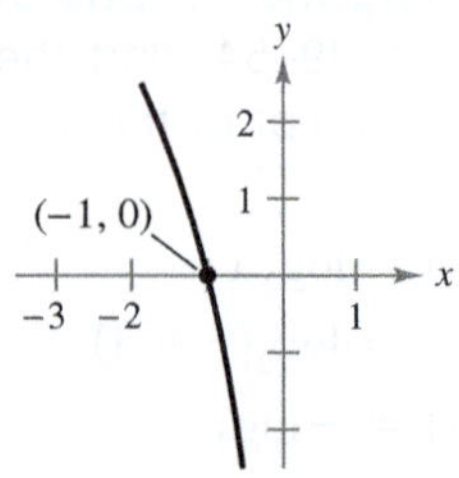

(c)

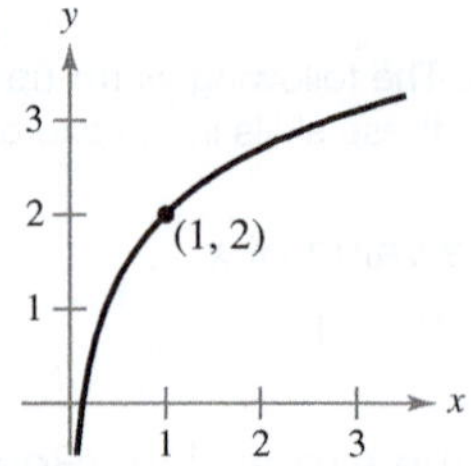

(d)

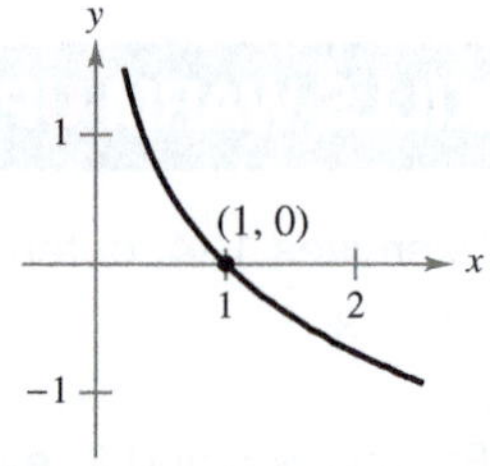

83. $f(x) = \ln x + 2$ **84.** $f(x) = -\ln x$

85. $f(x) = -3 \ln(x - 2)$ **86.** $f(x) = 4 \ln(-x)$

Graphing Logarithmic Functions In Exercises 87–92, use a graphing utility to graph the function. Be sure to use an appropriate viewing window.

87. $f(x) = \log(x + 1)$ **88.** $f(x) = \log(x - 1)$

89. $f(x) = \ln(x - 1)$ **90.** $f(x) = \ln(x + 2)$

91. $f(x) = \ln x + 1$ **92.** $f(x) = 3 \ln x - 1$

93. Population Growth The population of a town will double in

$$t = \frac{8 \ln 3}{\ln 63 - \ln 45} \text{ years.}$$

Find t.

94. Work The work W (in foot-pounds) done in compressing a volume of 9 cubic feet at a pressure of 15 pounds per square inch to a volume of 3 cubic feet is $W = 19{,}440(\ln 9 - \ln 3)$. Find W.

Skill Retention Model In Exercises 95 and 96, participants in an industrial psychology study were taught a simple mechanical task and tested monthly on this mechanical task for a period of 1 year. The average scores for the participants are given by the model $f(t) = 98 - 14 \log_{10}(t + 1)$, for $0 \le t \le 12$, where t is the time (in months).

95. Use a graphing utility to graph the function. Use the graph to discuss the domain and range of the function.

96. Think About It Based on the graph of f, do you think the study's participants practiced the simple mechanical task very often? Cite the behavior of the graph to justify your answer.

97. Human Memory Model Students in a seventh-grade class were given an exam. During the next 2 years, the same students were retested several times. The average score g can be approximated by the model

$$g(t) = 87 - 16 \log_{10}(t + 1), \quad 0 \le t \le 24$$

where t is the time (in months).

(a) What was the average score on the original exam?

(b) What was the average score after 6 months?

(c) When did the average score drop below 70?

98. **HOW DO YOU SEE IT?** The figure shows the graphs of $f(x) = 3^x$ and $g(x) = \log_3 x$. [The graphs are labeled m and n.]

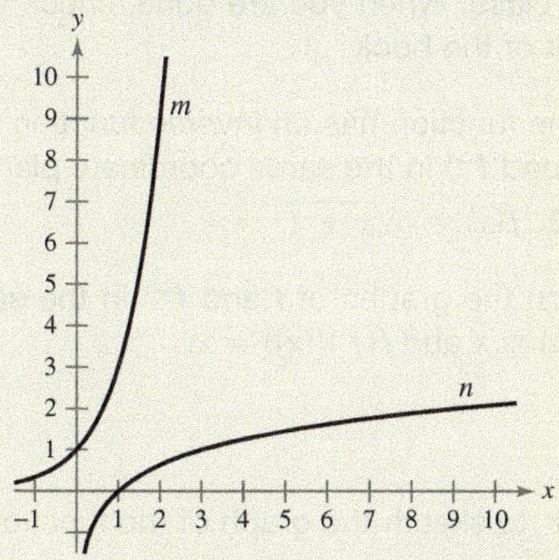

(a) Match each function with its graph.

(b) Given that $f(a) = b$, what is $g(b)$? Explain.

99. Monthly Payment The length t (in years) of a home mortgage of \$150,000 at 8% interest can be approximated by the model

$$t = 12.542 \ln\left(\frac{x}{x - 1000}\right), \quad x > 1000$$

where x represents the monthly payment (in dollars) (see figure).

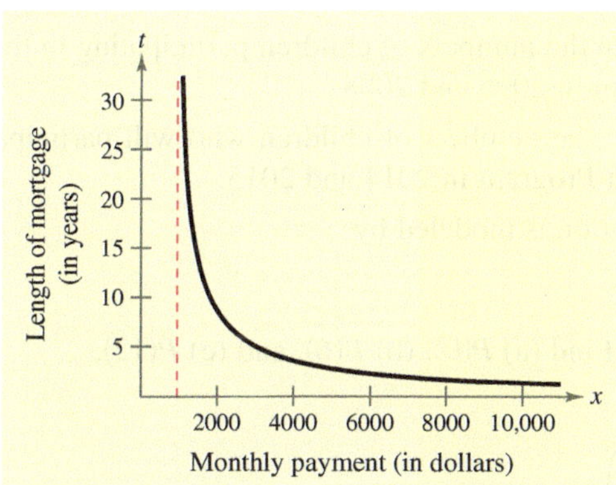

(a) Use the model to approximate the length of a \$150,000 mortgage at 8% interest when the monthly payment is \$1100.65 and when the monthly payment is \$1254.68.

(b) Approximate the total amount paid over the term of the mortgage with a monthly payment of \$1100.65 and with a monthly payment of \$1254.68.

(c) Approximate the total interest charge for a monthly payment of \$1100.65 and for a monthly payment of \$1254.68.

(d) What is the vertical asymptote of the model? Interpret its meaning in the context of the problem.

100. Median Age of U.S. Population The median age A of the U.S. population for each year from 1980 to 2050 can be approximated by the model

$$A = 16.15 - 0.036t + 5.956 \ln t, \quad 10 \le t \le 80$$

where t represents the year, with $t = 10$ corresponding to 1980 (see figure). *(Source: U.S. Census Bureau)*

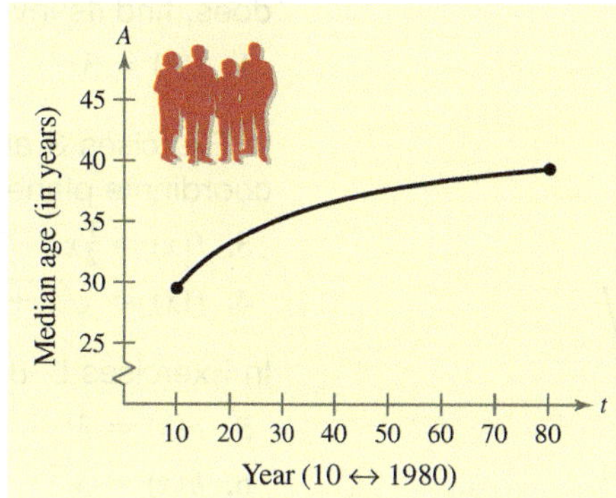

(a) Use the model to estimate the median age of the U.S. population in 1980.

(b) Use the model to estimate the median age of the U.S. population in 1990.

(c) Use the model to estimate the change in the median age of the U.S. population from 1980 to 2050.

(d) Use the model to project the change in the median age of the U.S. population from 1980 to 2050.

101. Investment Time A principal P, invested at 5.25% interest compounded continuously, increases to an amount that is K times the principal after t years, where t is given by

$$t = \frac{\ln K}{0.0525}.$$

(a) Complete the table.

K	1	2	4	6	8	10	12
t							

(b) Use the table in part (a) to graph the function.

102. Think About It The table of values was obtained by evaluating a function. Determine which of the statements may be true and which must be false.

x	1	2	8
y	0	1	3

(a) y is an exponential function of x.

(b) y is a logarithmic function of x.

(c) x is an exponential function of y.

(d) y is a linear function of x.

QUIZ YOURSELF

See www.CalcChat.com for worked-out solutions to odd-numbered exercises.

Take this quiz as you would take a quiz in class. When you are done, check your work against the answers given in the back of the book.

In Exercises 1 and 2, determine whether the function has an inverse function. If it does, find its inverse function and graph f and f^{-1} in the same coordinate plane.

1. $f(x) = 3x^2$ **2.** $f(x) = \sqrt[3]{x + 1}$

In Exercises 3 and 4, (a) find f^{-1}, (b) sketch the graphs of f and f^{-1} in the same coordinate plane, and (c) verify that $f^{-1}(f(x)) = x$ and $f(f^{-1}(x)) = x$.

3. $f(x) = \frac{1}{2}x - 3$

4. $f(x) = \sqrt{x + 1}$

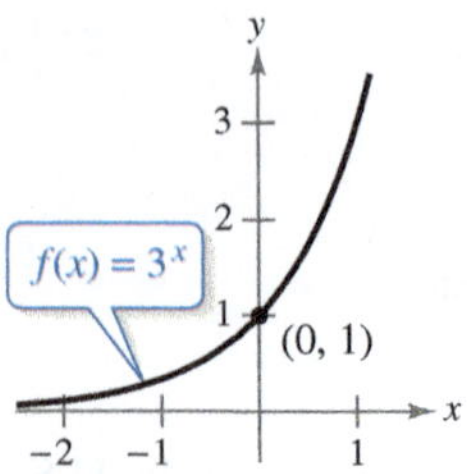

Figure for 5–8

In Exercises 5–8, use the graph of $f(x) = 3^x$ to sketch the graph of the function.

5. $g(x) = 3^x - 2$

6. $h(x) = 3^{-x}$

7. $k(x) = \log_3 x$

8. $j(x) = \log_3(x - 1)$

9. For $P = \$10{,}000$, $r = 5.15\%$, and $t = 8$ years, find the balance in an account when interest is compounded (a) monthly and (b) continuously.

10. The numbers of children C (in millions) participating in the Federal School Breakfast Program from 2000 through 2009 can be approximated by the model

$C = 7.50(1.04)^t, \quad 0 \le t \le 9$

where t represents the year, with $t = 0$ corresponding to 2000. *(Source: U.S. Department of Agriculture)*

(a) Use the model to estimate the numbers of children participating in the Federal School Breakfast Program in 2006 and 2008.

(b) Use the model to predict the numbers of children who will participate in the Federal School Breakfast Program in 2014 and 2015.

11. The size of a bacteria population is modeled by

$P(t) = 100e^{0.2154t}$

where t is the time in hours. Find (a) $P(0)$, (b) $P(6)$, and (c) $P(12)$.

12. Use the demand function

$$p = 4000\left(1 - \frac{8}{8 + e^{-0.003x}}\right)$$

to find the price for a demand of $x = 500$ MP3 players.

In Exercises 13–16, evaluate the expression without using a calculator.

13. $\log_{10} 100$

14. $\ln e^4$

15. $\log_4 \frac{1}{16}$

16. $\ln 1$

17. Sketch the graphs of $f(x) = 3^x$ and $g(x) = \log_3 x$ in the same coordinate plane. Identify the domains of f and g. Discuss the special relationship between f and g that is shown by their graphs.

4.4 Properties of Logarithms

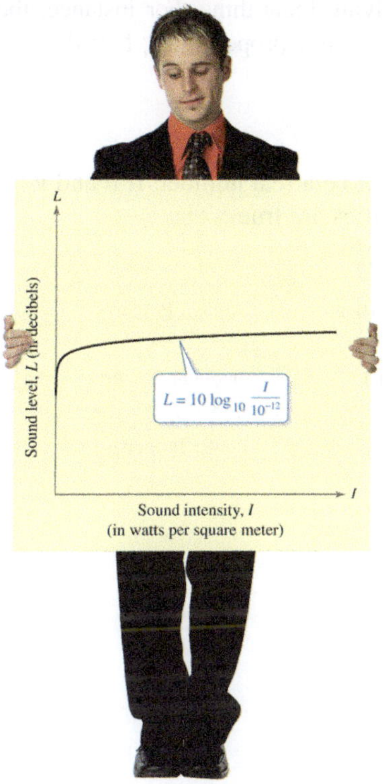

In Exercise 107 on page 376, you will use properties of logarithms to determine the difference in loudness between two sounds.

- Evaluate a logarithm using the change-of-base formula.
- Use properties of logarithms to evaluate or rewrite a logarithmic expression.
- Use properties of logarithms to expand or condense a logarithmic expression.
- Use logarithmic functions to model and solve real-life applications.

Change of Base

Most calculators have only two types of log keys, one for common logarithms (base 10) and one for natural logarithms (base e). In some situations, you must use logarithms with other bases. To evaluate logarithms with other bases, you can use the **change-of-base formula.**

Change-of-Base Formula

Let a, b, and x be positive real numbers such that $a \neq 1$ and $b \neq 1$. Then $\log_a x$ can be converted to a different base as shown.

Base b	*Base 10*	*Base e*
$\log_a x = \dfrac{\log_b x}{\log_b a}$	$\log_a x = \dfrac{\log_{10} x}{\log_{10} a}$	$\log_a x = \dfrac{\ln x}{\ln a}$

One way to look at the change-of-base formula is that logarithms to base a are simply *constant multiples* of logarithms to base b. The constant multiplier is

$$\frac{1}{\log_b a}.$$

Example 1 **Changing Bases Using Common Logarithms**

a. $\log_4 30 = \dfrac{\log_{10} 30}{\log_{10} 4} \approx \dfrac{1.47712}{0.60206} \approx 2.4534$

b. $\log_2 14 = \dfrac{\log_{10} 14}{\log_{10} 2} \approx \dfrac{1.14613}{0.30103} \approx 3.8074$

Example 2 **Changing Bases Using Natural Logarithms**

a. $\log_4 30 = \dfrac{\ln 30}{\ln 4} \approx \dfrac{3.40120}{1.386294} \approx 2.4534$

b. $\log_2 14 = \dfrac{\ln 14}{\ln 2} \approx \dfrac{2.63906}{0.693147} \approx 3.8074$

✓ Checkpoints 1 and 2

Evaluate $\log_8 56$ using common logarithms and natural logarithms. Round your results to three decimal places.

Notice in Examples 1 and 2 that the result is the same whether common logarithms or natural logarithms are used in the change-of-base formula.

Properties of Logarithms

You know from the preceding section that the logarithmic function with base a is the *inverse function* of the exponential function with base a. So, it makes sense that the properties of exponents should have corresponding properties involving logarithms. For instance, the exponential property $a^0 = 1$ has the corresponding logarithmic property $\log_a 1 = 0$.

STUDY TIP

There is no general property that can be used to rewrite $\log_a(u \pm v)$. Specifically, $\log_a(x + y)$ is *not* equal to $\log_a x + \log_a y$.

Properties of Logarithms

Let a be a positive number such that $a \neq 1$, and let n be a real number. If u and v are positive real numbers, then the following properties are true.

Logarithm with Base a	*Natural Logarithm*	
$\log_a(uv) = \log_a u + \log_a v$	$\ln(uv) = \ln u + \ln v$	Product Property
$\log_a \dfrac{u}{v} = \log_a u - \log_a v$	$\ln \dfrac{u}{v} = \ln u - \ln v$	Quotient Property
$\log_a u^n = n \log_a u$	$\ln u^n = n \ln u$	Power Property

Example 3 Using Properties of Logarithms

Write each logarithm in terms of ln 2 and ln 3.

a. $\ln 6$ **b.** $\ln \dfrac{2}{27}$

SOLUTION

a. $\ln 6 = \ln(2 \cdot 3)$ Rewrite 6 as $2 \cdot 3$.

$= \ln 2 + \ln 3$ Product Property

b. $\ln \dfrac{2}{27} = \ln 2 - \ln 27$ Quotient Property

$= \ln 2 - \ln 3^3$ Rewrite 27 as 3^3.

$= \ln 2 - 3 \ln 3$ Power Property

✓ Checkpoint 3

Write $\log_{10} \frac{25}{3}$ in terms of $\log_{10} 3$ and $\log_{10} 5$.

Example 4 Using Properties of Logarithms

Use the properties of logarithms to verify that $-\log_{10} \frac{1}{100} = \log_{10} 100$.

SOLUTION

$-\log_{10} \dfrac{1}{100} = -\log_{10}(100^{-1})$ Definition of negative exponent

$= -(-1) \log_{10} 100$ Power Property

$= \log_{10} 100$ Simplify.

✓ Checkpoint 4

Use the properties of logarithms to verify that $-\ln \dfrac{2}{e} = 1 - \ln 2$.

Rewriting Logarithmic Expressions

The properties of logarithms are useful for rewriting logarithmic expressions in forms that simplify the operations of algebra. This is true because these properties convert complicated products, quotients, and exponential forms into simpler sums, differences, and products, respectively.

Example 5 **Expanding Logarithmic Expressions**

Expand each logarithmic expression.

a. $\log_4 5x^3y$ **b.** $\ln \dfrac{\sqrt{3x-5}}{7}$

SOLUTION

a. $\log_4 5x^3y = \log_4 5 + \log_4 x^3 + \log_4 y$ Product Property

$= \log_4 5 + 3\log_4 x + \log_4 y$ Power Property

b. $\ln \dfrac{\sqrt{3x-5}}{7} = \ln \dfrac{(3x-5)^{1/2}}{7}$ Rewrite using rational exponent.

$= \ln(3x-5)^{1/2} - \ln 7$ Quotient Property

$= \dfrac{1}{2}\ln(3x-5) - \ln 7$ Power Property

STUDY TIP

Be careful when rewriting the logarithmic expression $\log ax^n$. It should be rewritten as

$\log a + n \log x$

instead of $n \log ax$.

Checkpoint 5

Expand the expression $\ln 2mn^2$.

Example 6 **Condensing Logarithmic Expressions**

Condense each logarithmic expression.

a. $\frac{1}{2}\log_{10} x + 3\log_{10}(x+1)$ **b.** $2\ln(x+2) - \ln x$ **c.** $\frac{1}{3}[\log_2 x + \log_2(x+1)]$

SOLUTION

a. $\frac{1}{2}\log_{10} x + 3\log_{10}(x+1) = \log_{10}x^{1/2} + \log_{10}(x+1)^3$ Power Property

$= \log_{10}\left[\sqrt{x}(x+1)^3\right]$ Product Property

b. $2\ln(x+2) - \ln x = \ln(x+2)^2 - \ln x$ Power Property

$= \ln \dfrac{(x+2)^2}{x}$ Quotient Property

c. $\frac{1}{3}[\log_2 x + \log_2(x+1)] = \frac{1}{3}\{\log_2[x(x+1)]\}$ Product Property

$= \log_2[x(x+1)]^{1/3}$ Power Property

$= \log_2 \sqrt[3]{x(x+1)}$ Rewrite with a radical.

STUDY TIP

Be careful when rewriting the logarithmic expression $\log x - \log y$. It should be rewritten as

$\log \dfrac{x}{y}$

instead of

$\dfrac{\log x}{\log y}$.

Checkpoint 6

Condense the expression $2\log_{10}(x+1) - 3\log_{10}(x-1)$.

When applying the properties of logarithms to a logarithmic function, be careful to check the domain of the function. For example, the domain of $f(x) = \ln x^2$ is all real $x \neq 0$, whereas the domain of $g(x) = 2\ln x$ is all real $x > 0$.

Applications

One method of determining how the x- and y-values of a set of nonlinear data are related begins by taking the natural logarithm of each of the x- and y-values. If you graph the points $(\ln x, \ln y)$ and they fall in a straight line, then you can determine that the x- and y-values are related by the equation

$$\ln y = m \ln x$$

where m is the slope of the straight line.

Example 7 Finding a Mathematical Model

The table shows the mean distance from the sun x and the period (the time it takes a planet to orbit the sun) y for each of the six planets that are closest to the sun. In the table, the mean distance is given in astronomical units (where Earth's mean distance is defined as 1.0), and the period is given in years. Find an equation that relates y and x.

Planet	Mean distance, x	Period, y
Mercury	0.387	0.241
Venus	0.723	0.615
Earth	1.000	1.000
Mars	1.524	1.881
Jupiter	5.203	11.863
Saturn	9.537	29.447

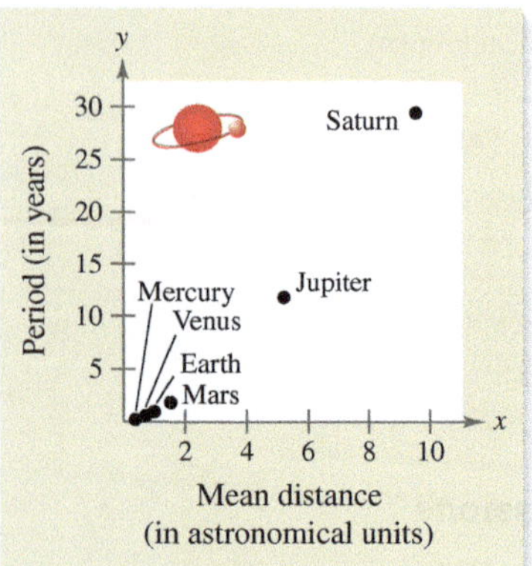

FIGURE 4.21

SOLUTION The points in the table are plotted in Figure 4.21. From this figure, it is not clear how to find an equation that relates y and x. To solve this problem, take the natural logarithm of each of the x- and y-values in the table. This produces the following results.

Planet	Mercury	Venus	Earth	Mars	Jupiter	Saturn
$\ln x$	−0.949	−0.324	0.000	0.421	1.649	2.255
$\ln y$	−1.423	−0.486	0.000	0.632	2.473	3.383

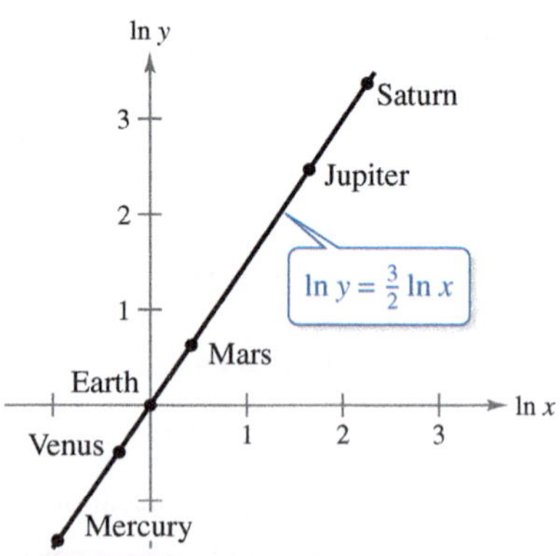

FIGURE 4.22

Now, by plotting the points in the second table, you can see that all six of the points appear to lie in a line (see Figure 4.22). Choose any two points to determine the slope of the line. Using the two points (0.421, 0.632) and (0, 0), you can determine that the slope of the line is

$$m = \frac{0.632 - 0}{0.421 - 0} \approx 1.5 = \frac{3}{2}.$$

By the point-slope form, the equation of the line is $Y = \frac{3}{2}X$, where $Y = \ln y$ and $X = \ln x$. You can therefore conclude that $\ln y = \frac{3}{2} \ln x$.

✓ Checkpoint 7

Find a logarithmic equation that relates y and x.

x	1	2	3	4
y	1	1.414	1.732	2

Example 8 Sound Intensity

The level of sound L (in decibels) with an intensity of I (in watts per square meter) is given by

$$L = 10 \log_{10} \frac{I}{I_0}$$

where I_0 represents the faintest sound that can be heard by the human ear, and is approximately equal to 10^{-12} watt per square meter. You and your roommate are playing your stereos at the same time and at the same intensity. How much louder is the music when both stereos are playing than when just one stereo is playing?

SOLUTION Let L_1 represent the level of sound when one stereo is playing and let L_2 represent the level of sound when both stereos are playing. Using the formula for level of sound, you can express L_1 as

$$L_1 = 10 \log_{10} \frac{I}{10^{-12}}.$$

For L_2, multiply I by 2 as shown below

$$L_2 = 10 \log_{10} \frac{2I}{10^{-12}}$$

because L_2 represents the level of sound when *two* stereos are playing at the same intensity I. To determine the increase in loudness, subtract L_1 from L_2 as shown.

$$\begin{aligned} L_2 - L_1 &= 10 \log_{10} \frac{2I}{10^{-12}} - 10 \log_{10} \frac{I}{10^{-12}} \\ &= 10\left(\log_{10} \frac{2I}{10^{-12}} - \log_{10} \frac{I}{10^{-12}}\right) \\ &= 10\left(\log_{10} 2 + \log_{10} \frac{I}{10^{-12}} - \log_{10} \frac{I}{10^{-12}}\right) \\ &= 10 \log_{10} 2 \approx 3 \end{aligned}$$

So, the music is about 3 decibels louder. Notice that the variable I drops out of the equation when it is simplified. This means that the loudness increases by 3 decibels when both stereos are playing at the same intensity, regardless of the individual intensities of the stereos.

✓ Checkpoint 8

Two sounds have intensities of $I_1 = 10^{-6}$ watt per square meter and $I_2 = 10^{-9}$ watt per square meter. Use the formula for the level of sound in Example 8 to find the difference in loudness between the two sounds.

SUMMARIZE (Section 4.4)

1. State the change-of-base formula *(page 369)*. For examples that use the change-of-base formula, see Examples 1 and 2.
2. Make a list of the properties of logarithms *(page 370)*. For examples that use properties of logarithms, see Examples 3 and 4.
3. Describe how the properties of logarithms are useful for rewriting logarithmic expressions *(page 371)*. For examples of expanding and condensing logarithmic expressions, see Examples 5 and 6.

Kurhan/Shutterstock.com

SKILLS WARM UP 4.4 The following warm-up exercises involve skills that were covered in earlier sections. You will use these skills in the exercise set for this section. For additional help, review Sections 0.3, 0.4, and 4.3.

In Exercises 1–4, evaluate the expression without using a calculator.

1. $\log_7 49$
2. $\log_2 \dfrac{1}{32}$
3. $\ln \dfrac{1}{e^2}$
4. $\log_{10} 0.001$

In Exercises 5–8, simplify the expression.

5. e^2e^3
6. $\dfrac{e^2}{e^3}$
7. $(e^2)^3$
8. $(e^2)^0$

In Exercises 9–12, rewrite the equation in exponential form.

9. $y = \dfrac{1}{x^2}$
10. $y = \sqrt{x}$
11. $\log_4 64 = 3$
12. $\log_{16} 4 = \dfrac{1}{2}$

Exercises 4.4

See www.CalcChat.com for worked-out solutions to odd-numbered exercises.

Changing the Base Using Common Logarithms In Exercises 1–4, evaluate the logarithm using common logarithms. Round your result to three decimal places. *See Example 1.*

1. $\log_5 8$
2. $\log_7 12$
3. $\log_4 30$
4. $\log_9 20$

Changing the Base Using Natural Logarithms In Exercises 5–8, evaluate the logarithm using natural logarithms. Round your result to three decimal places. *See Example 2.*

5. $\log_5 8$
6. $\log_7 12$
7. $\log_6 14$
8. $\log_8 18$

Changing the Base In Exercises 9–18, evaluate the logarithm. Round your result to three decimal places. *See Examples 1 and 2.*

9. $\log_2 6$
10. $\log_8 3$
11. $\log_{27} 35$
12. $\log_{19} 42$
13. $\log_{15} 1250$
14. $\log_{20} 1575$
15. $\log_5 \frac{1}{3}$
16. $\log_9 \frac{3}{5}$
17. $\log_{1/4} 10$
18. $\log_{1/3} 5$

Using Properties of Logarithms In Exercises 19–24, write the logarithm in terms of ln 2 and ln 5. *See Example 3.*

19. $\ln 10$
20. $\ln \frac{2}{5}$
21. $\ln \frac{5}{2}$
22. $\ln 20$
23. $\ln 100$
24. $\ln \frac{125}{16}$

Using Properties of Logarithms In Exercises 25–28, use the properties of logarithms to verify the statement. *See Example 4.*

25. $-\log_{10} \frac{11}{8} = \log_{10} \frac{8}{11}$
26. $-\ln \frac{3}{4} = \ln \frac{4}{3}$
27. $-2 \ln \dfrac{e}{7} = 2 \ln 7 - 2$
28. $-3 \log_{10} \frac{8}{100} = 6 - 3 \log_{10} 8$

Using Properties of Logarithms In Exercises 29–42, approximate the logarithm using the properties of logarithms, given $\log_b 2 \approx 0.3562$, $\log_b 3 \approx 0.5646$, and $\log_b 5 \approx 0.8271$.

29. $\log_b 10$
30. $\log_b 15$
31. $\log_b \frac{2}{3}$
32. $\log_b \frac{3}{5}$
33. $\log_b 8$
34. $\log_b 81$
35. $\log_b \sqrt{2}$
36. $\log_b \sqrt{5}$
37. $\log_b 40$
38. $\log_b 45$
39. $\log_b (2b)^{-2}$
40. $\log_b (3b^2)$
41. $\log_b \sqrt[3]{4b}$
42. $\log_b \sqrt[3]{3b}$

Using Properties of Logarithms In Exercises 43–50, use the properties of logarithms to simplify the logarithmic expression.

43. $\log_9 \frac{1}{18}$
44. $\log_5 \frac{1}{15}$
45. $\log_7 \sqrt{70}$
46. $\log_5 \sqrt{75}$
47. $\log_5 \frac{1}{250}$
48. $\log_{10} \frac{9}{300}$
49. $\ln(5e^6)$
50. $\ln \dfrac{6}{e^2}$

Error Analysis In Exercises 51–54, describe and correct the error.

51. $\log_2 5 = \dfrac{\ln 2}{\ln 5}$

52. $\ln 8x = \ln 8 \cdot \ln x$

53. $\ln 8 - \ln 2 = \dfrac{\ln 8}{\ln 2}$

54. $\log_{10} 5x^4 = 4 \log_{10} 5x$

Expanding a Logarithmic Expression In Exercises 55–76, use the properties of logarithms to expand the expression as a sum, difference, and/or multiple of logarithms. (Assume all variables are positive.) *See Example 5.*

55. $\log_2(4^3 \cdot 3^5)$

56. $\log_3(3^2 \cdot 4^2)$

57. $\log_3 4n$

58. $\log_6 6x$

59. $\log_5 \dfrac{x}{25}$

60. $\log_{10} \dfrac{y}{2}$

61. $\log_2 x^4$

62. $\log_2 z^{-3}$

63. $\ln \sqrt{z}$

64. $\ln \sqrt[3]{t}$

65. $\ln xyz$

66. $\ln \dfrac{xy}{z}$

67. $\ln \sqrt{a-1}, \quad a > 1$

68. $\ln \sqrt[3]{y-2}, \quad y > 2$

69. $\ln\left[\dfrac{(z-1)^2}{z}\right]$

70. $\ln\left(\dfrac{x}{\sqrt{x^2+1}}\right)$

71. $\ln \dfrac{yz}{\sqrt[3]{z+3}}$

72. $\log_9 \dfrac{\sqrt{y}}{7z^2}$

73. $\ln 5\sqrt[3]{\dfrac{x}{y}}$

74. $\ln 3\sqrt{\dfrac{x^2}{y^3}}$

75. $\ln \sqrt[4]{x^3(x^2+3)}$

76. $\ln \sqrt{x^2(x+2)}$

Condensing a Logarithmic Expression In Exercises 77–92, condense the expression to the logarithm of a single quantity. *See Example 6.*

77. $\log_3 x + \log_3 5$

78. $\log_5 y + \log_5 x$

79. $\log_4 8 - \log_4 x$

80. $\log_{10} 4 - \log_{10} z$

81. $2 \log_{10}(x + 4)$

82. $-4 \log_{10} 2x$

83. $-\ln x - 3 \ln 6$

84. $2 \ln 8 + 5 \ln z$

85. $\frac{1}{3} \ln 5x - \ln(x + 1)$

86. $\frac{3}{2} \ln(z - 2) + \ln z$

87. $\log_8(x - 2) - \log_8(x + 2)$

88. $3 \log_7 x + 2 \log_7 y - 4 \log_7 z$

89. $2 \ln 4 + 2 \ln 3 - \frac{1}{2} \ln(x^2 + 1)$

90. $2 \ln(x^2 - 2) + \frac{3}{2} \ln t^6 - \frac{3}{4} \ln t^4$

91. $2 \ln x - [\ln(x + 2) + \ln(x - 2)]$

92. $4 \ln(x + 1) + 2 \ln(x - 1) - 3 \ln x$

93. Graphical Analysis Use a graphing utility to graph

$$f(x) = \ln 5x \quad \text{and} \quad g(x) = \ln 5 + \ln x$$

in the same viewing window. How do the two graphs compare? What property of logarithms do the graphs demonstrate?

94. Think About It Consider the functions below.

$$f(x) = \ln \frac{x}{2}, \quad g(x) = \frac{\ln x}{\ln 2}, \quad h(x) = \ln x - \ln 2$$

Which two functions have identical graphs? Verify your answer by using a graphing utility to graph all three functions in the same viewing window.

95. Graphical Analysis Use a graphing utility to graph

$$f(x) = \log_{10} x^3 \quad \text{and} \quad g(x) = 3 \log_{10} x$$

in the same viewing window. How do the two graphs compare? What property of logarithms is shown?

96. HOW DO YOU SEE IT? The figure shows the graphs of $y = \ln x$, $y = \ln x^2$, $y = \ln 2x$, and $y = \ln 2$. Match each function with its graph. (The graphs are labeled A through D.) Explain your reasoning.

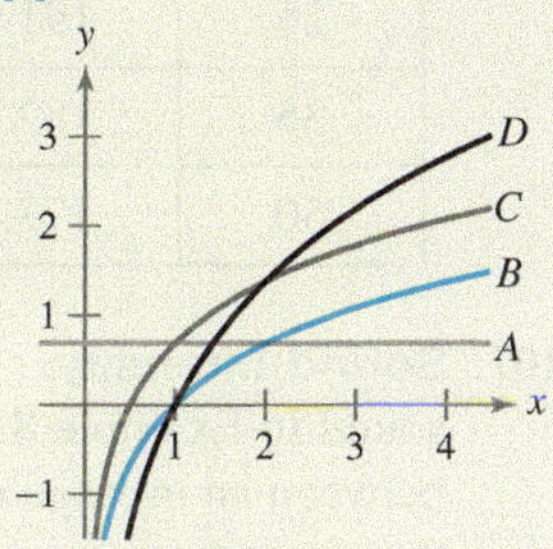

Using Inverse Properties In Exercises 97–102, find the exact value of the logarithmic expression without using a calculator.

97. $\log_4 \sqrt[3]{4}$

98. $\log_8 \sqrt[4]{8}$

99. $\ln \dfrac{1}{\sqrt{e}}$

100. $\ln \sqrt[4]{e^3}$

101. $\log_5 \frac{1}{125}$

102. $\log_7 \frac{49}{343}$

Finding a Mathematical Model In Exercises 103 and 104, find a logarithmic equation that relates y and x. Explain the steps used to find the equation. *See Example 7.*

103.

x	1	2	3	4	5	6
y	1	1.189	1.316	1.414	1.495	1.565

104.

x	1	2	3	4	5	6
y	1	1.587	2.080	2.520	2.924	3.302

105. Nail Length The approximate lengths and diameters (in inches) of common nails are shown in the table. Find a logarithmic equation that relates the diameter y of a common nail to its length x.

Length, x	Diameter, y	Length, x	Diameter, y
1	0.070	4	0.176
2	0.111	5	0.204
3	0.146	6	0.231

106. Galloping Speeds of Animals Four-legged animals run with two different types of motion: trotting and galloping. An animal that is trotting has at least one foot on the ground at all times, whereas an animal that is galloping has all four feet off the ground at some point in its stride. The number of strides per minute at which an animal breaks from a trot to a gallop depends on the weight of the animal. Use the table to find a logarithmic equation that relates an animal's lowest galloping speed y (in strides per minute) to its weight x (in pounds).

Weight, x	Galloping speed, y	Weight, x	Galloping speed, y
25	191.5	75	164.2
35	182.7	500	125.9
50	173.8	1000	114.2

107. Sound Intensity Use the equation for the level of sound in Example 8 to find the difference in loudness between an average office and a broadcast studio with the intensities given below.

Office: 1.26×10^{-7} watt per square meter

Broadcast studio: 3.16×10^{-10} watt per square meter

108. Sound Intensity Use the equation for the level of sound in Example 8 to find the difference in loudness between a bird singing and rustling leaves with the intensities given below.

Bird singing: 10^{-8} watt per square meter

Rustling leaves : 10^{-10} watt per square meter

109. Think About It For each property of logarithms given below, write a related property of exponents.

(a) $\log_a(uv) = \log_a u + \log_a v$

(b) $\log_a \dfrac{u}{v} = \log_a u - \log_a v$

(c) $\log_a u^n = n \log_a u$

110. Proof Complete the proof of the logarithmic property

$\log_a uv = \log_a u + \log_a v.$

Let $\log_a u = x$ and $\log_a v = y$.

$a^x = \square$ and $a^y = \square$ Rewrite in exponential form.

$u \cdot v = \square \cdot \square = a^{\square}$ Multiply and substitute for u and v.

$\square = x + y$ Rewrite in logarithmic form.

$\log_a uv = \square + \square$ Substitute for x and y.

111. Proof Complete the proof of the logarithmic property

$\log_a \dfrac{u}{v} = \log_a u - \log_a v.$

Let $\log_a u = x$ and $\log_a v = y$.

$a^x = \square$ and $a^y = \square$ Rewrite in exponential form.

$\dfrac{u}{v} = \dfrac{\square}{\square} = a^{\square}$ Divide and substitute for u and v.

$\square = x - y$ Rewrite in logarithmic form.

$\log_a \dfrac{u}{v} = \square - \square$ Substitute for x and y.

Courtesy of Maggie's Place

Business Capsule

Co-founded by five recent college graduates, Maggie's Place is a community of homes that provides hospitality for pregnant women who are alone or living on the streets. In homes shared with staff members, expectant mothers have their immediate needs met while being connected to community resources such as prenatal care, education programs, and low-cost housing. Maggie's Place opened its first home, the Magdalene House, on May 13, 2000 in Phoenix, Arizona. Today, there are five homes, and a resource and support center for women and children who have completed the Maggie's Place program.

112. Research Project Use your campus library, the Internet, or some other reference source to find information about a nonprofit group or company whose growth can be modeled by a logarithmic function. Write a brief report about the growth of the group or company.

4.5 Solving Exponential and Logarithmic Equations

- Understand the strategies for solving exponential and logarithmic equations.
- Solve an exponential equation.
- Solve a logarithmic equation.
- Use an exponential or a logarithmic model to solve an application problem.

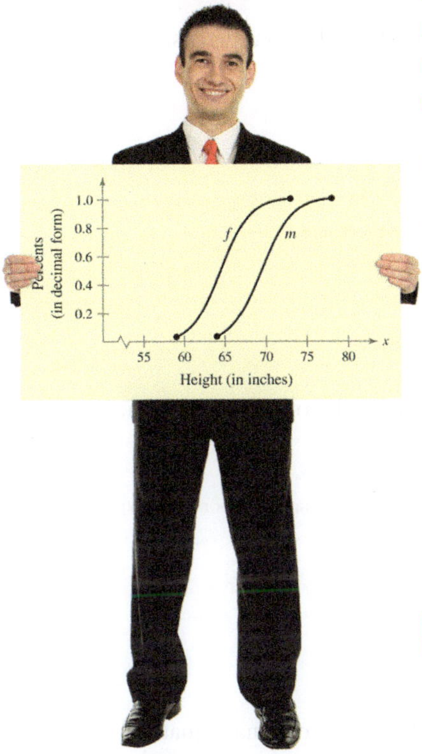

In Exercise 110 on page 386, you will use exponential models to find the median heights of American males and females between the ages of 20 and 29 years old.

Introduction

So far in this chapter, you have studied the definitions, graphs, and properties of exponential and logarithmic functions. In this section, you will study procedures for *solving equations* involving these exponential and logarithmic functions.

There are two basic strategies for solving exponential or logarithmic equations. The first is based on the One-to-One Properties and the second is based on the Inverse Properties. For $a > 0$ and $a \neq 1$, the following properties are true for all x and y for which $\log_a x$ and $\log_a y$ are defined.

One-to-One Properties	*Inverse Properties*
$a^x = a^y$ if and only if $x = y$.	$a^{\log_a x} = x$
$\log_a x = \log_a y$ if and only if $x = y$.	$\log_a a^x = x$

Example 1 Solving Simple Equations

	Original Equation	*Rewritten Equation*	*Solution*	*Property*
a.	$2^x = 32$	$2^x = 2^5$	$x = 5$	One-to-One
b.	$\ln x - \ln 3 = 0$	$\ln x = \ln 3$	$x = 3$	One-to-One
c.	$e^x = 7$	$\ln e^x = \ln 7$	$x = \ln 7$	Inverse
d.	$\ln x = -3$	$e^{\ln x} = e^{-3}$	$x = e^{-3}$	Inverse
e.	$\log_{10} x = -1$	$10^{\log_{10} x} = 10^{-1}$	$x = 10^{-1} = \frac{1}{10}$	Inverse

✓ **Checkpoint 1**

Solve each equation for x.

a. $3^x = 81$

b. $\log_6 x = 3$

c. $\ln 5 - \ln x = 0$

Strategies for Solving Exponential and Logarithmic Equations

1. Rewrite the original equation in a form that allows the use of the One-to-One Property of exponential or logarithmic functions.
2. Rewrite an *exponential* equation in logarithmic form and apply the Inverse Property of logarithmic functions.
3. Rewrite a *logarithmic* equation in exponential form and apply the Inverse Property of exponential functions.

Edyta Pawlowska /Shutterstock.com

Solving Exponential Equations

Example 2 Solving Exponential Equations

Solve each equation and approximate the result to three decimal places.

a. $4^x = 72$ **b.** $3(2^x) = 42$

SOLUTION

a. $4^x = 72$ — Write original equation.

$\log_4 4^x = \log_4 72$ — Take log (base 4) of each side.

$x = \log_4 72$ — Inverse Property

$x = \frac{\ln 72}{\ln 4} \approx 3.085$ — Change-of-base formula

The solution is $x = \log_4 72 \approx 3.085$. Check this in the original equation.

b. $3(2^x) = 42$ — Write original equation.

$2^x = 14$ — Divide each side by 3 to isolate the exponential expression.

$\log_2 2^x = \log_2 14$ — Take log (base 2) of each side.

$x = \log_2 14$ — Inverse Property

$x = \frac{\ln 14}{\ln 2} \approx 3.807$ — Change-of-base formula

The solution is $x = \log_2 14 \approx 3.807$. Check this in the original equation.

TECH TUTOR

When solving an exponential or logarithmic equation, you can check your solution graphically by "graphing the left and right sides separately" and using the *intersect* feature of your graphing utility to determine the point of intersection. For instance, to check the solution of the equation in Example 2(a), graph

$y_1 = 4^x$ and $y_2 = 72$

in the same viewing window, as shown below. Using the *intersect* feature, you can determine that the graphs intersect when $x \approx 3.085$, which confirms the solution found in Example 2(a).

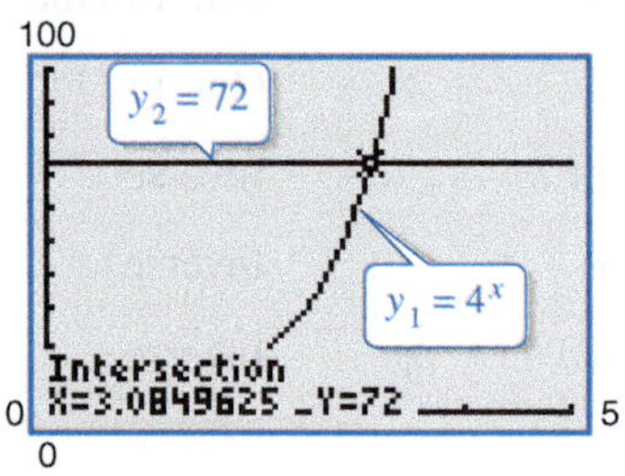

For instructions on how to use the *intersect* feature, see Appendix A. Consult the user's guide for your graphing utility for more information.

✓ Checkpoint 2

Solve each equation and approximate the result to three decimal places.

a. $6^x = 84$ **b.** $62 - 10^x = 24$ ■

In Example 2(a), the exact solution is $x = \log_4 72$ and the approximate solution is $x \approx 3.085$. An exact answer is preferred when the solution is an intermediate step in a larger problem. For a final answer, an approximate solution in decimal form is easier to comprehend.

Example 3 Solving an Exponential Equation

$e^x + 5 = 60$ — Original equation

$e^x = 55$ — Subtract 5 from each side to isolate the exponential expression.

$\ln e^x = \ln 55$ — Take natural log of each side.

$x = \ln 55$ — Inverse Property

$x \approx 4.007$ — Use a calculator.

The solution is $x = \ln 55 \approx 4.007$. Check this in the original equation.

✓ Checkpoint 3

Solve $2e^x - 7 = 23$ and approximate the result to three decimal places. ■

Example 4 Solving an Exponential Equation

Solve $2(3^{2t-5}) - 4 = 11$ and approximate the result to three decimal places.

SOLUTION

$2(3^{2t-5}) - 4 = 11$ Write original equation.

$2(3^{2t-5}) = 15$ Add 4 to each side.

$3^{2t-5} = \frac{15}{2}$ Divide each side by 2.

$\log_3 3^{2t-5} = \log_3 \frac{15}{2}$ Take log (base 3) of each side.

$2t - 5 = \log_3 \frac{15}{2}$ Inverse Property

$2t = 5 + \log_3 7.5$ Add 5 to each side.

$t = \frac{5}{2} + \frac{1}{2}\log_3 7.5$ Divide each side by 2.

$t \approx 3.417$ Use a calculator.

The solution is $t = \frac{5}{2} + \frac{1}{2}\log_3 7.5 \approx 3.417$. Check this in the original equation.

STUDY TIP

Remember that to evaluate a logarithm such as $\log_3 7.5$, you need to use the change-of-base formula as shown.

$$\log_3 7.5 = \frac{\log_{10} 7.5}{\log_{10} 3} \approx 1.834$$

Checkpoint 4

Solve $4(4^{2t-7}) + 14 = 110$ and approximate the result to three decimal places.

When an equation involves two or more exponential expressions, you can still use a procedure similar to that demonstrated in Examples 2, 3, and 4. However, the algebra is a bit more complicated.

Example 5 Solving an Exponential Equation of Quadratic Type

$e^{2x} - 3e^x + 2 = 0$ Original equation

$(e^x)^2 - 3e^x + 2 = 0$ Write in quadratic form.

$(e^x - 2)(e^x - 1) = 0$ Factor.

$e^x - 2 = 0 \Rightarrow x = \ln 2$ Set 1st factor equal to 0.

$e^x - 1 = 0 \Rightarrow x = 0$ Set 2nd factor equal to 0.

The solutions are $x = \ln 2$ and $x = 0$. Check these in the original equation. Or, check by graphing $y = e^{2x} - 3e^x + 2$ using a graphing utility. The graph should have two x-intercepts: $x = \ln 2$ and $x = 0$, as shown in Figure 4.23.

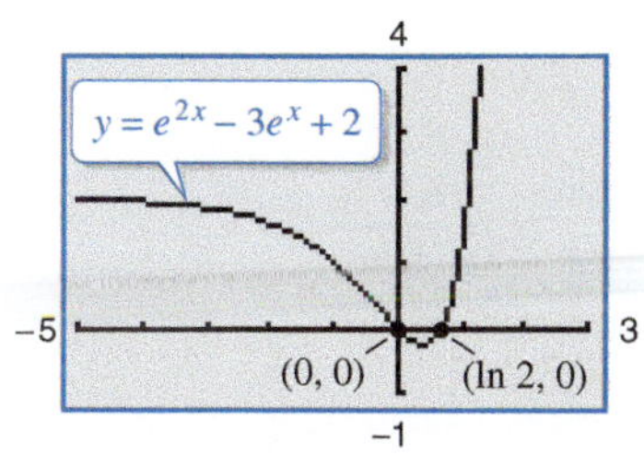

FIGURE 4.23

ALGEBRA TUTOR

For help with the algebra used to factor the left side of the equation in Example 5, see the *Chapter 4 Algebra Tutor* on page 401.

Checkpoint 5

Solve $e^{2x} - 7e^x + 12 = 0$.

Solving Logarithmic Equations

To solve a logarithmic equation such as

$\ln x = 3$ Logarithmic form

write the equation in exponential form as shown.

$e^{\ln x} = e^3$ Exponentiate each side.

$x = e^3$ Exponential form

This procedure is called **exponentiating** each side of an equation.

Example 6 Solving Logarithmic Equations

a. Solve $\ln(x + 5) = 1$. **b.** Solve $2 \log_5 3x = 4$.

SOLUTION

a. $\ln(x + 5) = 1$ Write original equation.

$e^{\ln(x+5)} = e^1$ Exponentiate each side.

$x + 5 = e$ Inverse Property

$x = e - 5$ Subtract 5 from each side.

The solution is $x = e - 5$. Check this in the original equation.

b. $2 \log_5 3x = 4$ Write original equation.

$\log_5 3x = 2$ Divide each side by 2.

$5^{\log_5 3x} = 5^2$ Exponentiate each side (base 5).

$3x = 25$ Inverse Property

$x = \frac{25}{3}$ Divide each side by 3.

The solution is $x = \frac{25}{3}$. Check this in the original equation.

✓Checkpoint 6

Solve each logarithmic equation.

a. $\ln(x - 4) = 0$ **b.** $\log_3 2x = 4$

Example 7 Solving a Logarithmic Equation

Solve $\log_3(5x - 1) = \log_3(x + 7)$.

SOLUTION

$\log_3(5x - 1) = \log_3(x + 7)$ Write original equation.

$5x - 1 = x + 7$ One-to-One Property

$4x = 8$ Add $-x$ and 1 to each side.

$x = 2$ Divide each side by 4.

The solution is $x = 2$. Check this in the original equation.

✓Checkpoint 7

Solve $\ln(3x + 2) = \ln(x + 8)$.

Ken Hurst/www.shutterstock.com

Example 8 Solving a Logarithmic Equation

Solve $5 + 2 \ln x = 4$ and approximate the result to three decimal places.

SOLUTION

$5 + 2 \ln x = 4$	Write original equation.
$2 \ln x = -1$	Subtract 5 from each side.
$\ln x = -\frac{1}{2}$	Divide each side by 2.
$e^{\ln x} = e^{-1/2}$	Exponentiate each side.
$x = e^{-1/2}$	Inverse Property
$x \approx 0.607$	Use a calculator.

The solution is $x = e^{-1/2} \approx 0.607$. Check this in the original equation.

✓ Checkpoint 8

Solve $4 + 3 \ln x = 16$ and approximate the result to three decimal places.

Because the domain of a logarithmic function generally does not include all real numbers, be sure to check for extraneous solutions of logarithmic equations.

Example 9 Checking for Extraneous Solutions

Solve $\log_{10} 5x + \log_{10}(x - 1) = 2$.

SOLUTION

$\log_{10} 5x + \log_{10}(x - 1) = 2$	Write original equation.
$\log_{10}[5x(x - 1)] = 2$	Product Property of logarithms
$10^{\log_{10}(5x^2 - 5x)} = 10^2$	Exponentiate each side (base 10).
$5x^2 - 5x = 100$	Inverse Property
$x^2 - x - 20 = 0$	Write in general form.
$(x - 5)(x + 4) = 0$	Factor.
$x - 5 = 0$	Set 1st factor equal to 0.
$x = 5$	Solution
$x + 4 = 0$	Set 2nd factor equal to 0.
$x = -4$	Solution

The solutions appear to be $x = 5$ and $x = -4$. However, when you check these in the original equation, you can see that $x = 5$ is the only solution.

✓ Checkpoint 9

Solve $\log_6 x + \log_6(x + 5) = 2$.

In Example 9, the domain of $\log_{10} 5x$ is $x > 0$ and the domain of $\log_{10}(x - 1)$ is $x > 1$, so the domain of the original equation is $x > 1$. Because the domain is all real numbers greater than 1, the solution $x = -4$ is extraneous.

TECH TUTOR

You can use a graphing utility to verify that the equation in Example 9 has $x = 5$ as its only solution. Graph

$$y_1 = \log_{10} 5x + \log_{10}(x - 1)$$

and

$$y_2 = 2$$

in the same viewing window, as shown below. It appears that the graphs intersect at one point. Use the *intersect* feature or the *zoom* and *trace* features to determine that $x = 5$ is the solution.

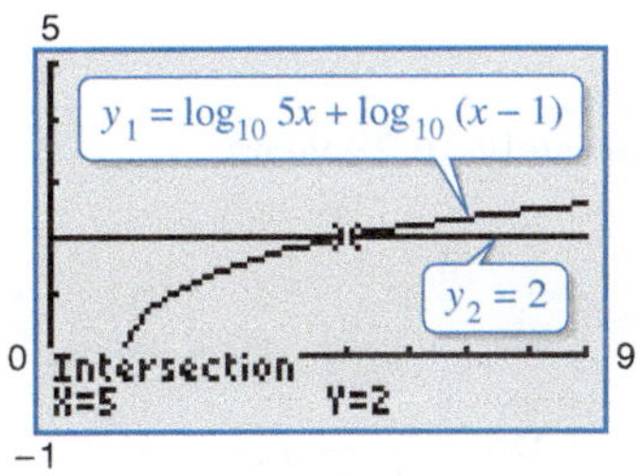

For instructions on how to use the *zoom* and *trace* features, see Appendix A. Consult the user's guide for your graphing utility for more information.

Applications

Example 10 Doubling and Tripling an Investment

You deposit \$500 in an account that pays 6.75% interest, compounded continuously.

a. How long will it take your money to double?

b. How long will it take your money to triple?

SOLUTION Using the formula for compound interest with continuous compounding, you can find that the balance in the account is given by

$$A = Pe^{rt} = 500e^{0.0675t}.$$

a. To find the time required for the balance to double, let $A = 1000$ and solve the resulting equation for t.

$500e^{0.0675t} = 1000$	Substitute 1000 for A.
$e^{0.0675t} = 2$	Divide each side by 500.
$\ln e^{0.0675t} = \ln 2$	Take natural log of each side.
$0.0675t = \ln 2$	Inverse Property
$t = \frac{1}{0.0675}\ln 2$	Divide each side by 0.0675.
$t \approx 10.27$	Use a calculator.

The balance in the account will double after approximately 10.27 years.

b. To find the time required for the balance to triple, let $A = 1500$ and solve the resulting equation for t.

$500e^{0.0675t} = 1500$	Substitute 1500 for A.
$e^{0.0675t} = 3$	Divide each side by 500.
$\ln e^{0.0675t} = \ln 3$	Take natural log of each side.
$0.0675t = \ln 3$	Inverse Property
$t = \frac{1}{0.0675}\ln 3$	Divide each side by 0.0675.
$t \approx 16.28$	Use a calculator.

The balance in the account will triple after approximately 16.28 years.

Notice that it took 10.27 years to earn the first \$500 and only 6.01 years to earn the second \$500. This result is graphically demonstrated in Figure 4.24.

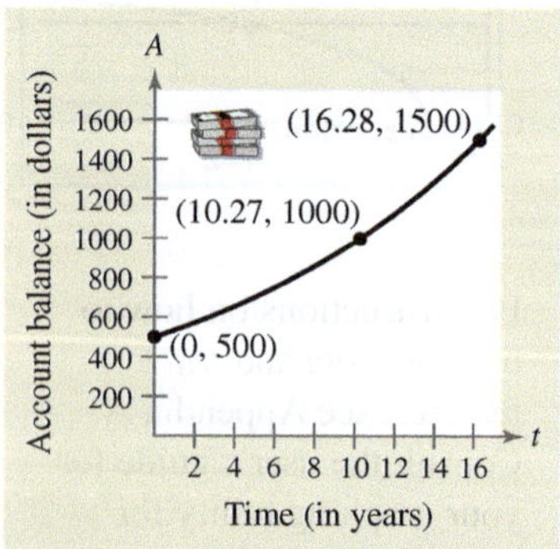

FIGURE 4.24

Checkpoint 10

In Example 10, how long will it take for the account balance to reach \$600? ■

Example 11 Hospital Expenses

From 1997 through 2008, the yearly expenses y (in billions of dollars) of hospitals in the United States can be approximated by

$$y = 208.05e^{0.067t}$$

where t represents the year, with $t = 7$ corresponding to 1997 (see Figure 4.25). Use the model to estimate the year in which the expenses of hospitals reached about \$610 billion. *(Source: Health Forum, An American Hospital Association Company)*

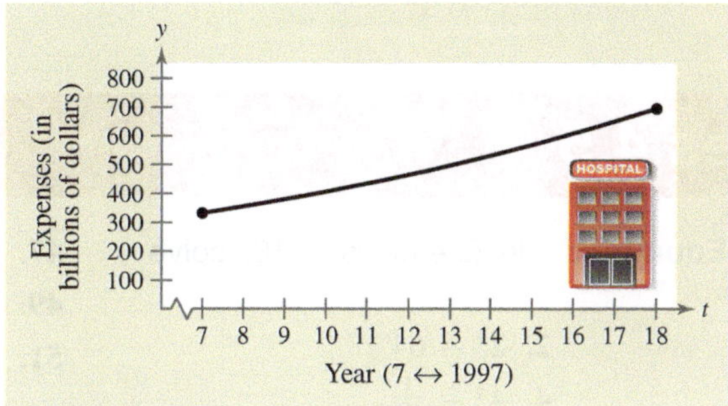

FIGURE 4.25

SOLUTION To solve this problem, let $y = 610$ and solve the resulting equation for t.

$208.05e^{0.067t} = 610$	Substitute 610 for y.
$e^{0.067t} \approx 2.932$	Divide each side by 208.05.
$\ln e^{0.067t} \approx \ln 2.932$	Take natural log of each side.
$0.067t \approx 1.076$	Inverse Property
$t \approx 16$	Divide each side by 0.067.

The solution is $t \approx 16$. Because $t = 7$ represents 1997, you can conclude that, according to the model, expenses were about \$610 billion in 2006.

✓Checkpoint 11

Use the model in Example 11 to estimate the year in which the expenses of hospitals reached about \$435 billion.

SUMMARIZE (Section 4.5)

1. State the One-to-One Properties and the Inverse Properties that can be used to solve simple exponential and logarithmic equations *(page 377)*. For an example of solving simple exponential and logarithmic equations, see Example 1.
2. Describe the strategies for solving exponential and logarithmic equations *(page 377)*. For examples of solving exponential equations, see Examples 2, 3, 4, and 5.
3. Describe what it means to exponentiate each side of an equation *(page 380)*. For examples of solving logarithmic equations, see Examples 6, 7, 8, and 9.
4. Describe a real-life example of how an exponential equation can be used to find the balance in an investment account *(page 382, Example 10)*.

SKILLS WARM UP 4.5

The following warm-up exercises involve skills that were covered in earlier sections. You will use these skills in the exercise set for this section. For additional help, review Sections 1.4 and 4.3.

In Exercises 1–6, solve for x.

1. $x \ln 2 = \ln 3$ **2.** $(x - 1) \ln 4 = 2$ **3.** $2xe^2 = e^3$

4. $4xe^{-1} = 8$ **5.** $x^2 - 4x + 5 = 0$ **6.** $2x^2 - 3x + 1 = 0$

In Exercises 7–10, simplify the expression.

7. $\log_{10} 10^x$ **8.** $\log_{10} 10^{2x}$ **9.** $\ln e^{2x}$ **10.** $\ln e^{-x^2}$

Exercises 4.5

See www.CalcChat.com for worked-out solutions to odd-numbered exercises.

Solving Simple Equations In Exercises 1–10, solve for x. *See Example 1.*

1. $5^x = 25$ **2.** $2^x = 64$

3. $7^x = \frac{1}{49}$ **4.** $4^x = \frac{1}{256}$

5. $4^{x-1} = 64$ **6.** $3^{x-1} = 27$

7. $\log_4 x = 3$ **8.** $\log_{10} x = -2$

9. $\log_5 5x = \log_5 2$ **10.** $\ln 2x = \ln 6$

Using Inverse Properties In Exercises 11–22, apply the Inverse Property of logarithmic or exponential functions to simplify the expression.

11. $\ln e^{x^2}$ **12.** $\ln e^{2x-1}$

13. $\log_{10} 10^x + 1$ **14.** $\log_{10} 10^{2x+3}$

15. $\log_5 5^{x^3} - 7$ **16.** $\log_8 8^{x^5} + 1$

17. $8 + e^{\ln x^3}$ **18.** $-1 + \ln e^{2x}$

19. $10^{\log_{10}(x+5)}$ **20.** $9^{\log_9(3x+7)}$

21. $2^{\log_2 x^2}$ **22.** $10^{\log_{10}(x^2+7x+10)}$

Solving Exponential Equations In Exercises 23–60, solve the exponential equation algebraically. Approximate the result to three decimal places. *See Examples 2, 3, 4, and 5.*

23. $3e^x = 9$ **24.** $5e^x = 20$

25. $2(3^x) = 16$ **26.** $3(4^x) = 81$

27. $e^x - 9 = 19$ **28.** $e^x + 4 = 18$

29. $3^{2x} = 80$ **30.** $6^{5x} = 3000$

31. $5^{-t/2} = 0.20$ **32.** $4^{-t/3} = 0.15$

33. $3^{x-1} = 28$ **34.** $2^{x-3} = 31$

35. $2^{3-x} = 565$ **36.** $8^{-2-x} = 431$

37. $8(10^{3x}) = 12$ **38.** $5(10^{2x}) = 7$

39. $3(5^{x-1}) = 21$ **40.** $8(3^{6-x}) = 40$

41. $e^{3x} = 12$ **42.** $e^{2x} = 50$

43. $500e^{-x} = 300$ **44.** $1000e^{-4x} = 75$

45. $7 - 2e^x = 6$ **46.** $-14 + 3e^x = 11$

47. $6(2^{3x-1}) - 7 = 9$ **48.** $8(4^{6-2x}) + 13 = 41$

49. $e^{2x} - 8e^x + 12 = 0$ **50.** $e^{2x} - 5e^x + 6 = 0$

51. $e^{2x} - 3e^x - 4 = 0$ **52.** $e^{2x} - 9e^x - 36 = 0$

53. $\dfrac{500}{100 - e^{x/2}} = 20$ **54.** $\dfrac{400}{1 + e^{-x}} = 350$

55. $\dfrac{3000}{2 + e^{2x}} = 2$ **56.** $\dfrac{119}{e^{6x} - 14} = 7$

57. $\left(1 + \dfrac{0.065}{365}\right)^{365t} = 4$ **58.** $\left(1 + \dfrac{0.075}{4}\right)^{4t} = 5$

59. $\left(1 + \dfrac{0.10}{12}\right)^{12t} = 2$ **60.** $\left(1 + \dfrac{0.0825}{26}\right)^{26t} = 9$

Solving Logarithmic Equations In Exercises 61–86, solve the logarithmic equation algebraically. Approximate the result to three decimal places. *See Examples 6, 7, 8, and 9.*

61. $\log_{10} x = 4$ **62.** $\log_{10} x = -5$

63. $\ln x = -3$ **64.** $\ln x = 6$

65. $\ln 2x = 2.4$ **66.** $\ln 4x = 1$

67. $\log_{10} 2x = 7$ **68.** $\log_{10} 3z = 2$

69. $4 \log_3(x + 1) = 12$ **70.** $5 \log_{10}(x - 2) = 10$

71. $3 \ln 5x = 10$ **72.** $2 \ln x = 7$

73. $\ln \sqrt{x + 2} = 1$ **74.** $\ln \sqrt{x - 8} = 5$

75. $7 + 3 \ln x = 5$ **76.** $2 - 6 \ln x = 10$

77. $\ln x - \ln(x + 1) = 2$ **78.** $\ln x - \ln(x + 2) = 3$

79. $\ln(x + 5) = \ln(x - 1) - \ln(x + 1)$

80. $\ln(x + 1) - \ln(x - 2) = \ln x$

81. $\log_2(2x - 3) = \log_2(x + 4)$

82. $\log_3(x + 8) = \log_3(3x + 2)$

83. $\log_{10}(x + 4) - \log_{10} x = \log_{10}(x + 2)$

84. $\log_{10} x + \log_{10}(x + 1) = \log_{10}(x + 3)$

85. $\log_4 x - \log_4(x - 1) = \frac{1}{2}$

86. $\log_3 x + \log_3(x - 8) = 2$

Rewriting Logarithmic Equations In Exercises 87–90, solve for y in terms of x.

87. $\ln y = \ln(2x + 1) + \ln 1$

88. $\ln y = 2 \ln x + \ln(x - 3)$

89. $\log_{10} y = 2 \log_{10}(x - 1) - \log_{10}(x + 2)$

90. $\log_{10}(y - 4) + \log_{10} x = 3 \log_{10} x$

Solving Equations Using a Graphing Utility In Exercises 91–94, use a graphing utility to solve the equation. Approximate the result to three decimal places. Verify your result algebraically.

91. $2^x - 7 = 0$

92. $500 - 1500e^{-x/2} = 0$

93. $3 - \ln x = 0$

94. $10 - 4 \ln(x - 2) = 0$

Doubling an Investment In Exercises 95 and 96, find the time required for a $1000 investment to double at interest rate r, compounded continuously. *See Example 10.*

95. $r = 0.0725$

96. $r = 0.065$

Tripling an Investment In Exercises 97 and 98, find the time required for a $1000 investment to triple at interest rate r, compounded continuously. *See Example 10.*

97. $r = 0.0825$

98. $r = 0.0775$

99. Suburban Wildlife The number V of varieties of suburban nondomesticated wildlife in a community is approximated by the model

$$V = 15 \cdot 10^{0.02x}, \quad 0 \le x \le 36$$

where x is the number of months since the development of the community was completed. Use this model to approximate the number of months since the development was completed when $V = 60$.

100. Native Prairie Grasses The number A of varieties of native prairie grasses per acre within a farming region is approximated by the model

$$A = 10.5 \cdot 10^{0.04x}, \quad 0 \le x \le 24$$

where x is the number of months since the farming region was plowed. Use this model to approximate the number of months since the region was plowed using a test acre for which $A = 80$.

101. Demand Function The demand function for a special limited edition coin set is given by

$$p = 1000\left(1 - \frac{5}{5 + e^{-0.001x}}\right).$$

(a) Find the demand x for a price of $p = \$139.50$.

(b) Find the demand x for a price of $p = \$99.99$.

(c) Use a graphing utility to confirm graphically the results found in parts (a) and (b).

102. Demand Function The demand function for a hot tub spa is given by

$$p = 105{,}000\left(1 - \frac{3}{3 + e^{-0.002x}}\right).$$

(a) Find the demand x for a price of $p = \$25{,}000$.

(b) Find the demand x for a price of $p = \$21{,}000$.

(c) Use a graphing utility to confirm graphically the results found in parts (a) and (b).

103. Forest Yield The yield V (in millions of cubic feet per acre) for a forest at age t years is given by

$$V = 6.7e^{-48.1/t}, \quad t > 0.$$

(a) Use a graphing utility to find the time necessary to obtain a yield of 1.3 million cubic feet per acre.

(b) Use the graphing utility to find the time necessary to obtain a yield of 2 million cubic feet per acre.

104. Human Memory Model In a group project on learning theory, a mathematical model for the percent P (in decimal form) of correct responses after n trials was found to be

$$P = \frac{0.98}{1 + e^{-0.3n}}, \quad n \ge 0.$$

(a) After how many trials will 80% of the responses be correct? (That is, for what value of n will $P = 0.8$?)

(b) Use a graphing utility to graph the memory model and confirm the result found in part (a).

(c) Write a paragraph describing the memory model.

105. Drug Prescriptions The numbers y (in millions) of mail-order drug prescriptions in the United States from 2002 through 2009 can be approximated by the model $y = 143.09 + 47.2 \ln t$, for $2 \le t \le 9$, where t represents the year, with $t = 2$ corresponding to 2002. *(Source: National Association of Chain Drug Stores)*

(a) Use a graphing utility to graph the model.

(b) Use the graphing utility to estimate the year when the number of mail-order drug prescriptions exceeded 200 million.

(c) Verify your answer to part (b) algebraically.

106. Air Transportation The numbers of employees y (in thousands) in air transportation in the United States from 2001 through 2009 can be approximated by the model $y = 608 - 64.2 \ln t$, for $1 \le t \le 9$, where t represents the year, with $t = 1$ corresponding to 2001. *(Source: U.S. Bureau of Labor Statistics)*

(a) Use a graphing utility to graph the model.

(b) Use the graphing utility to estimate the year in which the number of air transportation employees fell below 500,000.

(c) Verify your answer to part (b) algebraically.

107. Automobiles Automobiles are designed with crumple zones that help protect their occupants. The crumple zones allow the occupants to move short distances when the automobiles come to abrupt stops. The greater the distance moved, the fewer g's the crash victims experience. (One g is equal to the acceleration due to gravity. For very short periods of time, humans have withstood as much as 40 g's.) In crash tests with vehicles moving at 90 kilometers per hour, analysts measured the numbers of g's experienced during deceleration by crash dummies that were permitted to move x meters during impact. The data are shown in the table.

x	0.2	0.4	0.6	0.8	1.0
g's	158	80	53	40	32

A model for these data is given by

$$y = -3.00 + 11.88 \ln x + \frac{36.94}{x}$$

where y is the number of g's.

(a) Use the model to expand the table by adding a row of y-values that correspond to the given x-values.

(b) Use a graphing utility to graph the data points and the model in the same viewing window. How do they compare?

(c) Use the model to estimate the least distance traveled during impact for which the passenger does not experience more than 30 g's.

(d) Do you think it is practical to lower the number of g's experienced during impact to fewer than 23? Explain your reasoning.

108. HOW DO YOU SEE IT? Solving $\log_3 x + \log_3(x - 8) = 2$ algebraically, the solutions appear to be $x = 9$ and $x = -1$. Use the graph of $y = \log_3 x + \log_3(x - 8) - 2$ to determine whether each value is an actual solution of the equation. Explain your reasoning.

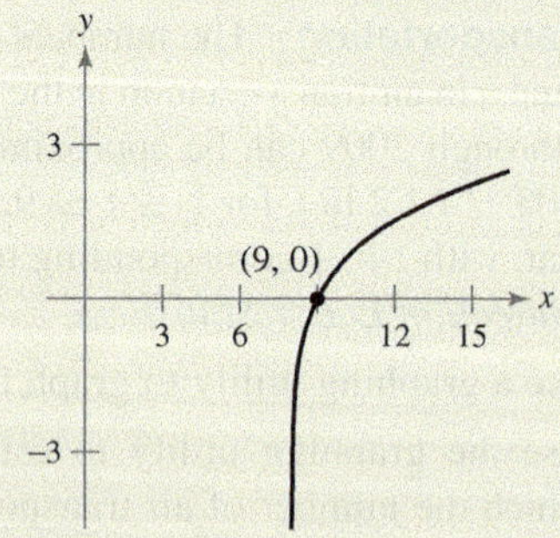

109. Think About It Is it possible for a logarithmic equation to have more than one extraneous solution? Explain.

110. Average Heights The percent m of American males (between 20 and 29 years old) who are less than x inches tall is approximated by

$$m = -0.018 + \frac{1.041}{1 + e^{-0.5325(x-69.44)}}, \quad 64 \le x \le 78$$

and the percent f of American females (between 20 and 29 years old) who are less than x inches tall is approximated by

$$f = -0.013 + \frac{1.031}{1 + e^{-0.5604(x-64.48)}}, \quad 59 \le x \le 73$$

where m and f are the percents (in decimal form) and x is the height (in inches) (see figure). *(Source: U.S. National Center for Health Statistics)*

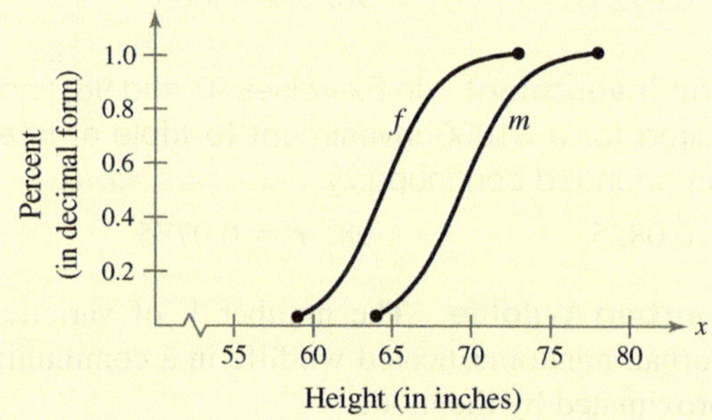

(a) What is the median height for each sex between 20 and 29 years old? (In other words, for what values of x are m and f equal to 0.5?)

(b) Write a paragraph describing each height model.

Writing Equations In Exercises 111–114, rewrite each verbal statement as an equation. Then decide whether the statement is true or false. Justify your answer.

111. The logarithm of the product of two numbers is equal to the sum of the logarithms of the numbers.

112. The logarithm of the sum of two numbers is equal to the product of the logarithms of the numbers.

113. The logarithm of the difference of two numbers is equal to the difference of the logarithms of the numbers.

114. The logarithm of the quotient of two numbers is equal to the difference of the logarithms of the numbers.

115. Think About It Are the times required for the investments in Exercises 95 and 96 to quadruple twice as long as the times for them to double? Give a reason for your answer and verify your answer algebraically.

4.6 Exponential and Logarithmic Models

- Understand the shapes of the graphs of the most common types of mathematical models involving exponential and logarithmic functions.
- Construct and use a model for exponential growth or exponential decay.
- Use a Gaussian model to solve an application problem.
- Use a logistic growth model to solve an application problem.
- Use a logarithmic model to solve an application problem.
- Choose an appropriate model involving exponential or logarithmic functions for a real-life situation.

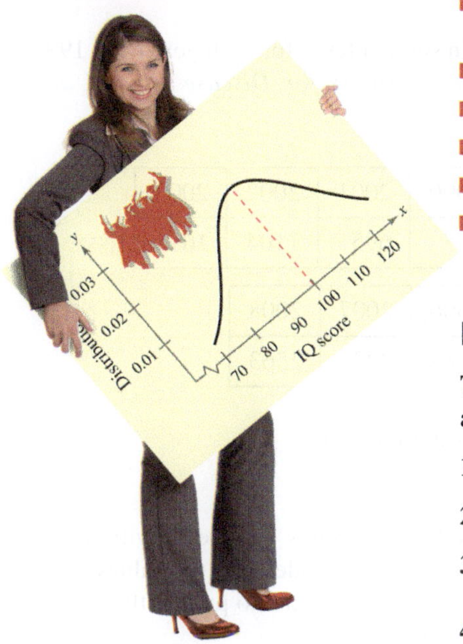

In Exercise 43 on page 397, you will use the graph of a Gaussian model to determine the average IQ score for the students at a college.

Introduction

The five most common types of mathematical models involving exponential functions and logarithmic functions are shown below.

1. Exponential growth model: $y = ae^{bx}, \quad b > 0$
2. Exponential decay model: $y = ae^{-bx}, \quad b > 0$
3. Gaussian model: $y = ae^{-(x-b)^2/c}$
4. Logistic growth model: $y = \dfrac{a}{1 + be^{-rx}}$
5. Logarithmic models: $y = a + b \ln x$

   $y = a + b \log_{10} x$

The basic shape of the graph of each of these models is shown in Figure 4.26.

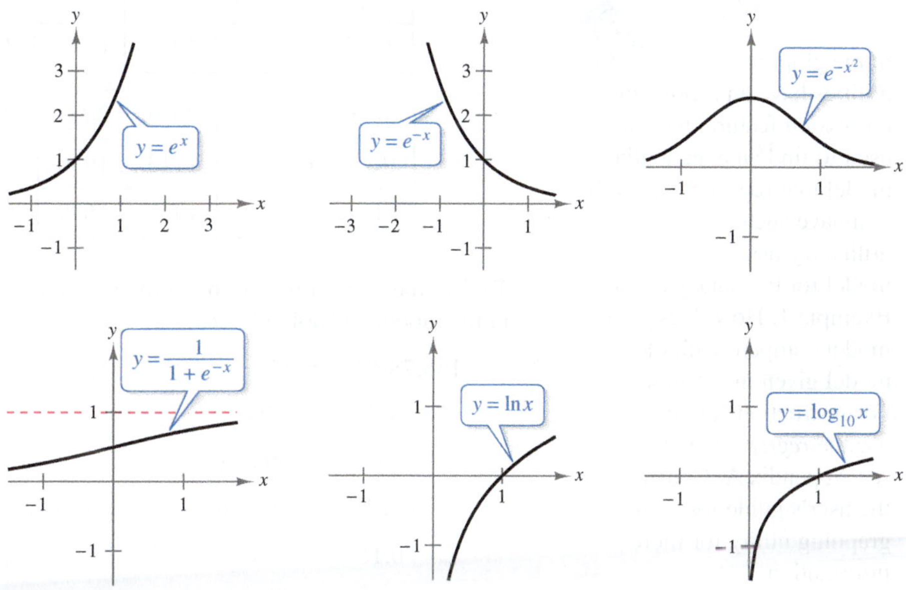

FIGURE 4.26

You can often gain quite a bit of insight into a situation modeled by an exponential or logarithmic function by identifying and interpreting the function's asymptotes. Use the graphs in Figure 4.26 to identify the asymptote(s) of the graph of each function.

Poznyakov/Shutterstock.com

Exponential Growth and Decay

Example 1 Rocky Mountain Spotted Fever

The number of reported cases of Rocky Mountain spotted fever for each year from 1998 through 2008 are shown in the table. *(Source: Centers for Disease Control and Prevention)*

Year	1998	1999	2000	2001	2002	2003
Reported cases	365	579	495	695	1104	1091

Year	2004	2005	2006	2007	2008
Reported cases	1713	1936	2288	2221	2563

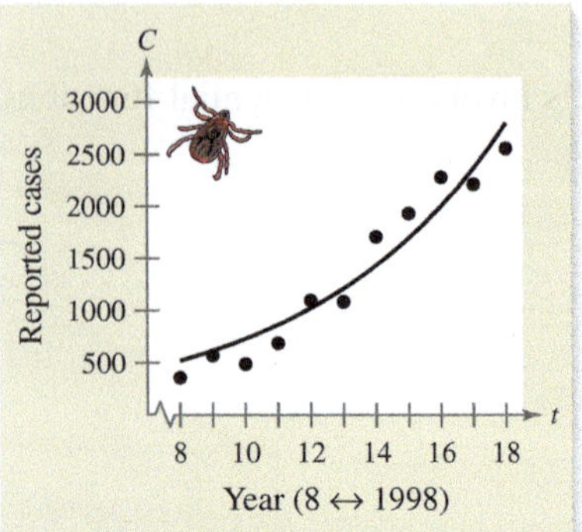

FIGURE 4.27

An exponential growth model that approximates these data is

$$C = 138.78e^{0.167t}, \quad 8 \le t \le 18$$

where C is the number of reported cases and t represents the year, with $t = 8$ corresponding to 1998. Compare the estimates given by the model with the values given by the Centers for Disease Control and Prevention. Use the model to predict the year in which the number of reported cases will reach 5500.

SOLUTION The following table compares the two sets of data. The graph of the model and the original data values are shown in Figure 4.27.

Year	1998	1999	2000	2001	2002	2003
Reported cases	365	579	495	695	1104	1091
Model	528	624	737	871	1030	1217

Year	2004	2005	2006	2007	2008
Reported cases	1713	1936	2288	2221	2563
Model	1438	1699	2008	2373	2804

To find the year in which the number of reported cases will reach 5500, let $C = 5500$ in the model and solve for t.

$138.78e^{0.167t} = C$ Write original model.

$138.78e^{0.167t} = 5500$ Substitute 5500 for C.

$e^{0.167t} \approx 39.631$ Divide each side by 138.78.

$\ln e^{0.167t} \approx \ln 39.631$ Take natural log of each side.

$0.167t \approx 3.680$ Inverse Property

$t \approx 22$ Divide each side by 0.167.

The solution is $t \approx 22$. Because $t = 8$ represents 1998, you can conclude that, according to the model, the number of reported cases will reach 5500 in 2012.

TECH TUTOR

Some graphing utilities have an *exponential regression* feature that can be used to find an exponential model that represents data. If you have such a graphing utility, try using it to find a model for the data given in Example 1. How does your model compare with the model given in Example 1? For instructions on how to use the *regression* feature, see Appendix A. Consult the user's guide for your graphing utility for more information.

✓ Checkpoint 1

Use the model in Example 1 to predict the year in which the number of reported cases will reach 7500.

The exponential model in Example 1 increases by the same percent each year. What is the annual percent increase for this exponential model?

In Example 1, you are given the exponential growth model. How could you find such a model if you are given only a set of data? To find an exponential growth model that fits a set of data, choose any two of the points and substitute them in the general exponential growth model $y = ae^{bx}$. This technique is demonstrated in Example 2.

Example 2 Finding an Exponential Growth Model

Find an exponential growth model whose graph passes through the points (0, 4453) and (7, 5024), as shown in Figure 4.28(a).

SOLUTION The general form of the model is

$$y = ae^{bx}.$$

From the fact that the graph passes through the point (0, 4453), you know that $y = 4453$ when $x = 0$. By substituting these values into the general model, you have

$$4453 = ae^0 \quad \Rightarrow \quad a = 4453.$$

In a similar way, from the fact that the graph passes through the point (7, 5024), you know that $y = 5024$ when $x = 7$. Substitute these values into the model $y = ae^{bx}$ and solve for b.

$$5024 = 4453e^{7b}$$ Substitute for y, a, and x.

$$1.12823 \approx e^{7b}$$ Divide each side by 4453.

$$\ln 1.12823 \approx \ln e^{7b}$$ Take natural log of each side.

$$0.12065 \approx 7b$$ Inverse Property

$$0.01724 \approx b$$ Divide each side by 7.

So, the exponential growth model is

$$y = 4453e^{0.01724x}.$$

The graph of the model is shown in Figure 4.28(b).

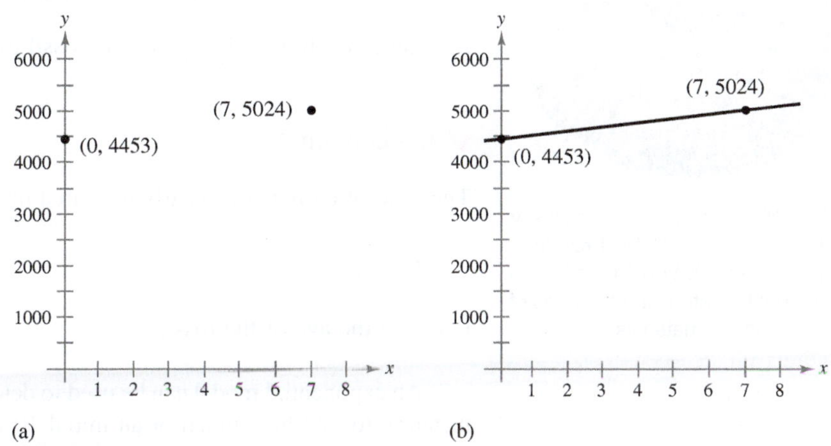

FIGURE 4.28

Checkpoint 2

Find an exponential growth model whose graph passes through the points (0, 3) and (5, 8).

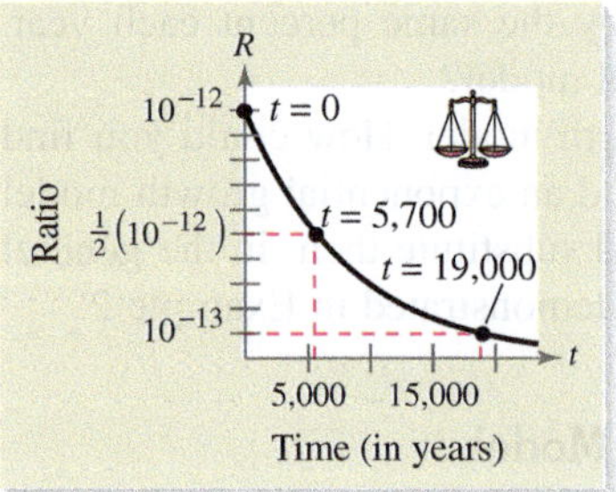

FIGURE 4.29

In living organic material, the ratio of the number of radioactive carbon isotopes (carbon 14) to the number of nonradioactive carbon isotopes (carbon 12) is about 1 to 10^{12}. When organic material dies, its carbon 12 content remains fixed, whereas its radioactive carbon 14 begins to decay with a half-life of about 5700 years. To estimate the age of dead organic material, scientists use the following formula, which denotes the ratio R of carbon 14 to carbon 12 present at any time t (in years).

$$R = \frac{1}{10^{12}}e^{-t/8223} \qquad \text{Carbon dating model}$$

In Figure 4.29, note that R decreases as the time t increases. Any material that is composed of carbon, such as wood, bone, hair, pottery, paper, and water, can be dated.

Example 3 Carbon Dating

The ratio of carbon 14 to carbon 12 in a newly discovered fossil is

$$R = \frac{1}{10^{13}}.$$

Estimate the age of the fossil.

SOLUTION In the carbon dating model, substitute for R and solve for t.

$$\frac{1}{10^{12}}e^{-t/8223} = R \qquad \text{Write original model.}$$

$$\frac{e^{-t/8223}}{10^{12}} = \frac{1}{10^{13}} \qquad \text{Substitute } \frac{1}{10^{13}} \text{ for } R.$$

$$e^{-t/8223} = \frac{1}{10} \qquad \text{Multiply each side by } 10^{12}.$$

$$\ln e^{-t/8223} = \ln\frac{1}{10} \qquad \text{Take natural log of each side.}$$

$$-\frac{t}{8223} \approx -2.3026 \qquad \text{Inverse Property}$$

$$t \approx 18{,}934 \qquad \text{Multiply each side by } -8223.$$

So, you can estimate the age of the fossil to be about 19,000 years.

In 1960, Willard Libby of the University of Chicago won the Nobel Prize for Chemistry for the carbon 14 method, a valuable tool for estimating the ages of ancient materials.

✓ Checkpoint 3

The ratio of carbon 14 to carbon 12 in a newly discovered fossil is

$$R = \frac{1}{9^{13}}.$$

Estimate the age of the fossil.

An exponential model can be used to determine the *decay* of radioactive isotopes. For instance, to find how much of an initial 10 grams of radioactive radium (^{226}Ra), with a half-life of 1599 years, is left after 500 years, use the exponential decay model, as shown.

$$y = ae^{-bt} \Rightarrow \frac{1}{2}(10) = 10e^{-b(1599)} \Rightarrow \ln\frac{1}{2} = -1599b \Rightarrow -\frac{\ln\frac{1}{2}}{1599} = b$$

Using the value of b found above, $a = 10$, and $t = 500$, the amount left is

$$y = 10e^{-[-\ln(1/2)/1599](500)} \approx 8.05 \text{ grams.}$$

Gaussian Models

As mentioned at the beginning of this section, Gaussian models are of the form

$$y = ae^{-(x-b)^2/c}.$$

This type of model is commonly used in probability and statistics to represent populations that are **normally distributed.** For *standard* normal distributions, the model takes the form

$$y = \frac{1}{\sqrt{2\pi}}e^{-x^2/2}.$$

The graph of a Gaussian model is called a **bell-shaped curve.** Try to sketch the standard normal distribution curve with a graphing utility. Can you see why it is called a bell-shaped curve?

The **average value** of a population can be found from the bell-shaped curve by observing where the maximum y-value of the function occurs. The x-value corresponding to the maximum y-value of the function represents the average value of the independent variable, x.

Example 4 ACT Scores

In 2010, the ACT science scores for high school graduates in the United States roughly followed a normal distribution given by

$$y = 0.0782e^{-(x-20.9)^2/52.02}, \quad 1 \le x \le 36$$

where x is the ACT science score. Sketch the graph of the function. From the graph, estimate the average ACT science score. *(Source: ACT, Inc.)*

SOLUTION The graph of the function is shown in Figure 4.30. On this bell-shaped curve, the x-value corresponding to the maximum value of the curve represents the average score. From the graph, you can estimate that the average ACT science score for high school graduates in 2010 was about 21.

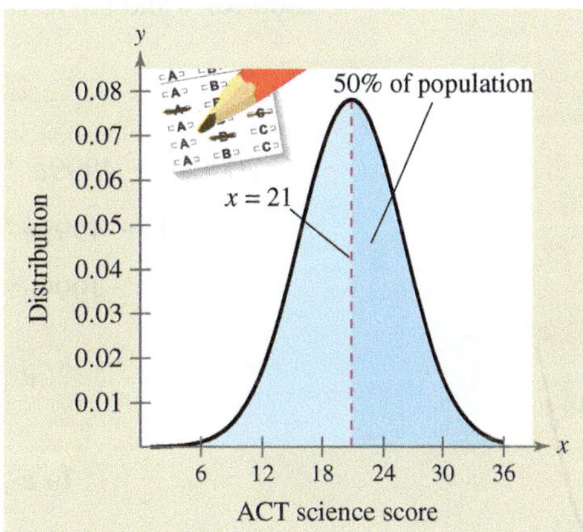

FIGURE 4.30

✓ Checkpoint 4

In 2010, the SAT critical reading scores for high school graduates in the United States roughly followed a normal distribution given by

$$y = 0.0036e^{-(x-501)^2/25{,}088}, \quad 200 \le x \le 800$$

where x is the SAT score for critical reading. Sketch the graph of the function. From the graph, estimate the average SAT critical reading score. *(Source: The College Board)*

Logistic Growth Models

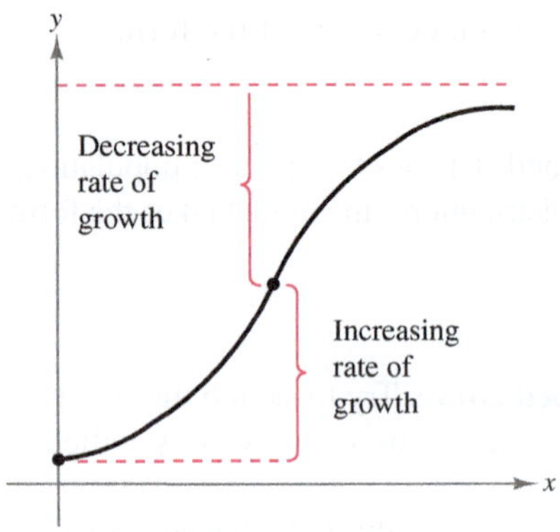

Logistic Curve
FIGURE 4.31

Some populations initially have rapid growth, followed by a declining rate of growth, as shown by the graph in Figure 4.31. One model for describing this type of growth pattern is the **logistic curve** given by the function

$$y = \frac{a}{1 + be^{-rx}}$$

where y is the population size and x is the time. An example is a bacteria culture that is initially allowed to grow under ideal conditions, followed by less favorable conditions that inhibit growth. A logistic growth curve is also called a **sigmoidal curve.**

Example 5 Spread of a Virus

On a college campus of 5000 students, one student returned from vacation with a contagious flu virus. The spread of the virus through the student population is given by

$$y = \frac{5000}{1 + 4999e^{-0.8t}}, \quad t \geq 0$$

where y is the total number of students infected after t days.

a. How many students are infected after 5 days?

b. The college will cancel classes when 40% or more of the students become infected. After how many days will the college cancel classes?

SOLUTION

a. After 5 days, the number of students infected is

$$y = \frac{5000}{1 + 4999e^{-0.8(5)}} = \frac{5000}{1 + 4999e^{-4}} \approx 54.$$

b. Classes are cancelled when the number infected is $(0.40)(5000) = 2000$. So, substitute 2000 for y and solve for t as shown.

$$2000 = \frac{5000}{1 + 4999e^{-0.8t}}$$

$$2000(1 + 4999e^{-0.8t}) = 5000$$

$$1 + 4999e^{-0.8t} = 2.5$$

$$4999e^{-0.8t} = 1.5$$

$$e^{-0.8t} = \frac{1.5}{4999}$$

$$\ln e^{-0.8t} = \ln \frac{1.5}{4999}$$

$$-0.8t = \ln \frac{1.5}{4999}$$

$$t \approx 10.1$$

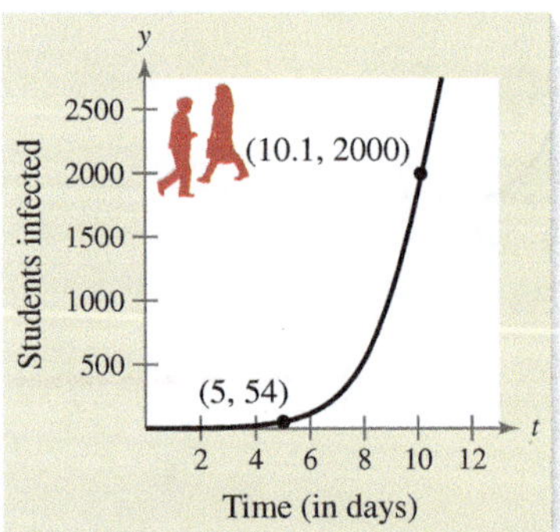

FIGURE 4.32

So, after about 10 days, 40% of the students will become infected, and the college will cancel classes. The graph of the function is shown in Figure 4.32.

✓ Checkpoint 5

In Example 5, how many days does it take for 25% of the students on campus to become infected?

Logarithmic Models

Example 6 Magnitudes of Earthquakes

The severity of the destruction caused by an earthquake depends on its magnitude and duration. Earthquakes can destroy buildings, and can cause landslides and tsunamis.

On the Richter scale, the magnitude R of an earthquake of intensity I per unit of area is given by $R = \log_{10}(I/I_0)$, where $I_0 = 1$ is the minimum intensity used for comparison. Find the intensity per unit of area for each earthquake. (Intensity is a measure of the wave energy of an earthquake.)

a. Chile, in 1960; $R = 9.5$

b. Southern Sumatra, Indonesia, in 2007; $R = 8.5$

SOLUTION

a. Because $I_0 = 1$ and $R = 9.5$,

$$9.5 = \log_{10} I$$

$$I = 10^{9.5} = 3{,}162{,}277{,}660.$$

b. For $R = 8.5$,

$$8.5 = \log_{10} I$$

$$I = 10^{8.5} = 316{,}227{,}766.$$

Note that an increase of 1 unit on the Richter scale (from 8.5 to 9.5) represents an intensity change by a factor of

$$\frac{3{,}162{,}277{,}660}{316{,}227{,}766} = 10.$$

In other words, the intensity of the earthquake in Chile in 1960 was 10 times greater than the intensity of the earthquake in Southern Sumatra, Indonesia in 2007.

✓ Checkpoint 6

In June 2010, an earthquake measuring 7.5 on the Richter scale occurred near the Nicobar Islands in the Indian Ocean. Find the intensity I per unit of area for this earthquake. (Let $I_0 = 1$.) How many times greater is the intensity of the 1960 Chile earthquake?

Example 7 pH Levels

Acidity, or pH level, is a measure of the hydrogen ion concentration $[H^+]$ (measured in moles of hydrogen per liter) of a solution. Use the model given by

$$\text{pH} = -\log_{10}[H^+]$$

to determine the hydrogen ion concentration of milk of magnesia, which has a pH of 10.5.

SOLUTION

$$\text{pH} = -\log_{10}[H^+] \quad \text{Write original model.}$$

$$10.5 = -\log_{10}[H^+] \quad \text{Substitute 10.5 for pH.}$$

$$-10.5 = \log_{10}[H^+] \quad \text{Multiply each side by } -1.$$

$$10^{-10.5} = 10^{\log_{10}[H^+]} \quad \text{Exponentiate each side (base 10).}$$

$$3.16 \times 10^{-11} = [H^+] \quad \text{Simplify.}$$

So, the hydrogen ion concentration is 3.16×10^{-11} mole of hydrogen per liter.

✓ Checkpoint 7

Use the model in Example 7 to determine the hydrogen ion concentration of coffee, which has a pH of 5.0.

Comparing Models

So far, you have been given the type of model to use for a data set. Now, you will use the general trends of the graphs of the five models presented in this section to choose appropriate models for real-life situations.

Example 8 Choosing an Appropriate Model

Decide whether to use an exponential growth model or a logistic growth model to represent each data set.

a.

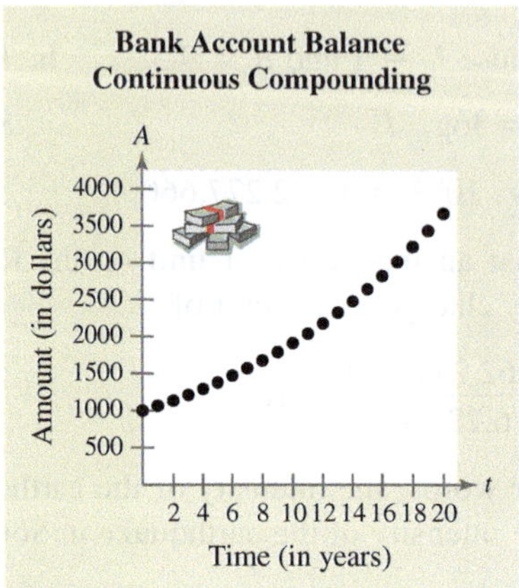

b.

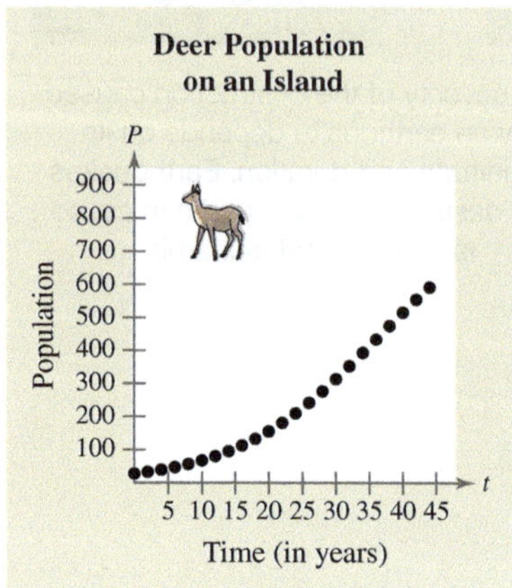

SOLUTION

a. As long as withdrawals and deposits are not made and the interest rate remains constant, the bank account balance will grow exponentially. So, an exponential growth model is an appropriate model.

b. The growth of the deer population will slow as the population approaches the carrying capacity of the island. So, a logistic growth model is an appropriate model.

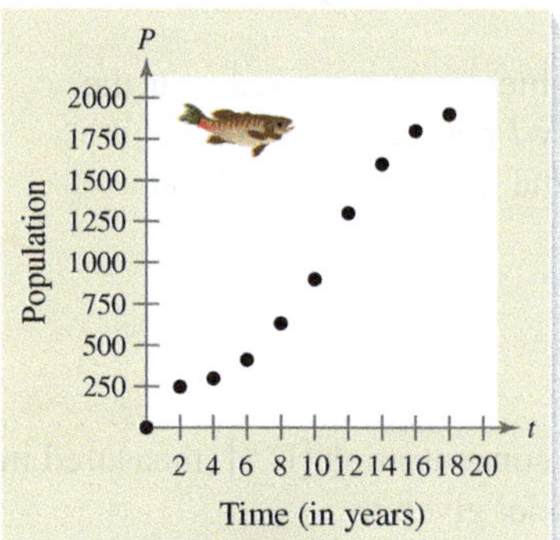

✓ Checkpoint 8

Decide whether to use an exponential growth model or a logistic growth model to represent the data for the fish population of a lake, shown in the figure at the left.

SUMMARIZE (Section 4.6)

1. State the functions for an exponential growth model and an exponential decay model *(page 387)*. For examples of finding and using exponential growth and decay models, see Examples 1, 2, and 3.
2. State the function for a Gaussian model and describe its shape *(page 391)*. For an example of using a Gaussian model, see Example 4.
3. State the function for a logistic growth model *(page 392)*. For an example of using a logistic growth model, see Example 5.
4. State the functions for logarithmic models *(page 387)*. For examples of using logarithmic models, see Examples 6 and 7.

SKILLS WARM UP 4.6

The following warm-up exercises involve skills that were covered in earlier sections. You will use these skills in the exercise set for this section. For additional help, review Sections 4.2, 4.3, and 4.5.

In Exercises 1–6, sketch the graph of the equation.

1. $y = e^{0.1x}$

2. $y = e^{-0.25x}$

3. $y = e^{-x^2/5}$

4. $y = \dfrac{2}{1 + e^{-x}}$

5. $y = \log_{10} 2x$

6. $y = \ln 4x$

In Exercises 7 and 8, solve the equation algebraically.

7. $3e^{2x} = 7$

8. $4 \ln 5x = 14$

In Exercises 9 and 10, solve the equation graphically.

9. $2e^{-0.2x} = 0.002$

10. $6 \ln 2x = 12$

Exercises 4.6

See www.CalcChat.com for worked-out solutions to odd-numbered exercises.

Matching In Exercises 1–6, match the function with its graph. [The graphs are labeled (a), (b), (c), (d), (e), and (f).]

(a)

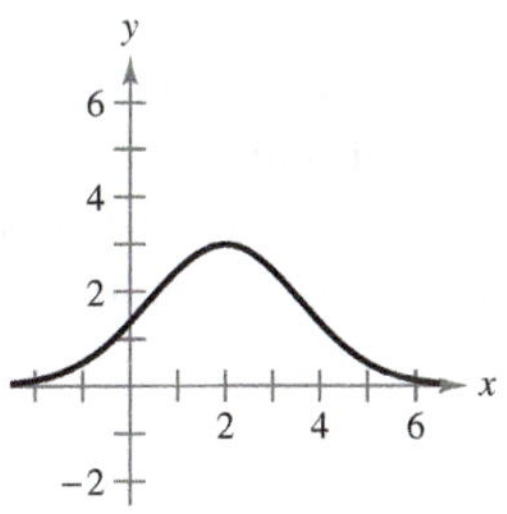

(b)

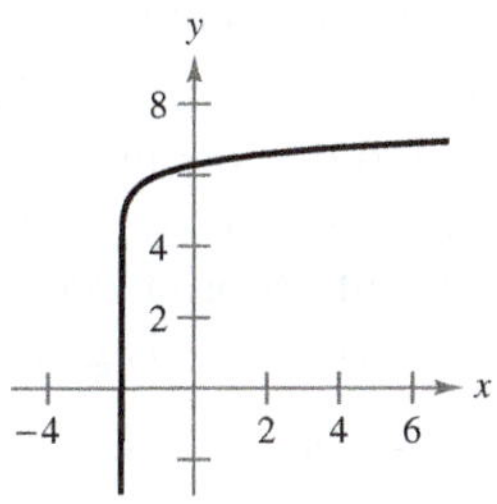

(c)

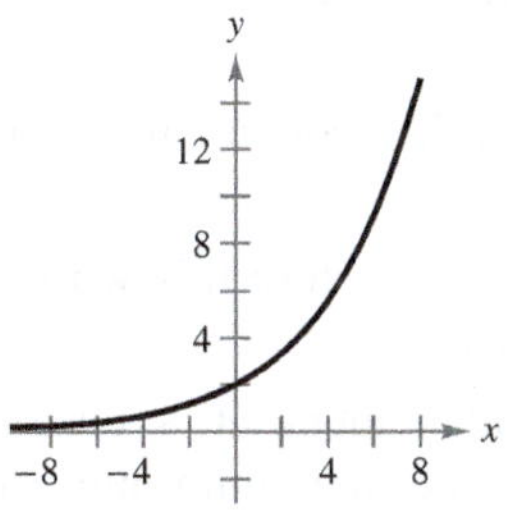

(d)

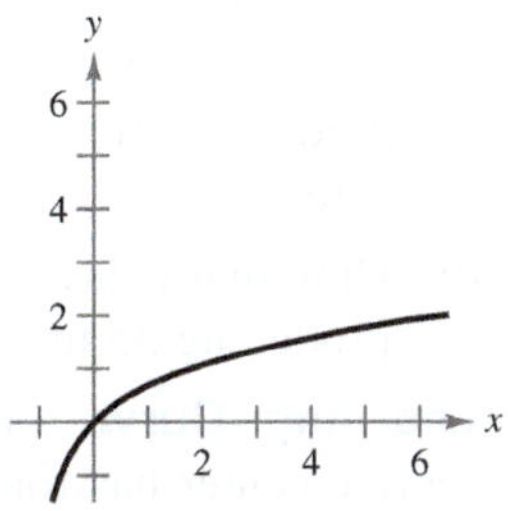

(e)

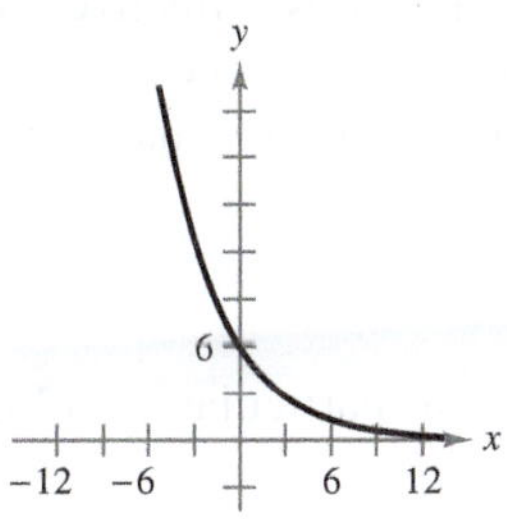

(f)

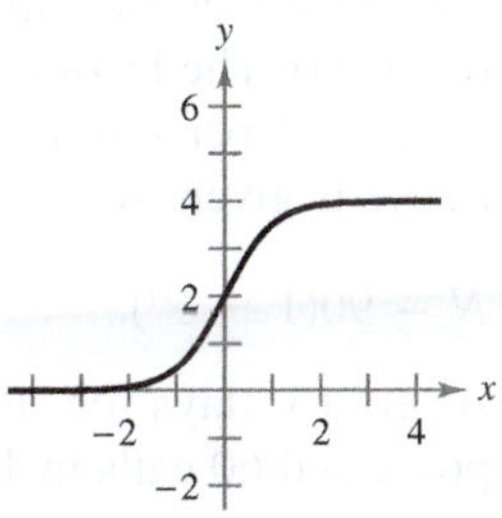

1. $y = 2e^{x/4}$

2. $y = 6e^{-x/4}$

3. $y = 6 + \log_{10}(x + 2)$

4. $y = 3e^{-(x-2)^2/5}$

5. $y = \ln(x + 1)$

6. $y = \dfrac{4}{1 + e^{-2x}}$

Classifying an Exponential Model In Exercises 7–10, classify the model as an exponential growth model or an exponential decay model.

7. $y = 3e^{0.5t}$

8. $y = 2e^{-0.6t}$

9. $y = 20e^{-1.5t}$

10. $y = 4e^{0.07t}$

11. Population The population P of a city is given by

$$P = 120{,}000e^{0.016t}$$

where t represents the year, with $t = 0$ corresponding to 2000. Sketch the graph of this equation. Use the model to predict the year in which the population of the city will reach 180,000.

12. Population The population P of a city is given by

$$P = 240{,}360e^{0.012t}$$

where t represents the year, with $t = 0$ corresponding to 2000. Sketch the graph of this equation. Use the model to predict the year in which the population of the city will reach 300,000.

Finding an Exponential Growth or Decay Model In Exercises 13–16, find the constants C and k such that the exponential function $y = Ce^{kt}$ passes through the points on the graph. *See Example 2.*

13.

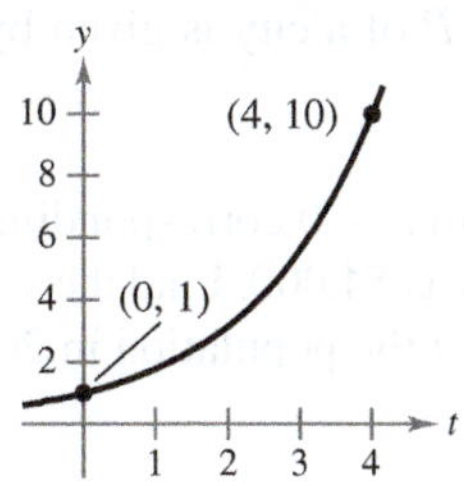

14.

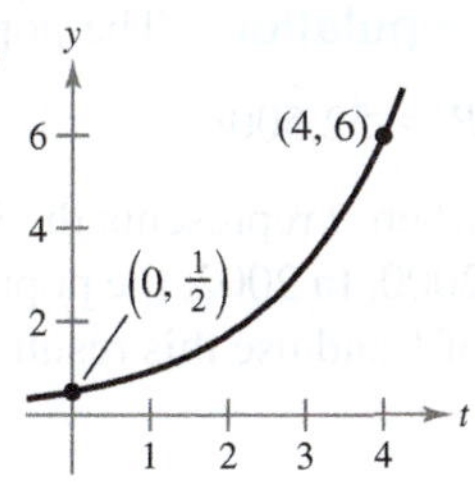

15. 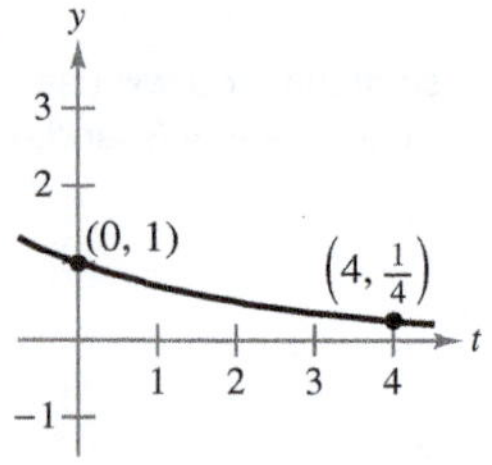

16. 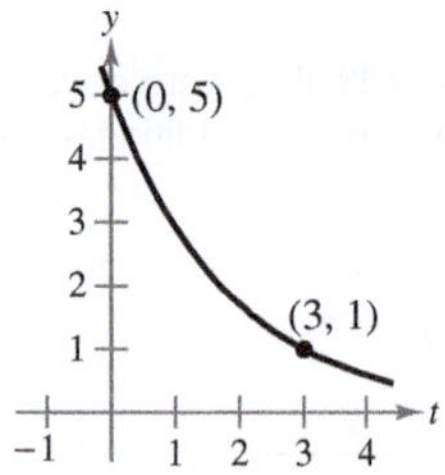

Compound Interest In Exercises 17–26, use the account balance formula $A = Pe^{rt}$ to complete the table for a savings account in which interest is compounded continuously.

	Initial Investment	Annual % Rate	Time to Double	Amount After 10 Years
17.	\$5000	7%		
18.	\$1000	$9\frac{1}{4}$%		
19.	\$500		10 yr	
20.	\$10,000		5 yr	
21.	\$1000			\$2281.88
22.	\$2000			\$3000
23.		11%		\$19,205
24.		8%		\$20,000
25.	\$5000			\$11,127.70
26.	\$250			\$600

Radioactive Decay In Exercises 27–32, use the exponential decay model $y = ae^{-bt}$ to complete the table for the radioactive isotope.

	Isotope	Half-Life (Years)	Initial Quantity	Amount After 1000 Years
27.	^{226}Ra	1599	4 g	
28.	^{226}Ra	1599		0.15 g
29.	^{14}C	5715		3.5 g
30.	^{14}C	5715	8 g	
31.	^{239}Pu	24,100		1.6 g
32.	^{239}Pu	24,100		0.38 g

33. Population The population P of a city is given by

$$P = 25{,}000e^{kt}$$

where t represents the year, with $t = 0$ corresponding to 2000. In 1980, the population was 15,000. Find the value of k and use this result to predict the population in 2015.

34. Population The population P of a city is given by

$$P = 52{,}500e^{kt}$$

where t represents the year, with $t = 0$ corresponding to 2000. In 2002, the population was 54,000. Find the value of k and use this result to predict the population in 2015.

35. Bacteria Growth The number N of bacteria in a culture is given by the model $N = 100e^{kt}$, where t is the time (in hours), with $t = 0$ corresponding to the time when $N = 100$. When $t = 6$, there are 140 bacteria. How long does it take the bacteria population to double in size? To triple in size?

36. Bacteria Growth The number N of bacteria in a culture is given by the model $N = 250e^{kt}$, where t is the time (in hours), with $t = 0$ corresponding to the time when $N = 250$. When $t = 10$, there are 320 bacteria. How long does it take the bacteria population to double in size? To triple in size?

37. Carbon Dating The ratio of carbon 14 to carbon 12 in a piece of wood discovered in a cave is $R = 1/8^{14}$. Estimate the age of the piece of wood.

38. Carbon Dating The ratio of carbon 14 to carbon 12 in a piece of paper buried in a tomb is $R = 1/13^{11}$. Estimate the age of the piece of paper.

39. Radioactive Decay What percent of a present amount of radioactive cesium (^{137}Cs) will remain after 100 years? Use the fact that radioactive cesium has a half-life of 30 years.

40. Radioactive Decay Find the half-life of radioactive iodine (^{131}I) if, after 20 days, 0.53 kilogram of an initial 3 kilograms remains.

41. Learning Curve The management at a factory has found that the maximum number of units a worker can produce in a day is 40. The learning curve for the number of units N produced per day after a new employee has worked t days is given by

$$N = 40(1 - e^{kt}).$$

After 20 days on the job, a particular worker produced 25 units in 1 day.

(a) Find the learning curve for this worker (first find the value of k).

(b) How many days should pass before this worker is producing 35 units per day?

42. Learning Curve The management at a customer service center has found that the maximum number of customer calls an employee can process effectively in a day is 90. The learning curve for the number N of calls processed per day after a new employee has worked t days is given by

$$N = 90(1 - e^{kt}).$$

After 15 days on the job, a particular employee processed 60 calls in 1 day.

(a) Find the learning curve for this worker (first find the value of k).

(b) How many days should pass before this employee will process 80 calls per day?

43. IQ Scores The IQ scores for the students at a college roughly follow the normal distribution

$$y = 0.0266e^{-(x-100)^2/450}, \quad 70 \le x \le 125$$

where x is the IQ score (see figure). Use the graph to estimate the average IQ score for the students.

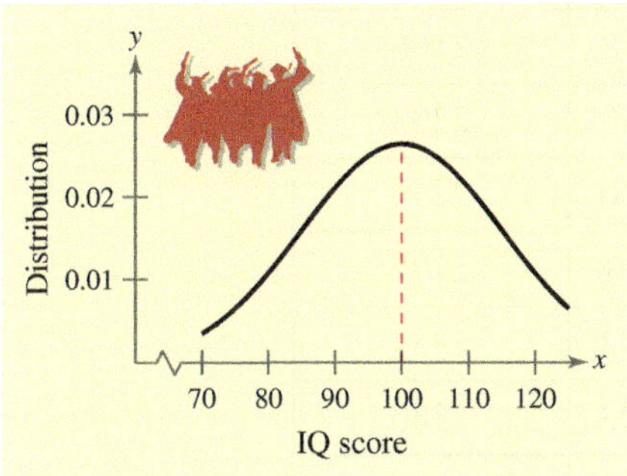

44. Education The amount of time (in hours per week) a student spends at a math-tutoring center roughly follows the normal distribution

$$y = 0.7979e^{-(x-5.4)^2/0.5}, \quad 4 \le x \le 7$$

where x is the number of hours.

(a) Use a graphing utility to graph the function.

(b) From the graph in part (a), estimate the average number of hours per week a student spends at the tutoring center.

45. Stocking a Lake with Fish A lake is stocked with 500 fish, and the fish population P increases according to the logistic curve

$$P = \frac{10{,}000}{1 + 19e^{-t/5}}, \quad t \ge 0$$

where t is the time (in months).

(a) Use a graphing utility to graph the logistic curve.

(b) Find the fish population after 5 months.

(c) After how many months will the fish population reach 2000?

46. Endangered Species A conservation organization releases 100 animals of an endangered species into a game preserve. The organization believes that the preserve has a carrying capacity of 1000 animals and that the growth of the herd will be modeled by the logistic curve

$$p = \frac{1000}{1 + 9e^{-kt}}, \quad t \ge 0$$

where p is the number of animals and t is the time (in years). The herd size is 134 after 2 years. Find k. Then find the population after 5 years.

47. Think About It Does the graph of a logistic growth model have a relative maximum value?

48. HOW DO YOU SEE IT? Match the description of each situation with its graph. Explain your reasoning. [The graphs are labeled (i), (ii), (iii), and (iv).]

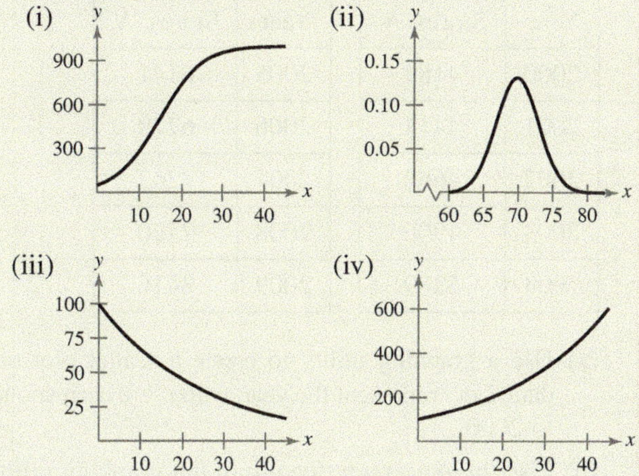

(a) The number of trout in a lake over time

(b) The amount of a radioactive substance over time

(c) The distribution of heights of men

(d) The balance of an interest-bearing account over time

49. Aged Population The table shows the projected U.S. populations P (in thousands) of people who are 85 years old or older for several years from 2020 through 2050. *(Source: U.S. Census Bureau)*

Year	Population, P
2020	6597
2025	7239
2030	8745
2035	11,450
2040	14,198
2045	16,985
2050	19,041

(a) Use a graphing utility to create a scatter plot of the data. Let t represent the year, with $t = 20$ corresponding to 2020.

(b) Use the *regression* feature of the graphing utility to find an exponential model for the data. Use the Inverse Property $b = e^{\ln b}$ to rewrite the model as an exponential model in base e.

(c) Use the graphing utility to graph the exponential model in base e.

(d) Use the exponential model in base e to estimate the populations of people who are 85 years old or older in 2022 and 2042.

50. Number of Stores The table shows the number N of stores operated by Wal-Mart each year from 2000 through 2009. *(Source: Wal-Mart Stores, Inc.)*

Year	Stores, N	Year	Stores, N
2000	4189	2005	6141
2001	4414	2006	6779
2002	4688	2007	7262
2003	4906	2008	7720
2004	5289	2009	8416

(a) Use a graphing utility to create a scatter plot of the data. Let t represent the year, with $t = 0$ corresponding to 2000.

(b) Use the *regression* feature of the graphing utility to find an exponential model for the data. Use the Inverse Property $b = e^{\ln b}$ to rewrite the model as an exponential model in base e.

(c) Use the graphing utility to graph the exponential model in base e.

(d) Use the exponential model in base e to predict the numbers of stores Wal-Mart will operate in 2010 and 2012.

51. Island Population The table shows the population P of an island town every five years for forty years.

Year	Population, P	Year	Population, P
0	280	25	2340
5	390	30	2830
10	650	35	3100
15	1160	40	3240
20	1730		

(a) Use a graphing utility to create a scatter plot of the data. Let t represent the year.

(b) Based on the scatter plot, do you think the data can be better represented by an exponential growth model or a logistic growth model? Explain.

(c) Use the *regression* feature of the graphing utility to find a model of the type you chose in part (b) for the data.

(d) Use the graphing utility to graph the model with the scatter plot. Does the model fit the data well?

(e) Use the model to predict the population of the island town in year 50.

52. Population The populations P of Utah (in thousands) from 1997 through 2009 are shown in the table. *(Source: U.S. Census Bureau)*

Year	Population, P
1997	2120
1998	2166
1999	2203
2000	2244
2001	2291
2002	2334
2003	2380
2004	2439
2005	2500
2006	2584
2007	2664
2008	2727
2009	2785

(a) Use a graphing utility to create a scatter plot of the data. Let t represent the year, with $t = 7$ corresponding to 1997.

(b) Use the *regression* feature of the graphing utility to find an exponential model for the data. Use the Inverse Property

$$b = e^{\ln b}$$

to rewrite the model as an exponential model in base e.

(c) Use the *regression* feature of the graphing utility to find a linear model and a quadratic model for the data.

(d) Use the graphing utility to graph the exponential model in base e and the models in part (c) with the scatter plot.

(e) Use each model to predict the populations in 2010, 2012, and 2014. Do all of the models give reasonable predictions? Explain.

Earthquake Magnitudes In Exercises 53 and 54, use the Richter scale for measuring the magnitudes of earthquakes. *See Example 6.*

53. Find the magnitude R (on the Richter scale) of an earthquake of intensity I. (Let $I_0 = 1$.)

(a) $I = 80{,}500{,}000$

(b) $I = 48{,}275{,}000$

54. Find the intensity I of an earthquake measuring R on the Richter scale. (Let $I_0 = 1$.)

(a) In 2011, an earthquake in the Vanuatu Islands had a magnitude of $R = 6.3$.

(b) In 2011, an earthquake near the coast of Japan had a magnitude of $R = 9.0$.

Intensity of Sound In Exercises 55 and 56, find the level of sound using the following information for determining sound intensity. The level of sound L (in decibels) of a sound with an intensity of I is given by

$$L = 10 \log_{10} \frac{I}{I_0}$$

where I_0 is an intensity of 10^{-12} watt per square meter, corresponding roughly to the faintest sound that can be heard by the human ear.

55. (a) $I = 10^{-10}$ watt per square meter (quiet room)

(b) $I = 10^{-5}$ watt per square meter (busy street corner)

56. (a) $I = 10^{-3}$ watt per square meter (loud car horn)

(b) $I \approx 10^{0}$ watt per square meter (threshold of pain)

pH Levels In Exercises 57–60, use the model $\text{pH} = -\log_{10}[H^+]$. *See Example 7.*

57. Compute the hydrogen ion concentration $[H^+]$ of a solution for which pH = 5.8.

58. Compute the hydrogen ion concentration $[H^+]$ of a solution for which pH = 7.3.

59. A grape has a pH of 3.5, and baking soda has a pH of 8.0. The hydrogen ion concentration of the grape is how many times that of the baking soda?

60. The pH of a solution is decreased by one unit. The hydrogen ion concentration is increased by what factor?

61. Estimating the Time of Death At 8:30 A.M., a coroner was called to the home of a person who had died during the night. The coroner assumed that the person had a normal body temperature of 98.6°F at death, and that the room temperature was a constant 70°F. From these two temperatures, the coroner was able to determine that the time elapsed since death and the body temperature are related by the formula

$$t = -10 \ln \frac{T - 70}{98.6 - 70}$$

where t is the time (in hours) elapsed since the person died, and T is the temperature (in degrees Fahrenheit) of the person's body. In order to estimate the time of death, the coroner took the person's temperature twice. At 9:00 A.M. the temperature was 85.7°F, and at 11:00 A.M. the temperature was 82.8°F. Use this formula to estimate the time of death of the person.

62. Thawing a Package of Steaks You take a three-pound package of steaks out of the freezer at 11 A.M. and place it in the refrigerator. Will the steaks be thawed in time to be grilled at 6 P.M.? Assume that the refrigerator temperature is 40°F and that the freezer temperature is 0°F. Use the formula for Newton's Law of Cooling

$$t = -5.05 \ln \frac{T - 40}{0 - 40}$$

where t is the time in hours (with $t = 0$ corresponding to 11 A.M.) and T is the temperature of the package of steaks (in degrees Fahrenheit).

63. Worker's Productivity The numbers n of units per day that a new worker can produce after t days on the job are listed in the table.

Days, t	5	10	15	20	25
Units, n	6	13	22	34	56

(a) Use a graphing utility to create a scatter plot of the data. Do the data fit an exponential model or a logarithmic model?

(b) Use the *regression* feature of the graphing utility to find the model of the type you chose in part (a).

(c) Use the graphing utility to graph the model with the data. Does the model fit the data well? Can you think of a better model to use for these data? Explain.

64. Chemical Reaction The table shows the yield y (in milligrams) of a chemical reaction after x minutes.

Minutes, x	Yield, y
1	1.5
2	7.4
3	10.2
4	13.4
5	15.8
6	16.3
7	18.2
8	18.3

(a) Use a graphing utility to create a scatter plot of the data. Do the data fit an exponential model or a logarithmic model?

(b) Use the *regression* feature of the graphing utility to find the model of the type you chose in part (a).

(c) Use the graphing utility to graph the model with the data. Does the model fit the data well?

ALGEBRA TUTOR

Using Properties of Exponents

Much of the algebra in Chapter 4 involves evaluating expressions by using the properties of exponents. On this page, you can review some of these properties.

Let a and b be real numbers, variables, or algebraic expressions, and let m and n be integers. (Assume all denominators and bases are nonzero.)

1. $a^m a^n = a^{m+n}$ Product of Powers

2. $\dfrac{a^m}{a^n} = a^{m-n}$ Quotient of Powers

3. $(ab)^m = a^m b^m$ Power of a Product

4. $\left(\dfrac{a}{b}\right)^m = \dfrac{a^m}{b^m}$ Power of a Quotient

5. $(a^m)^n = a^{mn}$ Power of a Power

6. $a^{-n} = \dfrac{1}{a^n}$ Definition of negative exponent

7. $a^0 = 1, \quad a \neq 0$ Definition of zero exponent

8. $\left(\dfrac{a}{b}\right)^{-n} = \left(\dfrac{b}{a}\right)^n, \quad a \neq 0, b \neq 0$

9. $|a^2| = |a|^2 = a^2$

Example 1 Using Properties of Exponents

a. $(2^4)^3 = 2^{4 \cdot 3}$ Power of a Power

$= 2^{12}$ Multiply exponents.

$= 4096$ Simplify.

b. $10^{12} \cdot \dfrac{1}{10^{13}} = \dfrac{10^{12}}{10^{13}}$ Multiply.

$= 10^{12-13}$ Quotient of Powers

$= 10^{-1}$ Subtract exponents.

$= \dfrac{1}{10}$ Definition of negative exponent

c. $8^{-2} \cdot 4^0 = \dfrac{1}{8^2} \cdot 4^0$ Definition of negative exponent

$= \dfrac{1}{8^2} \cdot 1$ Definition of zero exponent

$= \dfrac{1}{64}$ Simplify.

d. $\left(\dfrac{2}{3}\right)^{4(2)} = \left(\dfrac{2}{3}\right)^8$ Multiply exponents.

$= \dfrac{2^8}{3^8}$ Power of a Quotient

$= \dfrac{256}{6561}$ Simplify.

e. $\left(\dfrac{3}{4}\right)^{-2} = \left(\dfrac{4}{3}\right)^2$ Property 8

$= \dfrac{16}{9}$ Power of a Quotient

STUDY TIP

Be sure to remember that these properties apply for all integers m and n, not just positive integers.

Factoring Trinomials

To solve equations that are of quadratic type, you must be good at factoring trinomials. Some examples are presented on this page. Remember that to factor a trinomial of the form $ax^2 + bx + c$, use the following pattern.

Factors of a

$$ax^2 + bx + c = (\square x + \square)(\square x + \square)$$

Factors of c

The goal is to find a combination of factors of a and c such that the outer and inner products add up to the middle term bx.

Example 2 Factoring Trinomials

Factor each trinomial.

a. $x^2 + 7x + 12$ **b.** $x^2 - 6x + 5$ **c.** $x^2 - 14x + 45$ **d.** $2x^2 + 11x - 21$

SOLUTION

a. For $x^2 + 7x + 12$, you have $a = 1$, $b = 7$, and $c = 12$. Because b and c are positive, both factors of 12 must be positive. That is, $12 = (1)(12)$, $12 = (2)(6)$, or $12 = (3)(4)$. So, the possible factorizations of $x^2 + 7x + 12$ are

$$(x + 1)(x + 12), \quad (x + 2)(x + 6), \quad \text{and} \quad (x + 3)(x + 4).$$

Testing the middle term, you can find the correct factorization to be

$$x^2 + 7x + 12 = (x + 3)(x + 4).$$

b. For $x^2 - 6x + 5$, you have $a = 1$, $b = -6$, and $c = 5$. Because b is negative and c is positive, both factors of 5 must be negative. That is, $5 = (-1)(-5)$. So, the factorization is

$$x^2 - 6x + 5 = (x - 1)(x - 5).$$

c. For $x^2 - 14x + 45$, you have $a = 1$, $b = -14$, and $c = 45$. Because b is negative and c is positive, both factors of 45 must be negative. That is, $45 = (-1)(-45)$, $45 = (-3)(-15)$, or $45 = (-5)(-9)$. So, the possible factorizations of $x^2 - 14x + 45$ are

$$(x - 1)(x - 45), \quad (x - 3)(x - 15), \quad \text{and} \quad (x - 5)(x - 9).$$

Testing the middle term, you can find the correct factorization to be

$$x^2 - 14x + 45 = (x - 5)(x - 9).$$

d. For $2x^2 + 11x - 21$, you have $a = 2$ and $c = -21$, which means that the factors of -21 must have unlike signs. The eight possible factorizations are shown below.

$$\begin{array}{ll} (2x - 1)(x + 21) & (2x + 1)(x - 21) \\ (2x - 3)(x + 7) & (2x + 3)(x - 7) \\ (2x - 7)(x + 3) & (2x + 7)(x - 3) \\ (2x - 21)(x + 1) & (2x + 21)(x - 1) \end{array}$$

Testing the middle term, you can find the correct factorization to be

$$2x^2 + 11x - 21 = (2x - 3)(x + 7).$$

SUMMARY AND STUDY STRATEGIES

After studying this chapter, you should have acquired the following skills.
The exercise numbers are keyed to the Review Exercises that begin on page 404.
Answers to odd-numbered Review Exercises are given in the back of the text.*

Section 4.1	**Review Exercises**
■ Determine whether a function has an inverse function.	*1–6*
■ Find the inverse function of a function.	*3–6*
■ Graph a function and its inverse function.	*3–6*

Section 4.2	
■ Evaluate an exponential expression.	*7–12*
■ Sketch the graph of an exponential function.	*13–24*

Graph of $y = a^x$, $a > 1$
- Domain: $(-\infty, \infty)$, Range: $(0, \infty)$
- Intercept: $(0, 1)$
- Increasing
- x-axis is a horizontal asymptote: ($a^x \to 0$ as $x \to -\infty$)
- Continuous

Graph of $y = a^{-x}$, $a > 1$
- Domain: $(-\infty, \infty)$, Range: $(0, \infty)$
- Intercept: $(0, 1)$
- Decreasing
- x-axis is a horizontal asymptote: ($a^{-x} \to 0$ as $x \to \infty$)
- Continuous
- Reflection of graph of $y = a^x$ in y-axis

■ Use the compound interest formulas.	*25–29*

For n compoundings per year: $A = P(1 + r/n)^{nt}$

For continuous compounding: $A = Pe^{rt}$

■ Use an exponential model to solve an application problem.	*29, 30*

Section 4.3	
■ Recognize and evaluate a logarithmic function.	*31–44*

$y = \log_a x$ if and only if $x = a^y$ $\qquad$ $y = \log_e x = \ln x$

■ Sketch the graph of a logarithmic function.	*45–54*

Graph of $y = \log_a x$, $a > 1$
- Domain: $(0, \infty)$
- Range: $(-\infty, \infty)$
- Intercept: $(1, 0)$
- Increasing
- One-to-one; therefore has an inverse function
- y-axis is a vertical asymptote ($\log_a x \to -\infty$ as $x \to 0^+$)
- Continuous
- Reflection of graph of $y = a^x$ about the line $y = x$

■ Use a logarithmic model to solve an application problem.	*55–58*

* A wide range of valuable study aids are available to help you master the material in this chapter. The *Student Solutions Manual* includes step-by-step solutions to all odd-numbered exercises to help you review and prepare. The student website at *www.cengagebrain.com* offers algebra help and a *Graphing Technology Guide*, which contains step-by-step commands and instructions for a wide variety of graphing calculators.

Section 4.4

- Evaluate a logarithm using the change-of-base formula. *59–62*

$$\log_a x = \frac{\log_b x}{\log_b a}, \quad \log_a x = \frac{\log_{10} x}{\log_{10} a}, \quad \log_a x = \frac{\ln x}{\ln a}$$

- Use properties of logarithms to evaluate or rewrite a logarithmic expression. *63–70*

$$\log_a(uv) = \log_a u + \log_a v \qquad \ln(uv) = \ln u + \ln v$$

$$\log_a \frac{u}{v} = \log_a u - \log_a v \qquad \ln \frac{u}{v} = \ln u - \ln v$$

$$\log_a u^n = n \log_a u \qquad \ln u^n = n \ln u$$

- Use properties of logarithms to expand or condense a logarithmic expression. *71–82*
- Use logarithmic functions to model and solve real-life applications. *83, 84*

Section 4.5

- Solve an exponential equation. *85–90*
- Solve a logarithmic equation. *91–98*
- Use an exponential or a logarithmic model to solve an application problem. *99, 100*

Section 4.6

- Construct and use a model for exponential growth or exponential decay. *101–107*

$$y = ae^{bx}, \quad b > 0 \qquad y = ae^{-bx}, \quad b > 0$$

- Use a Gaussian model to solve an application problem. *108*

$$y = ae^{-(x-b)^2/c}$$

- Use a logistic growth model to solve an application problem. *109*

$$y = \frac{a}{1 + be^{-rx}}$$

- Use a logarithmic model to solve an application problem. *110, 111*

$$y = a + b \ln x, \quad y = a + b \log_{10} x$$

- Choose an appropriate model involving exponential or logarithmic functions for a real-life situation. *112*

Study Strategies

- **Solve Problems Algebraically or Graphically** When solving an exponential or logarithmic equation, you can use a variety of problem-solving strategies. For instance, you can solve the logarithmic equation $\ln(x + 4) - \ln x = 1$ *algebraically* or *graphically*.

  To solve the equation *algebraically*, you can use the properties of logarithms to rewrite the equation, exponentiate each side, use the Inverse Property, and solve the resulting equation to determine that $x \approx 2.328$.

  To solve the equation *graphically*, you can use a graphing utility to graph $y_1 = \ln(x + 4) - \ln x$ and $y_2 = 1$ in the same viewing window. Then use the *intersect* feature or the *zoom* and *trace* features to determine that the solution of the original equation is $x \approx 2.328$. (See figure.)

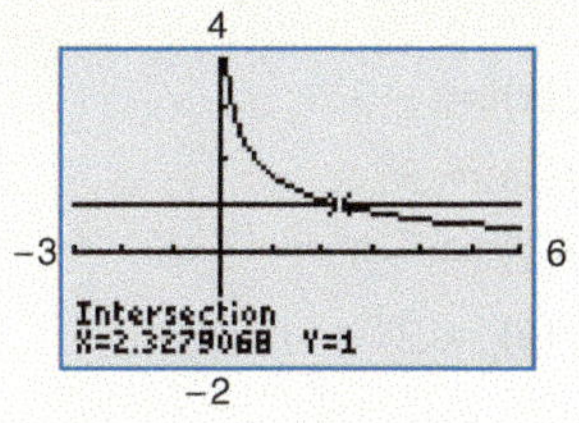

Review Exercises

See www.CalcChat.com for worked-out solutions to odd-numbered exercises.

Verifying Inverse Functions In Exercises 1 and 2, show that f and g are inverse functions of each other.

1. $f(x) = 3x + 5, \quad g(x) = \dfrac{x - 5}{3}$

2. $f(x) = \sqrt[3]{x - 3}, \quad g(x) = x^3 + 3$

Finding Inverse Functions In Exercises 3 and 4, determine whether the function has an inverse function. If it does, find its inverse function and graph f and f^{-1} in the same coordinate plane.

3. $f(x) = \dfrac{1}{x}$

4. $f(x) = \dfrac{x^2}{x^2 - 9}$

Finding and Sketching Inverse Functions In Exercises 5 and 6, (a) find f^{-1}, (b) sketch the graphs of f and f^{-1} in the same coordinate plane, and (c) verify that $f^{-1}(f(x)) = x$ and $f(f^{-1}(x)) = x$.

5. $f(x) = x^2, \quad x \geq 0$

6. $f(x) = \sqrt[3]{x - 1}$

Evaluating an Exponential Expression In Exercises 7–12, use a calculator to evaluate the expression. Round your result to three decimal places.

7. 4^{12}

8. $1.02^{-\pi}$

9. $2(100)^{-1/3}$

10. $7^{\sqrt{2}}$

11. $e^{3/4}$

12. $e^{-7.5}$

Matching In Exercises 13–16, match the function with its graph. [The graphs are labeled (a), (b), (c), and (d).]

(a)

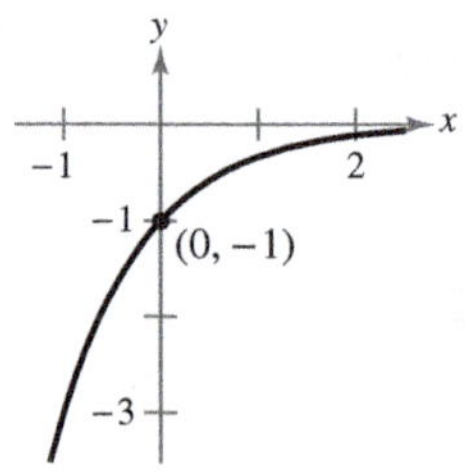

(b)

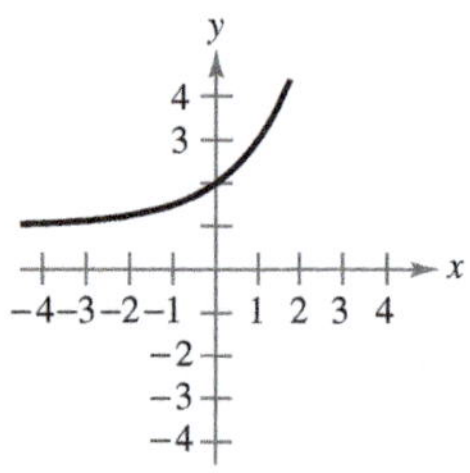

(c)

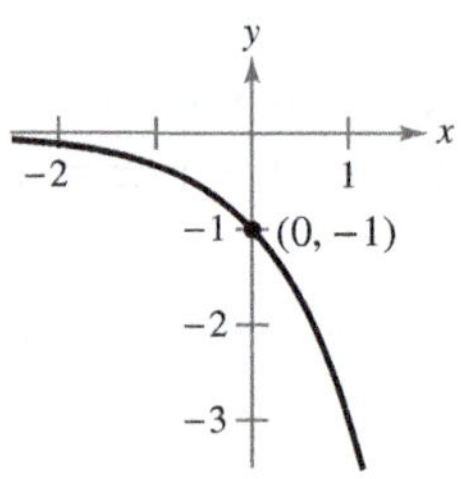

(d)

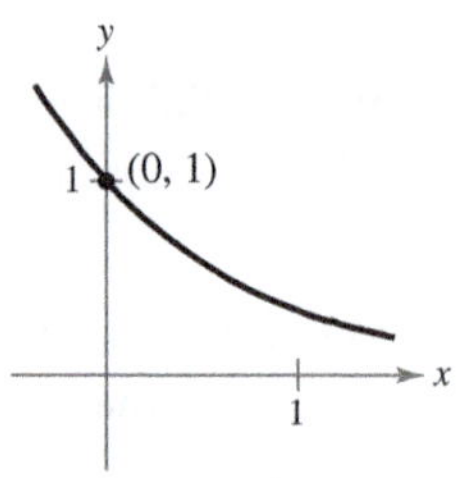

13. $f(x) = -3^x$

14. $f(x) = 3^{-x}$

15. $f(x) = -3^{-x}$

16. $f(x) = 2^x + 1$

Graphing Exponential Functions In Exercises 17–24, sketch the graph of the function.

17. $f(x) = 4^x$

18. $f(x) = 4^{x-1}$

19. $f(x) = \left(\frac{1}{2}\right)^x$

20. $f(x) = \left(\frac{1}{2}\right)^{x+1}$

21. $f(x) = 3e^{0.2x}$

22. $f(x) = 10e^{-0.1x}$

23. $f(x) = 3^{-x^2}$

24. $f(x) = 2^{1-x^2}$

Compound Interest In Exercises 25 and 26, complete the table to find the balance A for P dollars invested at rate r for t years, compounded n times per year.

n	1	2	4	12	365	Continuous
A						

25. $P = \$5000, \quad r = 8.5\%, \quad t = 12$ years

26. $P = \$8000, \quad r = 6.75\%, \quad t = 25$ years

Compound Interest In Exercises 27 and 28, complete the table to find the amount P that must be invested at rate r to obtain a balance of $A = \$200{,}000$ in t years.

t	1	10	20	30	40	50
P						

27. $r = 7.75\%$, compounded continuously

28. $r = 9.25\%$, compounded quarterly

29. Investment Plan You deposit \$6000 in a fund that yields 5.75% interest, compounded continuously. How much money will be in the fund after 6 years?

30. Population The population P of a town increases according to the model

$$P(t) = 15{,}000e^{0.025t}$$

where t is the time in years, with $t = 8$ corresponding to 2008. Use the model to approximate the population in 2009 and 2011.

Writing Logarithmic Equations In Exercises 31–34, write the exponential equation in logarithmic form.

31. $4^3 = 64$

32. $25^{3/2} = 125$

33. $e^2 = 7.3890 \ldots$

34. $e^x = 8$

Writing Exponential Equations In Exercises 35–38, write the logarithmic equation in exponential form.

35. $\log_3 81 = 4$

36. $\log_5 0.2 = -1$

37. $\ln 1 = 0$

38. $\ln 4 = 1.3862 \ldots$

Evaluating Logarithmic Expressions In Exercises 39–44, evaluate the expression without using a calculator.

39. $\log_2 32$

40. $\log_9 3$

41. $\ln e^7$

42. $\log_4 \frac{1}{4}$

43. $\ln e^{-1/2}$

44. $\ln 1$

Matching In Exercises 45–48, match the function with its graph. [The graphs are labeled (a), (b), (c), and (d).]

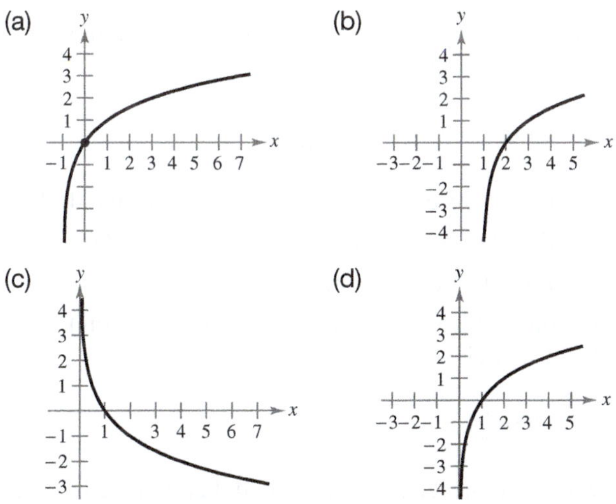

45. $f(x) = \log_2 x$ **46.** $f(x) = \log_2(x - 1)$

47. $f(x) = -\log_2 x$ **48.** $f(x) = \log_2(x + 1)$

Graphs of Exponential and Logarithmic Functions In Exercises 49 and 50, use the fact that f and g are inverse functions of each other to sketch their graphs in the same coordinate plane.

49. $f(x) = 10^x,\ g(x) = \log_{10} x$

50. $f(x) = e^x,\ g(x) = \ln x$

Sketching the Graphs of Logarithmic Functions In Exercises 51–54, find the domain, vertical asymptote, and x-intercept of the logarithmic function. Then sketch its graph.

51. $f(x) = \log_2(x - 3)$ **52.** $f(x) = 5 - 2\log_{10} x$

53. $g(x) = 2 \ln x$

54. $g(x) = \ln(4 - x)$

55. Human Memory Model Students in a sociology class were given an exam and then retested monthly for 6 months with an equivalent exam. The average score for the class is given by the human memory model

$$f(t) = 82 - 16\log_{10}(t + 1), \quad 0 \le t \le 6$$

where t is the time (in months). How did the average score change over the six-month period?

56. Investment Time A principal P, invested at 5.85% interest compounded continuously, increases to an amount that is K times the principal after t years, where t is given by

$$t = \frac{\ln K}{0.0585}.$$

Complete the table and describe the result.

K	1	2	3	4	6	8	10
t							

57. Antler Spread The antler spread a (in inches) and shoulder height h (in inches) of an adult American elk are related by the model

$$h = 116\log_{10}(a + 40) - 176.$$

(a) Approximate the shoulder height of an elk with an antler spread of 55 inches.

(b) Use a graphing utility to graph the model.

58. Snow Removal The number of miles s of roads cleared of snow is approximated by the model

$$s = 25 - \frac{13\ln(h/12)}{\ln 3}, \quad 2 \le h \le 15$$

where h is the depth of the snow in inches.

(a) Use the model to find s when $h = 10$ inches.

(b) Use a graphing utility to graph the model.

Changing Bases In Exercises 59–62, evaluate the logarithm using the change-of-base formula. Evaluate each problem twice, once with common logarithms and once with natural logarithms. (Round your answer to three decimal places.)

59. $\log_3 10$ **60.** $\log_{1/4} 7$

61. $\log_{12} 200$

62. $\log_3 0.28$

Using Properties of Logarithms In Exercises 63–66, approximate the logarithm using the properties of logarithms, given $\log_b 2 \approx 0.3562$, $\log_b 3 \approx 0.5646$, and $\log_b 5 \approx 0.8271$.

63. $\log_b 6$ **64.** $\log_b \frac{4}{25}$

65. $\log_b \sqrt{3}$ **66.** $\log_b 30$

Using Inverse Properties In Exercises 67–70, find the *exact* value of the logarithm.

67. $\log_7 49$

68. $\log_6 \frac{1}{36}$

69. $\ln e^{3.2}$

70. $\ln \sqrt[5]{e^3}$

Expanding Logarithmic Expressions In Exercises 71–76, use the properties of logarithms to expand the expression as a sum, difference, and/or multiple of logarithms. (Assume that all variables are positive.)

71. $\log_{10} \dfrac{x}{y}$

72. $\log_{10} \dfrac{xy^3}{z^2}$

73. $\ln\left(x\sqrt{x-3}\right)$

74. $\ln \sqrt[3]{\dfrac{x^3}{y^2}}$

75. $\log_5(y-3)^4$

76. $\log_2 2xy^2 z$

Condensing Logarithmic Expressions In Exercises 77–82, condense the expression to the logarithm of a single quantity.

77. $\log_4 2 + \log_4 3$

78. $\ln y + 2 \ln z$

79. $\frac{1}{2} \ln x$

80. $4 \log_3 x + \log_3 y - 2 \log_3 z$

81. $\ln x - \ln(x-3) - \ln(x+1)$

82. $\log_{10}(x+2) + 2 \log_{10} x - 3 \log_{10}(x+4)$

83. Finding a Mathematical Model Find a logarithmic equation that relates y and x (see figure). Explain the steps used to find the equation.

x	1	2	3	4	5	6
y	1	2.520	4.327	6.350	8.550	10.903

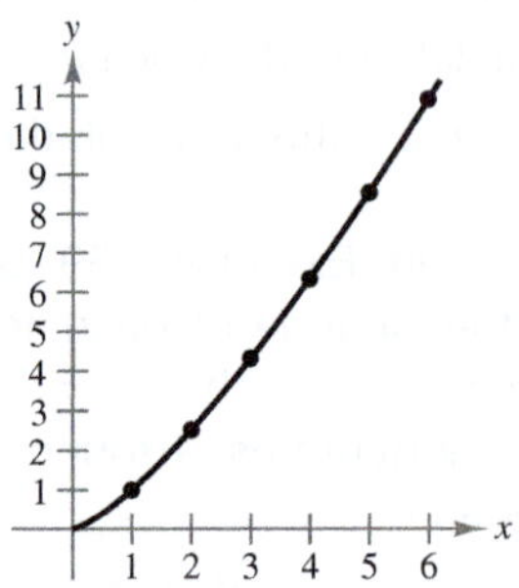

84. Human Memory Model Students in a learning theory study were given an exam and then retested monthly for 6 months with an equivalent exam. The average scores for the class are shown in the table, with $t = 1$ representing 1 month after the initial exam. Use the table to find a logarithmic equation that relates s and t.

Month, t	1	2	3	4	5	6
Score, s	87.9	79.7	74.8	71.3	68.6	66.5

Solving Exponential and Logarithmic Equations In Exercises 85–98, solve the equation. Approximate the result to three decimal places.

85. $e^x = 8$

86. $2e^{x+1} = 7$

87. $3^{4x-1} - 4 = 23$

88. $2^{3x+1} + 5 = 133$

89. $e^{2x} - 3e^x - 4 = 0$

90. $e^{2x} - 8e^x + 12 = 0$

91. $\ln 3x = 8.2$

92. $2 \log_3 4x = 15$

93. $-2 + \ln 5x = 0$

94. $\ln 4x^2 = 21$

95. $\ln x - \ln 3 = 2$

96. $\log_3 x - \log_3 4 = 5$

97. $\log_2 \sqrt[3]{x+1} = 1$

98. $\ln \sqrt{x+1} = 2$

99. Demand Function The demand x (in units) for a desk is given by

$$p = 6000\left(1 - \frac{5}{5 + e^{-0.004x}}\right)$$

where p represents the price (in dollars) of the desk.

(a) Find the demand x for a price of $p = \$500$.

(b) Find the demand x for a price of $p = \$400$.

100. Demand Function The demand x (in units) for a bicycle is given by

$$p = 4000\left(1 - \frac{3}{3 + e^{-0.004x}}\right)$$

where p represents the price (in dollars) of the bicycle.

(a) Find the demand x for a price of $p = \$700$.

(b) Find the demand x for a price of $p = \$400$.

Radioactive Decay In Exercises 101 and 102, complete the table for the radioactive isotope.

	Isotope	*Half-Life (Years)*	*Initial Quantity*	*Amount After 1000 Years*
101.	^{14}C	5715	12 g	
102.	^{239}Pu	24,100		3.1 g

103. Population The population P of a city is given by

$$P = 185{,}000e^{0.018t}$$

where t represents the year, with $t = 10$ corresponding to 2010.

(a) Use a graphing utility to graph this equation.

(b) Use the model to predict the year in which the population of the city will reach 270,000.

104. Population The population P of a city is given by

$$P = 50{,}000e^{kt}$$

where t represents the year, with $t = 0$ corresponding to 2000. In 1990, the population was 34,500.

(a) Find the value of k and use this result to predict the population in the year 2025.

(b) Use a graphing utility to confirm the result of part (a).

105. Bacteria Growth The number of bacteria N in a culture is given by the model $N = 250e^{kt}$, where t is the time (in hours), with $t = 0$ corresponding to the time when $N = 250$. When $t = 6$, there are 380 bacteria. How long does it take the bacteria population to double in size? To triple in size?

106. Bacteria Growth The number of bacteria N in a culture is given by the model

$$N = 200e^{kt}$$

where t is the time (in hours), with $t = 0$ corresponding to the time when $N = 200$. When $t = 5$, there are 325 bacteria. How long does it take for the bacteria population to double in size? To triple in size?

107. Learning Curve The management at a factory has found that the maximum number of units a worker can produce in a day is 50. The learning curve for the number of units N produced per day after a new employee has worked t days is given by

$$N = 50(1 - e^{kt}).$$

After 20 days on the job, a particular worker produced 31 units in 1 day.

(a) Find the learning curve for this worker.

(b) How many days should pass before this worker is producing 45 units per day?

108. Test Scores The scores on a general aptitude test roughly follow a normal distribution given by

$$y = 0.0040e^{-[(x-300)^2]/20{,}000}, \quad 100 \le x \le 500.$$

Sketch the graph of this function. Estimate the average score on this test.

109. Wildlife Management A state parks and wildlife department releases 100 deer into a wilderness area. The department believes that the carrying capacity of the area is 400 deer and that the growth of the herd will be modeled by the logistic curve

$$P = \frac{400}{1 + 3e^{-kt}}, \quad t \ge 0$$

where P is the number of deer and t is the time (in years).

(a) The herd size is 135 after 2 years. Find k.

(b) Find the populations after 5 years, after 10 years, and after 20 years.

110. Perrigo Sales The annual sales S (in millions of dollars) for the Perrigo Company from 2004 through 2010 are shown in the table. *(Source: Perrigo Company)*

Year	Sales, S
2004	898.2
2005	1024.1
2006	1366.8
2007	1447.4
2008	1822.1
2009	2006.9
2010	2268.9

(a) Use a graphing utility to create a scatter plot of the data. Let t represent the year, with $t = 4$ corresponding to 2004.

(b) Use the *regression* feature of the graphing utility to find an exponential model for the data. Use the Inverse Property $b = e^{\ln b}$ to rewrite the model as an exponential model in base e.

(c) Use the *regression* feature of the graphing utility to find a logarithmic model for the data.

(d) Use the exponential model in base e and the logarithmic model to predict sales in 2011. It is projected that sales in 2011 will be \$2740 million. Do the predictions from the two models agree with this projection? Explain.

111. Thawing a Package of Steaks You take a package of steaks out of a freezer at 10 A.M. and place it in the refrigerator. Will the steaks be thawed in time to be grilled at 6 P.M.? Assume that the refrigerator temperature is 40°F and the freezer temperature is 0°F. Use the formula

$$t = -3.95 \ln \frac{T - 40}{0 - 40}$$

where t is the time in hours (with $t = 0$ corresponding to 10 A.M.) and T is the temperature of the package of steaks (in degrees Fahrenheit).

112. Earthquake Magnitudes On the Richter scale, the magnitude R of an earthquake of intensity I is given by

$$R = \log_{10} \frac{I}{I_0}$$

where $I_0 = 1$ is the minimum intensity used for comparison. Find the intensity per unit of area for each value of R.

(a) $R = 8.4$ (b) $R = 6.85$ (c) $R = 9.1$

TEST YOURSELF

See www.CalcChat.com for worked-out solutions to odd-numbered exercises.

Take this test as you would take a test in class. When you are done, check your work against the answers given in the back of the book.

In Exercises 1 and 2, determine whether the function has an inverse function. If it does, find the inverse function and graph f and f^{-1} in the same coordinate plane.

1. $f(x) = 5x - 7$

2. $f(x) = x^2 + 1$

In Exercises 3–6, sketch the graph of the function.

3. $y = 2^x$

4. $y = e^{-2x}$

5. $y = \ln x$

6. $y = \log_3(x - 1)$

In Exercises 7 and 8, students in a psychology class were given an exam and then retested monthly with an equivalent exam. The average score $f(t)$ for the class is given by the human memory model

$$f(t) = 87 - 15 \log_{10}(t + 1), \quad 0 \le t \le 18$$

where t is the time (in months).

7. What was the average score on the original exam? After 2 months? After 4 months?

8. Find the average scores after 6 months, 12 months, and 18 months. What might these scores indicate about human memory?

In Exercises 9–12, expand the logarithmic expression. (Assume all variables are positive).

9. $\ln \dfrac{x^2y^3}{z}$

10. $\log_{10} 3xyz^2$

11. $\log_2\left(x\sqrt[3]{x - 2}\right)$

12. $\log_8 \sqrt[5]{x^2 + 1}$

In Exercises 13 and 14, condense the logarithmic expression.

13. $2 \ln x + 3 \ln y - \ln z$

14. $\frac{2}{3}(\log_{10} x + \log_{10} y)$

In Exercises 15–18, solve the equation. Approximate the result to three decimal places.

15. $2^{4x} = 21$

16. $e^{2x} - 8e^x + 12 = 0$

17. $\log_2(x + 1) - 7 = 0$

18. $\ln \sqrt{x + 2} = 3$

19. You deposit \$40,000 in a fund that pays 6.75% interest, compounded continuously. When will the balance be greater than \$120,000?

20. The population P of a city is given by

$$P = 85{,}000e^{0.025t}$$

where t represents the year, with $t = 8$ corresponding to 2008. When will the city have a population of 125,000? Explain.

21. The number of bacteria N in a culture is given by

$$N = 100e^{kt}$$

where t is the time (in hours), with $t = 0$ corresponding to the time when $N = 100$. When $t = 8$, $N = 175$. How long does it take the bacteria population to double?

22. Carbon 14 has a half-life of 5715 years. You have an initial quantity of 10 grams. How many grams will remain after 10,000 years? After 20,000 years?

23. Scientists release several bears on a small Alaskan island to start a bear population. Is the population likely to grow exponentially or logistically? Explain your reasoning.

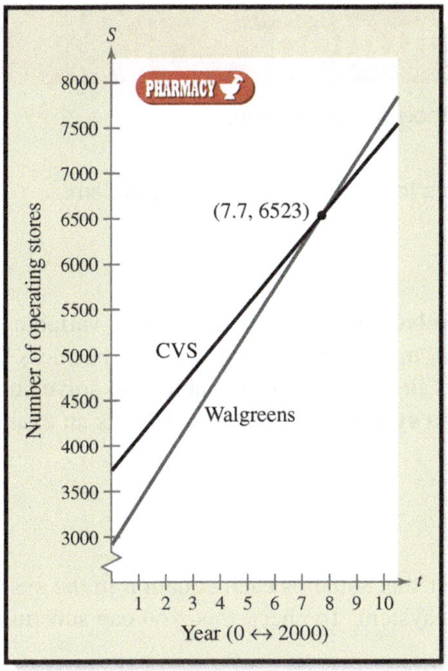

Example 7 on page 416 shows how a system of linear equations can be used to estimate the year when the number of operating stores of one pharmacy chain exceeded the number of stores of another chain.

5 Systems of Equations and Inequalities

5.1 Solving Systems Using Substitution

- Solve a system of equations by the method of substitution.
- Solve a system of equations graphically.
- Construct and use a system of equations to solve an application problem.

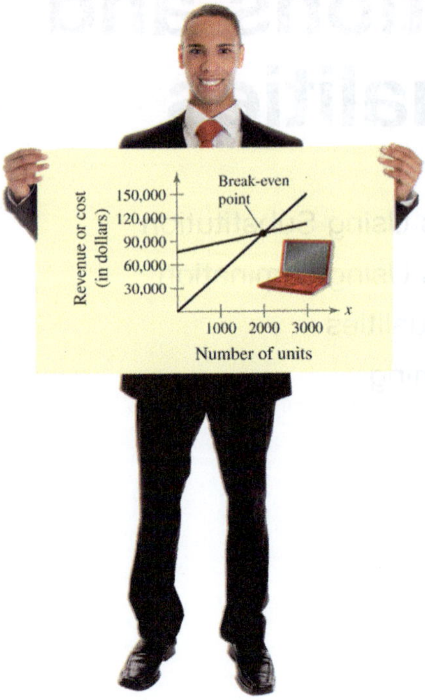

In Exercise 73 on page 419, you will use a system of equations to find the amount of sales a security software company needs to make in order to break even.

The Method of Substitution

So far, most problems in this text have involved either a function of one variable or a single equation in two variables. However, many problems in science, business, and engineering involve two or more equations in two or more variables. To solve such a problem, you need to find the solutions of a **system of equations.** Here is an example of a system of two equations in x and y.

$$\begin{cases} 2x + y = 5 & \text{Equation 1} \\ 3x - 2y = 4 & \text{Equation 2} \end{cases}$$

A **solution** of this system is an ordered pair that satisfies each equation in the system. The ordered pair $(2, 1)$ is a solution of this system. To check this, you can substitute 2 for x and 1 for y in *each* equation.

$$2x + y = 5 \qquad \text{Write Equation 1.}$$
$$2(2) + 1 \stackrel{?}{=} 5 \qquad \text{Substitute 2 for } x \text{ and 1 for } y.$$
$$4 + 1 \stackrel{?}{=} 5 \qquad \text{Multiply.}$$
$$5 = 5 \qquad \text{Solution checks in Equation 1. ✓}$$

$$3x - 2y = 4 \qquad \text{Write Equation 2.}$$
$$3(2) - 2(1) \stackrel{?}{=} 4 \qquad \text{Substitute 2 for } x \text{ and 1 for } y.$$
$$6 - 2 \stackrel{?}{=} 4 \qquad \text{Multiply.}$$
$$4 = 4 \qquad \text{Solution checks in Equation 2. ✓}$$

Finding the set of all solutions is called **solving the system of equations.** There are several different ways to solve systems of equations. In this chapter, you will study four of the most common techniques:

- *method of substitution,*
- *graphical approach,*
- *method of elimination,* and
- *Gaussian elimination.*

This section begins with the **method of substitution.**

Method of Substitution

1. *Solve* one of the equations for one variable in terms of the other.
2. *Substitute* the expression found in Step 1 into the other equation to obtain an equation in one variable.
3. *Solve* the equation obtained in Step 2.
4. *Back-substitute* the value found in Step 3 into the expression obtained in Step 1 to find the value of the other variable.
5. *Check* that the solution satisfies *each* of the original equations.

When using the method of substitution to solve a system of equations, it does not matter which variable you solve for first. You will obtain the same solution regardless. When making your choice, you should choose the variable that is easier to work with. For instance, solve for a variable that has a coefficient of 1 or -1 to avoid working with fractions.

Example 1 Solving a System of Two Equations by Substitution

Solve the system of equations.

$$\begin{cases} x + y = 4 & \text{Equation 1} \\ x - y = 2 & \text{Equation 2} \end{cases}$$

SOLUTION Begin by solving for y in Equation 1.

$$y = 4 - x \qquad \text{Revised Equation 1}$$

Next, substitute the expression $4 - x$ for y into Equation 2 and solve the resulting single-variable equation for x.

$$\begin{aligned} x - y &= 2 && \text{Write Equation 2.} \\ x - (4 - x) &= 2 && \text{Substitute } 4 - x \text{ for } y. \\ x - 4 + x &= 2 && \text{Distributive Property} \\ 2x &= 6 && \text{Combine like terms.} \\ x &= 3 && \text{Divide each side by 2.} \end{aligned}$$

Finally, you can solve for y by *back-substituting* $x = 3$ into the equation $y = 4 - x$.

$$\begin{aligned} y &= 4 - x && \text{Write revised Equation 1.} \\ y &= 4 - 3 && \text{Substitute 3 for } x. \\ y &= 1 && \text{Solve for } y. \end{aligned}$$

The solution is the ordered pair (3, 1).

TECH TUTOR

To check the solution in Example 1 with your graphing utility, graph $y_1 = -x + 4$ and $y_2 = x - 2$ in the same viewing window. Then use the *trace* feature to find the coordinates of the point of intersection. For instructions on how to use the *trace* feature, see Appendix A. Consult the user's guide for your graphing utility for more information.

CHECK

Substitute (3, 1) into Equation 1:

$$\begin{aligned} x + y &= 4 && \text{Write Equation 1.} \\ 3 + 1 &\stackrel{?}{=} 4 && \text{Substitute for } x \text{ and } y. \\ 4 &= 4 && \text{Solution checks in Equation 1.} \checkmark \end{aligned}$$

Substitute (3, 1) into Equation 2:

$$\begin{aligned} x - y &= 2 && \text{Write Equation 2.} \\ 3 - 1 &\stackrel{?}{=} 2 && \text{Substitute for } x \text{ and } y. \\ 2 &= 2 && \text{Solution checks in Equation 2.} \checkmark \end{aligned}$$

STUDY TIP

Because many steps are required to solve a system of equations, it is easy to make errors in arithmetic. So, you should always check your solution by substituting it into *each* equation in the original system.

✓ Checkpoint 1

Solve the system of equations.

$$\begin{cases} x + y = 6 \\ x - y = 4 \end{cases}$$

■

The term *back-substitution* implies that you work *backwards.* First you solve for one of the variables, and then you substitute that value *back* into one of the equations in the system to find the value of the other variable.

Example 2 Solving a System by Substitution

A total of \$10,500 is invested in two funds paying 3% and 5% simple interest. (Recall that the formula for simple interest is $I = Prt$, where P is the principal, r is the annual interest rate, and t is the time.) The total annual interest is \$465. How much is invested at each rate?

SOLUTION

Verbal Model:

Amount in 3% fund	+	Amount in 5% fund	=	Total investment
Interest for 3% fund	+	Interest for 5% fund	=	Total interest

Labels:

Amount in 3% fund $= x$	(dollars)
Interest for 3% fund $= 0.03x$	(dollars)
Amount in 5% fund $= y$	(dollars)
Interest for 5% fund $= 0.05y$	(dollars)
Total investment $= 10{,}500$	(dollars)
Total interest $= 465$	(dollars)

System: $\begin{cases} x + y = 10{,}500 & \text{Equation 1} \\ 0.03x + 0.05y = 465 & \text{Equation 2} \end{cases}$

To begin, it is convenient to multiply each side of Equation 2 by 100. This eliminates the need to work with decimals.

$$100(0.03x + 0.05y) = 100(465) \qquad \text{Multiply each side by 100.}$$

$$3x + 5y = 46{,}500 \qquad \text{Revised Equation 2}$$

To solve this system, you can begin by solving for x in Equation 1.

$$x = 10{,}500 - y \qquad \text{Revised Equation 1}$$

Then, substitute the expression $10{,}500 - y$ for x into revised Equation 2 and solve the resulting equation for y.

$$3x + 5y = 46{,}500 \qquad \text{Write revised Equation 2.}$$

$$3(10{,}500 - y) + 5y = 46{,}500 \qquad \text{Substitute } 10{,}500 - y \text{ for } x.$$

$$31{,}500 - 3y + 5y = 46{,}500 \qquad \text{Distributive Property}$$

$$2y = 15{,}000 \qquad \text{Combine like terms.}$$

$$y = 7500 \qquad \text{Divide each side by 2.}$$

Next, back-substitute $y = 7500$ into the equation $x = 10{,}500 - y$ to solve for x.

$$x = 10{,}500 - y \qquad \text{Write revised Equation 1.}$$

$$x = 10{,}500 - 7500 \qquad \text{Substitute 7500 for } y.$$

$$x = 3000 \qquad \text{Solve for } x.$$

The solution is (3000, 7500). So, \$3000 is invested at 3% and \$7500 is invested at 5%. Check this in the original system.

✓ Checkpoint 2

In Example 2, a total of \$15,000 is invested in the same two funds. The total annual interest is \$670. How much is invested at each rate?

The equations in Examples 1 and 2 are linear. The method of substitution can also be used to solve systems in which one or both of the equations are nonlinear.

Example 3 Substitution: Two-Solution Case

Solve the system of equations.

$$\begin{cases} x^2 - x - y = 1 & \text{Equation 1} \\ -x + y = -1 & \text{Equation 2} \end{cases}$$

STUDY TIP

Note that in Example 3, solving Equation 2 for y instead of x makes substitution into Equation 1 less tedious.

SOLUTION Begin by solving for y in Equation 2 to obtain $y = x - 1$. Next, substitute the expression $x - 1$ for y into Equation 1 and solve for x.

$$x^2 - x - y = 1 \quad \text{Write Equation 1.}$$

$$x^2 - x - (x - 1) = 1 \quad \text{Substitute for } y.$$

$$x^2 - 2x + 1 = 1 \quad \text{Simplify.}$$

$$x^2 - 2x = 0 \quad \text{General form}$$

$$x(x - 2) = 0 \quad \text{Factor.}$$

$$x = 0 \quad \text{Set 1st factor equal to 0.}$$

$$x - 2 = 0 \Rightarrow x = 2 \quad \text{Set 2nd factor equal to 0.}$$

Back-substituting these values of x to solve for the corresponding values of y produces the two solutions $(0, -1)$ and $(2, 1)$. Check these solutions in the original system.

When using the method of substitution, you may encounter an equation that has no solution, as shown in Example 4.

Example 4 Substitution: No-Real-Solution Case

Solve the system of equations.

$$\begin{cases} -x + y = 4 & \text{Equation 1} \\ x^2 + y = 3 & \text{Equation 2} \end{cases}$$

SOLUTION Begin by solving for y in Equation 1 to obtain $y = x + 4$. Next, substitute the expression $x + 4$ for y into Equation 2 and solve for x.

$$x^2 + y = 3 \quad \text{Write Equation 2.}$$

$$x^2 + (x + 4) = 3 \quad \text{Substitute } x + 4 \text{ for } y.$$

$$x^2 + x + 1 = 0 \quad \text{Simplify.}$$

$$x = \frac{-1 \pm \sqrt{1^2 - 4(1)(1)}}{2(1)} \quad \text{Use the Quadratic Formula.}$$

$$x = \frac{-1 \pm \sqrt{-3}}{2} \quad \text{Simplify.}$$

Because the discriminant is negative, the equation $x^2 + x + 1 = 0$ has no (real) solution. So, this system has no (real) solution.

✓ Checkpoints 3 and 4

Solve each system of equations.

a. $\begin{cases} x^2 + 4x - y = 7 \\ 2x - y = -1 \end{cases}$

b. $\begin{cases} 2x^2 - y = 1 \\ x + y = -2 \end{cases}$

Graphical Approach to Finding Solutions

From Examples 1, 3, and 4, you can see that a system of two equations in two unknowns can have exactly one solution, more than one solution, or no solution. In practice, you can gain insight about the location and number of solutions of a system of equations by graphing each of the equations in the same coordinate plane. The solution(s) of the system correspond to the **point(s) of intersection** of the graphs. For instance, the graph of the system in Example 1 is two lines with a *single point* of intersection, as shown in Figure 5.1(a). The graph of the system in Example 3 is a parabola and a line with *two points* of intersection, as shown in Figure 5.1(b). The graph of the system in Example 4 is a line and a parabola that have *no points* of intersection, as shown in Figure 5.1(c).

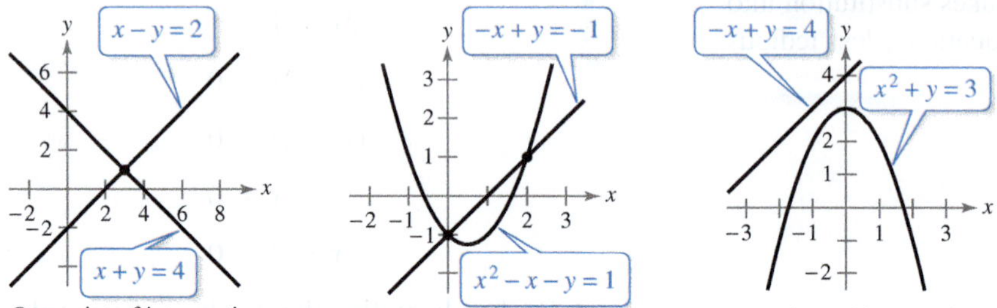

One point of intersection — (a) One solution

Two points of intersection — (b) Two solutions

No points of intersection — (c) No solution

FIGURE 5.1

Example 5 Solving a System of Equations Graphically

Solve the system of equations graphically.

$$\begin{cases} y = \ln x & \text{Equation 1} \\ x + y = 1 & \text{Equation 2} \end{cases}$$

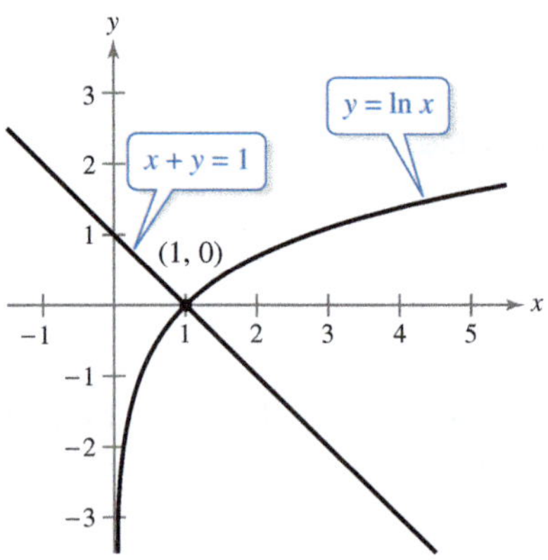

FIGURE 5.2

SOLUTION The graph of each equation is shown in Figure 5.2. From the graph, you can see that there is only one point of intersection. So, it appears that $(1, 0)$ is the solution point. You can confirm this by substituting 1 for x and 0 for y in *both* equations.

CHECK

Equation 1: $0 = \ln 1$ ✓

Equation 2: $1 + 0 = 1$ ✓

✓ Checkpoint 5

Solve the system of equations graphically.

$$\begin{cases} 2x + 2y = 8 \\ y = \ln \frac{1}{4}x \end{cases}$$

TECH TUTOR

Your graphing utility may have an *intersect* feature that approximates the point(s) of intersection of two graphs. Use the *intersect* feature to verify the solution of Example 5. For instructions on how to use the *intersect* feature, see Appendix A. Consult the user's guide for your graphing utility for more information.

Applications

The total cost C of producing x units of a product typically has two components—the initial cost and the cost per unit. When enough units have been sold so that the total revenue R equals the total cost C, the sales are said to have reached the **break-even point.** You will find that the break-even point corresponds to the point of intersection of the cost and revenue graphs.

In 2009, the average price for cross-training shoes in the United States was \$50.23. *(Source: National Sporting Goods Association)*

Example 6 Break-Even Analysis

A shoe company invests \$300,000 in equipment to produce cross-training shoes. Each pair of shoes costs \$10 to produce and is sold for \$75. How many pairs of shoes must be sold before the company breaks even?

SOLUTION The total cost of producing x units is

Total cost = Cost per unit · Number of units + Initial cost

$$C = 10x + 300{,}000.$$

The total revenue obtained by selling x units is

Total revenue = Price per unit · Number of units

$$R = 75x.$$

Because the break-even point occurs when $R = C$, you have $C = 75x$, and the system of equations to solve is

$$\begin{cases} C = 10x + 300{,}000 & \text{Equation 1} \\ C = 75x & \text{Equation 2} \end{cases}.$$

Now, you can solve by substitution.

$C = 10x + 300{,}000$ Write Equation 1.

$75x = 10x + 300{,}000$ Substitute $75x$ for C.

$65x = 300{,}000$ Subtract $10x$ from each side.

$x = \dfrac{300{,}000}{65}$ Divide each side by 65.

$x \approx 4615$ Use a calculator.

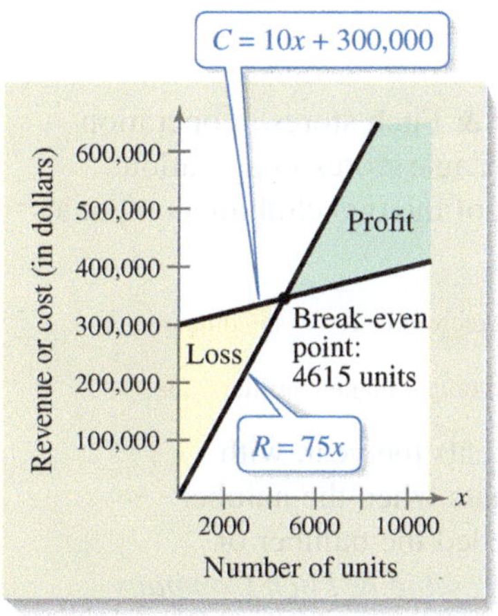

FIGURE 5.3

The company must sell about 4615 pairs of shoes to break even. Note in Figure 5.3 that sales less than the break-even point correspond to an overall loss, whereas sales greater than the break-even point correspond to a profit.

✓ Checkpoint 6

In Example 6, each pair of shoes costs \$12 to produce. How many pairs of shoes must be sold before the business breaks even?

Another way to view the solution in Example 6 is to consider the profit function

$P = R - C.$ Profit function

The break-even point occurs when the profit is 0, which is the same as saying that

$R = C.$ Profit is 0.

Rido/Shutterstock.com

Example 7 Operating Stores

From 2000 through 2010, the number of Walgreens stores in operation increased more quickly than the number of CVS stores in operation. Models that represent the numbers of operating stores of the two pharmacy chains are given by

$$S = 362.24t + 3728.0 \qquad \text{CVS Caremark Corporation}$$

$$S = 467.94t + 2912.5 \qquad \text{Walgreen Company}$$

where S is the number of operating stores and t represents the year, with $t = 0$ corresponding to 2000. Use the models to estimate when the number of Walgreens stores in operation first exceeded the number of CVS stores in operation. *(Source: CVS Caremark Corporation and Walgreen Company)*

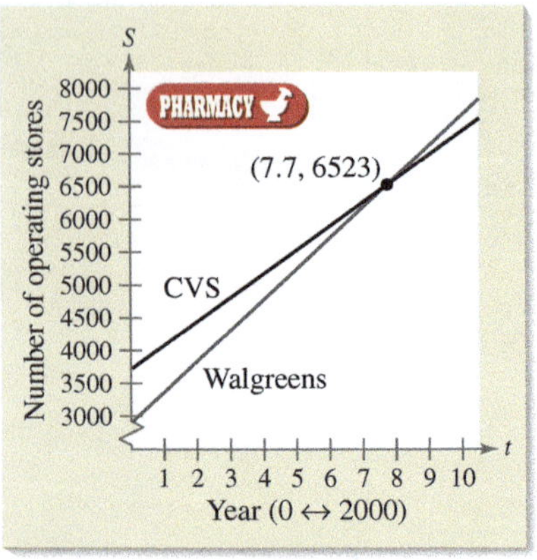

FIGURE 5.4

SOLUTION Both equations have already been solved for S in terms of t. So, substitute the value of S from the first equation into the second equation and solve for t.

$$S = 467.94t + 2912.5 \qquad \text{Write second equation.}$$

$$362.24t + 3728.0 = 467.94t + 2912.5 \qquad \text{Substitute } 362.24t + 3728.0 \text{ for } S.$$

$$-105.7t + 3728.0 = 2912.5 \qquad \text{Subtract } 467.94t \text{ from each side.}$$

$$-105.7t = -815.5 \qquad \text{Subtract 3728.0 from each side.}$$

$$t = \frac{-815.5}{-105.7} \qquad \text{Divide each side by } -105.7.$$

$$t \approx 7.7 \qquad \text{Use a calculator.}$$

Because the equations are equal when $t \approx 7.7$, you can conclude that the number of Walgreens stores first exceeded the number of CVS stores in 2007, as shown in Figure 5.4.

✓ Checkpoint 7

From 2002 through 2009, the number of Abercrombie & Fitch stores in operation increased more quickly than the number of American Eagle stores in operation. Models that represent the numbers of operating stores of the two clothing companies are given by

$$S = 76.81t + 469.5 \qquad \text{Abercrombie \& Fitch Company}$$

$$S = 52.14t + 634.7 \qquad \text{American Eagle Outfitters, Inc.}$$

where S is the number of operating stores and t represents the year, with $t = 2$ corresponding to 2002. Use the models to estimate when the number of Abercrombie & Fitch stores in operation first exceeded the number of American Eagle stores in operation. *(Source: Abercrombie & Fitch Company and American Eagle Outfitters, Inc.)*

SUMMARIZE (Section 5.1)

1. Describe how to solve a system of equations using the method of substitution *(page 410)*. For examples of solving systems of equations using the method of substitution, see Examples 1, 2, 3, and 4.
2. Describe how to solve a system of equations graphically *(page 414)*. For an example of solving a system of equations graphically, see Example 5.
3. State the definition of a break-even point *(page 415)*. For an example of using a system of equations to find a break-even point, see Example 6.

SKILLS WARM UP 5.1 The following warm-up exercises involve skills that were covered in earlier sections. You will use these skills in the exercise set for this section. For additional help, review Sections 0.5, 1.1, 1.3, and 2.1.

In Exercises 1–4, sketch the graph of the equation.

1. $y = -\frac{1}{3}x + 6$

2. $y = 2(x - 3)$

3. $x^2 + y^2 = 4$

4. $y = 5 - (x - 3)^2$

In Exercises 5–8, perform the indicated operations and simplify.

5. $(3x + 2y) - 2(x + y)$

6. $(-10u + 3v) + 5(2u - 8v)$

7. $x^2 + (x - 3)^2 + 6x$

8. $y^2 - (y + 1)^2 + 2y$

In Exercises 9 and 10, solve the equation.

9. $3x + (x - 5) = 15 + 4$

10. $y^2 + (y - 2)^2 = 2$

Exercises 5.1

See www.CalcChat.com for worked-out solutions to odd-numbered exercises.

Checking Solutions In Exercises 1–6, determine whether each ordered pair is a solution of the system of equations.

1. $\begin{cases} x + 4y = -3 \\ 5x - y = 6 \end{cases}$
(a) $(-1, -1)$
(b) $(1, -1)$

2. $\begin{cases} 2x - y = 2 \\ x + 3y = 8 \end{cases}$
(a) $(2, 1)$
(b) $(2, 2)$

3. $\begin{cases} 2x + 5y = -5 \\ 2x - y^2 = 1 \end{cases}$
(a) $(5, -3)$
(b) $(0, -1)$

4. $\begin{cases} 4x^2 + y = 3 \\ -x - y = 11 \end{cases}$
(a) $(-2, -9)$
(b) $(2, -13)$

5. $\begin{cases} y = -2e^x \\ 3x - y = 2 \end{cases}$
(a) $(-2, 0)$
(b) $(-1, 2)$

6. $\begin{cases} -\log_{10} x + 3 = y \\ \frac{1}{9}x + y = \frac{28}{9} \end{cases}$
(a) $(1, 3)$
(b) $\left(9, \frac{37}{9}\right)$

Solving a System by Substitution In Exercises 7–18, solve the system by the method of substitution. *See Example 1.*

7. $\begin{cases} 2x - y = -3 \\ -3x - 4y = -1 \end{cases}$

8. $\begin{cases} x + 2y = 1 \\ 5x - 4y = -23 \end{cases}$

9. $\begin{cases} 2x - y + 2 = 0 \\ 4x + y - 5 = 0 \end{cases}$

10. $\begin{cases} 6x - 3y - 4 = 0 \\ x + 2y - 4 = 0 \end{cases}$

11. $\begin{cases} x - y = 7 \\ 2x + y = 23 \end{cases}$

12. $\begin{cases} x - 2y = -2 \\ 3x - y = 6 \end{cases}$

13. $\begin{cases} 0.3x - 0.4y - 0.33 = 0 \\ 0.1x + 0.2y - 0.21 = 0 \end{cases}$

14. $\begin{cases} 1.5x + 0.8y = 2.3 \\ 0.3x - 0.2y = 0.1 \end{cases}$

15. $\begin{cases} \frac{1}{5}x + \frac{1}{2}y = 8 \\ x + y = 20 \end{cases}$

16. $\begin{cases} \frac{1}{2}x + \frac{3}{4}y = 10 \\ \frac{3}{2}x - y = 4 \end{cases}$

17. $\begin{cases} 6x + 5y = -3 \\ -x - \frac{5}{6}y = -7 \end{cases}$

18. $\begin{cases} -\frac{2}{3}x + y = 2 \\ 2x - 3y = 6 \end{cases}$

Solving a System by Substitution In Exercises 19–22, you are given the total annual interest earned from a total of \$12,000 invested in two funds paying the given rates of simple interest. Write and solve a system of equations to find the amount invested at each rate. *See Example 2.*

	Annual Interest	*Rate 1*	*Rate 2*
19.	\$500	3%	5%
20.	\$630	4%	6%
21.	\$396	2.8%	3.8%
22.	\$254	1.75%	2.25%

Solving a System with a Nonlinear Equation In Exercises 23–32, solve the system by the method of substitution. *See Examples 3 and 4.*

23. $\begin{cases} y = 2x \\ y = x^2 - 1 \end{cases}$

24. $\begin{cases} x + y = 4 \\ x^2 - y = 2 \end{cases}$

25. $\begin{cases} 3x - 7y + 6 = 0 \\ x^2 - y^2 = 4 \end{cases}$

26. $\begin{cases} x^2 + y^2 = 25 \\ 2x + y = 10 \end{cases}$

27. $\begin{cases} x - 2y = 4 \\ x^2 - y = 0 \end{cases}$

28. $\begin{cases} x^2 + y^2 = 9 \\ x - y = -5 \end{cases}$

29. $\begin{cases} y = x^4 - 2x^2 + 1 \\ y = 1 - x^2 \end{cases}$

30. $\begin{cases} y = x^3 - 2x^2 + x - 1 \\ y = -x^2 + 3x - 1 \end{cases}$

31. $\begin{cases} xy - 2 = 0 \\ y = \sqrt{x - 1} \end{cases}$

32. $\begin{cases} xy = 3 \\ y = \sqrt{x - 2} \end{cases}$

Solving a System by Substitution In Exercises 33–38, solve the system by the method of substitution. Then use the graph to confirm your solution.

33. $\begin{cases} x + y = -1 \\ -2x + y = -7 \end{cases}$

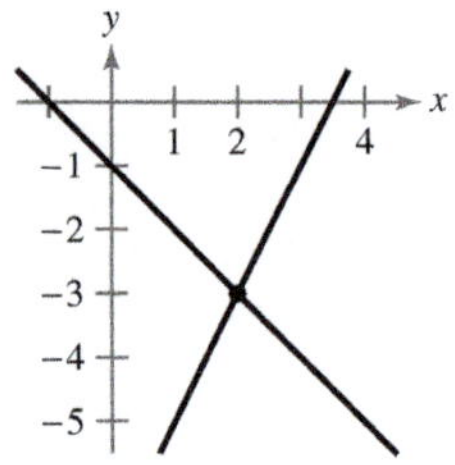

34. $\begin{cases} x - y = -5 \\ x + 2y = 4 \end{cases}$

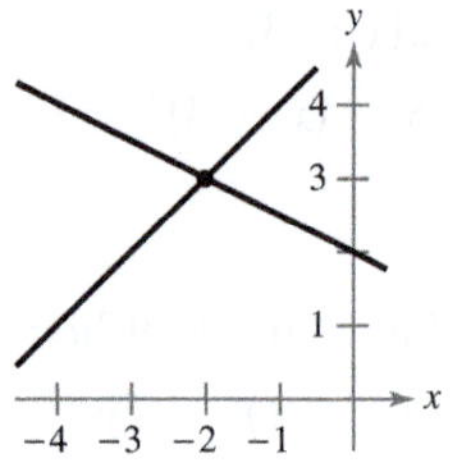

35. $\begin{cases} x - y = -3 \\ x^2 - y = -1 \end{cases}$

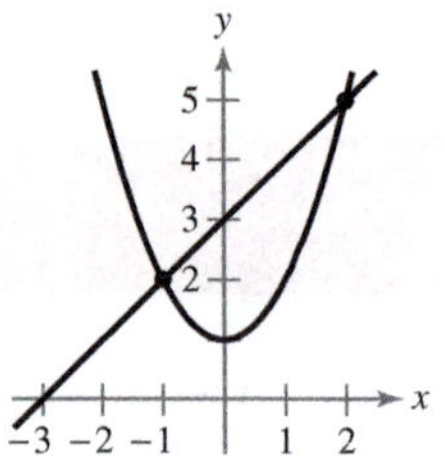

36. $\begin{cases} y = x^3 - 3x^2 + 3 \\ 2x + y = 3 \end{cases}$

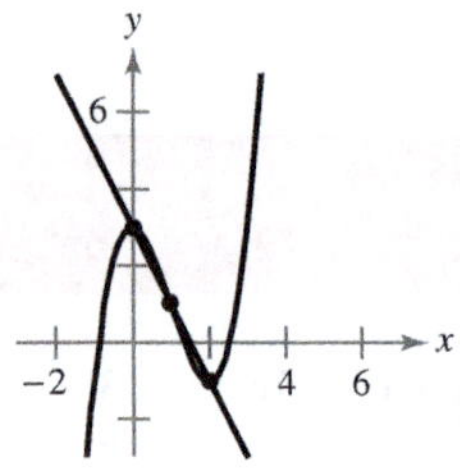

37. $\begin{cases} 3x + y = 4 \\ x^2 + y^2 = 16 \end{cases}$

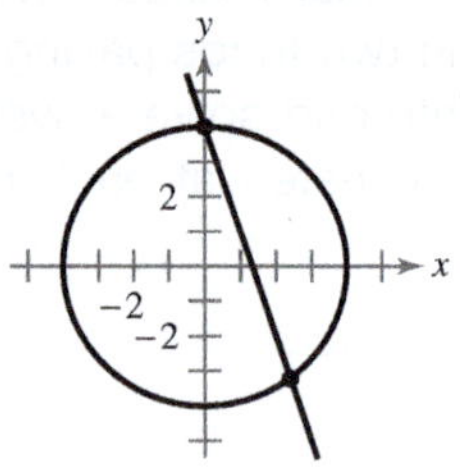

38. $\begin{cases} y = x^2 - 3x - 4 \\ y = -x^2 + 3x + 4 \end{cases}$

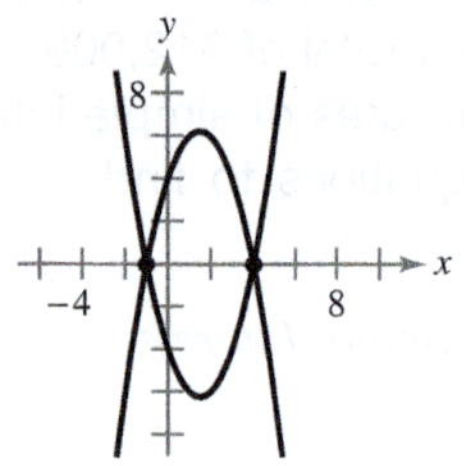

Solving a System of Equations Graphically In Exercises 39–50, solve the system graphically. *See Example 5.*

39. $\begin{cases} -x + 2y = 2 \\ 3x + y = 15 \end{cases}$

40. $\begin{cases} x + y = 0 \\ 3x - 2y = 10 \end{cases}$

41. $\begin{cases} x - 3y = -2 \\ 5x + 3y = 17 \end{cases}$

42. $\begin{cases} -x + 2y = 1 \\ x - y = 2 \end{cases}$

43. $\begin{cases} x + y = 4 \\ x^2 + y^2 - 4x = 0 \end{cases}$

44. $\begin{cases} -x + y = 3 \\ x^2 - 6x - 27 + y^2 = 0 \end{cases}$

45. $\begin{cases} x - y + 3 = 0 \\ x^2 - 4x + 7 = y \end{cases}$

46. $\begin{cases} y^2 - 4x + 11 = 0 \\ -\frac{1}{2}x + y = -\frac{1}{2} \end{cases}$

47. $\begin{cases} 7x + 8y = 24 \\ x - 8y = 8 \end{cases}$

48. $\begin{cases} x - y = 0 \\ 5x - 2y = 6 \end{cases}$

49. $\begin{cases} 3x - 2y = 0 \\ x^2 - y^2 = 4 \end{cases}$

50. $\begin{cases} 2x - y + 3 = 0 \\ x^2 + y^2 - 4x = 0 \end{cases}$

Finding the Number of Solutions In Exercises 51–56, use a graphing utility to determine whether the system of equations has *one solution, two solutions,* or *no solution.*

51. $\begin{cases} y = -5x + 1 \\ y = x + 3 \end{cases}$

52. $\begin{cases} -\frac{1}{2}x + y = -1 \\ 7x + y = 2 \end{cases}$

53. $\begin{cases} y = x^2 + 2x - 1 \\ y = 2x + 5 \end{cases}$

54. $\begin{cases} x^2 + 3x + y = 4 \\ 3x + y = -5 \end{cases}$

55. $\begin{cases} y = x^2 + 3x + 7 \\ y = -x^2 - 3x + 1 \end{cases}$

56. $\begin{cases} -10x + y = 2 \\ -10x + y = -3 \end{cases}$

Solving a System of Equations Graphically In Exercises 57–62, use a graphing utility to find the point(s) of intersection of the graphs. Then confirm your solution algebraically.

57. $\begin{cases} y = x^2 + 3x - 1 \\ y = -x^2 - 2x + 2 \end{cases}$

58. $\begin{cases} y = -2x^2 + x - 1 \\ y = x^2 - 2x - 1 \end{cases}$

59. $\begin{cases} x - y + 3 = 0 \\ x^2 - 4x + 7 = y \end{cases}$

60. $\begin{cases} x - y = 3 \\ x - y^2 = 1 \end{cases}$

61. $\begin{cases} y = e^x \\ x - y + 1 = 0 \end{cases}$

62. $\begin{cases} x^2 + y^2 = 8 \\ y = x^2 + 4 \end{cases}$

Choosing a Solution Method In Exercises 63–66, solve the system graphically or algebraically. Explain your choice of method.

63. $\begin{cases} 4x - y = 0 \\ x^2 - y = -3 \end{cases}$

64. $\begin{cases} x^2 + y^2 = 4 \\ x + y = 6 \end{cases}$

65. $\begin{cases} x^2 + y^2 = 25 \\ \frac{1}{2}x + y = \frac{5}{2} \end{cases}$

66. $\begin{cases} y = \sqrt{x + 7} \\ y = x + 1 \end{cases}$

67. Think About It Create systems of two equations in two variables that have (a) one solution, (b) two solutions, and (c) no solution. (There are many correct answers.)

68. HOW DO YOU SEE IT? The cost C of producing x units and the revenue R obtained by selling x units are shown in the figure.

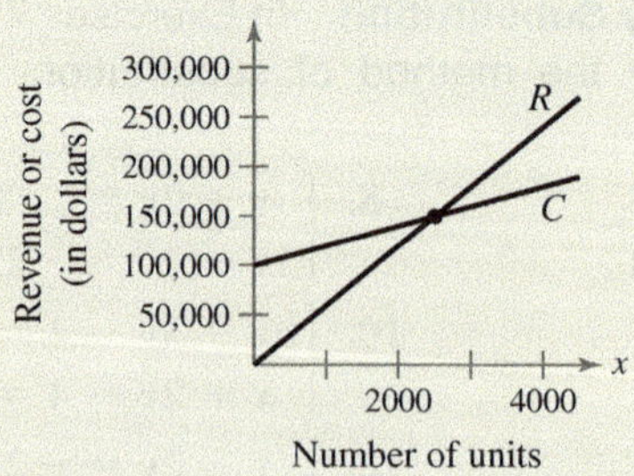

(a) Estimate the point of intersection. What does this point represent?

(b) Identify the area in the figure that corresponds to an overall loss. Identify the area that corresponds to a profit. Explain your reasoning.

Break-Even Analysis In Exercises 69–72, find the sales necessary to break even ($R = C$) for the cost C of producing x units and the revenue R obtained by selling x units. (Round your answer to the nearest whole unit.) *See Example 6.*

69. $C = 8650x + 250{,}000;\ R = 9950x$

70. $C = 5.5\sqrt{x} + 10{,}000;\ R = 3.29x$

71. $C = 2.65x + 350{,}000;\ R = 4.15x$

72. $C = 0.08x + 50{,}000;\ R = 0.25x$

73. Break-Even Analysis A security software company invests \$76,500 to produce software packages that will sell for \$49.95 each. Each software package can be produced for \$11.70. How many software packages must the company sell to break even?

74. Break-Even Analysis You invest \$3000 in a fishing lure business. A lure costs \$1.06 to produce and will be sold for \$5.86. How many lures must you sell to break even?

75. Comparing Populations From 2000 through 2009, the population of Alabama grew more slowly than that of Colorado. Models that represent the populations of the two states are given by

$$\begin{cases} P = 30.2t + 4420 & \text{Alabama} \\ P = 73.6t + 4332 & \text{Colorado} \end{cases}$$

where P is the population (in thousands) and t represents the year, with $t = 0$ corresponding to 2000. Use the models to estimate when the population of Colorado first exceeded the population of Alabama. *(Source: U.S. Census Bureau)*

76. Comparing Populations From 2000 through 2009, the population of Indiana grew more slowly than that of Arizona. Models that represent the populations of the two states are given by

$$\begin{cases} P = 37.6t + 6078 & \text{Indiana} \\ P = 168.5t + 5132 & \text{Arizona} \end{cases}$$

where P is the population (in thousands) and t represents the year, with $t = 0$ corresponding to 2000. Use the models to estimate when the population of Arizona first exceeded the population of Indiana. *(Source: U.S. Census Bureau)*

77. Body Mass Index Body mass index (BMI) is a measure of body fat based on height and weight. Models that represent the 85th percentile BMI for males and females, ages 9 to 20, are given by

$$\begin{cases} B = 0.77a + 11.7 & \text{Males} \\ B = 0.68a + 13.5 & \text{Females} \end{cases}$$

where B is the BMI (kg/m^2) and a represents the age, with $a = 9$ corresponding to 9 years old. Use a graphing utility to determine whether the BMI for males exceeds the BMI for females. *(Source: Centers for Disease Control and Prevention)*

78. Sales Per Share From 2000 through 2009, the sales per share of Ross Stores, Inc. grew faster than those of TJX Companies, Inc. Models that represent the sales per share of the two companies are given by

$$\begin{cases} S = 4.576t + 13.56 & \text{Ross Stores, Inc.} \\ S = 3.745t + 16.14 & \text{TJX Companies, Inc.} \end{cases}$$

where S is the sales per share (in dollars) and t represents the year, with $t = 0$ corresponding to 2000. Use a graphing utility to determine whether the sales per share of Ross Stores, Inc. will exceed the sales per share of TJX Companies, Inc. *(Source: Ross Stores, Inc. and TJX Companies, Inc.)*

79. Job Choices You are offered two jobs. Company A offers an annual salary of \$48,000 plus a year-end bonus of 2.5% of your total sales. Company B offers a salary of \$44,000 plus a year-end bonus of 6.5% of your total sales. What is the amount you must sell in one year to earn the same salary working for either company?

80. Camping You are choosing between camping outfitters. Outfitter A charges a reservation fee of \$150 plus a daily guide fee of \$70. Outfitter B charges a reservation fee of \$75 plus a daily guide fee of \$90. Estimate when the cost of Outfitter A equals the cost of Outfitter B.

81. Annual Sales The annual sales of food and beverage stores and general merchandise stores in the United States from 1992 through 2008 can be approximated by

$$\begin{cases} S = 0.026t^3 - 0.36t^2 + 10.3t + 348 & \text{Food \& beverage} \\ S = -0.024t^3 + 1.12t^2 + 8.2t + 231 & \text{General merchandise} \end{cases}$$

where S is the annual sales (in billions of dollars) and t represents the year, with $t = 2$ corresponding to 1992. Use a graphing utility to determine when the sales of general merchandise stores first exceeded the sales of food and beverage stores. Do you think these models will continue to be accurate? Explain your reasoning. *(Source: U.S. Census Bureau)*

82. Consumer Spending The amount spent per person per year on video games and newspapers in the United States from 1996 through 2008 can be approximated by

$$\begin{cases} A = 0.077t^3 - 2.63t^2 + 31.1t - 99 & \text{Video games} \\ A = -0.10t^2 + 2.0t + 44 & \text{Newspapers} \end{cases}$$

where A is the amount spent (in dollars) and t represents the year, with $t = 6$ corresponding to 1996. Use a graphing utility to determine when the amount spent per person per year on video games first exceeded the amount spent per person per year on newspapers. Do you think these models will continue to be accurate? Explain your reasoning. *(Source: Veronis Suhler Stevenson)*

5.2 Solving Systems Using Elimination

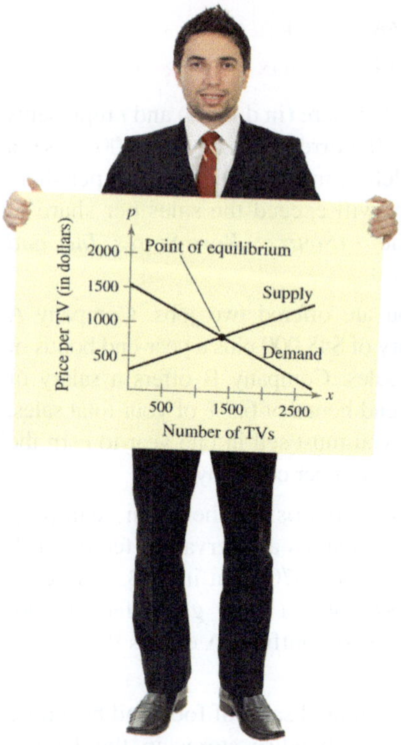

In Exercise 59 on page 430, you will use the method of elimination to solve a system of linear equations to find the equilibrium point that satisfies supply and demand equations.

- Solve a linear system by the method of elimination.
- Interpret the solution of a linear system graphically.
- Construct and use a linear system to solve an application problem.

The Method of Elimination

In Section 5.1, you studied two methods for solving a system of equations: substitution and graphing. In this section, you will study a third method called the **method of elimination.** The key step in the method of elimination is to obtain, for one of the variables, coefficients that differ only in sign, so that *adding* the two equations eliminates this variable. Here is an example.

$$\begin{array}{rll} 3x + 5y = & 7 & \text{Equation 1} \\ -3x - 2y = & -1 & \text{Equation 2} \\ \hline 3y = & 6 & \text{Add equations.} \end{array}$$

Note that by adding the two equations, you eliminate the variable x and obtain a single equation in y. Solving this equation for y produces $y = 2$, which you can then back-substitute into one of the original equations to solve for x.

Example 1 The Method of Elimination

Solve the system of linear equations.

$$\begin{cases} 3x + 2y = 4 & \text{Equation 1} \\ 5x - 2y = 8 & \text{Equation 2} \end{cases}$$

SOLUTION Because the coefficients of the y-terms differ only in sign, you can eliminate the y-terms by adding the two equations. This leaves you with a single equation in x.

$$\begin{array}{rll} 3x + 2y = & 4 & \text{Write Equation 1.} \\ 5x - 2y = & 8 & \text{Write Equation 2.} \\ \hline 8x \quad\;\; = & 12 & \text{Add equations.} \end{array}$$

Solving the equation $8x = 12$ produces $x = \frac{3}{2}$. By back-substituting $x = \frac{3}{2}$ into Equation 1, you can solve for y.

$$\begin{array}{rll} 3x + 2y = 4 & & \text{Write Equation 1.} \\ 3\left(\frac{3}{2}\right) + 2y = 4 & & \text{Substitute } \frac{3}{2} \text{ for } x. \\ \frac{9}{2} + 2y = 4 & & \text{Multiply.} \\ 2y = -\frac{1}{2} & & \text{Subtract } \frac{9}{2} \text{ from each side.} \\ y = -\frac{1}{4} & & \text{Divide each side by 2.} \end{array}$$

The solution is $\left(\frac{3}{2}, -\frac{1}{4}\right)$. Check this in the original system.

✓Checkpoint 1

Solve the system of linear equations.

$$\begin{cases} 2x - 3y = 5 \\ 5x + 3y = 9 \end{cases}$$

STUDY TIP

It is important that you understand both the method of substitution and the method of elimination. In some cases, it is more efficient to use one method instead of the other. For instance, when two equations are written such that one variable is represented in terms of the other variable, substitution is a more efficient solution method.

Blaj Gabriel/www.shutterstock.com

To obtain coefficients (for one of the variables) that differ only in sign, you may need to multiply one or both of the equations by a suitable constant or constants, as demonstrated in Example 2.

Example 2 The Method of Elimination

Solve the system of linear equations.

$$\begin{cases} 2x - 3y = -7 & \text{Equation 1} \\ 3x + y = -5 & \text{Equation 2} \end{cases}$$

SOLUTION For this system, you can obtain coefficients that differ only in sign by multiplying Equation 2 by 3. Then, by adding the two equations, you can eliminate the y-terms. This leaves you with a single equation in x.

$$\begin{array}{lcrl} 2x - 3y = -7 & \Rightarrow & 2x - 3y = -7 & \text{Write Equation 1.} \\ 3x + y = -5 & \Rightarrow & 9x + 3y = -15 & \text{Multiply Equation 2 by 3.} \\ & & 11x \quad = -22 & \text{Add equations.} \end{array}$$

Solving the equation $11x = -22$ produces $x = -2$. By back-substituting $x = -2$ into Equation 1, you can solve for y.

$$\begin{aligned} 2x - 3y &= -7 && \text{Write Equation 1.} \\ 2(-2) - 3y &= -7 && \text{Substitute } -2 \text{ for } x. \\ -4 - 3y &= -7 && \text{Multiply.} \\ -3y &= -3 && \text{Add 4 to each side.} \\ y &= 1 && \text{Divide each side by } -3. \end{aligned}$$

The solution is $(-2, 1)$.

CHECK

$$\begin{aligned} 2(-2) - 3(1) &\stackrel{?}{=} -7 \\ -4 - 3 &\stackrel{?}{=} -7 \\ -7 &= -7 \checkmark \end{aligned} \qquad \begin{aligned} 3(-2) + 1 &\stackrel{?}{=} -5 \\ -6 + 1 &\stackrel{?}{=} -5 \\ -5 &= -5 \checkmark \end{aligned}$$

✓ Checkpoint 2

Solve the system of linear equations.

$$\begin{cases} 3x - 5y = -1 \\ x - 2y = -1 \end{cases}$$

In Example 2, the two systems of linear equations

$$\begin{cases} 2x - 3y = -7 \\ 3x + y = -5 \end{cases} \quad \text{and} \quad \begin{cases} 2x - 3y = -7 \\ 9x + 3y = -15 \end{cases}$$

are called **equivalent systems** because they have precisely the same solution set. The operations that can be performed on a system of linear equations to produce an equivalent system are:

(1) interchanging any two equations,

(2) multiplying an equation by a nonzero constant, and

(3) adding a multiple of one equation to any other equation in the system.

The Method of Elimination

To use the **method of elimination** to solve a system of two linear equations in x and y, use the following steps.

1. Examine the system to determine which variable is easiest to eliminate.
2. Obtain coefficients of x (or y) that differ only in sign by multiplying all terms of one or both equations by suitably chosen constants.
3. Add the equations to eliminate one variable and solve the resulting equation.
4. Back-substitute the value obtained in Step 3 into either of the original equations and solve for the other variable.
5. Check your solution in both of the original equations.

Example 3 The Method of Elimination

Solve the system of linear equations.

$$\begin{cases} 5x + 3y = 9 & \text{Equation 1} \\ 2x - 4y = 14 & \text{Equation 2} \end{cases}$$

SOLUTION You can obtain coefficients of y that differ only in sign by multiplying Equation 1 by 4 and multiplying Equation 2 by 3.

$$5x + 3y = 9 \quad \Rightarrow \quad 20x + 12y = 36 \qquad \text{Multiply Equation 1 by 4.}$$

$$2x - 4y = 14 \quad \Rightarrow \quad 6x - 12y = 42 \qquad \text{Multiply Equation 2 by 3.}$$

$$26x \qquad = 78 \qquad \text{Add equations.}$$

Solving the equation $26x = 78$ produces $x = 3$. By back-substituting $x = 3$ into Equation 2, you can solve for y.

$$2x - 4y = 14 \qquad \text{Write Equation 2.}$$

$$2(3) - 4y = 14 \qquad \text{Substitute 3 for } x.$$

$$y = -2 \qquad \text{Solve for } y.$$

The solution is $(3, -2)$.

CHECK

$$5(3) + 3(-2) \stackrel{?}{=} 9 \qquad 2(3) - 4(-2) \stackrel{?}{=} 14$$

$$15 + (-6) \stackrel{?}{=} 9 \qquad 6 - (-8) \stackrel{?}{=} 14$$

$$9 = 9 \checkmark \qquad 14 = 14 \checkmark$$

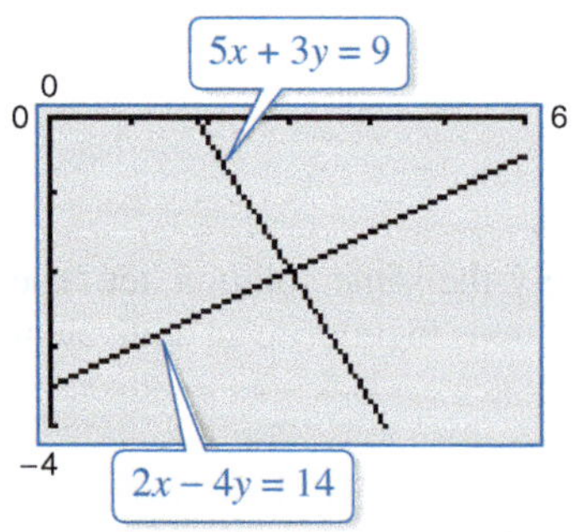

FIGURE 5.5

✓ Checkpoint 3

Solve the system of linear equations.

$$\begin{cases} -2x + 7y = -15 \\ 3x - 5y = 6 \end{cases}$$

Remember that you can check the solution of a system of equations graphically. For instance, to check the solution found in Example 3, graph both equations in the same viewing window, as shown in Figure 5.5. Notice that the two lines intersect at $(3, -2)$.

Example 4 illustrates a strategy for solving a system of linear equations that has decimal coefficients.

Example 4 A Linear System Having Decimal Coefficients

Solve the system of linear equations.

$$\begin{cases} 0.02x - 0.05y = -0.38 & \text{Equation 1} \\ 0.03x + 0.04y = 1.04 & \text{Equation 2} \end{cases}$$

SOLUTION Because the coefficients in this system have two decimal places, you can begin by multiplying each equation by 100. This produces a system in which the coefficients are all integers.

$$\begin{cases} 2x - 5y = -38 & \text{Revised Equation 1} \\ 3x + 4y = 104 & \text{Revised Equation 2} \end{cases}$$

Now, to obtain coefficients of x that differ only in sign, multiply revised Equation 1 by 3 and revised Equation 2 by -2.

$$\begin{aligned} 2x - 5y &= -38 &\Rightarrow\quad 6x - 15y &= -114 && \text{Multiply by 3.} \\ 3x + 4y &= 104 &\Rightarrow\quad -6x - 8y &= -208 && \text{Multiply by } -2. \\ & & -23y &= -322 && \text{Add equations.} \end{aligned}$$

Solving the equation $-23y = -322$ produces $y = 14$. Now, back-substitute $y = 14$ into any of the original or revised equations of the system that contain the variable y. By back-substituting $y = 14$ into revised Equation 2, you can solve for x.

$$\begin{aligned} 3x + 4y &= 104 && \text{Write revised Equation 2.} \\ 3x + 4(14) &= 104 && \text{Substitute 14 for } y. \\ 3x &= 48 && \text{Subtract 56 from each side.} \\ x &= 16 && \text{Divide each side by 3.} \end{aligned}$$

The solution is (16, 14). Check this in the original system.

✓ Checkpoint 4

Solve the system of linear equations.

$$\begin{cases} 0.03x + 0.04y = -0.13 \\ -0.04x + 0.05y = -0.24 \end{cases}$$

TECH TUTOR

The general solution of the linear system

$$\begin{cases} ax + by = c \\ dx + ey = f \end{cases}$$

is

$$x = (ce - bf)/(ae - bd) \quad \text{and} \quad y = (af - cd)/(ae - bd).$$

If $ae - bd = 0$, then the system does not have a unique solution. Several graphing utilities have programs for solving such a system. Try using one of these programs to check the solution of the system in Example 4.

Edyta Pawlowska/Shutterstock.com

Graphical Interpretation of Solutions

It is possible for a *general* system of equations to have

- exactly one solution,
- two or more solutions, or
- no solution.

If a system of *linear* equations has two different solutions, then it must have an *infinite* number of solutions. To see why this is true, consider the following graphical interpretations of systems of two linear equations in two variables. (Remember that the graph of a linear equation in two variables is a line.)

Graph	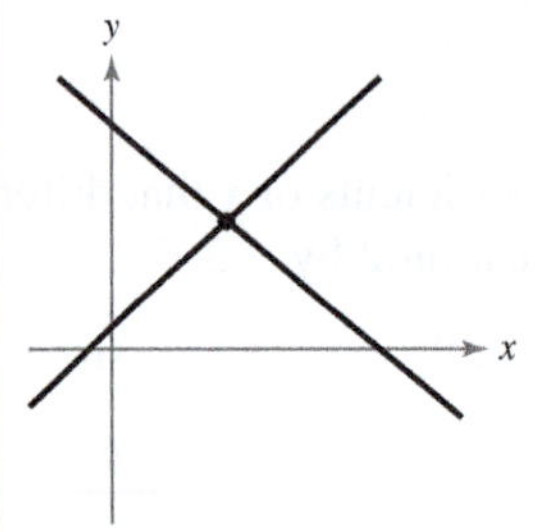	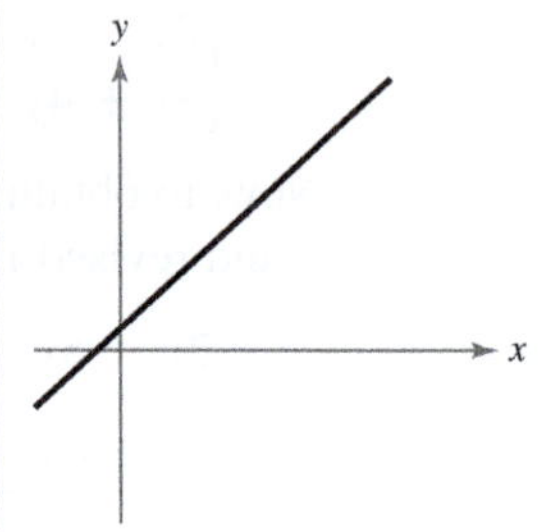	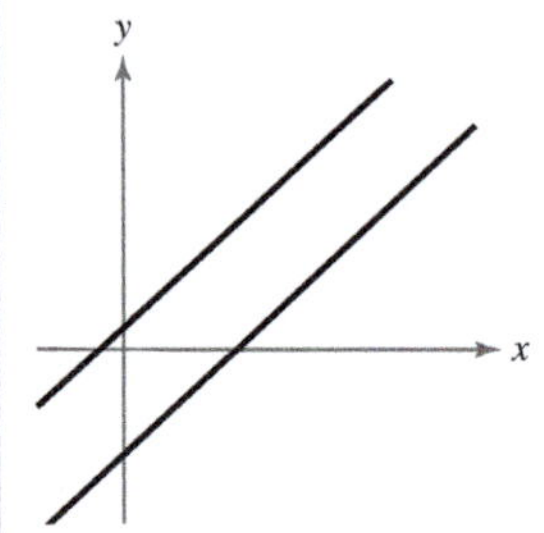
Graphical interpretation	The two lines intersect.	The two lines coincide (are identical).	The two lines are parallel.
Intersection	Single point of intersection	Infinitely many points of intersection	No point of intersection
Slopes of lines	Slopes are not equal.	Slopes are equal.	Slopes are equal.
Number of solutions	Exactly one solution	Infinitely many solutions	No solution
Type of system	**Independent (consistent) system**	**Dependent (consistent) system**	**Inconsistent system**

> **STUDY TIP**
>
> Keep in mind that the terminology and methods discussed in this section and the following section apply only to systems of *linear* equations.

A system of linear equations is *consistent* if it has at least one solution. A consistent system with exactly one solution is *independent*, whereas a consistent system with infinitely many solutions is *dependent*. A system is *inconsistent* if it has no solution.

From the graphs above, you can see that a comparison of the slopes and y-intercepts of two lines is helpful in determining the number of solutions of the corresponding system of equations. For instance:

- *Independent (consistent) systems* have lines with slopes that are not equal.
- *Dependent (consistent) systems* have lines with equal slopes and the same y-intercept.
- *Inconsistent systems* have lines with equal slopes, but different y-intercepts.

So, when solving a system of linear equations graphically, it is helpful to know the slope of each line. Writing each linear equation in the slope-intercept form

$$y = mx + b \qquad \text{Slope-intercept form}$$

enables you to identify the slopes quickly.

In Examples 5 and 6, note how you can use the method of elimination to determine when a linear system has no solution or infinitely many solutions.

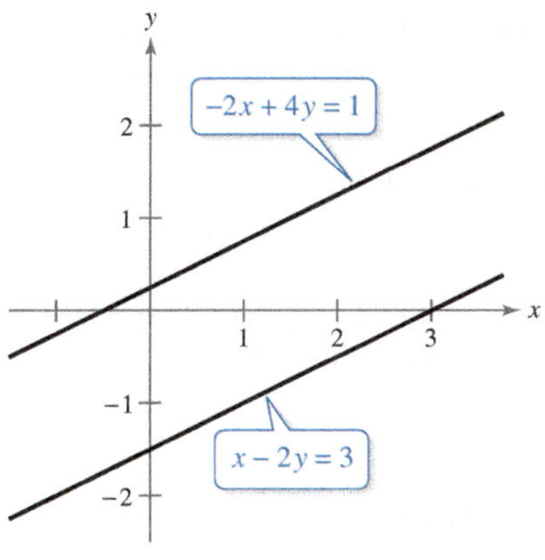

No Solution
FIGURE 5.6

Example 5 The Method of Elimination: No-Solution Case

Solve the system of linear equations.

$$\begin{cases} x - 2y = 3 & \text{Equation 1} \\ -2x + 4y = 1 & \text{Equation 2} \end{cases}$$

SOLUTION Obtain coefficients that differ only in sign.

$$\begin{array}{rcrl} x - 2y = 3 & \Rightarrow & 2x - 4y = 6 & \text{Multiply Equation 1 by 2.} \\ -2x + 4y = 1 & \Rightarrow & -2x + 4y = 1 & \text{Write Equation 2.} \\ \hline & & 0 = 7 & \text{False statement} \end{array}$$

Because there are no values of x and y for which $0 = 7$, you can conclude that the system is inconsistent and has no solution. The graphs of the equations are shown in Figure 5.6. Note that the two lines have equal slopes, but different y-intercepts. So, the lines are parallel and have no point of intersection.

✓ Checkpoint 5

Solve the system of linear equations.

$$\begin{cases} -3x + 6y = 5 \\ x - 2y = 2 \end{cases}$$

In Example 5, note that the occurrence of a false statement, such as $0 = 7$, indicates that the system has no solution. In the next example, note that the occurrence of a statement that is true for all values of the variables, such as $0 = 0$, indicates that the system has infinitely many solutions.

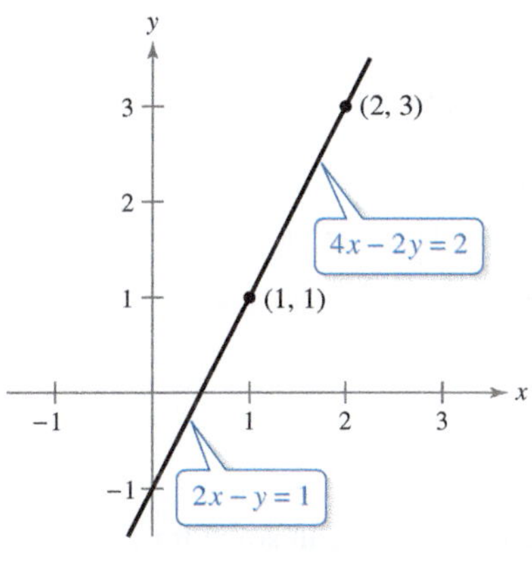

Infinitely Many Solutions
FIGURE 5.7

Example 6 The Method of Elimination: Many-Solutions Case

Solve the system of linear equations.

$$\begin{cases} 2x - y = 1 & \text{Equation 1} \\ 4x - 2y = 2 & \text{Equation 2} \end{cases}$$

SOLUTION Obtain coefficients that differ only in sign.

$$\begin{array}{rcrl} 2x - y = 1 & \Rightarrow & 2x - y = 1 & \text{Write Equation 1.} \\ 4x - 2y = 2 & \Rightarrow & -2x + y = -1 & \text{Multiply Equation 2 by } -\tfrac{1}{2}. \\ \hline & & 0 = 0 & \text{True statement} \end{array}$$

Because the two equations are equivalent (have the same solution set), you can conclude that the system is consistent and has infinitely many solutions. The solution set consists of all points (x, y) lying on the line

$$2x - y = 1$$

as shown in Figure 5.7. To represent the solution set as an ordered pair, let $x = a$, where a is any real number. Then $y = 2a - 1$, and the solution set can be written as

$$(a, 2a - 1).$$

✓ Checkpoint 6

Solve the system of linear equations.

$$\begin{cases} -x - y = 5 \\ 4x + 4y = -20 \end{cases}$$

Applications

At this point, you may be asking the question, "How can I tell which application problems can be solved using a system of linear equations?" The answer comes from the following considerations.

1. Does the problem involve more than one unknown quantity?
2. Are there two (or more) equations or conditions to be satisfied?

If the answer to one or both of these questions is "yes," then the appropriate model for the problem may be a system of linear equations. Example 7 shows how to construct such a model.

Example 7 An Application of a Linear System

An airplane flying into a headwind travels the 2000-mile flying distance between Wilmington, Delaware and Tucson, Arizona in 4 hours and 24 minutes. On the return flight, the same distance is traveled in 4 hours. Find the airspeed of the plane and the speed of the wind, assuming that both remain constant.

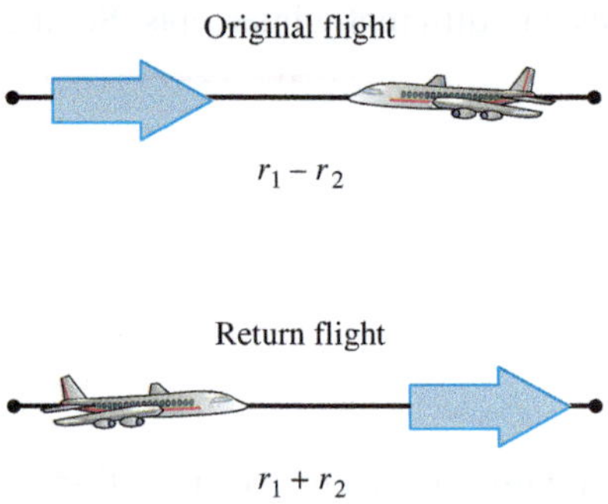

FIGURE 5.8

SOLUTION The two unknown quantities are the speeds of the wind and the plane. If r_1 is the airspeed of the plane and r_2 is the speed of the wind, then

$r_1 - r_2 =$ speed of the plane *against* the wind

$r_1 + r_2 =$ speed of the plane *with* the wind

as shown in Figure 5.8. Using the formula distance = (rate)(time) for these two speeds, you obtain the following equations.

$$2000 = (r_1 - r_2)\left(4 + \frac{24}{60}\right)$$

$$2000 = (r_1 + r_2)(4)$$

These two equations simplify as shown.

$$\begin{cases} 5000 = 11r_1 - 11r_2 & \text{Equation 1} \\ 500 = r_1 + r_2 & \text{Equation 2} \end{cases}$$

To solve this system by elimination, multiply Equation 2 by 11.

$$\begin{array}{lcll} 5000 = 11r_1 - 11r_2 & \Rightarrow & 5000 = 11r_1 - 11r_2 & \text{Write Equation 1.} \\ 500 = r_1 + r_2 & \Rightarrow & 5500 = 11r_1 + 11r_2 & \text{Multiply Equation 2 by 11.} \\ \hline & & 10{,}500 = 22r_1 & \text{Add equations.} \end{array}$$

Solving the equation $10{,}500 = 22r_1$ produces

$$r_1 = \frac{10{,}500}{22} = \frac{5250}{11} \approx 477.27.$$

By back-substituting $r_1 = 477.27$ into original Equation 2, you can solve for r_2.

$$500 = 477.27 + r_2 \qquad \text{Substitute 477.27 for } r_1.$$

$$22.73 = r_2 \qquad \text{Subtract 477.27 from each side.}$$

So, the airspeed of the plane is about 477.27 miles per hour and the speed of the wind is about 22.73 miles per hour. Check this solution in the original statement of the problem.

ALGEBRA TUTOR

For help with the algebra used to simplify the system of equations in Example 7, see the *Chapter 7 Algebra Tutor* on page 464.

✓ Checkpoint 7

In Example 7, the return flight takes 4 hours and 6 minutes. Find the airspeed of the plane and the speed of the wind, assuming that both remain constant.

In a free market, the demands for many products are related to the prices of the products. As the prices decrease, the demands by consumers increase and the amounts that producers are able or willing to supply decrease.

Example 8 Finding the Point of Equilibrium

The demand and supply equations for a video game console are given by

$$p = 350 - 0.00003x \qquad \text{Demand equation}$$

$$p = 200 + 0.00003x \qquad \text{Supply equation}$$

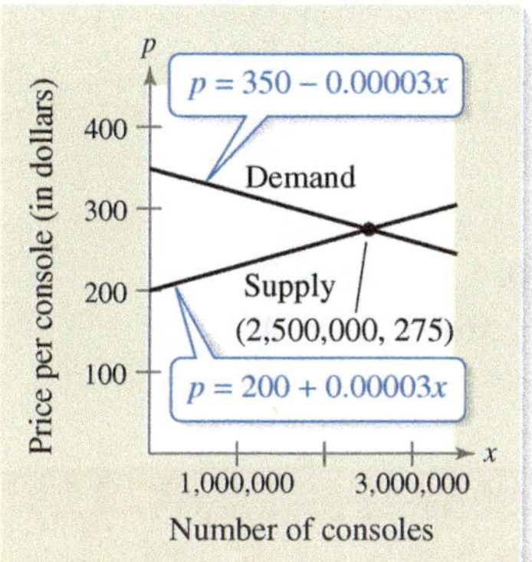

FIGURE 5.9

where p is the price (in dollars) and x is the number of video game consoles. For how many units will the quantity demanded equal the quantity supplied? What price corresponds to this value?

SOLUTION To obtain coefficients of p that differ only in sign, multiply the demand equation by -1.

$$\begin{array}{lcrl} p = 350 - 0.00003x & \Rightarrow & -p = -350 + 0.00003x & \text{Multiply demand equation by } -1. \\ \underline{p = 200 + 0.00003x} & \Rightarrow & \underline{p = 200 + 0.00003x} & \text{Write supply equation.} \\ & & 0 = -150 + 0.00006x & \text{Add equations.} \end{array}$$

By solving the equation $0 = -150 + 0.00006x$, you get $x = 2{,}500{,}000$. So, the quantity demanded equals the quantity supplied for 2,500,000 units (see Figure 5.9). The price that corresponds to this x-value is obtained by back-substituting $x = 2{,}500{,}000$ into either of the original equations. For instance, back-substituting into the demand equation produces

$$\begin{aligned} p &= 350 - 0.00003x && \text{Demand equation} \\ &= 350 - 0.00003(2{,}500{,}000) && \text{Substitute 2,500,000 for } x. \\ &= \$275. && \text{Simplify.} \end{aligned}$$

Back-substitute $x = 2{,}500{,}000$ into the supply equation to see that you obtain the same price. The solution (2,500,000, 275) is called the *point of equilibrium.* The **point of equilibrium** is the price p and the number of units x that satisfy both the demand and supply equations.

✓ Checkpoint 8

In Example 8, find the point of equilibrium when the supply equation is $p = 230 + 0.00003x$.

SUMMARIZE (Section 5.2)

1. Describe how to solve a system of linear equations using the method of elimination *(pages 420–422)*. For examples of solving systems of linear equations using the method of elimination, see Examples 1, 2, 3, and 4.
2. Explain what it means for a system of linear equations to be an independent system, a dependent system, or an inconsistent system *(page 424)*. For examples of solving these types of linear systems using the method of elimination, see Examples 5 and 6.
3. State the definition of the point of equilibrium *(page 427)*. For an example of solving a system of supply and demand equations to find the point of equilibrium, see Example 8.

SKILLS WARM UP 5.2

The following warm-up exercises involve skills that were covered in earlier sections. You will use these skills in the exercise set for this section. For additional help, review Section 2.2.

In Exercises 1 and 2, sketch the graph of the equation.

1. $2x + y = 4$

2. $5x - 2y = 3$

In Exercises 3 and 4, find an equation of the line passing through the two points.

3. $(-1, 3), (4, 8)$

4. $(2, 6), (5, 1)$

In Exercises 5 and 6, determine the slope of the line.

5. $3x + 6y = 4$

6. $7x - 4y = 10$

In Exercises 7–10, determine whether the lines represented by the pair of equations are parallel, perpendicular, or neither.

7. $\begin{aligned} 2x - 3y &= -10 \\ 3x + 2y &= 11 \end{aligned}$

8. $\begin{aligned} 4x - 12y &= 5 \\ -2x + 6y &= 3 \end{aligned}$

9. $\begin{aligned} 5x + y &= 2 \\ 3x + 2y &= 1 \end{aligned}$

10. $\begin{aligned} x - 3y &= 2 \\ 6x + 2y &= 4 \end{aligned}$

Exercises 5.2

See www.CalcChat.com for worked-out solutions to odd-numbered exercises.

Solving a System by Elimination In Exercises 1–8, solve the system by the method of elimination. Then use the graph to confirm your solution. Copy the graph and label each line with the appropriate equation. *See Examples 1, 2, 3, 5, and 6.*

1. $\begin{cases} 3x - 2y = 2 \\ x + 2y = 6 \end{cases}$

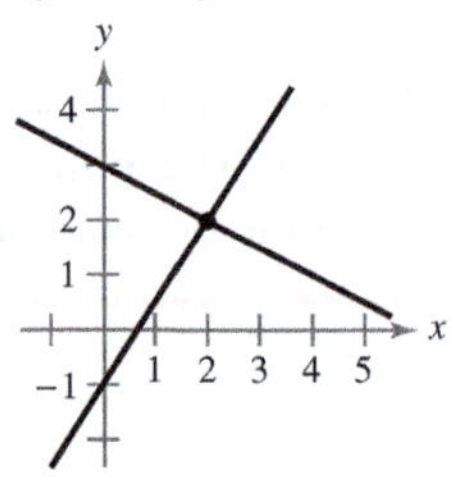

2. $\begin{cases} -x + 3y = 2 \\ x - 4y = -4 \end{cases}$

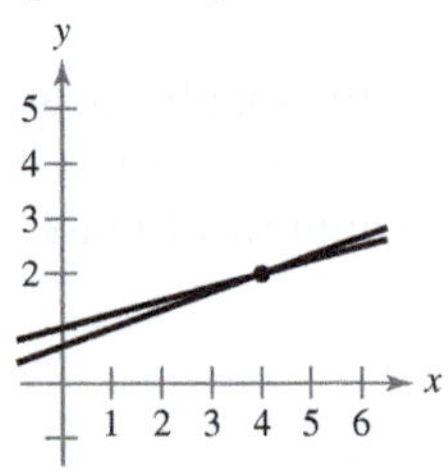

3. $\begin{cases} x - 4y = 2 \\ 2x + y = 4 \end{cases}$

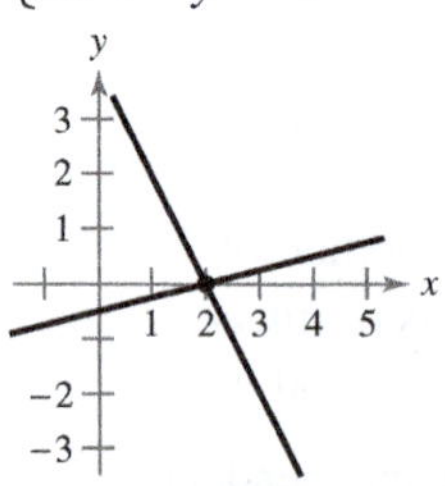

4. $\begin{cases} 2x - y = 2 \\ 4x + 3y = 24 \end{cases}$

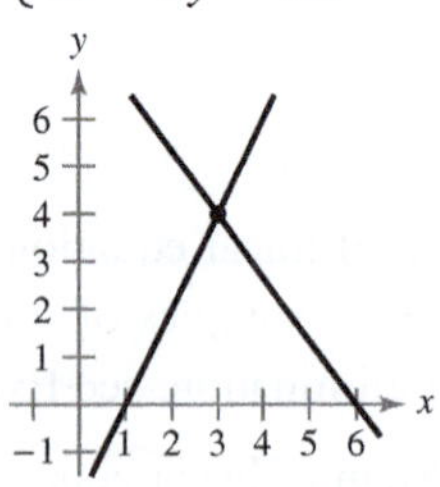

5. $\begin{cases} x - y = 1 \\ -2x + 2y = 5 \end{cases}$

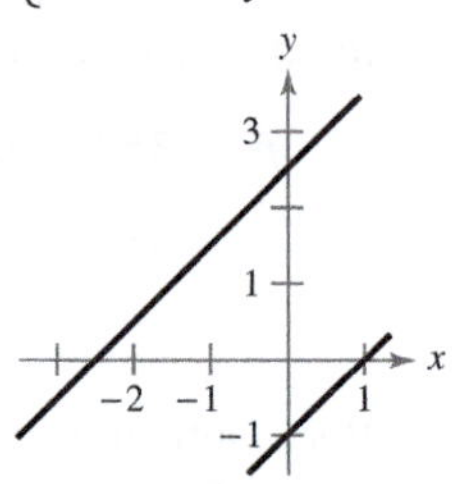

6. $\begin{cases} 3x + 2y = 2 \\ 6x + 4y = 14 \end{cases}$

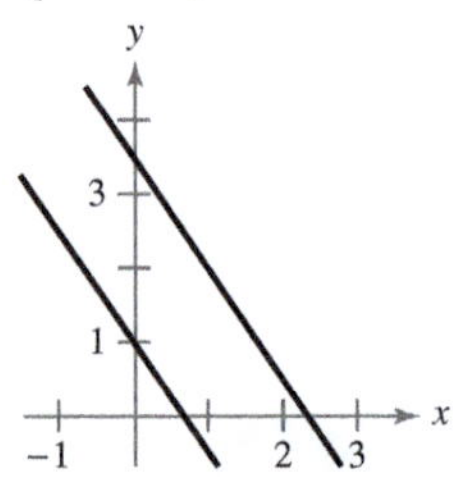

7. $\begin{cases} 3x - 2y = 6 \\ -6x + 4y = -12 \end{cases}$

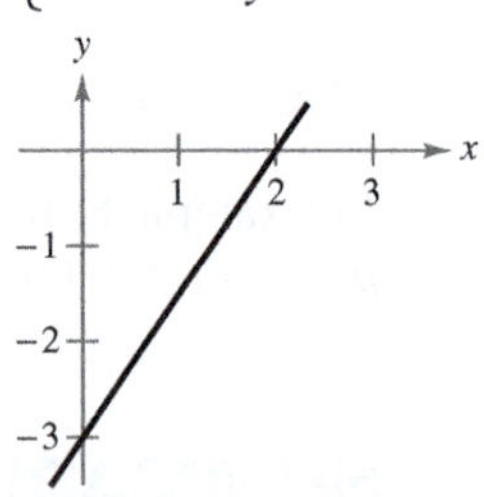

8. $\begin{cases} 2x + 4y = 8 \\ 6x + 12y = 24 \end{cases}$

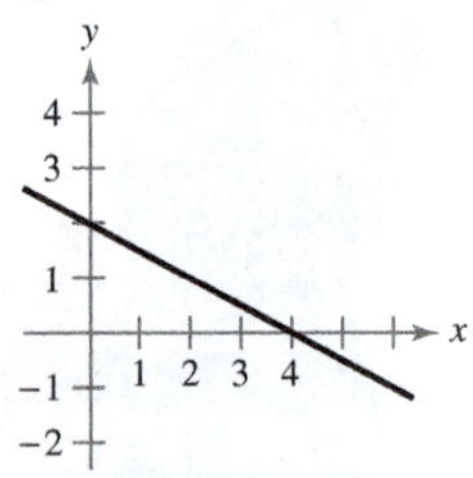

Solving a System by Elimination In Exercises 9–24, solve the system by the method of elimination. Then state whether the system is consistent or inconsistent. *See Examples 1, 2, 3, 5, and 6.*

9. $\begin{cases} x + 2y = 3 \\ x - 2y = 1 \end{cases}$

10. $\begin{cases} 2x - 3y = 4 \\ -2x - y = 4 \end{cases}$

11. $\begin{cases} 4x - 3y = 11 \\ -6x + 3y = 3 \end{cases}$

12. $\begin{cases} 3x - 5y = 2 \\ 2x + 5y = 13 \end{cases}$

13. $\begin{cases} 3x - y = 17 \\ 5x + 5y = -5 \end{cases}$

14. $\begin{cases} x + 7y = 12 \\ 3x - 5y = 10 \end{cases}$

15. $\begin{cases} 3x + 2y = 10 \\ 2x + 5y = 3 \end{cases}$ **16.** $\begin{cases} 8r + 16s = 20 \\ 16r + 50s = 55 \end{cases}$

17. $\begin{cases} 2u + v = 120 \\ u + 2v = 120 \end{cases}$ **18.** $\begin{cases} 5u + 6v = 24 \\ 3u + 5v = 18 \end{cases}$

19. $\begin{cases} 4b + 3m = 3 \\ 3b + 11m = 13 \end{cases}$ **20.** $\begin{cases} 3b + 3m = 7 \\ 3b + 5m = 3 \end{cases}$

21. $\begin{cases} \frac{x}{4} + \frac{y}{6} = 1 \\ x - y = 3 \end{cases}$ **22.** $\begin{cases} \frac{1}{6}x - \frac{2}{3}y = 3 \\ 3x + y = 15 \end{cases}$

23. $\begin{cases} \frac{x+3}{4} + \frac{y-1}{3} = 1 \\ x - y = 3 \end{cases}$ **24.** $\begin{cases} \frac{x-1}{2} + \frac{y+2}{3} = 4 \\ x - 2y = 5 \end{cases}$

Writing and Solving a System In Exercises 25–28, use the given statements to write a system of equations. Solve the system by the method of elimination.

25. The sum of a number x and a number y is 13. The difference of x and y is 3.

26. The sum of a number a and a number b is 43. The difference of a and b is -27.

27. The sum of twice a number r and a number s is 8. The difference of r and s is 7.

28. The difference of a number m and twice a number n is 1. The sum of two times m and n is 22.

Solving a System Having Decimal Coefficients In Exercises 29–34, solve the system by the method of elimination. Then state whether the system is consistent or inconsistent. *See Examples 4, 5, and 6.*

29. $\begin{cases} 6r - 5s = 3 \\ -1.2r + s = 0.5 \end{cases}$ **30.** $\begin{cases} 1.8x + 1.2y = 4 \\ 9x + 6y = 3 \end{cases}$

31. $\begin{cases} 2.5x - 3y = 1.5 \\ 10x - 12y = 6 \end{cases}$ **32.** $\begin{cases} 1.5x + 2y = 3.75 \\ 7.5x + 10y = 18.75 \end{cases}$

33. $\begin{cases} 0.05x - 0.03y = 0.21 \\ 0.07x + 0.02y = 0.16 \end{cases}$ **34.** $\begin{cases} 0.02x - 0.05y = -0.19 \\ 0.03x + 0.04y = 0.52 \end{cases}$

Choosing a Solution Method In Exercises 35–42, use any method to solve the system. Explain your choice of method.

35. $\begin{cases} 8x + 12y = 8 \\ 16x + 30y = 22 \end{cases}$ **36.** $\begin{cases} x - 9y = -7 \\ 5x + 9y = -35 \end{cases}$

37. $\begin{cases} -3x + 10y = -84 \\ y = x - 7 \end{cases}$ **38.** $\begin{cases} \frac{1}{8}x - \frac{3}{4}y = -3 \\ 4x + y = 29 \end{cases}$

39. $\begin{cases} y = e^x \\ 2x - 3y + 3 = 0 \end{cases}$ **40.** $\begin{cases} x + 5y = -11 \\ 8x - 6y = 50 \end{cases}$

41. $\begin{cases} 4x + 3y = 140 \\ 3x + 4y = 140 \end{cases}$ **42.** $\begin{cases} y = x^3 - 2x^2 \\ 10x - y = -7 \end{cases}$

43. Think About It Find a system of linear equations that has the given solution. (There are many correct answers.)

(a) $(6, 0)$ (b) $(-3, 4)$ (c) $\left(-\frac{3}{4}, -7\right)$

44. HOW DO YOU SEE IT? Use the graphs of the two equations shown below.

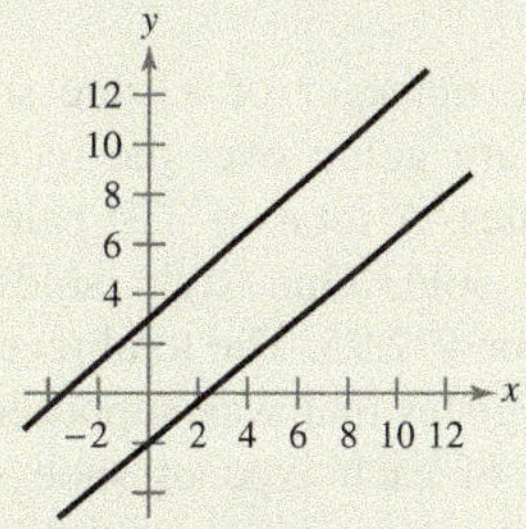

(a) Describe the graphs of the two equations.

(b) Can you conclude that the system of equations shown in the graph is inconsistent? Explain.

Recognizing a System of Equations In Exercises 45 and 46, the two lines appear to be parallel. Are they? Justify your answer by using the method of elimination to solve the system.

45. $\begin{cases} 200y - x = 200 \\ 199y - x = -198 \end{cases}$ **46.** $\begin{cases} 25x - 24y = 0 \\ 13x - 12y = 120 \end{cases}$

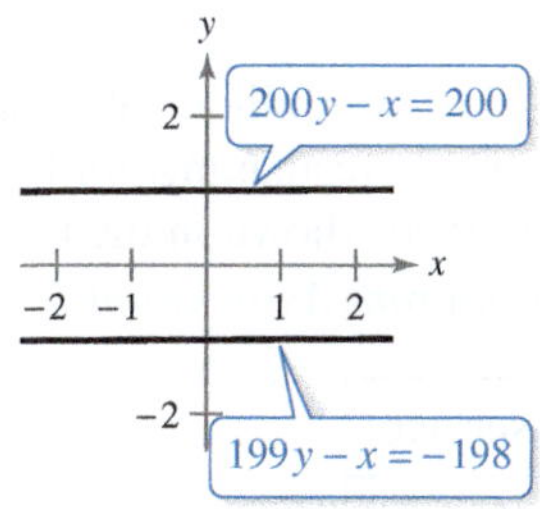

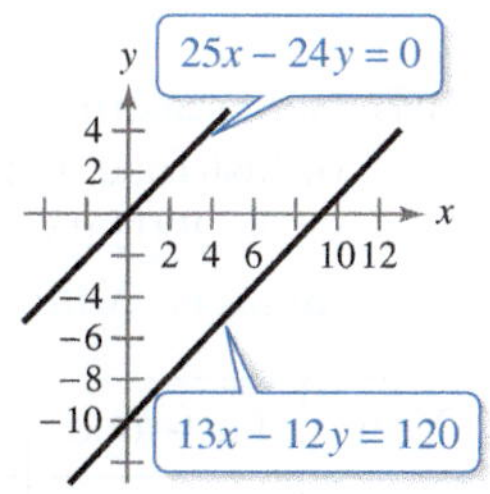

47. Airplane Speed An airplane flying into a headwind travels the 1800-mile flying distance between Los Angeles, California and South Bend, Indiana in 3 hours and 36 minutes. On the return flight, the distance is traveled in 3 hours. Find the airspeed of the plane and the speed of the wind, assuming that both remain constant.

48. Airplane Speed Two planes start from the same airport and fly in opposite directions. The second plane starts $\frac{1}{2}$ hour after the first plane, but its speed is 50 miles per hour faster. Find the airspeed of each plane if, 2 hours after the first plane departs, the planes are 2000 miles apart.

49. Acid Mixture Ten gallons of a 30% acid solution is obtained by mixing a 20% solution with a 50% solution. How much of each solution is required to obtain the specified concentration of the final mixture?

50. Fuel Mixture Five hundred gallons of 89-octane gasoline is obtained by mixing 87-octane gasoline with 92-octane gasoline. How much of each type of gasoline is required to obtain the specified mixture? (Octane ratings can be interpreted as percents.)

51. Investment Portfolio A total of $25,000 is invested in two corporate bonds that pay 4.75% and 5.5% simple interest. The total annual interest is $1315. How much is invested in each bond?

52. Shoe Sales You are the manager of a shoe store. On Sunday morning, you are going over the receipts for the previous week's sales. A total of 320 pairs of cross-training shoes were sold. One style sold for $56.95 and the other sold for $72.95. The total receipts were $21,024. The cash register that was supposed to keep track of the number of each type of shoe sold malfunctioned. Can you recover the information? If so, how many of each type were sold?

Finding the Point of Equilibrium In Exercises 53–56, find the point of equilibrium for the demand and supply equations. *See Example 8.*

	Demand	*Supply*
53.	$p = 56 - 0.0001x$	$p = 22 + 0.00001x$
54.	$p = 60 - 0.00001x$	$p = 15 + 0.00004x$
55.	$p = 140 - 0.00002x$	$p = 80 + 0.00001x$
56.	$p = 400 - 0.0002x$	$p = 225 + 0.0005x$

57. Restaurants The sales y (in billions of dollars) of fast-food and full-service restaurants for the years 2002 through 2009 are shown in the table. *(Source: National Restaurant Association)*

Year	Fast-food	Full-service
2002	115.1	141.9
2003	120.5	148.3
2004	129.5	156.9
2005	136.9	165.2
2006	142.9	172.8
2007	150.8	179.1
2008	157.7	189.4
2009	160.0	182.0

(a) Use a spreadsheet software program to create a scatter plot of the data for fast-food sales and use the *regression* feature to find a linear model. Let x represent the year, with $x = 2$ corresponding to 2002. Repeat the procedure for the data for full-service sales.

(b) Assuming that the amounts for the given 8 years are representative of future years, will fast-food sales ever equal full-service sales? Discuss the appropriateness of using these models to determine whether fast-food sales will ever equal full-service sales.

58. Prescriptions The numbers of prescriptions y (in thousands) filled at two pharmacies in the years 2007 through 2012 are shown in the table.

Year	Pharmacy A	Pharmacy B
2007	18.6	19.9
2008	19.2	20.4
2009	19.6	20.8
2010	20.0	21.1
2011	20.4	21.4
2012	21.3	22.0

(a) Use a spreadsheet software program to create a scatter plot of the data for pharmacy A and use the *regression* feature to find a linear model. Let x represent the year, with $x = 7$ corresponding to 2007. Repeat the procedure for the data for pharmacy B.

(b) Assuming that the amounts for the given 6 years are representative of future years, will the number of prescriptions filled at pharmacy A ever exceed the number of prescriptions filled at pharmacy B?

59. Supply and Demand The supply and demand equations for a small LCD television are given by

$$\begin{cases} p + 0.53x = 1542 & \text{Demand} \\ p - 0.37x = 300 & \text{Supply} \end{cases}$$

where p is the price (in dollars) and x represents the number of televisions. For how many units will the quantity demanded equal the quantity supplied? What price corresponds to this value?

60. Supply and Demand The supply and demand equations for a microscope are given by

$$\begin{cases} p + 0.85x = 650 & \text{Demand} \\ p - \quad 0.4x = \quad 75 & \text{Supply} \end{cases}$$

where p is the price (in dollars) and x represents the number of microscopes. For how many units will the quantity demanded equal the quantity supplied? What price corresponds to this value?

Fitting a Line to Data To find the least squares regression line $y = ax + b$ for a set of points (x_1, y_1), (x_2, y_2), . . ., (x_n, y_n), you can solve the following system for a and b. (If you are unfamiliar with summation notation, then look at the discussion in Section 7.1.)

$$\begin{cases} nb + \left(\sum_{i=1}^{n} x_i\right)a = \sum_{i=1}^{n} y_i \\ \left(\sum_{i=1}^{n} x_i\right)b + \left(\sum_{i=1}^{n} x_i^2\right)a = \sum_{i=1}^{n} x_i y_i \end{cases}$$

In Exercises 61–64, the sums have been evaluated. Solve the given system of linear equations for *a* and *b* to find the least squares regression line for the points. Use a graphing utility to confirm the result.

61. $\begin{cases} 5b + 10a = 20.2 \\ 10b + 30a = 50.1 \end{cases}$

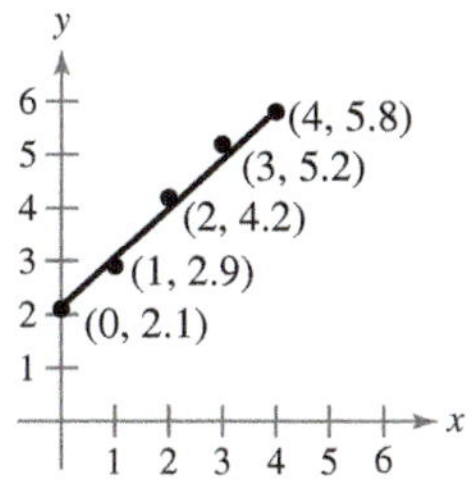

62. $\begin{cases} 5b + 10a = 11.7 \\ 10b + 30a = 25.6 \end{cases}$

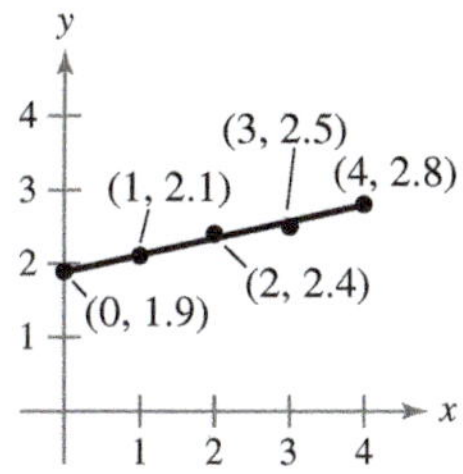

63. $\begin{cases} 7b + 21a = 35.1 \\ 21b + 91a = 114.2 \end{cases}$

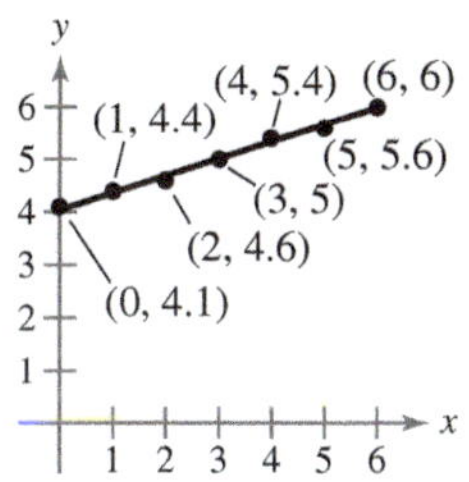

64. $\begin{cases} 6b + 15a = 23.6 \\ 15b + 55a = 48.8 \end{cases}$

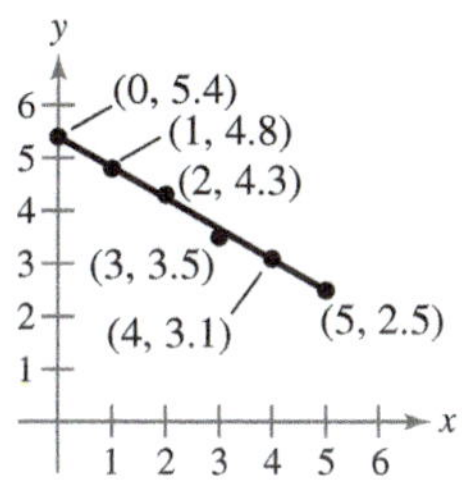

65. Atmosphere The concentration *y* (in parts per million) of carbon dioxide in the atmosphere is measured at the Mauna Loa Observatory in Hawaii. The greatest monthly carbon dioxide concentrations for the years 2006 through 2010 are shown in the table. *(Source: Scripps CO2 Program)*

Year	t	Concentration, y
2006	0	384.94
2007	1	386.43
2008	2	388.49
2009	3	390.18
2010	4	393.22

(a) Solve the following system for a and b to find the least squares regression line $y = at + b$ for the data. Let t represent the year, with $t = 0$ corresponding to 2006.

$$\begin{cases} 5b + 10a = 1943.26 \\ 10b + 30a = 3906.83 \end{cases}$$

(b) Use a graphing utility to graph the regression line and predict the greatest monthly carbon dioxide concentration in 2016.

(c) Use the *regression* feature of the graphing utility to find a linear model for the data. Compare this model with the one you found in part (a).

66. Stores in Operation The numbers of stores *y* operated by BJ's Wholesale Club, Inc. for the years 2005 through 2009 are shown in the table. *(Source: BJ's Wholesale Club, Inc.)*

Year	2005	2006	2007	2008	2009
t	0	1	2	3	4
Stores, y	165	172	177	180	187

(a) Solve the following system for a and b to find the least squares regression line $y = at + b$ for the data. Let t represent the year, with $t = 0$ corresponding to 2005.

$$\begin{cases} 5b + 10a = 881 \\ 10b + 30a = 1814 \end{cases}$$

(b) Use a graphing utility to graph the regression line and predict the number of stores in operation in 2011.

(c) Use the *regression* feature of the graphing utility to find a linear model for the data. Compare this model with the one you found in part (a).

Business Capsule

SAS is a leader in business software and services. Using SAS forecasting technologies, customers can accurately analyze and forecast processes that take place over time. SAS/ETS software contains popular forecasting methods such as regression analysis and trend extrapolation.

67. Research Project Use your campus library, the Internet, or some other reference source to find information about a company or small business that generates software which uses regression analysis to predict trends. Write a brief paper about the company or small business.

5.4 Systems of Inequalities

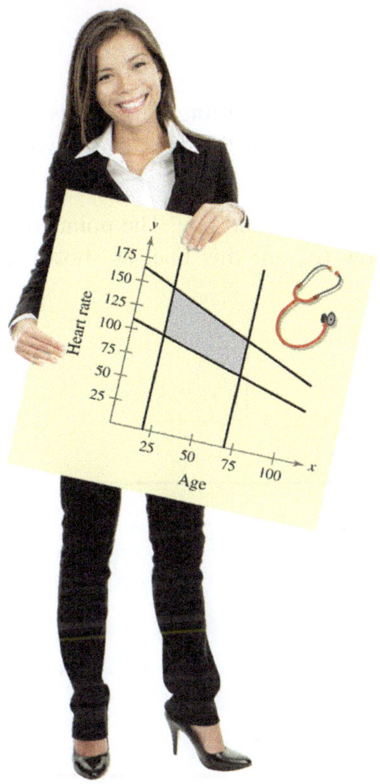

In Exercise 66 on page 453, you will use a graph to find the exercise target heart rate for individuals between the ages of 20 and 70 years old.

- Sketch the graph of an inequality in two variables.
- Solve a system of inequalities.
- Construct and use a system of inequalities to solve an application problem.

The Graph of an Inequality

The following statements are inequalities in two variables:

$$3x - 2y < 6 \quad \text{and} \quad 2x^2 + 3y^2 \geq 6.$$

An ordered pair (a, b) is a **solution of an inequality** in x and y if the inequality is true when a and b are substituted for x and y, respectively. The **graph of an inequality** is the collection of all solutions of the inequality. To sketch the graph of an inequality, begin by sketching the graph of the *corresponding equation*. The graph of the equation will normally separate the plane into two or more regions. In each such region, one of the following must be true.

1. *All* points in the region are solutions of the inequality.
2. *No* point in the region is a solution of the inequality.

So, you can determine whether the points in an entire region satisfy the inequality simply by testing *one* point in the region. When possible, use test points that are convenient to substitute into the inequality, such as $(0, 0)$.

Sketching the Graph of an Inequality in Two Variables

1. Replace the inequality sign with an equal sign, and sketch the graph of the resulting equation. (Use a dashed line for $<$ or $>$ and a solid line for $\leq$ or $\geq$.)
2. Test one point in each of the regions formed by the graph in Step 1. If the point satisfies the inequality, then shade the entire region to denote that every point in the region satisfies the inequality.

Example 1 Sketching the Graph of an Inequality

Sketch the graph of the inequality $y \geq x^2 - 1$.

SOLUTION The graph of the corresponding *equation* $y = x^2 - 1$ is a parabola, as shown in Figure 5.12. Test a point *above* the parabola $(0, 0)$ and a point *below* the parabola $(0, -2)$.

The points that satisfy the inequality are those lying above (or on) the parabola.

$(0, 0)$: $0 \overset{?}{\geq} 0^2 - 1$ $\qquad$ $(0, -2)$: $-2 \overset{?}{\geq} 0^2 - 1$

$0 \geq -1$ $\quad$ (0, 0) is a solution. $\qquad$ $-2 \ngeq -1$ $\quad$ (0, −2) is not a solution.

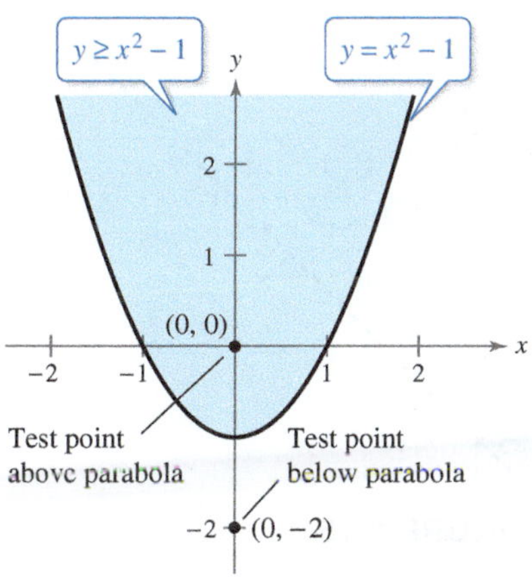

FIGURE 5.12

Checkpoint 1

Sketch the graph of $y < x^2 + 2$.

The inequality in Example 1 is a nonlinear inequality in two variables. Most of the following examples involve **linear inequalities** such as $ax + by < c$ (a and b are not both zero). The graph of a linear inequality is a half-plane lying on one side of the line $ax + by = c$.

TECH TUTOR

A graphing utility can be used to graph an inequality or a system of inequalities. For instance, to graph $y \geq x^2 - 2$, enter $y = x^2 - 2$ and use the *shade* feature of the graphing utility to shade the part of the graph consisting of the points that satisfy the inequality. You should obtain the graph shown below. Consult the user's guide for your graphing utility for more information.

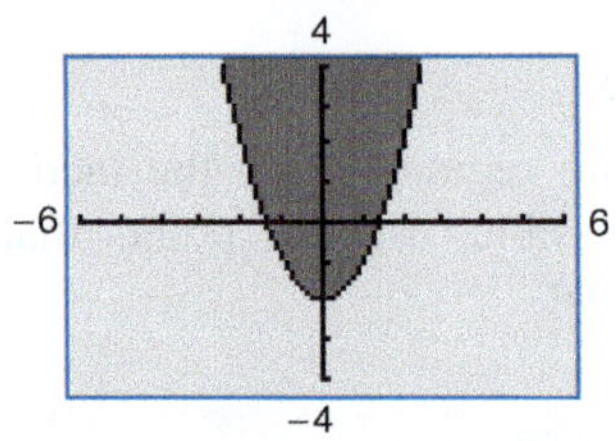

Example 2 Sketching the Graph of a Linear Inequality

Sketch the graph of each linear inequality.

a. $x > -2$ **b.** $y \leq 3$

SOLUTION

a. The graph of the corresponding equation $x = -2$ is a vertical line. The points that satisfy the inequality $x > -2$ are those lying to the right of this line, as shown in Figure 5.13.

b. The graph of the corresponding equation $y = 3$ is a horizontal line. The points that satisfy the inequality $y \leq 3$ are those lying below (or on) this line, as shown in Figure 5.14.

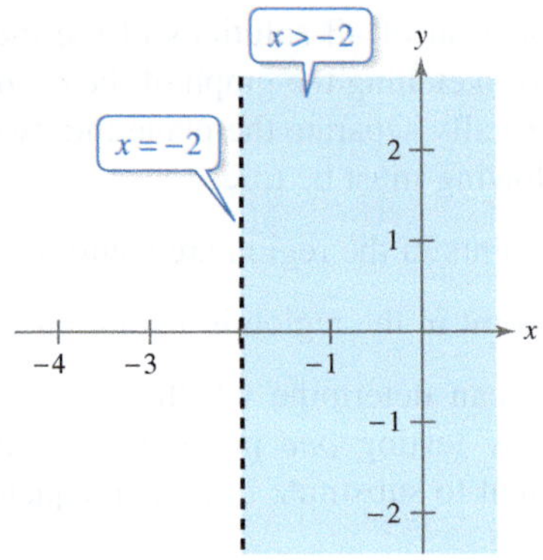

FIGURE 5.13

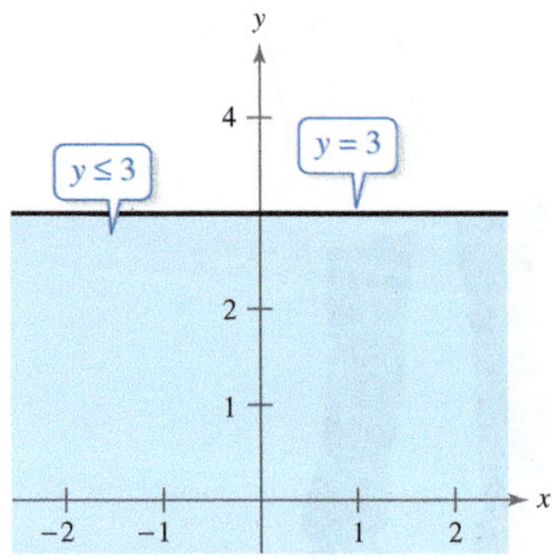

FIGURE 5.14

Checkpoint 2

Sketch the graph of $x \geq -1$.

Example 3 Sketching the Graph of a Linear Inequality

Sketch the graph of $x - y < 2$.

SOLUTION The graph of the corresponding equation $x - y = 2$ is a line, as shown in Figure 5.15. Because the origin $(0, 0)$ satisfies the inequality, the graph consists of the half-plane lying above the line. (Try checking a point below the line. Regardless of which point you choose, you will see that it does not satisfy the inequality.)

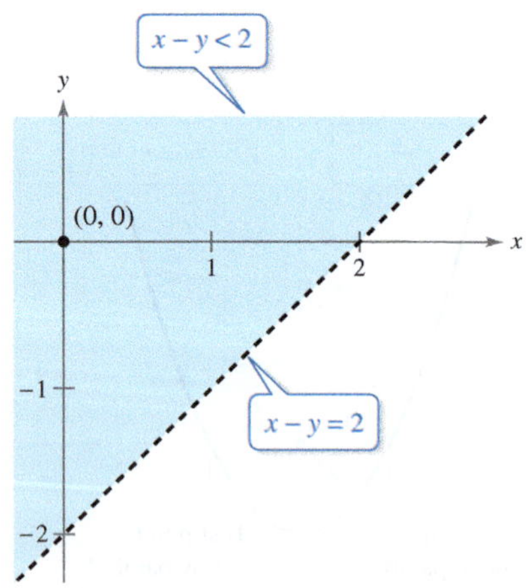

FIGURE 5.15

STUDY TIP

To graph a linear inequality, it can help to write the inequality in slope-intercept form. For instance, by writing $x - y < 2$ in the form

$$y > x - 2$$

you can see that the solution points lie above the line $x - y = 2$ (or $y = x - 2$), as shown in Figure 5.15.

Checkpoint 3

Sketch the graph of $x + y \leq 1$.

Systems of Inequalities

A **solution** of a system of inequalities in x and y is a point (x, y) that satisfies each inequality in the system. To solve a system of inequalities in two variables, first sketch the graph of each individual inequality (on the same coordinate system) and then find the region that is *common* to every graph in the system. This region represents the **solution set** of the system. For systems of *linear* inequalities, it is helpful to find the vertices of the solution region.

Example 4 Solving a System of Inequalities

Solve the system of linear inequalities.

$$\begin{cases} x - y < 2 & \text{Inequality 1} \\ x > -2 & \text{Inequality 2} \\ y \le 3 & \text{Inequality 3} \end{cases}$$

SOLUTION To solve the system, sketch the graph of the solution set. The graphs of these inequalities are shown in Figures 5.15, 5.13, and 5.14 on page 446. The triangular region common to all three graphs can be found by superimposing the graphs on the same coordinate plane, as shown in Figure 5.16. To find the vertices of the region, solve the three systems of equations obtained by taking the *pairs* of equations representing the boundaries of the individual regions.

Vertex A: $(-2, -4)$
Obtained by solving the system
$$\begin{cases} x - y = 2 \\ x = -2 \end{cases}$$

Vertex B: $(5, 3)$
Obtained by solving the system
$$\begin{cases} x - y = 2 \\ y = 3 \end{cases}$$

Vertex C: $(-2, 3)$
Obtained by solving the system
$$\begin{cases} x = -2 \\ y = 3 \end{cases}$$

STUDY TIP

Using pencils of different colors to shade the solutions of the inequalities in a system makes identifying the solution of the system of inequalities easier. The region common to every graph in the system is where all the shaded regions overlap.

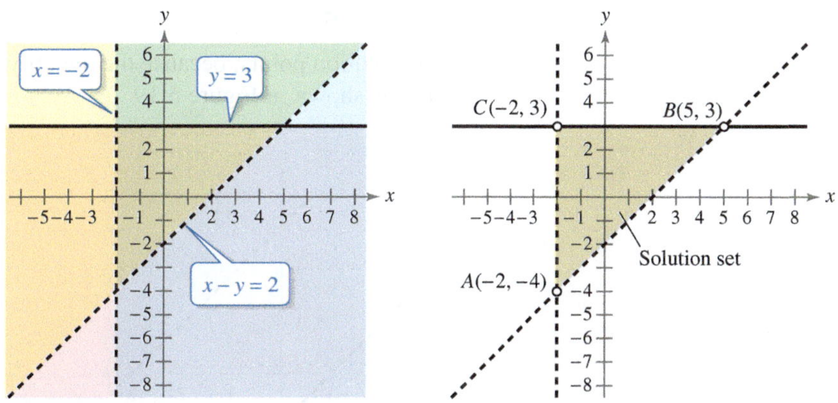

FIGURE 5.16

Checkpoint 4

Solve the system of linear inequalities.

$$\begin{cases} 2x - y > -3 \\ x \le 2 \\ y > -2 \end{cases}$$

For the triangular region shown in Figure 5.16, each point of intersection of a pair of boundary lines corresponds to a vertex. With more complicated regions, two border lines can sometimes intersect at a point that is *not* a vertex of the region, as shown in Figure 5.17. To determine which points of intersection are actually vertices of the region, sketch the region and refer to your sketch as you find each point of intersection.

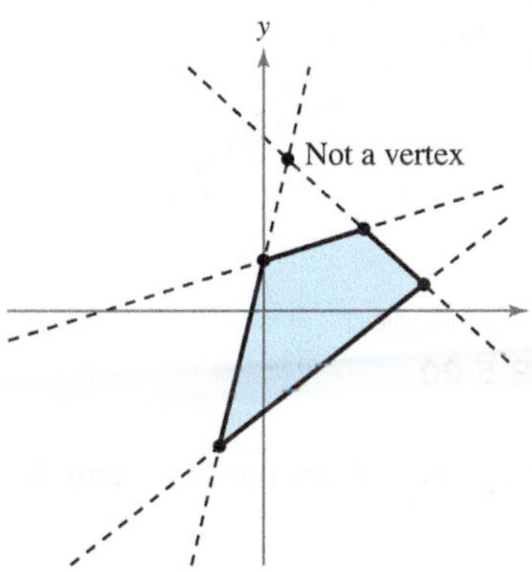

Boundary lines can intersect at a point that is not a vertex.
FIGURE 5.17

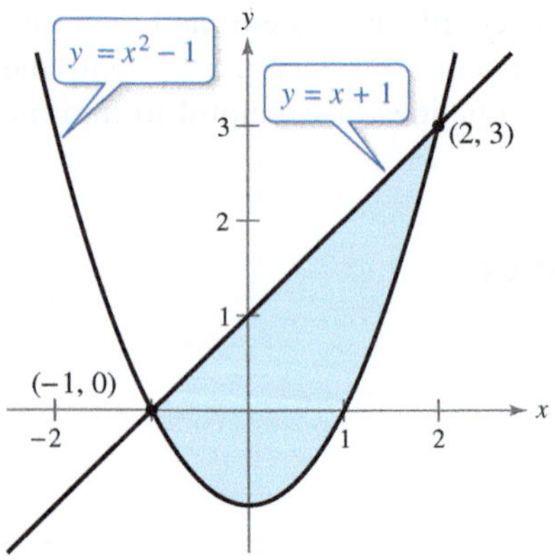

FIGURE 5.18

Example 5 Solving a System of Inequalities

Solve the system of inequalities.

$$\begin{cases} x^2 - y \le 1 & \text{Inequality 1} \\ -x + y \le 1 & \text{Inequality 2} \end{cases}$$

SOLUTION To solve the system, sketch the graph of the solution set. As shown in Figure 5.18, the points that satisfy the inequality $x^2 - y \le 1$ are the points lying above (or on) the parabola given by $y = x^2 - 1$. The points that satisfy the inequality $-x + y \le 1$ are the points lying below (or on) the line given by $y = x + 1$. To find the points of intersection of the parabola and the line, solve the system of corresponding equations.

$$\begin{cases} x^2 - y = 1 \\ -x + y = 1 \end{cases}$$

Using the method of substitution, you can find the points of intersection to be $(-1, 0)$ and $(2, 3)$. The graph of the solution set of the system is shown in Figure 5.18.

Checkpoint 5

Solve the system of inequalities.

$$\begin{cases} x^2 + y < 3 \\ x + y > -3 \end{cases}$$

When solving a system of inequalities, you should be aware that the system might have no solution. For instance, the system

$$\begin{cases} x + y > 3 \\ x + y < -1 \end{cases}$$

has no solution points, because the quantity $(x + y)$ cannot be less than -1 *and* greater than 3, as shown in Figure 5.19.

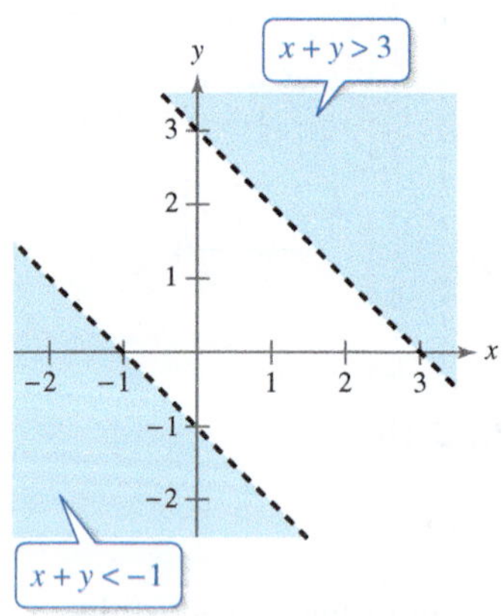

No Solution
FIGURE 5.19

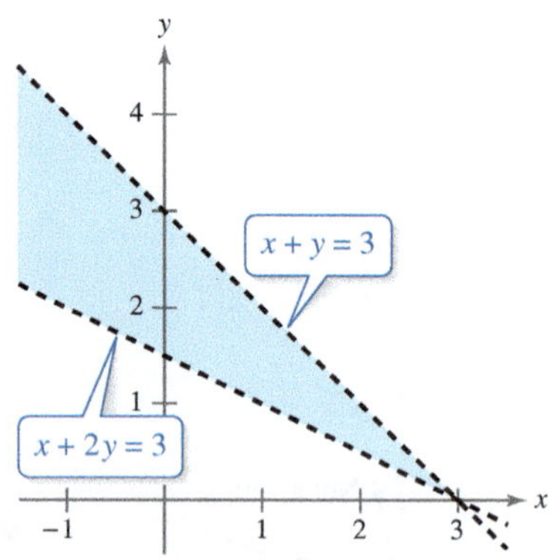

Unbounded Region
FIGURE 5.20

Another possibility is that the solution set of a system of inequalities can be unbounded. For instance, the solution set of

$$\begin{cases} x + y < 3 \\ x + 2y > 3 \end{cases}$$

forms an *infinite wedge*, as shown in Figure 5.20.

Applications

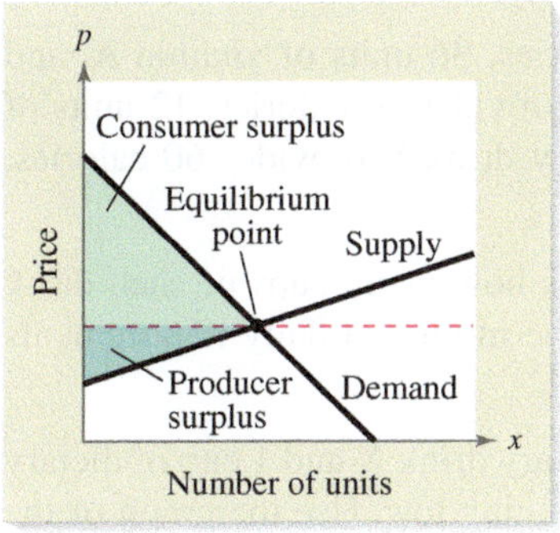

FIGURE 5.21

Example 8 in Section 5.2 discussed the *point of equilibrium* for a demand function and a supply function. The next example discusses two related concepts that economists call **consumer surplus** and **producer surplus.** As shown in Figure 5.21, the consumer surplus is defined as the area of the region that lies *below* the demand graph, *above* the horizontal line passing through the equilibrium point, and to the right of the p-axis. Similarly, the producer surplus is defined as the area of the region that lies *above* the supply graph, *below* the horizontal line passing through the equilibrium point, and to the right of the p-axis. The consumer surplus is a measure of the amount that consumers would have been willing to pay *above what they actually paid*, whereas the producer surplus is a measure of the amount that producers would have been willing to receive *below what they actually received.*

Example 6 Consumer and Producer Surpluses

The demand and supply equations for a video game console are given by

$$\begin{cases} p = 350 - 0.00003x & \text{Demand equation} \\ p = 200 + 0.00003x & \text{Supply equation} \end{cases}$$

where p is the price (in dollars) and x is the number of video game consoles. Find the consumer surplus and producer surplus for these two equations.

SOLUTION In Example 8 in Section 5.2, you saw that the point of equilibrium for these equations is

$$(2{,}500{,}000,\ 275).$$

So, the horizontal line passing through this point is $p = 275$. Now you can determine that the consumer surplus and producer surplus are the areas of the triangular regions given by the following systems of inequalities, respectively.

Consumer Surplus

$$\begin{cases} p \le 350 - 0.00003x \\ p \ge 275 \\ x \ge 0 \end{cases}$$

Producer Surplus

$$\begin{cases} p \ge 200 + 0.00003x \\ p \le 275 \\ x \ge 0 \end{cases}$$

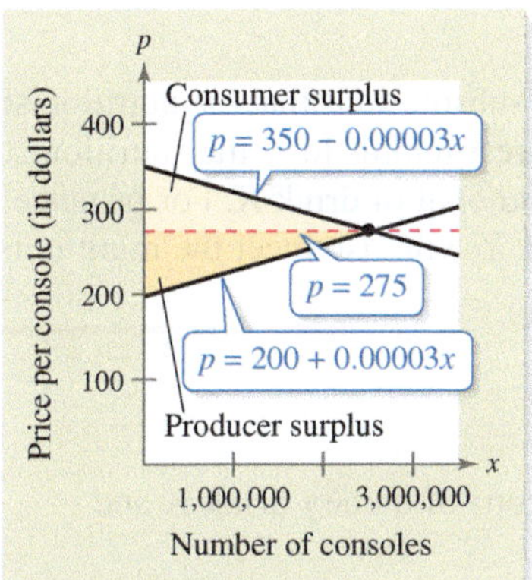

FIGURE 5.22

In Figure 5.22, you can see that the consumer and producer surpluses are defined as the areas of the shaded triangles. The base of the triangle representing the consumer surplus is 2,500,000 because the x-value of the point of equilibrium is 2,500,000. To find the height of this triangle, subtract the p-value of the point of equilibrium, 275, from the p-intercept of the demand equation, 350, to obtain 75. You can find the base and height of the triangle representing the producer surplus in a similar manner.

$$\begin{aligned} \text{Consumer surplus} &= \tfrac{1}{2}(\text{base})(\text{height}) \\ &= \tfrac{1}{2}(2{,}500{,}000)(75) \\ &= \$93{,}750{,}000 \end{aligned}$$

$$\begin{aligned} \text{Producer surplus} &= \tfrac{1}{2}(\text{base})(\text{height}) \\ &= \tfrac{1}{2}(2{,}500{,}000)(75) \\ &= \$93{,}750{,}000 \end{aligned}$$

✓ Checkpoint 6

In Example 6, the supply equation is given by $p = 230 + 0.00003x$ and the new point of equilibrium is (2,000,000, 290). Find the consumer surplus and producer surplus.

Example 7 Nutrition

The liquid portion of a diet requires at least 300 calories, 36 units of vitamin A, and 90 units of vitamin C daily. A cup of dietary drink X provides 60 calories, 12 units of vitamin A, and 10 units of vitamin C. A cup of dietary drink Y provides 60 calories, 6 units of vitamin A, and 30 units of vitamin C.

a. Set up a system of linear inequalities that describes how many cups of each drink should be consumed each day to meet or exceed the minimum daily requirements for calories and vitamins.

b. A nutritionist normally gives a patient 6 cups of dietary drink X and 1 cup of dietary drink Y per day. Supplies of dietary drink X are running low. Use the graph of the system of linear inequalities to determine other combinations of drinks X and Y that can be given that will meet the minimum daily requirements.

SOLUTION

a. Begin by letting x represent the number of cups of drink X and letting y represent the number of cups of drink Y. To meet or exceed the minimum daily requirements, the following inequalities must be satisfied.

$$\begin{cases} 60x + 60y \geq 300 & \text{Calories} \\ 12x + 6y \geq 36 & \text{Vitamin A} \\ 10x + 30y \geq 90 & \text{Vitamin C} \\ x \geq 0 & \\ y \geq 0 & \end{cases}$$

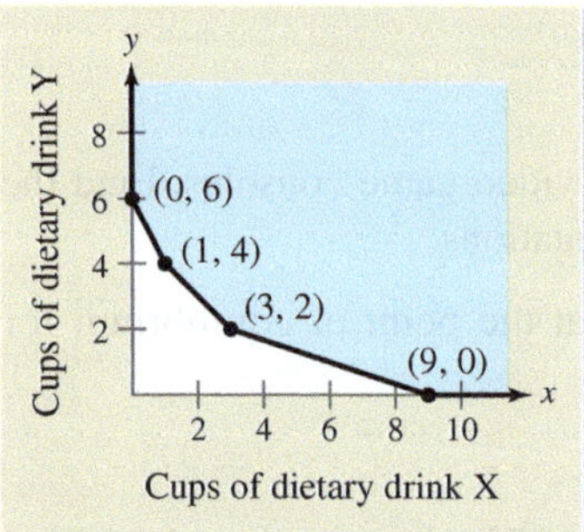

FIGURE 5.23

The last two inequalities are included because x and y cannot be negative. The graph of this system of inequalities is shown in Figure 5.23. (This application is discussed again in Example 5 in Section 5.5.)

b. From Figure 5.23, there are many different possible substitutions that the nutritionist can make. Because supplies of dietary drink X are running low, the nutritionist should choose a combination that contains a small amount of drink X. For instance, 1 cup of dietary drink X and 4 cups of dietary drink Y will also meet the minimum daily requirements.

✓ Checkpoint 7

In Example 7, should the nutritionist give a patient 4 cups of dietary drink X and 1 cup of dietary drink Y? Explain.

SUMMARIZE (Section 5.4)

1. Describe how to sketch the graph of an inequality in two variables *(page 445)*. For examples of sketching the graph of an inequality, see Examples 1, 2, and 3.
2. Describe how to solve a system of inequalities *(page 447)*. For examples of solving a system of inequalities, see Examples 4 and 5.
3. State the definitions of consumer surplus and producer surplus *(page 449)*. For an example of finding consumer surplus and producer surplus, see Example 6.
4. Describe a real-life example of how a system of linear inequalities can be used to determine the amounts of dietary drinks that should be consumed each day to meet or exceed the minimum daily requirements *(page 450, Example 7)*.

David Gilder/Shutterstock.com

SKILLS WARM UP 5.4

The following warm-up exercises involve skills that were covered in earlier sections. You will use these skills in the exercise set for this section. For additional help, review Sections 2.1, 2.2, 3.1, 5.1, and 5.2.

In Exercises 1–6, classify the graph of the equation as a line, a parabola, or a circle.

1. $x + y = 3$ **2.** $y = x^2 - 4$ **3.** $x^2 + y^2 = 9$

4. $y = -x^2 + 1$ **5.** $4x - y = 8$ **6.** $y^2 = 16 - x^2$

In Exercises 7–10, solve the system of equations.

7. $\begin{cases} x + 2y = 3 \\ 4x - 7y = -3 \end{cases}$ **8.** $\begin{cases} 2x - 3y = 4 \\ x + 5y = 2 \end{cases}$ **9.** $\begin{cases} x^2 + y = 5 \\ 2x - 4y = 0 \end{cases}$ **10.** $\begin{cases} x^2 + y^2 = 13 \\ x + y = 5 \end{cases}$

Exercises 5.4

See www.CalcChat.com for worked-out solutions to odd-numbered exercises.

Matching In Exercises 1–6, match the inequality with its graph. [The graphs are labeled (a), (b), (c), (d), (e), and (f).]

(a)

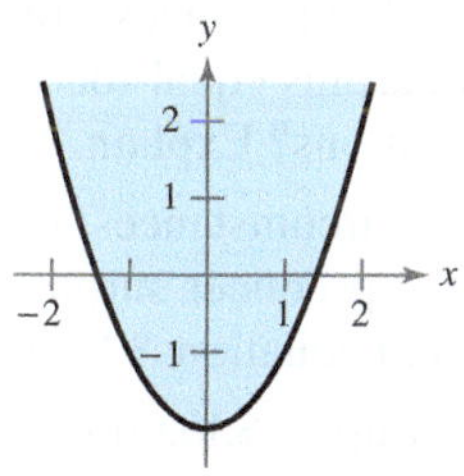

(b)

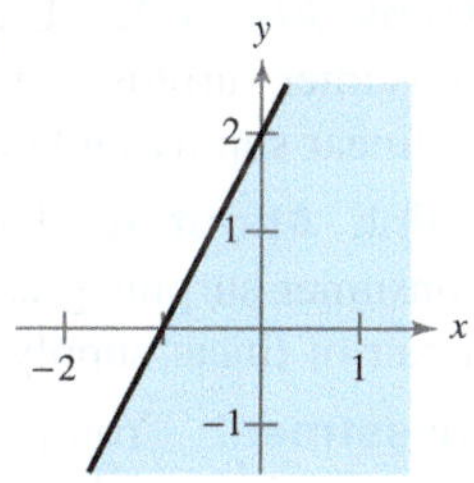

(c)

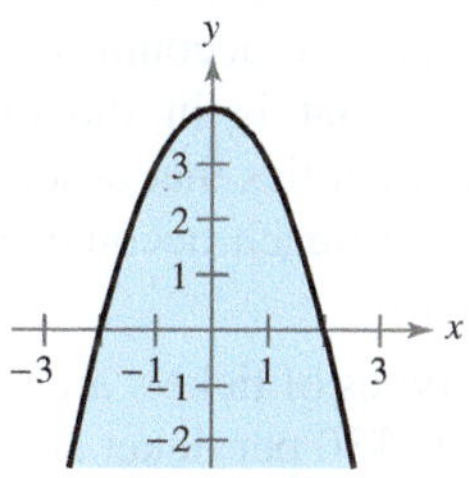

(d)

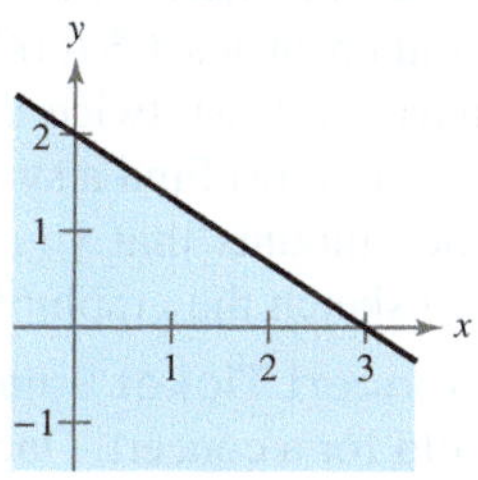

(e)

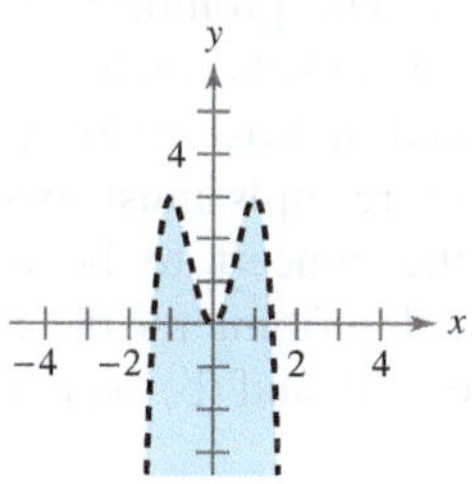

(f)

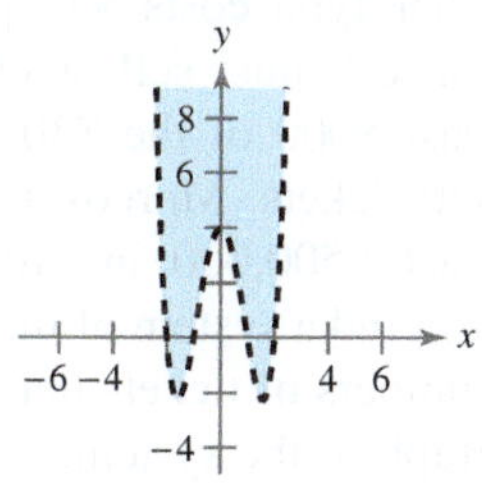

1. $2x + 3y \le 6$ **2.** $2x - y \ge -2$

3. $x^2 - y \le 2$ **4.** $y \le 4 - x^2$

5. $y > x^4 - 5x^2 + 4$ **6.** $3x^4 + y < 6x^2$

Sketching the Graph of an Inequality In Exercises 7–22, sketch the graph of the inequality. *See Examples 1, 2, and 3.*

7. $x \ge 2$ **8.** $x < 4$

9. $y + 2x^2 > 0$ **10.** $y^2 - x < 0$

11. $y > -1$ **12.** $y \le 3$

13. $y < 2 - x$ **14.** $y > 2x - 4$

15. $2y - x \ge 4$ **16.** $5x + 3y \ge -15$

17. $y < \ln x$ **18.** $y \ge -\ln x + 1$

19. $x^2 + y^2 \le 4$ **20.** $x^2 + y^2 > 4$

21. $y - (x + 2)^3 \le 0$ **22.** $y - (x - 3)^3 \ge 0$

Solving a System of Inequalities In Exercises 23–44, graph the solution set of the system of inequalities. *See Examples 4 and 5.*

23. $\begin{cases} x + y \le 2 \\ -x + y \le 2 \\ y \ge 0 \end{cases}$ **24.** $\begin{cases} 3x + 2y < 6 \\ x > 1 \\ y > 0 \end{cases}$

25. $\begin{cases} x + y \le 5 \\ x \ge 2 \\ y \ge 0 \end{cases}$ **26.** $\begin{cases} 2x + y \ge 2 \\ x \le 2 \\ y \le 1 \end{cases}$

27. $\begin{cases} -3x + 2y < 6 \\ x + 4y < -2 \\ 2x + y < 3 \end{cases}$ **28.** $\begin{cases} x - 7y > -36 \\ 5x + 2y > 5 \\ 6x - 5y > 6 \end{cases}$

29. $\begin{cases} x^2 + y \le 6 \\ x \ge -1 \\ y \ge 0 \end{cases}$ **30.** $\begin{cases} 2x^2 + y > 4 \\ x < -1 \\ y > -2 \end{cases}$

31. $\begin{cases} 2x + y > 2 \\ 6x + 3y < 2 \end{cases}$ **32.** $\begin{cases} 5x - 3y > -6 \\ 5x - 3y < -9 \end{cases}$

33. $\begin{cases} y \ge -3 \\ y \le 1 - x^2 \end{cases}$ **34.** $\begin{cases} y < -(x - 1)^2 \\ y > (x - 3)^2 - 4 \end{cases}$

35. $\begin{cases} x^2 + y^2 \le 16 \\ x^2 + y^2 < 1 \end{cases}$ **36.** $\begin{cases} x^2 + y^2 \le 25 \\ x^2 + y^2 \ge 9 \end{cases}$

37. $\begin{cases} x > y^2 \\ x < y + 2 \end{cases}$ **38.** $\begin{cases} x < 2y - y^2 \\ 0 < x + y \end{cases}$

39. $\begin{cases} y \le \sqrt{3x} + 1 \\ y \ge x + 1 \end{cases}$ **40.** $\begin{cases} y < \sqrt{2x} + 3 \\ y > x + 3 \end{cases}$

41. $\begin{cases} y < x^3 - 2x + 1 \\ y > -2x \\ x \leq 1 \end{cases}$

42. $\begin{cases} x^2 + y \leq 4 \\ y \geq 2x \\ x \geq -1 \end{cases}$

43. $\begin{cases} y \leq e^x \\ y \geq \ln x \\ x \geq \frac{1}{2} \\ x \leq 2 \end{cases}$

44. $\begin{cases} y \leq e^{-x^2/2} \\ y \geq 0 \\ x \geq -1 \\ x \leq 0 \end{cases}$

Geometry In Exercises 45–50, write a system of inequalities that corresponds to the solution set shown in the graph.

45. Parallelogram

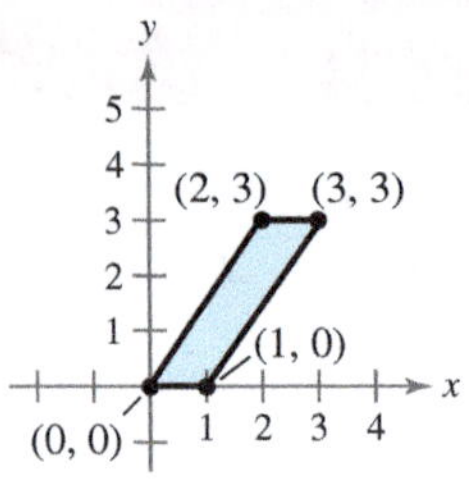

46. Rectangle

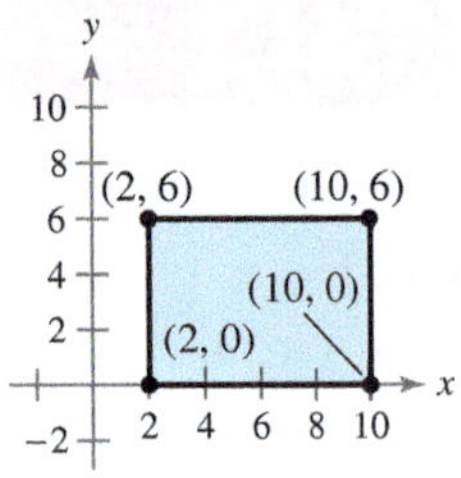

47. Triangle

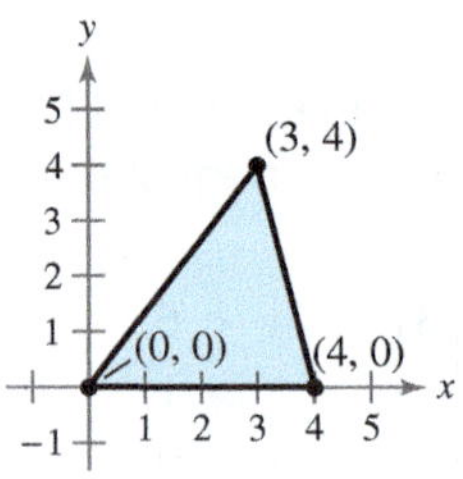

48. Triangle

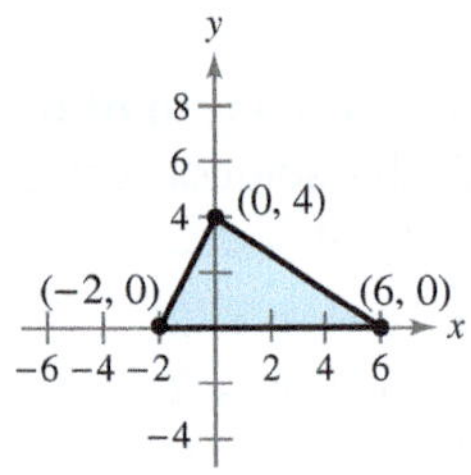

49. Sector of a circle

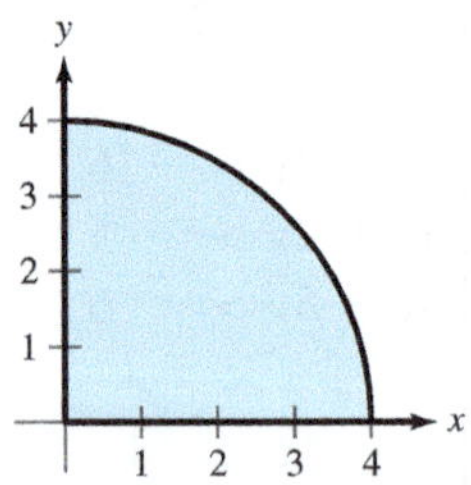

50. Sector of a circle

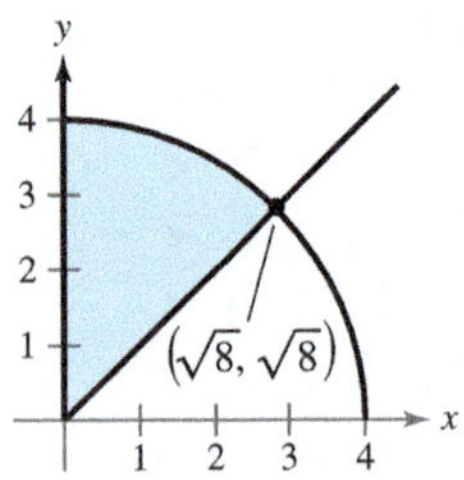

51. Furniture Production A furniture company produces tables and chairs. Each table requires 2 hours in the assembly center and $1\frac{1}{2}$ hours in the finishing center. Each chair requires $1\frac{1}{2}$ hours in the assembly center and $\frac{3}{4}$ hour in the finishing center. The company's assembly center is available 18 hours per day, and its finishing center is available 12 hours per day. Let x and y be the numbers of tables and chairs produced per day, respectively. (a) Find a system of inequalities describing all possible production levels, and (b) sketch the graph of the system.

52. Kayak Inventory A store sells two models of kayaks. Because of the demand, it is necessary to stock at least twice as many units of model A as units of model B. The costs to the store for the two models are \$500 and \$700, respectively. The management does not want more than \$25,000 in kayak inventory at any one time, and it wants at least six model A kayaks and three model B kayaks in inventory at all times. (a) Find a system of inequalities describing all possible inventory levels, and (b) sketch the graph of the system.

Consumer and Producer Surpluses In Exercises 53–56, find the consumer surplus and producer surplus for the demand and supply equations. *See Example 6.*

	Demand	*Supply*
53.	$p = 56 - 0.0001x$	$p = 22 + 0.00001x$
54.	$p = 60 - 0.00001x$	$p = 15 + 0.00004x$
55.	$p = 140 - 0.00002x$	$p = 80 + 0.00001x$
56.	$p = 600 - 0.0002x$	$p = 125 + 0.0006x$

57. Think About It Under what circumstances are the consumer surplus and producer surplus equal for a pair of linear supply and demand equations? Explain.

58. Think About It Under what circumstances is the consumer surplus greater than the producer surplus for a pair of linear supply and demand equations? Explain.

59. Investment You plan to invest up to \$30,000 in two different interest-bearing accounts. Each account is to contain at least \$7000. Moreover, one account should have at least twice the amount that is in the other account. (a) Find a system of inequalities that describes the amounts that you can invest in each account, and (b) sketch the graph of the system.

60. Concert Ticket Sales Two types of tickets are to be sold for a concert. One type costs \$30 per ticket and the other type costs \$40 per ticket. The promoter of the concert must sell at least 15,000 tickets, including at least 8000 of the \$30 tickets and at least 4000 of the \$40 tickets. Moreover, the gross receipts must total at least \$500,000 in order for the concert to be held. (a) Find a system of inequalities describing the different numbers of tickets that must be sold, and (b) sketch the graph of the system.

61. Diet Supplement A dietitian designs a special diet supplement using two different foods. Each ounce of food X contains 20 units of calcium, 10 units of iron, and 15 units of vitamin B. Each ounce of food Y contains 15 units of calcium, 20 units of iron, and 20 units of vitamin B. The minimum daily requirements for the diet are 400 units of calcium, 250 units of iron, and 220 units of vitamin B. (a) Find a system of inequalities describing the different amounts of food X and food Y that the dietitian can use in the diet, and (b) sketch the graph of the system.

62. Diet Supplement A dietitian designs a special diet supplement using two different foods. Each ounce of food X contains 12 units of calcium, 10 units of iron, and 20 units of vitamin B. Each ounce of food Y contains 15 units of calcium, 20 units of iron, and 12 units of vitamin B. The minimum daily requirements for the diet are 300 units of calcium, 280 units of iron, and 300 units of vitamin B.

(a) Find a system of inequalities describing the different amounts of food X and food Y that the dietitian can use in the diet.

(b) Sketch the graph of the system.

(c) A nutritionist normally gives a patient 10 ounces of food X and 12 ounces of food Y per day. Supplies of food Y are running low. What other combinations of foods X and Y can be given to the patient to meet the minimum daily requirements?

63. Nesting Falcons The numbers of nesting pairs y of falcons in a nature conservancy from 2006 through 2010 can be approximated by the linear model

$$y = 2.8t + 1.8, \quad 6 \le t \le 10$$

where t represents the year, with $t = 6$ corresponding to 2006.

(a) The *total* number of nesting pairs during this five-year period can be approximated by finding the area of the trapezoid represented by the system below.

$$\begin{cases} y \le 2.8t + 1.8 \\ y \ge 0 \\ t \ge 5.5 \\ t \le 10.5 \end{cases}$$

Graph this region using a graphing utility.

(b) Use the formula for the area of a trapezoid to approximate the total number of nesting pairs.

64. Sales The annual sales y (in millions of dollars) for WebMD Health from 2005 through 2010 can be approximated by the linear model

$$y = 69.50t - 169.5, \quad 5 \le t \le 10$$

where t represents the year, with $t = 5$ corresponding to 2005. *(Source: WebMD Health Corporation)*

(a) The *total* sales during this six-year period can be approximated by finding the area of the trapezoid represented by the system below.

$$\begin{cases} y \le 69.50t - 169.5 \\ y \ge 0 \\ t \ge 4.5 \\ t \le 10.5 \end{cases}$$

Graph this region using a graphing utility.

(b) Use the formula for the area of a trapezoid to approximate the total sales.

65. Writing Explain the difference between the graphs of the inequality $x \le 4$ on the real number line and on the rectangular coordinate system.

66. HOW DO YOU SEE IT? A person's maximum heart rate is

$$220 - x$$

where x is the person's age in years for $20 \le x \le 70$. When a person exercises, it is recommended that the person strive for a heart rate that is at least 50% of the maximum and at most 75% of the maximum. The exercise target heart rate region can be approximated by the system of inequalities below, where y represents a person's heart rate. *(Source: American Heart Association)*

$$\begin{cases} y \ge 0.5(220 - x) \\ y \le 0.75(220 - x) \\ x \ge 20 \\ x \le 70 \end{cases}$$

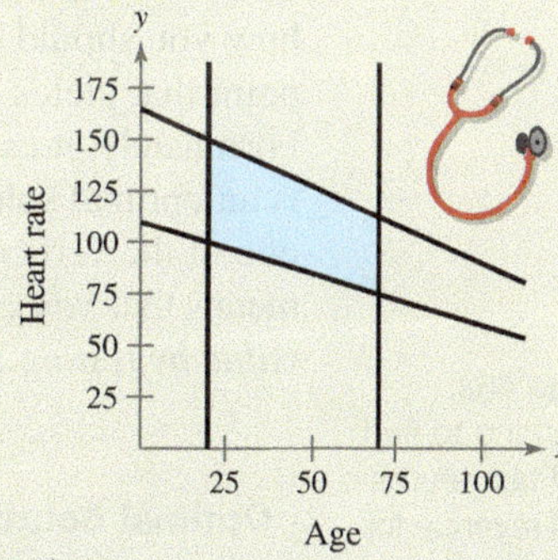

Using the graph of the system of inequalities above, find two solutions of the system and interpret the meanings of the solutions in the context of the problem.

67. Geometry Write a system of inequalities whose graphed solution set is a rectangle.

68. Geometry Write a system of inequalities whose graphed solution set is an isosceles triangle.

69. Graphical Reasoning Two concentric circles have radii x and y, where $y > x$. The area between the circles must be at least 10 square units.

(a) Find a system of inequalities describing the constraints on the circles.

(b) Use a graphing utility to graph the system of inequalities in part (a). Graph the line $y = x$ in the same viewing window.

(c) Identify the graph of the line in relation to the boundary of the inequality. Explain its meaning in the context of the problem.

5.5 Linear Programming

- Use linear programming to minimize or maximize an objective function.
- Use linear programming to optimize an application.

In Exercise 43 on page 462, you will use linear programming to find the numbers of audits and tax returns that will bring in an optimal revenue for an accounting firm.

Linear Programming: A Graphical Approach

Many applications in business and economics involve a process called **optimization,** in which you are asked to find the minimum cost, the maximum profit, or the minimum use of resources. In this section, you will study an optimization strategy that applies systems of linear inequalities, called **linear programming.**

A two-dimensional linear programming problem consists of a linear **objective function** and a system of linear inequalities called **constraints.** The objective function gives the quantity that is to be maximized (or minimized), and the constraints determine the set of **feasible solutions.** For example, suppose you are asked to maximize the value of

$$z = ax + by \qquad \text{Objective function}$$

subject to a set of constraints that determines the region in Figure 5.24. Because every point in the region satisfies each constraint, it is not clear how you should go about finding the point that yields a maximum value of z. Fortunately, it can be shown that if there is an optimal solution, it must occur at one of the vertices of the region. This means that *you can find the maximum value by testing z at each of the vertices.*

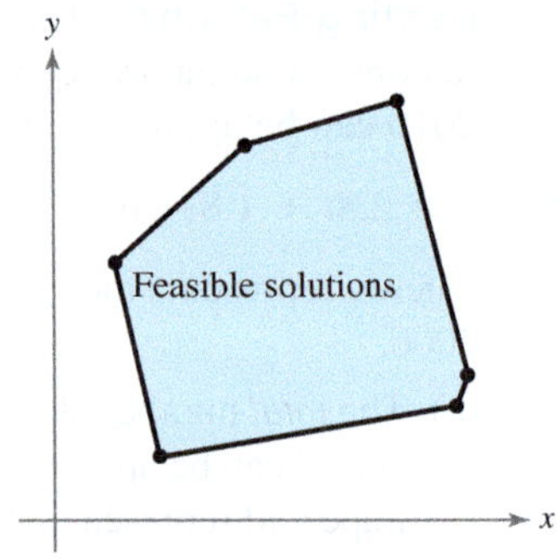

FIGURE 5.24

Optimal Solution of a Linear Programming Problem

If a linear programming problem has a solution, then it must occur at a vertex of the set of feasible solutions. If the problem has more than one solution, then at least one solution must occur at a vertex of the set of feasible solutions. In either case, the value of the objective function is unique.

A linear programming problem can include hundreds, and sometimes even thousands, of variables. However, in this section, you will solve linear programming problems that involve only two variables. The guidelines for solving a linear programming problem in two variables are listed below.

Guidelines for Solving a Linear Programming Problem

1. Sketch the region corresponding to the system of constraints. (The points inside or on the boundary of the region are *feasible solutions.*)
2. Find the vertices of the region.
3. Test the objective function at each of the vertices and select the values of the variables that optimize the objective function. For a bounded region, both a minimum and a maximum value will exist. (For an unbounded region, *if* an optimal solution exists, then it will occur at a vertex.)

Example 1 Solving a Linear Programming Problem

Find the maximum value of

$$z = 3x + 2y \qquad \text{Objective function}$$

subject to the following constraints.

$$\left.\begin{aligned} x &\ge 0 \\ y &\ge 0 \\ x + 2y &\le 4 \\ x - y &\le 1 \end{aligned}\right\} \qquad \text{Constraints}$$

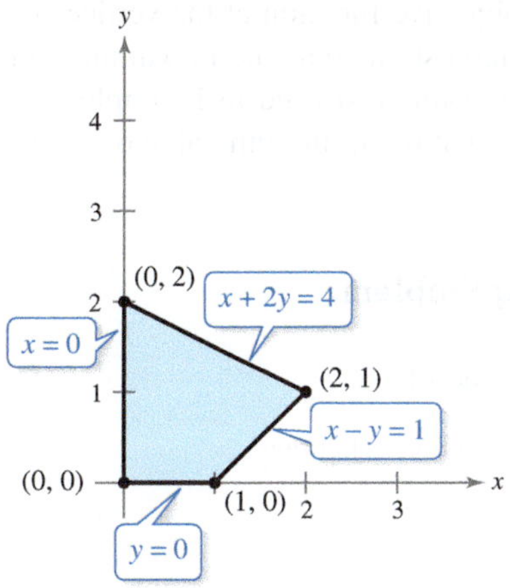

FIGURE 5.25

SOLUTION The constraints form the region shown in Figure 5.25. At the four vertices of this region, the objective function has the values shown below.

At (0, 0): $z = 3(0) + 2(0) = 0$

At (1, 0): $z = 3(1) + 2(0) = 3$

At (2, 1): $z = 3(2) + 2(1) = 8$ Maximum value of z

At (0, 2): $z = 3(0) + 2(2) = 4$

So, the maximum value of z is 8, which occurs when $x = 2$ and $y = 1$.

Checkpoint 1

Find the maximum value of

$$z = 2x + 3y$$

subject to the following constraints.

$$\left.\begin{aligned} x &\ge 0 \\ y &\ge 0 \\ x + y &\le 3 \\ x - y &\le 2 \end{aligned}\right\}$$

STUDY TIP

Remember that a vertex of a region can be found using a system of linear equations. The system will consist of the equations of the lines passing through the vertex.

In Example 1, try testing some of the *interior* points of the region. You will see that the corresponding values of z are less than 8. Here are some examples.

At (1, 1): $z = 3(1) + 2(1) = 5$

At $\left(1, \frac{1}{2}\right)$: $z = 3(1) + 2\left(\frac{1}{2}\right) = 4$

At $\left(\frac{1}{2}, \frac{3}{2}\right)$: $z = 3\left(\frac{1}{2}\right) + 2\left(\frac{3}{2}\right) = \frac{9}{2}$

At $\left(\frac{3}{2}, 1\right)$: $z = 3\left(\frac{3}{2}\right) + 2(1) = \frac{13}{2}$

To see why the maximum value of the objective function in Example 1 must occur at a vertex, consider writing the objective function in slope-intercept form

$$y = -\frac{3}{2}x + \frac{z}{2} \qquad \text{Family of lines}$$

where

$$\frac{z}{2}$$

is the y-intercept of the objective function. This equation represents a family of lines, each of slope $-\frac{3}{2}$. Of these infinitely many lines, you want the one that has the largest z-value while still intersecting the region determined by the constraints. In other words, of all the lines whose slope is $-\frac{3}{2}$, you want the one that has the largest y-intercept and intersects the given region, as shown in Figure 5.26. It should be clear that such a line will pass through one (or more) of the vertices of the region.

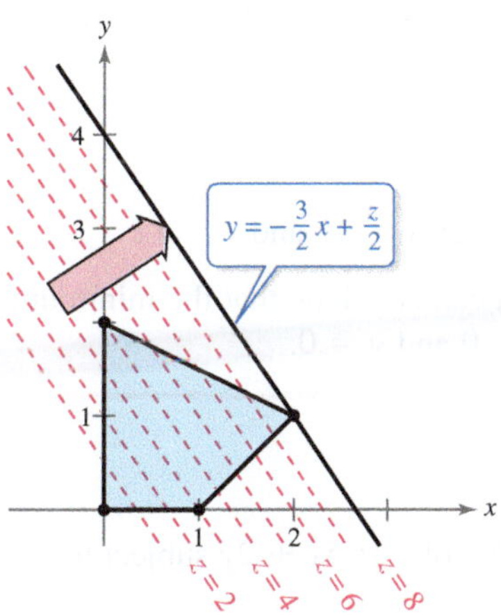

FIGURE 5.26

The guidelines for solving a linear programming problem will work whether the objective function is to be maximized or minimized. The steps used are precisely the same. In other words, once you have evaluated the objective function at the vertices of the set of feasible solutions, you simply choose the largest value as the maximum and the smallest value as the minimum. For instance, the same test used in Example 1 to find the maximum value of z can be used to conclude that the minimum value of z is 0, and that this value occurs at the vertex (0, 0).

Example 2 Solving a Linear Programming Problem

Find (a) the maximum value and (b) the minimum value of

$$z = 4x + 6y \qquad \text{Objective function}$$

subject to the following constraints.

$$\left.\begin{aligned} x &\geq 0 \\ y &\geq 0 \\ -x + y &\leq 11 \\ x + y &\leq 27 \\ 2x + 5y &\leq 90 \end{aligned}\right\} \qquad \text{Constraints}$$

SOLUTION

a. The region bounded by the constraints is shown in Figure 5.27.

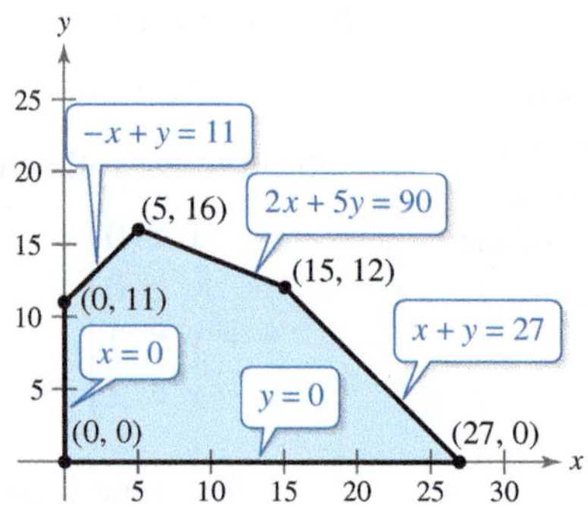

FIGURE 5.27

By testing the objective function at each vertex, you obtain the values shown below.

At (0, 0): $z = 4(0) + 6(0) = 0$

At (0, 11): $z = 4(0) + 6(11) = 66$

At (5, 16): $z = 4(5) + 6(16) = 116$

At (15, 12): $z = 4(15) + 6(12) = 132$ Maximum value of z

At (27, 0): $z = 4(27) + 6(0) = 108$

So, the maximum value of z is 132, which occurs when $x = 15$ and $y = 12$.

b. Using the values of z at the vertices in part (a), you can conclude that the minimum value of z is 0, and that this value occurs when $x = 0$ and $y = 0$.

ALGEBRA TUTOR

For help with the algebra used to find the vertices of the region in Example 2, see the *Chapter 5 Algebra Tutor* on page 465.

✓ Checkpoint 2

Find (a) the maximum value and (b) the minimum value of $z = 5x + 2y$ subject to the same constraints as in Example 2.

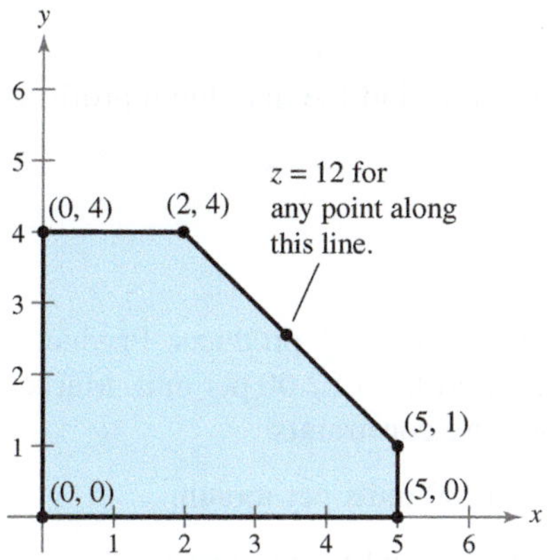

FIGURE 5.28

It is possible for the maximum (or minimum) value in a linear programming problem to occur at *two* different vertices. For instance, at the vertices of the region shown in Figure 5.28, the objective function

$z = 2x + 2y$ Objective function

has the values shown below.

At $(0, 0)$: $z = 2(0) + 2(0) = 0$

At $(0, 4)$: $z = 2(0) + 2(4) = 8$

At $(2, 4)$: $z = 2(2) + 2(4) = 12$ Maximum value of z

At $(5, 1)$: $z = 2(5) + 2(1) = 12$ Maximum value of z

At $(5, 0)$: $z = 2(5) + 2(0) = 10$

In this case, you can conclude that the objective function has a maximum value not only at the vertices $(2, 4)$ and $(5, 1)$; it also has a maximum value (of 12) at *any point on the line segment connecting these two vertices*. Note that the objective function

$$y = -x + \frac{1}{2}z$$

has the same slope as the line through the vertices $(2, 4)$ and $(5, 1)$.

Some linear programming problems have no optimal solution. This can occur when the region determined by the constraints is *unbounded*. This is shown in Example 3.

Example 3 An Unbounded Region

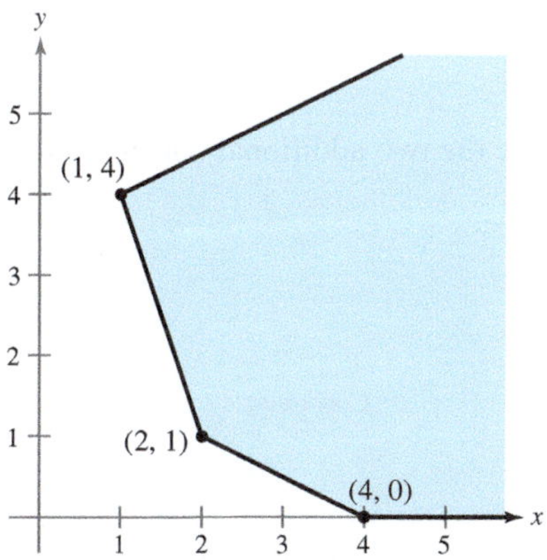

FIGURE 5.29

Find the maximum value of

$z = 4x + 2y$ Objective function

where $x \geq 0$ and $y \geq 0$, subject to the following constraints.

$$\left.\begin{aligned} x + 2y &\geq 4 \\ 3x + y &\geq 7 \\ -x + 2y &\leq 7 \end{aligned}\right\}$$ Constraints

SOLUTION The region determined by the constraints is shown in Figure 5.29. For this unbounded region, there is no maximum value of z. To see this, note that the point $(x, 0)$ lies in the region for all values of $x \geq 4$. By choosing x to be large, you can obtain values of

$$\begin{aligned} z &= 4(x) + 2(0) \\ &= 4x \end{aligned}$$

that are as large as you want. So, there is no maximum value of z. For this problem, there *is* a minimum value of z, $z = 10$, which occurs at the vertex $(2, 1)$, as shown below.

At $(1, 4)$: $z = 4(1) + 2(4) = 12$

At $(2, 1)$: $z = 4(2) + 2(1) = 10$ Minimum value of z

At $(4, 0)$: $z = 4(4) + 2(0) = 16$

Checkpoint 3

Find the maximum value of the objective function

$z = x + 8y$

where $x \geq 0$ and $y \geq 0$, subject to the same constraints as in Example 3.

Applications

Example 4 shows how linear programming can be used to find the maximum profit in a business application.

Example 4 Optimal Profit

A manufacturer wants to maximize the profit for two laboratory products. Product I yields a profit of \$1.50 per unit, and product II yields a profit of \$2.00 per unit. Market tests and available resources have indicated the following constraints.

1. The combined production level should not exceed 1200 units per month.
2. The demand for product II is no more than half the demand for product I.
3. The production level of product I is less than or equal to 600 units plus three times the production level of product II.

What is the optimal production level for each product?

SOLUTION Let x be the number of units of product I and let y be the number of units of product II. The objective function (for the combined profit) is given by

$$P = 1.5x + 2y. \qquad \text{Objective function}$$

The three constraints translate into the following linear inequalities.

1. $x + y \le 1200 \quad \Rightarrow \quad x + y \le 1200$

2. $y \le \frac{1}{2}x \quad \Rightarrow \quad -x + 2y \le 0$

3. $x \le 3y + 600 \quad \Rightarrow \quad x - 3y \le 600$

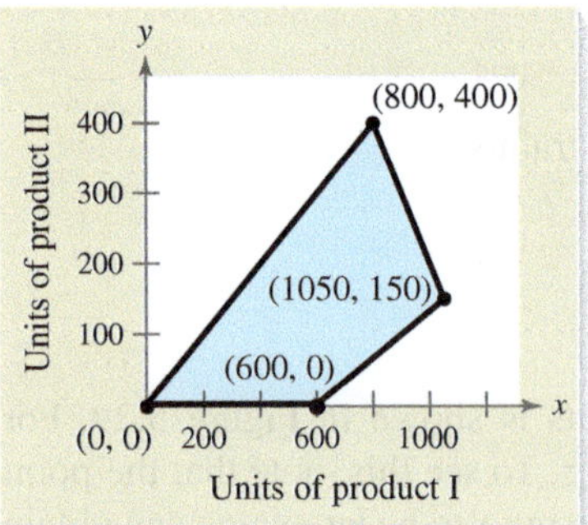

FIGURE 5.30

Because neither x nor y can be negative, you also have the two additional constraints of $x \ge 0$ and $y \ge 0$.

So, the constraints are as shown.

$$\left.\begin{aligned} x &\ge 0 \\ y &\ge 0 \\ x + y &\le 1200 \\ -x + 2y &\le 0 \\ x - 3y &\le 600 \end{aligned}\right\} \qquad \text{Constraints}$$

Figure 5.30 shows the region determined by the constraints. To find the maximum profit, test the value of P at each vertex of the region.

$$\begin{aligned} &\text{At } (0, 0): & P &= 1.5(0) + 2(0) = 0 \\ &\text{At } (800, 400): & P &= 1.5(800) + 2(400) = 2000 \qquad \text{Maximum profit} \\ &\text{At } (1050, 150): & P &= 1.5(1050) + 2(150) = 1875 \\ &\text{At } (600, 0): & P &= 1.5(600) + 2(0) = 900 \end{aligned}$$

So, the maximum profit is \$2000, and it occurs when the monthly production levels are 800 units of product I and 400 units of product II.

Checkpoint 4

In Example 4, the manufacturer improves the production of product I so that it yields a profit of \$2.50 per unit. How would this improvement affect the optimal number of units the manufacturer should sell in order to obtain a maximum profit?

Example 5 shows how linear programming can be used to find the optimal cost in a real-life application.

Example 5 Optimal Cost

The liquid portion of a diet requires at least 300 calories, 36 units of vitamin A, and 90 units of vitamin C daily. A cup of dietary drink X costs \$0.12 and provides 60 calories, 12 units of vitamin A, and 10 units of vitamin C. A cup of dietary drink Y costs \$0.15 and provides 60 calories, 6 units of vitamin A, and 30 units of vitamin C. How many cups of each drink should be consumed each day to obtain an optimal cost and still meet the daily requirements?

SOLUTION As in Example 7 on page 450, let x be the number of cups of dietary drink X and let y be the number of cups of dietary drink Y.

$$\left.\begin{array}{ll} \text{For calories:} & 60x + 60y \geq 300 \\ \text{For vitamin A:} & 12x + 6y \geq 36 \\ \text{For vitamin C:} & 10x + 30y \geq 90 \\ & x \geq 0 \\ & y \geq 0 \end{array}\right\} \quad \text{Constraints}$$

The cost C is given by

$$C = 0.12x + 0.15y. \qquad \text{Objective function}$$

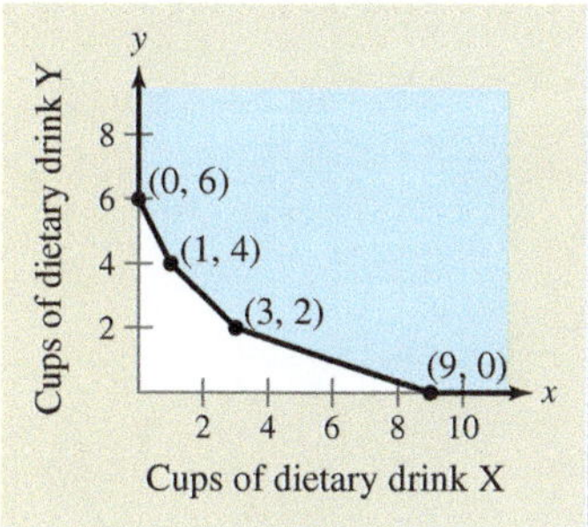

FIGURE 5.31

The graph of the region corresponding to the constraints is shown in Figure 5.31. Because you want to incur as little cost as possible, you want to determine the *minimum* cost. To determine the minimum cost, test C at each vertex of the region, as shown below.

At $(0, 6)$: $C = 0.12(0) + 0.15(6) = 0.90$

At $(1, 4)$: $C = 0.12(1) + 0.15(4) = 0.72$

At $(3, 2)$: $C = 0.12(3) + 0.15(2) = 0.66$ Minimum value of C

At $(9, 0)$: $C = 0.12(9) + 0.15(0) = 1.08$

So, the minimum cost is \$0.66 per day, and this cost occurs when 3 cups of dietary drink X and 2 cups of dietary drink Y are consumed each day.

✓ Checkpoint 5

In Example 5, a cup of dietary drink Y costs \$0.11. How would this affect the number of cups of each drink that should be consumed each day to obtain an optimal cost and still meet the daily requirements?

SUMMARIZE (Section 5.5)

1. Describe the guidelines for solving a linear programming problem *(page 454)*. For an example of solving a linear programming problem, see Example 1.
2. Describe how to find the maximum and minimum values of an objective function of a linear programming problem *(page 456)*. For an example of maximizing and minimizing the objective function of a linear programming problem, see Example 2.
3. Describe a situation in which a linear programming problem has no optimal solution *(page 457)*. For an example of solving a linear programming problem that has no optimal solution, see Example 3.
4. Describe a real-life example of how linear programming can be used to find the maximum profit in a business application *(page 458, Example 4)*.

SKILLS WARM UP 5.5

The following warm-up exercises involve skills that were covered in earlier sections. You will use these skills in the exercise set for this section. For additional help, review Sections 2.2, 5.1, and 5.4.

In Exercises 1–4, sketch the graph of the linear equation.

1. $y + x = 3$

2. $y - x = 12$

3. $x = 0$

4. $y = 4$

In Exercises 5–8, find the point of intersection of the two lines.

5. $x + y = 4$
$x = 0$

6. $x + 2y = 12$
$y = 0$

7. $x + y = 4$
$2x + 3y = 9$

8. $x + 2y = 12$
$2x + y = 9$

In Exercises 9 and 10, sketch the graph of the inequality.

9. $2x + 3y \geq 18$

10. $4x + 3y \geq 12$

Exercises 5.5

See www.CalcChat.com for worked-out solutions to odd-numbered exercises.

Solving a Linear Programming Problem In Exercises 1–8, find the minimum and maximum values of the objective function and where they occur, subject to the indicated constraints. (For each exercise, the graph of the region determined by the constraints is provided.) *See Examples 1 and 2.*

1. *Objective function:*

$z = 3x + 4y$

Constraints:

$x \geq 0$

$y \geq 0$

$x + y \leq 5$

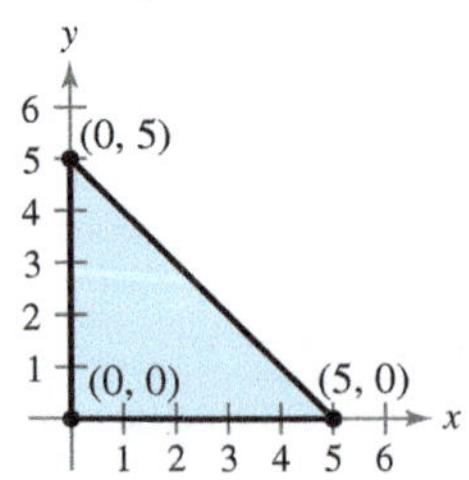

2. *Objective function:*

$z = 2x + 8y$

Constraints:

$x \geq 0$

$y \geq 0$

$2x + y \leq 4$

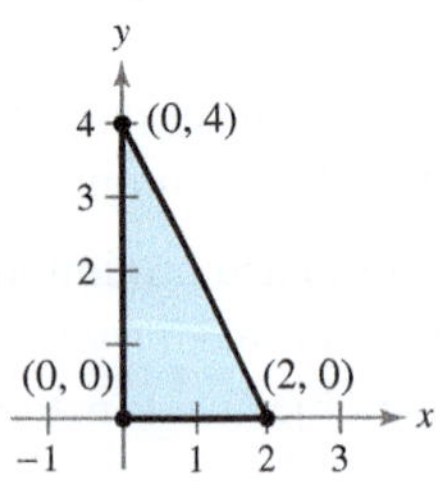

3. *Objective function:*

$z = 9x + 6y$

Constraints:

See Exercise 1.

4. *Objective function:*

$z = 7x + 3y$

Constraints:

See Exercise 2.

5. *Objective function:*

$z = 6x + 2y$

Constraints:

$x \geq 0$

$y \geq 0$

$x + 3y \leq 15$

$4x + y \leq 16$

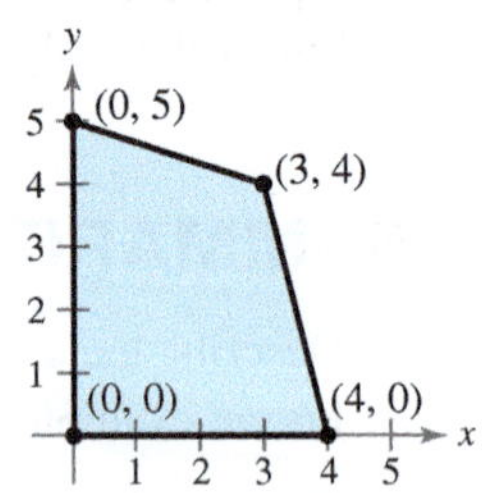

6. *Objective function:*

$z = 5x + 4y$

Constraints:

$x \geq 0$

$y \geq 0$

$2x + 3y \geq 6$

$3x - 2y \leq 9$

$x + 5y \leq 20$

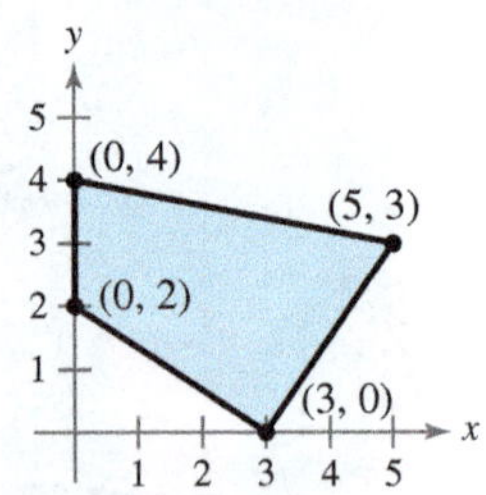

7. *Objective function:*

$z = 5x + 0.5y$

Constraints:

See Exercise 5.

8. *Objective function:*

$z = x + 6y$

Constraints:

See Exercise 6.

Solving a Linear Programming Problem In Exercises 9–20, sketch the region determined by the indicated constraints. Then find the minimum and maximum values of the objective function and where they occur, subject to the constraints. *See Examples 1, 2, and 3.*

9. *Objective function:*

$z = 6x + 10y$

Constraints:

$x \geq 0$

$y \geq 0$

$3x + 5y \leq 15$

10. *Objective function:*

$z = 7x + 8y$

Constraints:

$x \geq 0$

$y \geq 0$

$x + 2y \leq 8$

11. *Objective function:*

$z = 8x + 4y$

Constraints:

See Exercise 9.

12. *Objective function:*

$z = 10x + 3y$

Constraints:

See Exercise 10.

13. *Objective function:*

$z = 4x + 5y$

Constraints:

$x \geq 0$

$y \geq 0$

$x + y \geq 8$

$3x + 5y \geq 30$

14. *Objective function:*

$z = 4x + 5y$

Constraints:

$x \geq 0$

$y \geq 0$

$x + y \leq 5$

$x + 2y \leq 6$

15. *Objective function:*

$z = 2x + 7y$

Constraints:

See Exercise 13.

16. *Objective function:*

$z = 2x - y$

Constraints:

See Exercise 14.

17. *Objective function:*

$z = 3x + 6y$

Constraints:

$x \geq 0$

$y \geq 0$

$x + 2y \leq 40$

$x + y \leq 30$

$2x + 3y \leq 65$

18. *Objective function:*

$z = x$

Constraints:

$x \geq 0$

$y \geq 0$

$2x + 3y \leq 60$

$2x + y \leq 28$

$4x + y \leq 48$

19. *Objective function:*

$z = x + y$

Constraints:

See Exercise 17.

20. *Objective function:*

$z = y$

Constraints:

See Exercise 18.

Maximizing the Objective Function In Exercises 21–24, maximize the objective function subject to the constraints $3x + y \leq 15$, $4x + 3y \leq 30$, $x \geq 0$, and $y \geq 0$.

21. $z = 2x + y$

22. $z = 5x + y$

23. $z = 4x + 3y$

24. $z = 3x + y$

Maximizing the Objective Function In Exercises 25–28, maximize the objective function subject to the constraints $x + 5y \leq 25$, $x + y \leq 9$, $4x + 2y \leq 32$, $x \geq 0$, and $y \geq 0$.

25. $z = 6x + 7y$

26. $z = 10x + 8y$

27. $z = 15x + y$

28. $z = 5x + 11y$

Think About It In Exercises 29–36, find an objective function that has a maximum or minimum value at the indicated vertex of the constraint region shown. (There are many correct answers.)

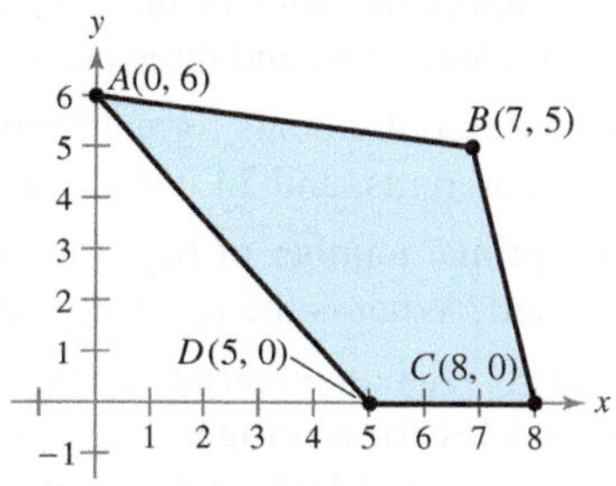

29. The maximum occurs at vertex A.

30. The maximum occurs at vertex B.

31. The minimum occurs at vertex C.

32. The minimum occurs at vertex D.

33. The maximum occurs at vertices A and B.

34. The maximum occurs at vertices B and C.

35. The minimum occurs at vertices A and D.

36. The minimum occurs at vertices C and D.

37. Optimal Profit A fruit grower raises crops A and B. The profit is \$185 per acre for crop A and \$245 per acre for crop B. Research and available resources indicate the following constraints.

- The fruit grower has 150 acres of land for raising the crops.
- It takes 1 day to trim an acre of crop A and 2 days to trim an acre of crop B, and there are 240 days per year available for trimming.
- It takes 0.3 day to pick an acre of crop A and 0.1 day to pick an acre of crop B, and there are 30 days per year available for picking.

What is the optimal acreage for each fruit? What is the optimal profit?

38. Optimal Profit The costs to a store for two models of Global Positioning System (GPS) receivers are \$80 and \$100. The \$80 model yields a profit of \$25 and the \$100 model yields a profit of \$30. Market tests and available resources indicate the following constraints.

- The merchant estimates that the total monthly demand will not exceed 200 units.
- The merchant does not want to invest more than \$18,000 in GPS receiver inventory.

What is the optimal inventory level for each model? What is the optimal profit?

39. Optimal Cost A farming cooperative mixes two brands of cattle feed. Brand X costs $30 per bag, and brand Y costs $25 per bag. Research and available resources have indicated the following constraints.

- Brand X contains two units of nutritional element A, two units of element B, and two units of element C.
- Brand Y contains one unit of nutritional element A, nine units of element B, and three units of element C.
- The minimum requirements for nutrients A, B, and C are 12 units, 36 units, and 24 units, respectively.

What is the optimal number of bags of each brand that should be mixed? What is the optimal cost?

40. Optimal Cost A humanitarian agency can use two models of vehicles for a refugee rescue mission. Each model A vehicle costs $1000 and each model B vehicle costs $1500. Mission strategies and objectives indicate the following constraints.

- A total of at least 20 vehicles must be used.
- A model A vehicle can hold 45 boxes of supplies. A model B vehicle can hold 30 boxes of supplies. The agency must deliver at least 690 boxes of supplies to the refugee camp.
- A model A vehicle can hold 20 refugees. A model B vehicle can hold 32 refugees. The agency must rescue at least 520 refugees.

What is the optimal number of vehicles of each model that should be used? What is the optimal cost?

41. Optimal Profit A manufacturer produces two models of bicycles. The times (in hours) required for assembling, painting, and packaging each model are shown in the table.

Process	Model A	Model B
Assembling	2	2.5
Painting	4	1
Packaging	1	0.75

The total times available for assembling, painting, and packaging are 4000 hours, 4800 hours, and 1500 hours, respectively. The profits per unit are $50 for model A and $75 for model B. What is the optimal production level for each model? What is the optimal profit?

42. Optimal Profit A company makes two models of doghouses. The times (in hours) required for assembling, painting, and packaging are shown in the table.

Process	Model A	Model B
Assembling	2.5	3
Painting	2	1
Packaging	0.75	1.25

The total times available for assembling, painting, and packaging are 4000 hours, 2500 hours, and 1500 hours, respectively. The profits per unit are $60 for model A and $75 for model B. What is the optimal production level for each model? What is the optimal profit?

43. Optimal Revenue An accounting firm charges $2500 for an audit and $350 for a tax return. Research and available resources have indicated the following constraints.

- The firm has 900 hours of staff time available each week.
- The firm has 155 hours of review time available each week.
- Each audit requires 75 hours of staff time and 10 hours of review time.
- Each tax return requires 12.5 hours of staff time and 2.5 hours of review time.

What numbers of audits and tax returns will bring in an optimal revenue?

44. Optimal Revenue The accounting firm in Exercise 43 lowers its charge for an audit to $2000. What numbers of audits and tax returns will bring in an optimal revenue?

45. Media Selection A company has budgeted a maximum of $4,800,000 for national advertising of an allergy medication. Each minute of television time costs $400,000 and each one-page newspaper ad costs $80,000. Each television ad is expected to be viewed by 20 million viewers, and each newspaper ad is expected to be seen by 5 million readers. The company's market research department recommends that at most 80% of the advertising budget be spent on television ads. What is the optimal amount that should be spent on each type of ad? What is the optimal total audience?

46. Optimal Profit A fruit juice company makes two drinks by blending apple and pineapple juices. The percents of apple juice and pineapple juice in each drink are shown in the table.

Mixture	Drink A	Drink B
Apple juice	30%	60%
Pineapple juice	70%	40%

There are 1000 liters of apple juice and 1500 liters of pineapple juice available. The profit for drink A is $0.70 per liter and the profit for drink B is $0.60 per liter. What is the optimal production level for each type of drink? What is the optimal profit?

47. **Investments** An investor has up to \$250,000 to invest in two types of investments. Type A investments pay 7% annually and type B pay 12% annually. To have a well-balanced portfolio, the investor imposes the following conditions. At least one-fourth of the total portfolio is to be allocated to type A investments and at least one-fourth is to be allocated to type B investments. What is the optimal amount that should be invested in each type of investment? What is the optimal return?

48. **Investments** An investor has up to \$450,000 to invest in two types of investments. Type A investments pay 8% annually and type B pay 14% annually. To have a well-balanced portfolio, the investor imposes the following conditions. At least one-half of the total portfolio is to be allocated to type A investments and at least one-fourth is to be allocated to type B investments. What is the optimal amount that should be invested in each type of investment? What is the optimal return?

49. **Optimal Profit** A company makes two models of a patio furniture set. The times for assembling, finishing, and packaging model A are 3 hours, 2.5 hours, and 0.6 hour, respectively. The times for model B are 2.75 hours, 1 hour, and 1.25 hours. The total times available for assembling, finishing, and packaging are 3000 hours, 2400 hours, and 1200 hours, respectively. The profit per unit for model A is \$100 and the profit per unit for model B is \$85. What is the optimal production level for each model? What is the optimal profit?

50. **Optimal Profit** A manufacturer produces two models of elliptical cross-training exercise machines. The times for assembling, finishing, and packaging model A are 3 hours, 3 hours, and 0.8 hour, respectively. The times for model B are 4 hours, 2.5 hours, and 0.4 hour. The total times available for assembling, finishing, and packaging are 6000 hours, 4200 hours, and 950 hours, respectively. The profits per unit are \$300 for model A and \$375 for model B. What is the optimal production level for each model? What is the optimal profit?

Recognizing Unusual Characteristics In Exercises 51–56, the given linear programming problem has an unusual characteristic. Sketch a graph of the solution region for the problem and describe the unusual characteristic. Find the maximum value of the objective function and where it occurs.

51. *Objective function:*

$z = 2.5x + y$

Constraints:

$$x \geq 0$$
$$y \geq 0$$
$$3x + 5y \leq 15$$
$$5x + 2y \leq 10$$

52. *Objective function:*

$z = x + y$

Constraints:

$$x \geq 0$$
$$y \geq 0$$
$$-x + y \leq 1$$
$$-x + 2y \leq 4$$

53. *Objective function:*

$z = -x + 2y$

Constraints:

$$x \geq 0$$
$$y \geq 0$$
$$x \leq 10$$
$$x + y \leq 7$$

54. *Objective function:*

$z = x + y$

Constraints:

$$x \geq 0$$
$$y \geq 0$$
$$-x + y \leq 1$$
$$-3x + y \geq 3$$

55. *Objective function:* $z = 3x + 4y$

Constraints:

$$x \geq 0$$
$$y \geq 0$$
$$x + y \leq 1$$
$$2x + y \leq 4$$

56. *Objective function:* $z = x + 2y$

Constraints:

$$x \geq 0$$
$$y \geq 0$$
$$x + 2y \leq 4$$
$$2x + y \leq 4$$

57. **Reasoning** An objective function has a minimum value at the vertex (20, 0). Can you conclude that it also has a minimum value at the point (0, 0)?

58. **HOW DO YOU SEE IT?** A company trying to determine an optimal profit finds that the objective function has a maximum value at the vertices shown in the graph.

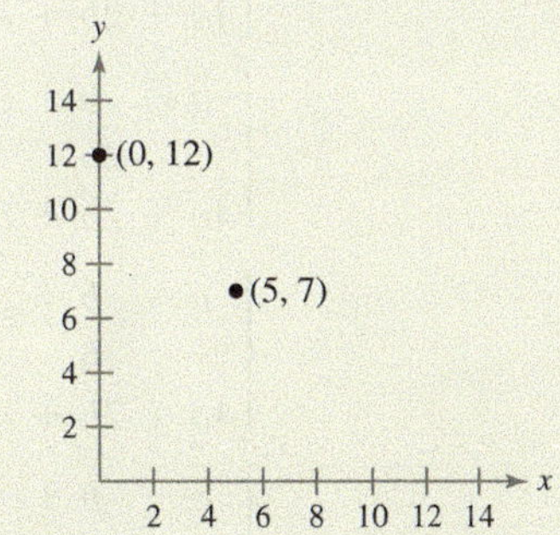

(a) Can you conclude that it also has a maximum value at the point (3, 9)? Explain.

(b) Can you conclude that it also has a maximum value at the point (6, 6)? Explain.

(c) Find two additional points that maximize the objective function.

59. **Reasoning** When solving a linear programming problem, you find that the objective function has a maximum value at more than one vertex. Can you assume that there are an infinite number of points that will produce the maximum value? Explain your reasoning.

ALGEBRA TUTOR

Simplifying and Solving Systems of Equations

On this page, you can review the algebra used in solving some of the examples in Chapter 5.

Example 1 Simplifying a System of Equations

Example 7, page 426

$$2000 = (r_1 - r_2)\left(4 + \frac{24}{60}\right)$$ Equation 1

$$2000 = (r_1 - r_2)\left(\frac{264}{60}\right)$$ Add.

$$\frac{5000}{11} = (r_1 - r_2)$$ Multiply each side by $\frac{60}{264}$.

$$5000 = 11(r_1 - r_2)$$ Multiply each side by 11.

$$5000 = 11r_1 - 11r_2$$ Distributive Property

$$2000 = (r_1 + r_2)(4)$$ Equation 2

$$500 = r_1 + r_2$$ Divide each side by 4.

Example 2 Solving a System of Equations

Example 7, page 438

$$\begin{cases} a - b + c = 3 & \text{Equation 1} \\ a + b + c = 1 & \text{Equation 2} \\ 4a + 2b + c = 6 & \text{Equation 3} \end{cases}$$

$$\begin{cases} a - b + c = 3 \\ 2b = -2 \\ 4a + 2b + c = 6 \end{cases}$$

Adding -1 times the first equation to the second equation produces a new second equation.

$$\begin{cases} a - b + c = 3 \\ b = -1 \\ 4a + 2b + c = 6 \end{cases}$$

Multiplying the second equation by $\frac{1}{2}$ produces a new second equation.

$$\begin{cases} a - b + c = 3 \\ b = -1 \\ 6b - 3c = -6 \end{cases}$$

Adding -4 times the first equation to the third equation produces a new third equation.

$$\begin{cases} a - b + c = 3 \\ b = -1 \\ -3c = 0 \end{cases}$$

Adding -6 times the second equation to the third equation produces a new third equation.

$$\begin{cases} a - b + c = 3 \\ b = -1 \\ c = 0 \end{cases}$$

Multiplying the third equation by $-\frac{1}{3}$ produces a new third equation.

To solve for a, substitute $b = -1$ and $c = 0$ into Equation 1 to obtain

$$a - (-1) + 0 = 3 \quad \Rightarrow \quad a = 2.$$

So, the solution is $a = 2$, $b = -1$, and $c = 0$.

Finding Vertices of a Region

On this page, you can review the algebra used in finding the vertices of the region bounded by the constraints in Example 2 on page 456.

Example 3 Finding Vertices of a Region

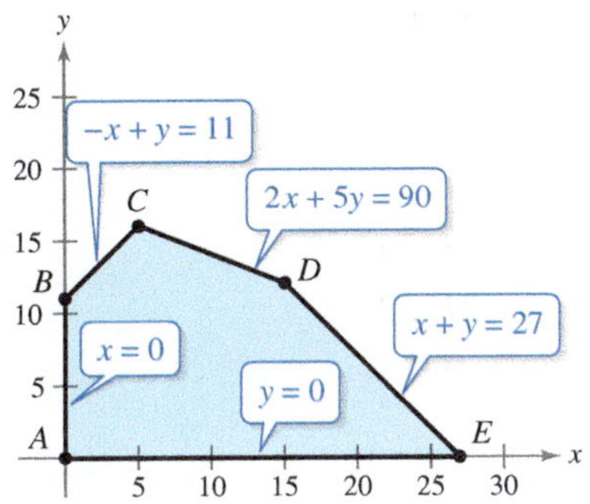

Find the vertices of the region bounded by the constraints shown in the figure.

SOLUTION

First, you know that A is at (0, 0) because $x = 0$ and $y = 0$ intersect at the origin. To find vertex B, solve the system

$$\begin{cases} -x + y = 11 & \text{Equation 1} \\ x = 0 & \text{Equation 2} \end{cases}$$

by substituting $x = 0$ into Equation 1 and solving for y.

$$0 + y = 11 \quad \Rightarrow \quad y = 11$$

So, B is at (0, 11). To find vertex C, solve the system

$$\begin{cases} -x + y = 11 & \text{Equation 1} \\ 2x + 5y = 90 & \text{Equation 2} \end{cases}$$

$$\begin{cases} -x + y = 11 \\ 7y = 112 \end{cases}$$

Adding 2 times the first equation to the second equation produces a new second equation.

So, $y = 16$. By back-substituting $y = 16$ into Equation 1, you obtain

$$-x + 16 = 11 \quad \Rightarrow \quad x = 5.$$

So, C is at (5, 16). To find vertex D, solve the system

$$\begin{cases} 2x + 5y = 90 & \text{Equation 1} \\ x + y = 27 & \text{Equation 2} \end{cases}$$

First, solve for y in Equation 2 to get $y = 27 - x$. Then substitute the expression $27 - x$ for y into Equation 1 and solve for x.

$$2x + 5(27 - x) = 90 \qquad \text{Substitute } 27 - x \text{ for } y.$$

$$2x + 135 - 5x = 90 \qquad \text{Distributive Property}$$

$$-3x = -45 \qquad \text{Combine like terms.}$$

$$x = 15 \qquad \text{Divide each side by } -3.$$

By back-substituting $x = 15$ into Equation 2, you obtain

$$15 + y = 27 \quad \Rightarrow \quad y = 12.$$

So, D is at (15, 12). To find vertex E, solve the system

$$\begin{cases} x + y = 27 & \text{Equation 1} \\ y = 0 & \text{Equation 2} \end{cases}$$

by substituting $y = 0$ into Equation 1 and solving for x. You obtain

$$x + 0 = 27 \quad \Rightarrow \quad x = 27.$$

So, E is at (27, 0).

SUMMARY AND STUDY STRATEGIES

After studying this chapter, you should have acquired the following skills.
The exercise numbers are keyed to the Review Exercises that begin on page 467.
Answers to odd-numbered Review Exercises are given in the back of the text.*

Section 5.1	Review Exercises
■ Solve a system of equations by the method of substitution.	*1–6*
■ Solve a system of equations graphically.	*7, 8, 12*
■ Construct and use a system of equations to solve an application problem.	*9–12*

Section 5.2	
■ Solve a linear system by the method of elimination.	*13–20*
■ Interpret the solution of a linear system graphically.	*21, 22*
■ Construct and use a linear system to solve an application problem.	*23–28*

Section 5.3	
■ Solve a linear system in row-echelon form using back-substitution.	*29, 30*
■ Use Gaussian elimination to solve a linear system.	*31–36*
■ Solve a nonsquare linear system.	*33, 34*
■ Find the equation of a parabola or a circle using a linear system in three or more variables.	*37–40*
■ Construct and use a linear system in three or more variables to solve an application problem.	*41–44*

Section 5.4	
■ Sketch the graph of an inequality in two variables.	*45–50*
■ Solve a system of inequalities.	*51–58*
■ Construct and use a system of inequalities to solve an application problem.	*59–64*

Section 5.5	
■ Use linear programming to minimize or maximize an objective function.	*65–72*
■ Use linear programming to optimize an application.	*73–77*

Study Strategies

- **Units of Variables in Applied Problems** When using systems of equations to solve real-life applications, be sure to keep track of the unit(s) assigned to each variable. This will allow you to write correctly each equation of the system based on the constraints given in the application.

---

* A wide range of valuable study aids are available to help you master the material in this chapter. The *Student Solutions Manual* includes step-by-step solutions to all odd-numbered exercises to help you review and prepare. The student website at *www.cengagebrain.com* offers algebra help and a *Graphing Technology Guide*, which contains step-by-step commands and instructions for a wide variety of graphing calculators.

Review Exercises

See www.CalcChat.com for worked-out solutions to odd-numbered exercises.

Solving a System by Substitution In Exercises 1–6, solve the system by the method of substitution.

1. $\begin{cases} x + 3y = 10 \\ 4x - 5y = -28 \end{cases}$

2. $\begin{cases} 3x - y - 13 = 0 \\ 4x + 3y - 26 = 0 \end{cases}$

3. $\begin{cases} \frac{1}{2}x + \frac{3}{5}y = -2 \\ 2x + y = 6 \end{cases}$

4. $\begin{cases} 1.3x + 0.9y = 7.5 \\ 0.4x - 0.5y = -0.8 \end{cases}$

5. $\begin{cases} x^2 + y^2 = 100 \\ x + 2y = 20 \end{cases}$

6. $\begin{cases} y = x^3 - 2x^2 - 2x - 3 \\ y = -x^2 + 4x - 3 \end{cases}$

Solving a System of Equations Graphically In Exercises 7 and 8, use a graphing utility to find the point(s) of intersection of the graphs.

7. $\begin{cases} y = x^2 - 3x + 11 \\ y = -x^2 + 2x + 8 \end{cases}$

8. $\begin{cases} y = \sqrt{9 - x^2} \\ y = e^x + 1 \end{cases}$

9. **Break-Even Analysis** You invest \$5000 in a greenhouse. The planter, potting soil, and seed for each plant costs \$6.43, and the selling price is \$12.68. How many of the potted plants do you need to sell to break even?

10. **Break-Even Analysis** You invest \$50,000 to open a temporary employment agency. Each hour of labor that you contract costs you \$13.72 and brings in revenue of \$22.26. How many hours of labor must you contract to break even?

11. **Choice of Newscasts** Television stations A and B are competing for the 6 P.M. newscast audience. Station A is implementing a new newscast format for the 6 P.M. audience. Models that represent the numbers of 6 P.M. viewers each month for the two stations are given by

$$\begin{cases} y = 950x + 10{,}000 & \text{Station A (new format)} \\ y = -875x + 18{,}000 & \text{Station B} \end{cases}$$

where y represents the number of viewers and x represents the month, with $x = 1$ corresponding to the first month of station A's new format. Use the models to estimate when the number of viewers for station A's 6 P.M. newscast will exceed the number of viewers for station B's 6 P.M. newscast.

12. **Comparing Populations** From 2000 through 2009, the population of Kentucky grew more slowly than that of South Carolina. Models that represent the populations of the two states are given by

$$\begin{cases} P = 30.8t + 4035 & \text{Kentucky} \\ P = 61.6t + 3985 & \text{South Carolina} \end{cases}$$

where P represents the population (in thousands) and t represents the year, with $t = 0$ corresponding to 2000. Use a graphing utility to estimate the year when the population of South Carolina first exceeded that of Kentucky. *(Source: U.S. Census Bureau)*

Solving a System by Elimination In Exercises 13–20, solve the system by the method of elimination.

13. $\begin{cases} 2x - 3y = 21 \\ 3x + y = 4 \end{cases}$

14. $\begin{cases} 3u + 5v = 9 \\ 12u + 10v = 22 \end{cases}$

15. $\begin{cases} 4x - 3y = 10 \\ 8x - 6y = 20 \end{cases}$

16. $\begin{cases} 3x + 4y = 18 \\ 6x + 8y = 18 \end{cases}$

17. $\begin{cases} 1.5x + 2.5y = 8.5 \\ 6x + 10y = 24 \end{cases}$

18. $\begin{cases} 1.25x - 2y = 3.5 \\ 5x - 8y = 14 \end{cases}$

19. $\begin{cases} \dfrac{x - 2}{3} + \dfrac{y + 3}{4} = 5 \\ 2x - y = 7 \end{cases}$

20. $\begin{cases} \dfrac{3}{5}x + \dfrac{2}{7}y = 10 \\ x + 2y = 38 \end{cases}$

Interpreting the Solution Graphically In Exercises 21 and 22, describe the graph of the solution of the linear system.

21. $\begin{cases} 2x + y = -1 \\ 3x - 2y = -5 \end{cases}$

22. $\begin{cases} x - 2y = -1 \\ -2x + 4y = 2 \end{cases}$

23. **Acid Mixture** Twelve gallons of a 25% acid solution is obtained by mixing a 10% solution with a 50% solution.
    (a) Write a system of equations that represents the problem and use a graphing utility to graph the equations in the same viewing window.
    (b) How much of each solution is required to obtain the specified concentration of the final mixture?

24. Acid Mixture Twenty gallons of a 30% acid solution is obtained by mixing a 12% solution with a 60% solution.

(a) Write a system of equations that represents the problem and use a graphing utility to graph the equations in the same viewing window.

(b) How much of each solution is required to obtain the specified concentration of the final mixture?

25. Fitting a Line to Data Find the least squares regression line $y = ax + b$ for the points shown in the graph below by solving the following system of linear equations for a and b.

$$\begin{cases} 5b + 10a = 17.8 \\ 10b + 30a = 45.7 \end{cases}$$

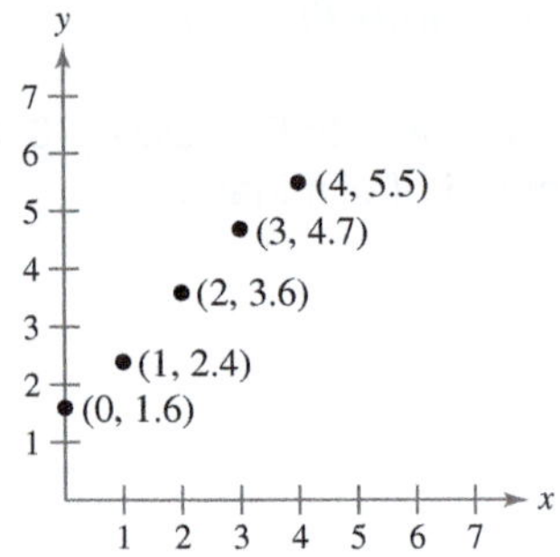

26. Ethanol Production The table shows the amount of corn y (in billions of bushels) used in the production of ethanol each year from 2005 through 2009. *(Source: U.S. Department of Agriculture)*

Year	2005	2006	2007	2008	2009
t	0	1	2	3	4
Amount, y	1.6	2.1	3.0	3.7	4.3

(a) Solve the system below for a and b to find the least squares regression line $y = at + b$ for the data. Let t represent the year, with $t = 0$ corresponding to 2005.

$$\begin{cases} 5b + 10a = 14.7 \\ 10b + 30a = 36.4 \end{cases}$$

(b) Use a graphing utility to graph the regression line and predict the amount of corn used to produce ethanol in 2010.

(c) Use the *regression* feature of the graphing utility to find a linear model for the data. Compare this model with the one you found in part (a).

Finding the Point of Equilibrium In Exercises 27 and 28, find the point of equilibrium for the demand and supply equations.

	Demand	*Supply*
27.	$p = 37 - 0.0002x$	$p = 22 + 0.00001x$
28.	$p = 120 - 0.0001x$	$p = 45 + 0.0002x$

Using Back-Substitution In Exercises 29 and 30, use back-substitution to solve the system of linear equations.

29. $\begin{cases} x - 3y + 2z = 1 \\ y - 2z = 1 \\ z = 2 \end{cases}$

30. $\begin{cases} x + y - 4z = 6 \\ y + 2z = 2 \\ z = -4 \end{cases}$

Solving a System of Linear Equations In Exercises 31–36, solve the system of linear equations.

31. $\begin{cases} 2x + y + z = 6 \\ x - 4y - z = 3 \\ x + y + z = 4 \end{cases}$

32. $\begin{cases} x + 3y - z = 13 \\ 2x - 5z = 23 \\ 4x - y - 2z = 4 \end{cases}$

33. $\begin{cases} x + y + z = 10 \\ -2x + 3y + 4z = 22 \end{cases}$

34. $\begin{cases} 5x - 12y + 7z = 16 \\ 3x - 7y + 4z = 9 \end{cases}$

35. $\begin{cases} 2x + 6y - z = 1 \\ x - 3y + z = 2 \\ \frac{3}{2}x + \frac{3}{2}y = 6 \end{cases}$

36. $\begin{cases} x + y + z + w = 8 \\ 4y + 5z - 2w = 3 \\ 2x + 3y - z = -2 \\ 3x + 2y - 4w = -20 \end{cases}$

Data Analysis: Curve Fitting In Exercises 37 and 38, find the quadratic equation $y = ax^2 + bx + c$ whose graph passes through the points.

37.

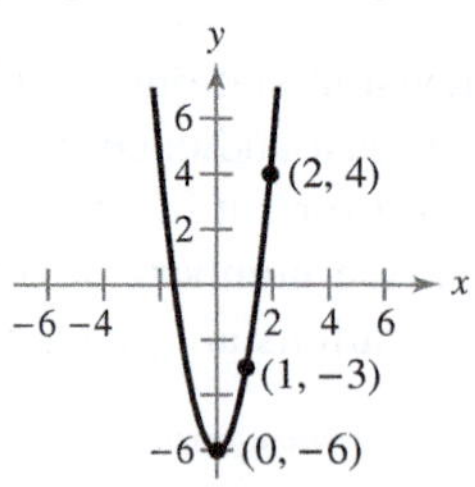

38.

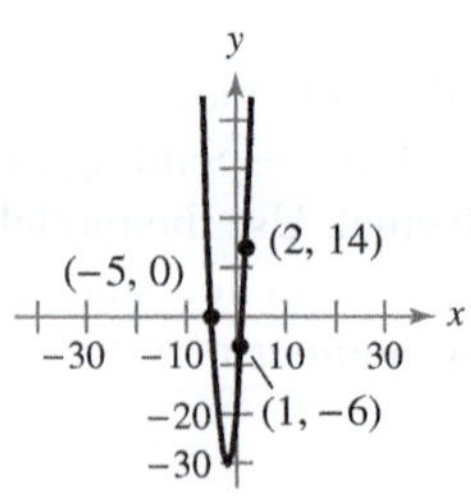

Data Analysis: Curve Fitting In Exercises 39 and 40, find the equation $x^2 + y^2 + Dx + Ey + F = 0$ of the circle that passes through the points.

39.

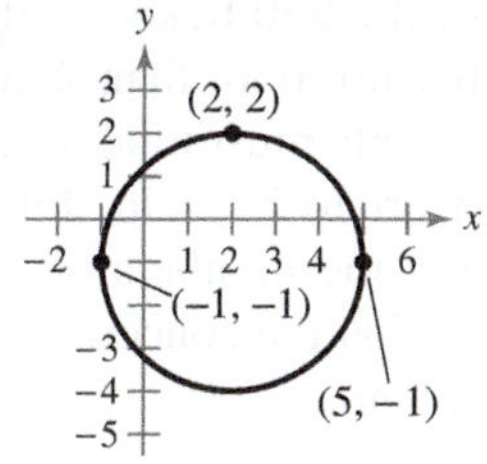

40.

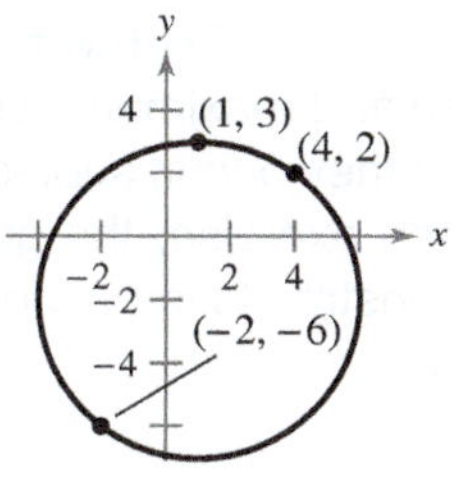

41. Investment Portfolio You are creating an investment portfolio by investing a total of \$500,000 in certificates of deposit, municipal bonds, blue-chip stocks, and growth stocks. The certificates of deposit pay 5% simple annual interest, and the municipal bonds pay 8% simple annual interest. Over a five-year period, you expect the blue-chip stocks to return 10% simple annual interest and the growth stocks to return 15% simple annual interest. You want to have two-fifths of your portfolio in stocks, and twice as much in certificates of deposit as in municipal bonds. How much should you allocate to each type of investment to have a combined return of 9.45%?

42. Investment You receive \$8580 a year in simple annual interest from three investments. The interest rates for the three investments are 6%, 8%, and 10%. The value of the 10% investment is two times that of the 6% investment, and the 8% investment is \$1000 more than the 6% investment. What is the amount of each investment?

43. Fitting a Parabola to Data Find the least squares regression parabola $y = ax^2 + bx + c$ for the points shown in the graph below by solving the following system of linear equations for a, b, and c.

$$\begin{cases} 5c + 10a = 9.1 \\ 10b = 8.0 \\ 10c + 34a = 19.8 \end{cases}$$

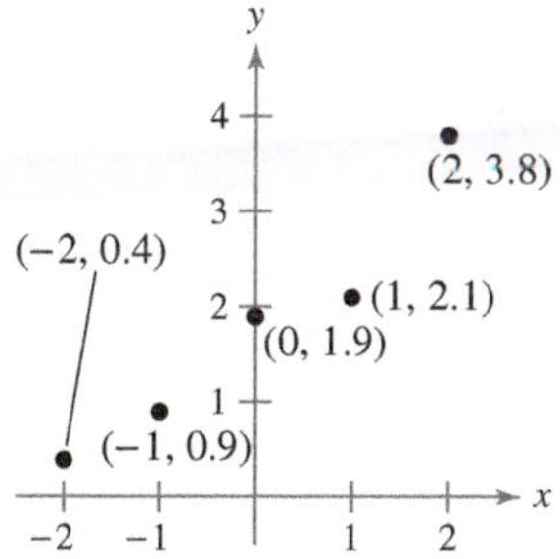

44. Sailboat Sales The table shows the number y (in thousands) of new sailboats sold in the United States each year from 2005 through 2009. In the table, t represents the year, with $t = -1$ corresponding to 2005. *(Source: The Sailing Company's Annual Sailing Business Review)*

Year, t	Number, y
−1	14.4
0	12.9
1	11.8
2	9.3
3	5.4

(a) Use a graphing utility to create a scatter plot of the data.

(b) Solve the following system for a and b to find the least squares regression line $y = at + b$ for the data.

$$\begin{cases} 5b + 5a = 53.8 \\ 5b + 15a = 32.2 \end{cases}$$

(c) Solve the following system for a, b, and c to find the least squares regression parabola $y = at^2 + bt + c$ for the data.

$$\begin{cases} 5c + 5b + 15a = 53.8 \\ 5c + 15b + 35a = 32.2 \\ 15c + 35b + 99a = 112 \end{cases}$$

(d) Use the *regression* feature of the graphing utility to find linear and quadratic models for the data. Compare the models with the least squares regression models found in parts (b) and (c).

(e) Use the graphing utility to graph the linear and quadratic models with the scatter plot of the data. Use the models to predict the numbers of sailboats sold in 2010 and 2011. Discuss your results in the context of the situation.

Sketching the Graph of an Inequality In Exercises 45–50, sketch the graph of the inequality.

45. $x \geq -5$

46. $y < -2$

47. $y \leq 5 + \frac{1}{2}x$

48. $3y - x \geq 7$

49. $y - 4x^2 > -1$

50. $y \leq \dfrac{3}{x^2 + 2}$

Solving a System of Inequalities In Exercises 51–58, graph the solution set of the system of inequalities.

51. $\begin{cases} x + 4y < 8 \\ x > 0 \\ y > 0 \end{cases}$

52. $\begin{cases} 2x - y > 6 \\ x < 6 \\ y \geq -2 \end{cases}$

53. $\begin{cases} 3x - y > -4 \\ 2x + y > -1 \\ 7x + y < 4 \end{cases}$

54. $\begin{cases} x + y > 4 \\ 3x + y > 10 \\ x - y \geq 0 \end{cases}$

55. $\begin{cases} x^2 + y^2 \leq 9 \\ x^2 - x - 2 \leq y \end{cases}$

56. $\begin{cases} x^2 + y^2 \leq 4 \\ -2x^2 + 2 < y \end{cases}$

57. $\begin{cases} \ln x < y \\ y > -1 \\ x < 4 \end{cases}$

58. $\begin{cases} \ln x \geq y \\ -x + y < -2 \\ x > 2 \end{cases}$

Geometry In Exercises 59 and 60, write a system of inequalities that corresponds to the solution set shown in the graph.

59. Parallelogram

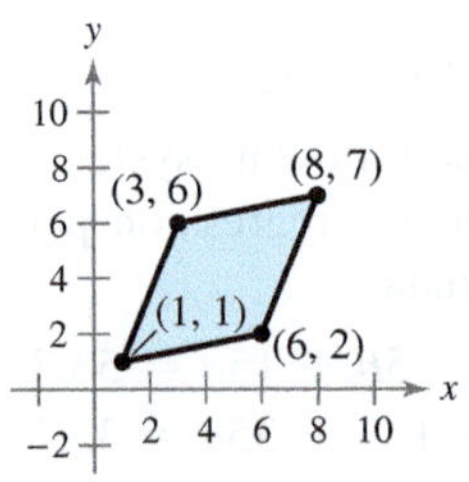

60. Triangle

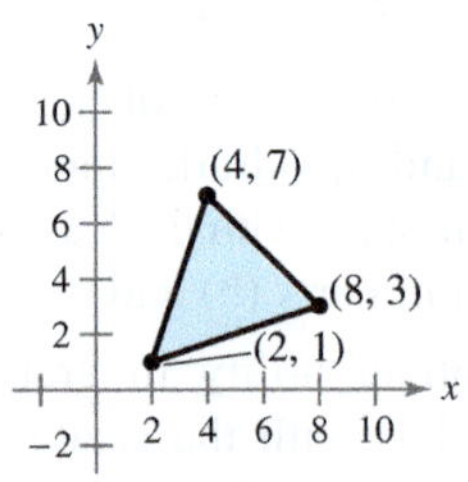

Consumer and Producer Surpluses In Exercises 61 and 62, find the consumer surplus and producer surplus for the demand and supply equations.

	Demand	*Supply*
61.	$p = 160 - 0.0001x$	$p = 70 + 0.0002x$
62.	$p = 130 - 0.0002x$	$p = 30 + 0.0003x$

63. Movie Player Inventory A store sells two models of Blu-ray Disc™ players (BDPs). Because of the demand, it is necessary to stock at least twice as many units of model Y as units of model Z. The costs to the store for the two models are \$100 and \$150, respectively. The management does not want more than \$4000 in BDP inventory at any one time, and it wants at least four model Y BDPs and two model Z BDPs in inventory at all times. Find a system of inequalities that describes all possible inventory levels. Sketch the graph of the system.

64. Concert Ticket Sales Two types of tickets are to be sold for a concert. One type costs \$30 per ticket and the other type costs \$50 per ticket. The promoter of the concert can sell up to 10,000 of the \$50 tickets and up to 18,000 of the \$30 tickets, but no more than 24,000 tickets in all. Also, the gross receipts must total at least \$550,000 in order for the concert to be held. Find a system of inequalities describing the numbers of each type of ticket that can be sold so that the concert can be held. Sketch the graph of the system.

Solving a Linear Programming Problem In Exercises 65–68, find the minimum and maximum values of the objective function and where they occur, subject to the indicated constraints. (For each exercise, the graph of the region determined by the constraints is provided.)

65. *Objective function:* $z = 5x + 6y$

Constraints: $x \geq 0$
$y \geq 0$
$x + y \leq 8$

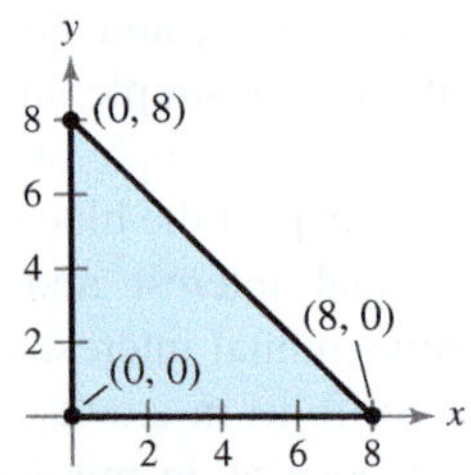

66. *Objective function:* $z = 15x + 12y$

Constraints: $x \geq 0$
$y \geq 0$
$x + 3y \leq 12$
$3x + 2y \leq 15$

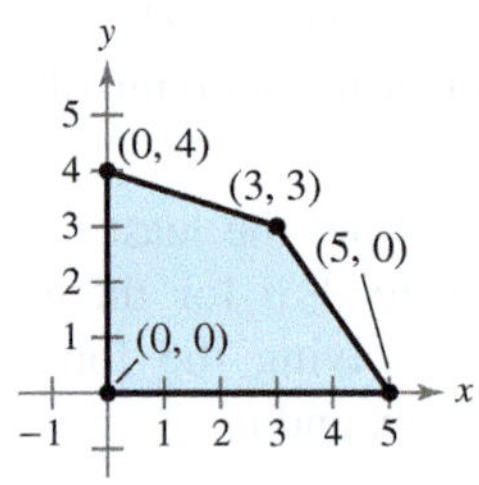

67. *Objective function:* $z = 8x + 10y$

Constraints: $0 \leq x \leq 50$
$0 \leq y \leq 35$
$4x + 5y \leq 275$

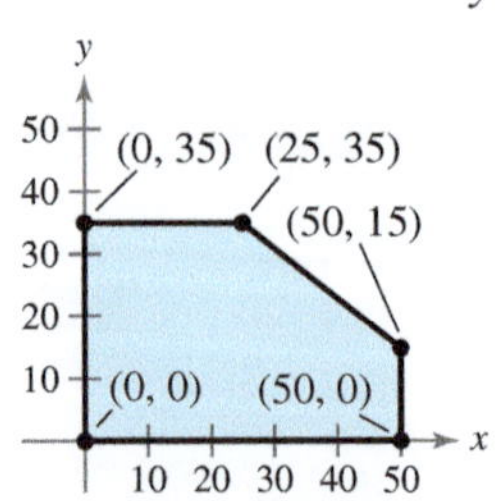

68. *Objective function:* $z = 50x + 60y$

Constraints:
$$x \geq 0$$
$$y \geq 0$$
$$3x + 4y \geq 1200$$
$$5x + 6y \leq 3000$$

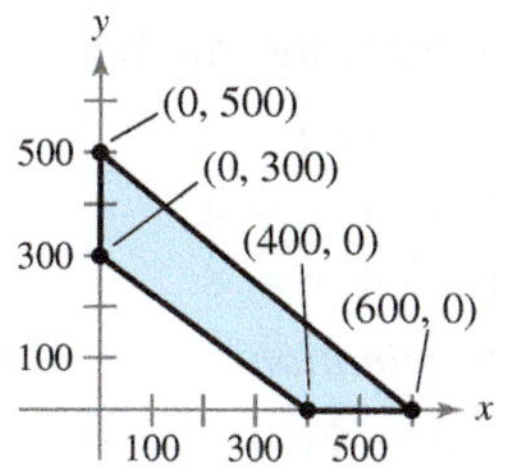

Solving a Linear Programming Problem In Exercises 69–72, sketch the region determined by the indicated constraints. Then find the minimum and maximum values of the objective function and where they occur, subject to the constraints.

69. *Objective function:* $z = 6x + 8y$

Constraints:
$$x \geq 0$$
$$y \geq 0$$
$$x + 4y \leq 16$$
$$3x + 2y \leq 18$$

70. *Objective function:* $z = 5x + 8y$

Constraints:
$$0 \leq x \leq 5$$
$$y \geq 0$$
$$x + 2y \leq 12$$
$$2x + 3y \leq 19$$

71. *Objective function:* $z = 8x + 3y$

Constraints:
$$0 \leq x \leq 5$$
$$0 \leq y \leq 7$$
$$x + y \leq 9$$
$$3x + y \leq 17$$

72. *Objective function:* $z = 10x + 11y$

Constraints:
$$x \geq 0$$
$$y \geq 0$$
$$2x + 5y \leq 30$$
$$x + y \geq 3$$
$$2x + y \leq 14$$

73. Optimal Profit A company makes two models of desks. The times (in hours) required for assembling, finishing, and packaging each model are shown in the table.

Process	Model A	Model B
Assembling	3.5	8
Finishing	2.5	2
Packaging	1.3	0.7

The total times available each month for assembling, finishing, and packaging are 5600 hours, 2000 hours, and 910 hours, respectively. The profits per unit are \$100 for model A and \$150 for model B. What is the optimal monthly production level for each model? What is the optimal monthly profit?

74. Optimal Profit A factory manufactures two television set models: an LCD model that yields \$100 profit and a plasma model that yields a profit of \$180. The times (in hours) required for assembling, testing, and packaging each model are shown in the table.

Process	LCD model	Plasma model
Assembling	2	5
Testing	1	2
Packaging	1	1

The total times available each month for assembling, testing, and packaging are 3000 hours, 1300 hours, and 1000 hours, respectively. What is the optimal monthly production level for each model? What is the optimal monthly profit?

75. Optimal Profit The costs to a merchant for two models of digital camcorders are \$525 and \$675. The \$525 model yields a profit of \$75 and the \$675 model yields a profit of \$125. The merchant estimates that the total monthly demand will not exceed 350 units. There should be no more than \$206,250 in digital camcorder inventory. Find the number of units of each model that should be stocked in order to optimize profit. What is the optimal profit?

76. Optimal Profit The costs to a merchant for two models of home theater systems are \$270 and \$455. The \$270 model yields a profit of \$30 and the \$455 model yields a profit of \$45. The merchant estimates that the total monthly demand will not exceed 100 units. There should be no more than \$36,250 in home theater system inventory. Find the number of units of each model that should be stocked in order to optimize profit. What is the optimal profit?

77. Optimal Revenue An accounting firm has 800 hours of staff time and 90 hours of review time available each week. The firm charges \$2500 for an audit and \$200 for a tax return. Each audit requires 100 hours of staff time and 10 hours of review time. Each tax return requires 10 hours of staff time and 2 hours of review time. What numbers of audits and tax returns will bring in an optimal revenue? What is the optimal revenue?

TEST YOURSELF

See www.CalcChat.com for worked-out solutions to odd-numbered exercises.

Take this test as you would take a test in class. When you are done, check your work against the answers given in the back of the book.

In Exercises 1–6, solve the system of equations using the indicated method.

1. *Substitution*

$$\begin{cases} 5x - 7y = -18 \\ 4x + 3y = 20 \end{cases}$$

2. *Substitution*

$$\begin{cases} x + y = 3 \\ x^2 + y = 9 \end{cases}$$

3. *Graphing*

$$\begin{cases} 2x + y = 4 \\ 2x^2 + y = 8 \end{cases}$$

4. *Graphing*

$$\begin{cases} 1.5x - 2.25y = 8 \\ 2.5x + 2y = 5.75 \end{cases}$$

5. *Elimination*

$$\begin{cases} 2x - 4y + z = 11 \\ x + 2y + 3z = 9 \\ 3y + 5z = 12 \end{cases}$$

6. *Elimination*

$$\begin{cases} 3x - 2y + z = 16 \\ 5x - z = 6 \\ 2x - y - z = 3 \end{cases}$$

7. A total of \$80,000 is invested in two funds paying 2.3% and 3.1% simple interest. The total annual interest is \$2000. How much is invested in each fund?

8. Find the point of equilibrium for a system that has a demand equation of $p = 49 - 0.0003x$ and a supply equation of $p = 33 + 0.00002x$.

9. The numbers y of adults (in millions) who participated in baking as a leisure activity in the years 2005 through 2009 are shown in the table at the left. In the table, x represents the year, with $x = 0$ corresponding to 2007. *(Source: Mediamark Research & Intelligence)*

Year, x	Number, y
−2	39.6
−1	41.3
0	47.6
1	52.2
2	58.9

Table for 9

(a) Find the least squares regression parabola $y = ax^2 + bx + c$ for the data by solving the system below.

$$\begin{cases} 5c + 10a = 239.6 \\ 10b = 49.5 \\ 10c + 34a = 487.5 \end{cases}$$

(b) Use the model found in part (a) to predict the number of adults who participated in baking as a leisure activity in 2010.

In Exercises 10–13, sketch the graph of the inequality.

10. $x \geq 0$

11. $y \geq 0$

12. $x + 3y \leq 12$

13. $3x + 2y \leq 15$

14. Sketch the solution set of the system of inequalities composed of the inequalities in Exercises 10–13.

15. Find the minimum and maximum values of the objective function $z = 6x + 7y$, subject to the constraints given in Exercises 10–13.

16. A manufacturer produces two models of exercise equipment. The times required for assembling, painting, and packaging each model are as follows.

- Assembling: 3.5 hours for model A; 8 hours for model B
- Painting: 2.5 hours for model A; 2 hours for model B
- Packaging: 1.3 hours for model A; 0.9 hour for model B

The total times available for assembling, painting, and packaging are 5600 hours, 2000 hours, and 900 hours, respectively. The profits per unit are \$200 for model A and \$275 for model B. What is the optimal production level for each model? What is the optimal profit? Explain your reasoning.

Answers to Selected Exercises

Chapter 1

Section 1.1 (page 9)

Skills Warm Up (page 9)

1. $-3x - 10$ **2.** $5x - 12$ **3.** x **4.** $x + 26$
5. $\frac{9x}{14}$ **6.** $\frac{3x}{4}$ **7.** $-\frac{1}{x(x+1)}$ **8.** $\frac{8}{x}$
9. $\frac{7x-8}{x(x-2)}$ **10.** $-\frac{2}{x^2-1}$

1. Identity **3.** Conditional equation
5. Conditional equation
7. (a) No (b) No (c) Yes (d) No
9. (a) Yes (b) Yes (c) No (d) No
11. (a) Yes (b) No (c) No (d) No
13. (a) No (b) No (c) No (d) Yes
15. (a) Yes (b) No (c) No (d) No
17. 5 **19.** -4 **21.** 3 **23.** No solution
25. Infinitely many solutions **27.** 9 **29.** -26
31. No solution **33.** -4 **35.** $z = -\frac{6}{5}$ **37.** 10
39. 4 **41.** 3 **43.** 5 **45.** No solution **47.** $x = 1$
49. 9 **51.** 0 **53.** All real numbers **55.** No solution
57. Because substituting 2 for x in the equation produces division by zero, $x = 2$ cannot be a solution to the equation.
59. Extraneous solutions may arise when a fractional expression is multiplied by factors involving the variable.
61. Equivalent equations have the same solutions.
Example: $2x - 6 = 0$ and $x - 3 = 0$ both have the solution $x = 3$.
63. $x \approx 138.889$ **65.** $x \approx 62.372$ **67.** $x \approx 19.993$
69. Use the *table* feature in ASK mode, or use the scientific calculator part of the graphing utility.
71. (a) 6.46 (b) 6.41
(c) There is a roundoff error because decimals in the original problem have 5 decimal places.
73. (a) 56.09 (b) 56.13
(c) There is a roundoff error because the fractions in the numerator and denominator have infinitely many decimal places.
75. 2009 **77.** 65.8 in. **79.** (a) 2007 (b) 2010

Section 1.2 (page 20)

Skills Warm Up (page 20)

1. 14 **2.** 4 **3.** -3 **4.** 4 **5.** -2
6. 1 **7.** $\frac{2}{5}$ **8.** $\frac{10}{3}$ **9.** 6 **10.** $-\frac{11}{5}$

1. (a) One natural number + Next consecutive natural number
(b) One natural number $= x$
Next consecutive natural number $= x + 1$
(c) $x + (x + 1) = 2x + 1$
3. (a) Distance $= t$ hours $\times$ Car speed
(b) Hours $= t$
Car speed = 50 mi/h
(c) $50t$
5. (a) Perimeter = 2(Width) + 2(Length)
(b) Width $= x$
Length $= 2x$
(c) $2x + 2(2x) = 6x$
7. $5 + x = 8$ **9.** $\frac{r}{2} = 9$ **11.** $60{,}000 + 0.08(60{,}000)$
13. (a) Sum = One number + Next consecutive number
(b) One number $= x$
Next consecutive number $= x + 1$
Sum = 525
(c) $525 = x + (x + 1)$ (d) 262, 263
15. (a) 148 = One positive number − Another positive number
(b) One positive number $= x$
Another positive number $= 5x$
(c) $148 = 5x - x$ (d) 37, 185
17. (a) Raise = Percent × Old wage
(b) Old wage = \$440
Raise = \$66
Percent $= r$
(c) $66 = 440r$ (d) 15%
19. (a) Change in price = Percent × Original price
(b) Change in price = \$30
Percent $= r$
Original price = \$149
(c) $30 = 149r$ (d) ≈ 20.13%
21. (a) Total paycheck = Your weekly paycheck + Your coworker's paycheck
(b) Total paycheck = \$848
Your weekly paycheck $= 1.12x$
Your coworker's paycheck $= x$
(c) $848 = 2.12x$
(d) \$400, \$448
23. ≈ 17.5% decrease **25.** ≈ 16.2% increase
27. ≈ 3.4% increase **29.** ≈ 128.57%
31. ≈ 54.29% **33.** ≈ 56.41% decrease
35. (a) \$37,800 (b) \$40,748.40 (c) \$44,578.75
37. (a) 46.6%
(b) Cars: 5,459,515
Light trucks: 4,940,066
Other trucks: 201,419
(c) ≈ 96.1%
39. ≈ 0.57 ft × 0.93 ft **41.** ≈ 8.9 years
43. 97 or greater **45.** \$1411.76 **47.** \$18
49. (a) \$361.25
(b) Percent increase needed: ≈ 17.65%
A higher percent increase is needed because you are taking a percentage of a smaller number.

51. $2\frac{1}{3}$ hours **53.** 62.5 feet **55.** 1.28 seconds
57. \$10,500 at 6.5% and \$4500 at 7.5%
59. Stock A: \$2200, Stock B: \$2800
61. 8571 units per month **63.** $h = \frac{2A}{b}$ **65.** $l = \frac{P - 2w}{2}$
67. $L = \frac{S}{1 - R}$ **69.** $P = \frac{A}{\left(1 + \frac{r}{n}\right)^{nt}}$
71. $\approx$ 48 ft **73.** $\approx$ 32.1 gallons
75. Williams: $\approx$ \$24,933
Gonzalez: $\approx$ \$27,833
Walters: $\approx$ \$25,633
Gilbert: $\approx$ \$26,100
Hart: $\approx$ \$24,433
Team average: April: \$25,600, May: \$25,920, June: \$25,840
77. "takes 30 minutes"; "from a depth of 150 feet"

Section 1.3 *(page 32)*

Skills Warm Up *(page 32)*

1. $\frac{\sqrt{14}}{10}$ **2.** $4\sqrt{2}$ **3.** 14 **4.** $\frac{\sqrt{10}}{4}$
5. $x(3x + 5)$ **6.** $(2x - 5)(2x + 5)$
7. $-(x - 7)(x - 15)$ **8.** $(x - 2)(x + 9)$
9. $(5x - 1)(2x + 3)$ **10.** $(6x - 1)(x - 12)$

1. $2x^2 + 5x - 3 = 0$ **3.** $x^2 - 19x = 0$
5. $x^2 - 6x + 4 = 0$ **7.** $2x^2 - 2x + 1 = 0$
9. $3x^2 - 60x - 10 = 0$ **11.** $4, -2$ **13.** $0, -\frac{1}{2}$
15. -5 **17.** $3, -\frac{1}{2}$ **19.** $2, -6$ **21.** $-2, -5$
23. ± 4 **25.** $\pm\sqrt{7} \approx \pm 2.65$ **27.** $\pm 2\sqrt{3} \approx \pm 3.46$
29. 0, 24
31. $-2 + 2\sqrt{3} \approx 1.46$
$-2 - 2\sqrt{3} \approx -5.46$
33. ± 5 **35.** $\pm\sqrt{38} \approx \pm 6.16$ **37.** $\pm\frac{\sqrt{115}}{5} \approx \pm 2.14$
39. $\pm\frac{\sqrt{78}}{3} \approx \pm 2.94$ **41.** ± 8 **43.** 1 **45.** $\pm\frac{3}{4}$
47. $\frac{3}{2}$ **49.** $3, -11$ **51.** $\frac{3}{2}, -\frac{1}{2}$ **53.** $5, -\frac{10}{3}$
55. 9, 3 **57.** $\frac{1}{5}, 1$ **59.** $-1, -5$ **61.** $-\frac{1}{2}$
63. Algebra argument:

$(x + 2)^2 = (x + 2)(x + 2)$	Definition of exponent
$= x^2 + 2x + 2x + 4$	FOIL
$= x^2 + 4x + 4$	Combine like terms.

So, $(x + 2)^2 \neq x^2 + 4$.
Graphing utility argument:
(1) Let $y_1 = (x + 2)^2$ and $y_2 = x^2 + 4$. Use the table feature with an arbitrary value of x (but not $x = 0$). The table will show that the values of y_1 are not the same as the values of y_2.
(2) Use the scientific calculator portion of the graphing utility to show that if $x = 5$, $(5 + 2)^2 = 49$ and $5^2 + 4 = 29$. So, $(x + 2)^2$ is not equal to $x^2 + 4$.

65. 34 feet × 48 feet
67. Base: $2\sqrt{2}$ feet
Height: $2\sqrt{2}$ feet
69. 5 feet **71.** $\approx$ 3.54 seconds **73.** $\approx$ 1.43 seconds
75. 42 seconds faster **77.** $\approx$ 4.24 centimeters
79. 976 miles **81.** $\approx$ 494.97 meters **83.** 60,000 units
85. 2019 ($t \approx 8.85$)
87. (a) 1987 ($t \approx 18.74$)
(b) Yes; the model is a good representation through 1890.
(c) 2010: 312,493.19 thousand
They are close.
(d) Yes
89. 2019 ($t \approx 9.13$)
91. 1 P.M. ($t \approx 12.96$)
No; For 7 P.M. ($t = 19$), the model yields a value of $T = 144.81$°F, which is extremely high and unreasonable.

Section 1.4 *(page 42)*

Skills Warm Up *(page 42)*

1. 13 **2.** 5 **3.** $4\sqrt{6}$ **4.** $3\sqrt{73}$
5. $2, -1$ **6.** $\frac{3}{2}, -3$ **7.** $5, -1$ **8.** $\frac{1}{2}, -7$
9. 5, 2 **10.** $4, -1$

1. One real solution **3.** Two real solutions
5. No real solutions **7.** Two real solutions
9. $\frac{1}{2}, -1$ **11.** $\frac{1}{4}, -\frac{3}{4}$ **13.** $-3 \pm \sqrt{13}$ **15.** $5 \pm \sqrt{3}$
17. $-4 \pm 2\sqrt{5}$ **19.** $\frac{2}{3} \pm \frac{\sqrt{7}}{3}$ **21.** $-\frac{1}{3} \pm \frac{\sqrt{11}}{6}$
23. $-\frac{1}{2} \pm \sqrt{2}$ **25.** $\frac{2}{7}$ **27.** $2 \pm \frac{\sqrt{6}}{2}$ **29.** $6 \pm \sqrt{11}$
31. $x \approx 0.976, -0.643$ **33.** $-0.290, -2.200$
35. No real solution **37.** -11 **39.** $\pm\sqrt{10}$
41. $1 \pm \sqrt{2}$ **43.** $-2, 4$
45. $\frac{-1 \pm \sqrt{37}}{6}$ **47.** 50, 50; Real-life problems will vary.
49. 7, 8, or $-8, -7$; Real-life problems will vary.
51. 200 units **53.** 653 units
55. 6 rows, 18 chairs per row
57. 14 in. × 14 in. **59.** $\approx$ 7.2 seconds
61. Moon: $\approx$ 14.9 seconds **63.** Moon
Earth: $\approx$ 2.6 seconds
65. Oklahoma City to Austin: 354.45 mi
Austin to New Orleans: 433.55 mi
67. 4:00 P.M. ($t \approx 3.9$)
69. Southbound: $\approx$ 550 miles per hour
Eastbound: $\approx$ 600 miles per hour
71. (a) 2011 ($t \approx 11$) (b) 2008 ($t \approx 8$)
(c) 2016 ($t \approx 16.6$); yes
73. (a) $\approx$ 16.8°C (b) $\approx$ 2.5
75. 5279 units or 94,721 units
77. No; the equation that has a discriminant of 5 has two distinct real solutions, while the equation that has a discriminant of -10 has no real solutions.
79. Answers will vary.

Quiz Yourself *(page 46)*

1. $x = -6$ **2.** $x = 6$ **3.** $x = -2$ **4.** No solution

5. Use the *table* feature in ASK mode or the scientific calculator portion of the graphing utility.

6. 328.954 **7.** 431.398

8. $8.50x + 30{,}000 = 200{,}000$; 20,000 units

9. $300{,}000 = x(75 - 0.0002x)$; 4044 units or 370,956 units

10. $x = \frac{2}{3}, -5$ **11.** $x = \pm\sqrt{5}; x \approx \pm 2.24$

12. $x = -3 \pm \sqrt{17}; x \approx -7.12, 1.12$

13. $x = -1 \pm \sqrt{7}; x \approx -3.65, 1.65$

14. $x = \dfrac{-7 \pm \sqrt{73}}{6}; x \approx -2.59, 0.26$

15. $x \approx 1.568, -0.068$ **16.** No real solutions

17. One real solution

18. Answers will vary. Sample answer: Use the FOIL method $[(x + 3)^2 = (x + 3)(x + 3) = x^2 + 6x + 9]$, use the *table* feature of your graphing utility, or use the scientific calculator portion of your graphing utility to evaluate the solution.

19. 4.5 seconds **20.** 9 in. × 9 in. × 6 in.

Section 1.6 *(page 56)*

Skills Warm Up *(page 56)*

1. $-\frac{1}{2}$ **2.** $-\frac{1}{6}$ **3.** -3 **4.** -6 **5.** $x \geq 0$

6. $-3 < z < 10$ **7.** $P \leq 2$ **8.** $W \geq 200$

9. 2, 7 **10.** 0, 1

1. $-1 \leq x \leq 5$; Bounded **3.** $x > 11$; Unbounded

5. $x < -2$; Unbounded

7. c **8.** h **9.** f **10.** e

11. g **12.** a **13.** b **14.** d

15. (a) Yes (b) No (c) Yes (d) No

17. (a) Yes (b) No (c) No (d) Yes

19. (a) No (b) Yes (c) Yes (d) Yes

21. (a) No (b) Yes (c) No (d) Yes

23. If $2x > 6$, then $x > 3$. **25.** If $2x \leq -8$, then $x \leq -4$.

27. If $2 - 4x > -10$, then $x < 3$.

29. If $-\frac{2}{3}x \geq -6$, then $x \leq 9$.

31. $x \geq 6$

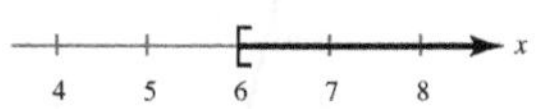

33. $x > -4$

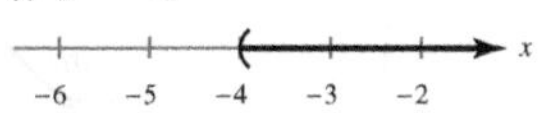

35. $x < 25$

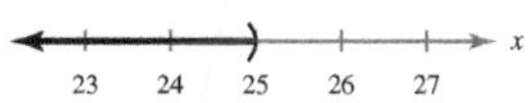

37. $x > 2$

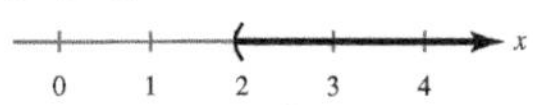

39. $x \leq -\frac{2}{3}$

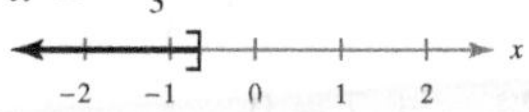

41. $x < -18$

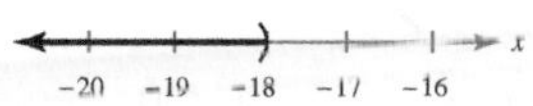

43. $x > \frac{2}{5}$

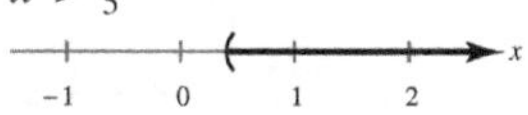

45. $2 \leq x < 4$

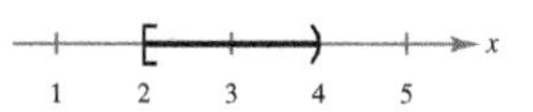

47. $-4 \leq x < 2$

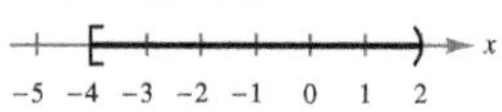

49. $-9 < x < 15$

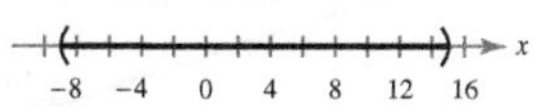

51. $-\frac{3}{4} < x < -\frac{1}{4}$

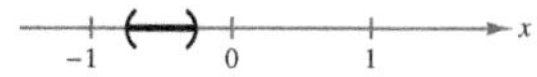

53. $-6 < x < 6$

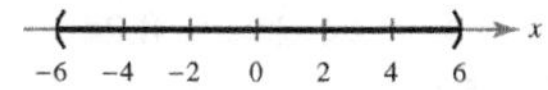

55. $x < -6$ or $x > 6$

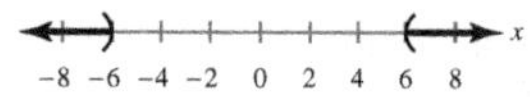

57. $-8 < x < 2$

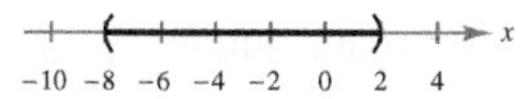

59. No solution

61. $x < -\frac{1}{2}$ or $x > \frac{11}{2}$

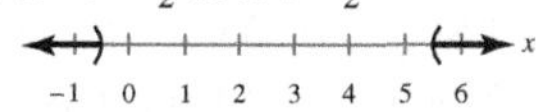

63. $x \leq -\frac{29}{2}$ or $x \geq -\frac{11}{2}$

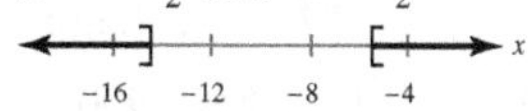

65. $x \leq -7$ or $x \geq 13$

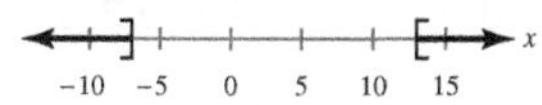

67. $8 < x < 10$

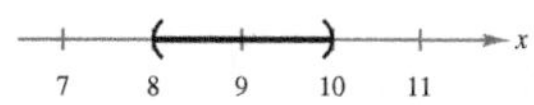

69. $x < 0$ or $x > 5$

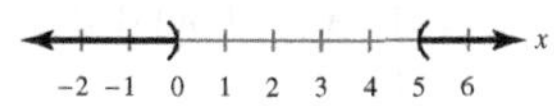

71. $|x| \leq 2$ **73.** $|x - 9| \geq 3$ **75.** $|x - 12| \leq 10$

77. $|x + 3| > 5$ **79.** Less than 18,269.23 miles

81. 2008 ($t > 7.9$) **83.** More than 250 miles

85. (a) 1998 to 2005 ($t < 15.23$); yes; According to the bar graph, the average salary was less than \$1.5 million from 1998 to 2005.

(b) 2011; Yes; The average salary has been increasing.

87. $33\frac{1}{3}$ weeks **89.** ≤ 15 inches

91. $\geq 10{,}417$ units **93.** Greater than 4%

95. $[\approx 106.864, \approx 109.464]$

97. [65.8, 71.2]

99. Minimum = 20%; Maximum = 80%

101. ≈ \$0.28

Review Exercises *(page 84)*

1. Conditional equation **3.** Identity

5. (a) No (b) No (c) Yes (d) No

7. (a) No (b) No (c) No (d) No

9. 13 **11.** $-\frac{1}{2}$ **13.** No solution

15. All real numbers **17.** $-\frac{5}{3}$ **19.** $-\frac{2}{3}$

21. 377.778 **23.** 0.033 **25.** 2009

27. (a) Sum = First even integer + Second even integer + Third even integer

(b) Sum = 42

First even integer = x

Second even integer = $x + 2$

Third even integer = $x + 4$

(c) $x + (x + 2) + (x + 4) = 42$

(d) 12

29. (a) New salary = 7%(Annual salary) + Annual salary

(b) New salary = S (dollars)

Annual salary = 28,900 (dollars)

Percent = 0.07 (percent in decimal form)

(c) $S = 0.07(28{,}900) + 28{,}900$

(d) \$30,923

31. \$163.53 **33.** 2 hours **35.** ≈ 26.58%

37. $2\frac{2}{9}$ quarts **39.** 29.5 feet × 59 feet **41.** \$20

43. $-\frac{1}{2}, \frac{4}{3}$ **45.** 3, 8 **47.** $\pm\sqrt{11}, \approx \pm 3.32$

49. $-4 + 3\sqrt{2} \approx 0.24$
$-4 - 3\sqrt{2} \approx -8.24$
51. 15 feet × 27 feet
53. 200,000 units or 400,000 units
55. Two real solutions **57.** No real solutions
59. $6 \pm \sqrt{6}$ **61.** $\dfrac{-19 \pm \sqrt{165}}{2}$ **63.** $-3 \pm 2\sqrt{3}$
65. $1.866, -0.283$ **67.** 8.544, 0.162
69. Moon: ≈ 8.61 seconds
Earth: ≈ 3.54 seconds
71. $0, -1, 4$ **73.** $\pm 2, \pm 1$ **75.** $\frac{25}{4}$ **77.** No solution
79. $\pm 4\sqrt{2}$ **81.** $-3, \frac{7}{5}$ **83.** $2 \pm \sqrt{19}$ **85.** \$900
87. ≈ 313,251 units **89.** $-4 \le x \le 7$; bounded
91. $x \le 1000$; unbounded
93. $x < 11$

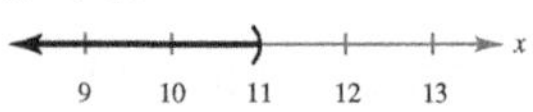

95. $-\frac{13}{2} < x < \frac{11}{2}$

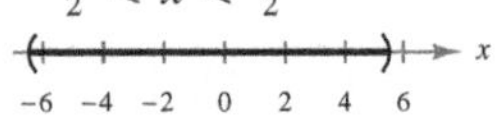

97. $-12 < x < -8$

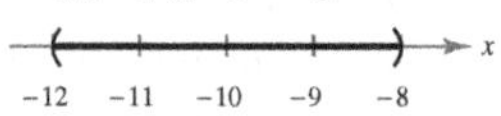

99. $x > 45$ units

101. $(-\infty, -1), (-1, 1), (1, \infty)$
103. $(-\infty, 1), (1, 5), (5, \infty)$
105. $(-1, 3)$

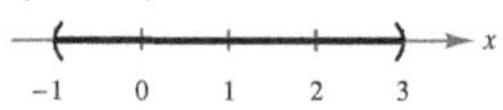

107. $(-\infty, -3) \cup (0, 3)$

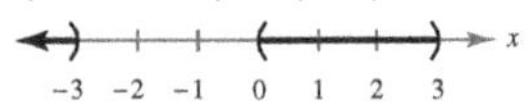

109. $\left(-\infty, \frac{6}{5}\right) \cup (4, \infty)$

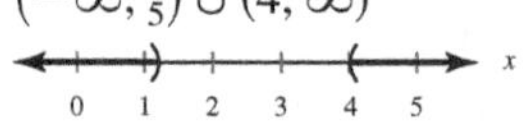

111. $(-1.69, 1.69)$ **113.** $(1.65, 1.74)$ **115.** $[10, \infty)$
117. $(-\infty, 6] \cup [9, \infty)$
119. Between 2.5 and 3.75 seconds
121. Between about 15.5 meters and about 24.5 meters
123. Greater than 4.25%
125. Between about \$41.34 and about \$58.66

Test Yourself *(page 88)*

1. $\frac{17}{23}$ **2.** (a) All real numbers (b) $-3 \le x \le 3$
3. April: \$325,786.00 **4.** $-\frac{5}{3}, \frac{1}{2}$ **5.** $4, -\frac{3}{2}$
May: \$299,723.12
6. $\pm\sqrt{15}$ **7.** $\dfrac{-13 \pm \sqrt{69}}{2}$ **8.** $\dfrac{11 \pm \sqrt{145}}{6}$
9. $1.038, -0.446$ **10.** $-\frac{7}{2}, \frac{13}{2}$
11. 4 (7 is extraneous.)
12. $-1, 1, -3, 3$ **13.** $-6, 6$
14. Selling either 341,421 units or 58,579 units will produce a revenue of \$2,000,000.
15. $x < 3$

16. $x \le -4$ or $x \ge \frac{28}{5}$

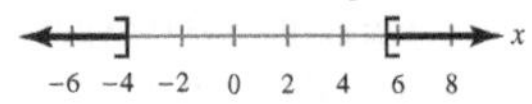

17. $(-11, -7)$

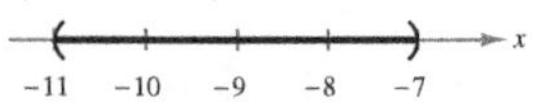

18. $(-\infty, -2] \cup [0, 2]$

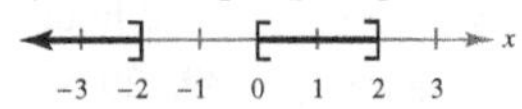

19. Between 19,189 units and 143,311 units **20.** 2012

Chapter 2

Section 2.2 *(page 74)*

Skills Warm Up *(page 74)*

1. -4 **2.** $-\frac{13}{3}$ **3.** $-\frac{5}{4}$ **4.** $\frac{1}{3}$ **5.** $y = \frac{2}{3}x - 2$
6. $y = -2x$ **7.** $y = 3x - 1$ **8.** $y = \frac{2}{3}x + 5$
9. $y = -2x + 7$ **10.** $y = x + 3$

1. $\frac{1}{2}$ **3.** -2 **5.** (a) L_2 (b) L_3 (c) L_1
7.

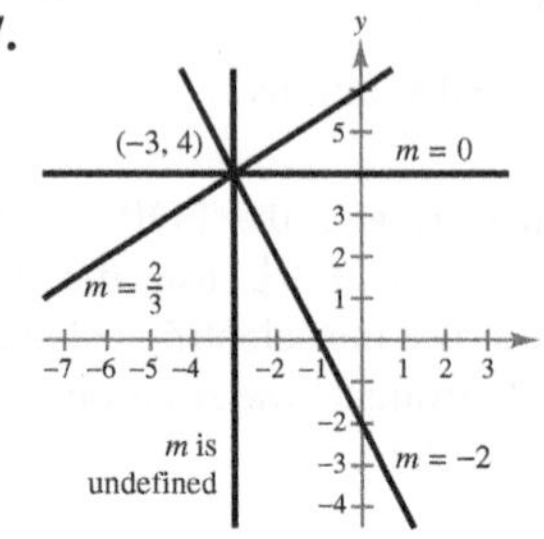

9.

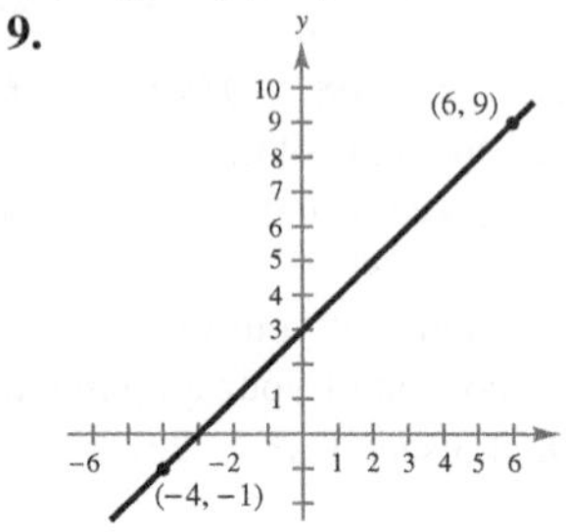

$m = 1$

11.

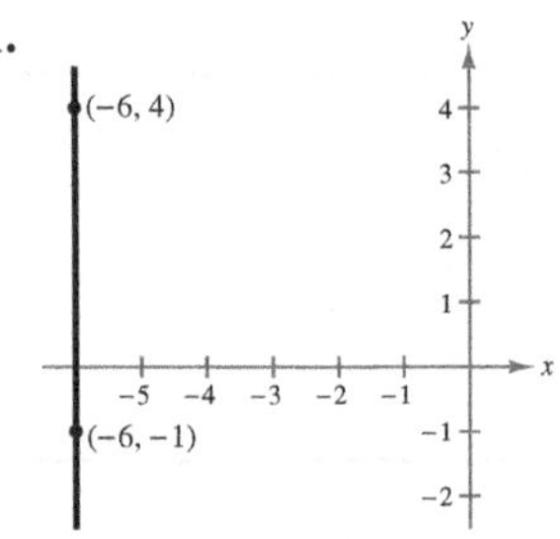

m is undefined.

13.

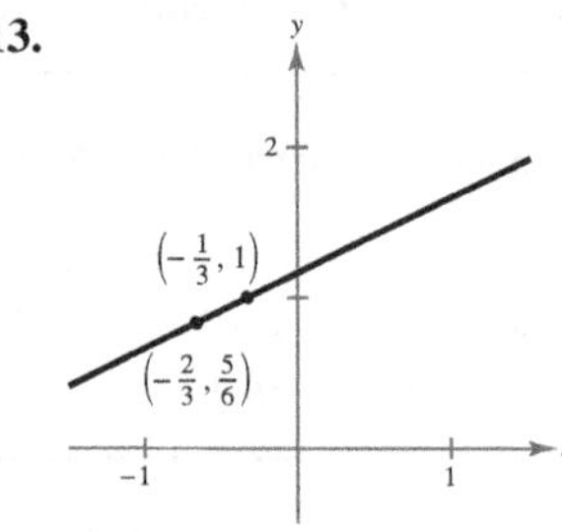

$m = \frac{1}{2}$

15. Answers will vary.
Sample answer: $(3, -2), (-1, -2), (0, -2)$
17. Answers will vary.
Sample answer: $(2, -3), (2, -7), (2, 9)$
19. Answers will vary.
Sample answer: $(6, -5), (7, -4), (8, -3)$
21. Answers will vary.
Sample answer: $(2, 3), (-4, 0), (4, 4)$
23. $x - y - 7 = 0$

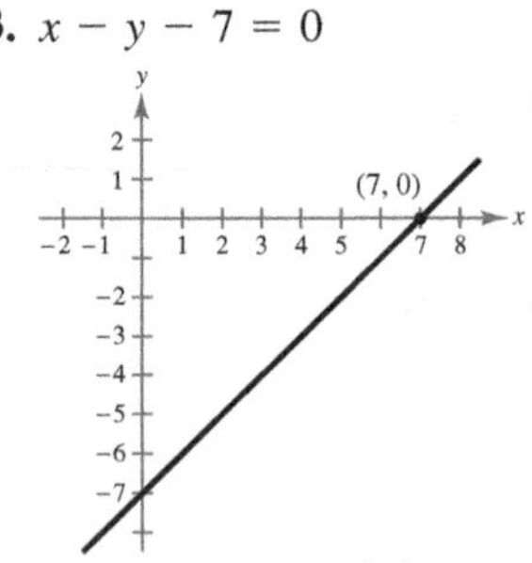

25. $4x + y + 8 = 0$

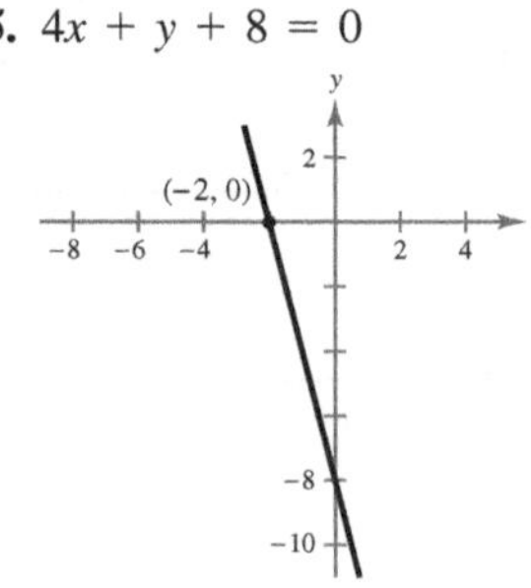

27. $x + 3y - 4 = 0$

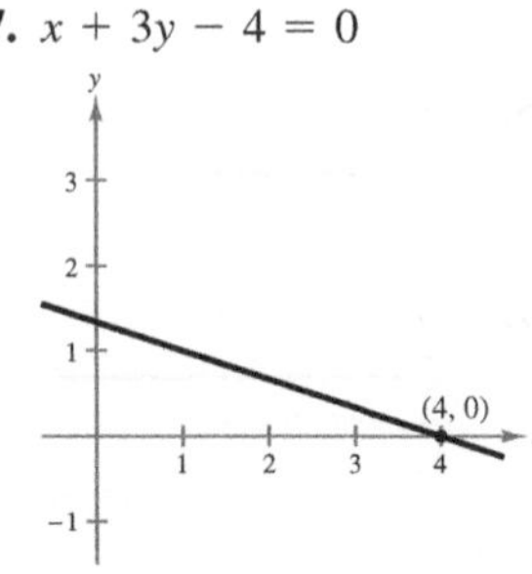

29. $y + 7 = 0$

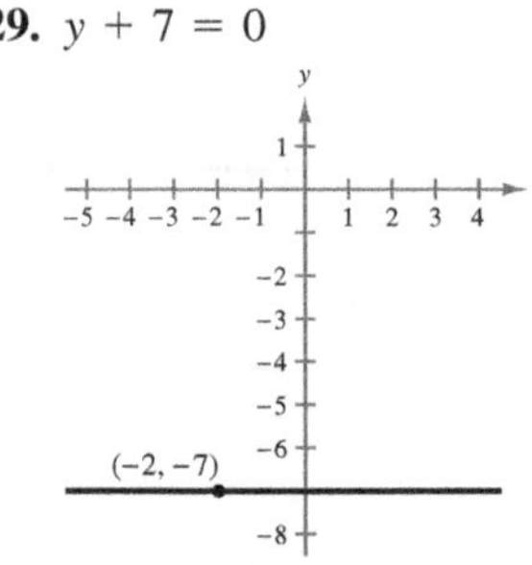

31. $8x - 6y - 17 = 0$

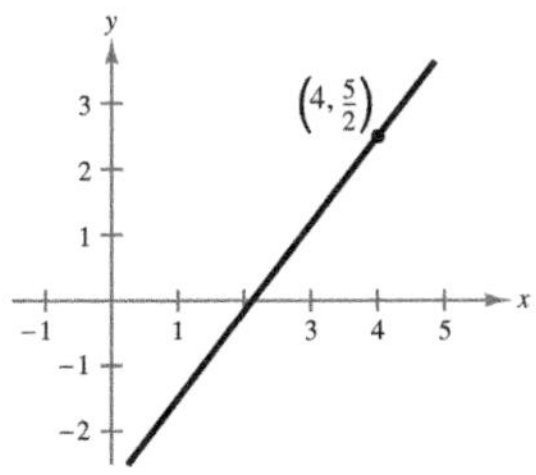

33. $y = 4x$

35. $x + 2y + 1 = 0$ **37.** $x + 9 = 0$ **39.** $y - 7 = 0$

41. $x + 2y - 3 = 0$ **43.** $2x - 5y + 1 = 0$

45. $(0, 2)$

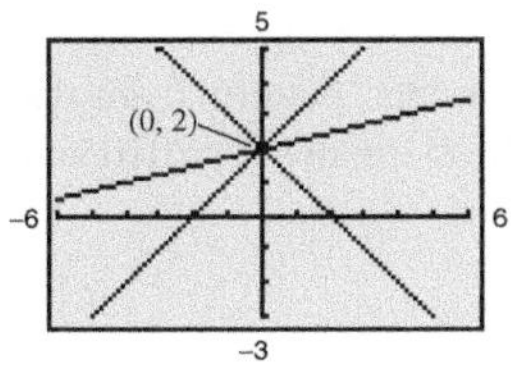

47. $m = 2, (0, -1)$

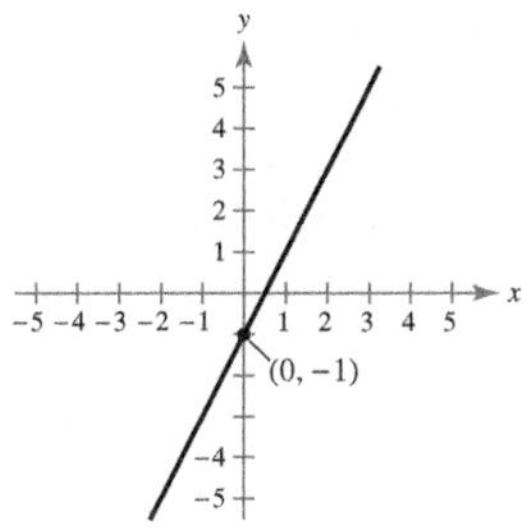

49. $m = 4, (0, -6)$

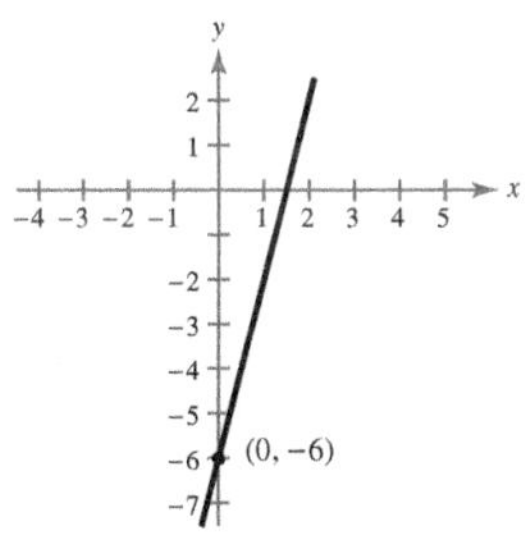

51. m is undefined; no y-intercept

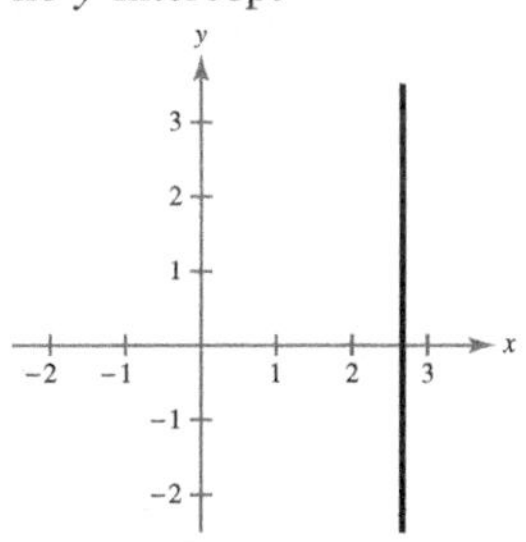

53. $m = -\frac{7}{6}, (0, 5)$

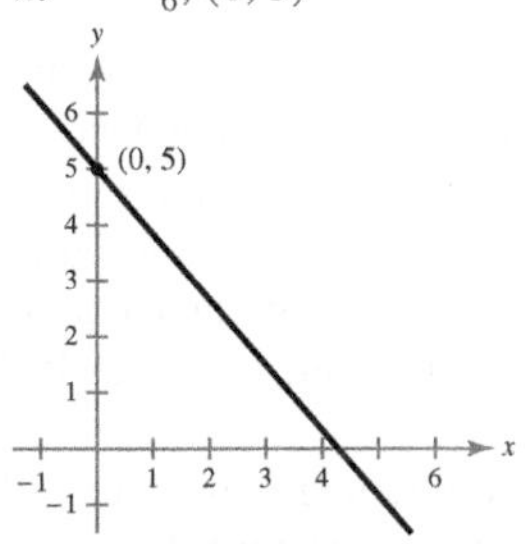

55. $m = 0, \left(0, \frac{7}{2}\right)$

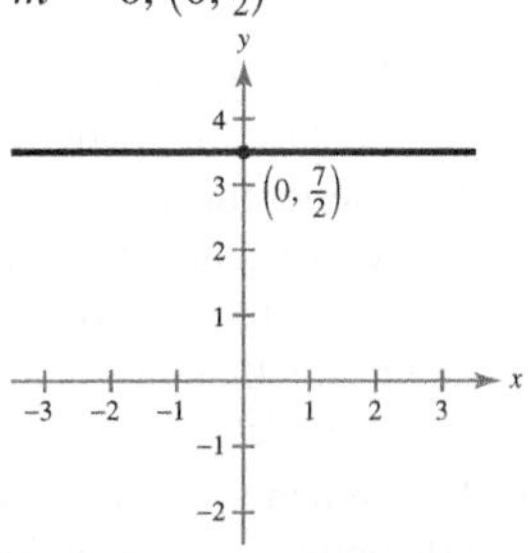

57. Answers will vary. Sample answer: You could graph a vertical line and pick two convenient points on the line to find the slope. Regardless of the points selected, the slope will have zero in the denominator. Division by zero is not possible, so the slope does not exist.

59. $4x - y - 4 = 0$ **61.** $x + y + 2 = 0$

63. $12x + 3y + 2 = 0$

65. Changing the viewing window will affect the appearance of the slope of the line. Answers will vary.

67. Neither **69.** Perpendicular **71.** Parallel

73. Perpendicular **75.** Parallel **77.** Perpendicular

79. Neither **81.** Parallel

83. (a) $2x - y - 10 = 0$ (b) $x + 2y - 10 = 0$

85. (a) $4x - 6y - 5 = 0$ (b) $36x + 24y + 7 = 0$

87. (a) $y = 0$ (b) $x + 1 = 0$ **89.** Yes $\left(m = \frac{17}{180}\right)$

91. $F = \frac{9}{5}C + 32$ **93.** $y = 7000x + 151{,}000$; \$179,000; No

95. $p = \frac{1}{33}d + 1$; $\frac{1}{33}$ atmosphere per foot

Section 2.3 *(page 84)*

Skills Warm Up *(page 84)*

1.

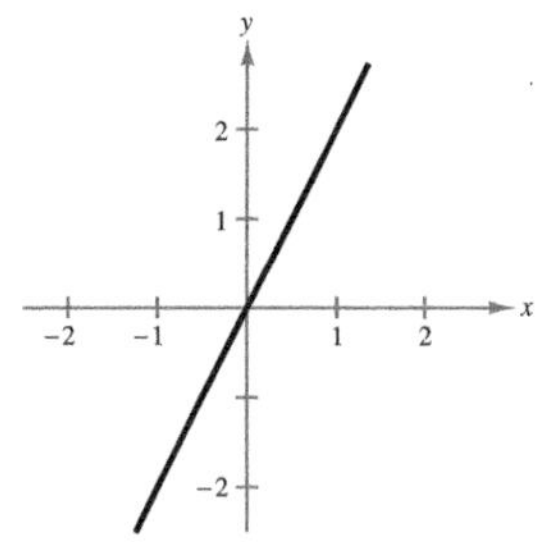

2.

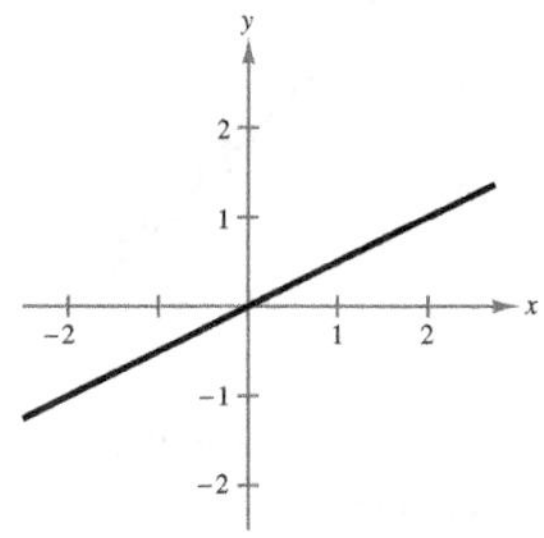

3.

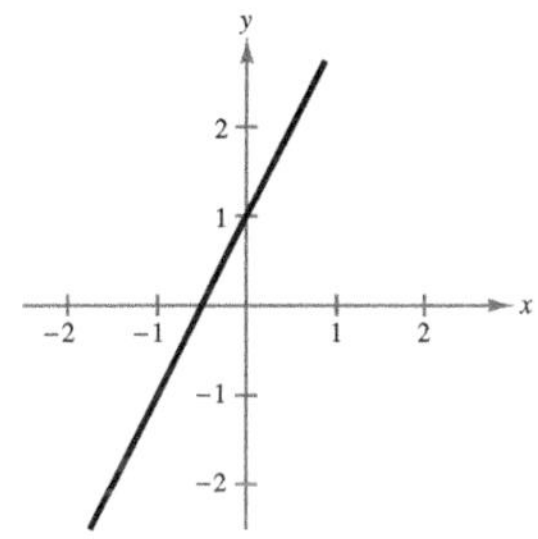

4.

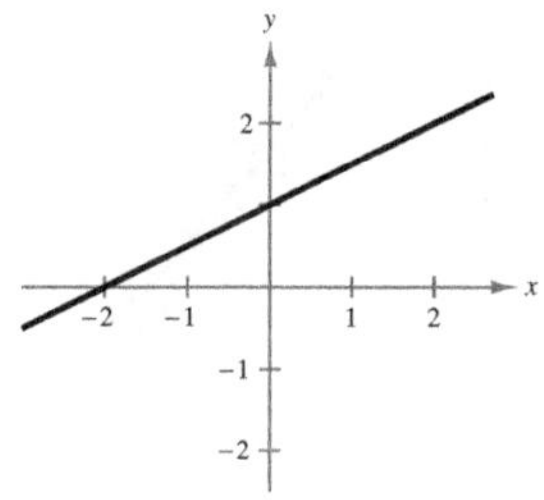

5. $y = x + 2$ **6.** $y = \frac{3}{2}x + 3$ **7.** $y = x + 2$

8. $y = \frac{6}{7}x + 4$ **9.** $5x + 40y - 213 = 0$

10. $29x + 60y - 448 = 0$

1.

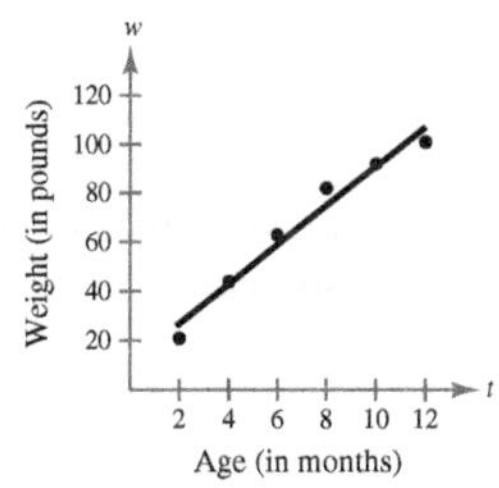

The model is a good fit for the actual data.

3.

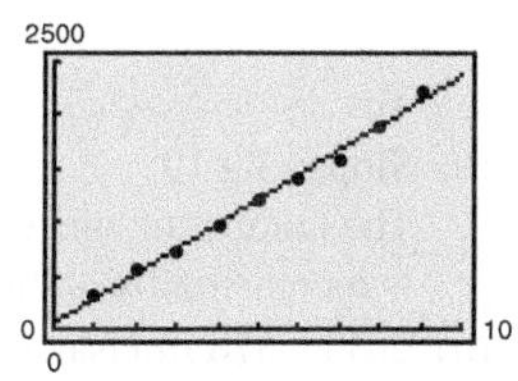

$y = 231.1x + 53.256$

The model is a good fit for the actual data.

5. $y = \dfrac{3}{8}x$ **7.** $y = 20x$ **9.** $y = \dfrac{3.2}{7}x$ or $y = \dfrac{16}{35}x$

11. $H = 3p$ **13.** $c = \frac{3}{5}d$ **15.** $I = 0.06P$

17. (a) $y = 0.0134x$ (b) \$2479

19. (a) $M = \frac{64}{103}K$

(b)

km/h	40	60	80	100	120
mi/h	24.85	37.28	49.71	62.14	74.56

21. $V = 140t + 860,\ 12 \le t \le 17$

23. $V = -2142t + 46{,}104,\ 12 \le t \le 17$

25. $V = 10{,}780t + 24{,}640,\ 12 \le t \le 17$

27. (a) $h = 7000 - 20t$ (b) 2:13:50 P.M.

29. $V = 875 - 175t,\ \ 0 \le t \le 5$

31. $W = 0.75x + 11.50$ **33.** (a) $H = 0.5t + 4$ (b) 13 in.

35. Yes; Answers will vary.
Sample answer:

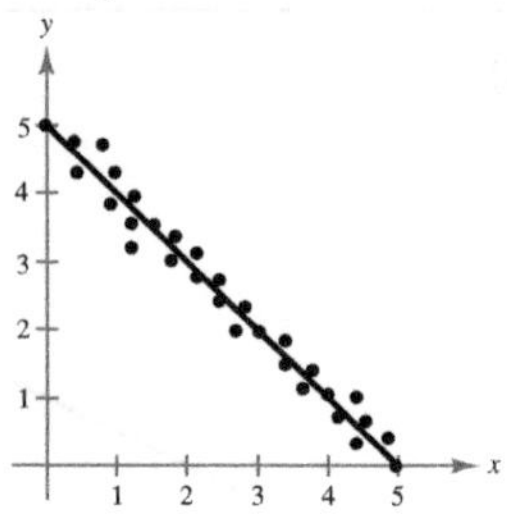

$y = 5 - x$

37. No

39. Yes; Answers will vary.
Sample answer:

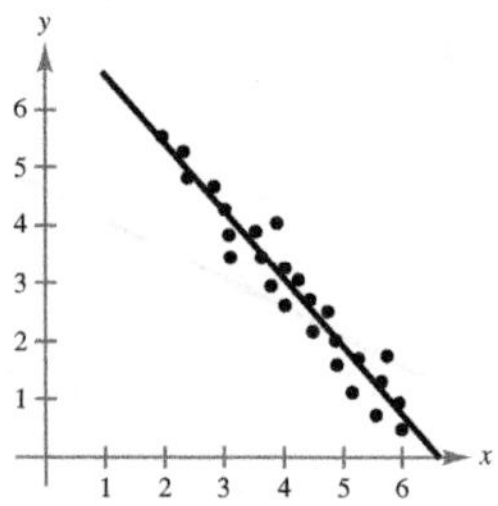

$y = -1.18x + 7.76$

41. (a)

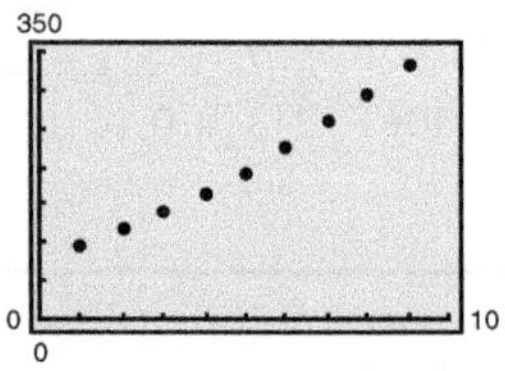

Yes

(b) $C = 29.19t + 57.1$

(c) Slope: 29.19
The number of autistic children receiving disability services is increasing by about 29 thousand children per year.

(d) 2011: 378 thousand children
2012: 407 thousand children
The estimates are reasonable.

43. (a)

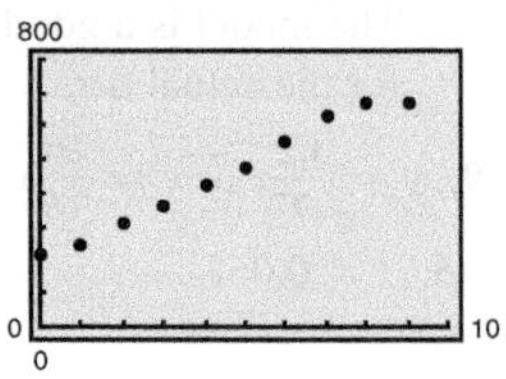

(b) Answers will vary.
Sample answer:
$y = 58.1t + 210.8$

(c) $y = 56.69t + 200.7$
Using linear regression equation:
2010: \$767.6 million
2011: \$824.29 million
Using equation from part (b):
2010: \$791.8 million
2011: \$849.9 million

(d) The projections made by California Pizza Kitchen are lower than the predictions given by the models.

(e) Yes; Using the linear regression equation, the yearly revenue is expected to reach \$937.67 million by 2013. Using the equation from part (b), the yearly revenue is expected to reach \$966.1 million by 2013.

45. (a)

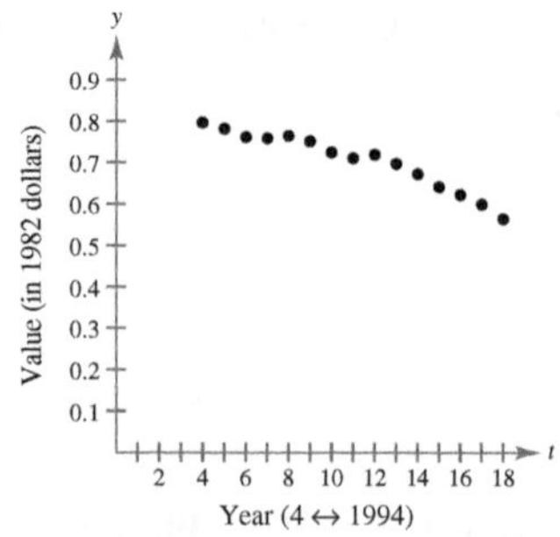

The data appear to be approximately linear.

(b) $y = -0.0152t + 0.873$

(c) 2010: 0.569; 2011: 0.5538
Because the data follow a linear pattern from 1994 to 2008, you can assume that the estimates for 2010 and 2011 are reliable.

47. (a)

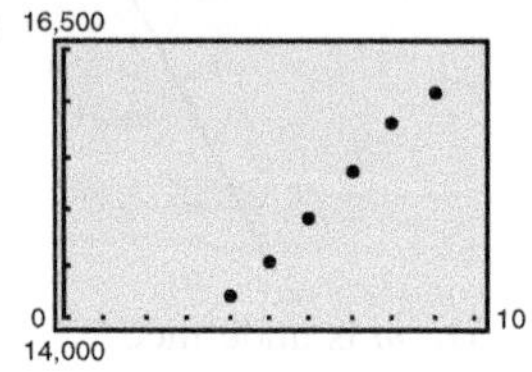

The data appear to be approximately linear.

(b) $E = 394.17t + 12{,}592.9$

(c) 2011: 16,928,770 employees
2013: 17,717,110 employees

(d)

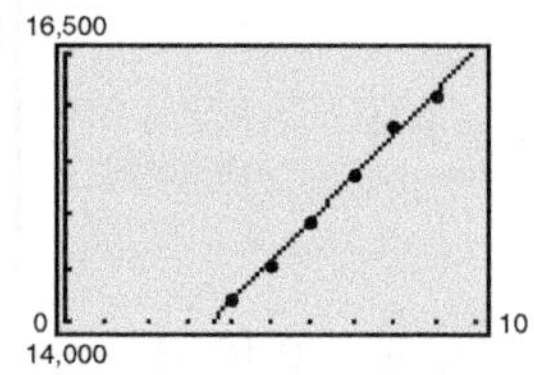

The predictions are most likely going to be just about right because the model is a good fit for the actual data.

49. The model for population, because population tends to change at a consistent rate, whereas snowfall can be quite different from year to year. You should use more than three data points to ensure that the data can be represented accurately by a linear model.

51. Yes; When you start, you have zero wrong moves $(x = 0)$ and 800 points $(y = 800)$. So, 800 is the y-intercept. Then you lose 50 points for each wrong move. This is a slope of -50.

53. A known rate of change corresponds to the slope of a linear model, so determine whether the rate of change of y with respect to x is constant. The data points show a 10-unit increase in y for each 1-unit increase in x, so the rate of change is constant and the data points are linear.

Section 2.4 *(page 97)*

Skills Warm Up *(page 9 7)*

1. -73 **2.** 13 **3.** $2(x + 2)$ **4.** $-8(x - 2)$
5. $y = \frac{7}{5} - \frac{2}{5}x$ **6.** $y = \pm x$ **7.** $x \le -2, x \ge 2$
8. $-3 \le x \le 3$ **9.** All real numbers
10. $x \le 1, x \ge 2$

1. This is a function from A to B because each element of A is matched with an element of B.
3. Not a function; The relationship does not match the element b of A with an element of B.
5. This is a function from A to B because each element of A is matched with an element of B.
7. This is a function from A to B because each element of A is matched with an element of B.
9. Not a function; The relationship assigns two elements of B to the element c of A.
11. Not a function from A to B; The relationship defines a function from B to A.
13. This is a function from A to B because each element of A is matched with an element of B.
15. Not a function **17.** Function **19.** Function
21. Not a function **23.** Function
25. (a) -6 (b) 34 (c) $6 - 4t$ (d) $2 - 4c$
27. (a) -1 (b) $\frac{1}{15}$ (c) $\frac{1}{t^2 - 2t}$ (d) $\frac{1}{t^2 - 1}$
29. (a) -1 (b) -9 (c) $2x - 5$ (d) $-\frac{5}{2}$
31. (a) 0 (b) 3 (c) $x^2 + 2x$ (d) -0.75
33. (a) 36π (b) 0 (c) $\frac{9\pi}{2}$ (d) $\frac{32\pi r^3}{3}$
35. (a) 1 (b) -7 (c) $3 - 2|x|$ (d) 2.5
37. (a) Undefined (b) $-\frac{1}{16}$
(c) $\frac{1}{y^2 + 4y - 12}$, $y \neq -6, 2$
(d) $\frac{1}{y^2 - 4y - 12}$, $y \neq -2, 6$
39. (a) 1 (b) -1 (c) 1 (d) $\frac{|x - 1|}{x - 1}$
41. (a) -4 (b) 3 (c) -7 (d) 7
43. (a) 2 (b) -1 (c) 3 (d) 3 **45.** 5 **47.** ± 3
49. $0, 1, -1$ **51.** $\frac{10}{7}$ **53.** $\{0, 3, 5, 8, 9\}$
55. $\{-2, -1, 0, 1, 2\}$ **57.** All real numbers x
59. All real numbers except $t = 0$ **61.** All real numbers y
63. $-1 \le x \le 1$ **65.** All real numbers except $x = 0, -2$
67. All real numbers $x \ge -1$ except $x = 2$ **69.** $x > 0$
71. $\{(-2, 4), (-1, 1), (0, 0), (1, 1), (2, 4)\}$
73. $\{(-2, 0)\ (-1, 1)\ (0, \sqrt{2})\ (1, \sqrt{3})\ (2, 2)\}$
75. The domain of $f(x) = \sqrt{x - 2}$ is all real numbers $x \ge 2$ because an even root of a negative number is not a real number. The domain of $g(x) = \sqrt[3]{x - 2}$ is all real numbers. f and g have different domains because an odd root of a negative number is a real number, but an even root of a negative number is not a real number.
77. n ordered pairs
79. (a) $V = x(18 - 2x)^2$ (b) Domain: $0 < x < 9$
(c) 400 cubic inches
81. Yes $[y(30) = 6]$
83. 2000: 945 million acres; 2009: 917.98 million acres
85. (a) $C = 57{,}000 + 2.05x$ (b) $\overline{C} = \frac{57{,}000}{x} + 2.05$
(c)

x	100	1000	10,000	100,000
$\overline{C}$	572.05	59.05	7.75	2.62

(d) Answers will vary. Sample answer: The average cost per unit decreases as x gets larger.

87. (a)

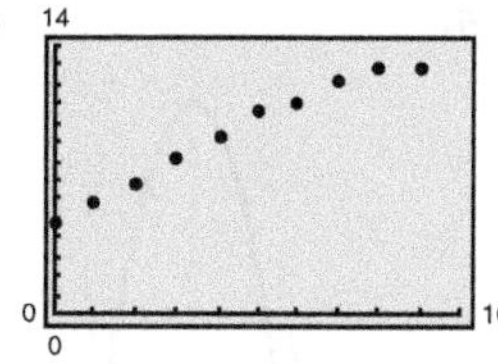

(b) Linear model: $P = 0.965t + 5.11$
Quadratic model: $P = -0.0456t^2 + 1.375t + 4.56$
(c)

Year	P (Actual)	P (Linear)	P (Quadratic)
2000	4.80	5.11	4.56
2001	5.89	6.08	5.89
2002	6.81	7.04	7.13
2003	8.10	8.01	8.27
2004	9.30	8.97	9.33
2005	10.55	9.94	10.30
2006	11.13	10.90	11.17
2007	12.09	11.87	11.95
2008	12.95	12.83	12.64
2009	12.91	13.80	13.24

The quadratic model is a better fit because its values of P tend to be closer to the actual values than those of the linear model.

89. (a) $r(t) = 0.75t$ (b) $A = 0.5625\pi t^2$

Time, t	1	2	3	4	5
Radius, r (in feet)	0.75	1.5	2.25	3	3.75
Area, A (in square feet)	1.767	7.069	15.904	28.274	44.179

(c) $\frac{A(2)}{A(1)} = \frac{7.069}{1.767} = 4.000$, $\frac{A(4)}{A(2)} = \frac{28.274}{7.069} = 4.000$
Predicted area when $t = 8$: 113.096 square feet
Calculated area when $t = 8$: 113.097 square feet
91. (a) Correct; Each price has exactly one tax amount.
(b) Incorrect; For any given value of education, there is not exactly one specific value of intelligence.
93. Answers will vary.

Quiz Yourself *(page 102)*

1. (a)

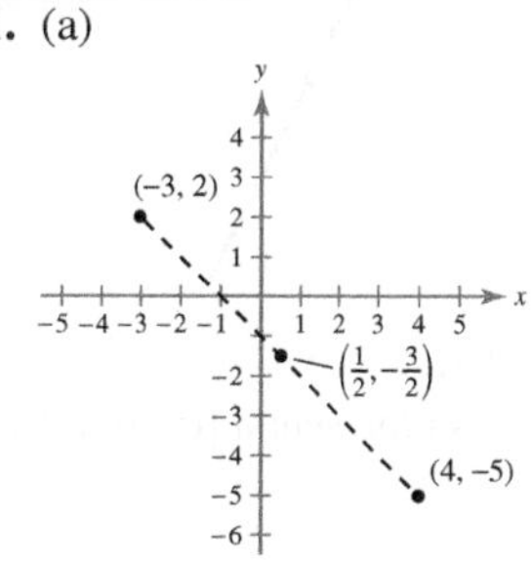

(b) $7\sqrt{2}$
(c) $\left(\frac{1}{2}, -\frac{3}{2}\right)$

2. (a)

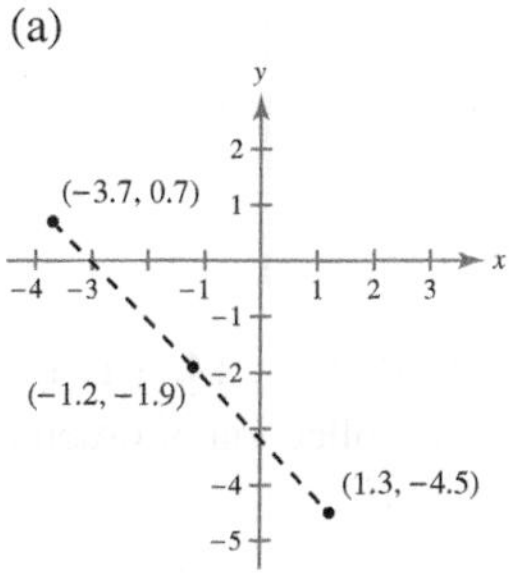

(b) $\sqrt{52.04}$
(c) $(-1.2, -1.9)$

CHAPTER 2

3. (a)

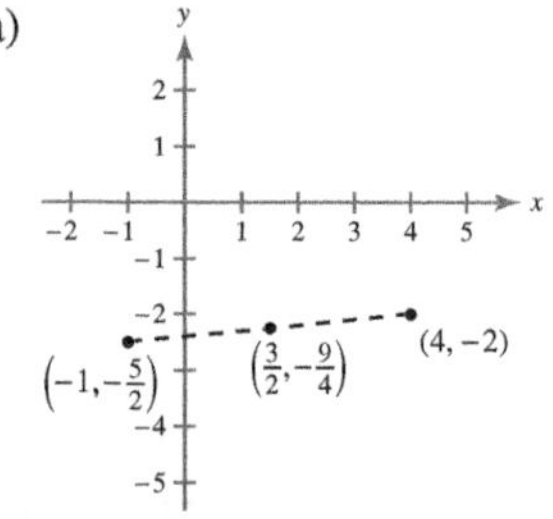

(b) $\dfrac{\sqrt{101}}{2}$ (c) $\left(\dfrac{3}{2}, -\dfrac{9}{4}\right)$

4.

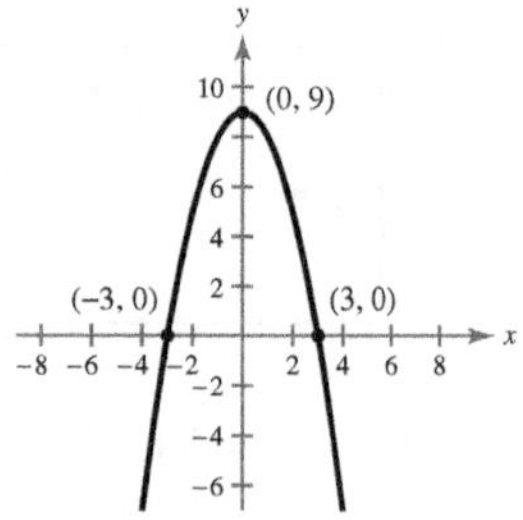

Intercepts: $(-3, 0), (3, 0), (0, 9)$

Symmetry: y-axis

5.

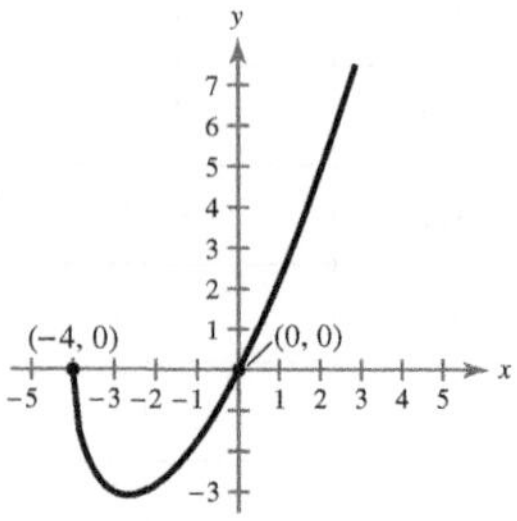

Intercepts: $(-4, 0), (0, 0)$

Symmetry: none

6.

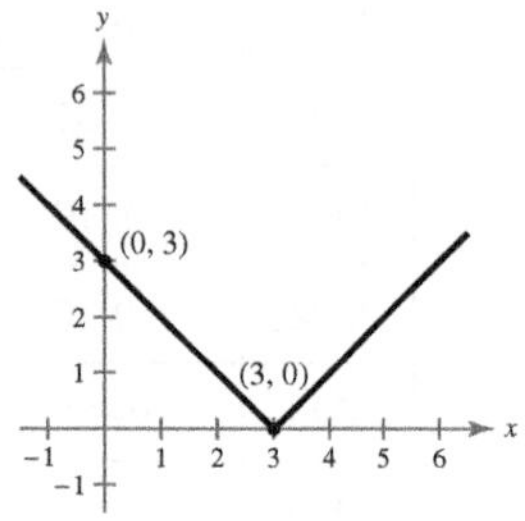

Intercepts: $(3, 0), (0, 3)$

Symmetry: none

7. $(x - 4)^2 + (y - 5)^2 = 20$

8. $(x - 1)^2 + (y + 2)^2 = 9$

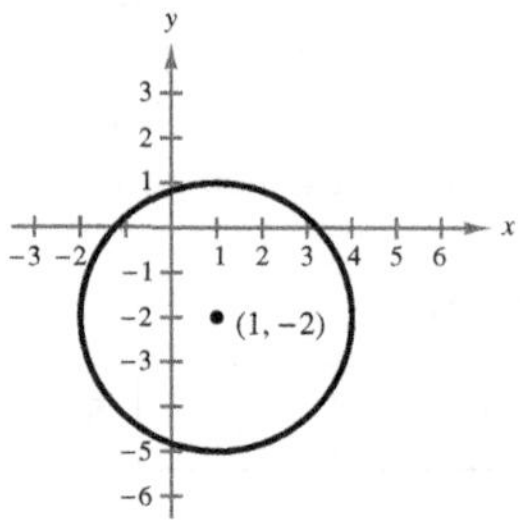

9. $2x - 3y + 9 = 0$

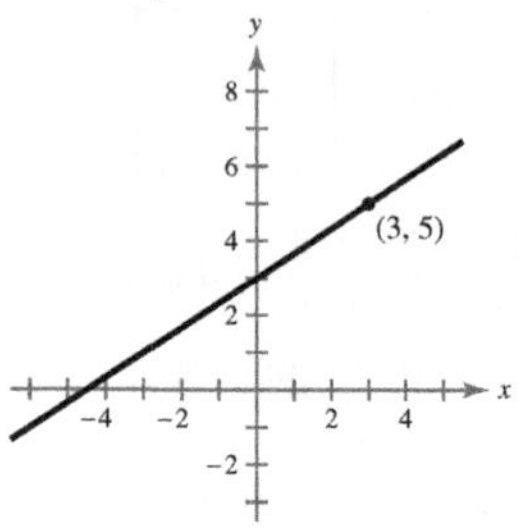

10. $y - 4 = 0$

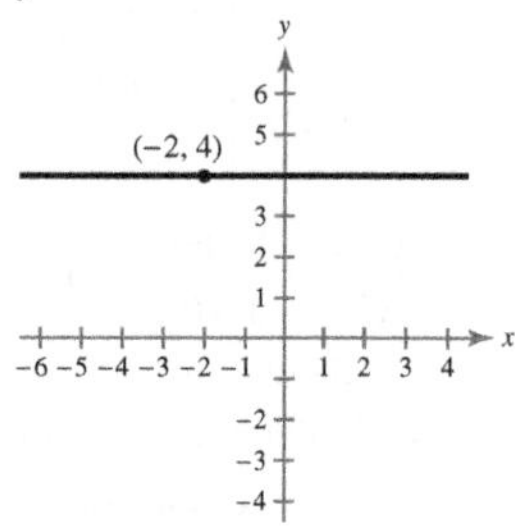

11. $x - 2 = 0$

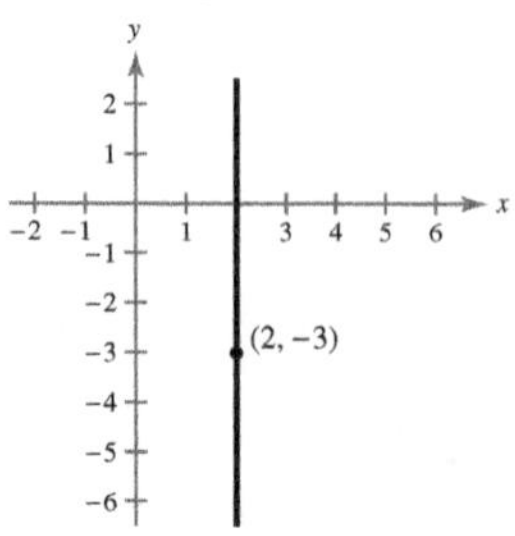

12. $2x + y + 9 = 0$

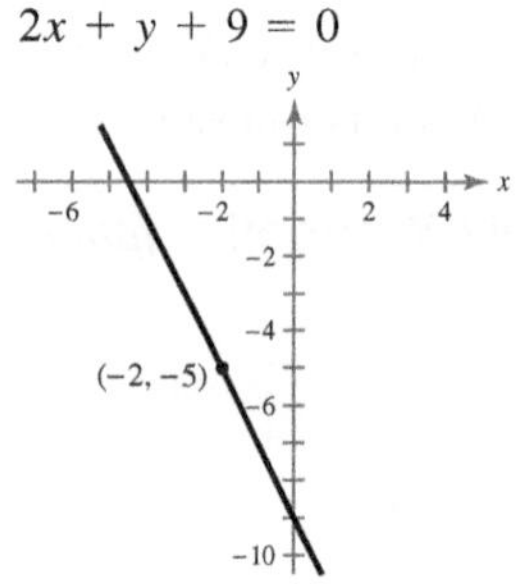

13. 261,019 **14.** y is a function of x.

15. The collection of ordered pairs does represent a function from A to B.

16. The collection of ordered pairs does not represent a function from A to B.

17. (a) 2 (b) -7 **18.** (a) 1 (b) -20

19. $x \geq 4$ **20.** All real numbers except $x = -2$

21.

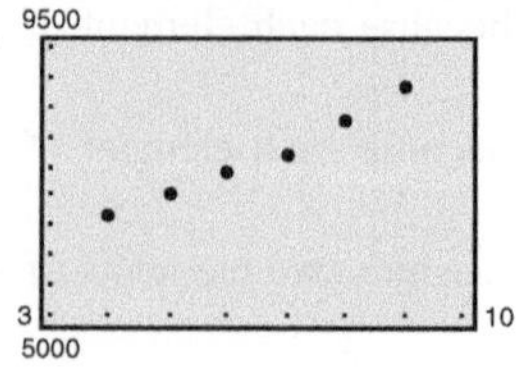

Linear: $y = 428.5t + 4873$

Quadratic: $y = 35.29t^2 - 30.2t + 6261$

22. Linear:
2010: $9158 million
2011: $9586.5 million

Quadratic:
2010: $9488 million
2011: $10,198.89 million

23. $A = \dfrac{C^2}{4\pi}$

Section 2.5 *(page 110)*

Skills Warm Up *(page 110)*

1. 2 **2.** 0 **3.** $-\dfrac{3}{x}$ **4.** $x^2 + 3$ **5.** $0, \pm 4$

6. $\frac{1}{2}, 1$ **7.** All real numbers except $x = 4$

8. All real numbers except $x = 4, 5$ **9.** $t \leq \frac{5}{3}$

10. All real numbers

1. Domain: $[1, \infty)$; Range: $[0, \infty)$; 0

3. Domain: $(-\infty, \infty)$; Range: $(-\infty, 4]$; 4

5. Domain: $(-\infty, \infty)$; Range: $(-\infty, \infty)$; -1

7. Domain: $[-5, 5]$; Range: $[0, 5]$; 5

9. Function **11.** Not a function **13.** Function

15. Not a function **17.** Increasing on $(-\infty, \infty)$; No change

19. Increasing on $(-\infty, 0)$ and $(2, \infty)$, decreasing on $(0, 2)$; behavior changes at $(0, 0)$ and $(2, -4)$.

21. Increasing on $(-1, 0)$ and $(1, \infty)$, decreasing on $(-\infty, -1)$ and $(0, 1)$; behavior changes at $(-1, -3)$, $(0, 0)$, and $(1, -3)$.

23. Increasing on $(-2, \infty)$, decreasing on $(-3, -2)$; behavior changes at $(-2, -2)$.

25. Increasing on $(2, \infty)$, decreasing on $(-\infty, 2)$; behavior changes at $(2, -4)$.

27. Increasing on $(-\infty, 0)$ and $(2, \infty)$, decreasing on $(0, 2)$; behavior changes at $(0, 2)$ and $(2, -2)$.

29.

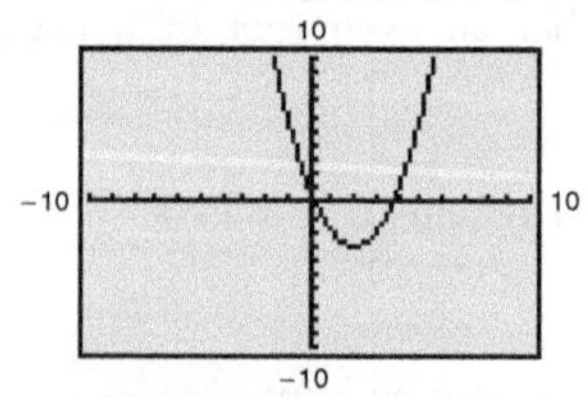

Minimum: $(2, -3)$
Increasing: $(2, \infty)$
Decreasing: $(-\infty, 2)$

31.

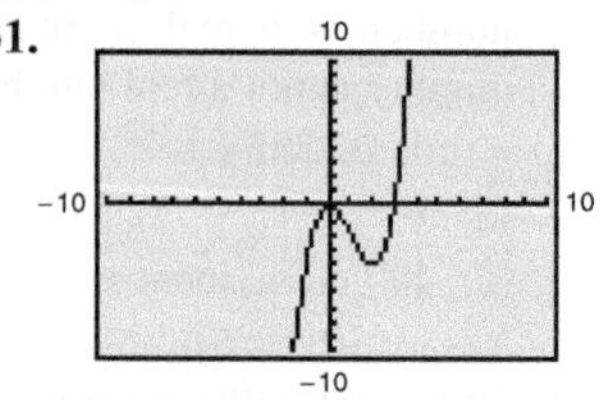

Relative maximum: $(0, 0)$
Relative minimum: $(2, -4)$
Increasing: $(-\infty, 0), (2, \infty)$
Decreasing: $(0, 2)$

33.

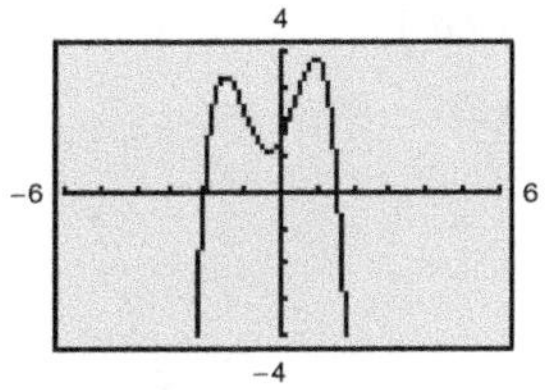

Relative maxima: $(-1.54, 3.29)$, $(0.95, 3.77)$
Relative minimum: $(-0.34, 1.14)$
Increasing: $(-\infty, -1.54)$, $(-0.34, 0.95)$
Decreasing: $(-1.54, -0.34)$, $(0.95, \infty)$

35. (a) 2 (b) 2 (c) -3 (d) -4

37. (a) 2 (b) 1 (c) -8 (d) -9

39. Even **41.** Odd **43.** Odd

45.

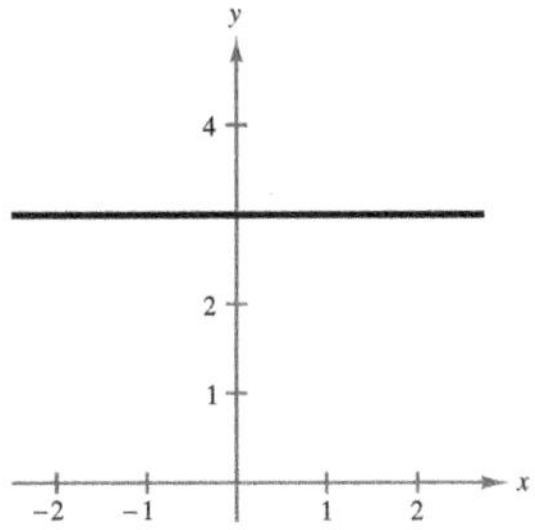

Even

47.

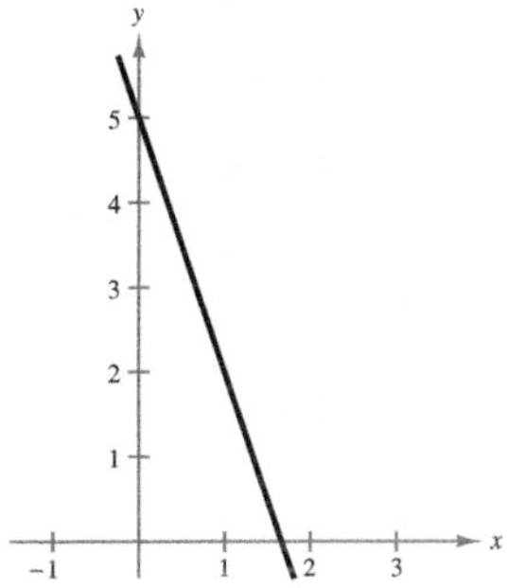

Neither even nor odd

49.

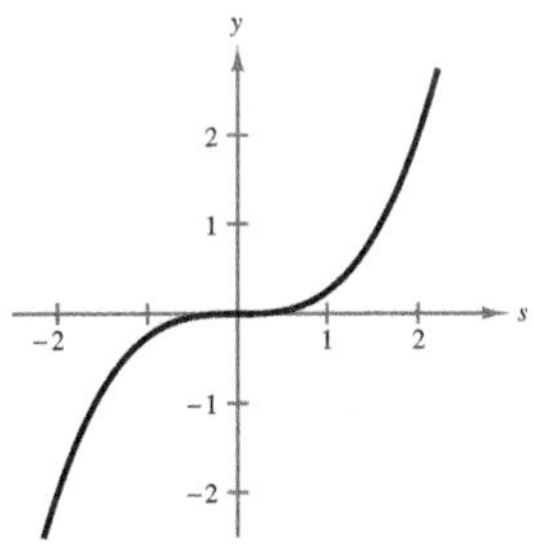

Odd

51.

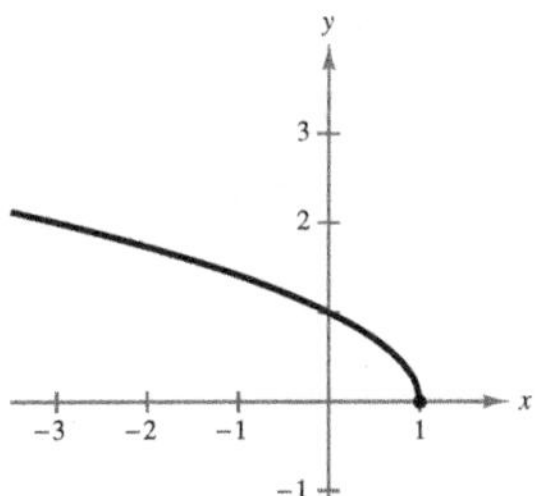

Neither even nor odd

53.

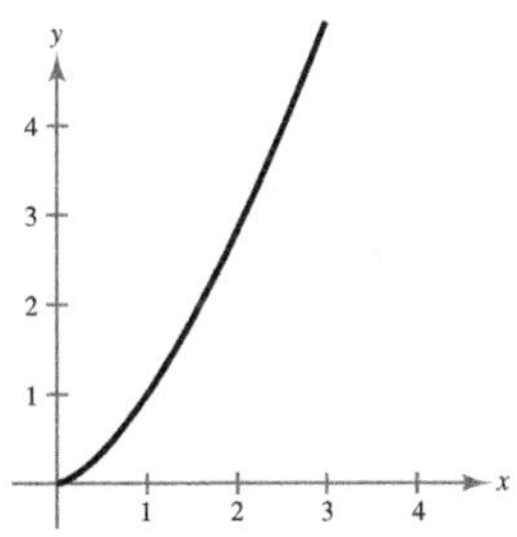

Neither even nor odd

55.

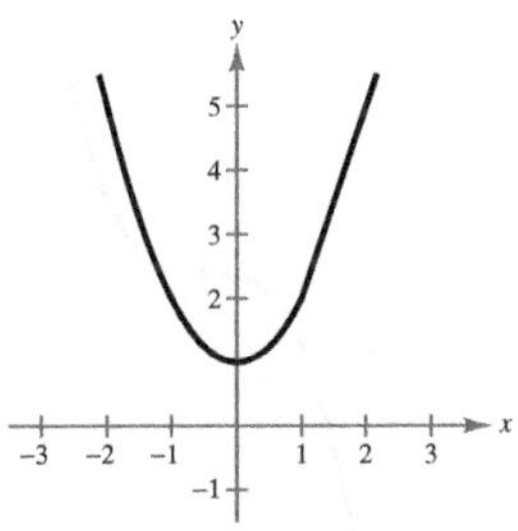

Neither even nor odd

57.

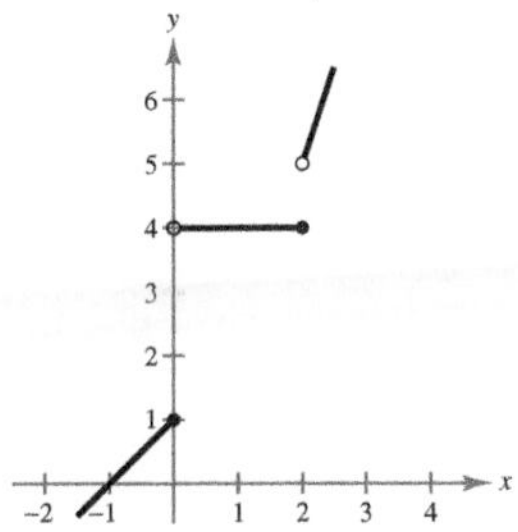

Neither even nor odd

59.

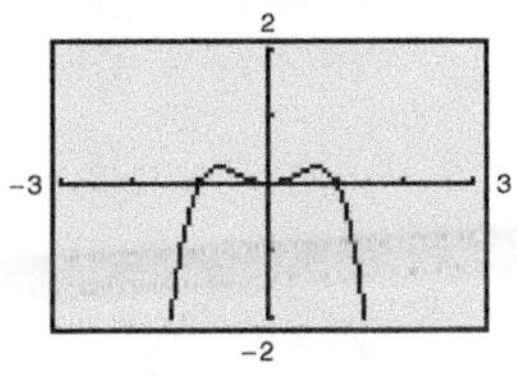

Even

61.

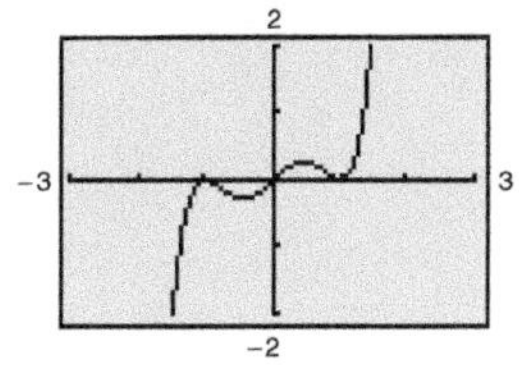

Odd

63.

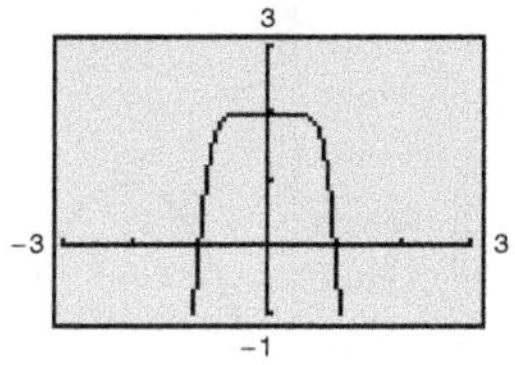

Even

65.

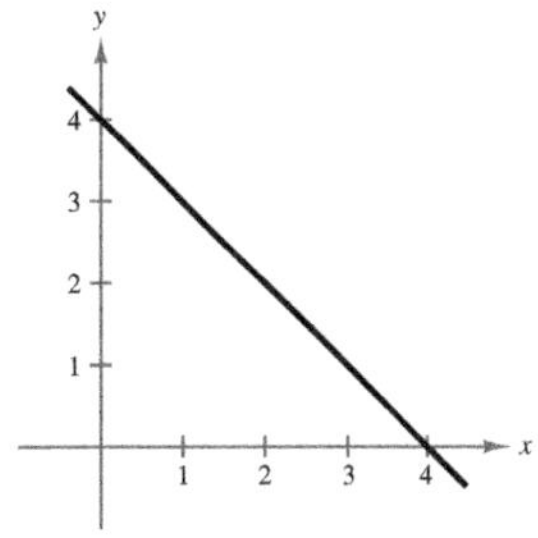

67.

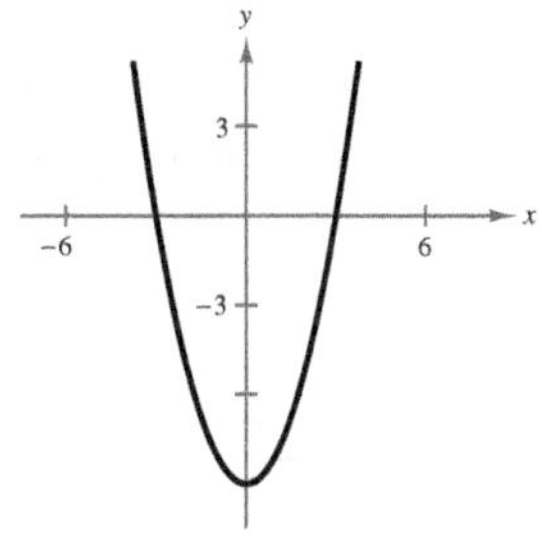

69.

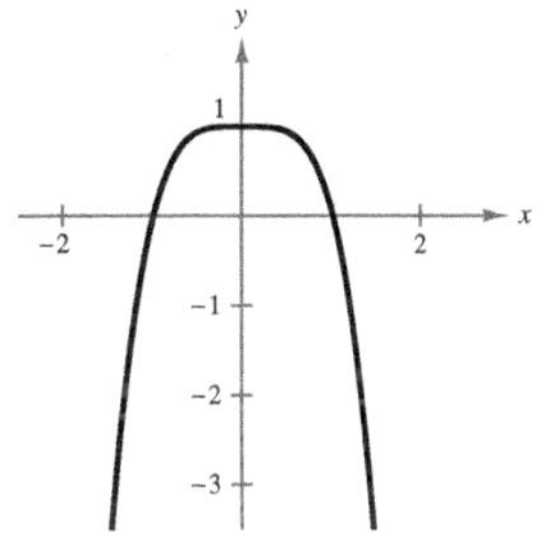

71.

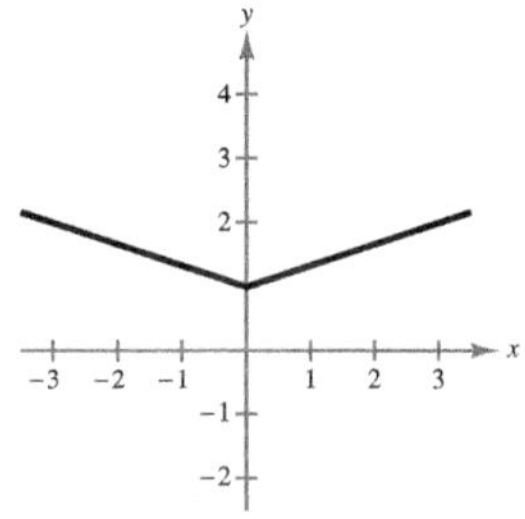

73.

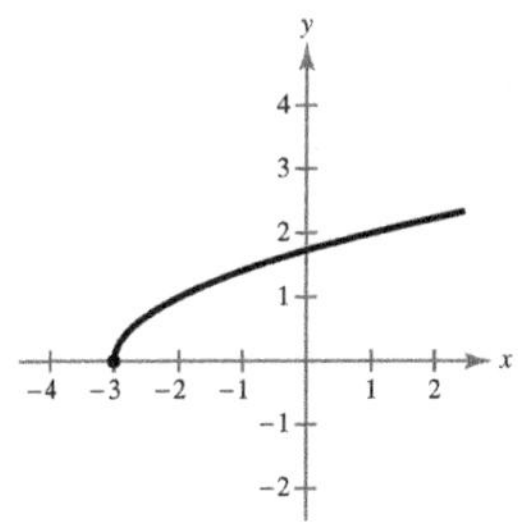

75.

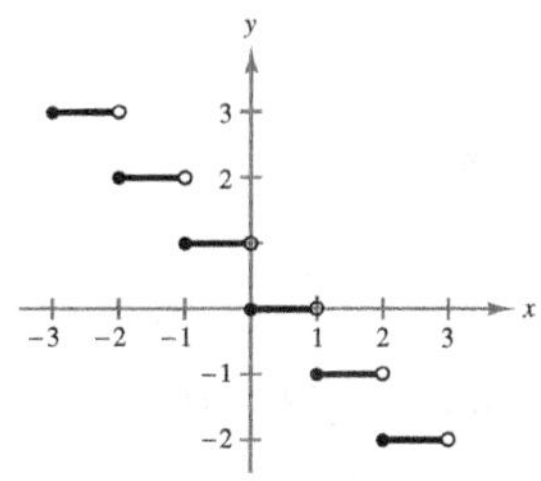

77.

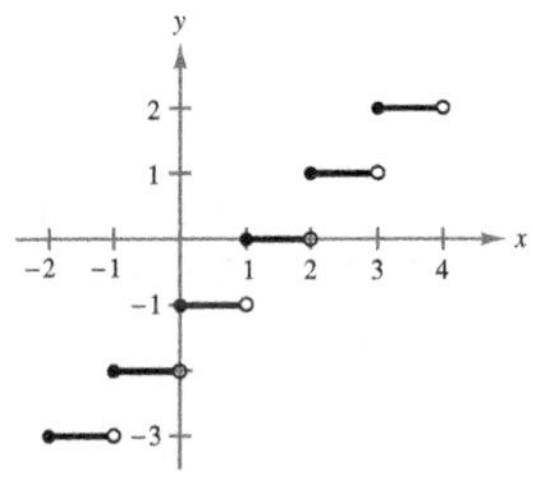

79. (a) Maximum: \$1212.62
(b) Decreasing: Price did not decrease during 2000–2010.
Increasing: 2000–2010
(c) The slope of the graph seems to increase quite rapidly in future years. The price of gold probably will not increase that quickly.

81. (a)

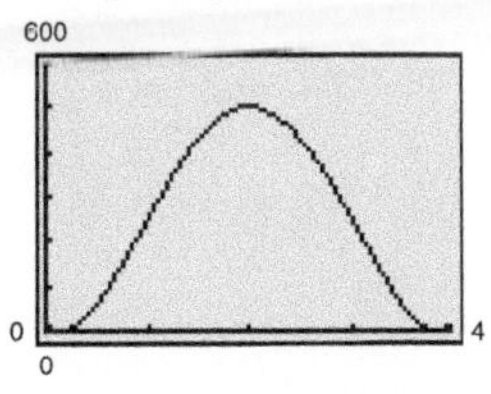

Increasing: 0 seconds to 2 seconds
Decreasing: 2 seconds to 4 seconds
(b) Maximum change in volume: ≈ 501.9 milliliters

CHAPTER 2

83.

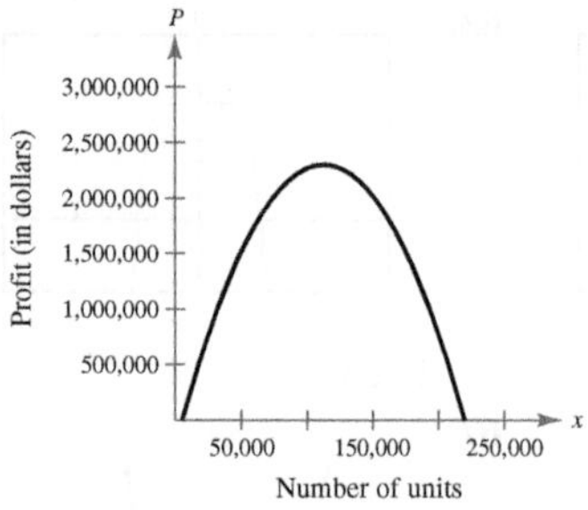

Approximately 112,500 units

85.

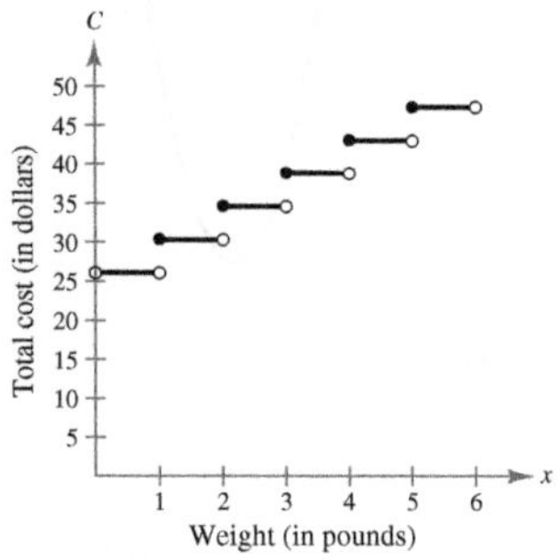

87.

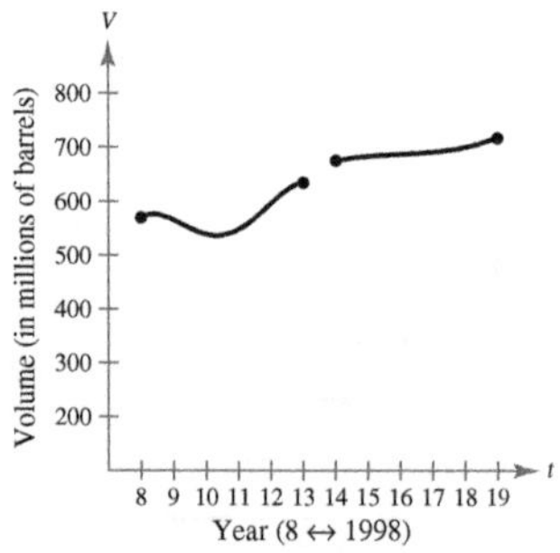

89. (a) $y = 425.71t + 705.54$
Domain: $(0, 4)$
Range: $(705.54, 2408.38)$
(b) $y = -202.95t^2 + 3304.67t - 7311.8$
Domain: $(5, 9)$
Range: $(4137.8, 6140.3)$
(c) $y = \begin{cases} 425.71t + 705.54, & 0 \leq t \leq 4 \\ -202.95t^2 + 3304.67t - 7311.8, & 5 \leq t \leq 9 \end{cases}$

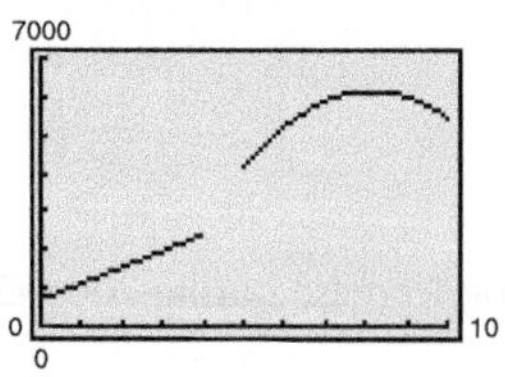

(d) The revenues increased from 2000 to 2008 and decreased from 2008 to 2009.

91. (a) $\left(\frac{3}{2}, 4\right)$ (b) $\left(\frac{3}{2}, -4\right)$

93. (a) $(-4, 9)$ (b) $(-4, -9)$

Review Exercises *(page 114)*

1. (a)

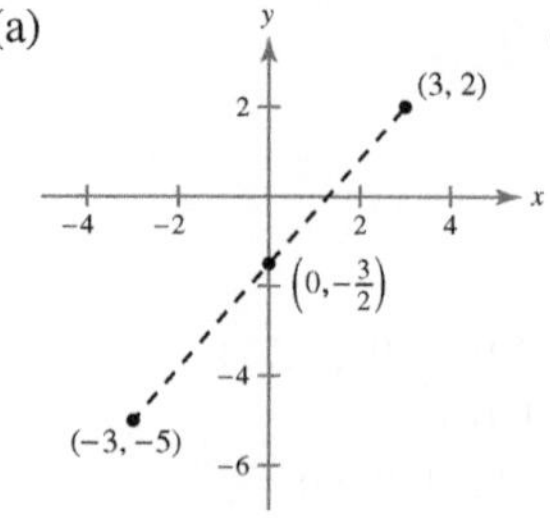

(b) $\sqrt{85} \approx 9.22$
(c) $\left(0, -\frac{3}{2}\right)$

3. (a)

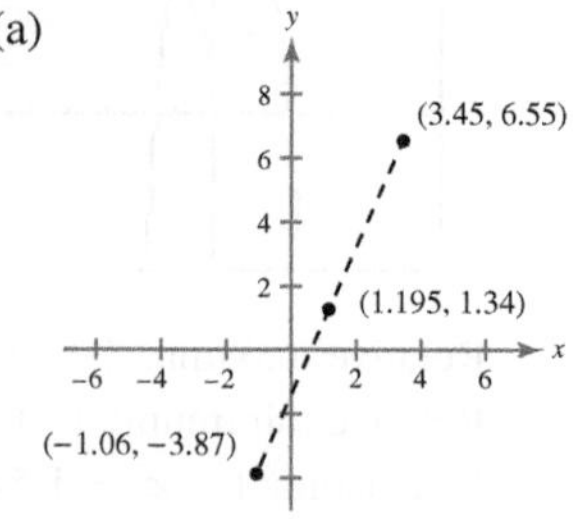

(b) $\sqrt{128.9165} \approx 11.35$
(c) $(1.195, 1.34)$

5. $x = -10$ or 30

7. (a) Yes (b) Yes

9.

x	y
-2	3
0	2
2	1
3	$\frac{1}{2}$
4	0

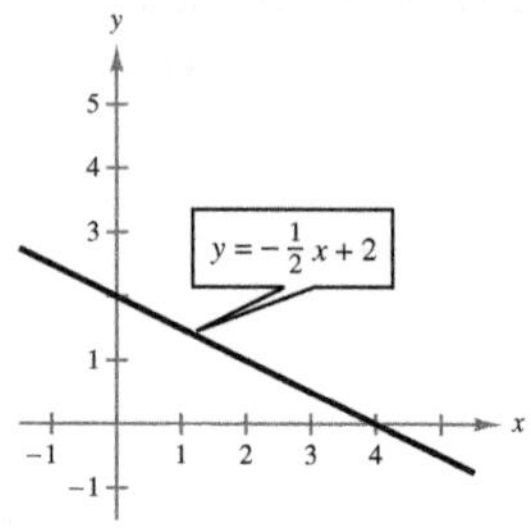

11.

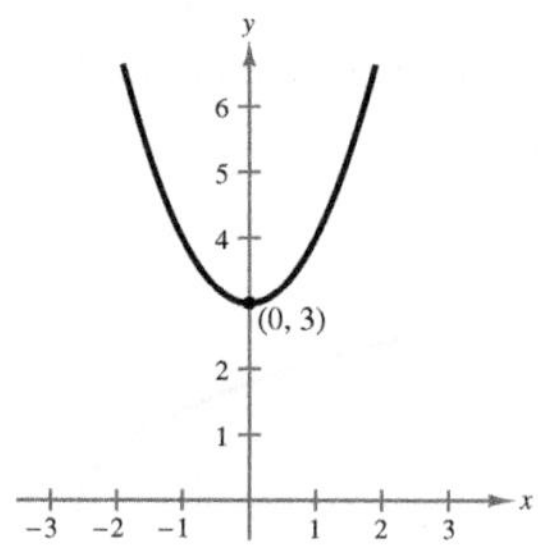

y-intercept: $(0, 3)$
Symmetry: y-axis

13.

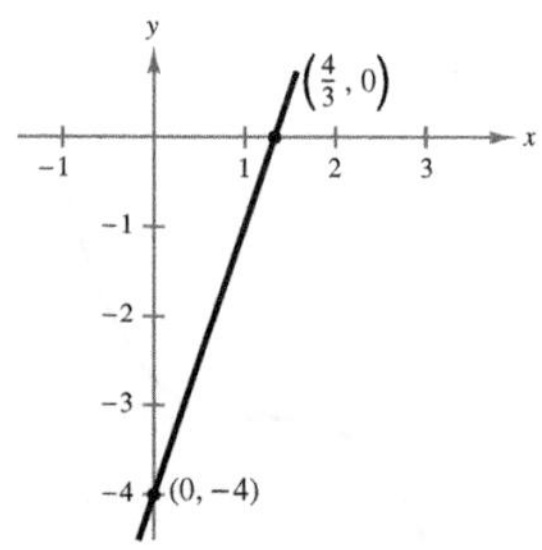

x-intercept: $\left(\frac{4}{3}, 0\right)$
y-intercept: $(0, -4)$
Symmetry: none

15.

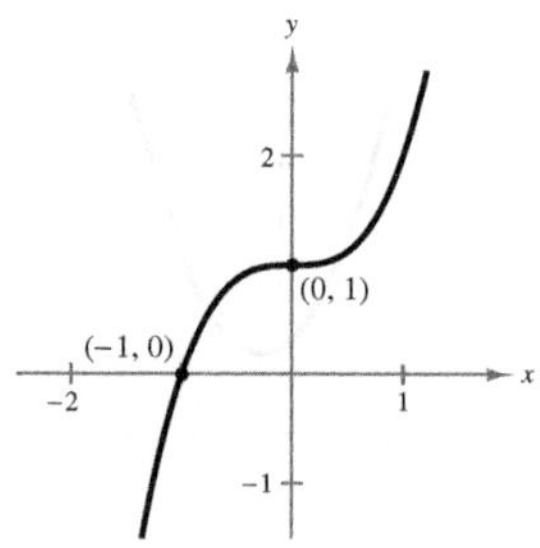

x-intercept: $(-1, 0)$
y-intercept: $(0, 1)$
Symmetry: none

17. $(x + 1)^2 + (y - 2)^2 = 36$

19. $(x - 2)^2 + (y + 3)^2 = 25$

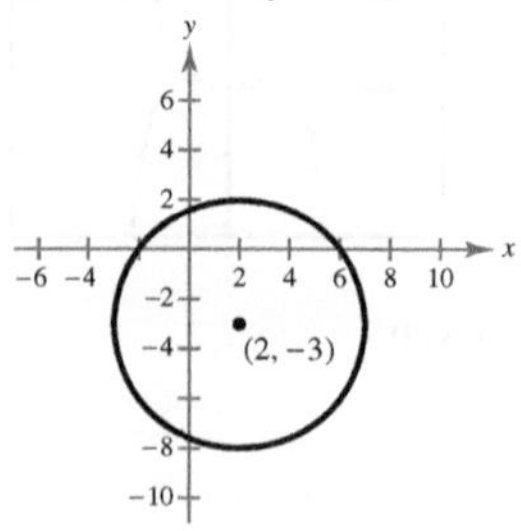

21.

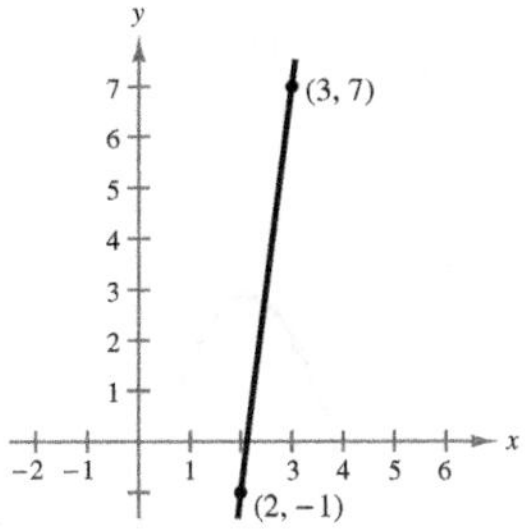

$m = 8$

23.

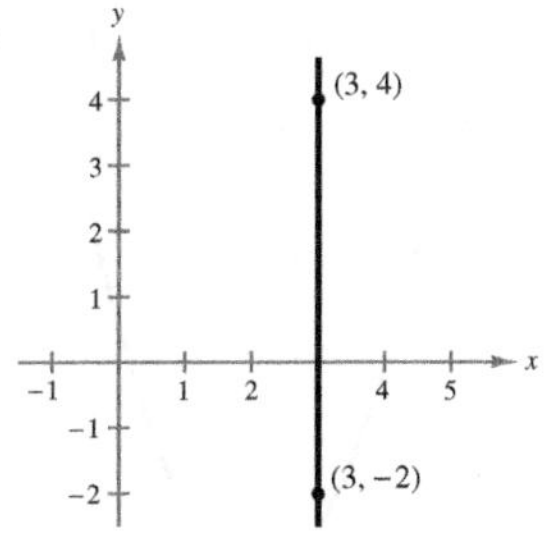

m is undefined.

25. $3x - 2y - 10 = 0$

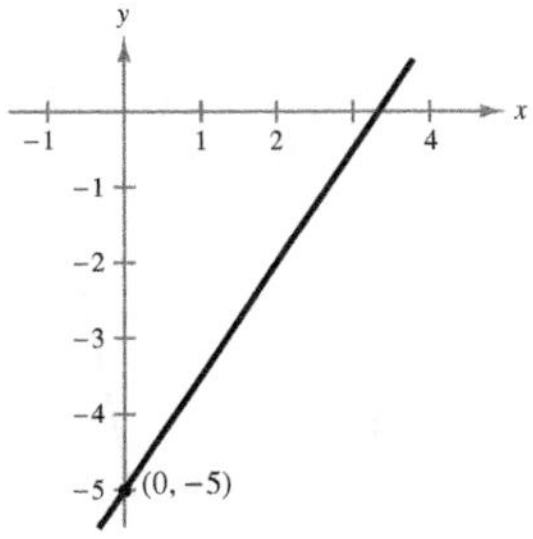

27. $y = 6$

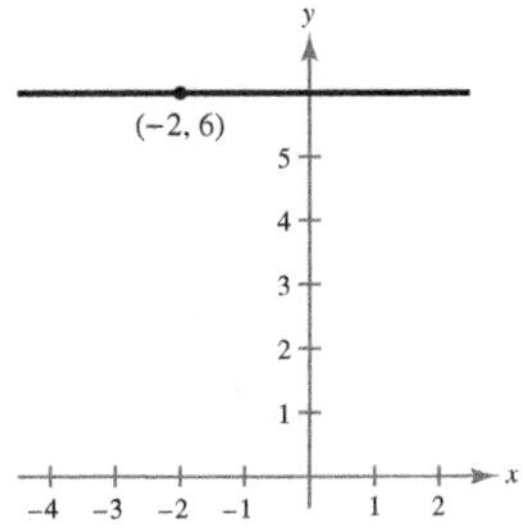

29. Slope: 2
y-intercept: $(0, 4)$

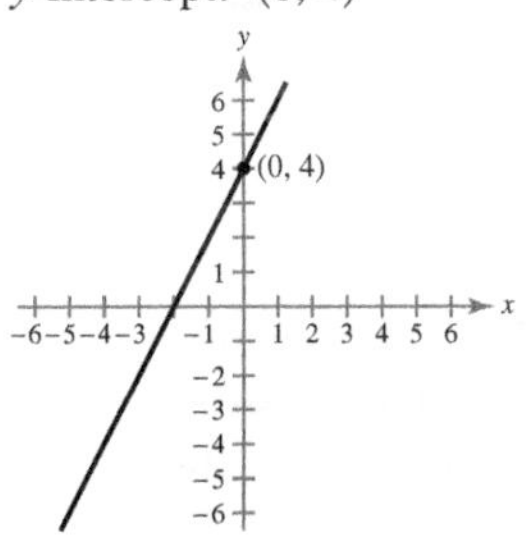

31. Slope: undefined
y-intercept: none

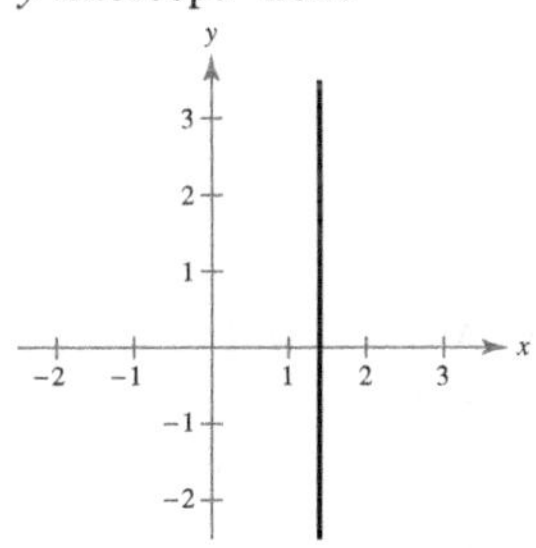

33. Parallel **35.** Neither

37. (a) $y = \frac{1}{2}x + 1$ (b) $y = -2x + 11$

39. (a) $5x - 4y - 23 = 0$ (b) $4x + 5y - 2 = 0$

41. (a) $y = -2$ (b) $x = -1$

43. $y = \frac{7}{3}x$ **45.** $y = 348x$ **47.** $A = 5r$ **49.** $a = \frac{3}{4}b$

51. $y = 0.0084x$; \$1470 **53.** \$3,350,000

55. $V = -12{,}950t + 135{,}000,\ 0 \le t \le 10$

57. y is a function of x. **59.** y *is not* a function of x.

61. Function; Every element of A is assigned to an element of B.

63. (a) -2 (b) -3 (c) -5 (d) $\sqrt{x + 7} - 5$

65. All real numbers x **67.** $x \ge -5$

69. $1 \le t < 4,\ t > 4$

71. The domain of $h(x)$ is all real numbers except $x = 0$, because division by zero is undefined. The domain of $k(x)$ is all real numbers except $x = -2$ and $x = 2$, because if $x = 2$ or $x = -2$, then $x^2 - 4$ equals zero, and division by zero is undefined. When a graphing utility and the *table* feature are used, $h(0)$ results in an error, and $k(-2)$ and $k(2)$ also result in errors.

73. (a) $B = 6500\left(1 + \dfrac{0.0575}{4}\right)^{4t}$ (b) $t \ge 0$

75. (a) Domain: all real numbers
Range: $[1, \infty)$
(b) Decreasing: $(-\infty, 0)$
Increasing: $(0, \infty)$
(c) Even (d) Minimum: $(0, 1)$

77. (a) Domain: all real numbers
Range: all real numbers
(b) Decreasing: $\left(0, \frac{8}{3}\right)$
Increasing: $(-\infty, 0) \cup \left(\frac{8}{3}, \infty\right)$
(c) Neither
(d) Relative minimum: $\left(\frac{8}{3}, -\frac{256}{27}\right)$
Relative maximum: $(0, 0)$

79. y is a function of x. **81.** y is not a function of x.

83.

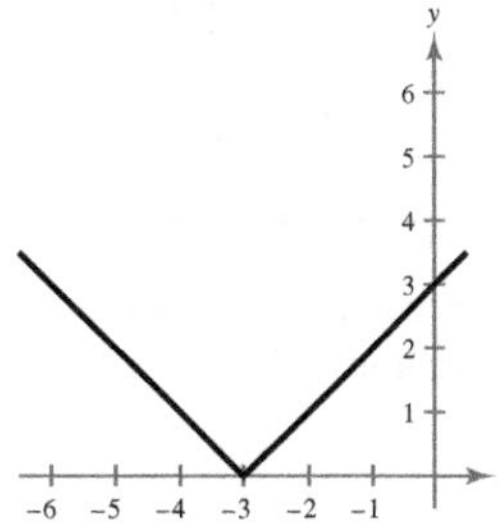

85.

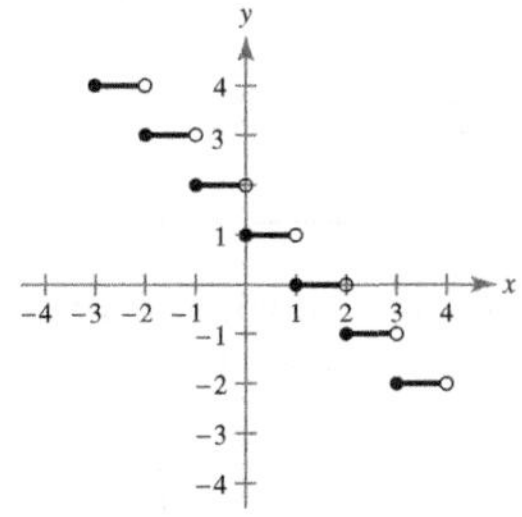

87.

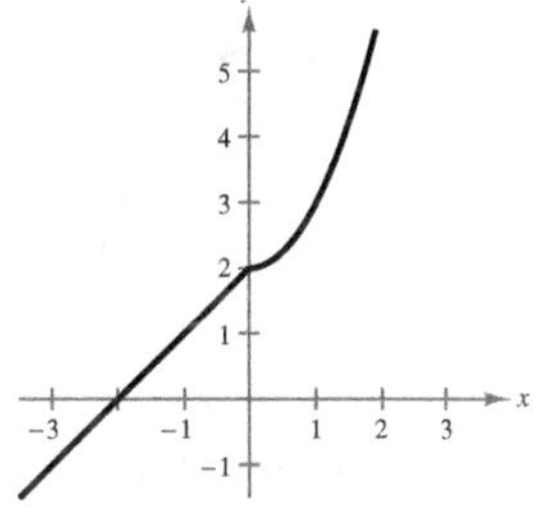

89.

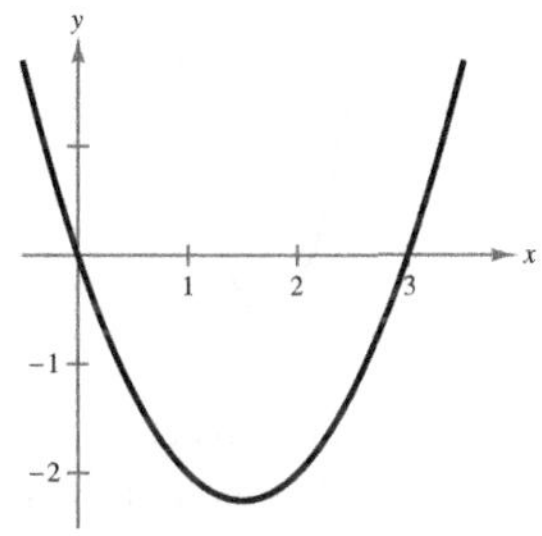

91. Reflected about the x-axis and shifted two units downward and one unit to the right

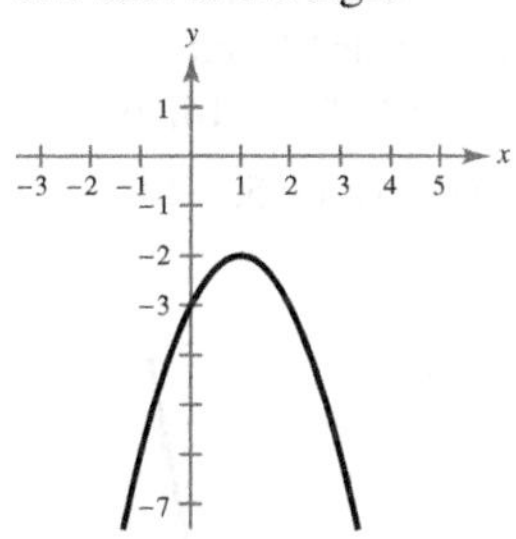

93. Shifted two units to the right

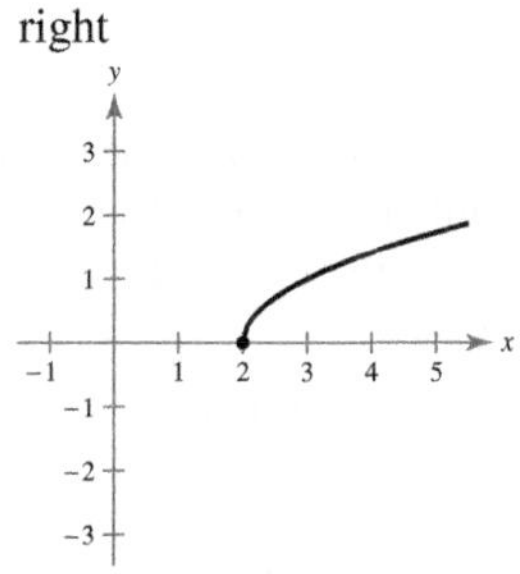

95. Shifted two units to the left

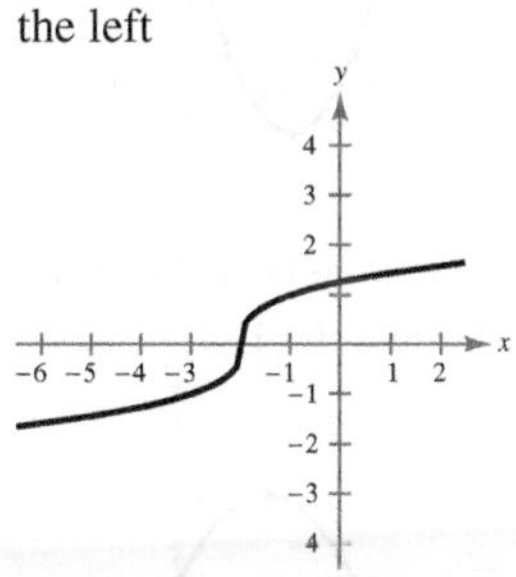

97. Common function: $y = \sqrt{x}$
Transformation: reflected about the y-axis and shifted three units to the right
Equation: $y = \sqrt{3 - x}$ or
$y = \sqrt{-(x - 3)}$

99. (a) $(f + g)(x) = x^2 + 5x - 1$
(b) $(f - g)(x) = -x^2 + x - 1$
(c) $(fg)(x) = 3x^3 + 5x^2 - 2x$
(d) $(f/g)(x) = \dfrac{3x - 1}{x^2 + 2x}$
Domain of f/g: $x < -2,\ -2 < x < 0,\ x > 0$, or $(-\infty, -2) \cup (-2, 0) \cup (0, \infty)$

Chapter 3

Section 3.1 *(page 128)*

Skills Warm Up *(page 128)*

1. $\frac{1}{2}, -6$ **2.** $-\frac{3}{5}, 3$ **3.** $\frac{3}{2}, -1$ **4.** -10

5. $3 \pm \sqrt{5}$ **6.** $-2 \pm \sqrt{3}$ **7.** $4 \pm \frac{\sqrt{14}}{2}$

8. $-5 \pm \frac{\sqrt{3}}{3}$ **9.** $-\frac{3}{2} \pm \frac{\sqrt{5}}{2}$ **10.** $-\frac{3}{2} \pm \frac{\sqrt{21}}{2}$

1. Compared with the graph of $y = x^2$, each output of $f(x) = 5x^2$ vertically stretches the graph by a factor of 5.

3.

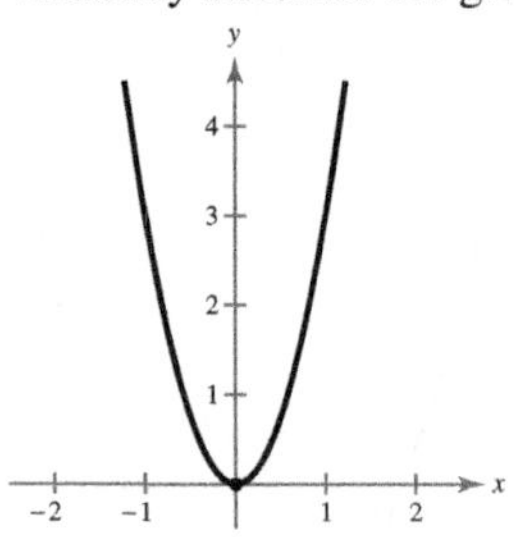

The graph of $f(x) = 3x^2$ is a vertical stretch of the graph of $y = x^2$ by a factor of 3.

5.

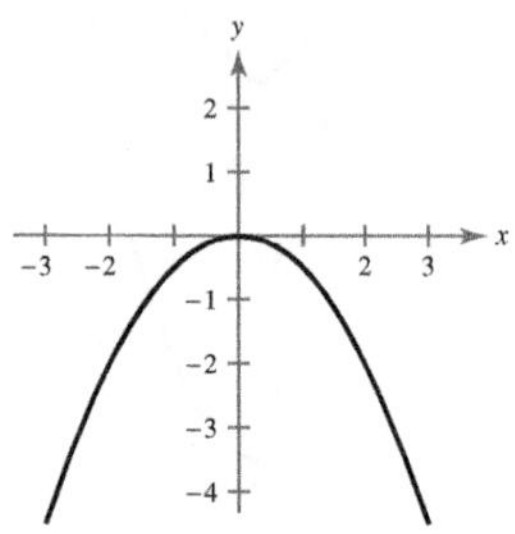

The graph of $g(x) = -\frac{1}{2}x^2$ is a reflection in the x-axis of the graph of $y = x^2$ and a vertical shrink by a factor of $\frac{1}{2}$.

7. The graph of f is the graph of $y = x^2$ reflected in the x-axis, shifted to the left 1 unit and shifted upward 1 unit.

9. e **10.** c **11.** a **12.** d **13.** f **14.** b

15. $f(x) = -x^2 + 16$
Vertex: $(0, 16)$

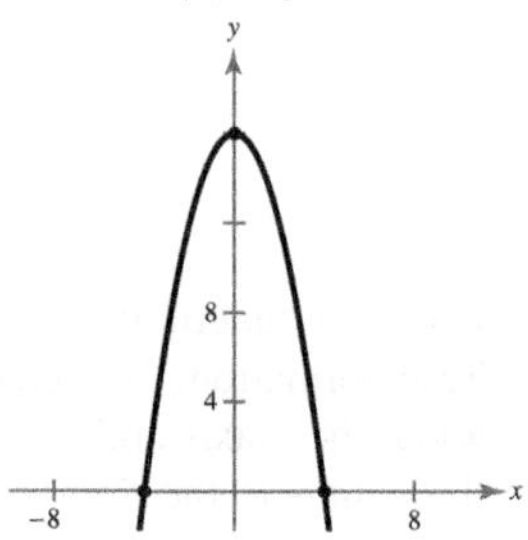

17. Vertex: $(-5, -6)$

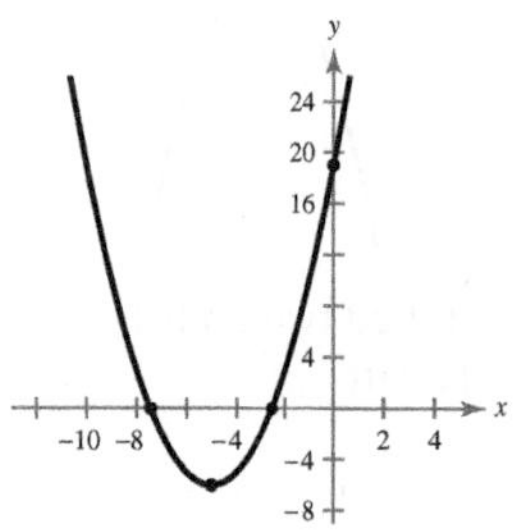

19. $g(x) = (x + 1)^2$
Vertex: $(-1, 0)$

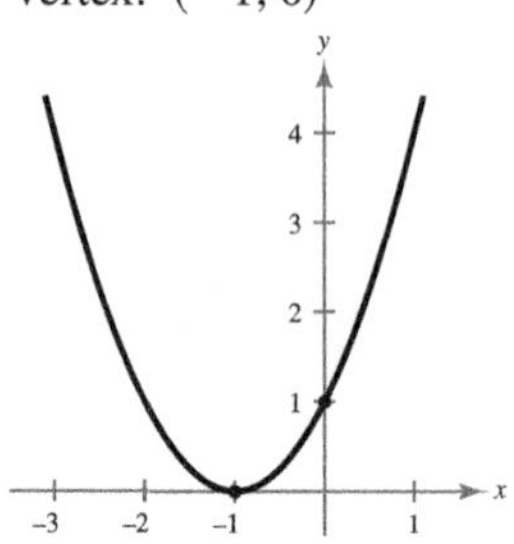

21. $f(x) = -(x + 1)^2 + 4$
Vertex: $(-1, 4)$

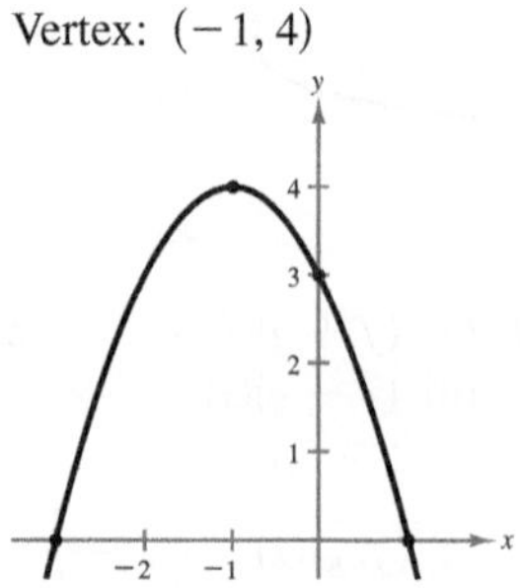

23. $f(x) = \left(x - \frac{1}{2}\right)^2 + 1$
Vertex: $\left(\frac{1}{2}, 1\right)$

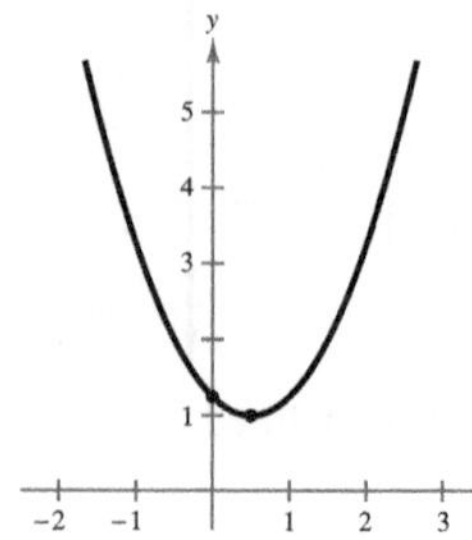

25. $f(x) = -(x - 1)^2 + 6$
Vertex: $(1, 6)$

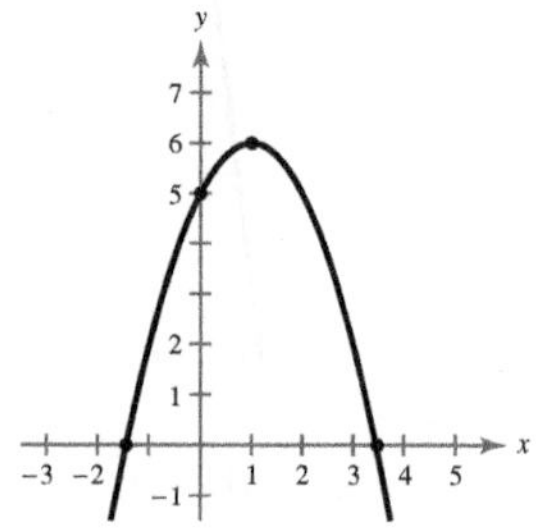

27. $h(x) = 4\left(x - \frac{1}{2}\right)^2 + 20$
Vertex: $\left(\frac{1}{2}, 20\right)$

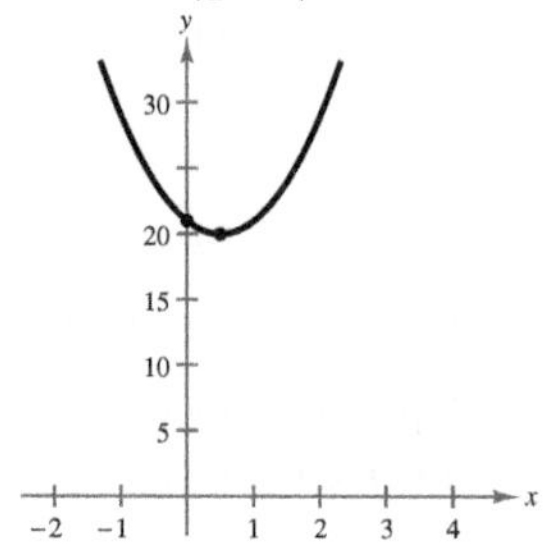

29. $f(x) = \frac{1}{4}(x - 8)^2 - 8$
Vertex: $(8, -8)$

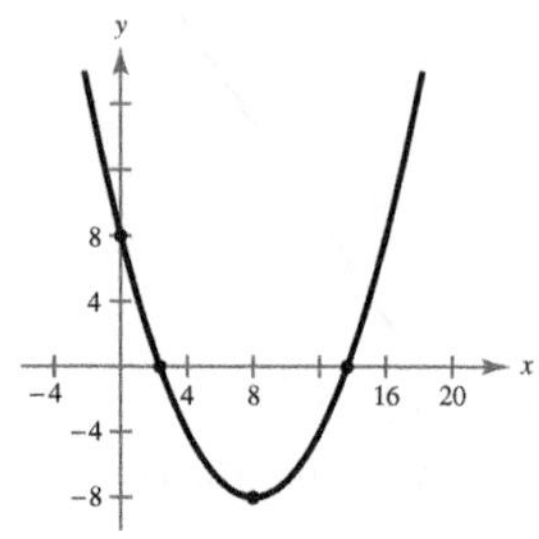

31. $y = -(x + 2)^2$ **33.** $y = (x - 3)^2 - 9$

35. $y = -2(x + 3)^2 + 3$

37. $y = -\frac{1}{2}(x - 2)^2 - 1$ **39.** $y = \frac{3}{4}(x - 5)^2 + 12$

41. Answers will vary.
Sample answer:
$f(x) = x^2 - x - 2$
$g(x) = -x^2 + x + 2$

43. Answers will vary.
Sample answer:
$f(x) = x^2 - 10x$
$g(x) = -x^2 + 10x$

45. Answers will vary.
Sample answer:
$f(x) = 2x^2 + 7x + 3$
$g(x) = -2x^2 - 7x - 3$

47. $A = 100x - x^2$; $(50, 2500)$; The rectangle has the greatest area ($A = 2500$ square feet) when its width is 50 feet.

49. $x = 150$ feet, $y = 200$ feet; 300 feet × 200 feet

51. 25,000 units **53.** 20 fixtures **55.** 14 feet

57. (a) and (b)

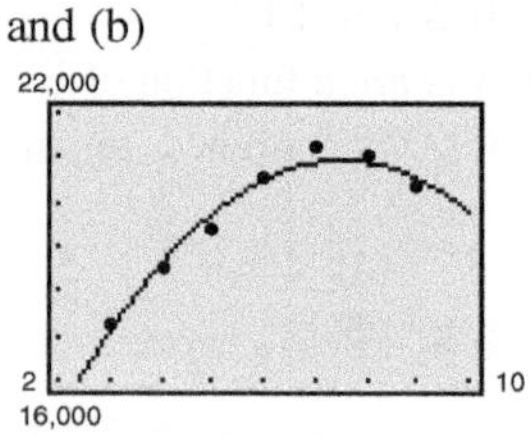

$R = -186.43t^2 + 2811.9t + 10{,}329$

(c) 2007 ($t \approx 7.5$); Yes, this agrees with the actual data.

59. (a)

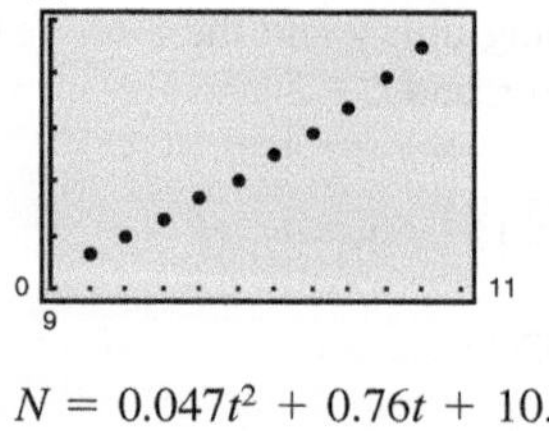

(b) $N = 0.047t^2 + 0.76t + 10.1$

(c) 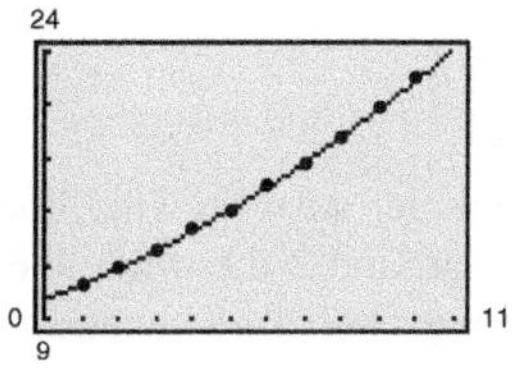

(d) 25,988 physicians

61. (a) 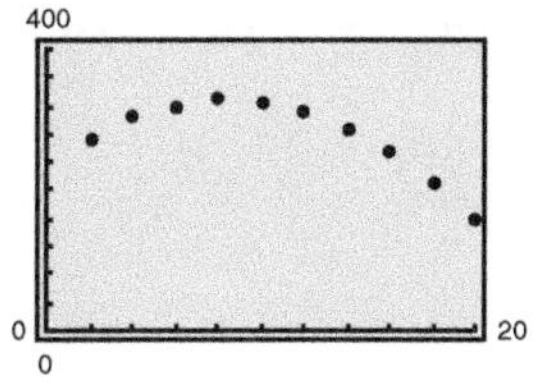

(b) $p(x) = -1.276x^2 + 21.63x + 236.5$

(c) 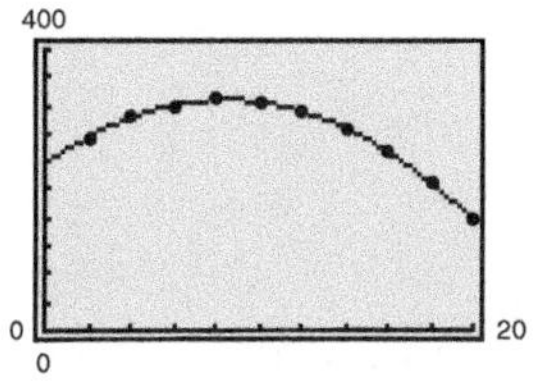

(d) $\approx (8.5, 328.2)$; Producing about 85,000 units yields the maximum profit, about $32,820,000.

(e) Sample answer: Production costs may be growing faster than revenue, so profit decreases.

63. $f(x) = a\left(x + \dfrac{b}{2a}\right)^2 + \dfrac{4ac - b^2}{4a}$

Review Exercises *(page 132)*

1. 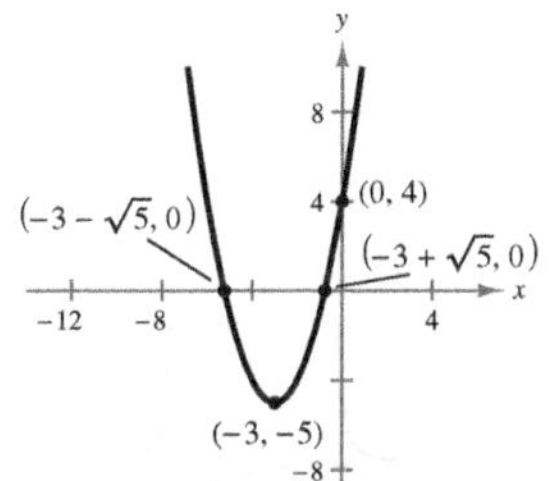

Vertex: $(-3, -5)$
Intercepts:
$(0, 4), (-3 \pm \sqrt{5}, 0)$

3. $h(x) = 3(x - 2)^2 - 1$

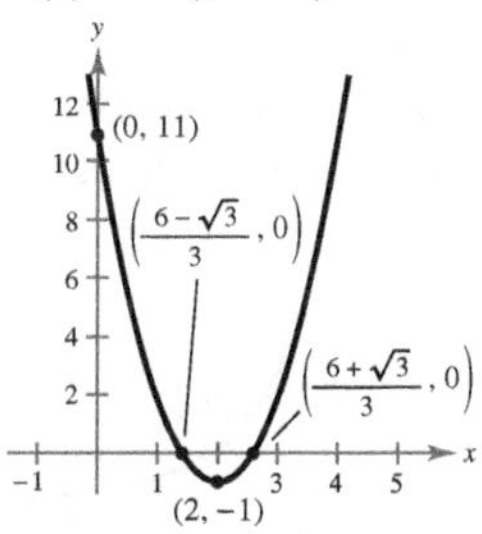

Vertex: $(2, -1)$
Intercepts:
$(0, 11), \left(\dfrac{6 \pm \sqrt{3}}{3}, 0\right)$

5. $f(x) = \frac{7}{9}(x + 5)^2 - 1$

7. $A(x) = 250x - x^2$; 125 feet × 125 feet

9. (a) 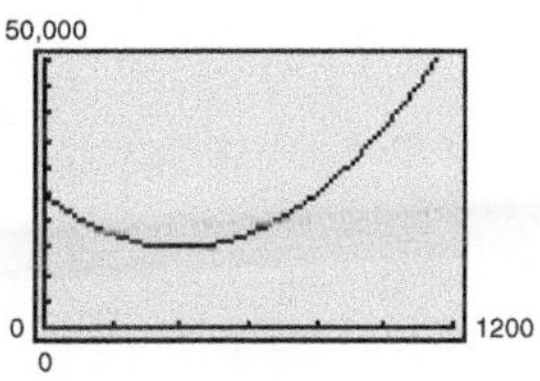

(b) About 385 units

(c) Write the equation of the quadratic function in standard form. The vertex is the minimum cost.

11. (a) 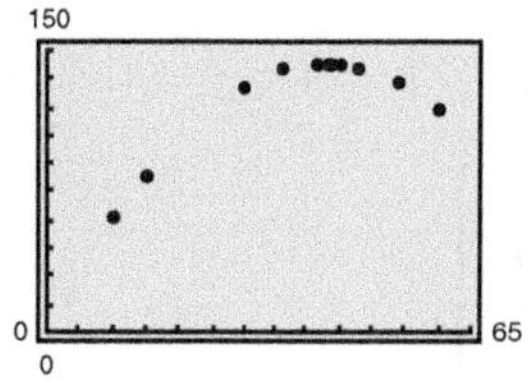

(b) $d(x) = -0.077x^2 + 6.59x + 2.4$

(c) 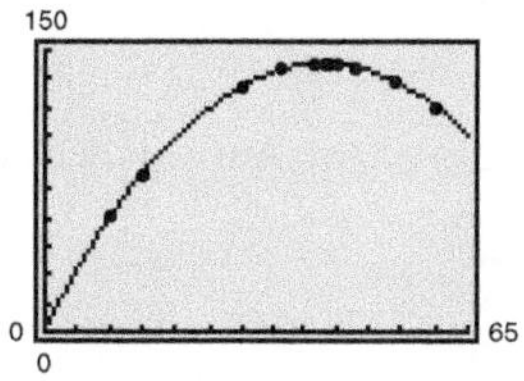

(d) (42.8, 143.4); The vertex gives the angle 42.8°, which results in the greatest distance of 143.4 meters.

13. 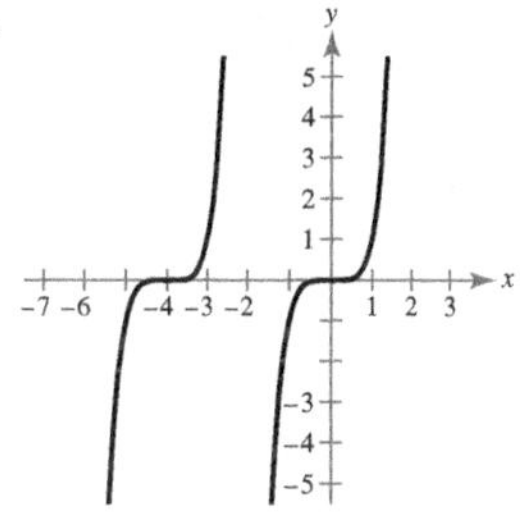

15. 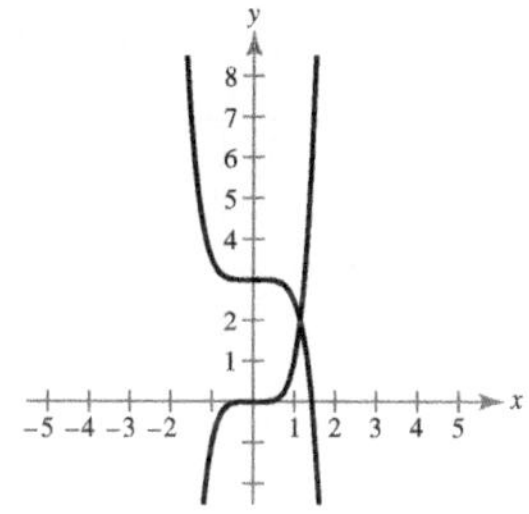

17. 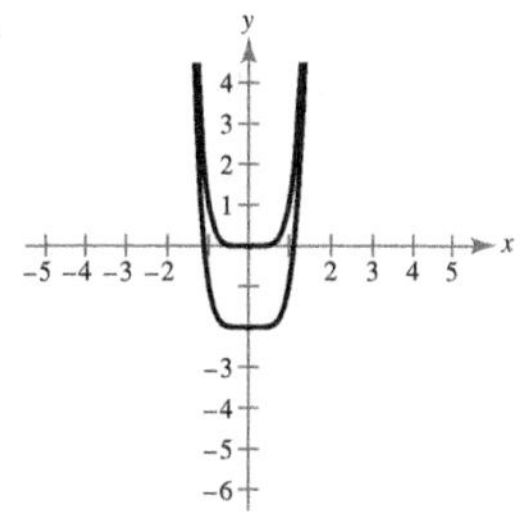

19. 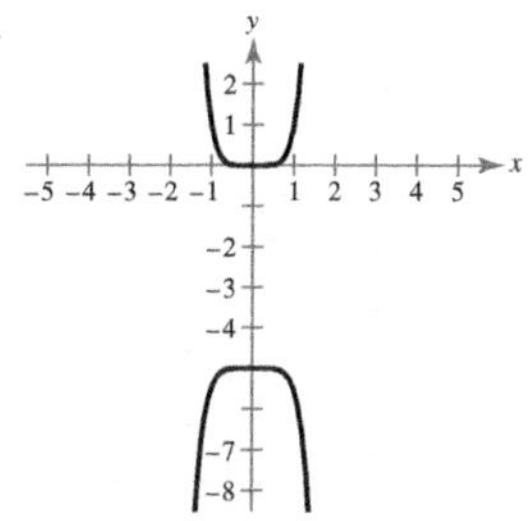

21. Falls to the left
Rises to the right

23. Falls to the left
Falls to the right

25. ± 4

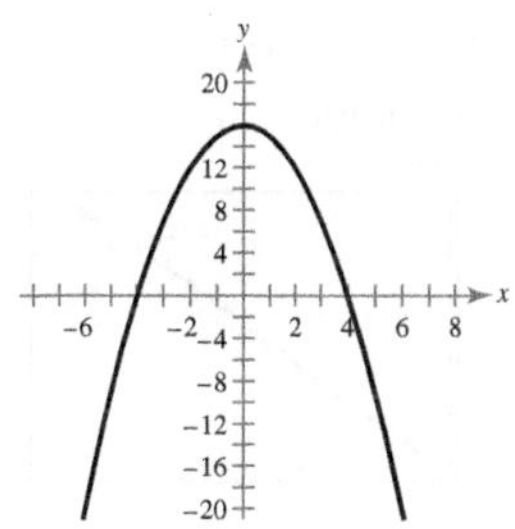

27. 0, 2, 5

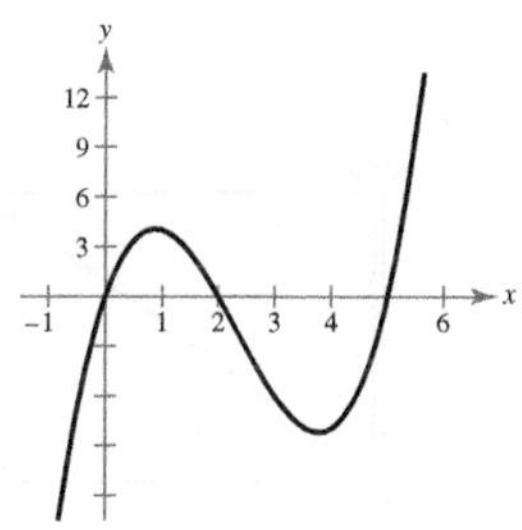

29. $x^2 - 3x + 1 - \dfrac{1}{2x + 1}$

31. $x^2 - 3x + 3$

33. (a) -10 (b) 11

CHAPTER 3

35. (a) Proof (b) $x - 2$
(c) $(x - 5)(x + 3)(x - 2)$ (d) $-3, 2, 5$
(e)

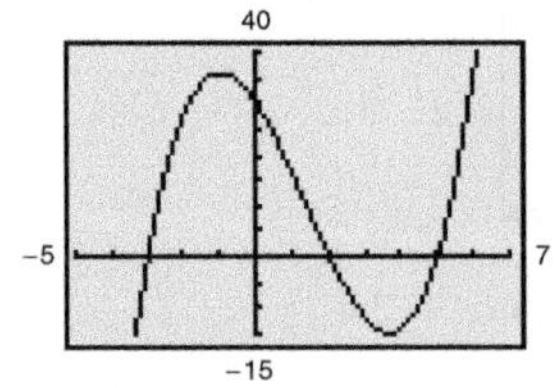

37. $x^2 + 11x + 24$

39. (a)

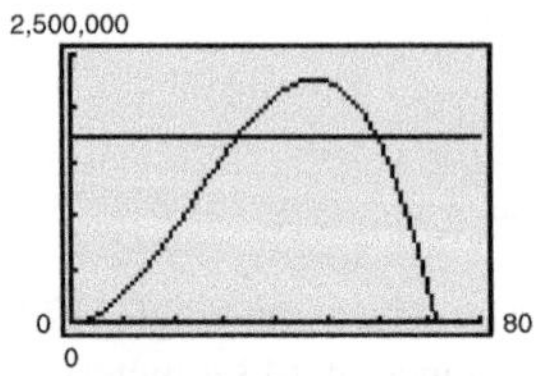

$\approx \$325,167$

(b) Answers will vary.

41. $\pm1, \pm3, \pm5, \pm15, \pm\frac{1}{2}, \pm\frac{3}{2}, \pm\frac{5}{2}, \pm\frac{15}{2}, \pm\frac{1}{4}, \pm\frac{3}{4}, \pm\frac{5}{4}, \pm\frac{15}{4}$

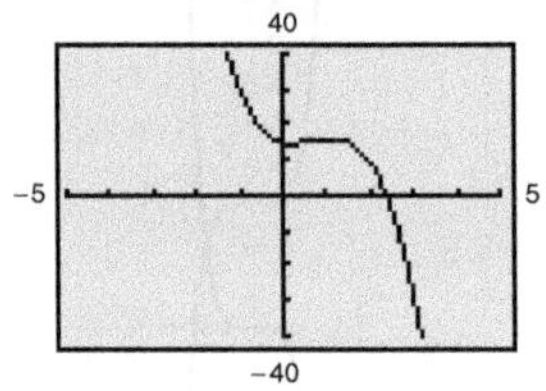

From the graph: $x \approx 2.357$, so the zero is not rational.

43. $-3, -1, 2$ **45.** $\pm2, \pm\sqrt{5}$ **47.** $-\frac{1}{3}, \frac{3}{2}, 2$

49. $-1, 2$ **51.** $x \approx -2.3$ **53.** $-1.321, -0.283, 1.604$

55. (a)

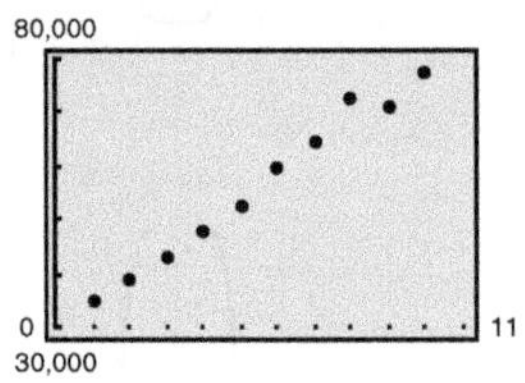

(b) Linear: $R = 4996.7t + 28,927$
Quadratic: $R = -36.14t^2 + 5394.3t + 28,132$
Quartic: $R = 3.4725t^4 - 134.788t^3 + 1486.42t^2 - 577.4t + 34,333$

(c) Linear

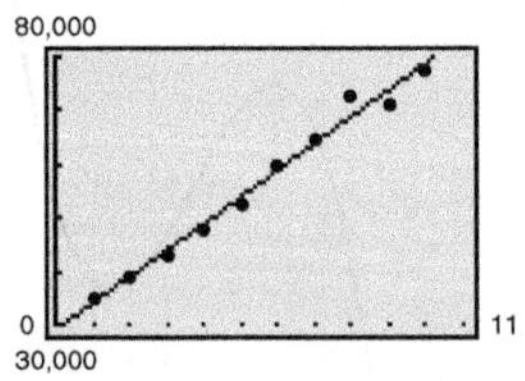

Quadratic

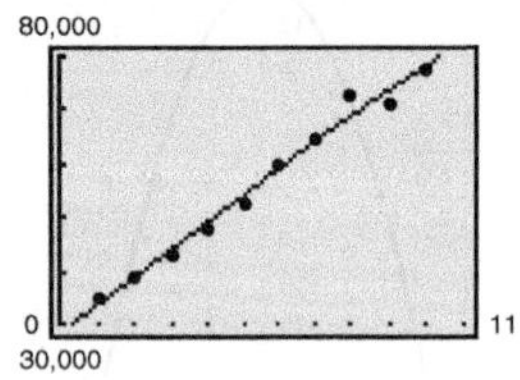

Quartic

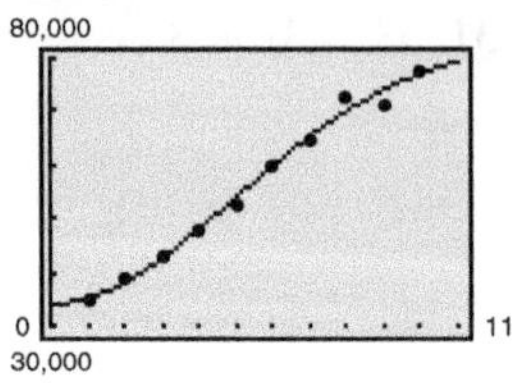

Each model could be considered a good fit for the data.

(d) Linear: 2011 ($t \approx 11.6$)
Quadratic: 2011 ($t \approx 11.9$)
Quartic: 2018 ($t \approx 18.8$)
Answers will vary. Sample answer: The linear and quadratic predictions are around 2011, while the quartic model levels off for a few years before increasing.

57. $4\sqrt{2}i; -4\sqrt{2}i$ **59.** $-3 + 4i; -3 - 4i$ **61.** $5 + 2i$

63. 89 **65.** $-10 - 8i$ **67.** $-7 + 24i$ **69.** 10

71. $3 - 2i$ **73.** $-3 - 4i$

75. $\dfrac{1 \pm \sqrt{23}i}{4}$ **77.** $\dfrac{-11 \pm \sqrt{73}}{8}$

79.

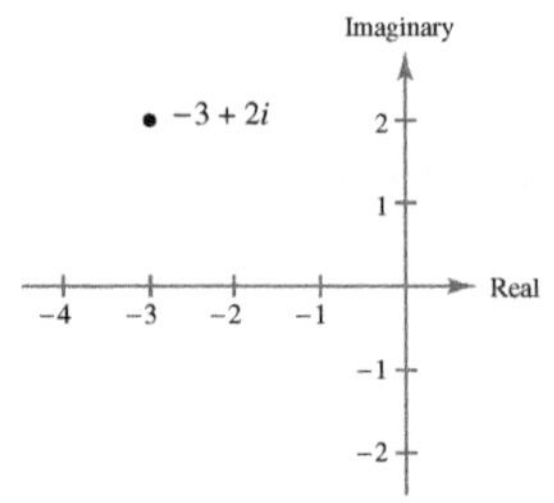

81. $\pm3, \pm3i; (x - 3)(x + 3)(x - 3i)(x + 3i)$

83. $-5, \pm\sqrt{3}i; (t + 5)(t - \sqrt{3}i)(t + \sqrt{3}i)$

85. $2, \pm\frac{3i}{2}; (x - 2)(2x - 3i)(2x + 3i)$

87. Answers will vary. Sample answer: $x^3 - x^2 + 9x - 9$

89. (a) $(x^2 + 8)(x^2 - 3)$ (b) $(x^2 + 8)(x - \sqrt{3})(x + \sqrt{3})$
(c) $(x + 2\sqrt{2}i)(x - 2\sqrt{2}i)(x - \sqrt{3})(x + \sqrt{3})$

91. $\pm4i, \frac{1}{4}$ **93.** $-1 \pm 3i, -1, -4$

95. Domain: All $x \neq -2$
Vertical asymptote: $x = -2$
Horizontal asymptote: $y = 0$

97. Domain: All $x \neq \pm3$
Vertical asymptotes: $x = -3, x = 3$
Horizontal asymptote: $y = 2$

99.

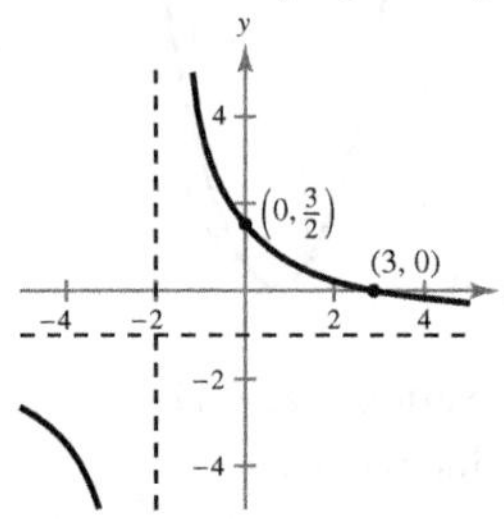

101.

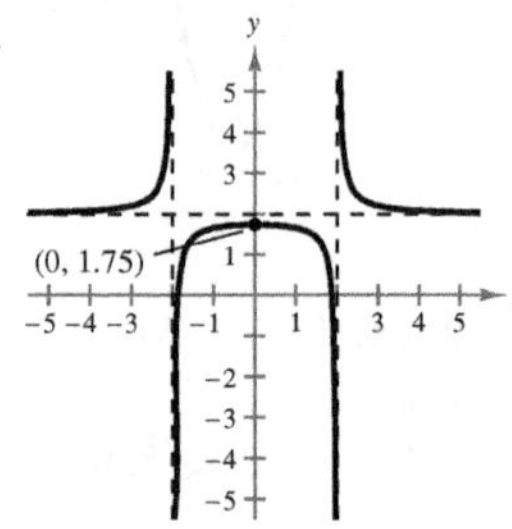

103. None

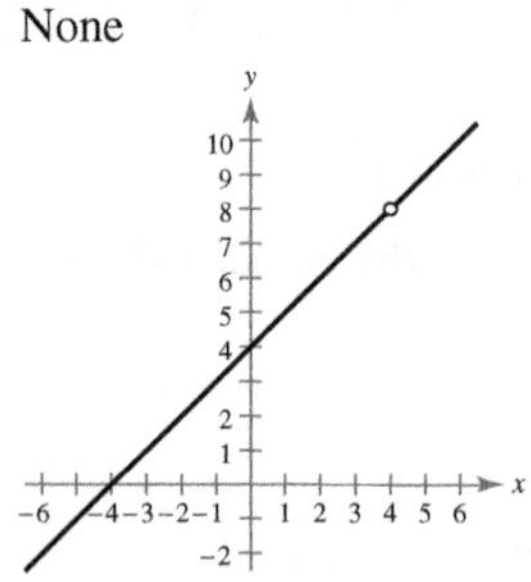

105. (a) 430,769 fish; 662,500 fish; 1,024,000 fish
(b) 1,666,667 fish

107. (a) \$35,000 (b) \$157,500

Test Yourself *(page 136)*

1. 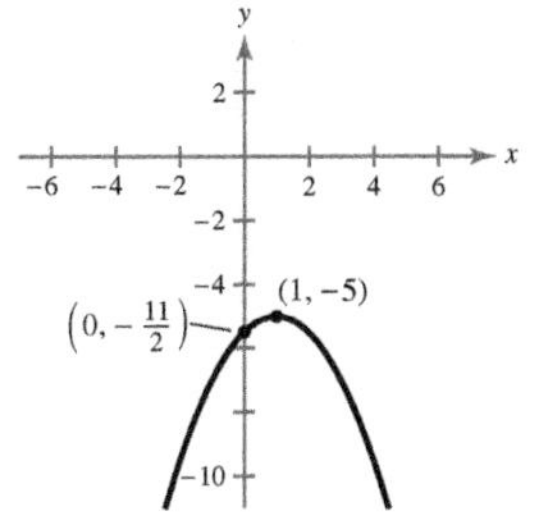

Vertex: $(1, -5)$
Intercept: $\left(0, -\frac{11}{2}\right)$

2. (a) Falls to the left; Rises to the right (b) Rises to the left; Rises to the right

3. $x^2 + 7x + 12$

4. $\pm 1, \pm 3, \pm 9, \pm 27, \pm\frac{1}{4}, \pm\frac{1}{2}, \pm\frac{3}{4}, \pm\frac{3}{2}, \pm\frac{9}{4}, \pm\frac{9}{2}, \pm\frac{27}{4}, \pm\frac{27}{2}$
$f\left(\frac{3}{2}\right) = 0$
$f\left(-\frac{3}{2}\right) = 0$
$f(x) = \left(x - \frac{3}{2}\right)\left(x + \frac{3}{2}\right)(x - 1)(x - 3)$

5. (a) 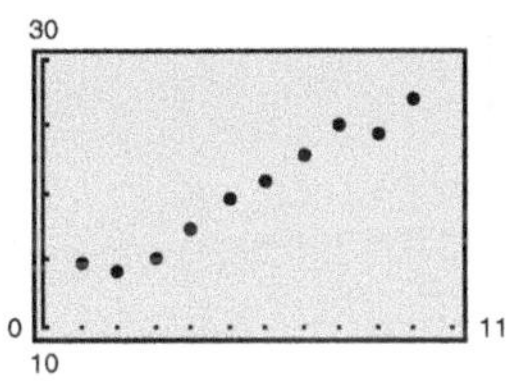

(b) Linear: $S = 1.517t + 11.81$
Quadratic: $S = 0.024t^2 + 1.253t + 12.34$
Quartic:
$S = 0.010739t^4 - 0.26789t^3 + 2.2749t^2 - 5.879t + 18.74$

(c) Linear

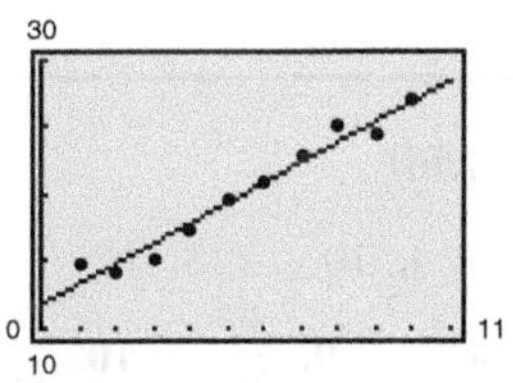

Quadratic

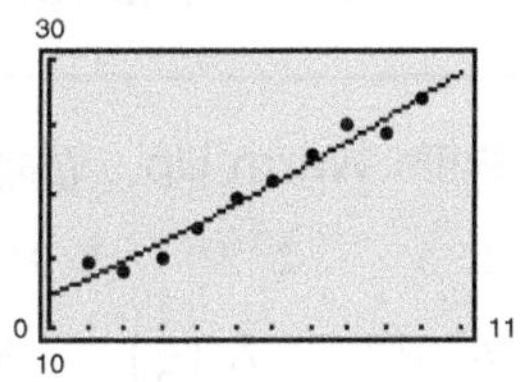

Quartic

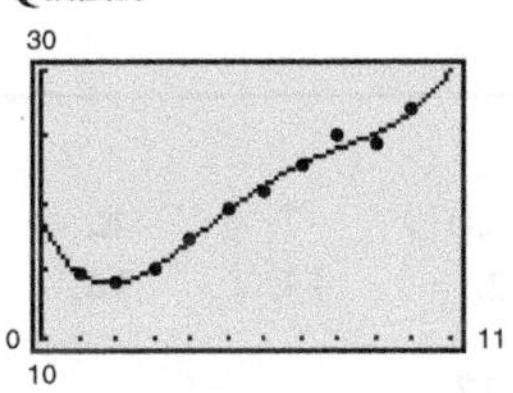

Each model could be considered a good fit for the data.

(d) Linear: 2015 ($t \approx 15.29$)
Quadratic: 2014 ($t \approx 14.21$)
Quartic: 2011 ($t \approx 11.9$)
Answers will vary. Sample answer: All three models give predictions that reflect the increasing data. The quartic model increases most rapidly, while the linear and quadratic models increase at slower rates.

6. $16 - 2i$ 7. $7 - 10i$ 8. $27 - 4\sqrt{3}i$ 9. $23 - 14i$

Chapter 4

Section 4.1 *(page 145)*

Skills Warm Up *(page 145)*

1. All real numbers 2. $[-1, \infty)$
3. All real numbers except $x = 0, 2$
4. All real numbers except $x = -\frac{5}{3}$ 5. x 6. x
7. x 8. x 9. $x = \frac{3}{2}y + 3$ 10. $x = \frac{y^3}{2} + 2$

1. $f^{-1} = \{(4, 1), (5, 2), (6, 3), (7, 4)\}$
3. $f^{-1} = \{(1, -1), (2, -2), (3, -3), (4, -4)\}$
5. $f^{-1}(x) = \frac{1}{2}x$ 7. $f^{-1}(x) = x + 5$
9. $f(g(x)) = 5\left(\frac{x - 1}{5}\right) + 1 = x;\ g(f(x)) = \frac{(5x + 1) - 1}{5} = x$
11. $f(g(x)) = \left(\sqrt[3]{x}\right)^3 = x;\ g(f(x)) = \sqrt[3]{x^3} = x$
13. $f(g(x)) = \sqrt{(x^2 + 4) - 4} = x;\ g(f(x)) = \left(\sqrt{x - 4}\right)^2 + 4 = x$
15. $f(g(x)) = 1 - \left(\sqrt[3]{1 - x}\right)^3 = x;\ g(f(x)) = \sqrt[3]{1 - (1 - x^3)} = x$

17.

x	0	1	2	3	4
$f^{-1}(x)$	−2	0	1	2	4

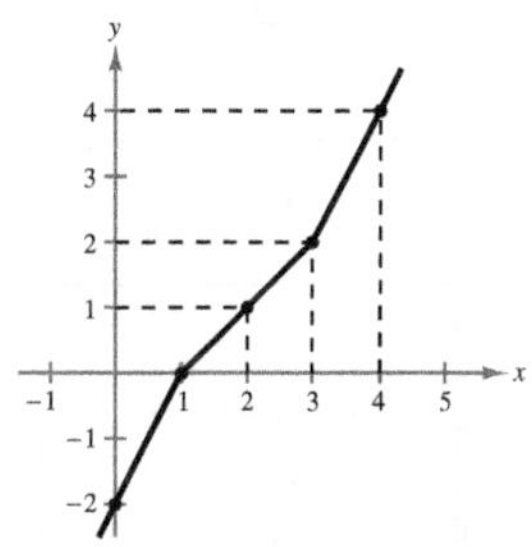

19. f doesn't have an inverse function. 21. $g^{-1}(x) = 8x$
23. p doesn't have an inverse function.
25. $f^{-1}(x) = \sqrt{x} - 3,\ x \geq 0$
27. $h^{-1}(x) = \frac{1}{x}$ 29. $f^{-1}(x) = \frac{x^2 - 3}{2},\ x \geq 0$
31. g doesn't have an inverse function.
33. $f^{-1}(x) = -\sqrt{25 - x},\ x \leq 25$

35. 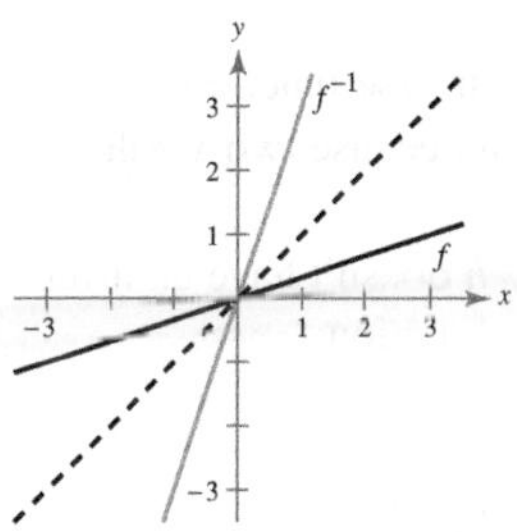

37. 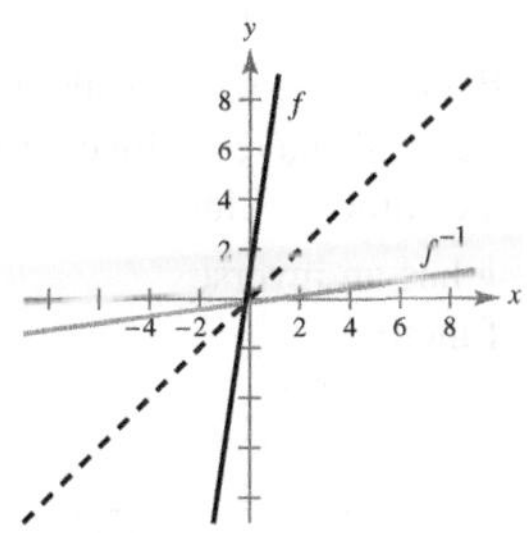

CHAPTER 3

x	-3	0	3
$f(x)$	-1	0	1

x	-1	0	1
$f^{-1}(x)$	-3	0	3

x	-1	0	1
$f(x)$	-6	1	8

x	-6	1	8
$f^{-1}(x)$	-1	0	1

39.

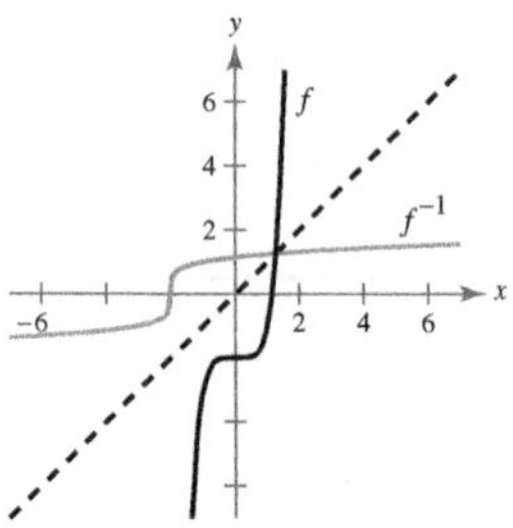

x	-1	0	1
$f(x)$	-3	-2	-1

x	-3	-2	-1
$f^{-1}(x)$	-1	0	1

41. Error: f^{-1} does not mean to take the reciprocal of $f(x)$.
$f^{-1}(x) = \dfrac{x^2 + 5}{2}$

43. $f^{-1}(x) = \dfrac{x + 3}{2}$

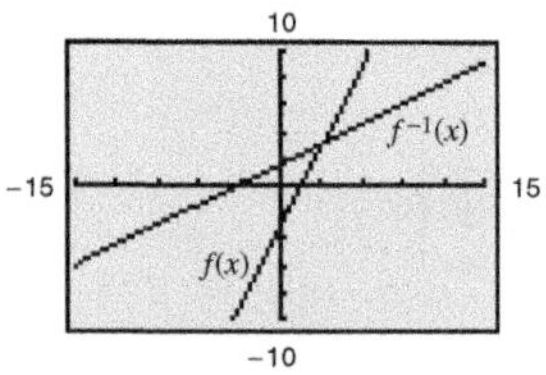

45. $f^{-1}(x) = \sqrt[5]{x}$

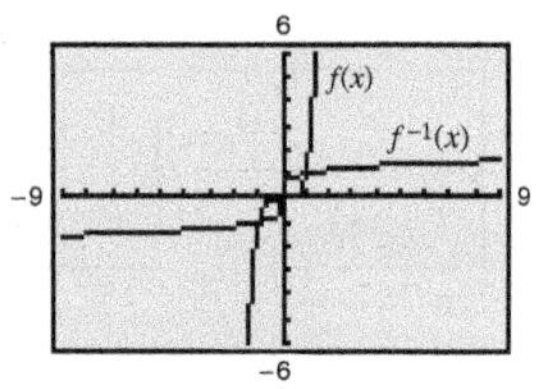

47. $f^{-1}(x) = x^2$, $x \geq 0$

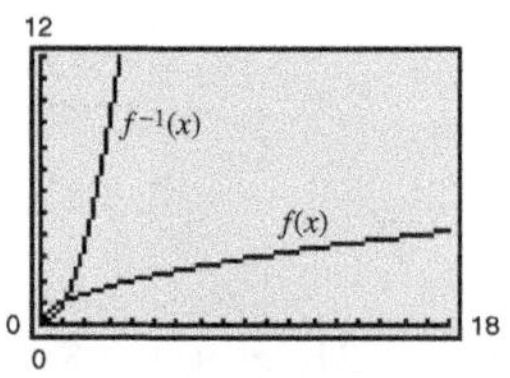

49. $f^{-1}(x) = \sqrt{16 - x^2}$, $0 \leq x \leq 4$

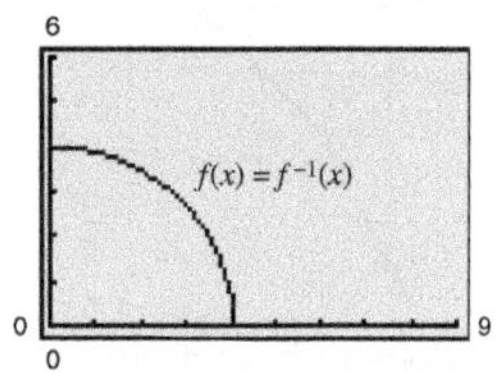

51. $f^{-1}(x) = x^3 - 2$

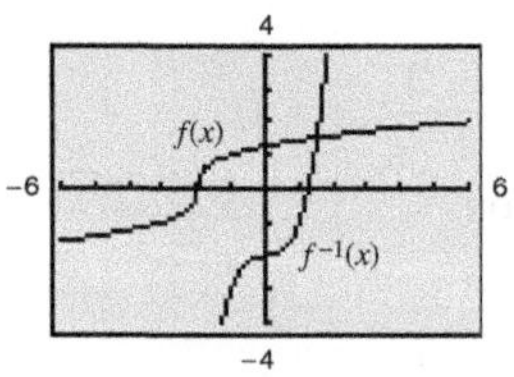

53. Because f is one-to-one, f has an inverse function.

55. f doesn't have an inverse function because two x-values share the same y-value.

57. g has an inverse function.

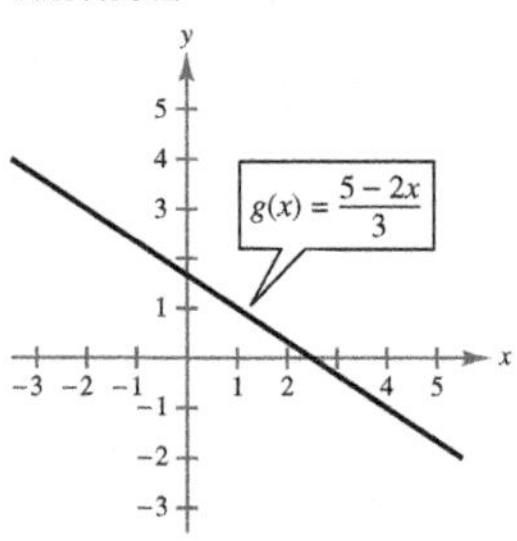

59. h doesn't have an inverse function.

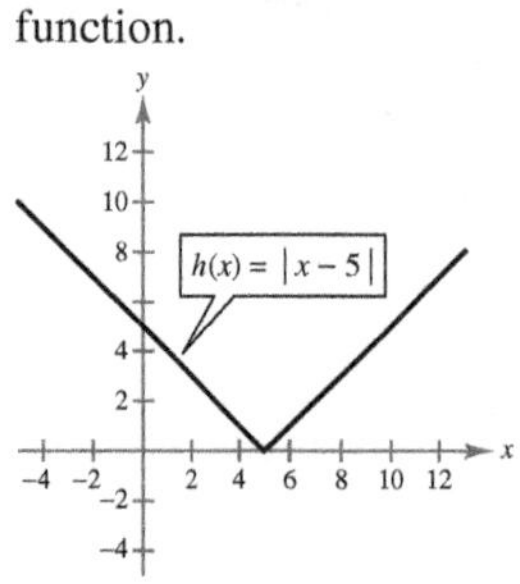

61. f doesn't have an inverse function.

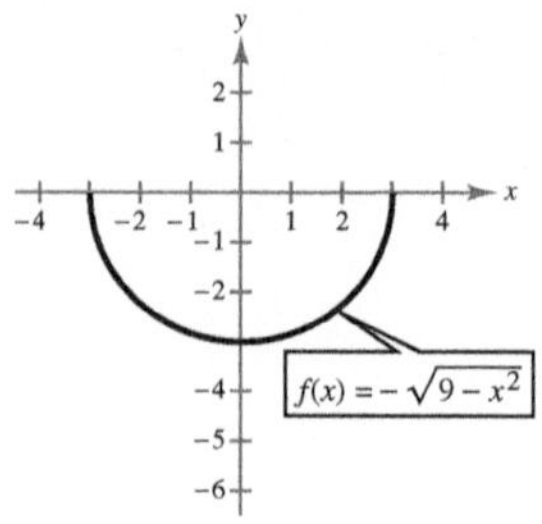

63. $(g^{-1} \circ f^{-1})(x) = \dfrac{x + 2}{2}$ **65.** $(f \circ g)^{-1}(x) = \dfrac{x + 2}{2}$

67. 32 **69.** 600

71. $P^{-1}(x) = \dfrac{x + 5736}{47}$; $P^{-1}(x)$ computes the number of units that must be sold to obtain a profit of x.
Domain of P: $[0, \infty)$
Domain of P^{-1}: $[-5736, \infty)$

73. (a)

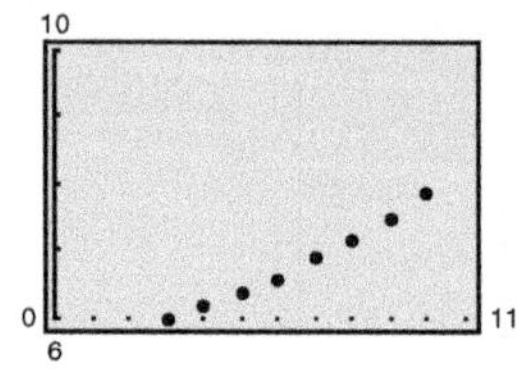

(b) $y = 0.263t + 5.12$

(c) $y^{-1} = \dfrac{t - 5.12}{0.263}$; y^{-1} represents the year in which the average price is t dollars.

(d) 2014 ($y^{-1} \approx 14.75$)

75. $f^{-1}(x) = \sqrt[3]{\dfrac{x^2 + 0.021}{0.0209}}$; \$3.49

Section 4.2 *(page 156)*

Skills Warm Up *(page 156)*

1. 5^x **2.** 3^{2x} **3.** 4^{3x} **4.** 10^x **5.** 4^{2x}
6. 4^{10x} **7.** $\left(\dfrac{3}{2}\right)^x$ **8.** 4^{3x} **9.** $\dfrac{1}{2^x}$ **10.** $\dfrac{3^x}{5^x}$
11. 2^x **12.** 3^x

1. 3.463 **3.** 94.818 **5.** 0.079 **7.** g **8.** e
9. b **10.** h **11.** d **12.** a **13.** f **14.** c

15.

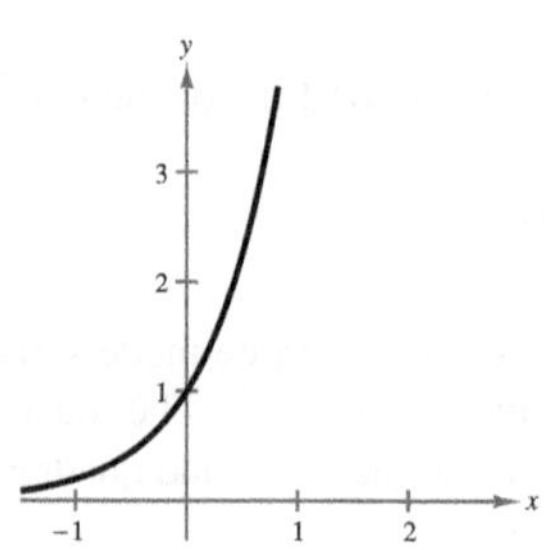

17.

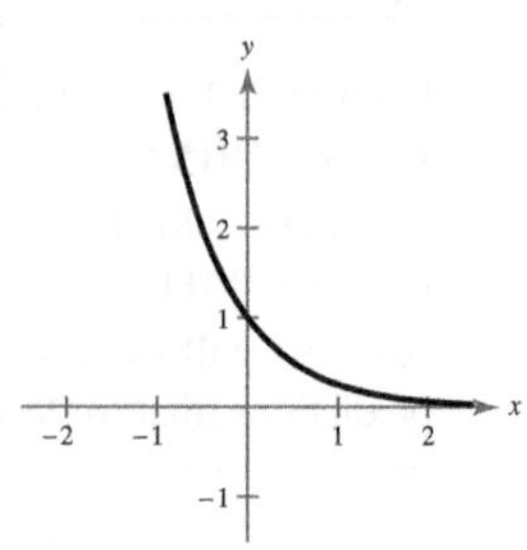

19.

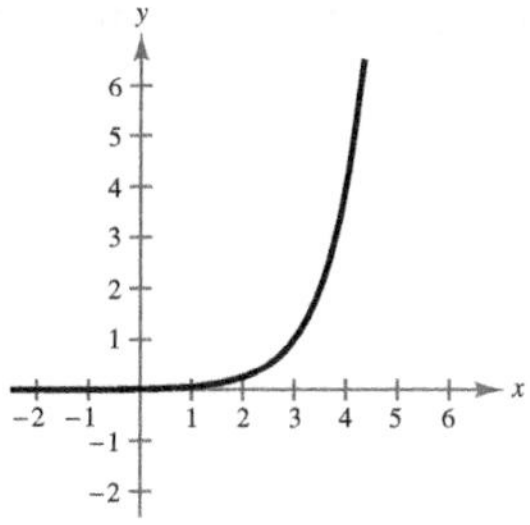

21.

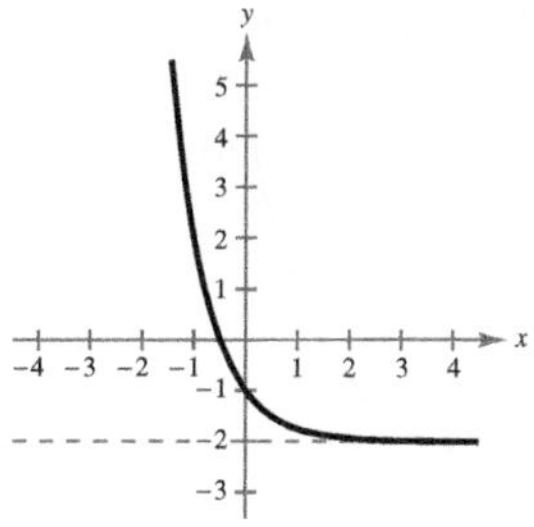

23.

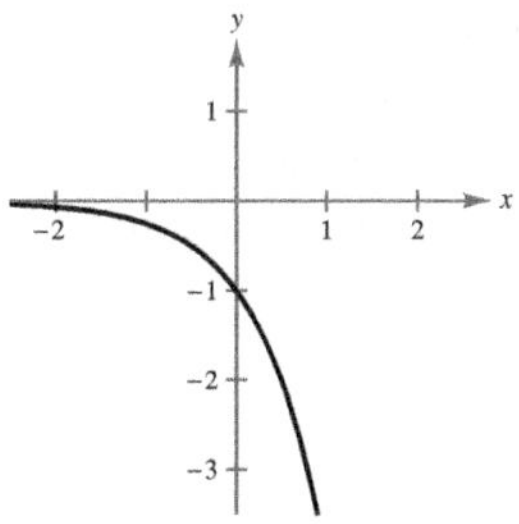

25.

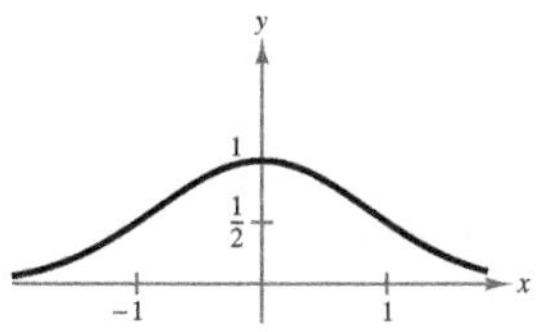

27. 54.598 **29.** 0.067

31.

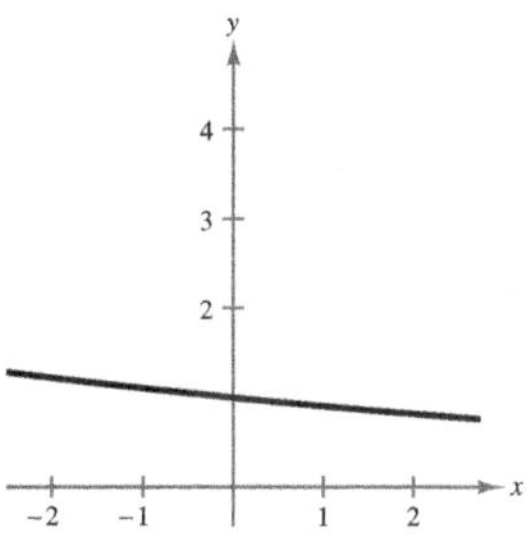

33.

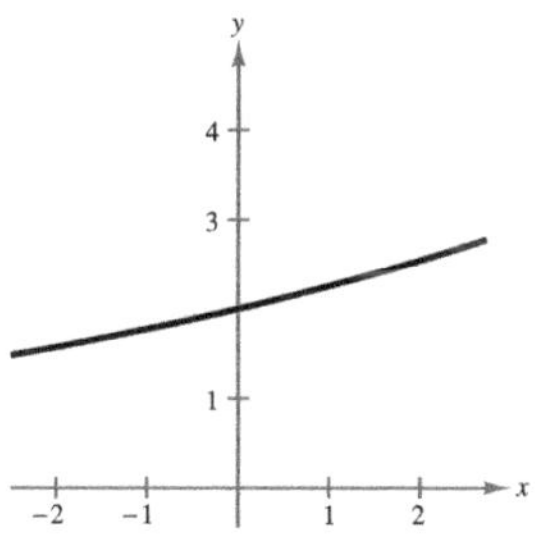

35.

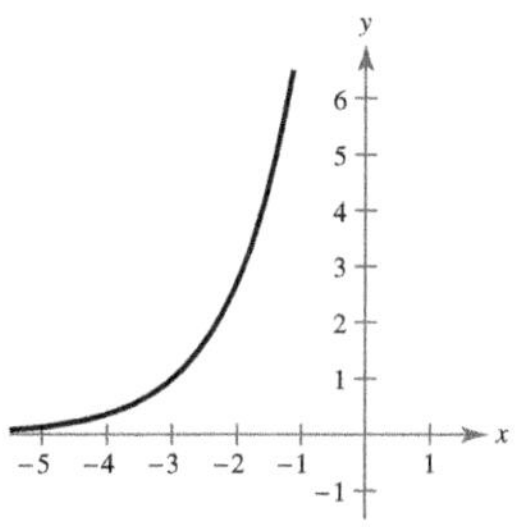

37.

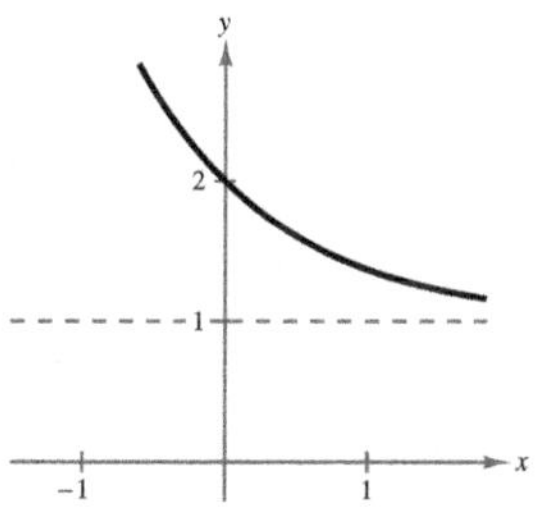

39. Yes. For the natural exponential function, $f(x) = e^x$, the base is $a = e$.

41.

n	1	2	4
A	\$7346.64	\$7401.22	\$7429.74

n	12	365	Continuous
A	\$7449.23	\$7458.80	\$7459.12

43.

n	1	2	4
A	\$6633.24	\$6712.66	\$6753.71

n	12	365	Continuous
A	\$6781.60	\$6795.24	\$6795.70

45. The account paying 5% interest compounded quarterly earns more money. Even though the interest is compounded less frequently, the higher interest rate yields a higher return.

47. \$7424.70 **49.** \$12,434.43

51.

t	1	10	20
P	\$90,521.24	\$36,940.70	\$13,646.15

t	30	40	50
P	\$5040.98	\$1862.17	\$687.90

53.

t	1	10	20
P	\$91,393.12	\$40,656.97	\$16,529.89

t	30	40	50
P	\$6720.55	\$2732.37	\$1110.90

55. \$19,691.17 **57.** \$147,683.76

59. \$20,700.76 **61.** \$155,255.66

63. (a) \$182.91 (b) \$29.58

(c)

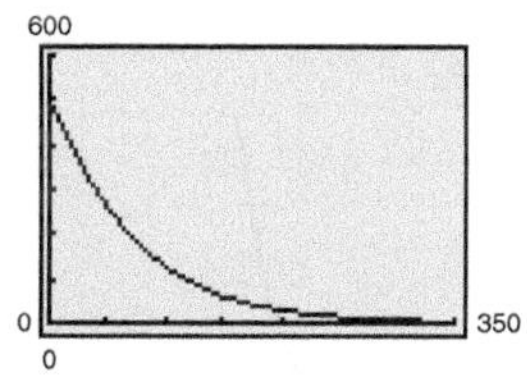

(d) ≈ 117 books

65. (a) 100 (b) ≈ 110 (c) ≈ 121 (d) ≈ 158

67. (a) 5907 (b) 6237 (c) 6767 (d) 7753

69. ≈ 30.42 kilograms

71. (a)

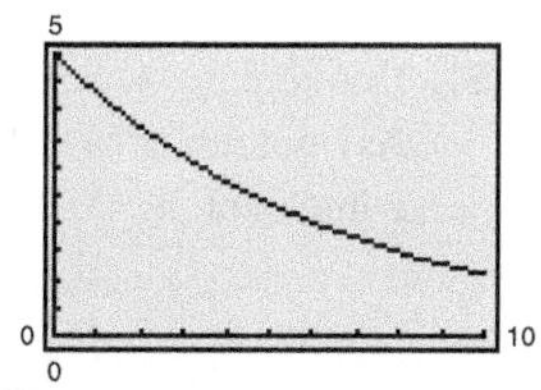

(b) ≈ 1.11 pounds

(c) On the graph, when $P = 2.5$, $t \approx 4.6$ months.

73. (a) 9000 screens, 39,000 screens

(b) 9059 screens, 39,186 screens

75. (a) 22, 24, 25, 26 (b) 22.1, 24.0, 25.1, 25.9

77. (a)

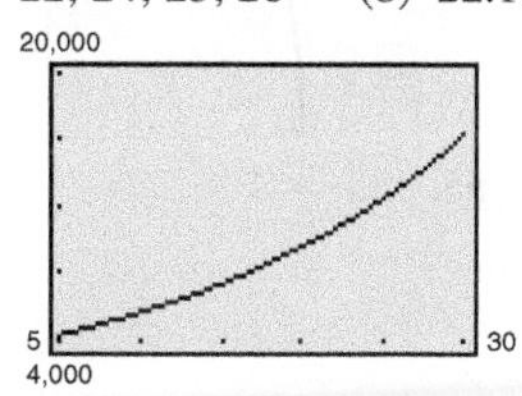

(b) \$5682.8 billion, \$9624.6 billion, \$15,463.8 billion

79. Answers will vary.

CHAPTER 4

Section 4.3 *(page 167)*

Skills Warm Up *(page 167)*

1. 3 **2.** 0 **3.** -1 **4.** 1
5. 7.389 **6.** 0.368
7. The graph of g is the graph of f shifted two units to the left.
8. The graph of g is the graph of f reflected about the x-axis.
9. The graph of g is the graph of f shifted downward one unit.
10. The graph of g is the graph of f reflected about the y-axis.

1. c **2.** f **3.** b **4.** d **5.** a **6.** e
7. $\log_4 256 = 4$ **9.** $\log_{81} 3 = \frac{1}{4}$ **11.** $\log_6 \frac{1}{36} = -2$
13. $4^2 = 16$ **15.** $2^{-1} = \frac{1}{2}$ **17.** $5^{-1} = 0.2$
19. $27^{1/3} = 3$ **21.** 2 **23.** -4 **25.** $\frac{1}{3}$ **27.** 1
29. -4 **31.** 2.538 **33.** -0.097 **35.** 0.452 **37.** 1
39. -1 **41.** 3 **43.** 7

45.
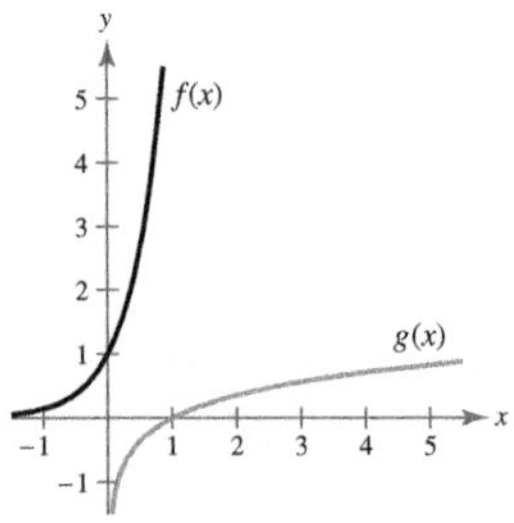

47.
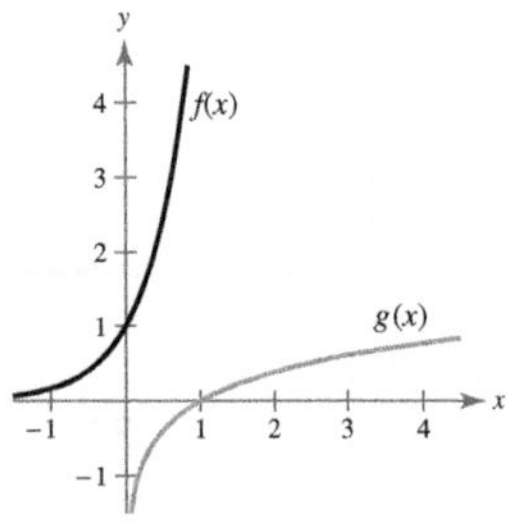

49. Domain: $(0, \infty)$
Asymptote: $x = 0$
x-intercept: $(1, 0)$

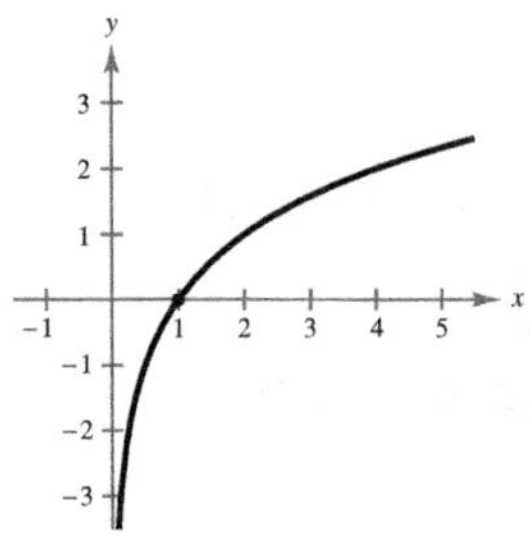

51. Domain: $(-4, \infty)$
Asymptote: $x = -4$
x-intercept: $(-3, 0)$

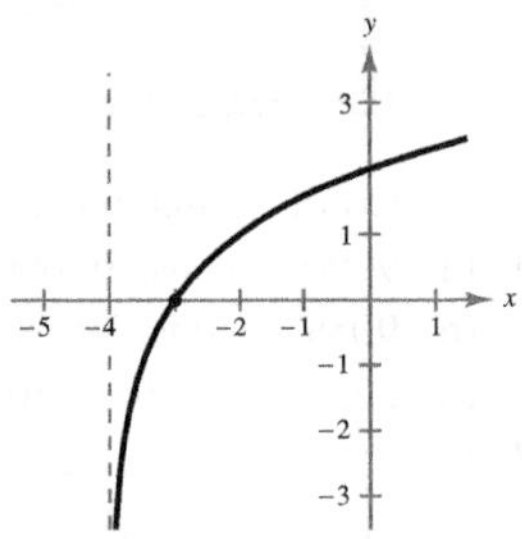

53. Domain: $(0, \infty)$
Asymptote: $x = 0$
x-intercept: $(1, 0)$

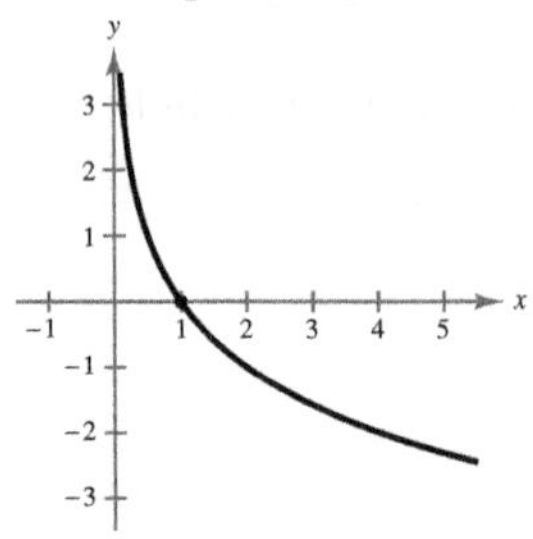

55. $\ln e = 1$

57. $\ln 4 = x$ **59.** $e^1 = e$ **61.** $e^0 = 1$ **63.** 1 **65.** -4

67. 1.946 **69.** 2.913 **71.** 0.896 **73.** 4 **75.** ± 5

77.
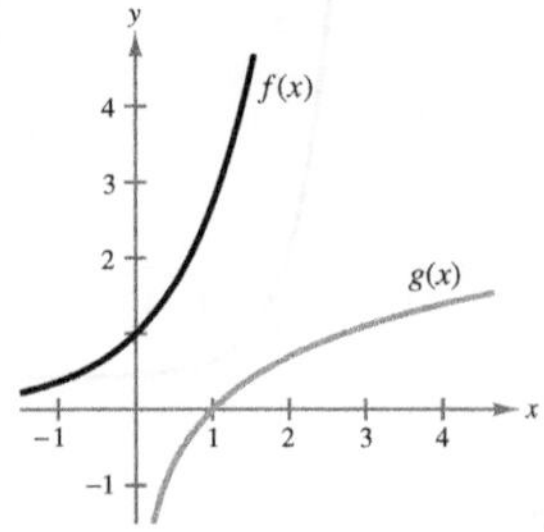

79. Domain: $(-\infty, 0)$
Asymptote: $x = 0$
x-intercept: $(-1, 0)$

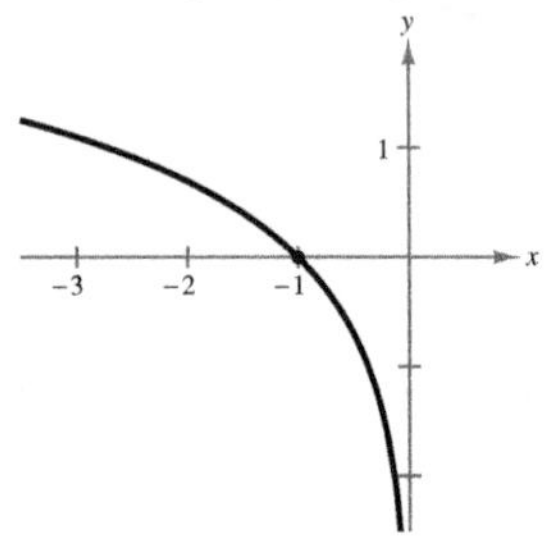

81. Domain: $(-1, \infty)$
Asymptote: $x = -1$
x-intercept: $(0, 0)$

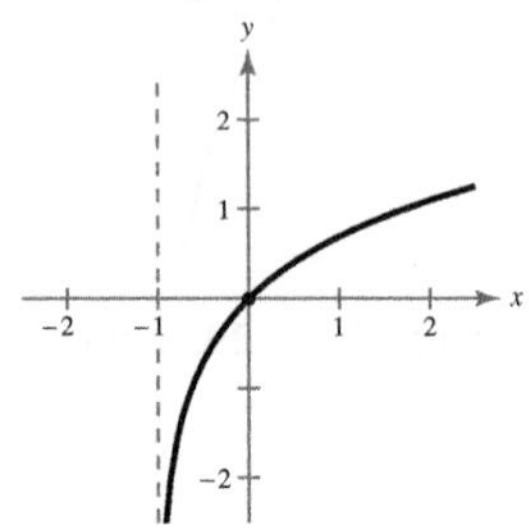

83. c **84.** d **85.** a **86.** b

87.
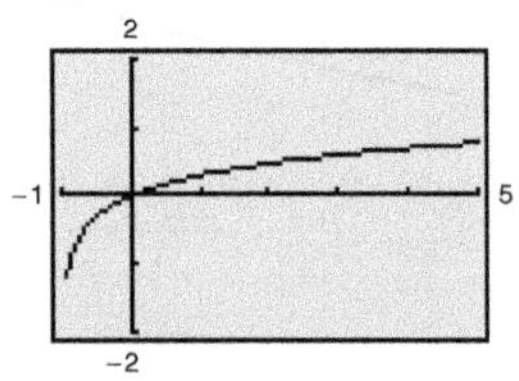

89.
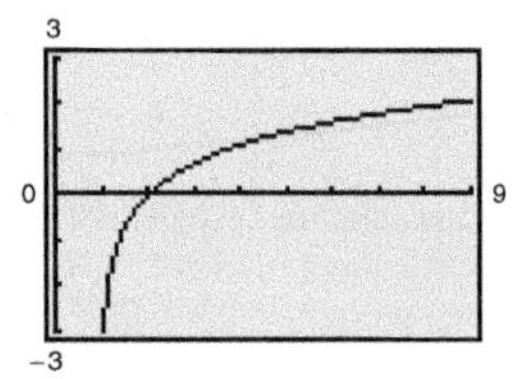

91.
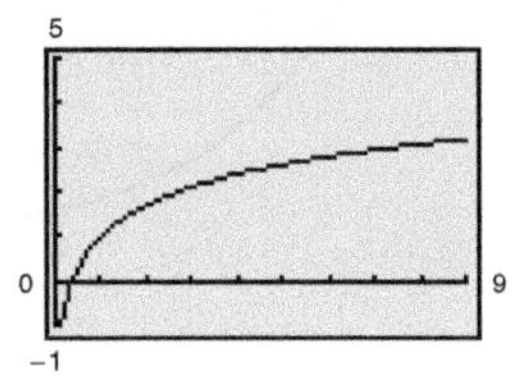

93. $t \approx 26.1$ years

95.
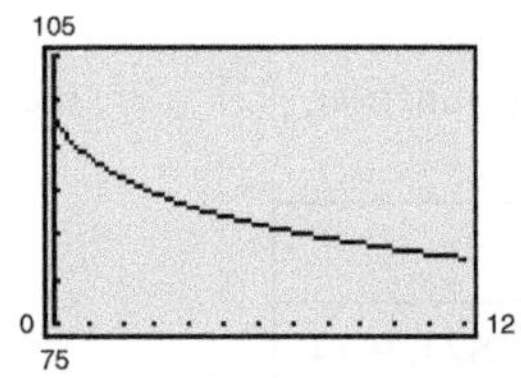

The domain $0 \le t \le 12$ covers the period of 12 months or 1 year. The scores range from the average score on the original exam to the average score after 12 months, so the range is $82.40 \le f(t) \le 98$

97. (a) 87 (b) ≈ 73.5 (c) In 11 months ($t \approx 10.6$)

99. (a) 30 years; 20 years
(b) \$396,234; \$301,123
(c) \$246,234; \$151,123
(d) $x = 1000$; As the monthly payment approaches \$1000, the length of time it takes to pay off the loan increases without bound.

101. (a)

K	1	2	4	6	8	10	12
t	0	13.2	26.4	34.1	39.6	43.9	47.3

(b)

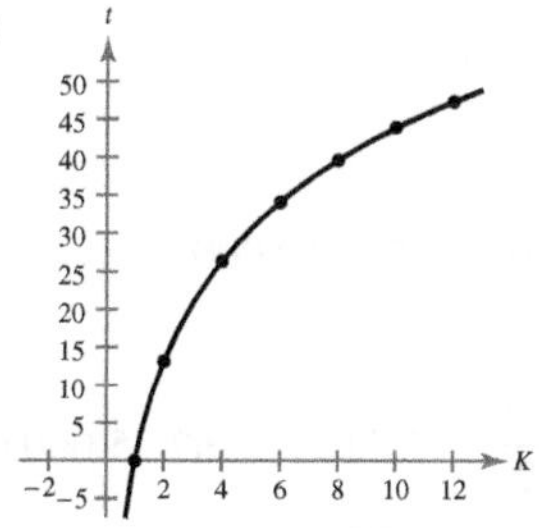

Quiz Yourself *(page 170)*

1. f does not have an inverse function.

2. Yes; $f^{-1}(x) = x^3 - 1$

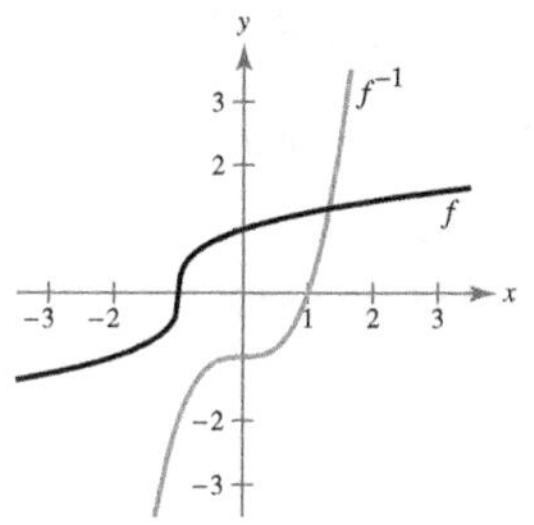

3. (a) $f^{-1}(x) = 2(x + 3)$

(b)

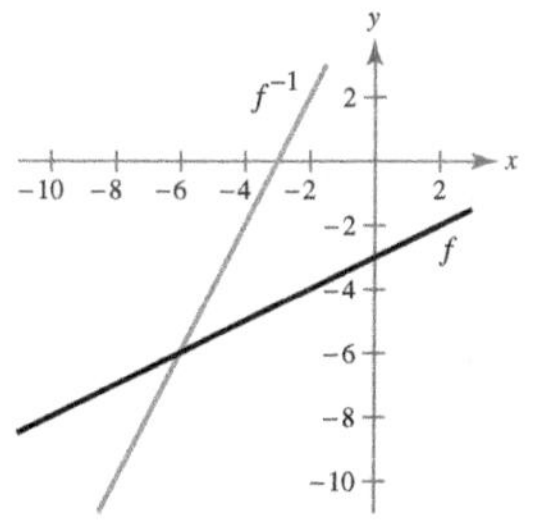

(c) $f^{-1}(f(x)) = 2(\frac{1}{2}x - 3 + 3) = x$

$f(f^{-1}(x)) = \frac{1}{2}[2(x + 3)] - 3 = x$

4. (a) $f^{-1}(x) = x^2 - 1,\ x \geq 0$

(b)

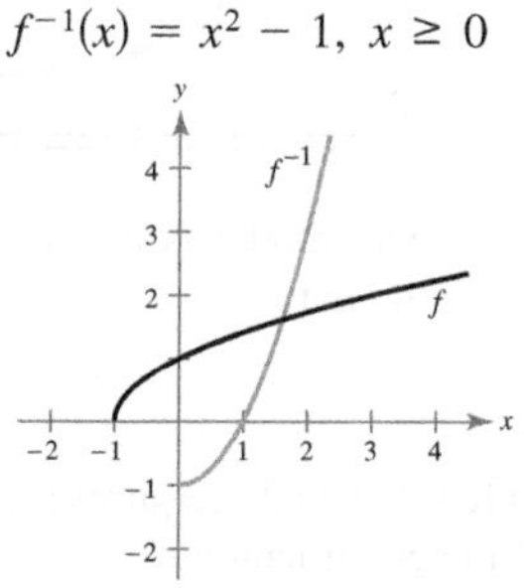

(c) $f^{-1}(f(x)) = (\sqrt{x + 1})^2 - 1 = x$

$f(f^{-1}(x)) = \sqrt{x^2 - 1 + 1} = x,\ x \geq 0$

5.

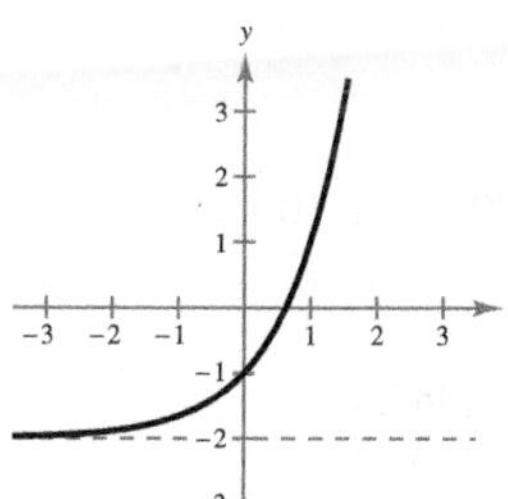

6.

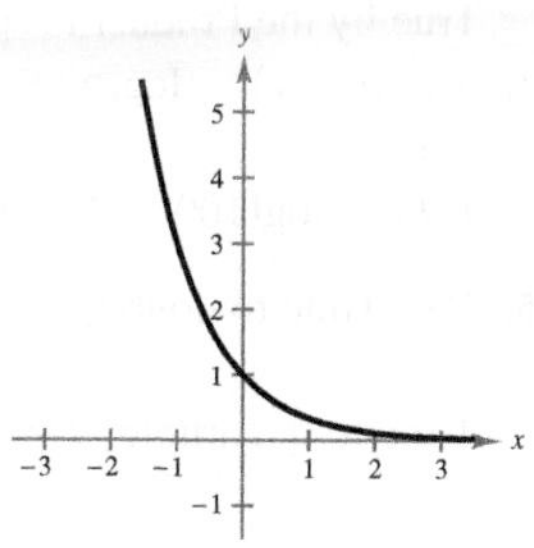

7.

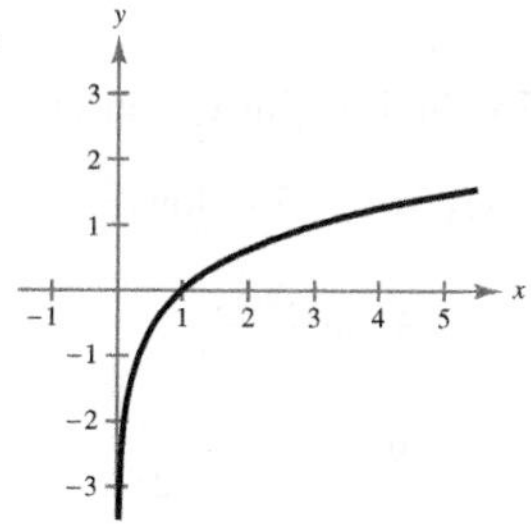

8.

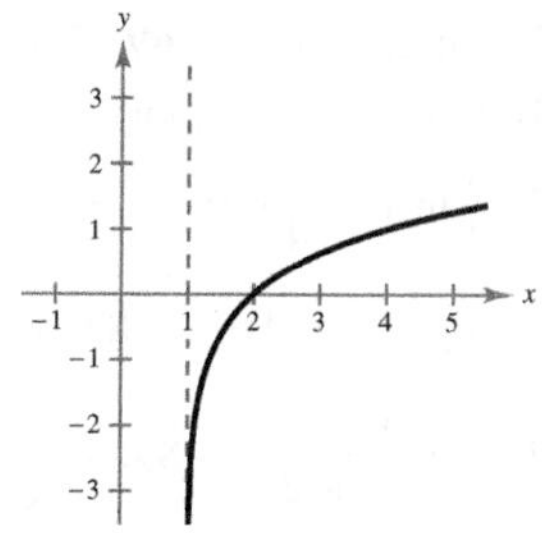

9. (a) Monthly: \$15,085.04 (b) Continuously: \$15,098.34

10. (a) 2006: 9.49 million; 2008: 10.26 million

(b) 2014: 12.99 million; 2015: 13.51 million

11. (a) 100 (b) ≈ 364 (c) ≈ 1326

12. \$108.54 **13.** 2 **14.** 4 **15.** -2 **16.** 0

17.

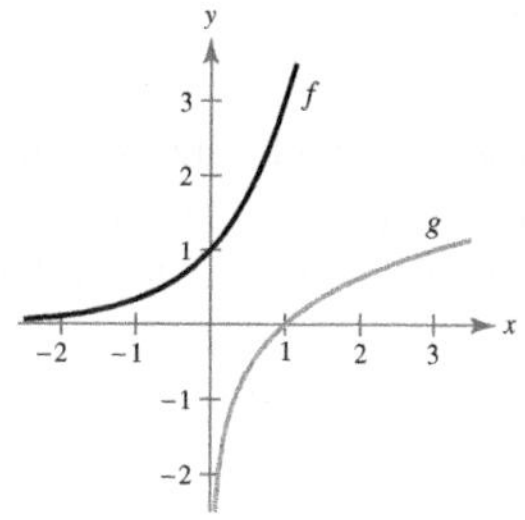

Domain of $f(x)$: $(-\infty, \infty)$

Domain of $g(x)$: $(0, \infty)$

The graphs are reflections of each other about the line $y = x$.

Section 4.4 *(page 176)*

Skills Warm Up *(page 176)*

1. 2 **2.** -5 **3.** -2 **4.** -3 **5.** e^5

6. $\dfrac{1}{e}$ **7.** e^6 **8.** 1 **9.** $y = x^{-2}$

10. $y = x^{1/2}$ **11.** $4^3 = 64$ **12.** $16^{1/2} = 4$

1. $\dfrac{\log_{10} 8}{\log_{10} 5} \approx 1.292$ **3.** $\dfrac{\log_{10} 30}{\log_{10} 4} \approx 2.453$

5. $\dfrac{\ln 8}{\ln 5} \approx 1.292$ **7.** $\dfrac{\ln 14}{\ln 6} \approx 1.473$ **9.** 2.585

11. 1.079 **13.** 2.633 **15.** -0.683 **17.** -1.661

19. $\ln 5 + \ln 2$ **21.** $\ln 5 - \ln 2$ **23.** $2(\ln 5 + \ln 2)$

25. $-\log_{10} \frac{11}{8} = \log_{10}\left(\frac{11}{8}\right)^{-1} = \log_{10} \frac{8}{11}$

27. $-2 \ln \dfrac{e}{7} = \ln\left(\dfrac{e}{7}\right)^{-2} = \ln\left(\dfrac{7}{e}\right)^2 = 2(\ln 7 - \ln e) = 2 \ln 7 - 2$

29. 1.1833 **31.** -0.2084 **33.** 1.0686 **35.** 0.1781

37. 1.8957 **39.** -2.7124 **41.** 0.5708

43. $-1 - \log_9 2$ **45.** $\frac{1}{2} + \frac{1}{2}\log_7 10$

47. $-3 - \log_5 2$ **49.** $6 + \ln 5$

51. The change-of-base formula was used incorrectly. The formula is $\log_a x = \dfrac{\ln x}{\ln a}$. So, $\log_2 5 = \dfrac{\ln 5}{\ln 2}$.

53. The Quotient Property was used incorrectly. The Quotient Property is $\ln \dfrac{u}{v} = \ln u - \ln v$. So, $\ln 8 - \ln 2 = \ln \dfrac{8}{2} = \ln 4$.

55. $6 + 5 \log_2 3$ **57.** $\log_3 4 + \log_3 n$ **59.** $\log_5 x - 2$

61. $4 \log_2 x$ **63.** $\frac{1}{2} \ln z$ **65.** $\ln x + \ln y + \ln z$

67. $\frac{1}{2}\ln(a-1)$ **69.** $2\ln(z-1)-\ln z$
71. $\ln y+\ln z-\frac{1}{3}\ln(z+3)$ **73.** $\ln 5+\frac{1}{3}(\ln x-\ln y)$
75. $\frac{3}{4}\ln x+\frac{1}{4}\ln(x^2+3)$ **77.** $\log_3 5x$ **79.** $\log_4\frac{8}{x}$
81. $\log_{10}(x+4)^2$ **83.** $\ln\frac{1}{216x}$ **85.** $\ln\frac{\sqrt[3]{5x}}{x+1}$
87. $\log_8\frac{x-2}{x+2}$ **89.** $\ln\frac{144}{\sqrt{x^2+1}}$ **91.** $\ln\frac{x^2}{(x+2)(x-2)}$

93.

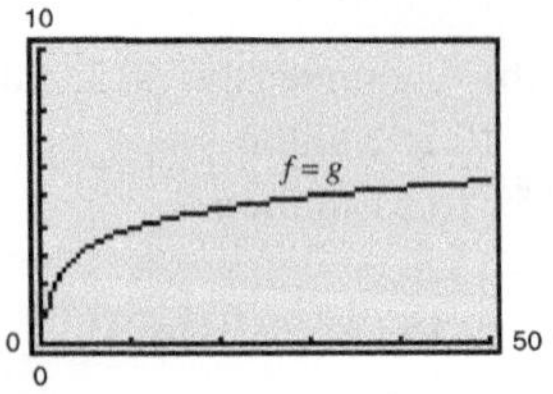

The two graphs are the same. The property is $\log_a(uv)=\log_a u+\log_a v$.

95.

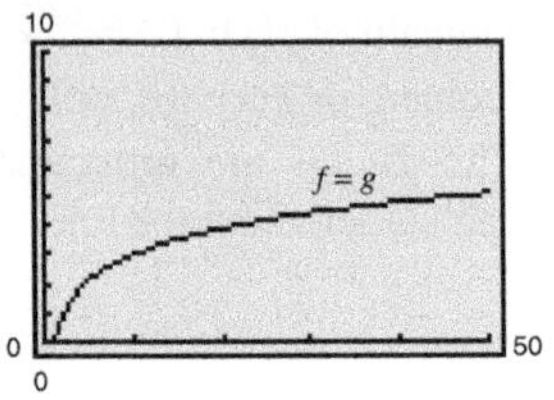

The two graphs are the same. The graphs demonstrate the Power Property for logarithms.

97. $\frac{1}{3}$ **99.** $-\frac{1}{2}$ **101.** -3 **103.** $\ln y=\frac{1}{4}\ln x$
105. $\ln y=\frac{2}{3}\ln x+\ln 0.070$ **107.** $\approx$ 26 decibels
109. (a) $a^m a^n=a^{m+n}$ (b) $\frac{a^m}{a^n}=a^{m-n}$ (c) $(a^m)^n=a^{mn}$
111. Let $\log_a u=x$ and $\log_a v=y$.
$a^x=u$ and $a^y=v$
$\frac{u}{v}=\frac{a^x}{a^y}=a^{x-y}$
$\log_a a^{x-y}=x-y$
$\log_a\frac{u}{v}=\log_a u-\log_a v$

Section 4.5 *(page 186)*

Skills Warm Up *(page 186)*

1. $\frac{\ln 3}{\ln 2}$ **2.** $1+\frac{2}{\ln 4}$ **3.** $\frac{e}{2}$ **4.** $2e$
5. $2\pm i$ **6.** $\frac{1}{2}, 1$ **7.** x **8.** $2x$
9. $2x$ **10.** $-x^2$

1. 2 **3.** -2 **5.** 4 **7.** 64 **9.** $\frac{2}{5}$ **11.** x^2
13. $x+1$ **15.** x^3-7 **17.** $8+x^3$ **19.** $x+5$
21. x^2 **23.** $\ln 3\approx 1.099$ **25.** $\log_3 8\approx 1.893$
27. $\ln 28\approx 3.332$ **29.** $\frac{1}{2}\log_3 80\approx 1.994$ **31.** 2
33. $\log_3 28+1\approx 4.033$ **35.** $3-\log_2 565\approx -6.142$
37. $\frac{1}{3}\log_{10}\frac{3}{2}\approx 0.059$ **39.** $\log_5 7+1\approx 2.209$
41. $\frac{1}{3}\ln 12\approx 0.828$ **43.** $\ln\frac{5}{3}\approx 0.511$
45. $\ln\frac{1}{2}\approx -0.693$ **47.** $\frac{1}{3}+\frac{1}{3}\log_2\frac{8}{3}\approx 0.805$
49. $\ln 6\approx 1.792, \ln 2\approx 0.693$ **51.** $\ln 4\approx 1.386$
53. $2\ln 75\approx 8.635$ **55.** $\frac{1}{2}\ln 1498\approx 3.656$

57. $\frac{\ln 4}{365\ln\left(1+\frac{0.065}{365}\right)}\approx 21.330$
59. $\frac{\ln 2}{12\ln\left(1+\frac{0.10}{12}\right)}\approx 6.960$ **61.** 10,000
63. $e^{-3}\approx 0.050$ **65.** $\frac{e^{2.4}}{2}\approx 5.512$ **67.** 5,000,000
69. 26 **71.** $\frac{1}{5}e^{10/3}\approx 5.606$ **73.** $e^2-2\approx 5.389$
75. $e^{-2/3}\approx 0.513$ **77.** No solution **79.** No solution
81. 7 **83.** $\frac{-1+\sqrt{17}}{2}\approx 1.562$ **85.** 2
87. $y=2x+1, x>-\frac{1}{2}$ **89.** $y=\frac{(x-1)^2}{x+2}, x\geq 1$
91. 2.807 **93.** 20.086 **95.** $\approx$ 9.56 years
97. $\approx$ 13.32 years **99.** 30 months
101. (a) $\approx$ 210 coin sets (b) $\approx$ 588 coin sets
(c)

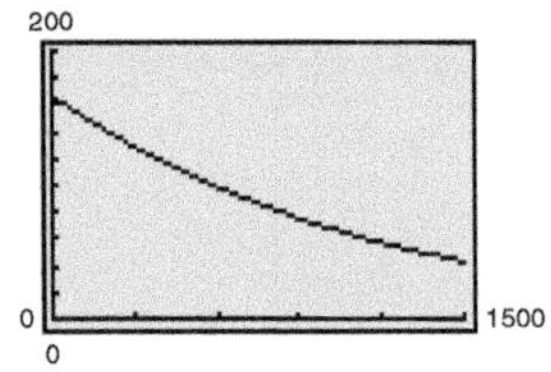

103. (a) $\approx$ 29.3 years (b) $\approx$ 39.8 years
105. (a)

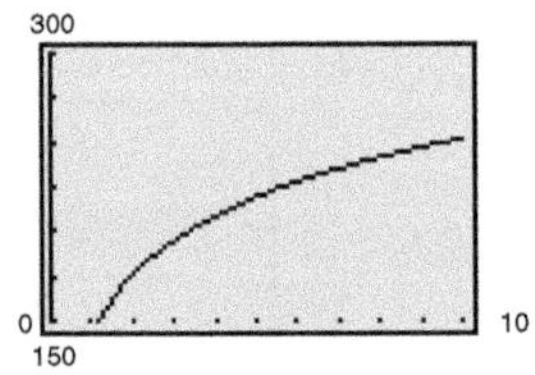

(b) and (c) 2003

107. (a)

x	y
0.2	162.6
0.4	78.5
0.6	52.5
0.8	40.5
1.0	33.9

(b)

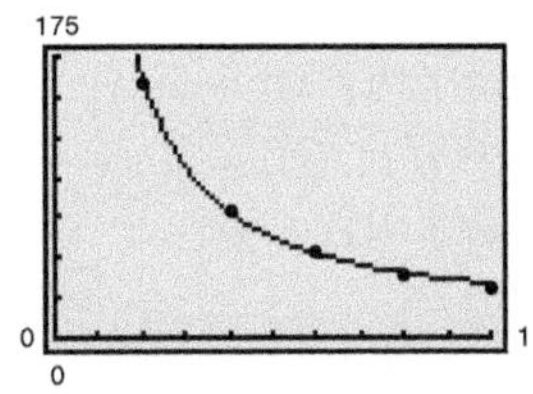

The model is a good fit for the data.

(c) 1.197 meters
(d) No. To reduce the g's to fewer than 23 requires a crumple zone of more than 2.27 meters, a length that exceeds the front width of most cars.

109. Yes. See Exercise 79.
111. $\log_b uv=\log_b u+\log_b v$
True by the Product Property in Section 4.4.
113. $\log_b(u-v)=\log_b u-\log_b v$
False.
$1.95\approx\log(100-10)\neq\log 100-\log 10=1$
115. Yes. Time to double: $t=\frac{\ln 2}{r}$.
Time to quadruple: $t=\frac{\ln 4}{r}=2\left(\frac{\ln 2}{r}\right)$

Section 4.6 *(page 197)*

Skills Warm Up *(page 197)*

1.
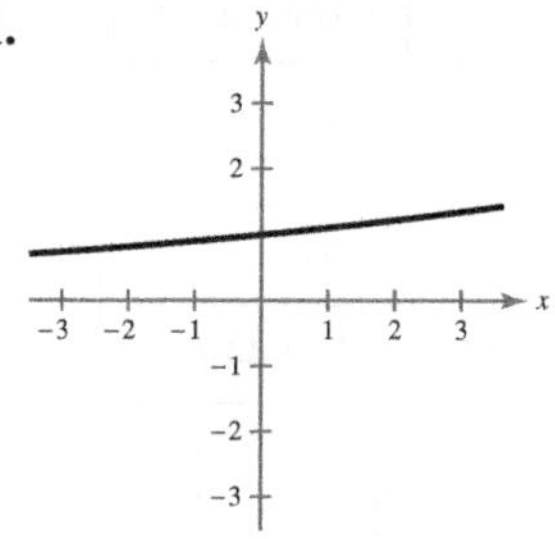

2.
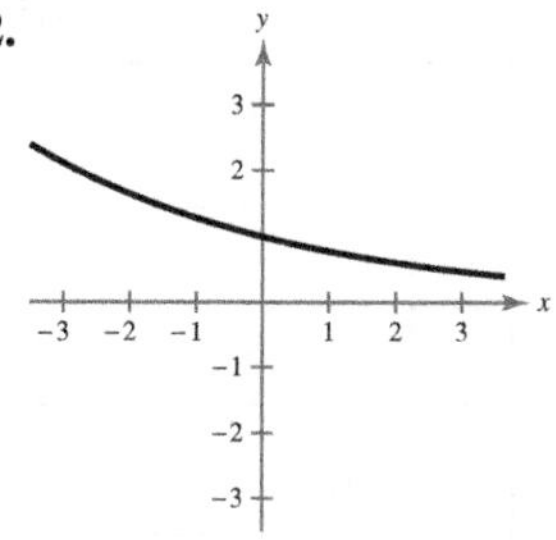

3.
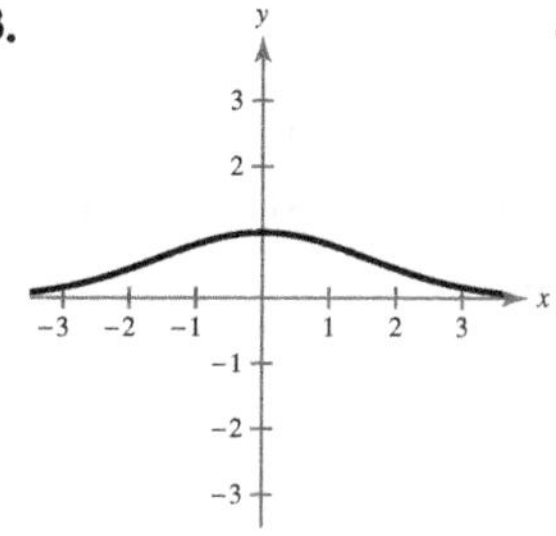

4.
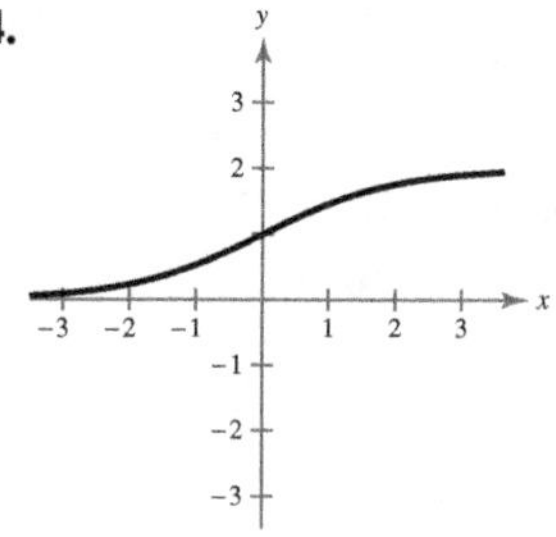

5.
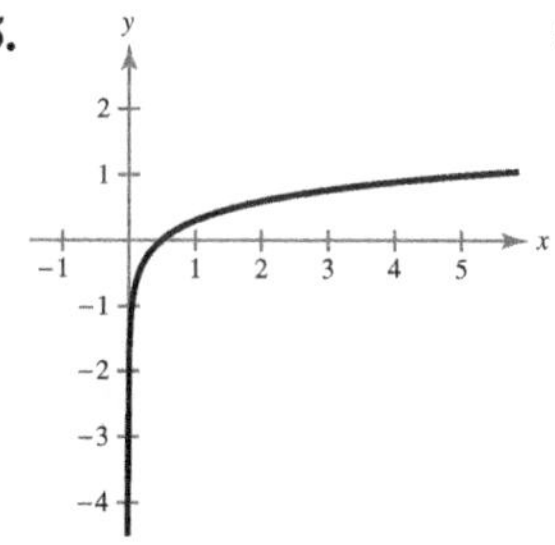

6.
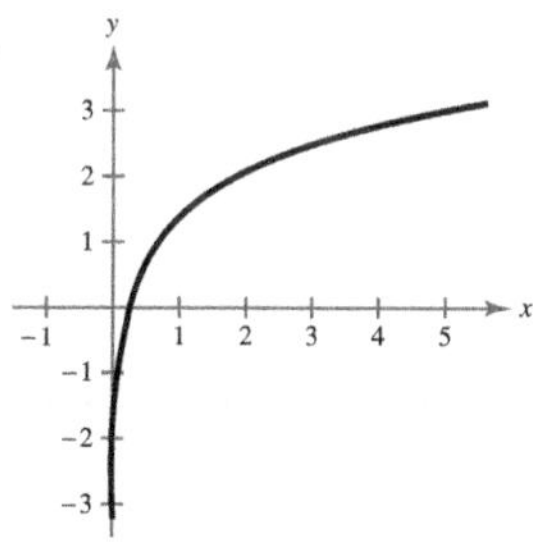

7. $\frac{1}{2}\ln\frac{7}{3}$ **8.** $\frac{1}{5}e^{7/2}$ **9.** ≈ 34.539 **10.** ≈ 3.695

1. c **2.** e **3.** b **4.** a **5.** d **6.** f
7. Exponential growth **9.** Exponential decay

11.
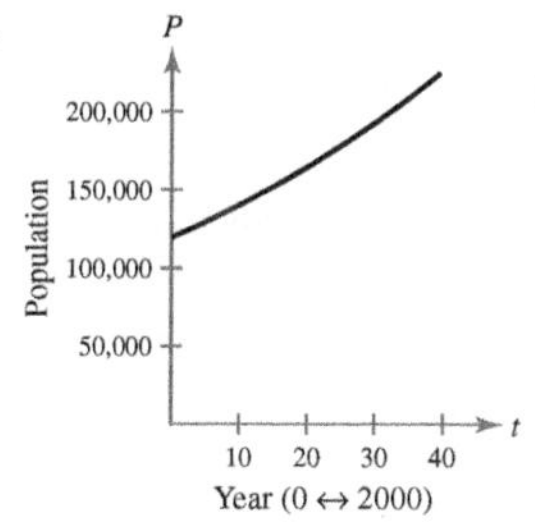

2025

13. $C = 1, k = \frac{1}{4}\ln 10$ **15.** $C = 1, k = \frac{1}{4}\ln\frac{1}{4}$

	Initial Investment	Annual % Rate	Time to Double	Amount After 10 Years
17.	\$5000	7%	9.90 years	\$10,068.76
19.	\$500	6.93%	10 years	\$1000.00
21.	\$1000	8.25%	8.40 years	\$2281.88
23.	\$6392.79	11%	6.30 years	\$19,205.00
25.	\$5000	8%	8.66 years	\$11,127.70

	Isotope	Half-Life (Years)	Initial Quantity	Amount After 1000 Years
27.	^{226}Ra	1599	4 g	2.59 g
29.	^{14}C	5715	3.95 g	3.5 g
31.	^{239}Pu	24,100	1.65 g	1.6 g

33. $k = -\frac{1}{20}\ln\frac{3}{5}$; $\approx$36,671 people
35. $\approx$ 12.36 hours to double
$\approx$ 19.59 hours to triple
37. 12,180 years **39.** $\approx$ 9.92%
41. (a) $N = 40(1 - e^{-0.049t})$ (b) $\approx$ 42 days **43.** 100
45. (a)
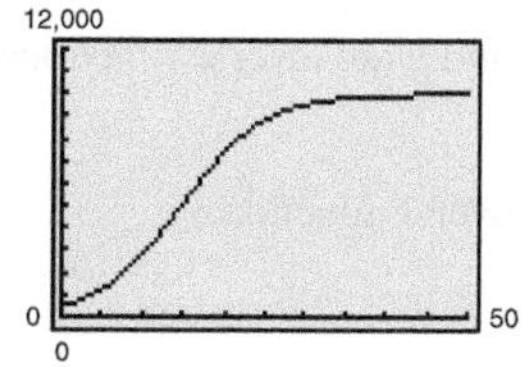

(b) $\approx$ 1252 fish
(c) $\approx$ 7.8 months

47. No, because the function is always increasing.
49. (a)
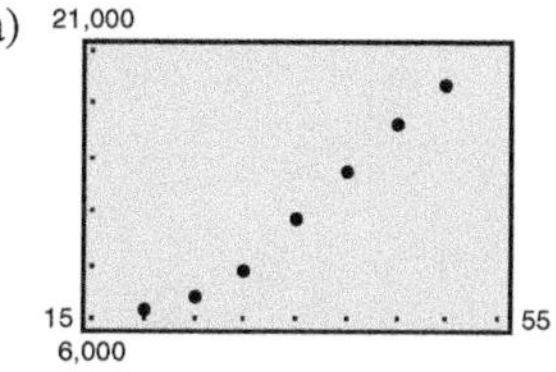

(b) $P = 2922.32(1.0391)^t$; $P = 2922.32e^{0.0384t}$
(c)
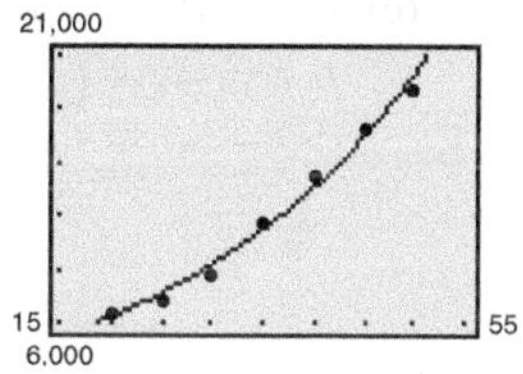

(d) 2022: $\approx$6,801,736 people
2042: $\approx$14,660,808 people

51. (a)
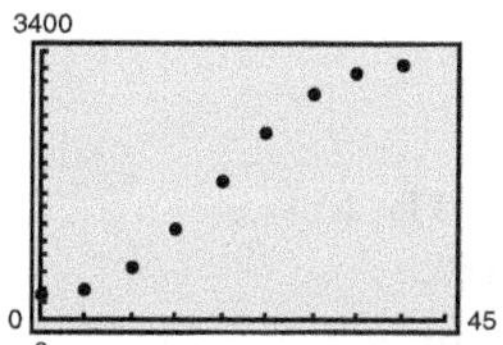

(b) Logistic growth model; The data points seem to be leveling off in the later years as though approaching an asymptote.
(c) $P = \dfrac{3455}{1 + 16.51e^{-0.1415t}}$
(d)
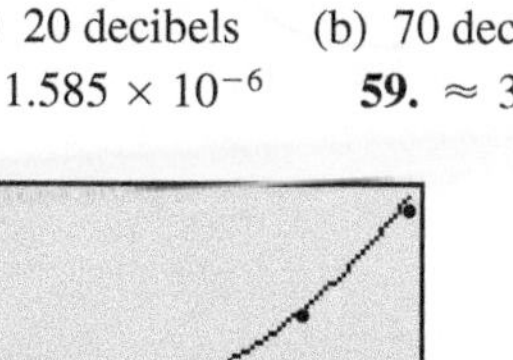

(e) 3407

The model is a good fit.

53. (a) $\approx$ 7.906 (b) $\approx$ 7.684
55. (a) 20 decibels (b) 70 decibels
57. $\approx 1.585 \times 10^{-6}$ **59.** $\approx$ 31,623 **61.** 3:00 A.M.
63.
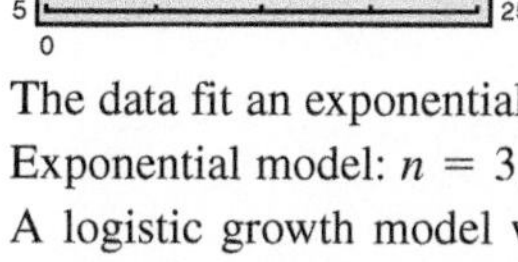

The data fit an exponential model.
Exponential model: $n = 3.9405e^{0.1086t}$
A logistic growth model would be more appropriate for this data because after the initial rapid growth in productivity, the worker's production rate will eventually level off.

CHAPTER 4

Review Exercises *(page 206)*

1. $f(g(x)) = 3\left(\frac{x-5}{3}\right) + 5 = x$

$g(f(x)) = \frac{3x + 5 - 5}{3} = x$

$f(g(x)) = x = g(f(x))$, so f and g are inverse functions of each other.

3. $f^{-1}(x) = \frac{1}{x}$ (f is its own inverse function.)

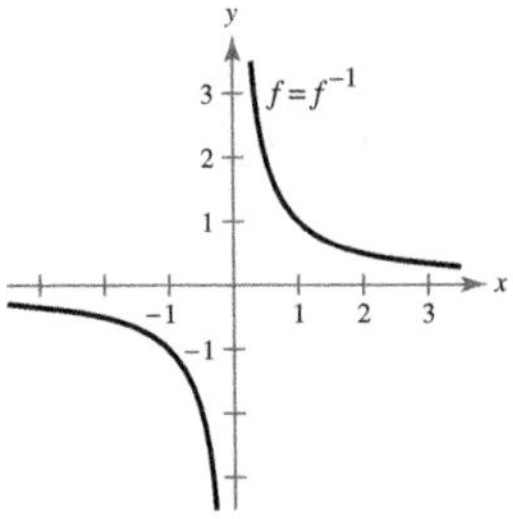

5. (a) $f^{-1}(x) = \sqrt{x},\ x \geq 0$

(b)

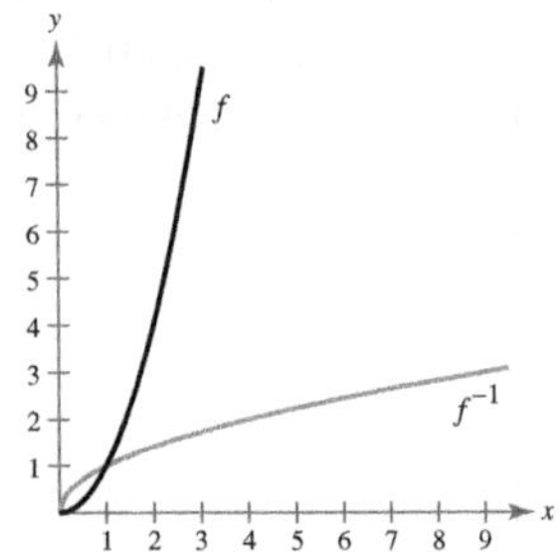

(c) $f^{-1}(f(x)) = \sqrt{x^2} = x$

$f(f^{-1}(x)) = (\sqrt{x})^2 = x$

7. 16,777,216 **9.** 0.431 **11.** 2.117

13. c **14.** d **15.** a **16.** b

17.

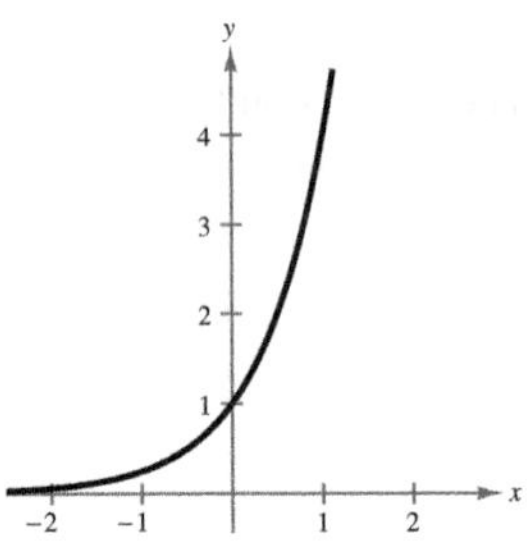

19.

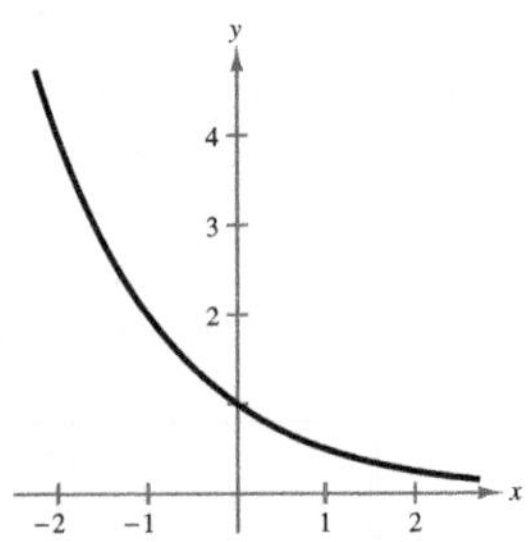

21.

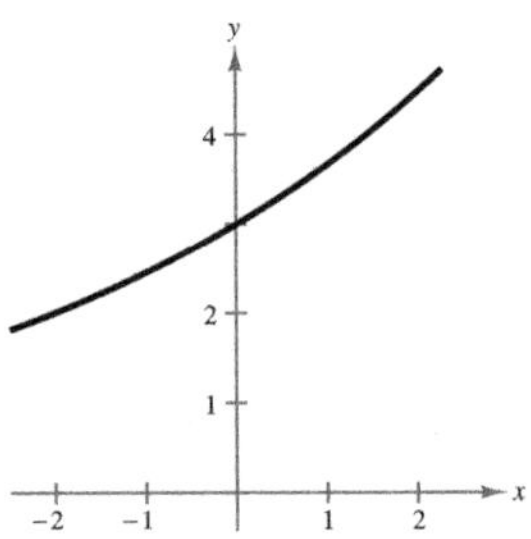

23.

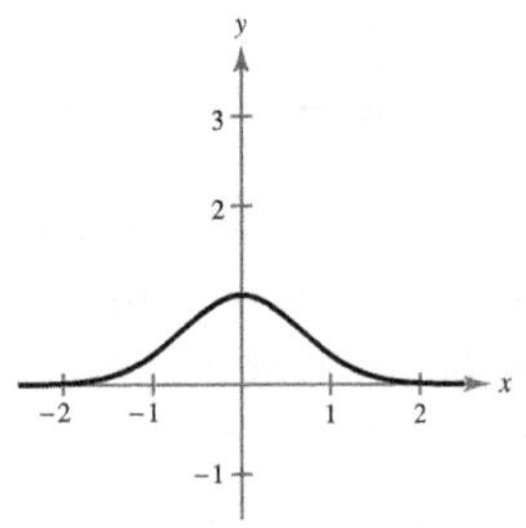

25.

n	1	2	4
A	\$13,308.43	\$13,576.74	\$13,718.59

n	12	365	Continuous
A	\$13,816.21	\$13,864.33	\$13,865.97

27.

t	1	10	20
P	\$185,085.40	\$92,140.76	\$42,449.59

t	30	40	50
P	\$19,556.69	\$9009.84	\$4150.87

29. \$8471.94 **31.** $\log_4 64 = 3$ **33.** $\ln 7.3890\ldots = 2$

35. $3^4 = 81$ **37.** $e^0 = 1$ **39.** 5 **41.** 7 **43.** $-\frac{1}{2}$

45. d **46.** b **47.** c **48.** a

49.

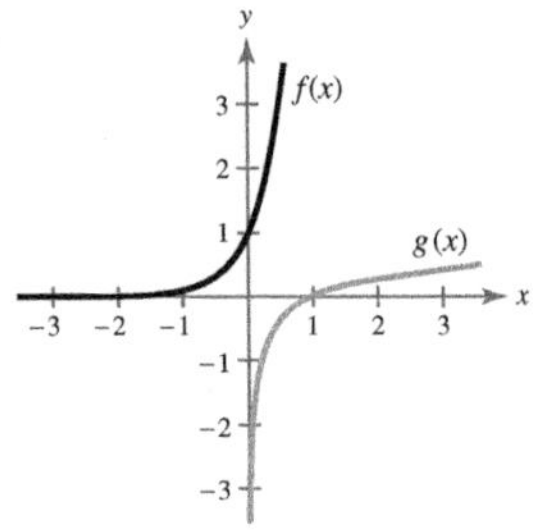

51. Domain: $(3, \infty)$
Vertical asymptote: $x = 3$
x-intercept: $(4, 0)$

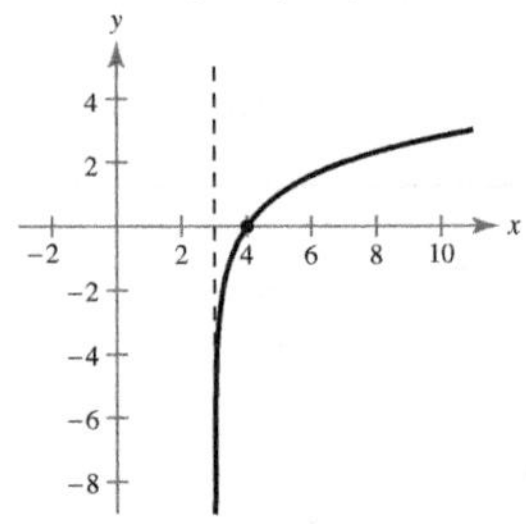

53. Domain: $(0, \infty)$
Vertical asymptote: $x = 0$
x-intercept: $(1, 0)$

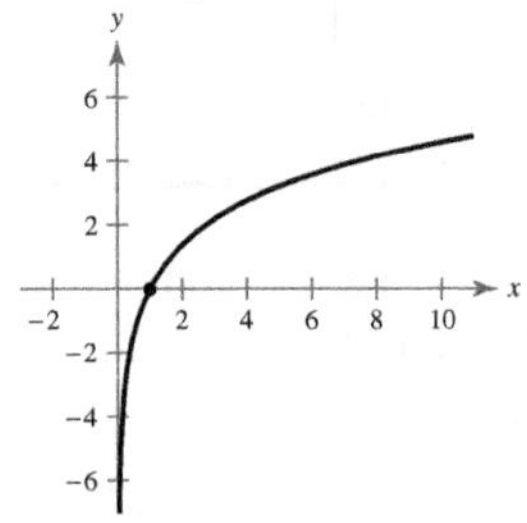

55. The average score decreased from 82 to about 68.

57. (a) 53.42 inches

(b)

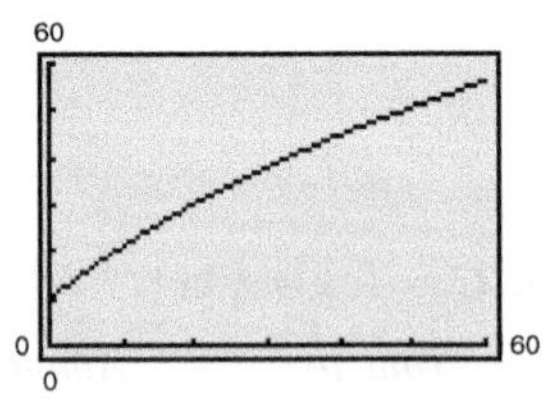

59. 2.096 **61.** 2.132 **63.** 0.9208 **65.** 0.2823

67. 2 **69.** 3.2 **71.** $\log_{10} x - \log_{10} y$

73. $\ln x + \frac{1}{2}\ln(x - 3)$ **75.** $4 \log_5(y - 3)$ **77.** $\log_4 6$

79. $\ln \sqrt{x}$ **81.** $\ln \dfrac{x}{(x-3)(x+1)}$ **83.** $\ln y = \dfrac{4}{3} \ln x$

85. $\ln 8 \approx 2.079$ **87.** 1 **89.** $\ln 4 \approx 1.386$

91. $\frac{1}{3}e^{8.2} \approx 1213.650$ **93.** $\frac{1}{5}e^2 \approx 1.478$

95. $3e^2 \approx 22.167$ **97.** 7

99. (a) $\approx$ 197 desks (b) $\approx$ 257 desks **101.** 10.63 g

103. (a)

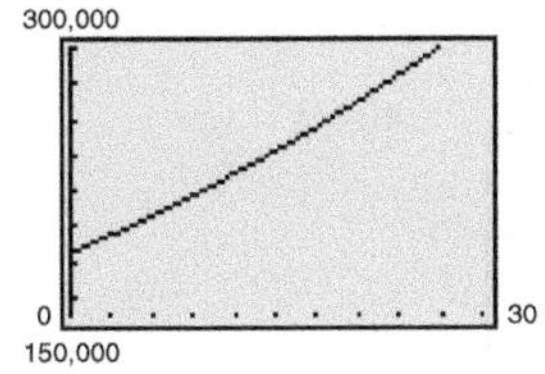

(b) 2021 ($t \approx 21.00$)

105. $\approx$ 9.93 hours to double
$\approx$ 15.74 hours to triple

107. (a) $N = 50(1 - e^{-0.04838t})$ (b) 48 days

109. (a) $k \approx 0.2121$
(b) $\approx$ 196 deer; $\approx$ 294 deer; $\approx$ 383 deer

111. Yes

Test Yourself *(page 210)*

1. Yes; $f^{-1}(x) = \dfrac{x+7}{5}$

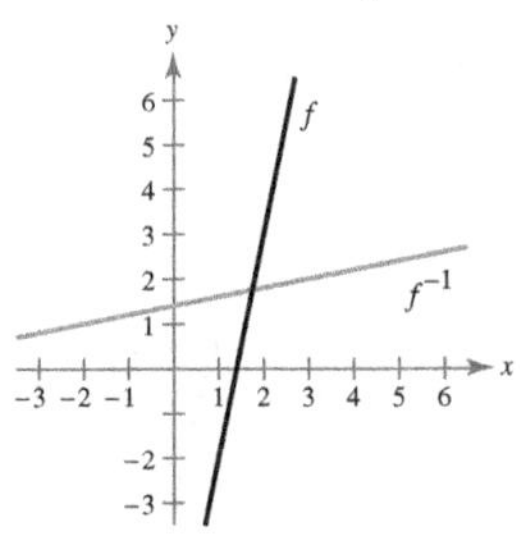

2. The function does not have an inverse function.

3.

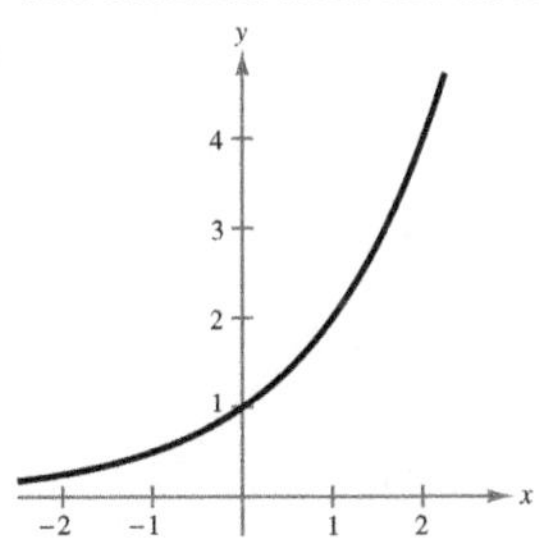

4.

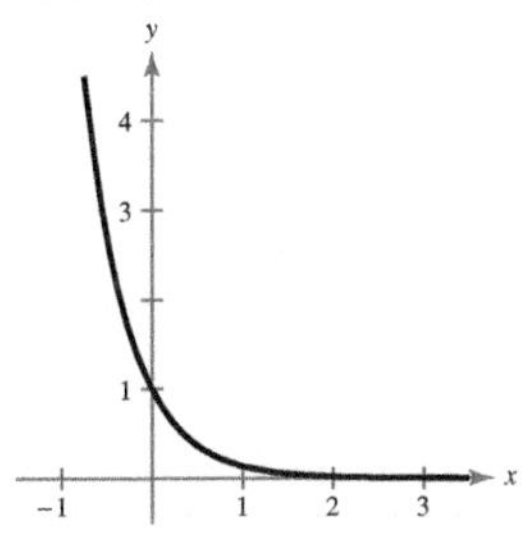

5.

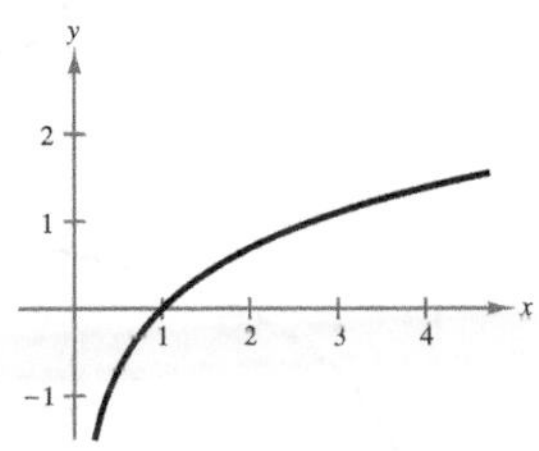

6.

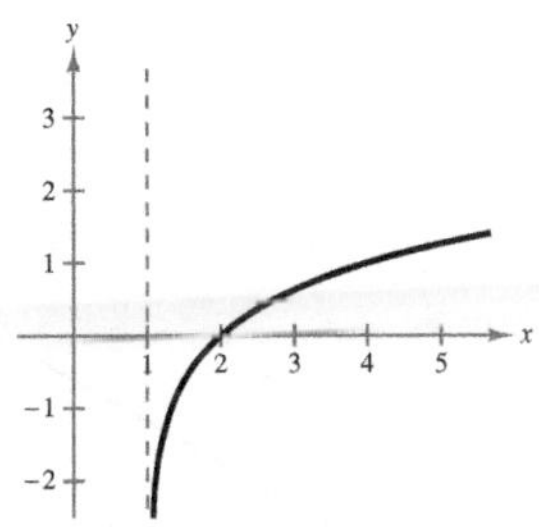

7. 87; $\approx$ 79.8; $\approx$ 76.5

Chapter 5

Section 5.1 *(page 219)*

Skills Warm Up *(page 219)*

1.

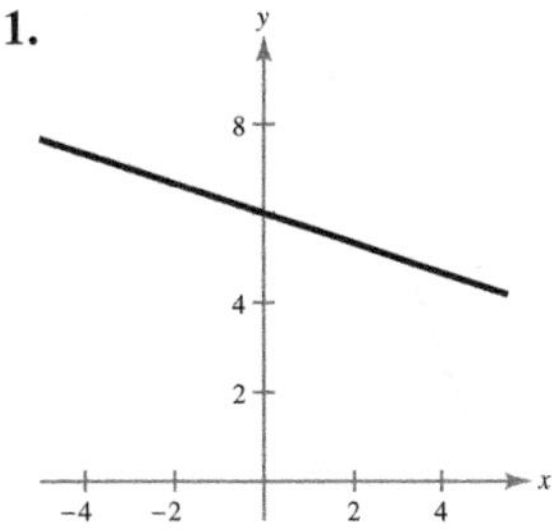

2.

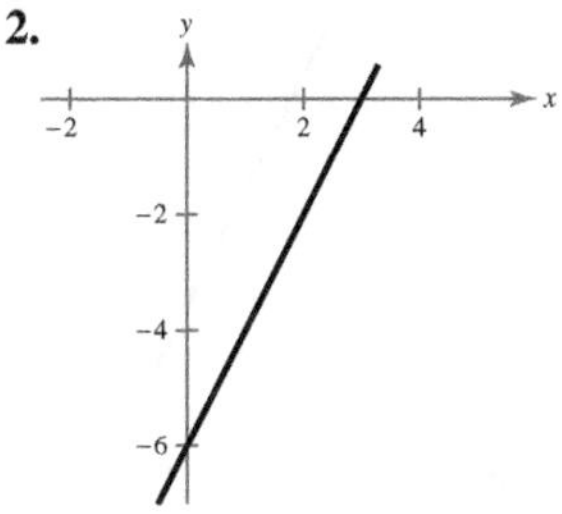

3.

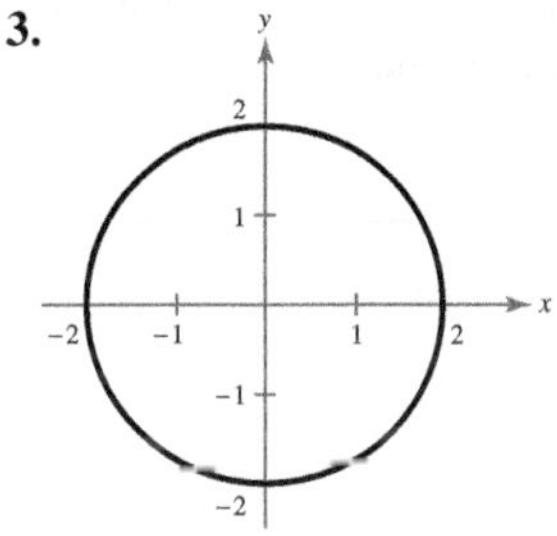

4.

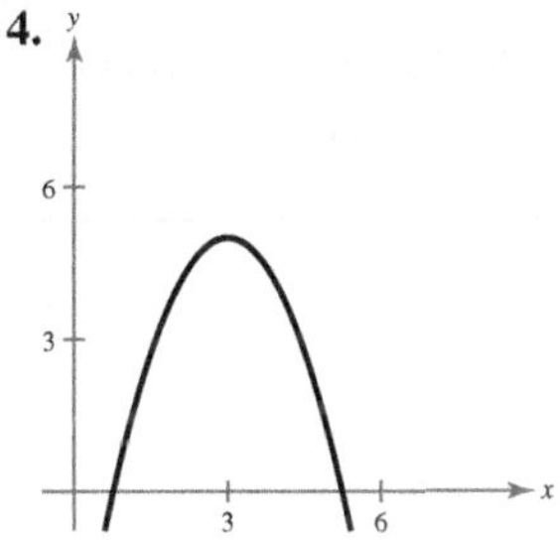

5. x **6.** $-37v$ **7.** $2x^2 + 9$ **8.** -1

9. $x = 6$ **10.** $y = 1$

1. (a) No (b) Yes **3.** (a) Yes (b) No

5. (a) No (b) No **7.** $(-1, 1)$ **9.** $\left(\frac{1}{2}, 3\right)$ **11.** $(10, 3)$

13. $(1.5, 0.3)$ **15.** $\left(\frac{20}{3}, \frac{40}{3}\right)$ **17.** No solution

19. \$5000 at 3%, \$7000 at 5%

21. \$6000 at 2.8%, \$6000 at 3.8%

23. $\left(1 + \sqrt{2}, 2 + 2\sqrt{2}\right), \left(1 - \sqrt{2}, 2 - 2\sqrt{2}\right)$

25. $\left(\frac{29}{10}, \frac{21}{10}\right), (-2, 0)$ **27.** No solution **29.** $(0, 1), (\pm 1, 0)$

31. $(2, 1)$ **33.** $(2, -3)$ **35.** $(-1, 2), (2, 5)$

37. $(0, 4), \left(\frac{12}{5}, -\frac{16}{5}\right)$ **39.** $(4, 3)$ **41.** $\left(\frac{5}{2}, \frac{3}{2}\right)$

43. $(2, 2), (4, 0)$ **45.** $(1, 4), (4, 7)$ **47.** $\left(4, -\frac{1}{2}\right)$

49. No solution **51.** One solution **53.** Two solutions

55. No solution **57.** $\left(\frac{1}{2}, \frac{3}{4}\right), (-3, -1)$ **59.** $(1, 4), (4, 7)$

61. $(0, 1)$ **63.** $(1, 4), (3, 12)$ **65.** $(-3, 4), (5, 0)$

67. Answers will vary. Sample answers:

(a) $\begin{cases} x + 2y = 0 \\ 2x - 4y = 3 \end{cases}$ (b) $\begin{cases} 2x^2 - y = -1 \\ x + y = 4 \end{cases}$

(c) $\begin{cases} 3x + y = 8 \\ 3x + y = -2 \end{cases}$

69. 192 units **71.** 233,333 units

73. 2000 software packages **75.** 2002 ($t \approx 2.03$)

77. Yes, at age 20. **79.** \$100,000

81. According to the graphs, sales of general merchandise stores first exceeded sales of food and beverage stores in 2003 ($t \approx 13.37$). Both models eventually decrease and become negative. So, it is unlikely that these models will continue to be accurate.

Section 5.2 *(page 230)*

Skills Warm Up *(page 230)*

1.

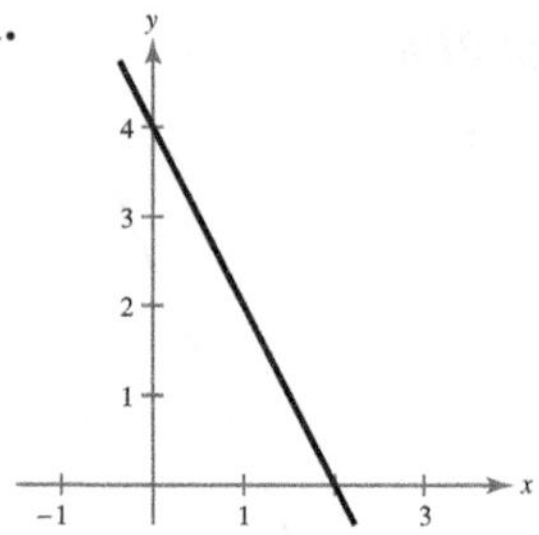

2.

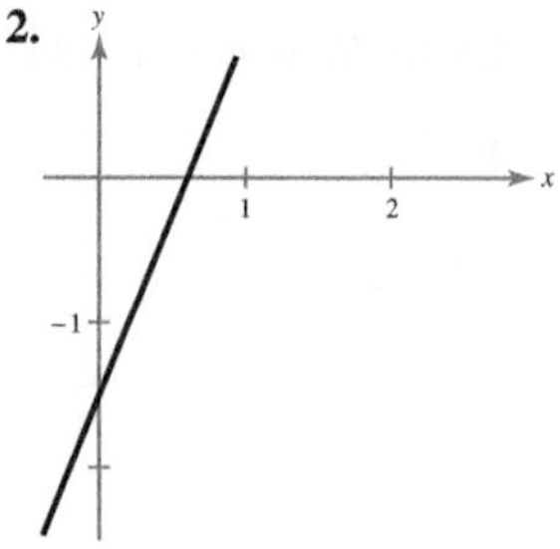

3. $x - y + 4 = 0$ **4.** $5x + 3y - 28 = 0$

5. $-\frac{1}{2}$ **6.** $\frac{7}{4}$ **7.** Perpendicular **8.** Parallel

9. Neither parallel nor perpendicular

10. Perpendicular

1. $(2, 2)$ **3.** $(2, 0)$ **5.** Inconsistent

7. $(2a, 3a - 3)$, where a is any real number

9. $\left(2, \frac{1}{2}\right)$; consistent **11.** $(-7, -13)$; consistent

13. $(4, -5)$; consistent **15.** $(4, -1)$; consistent

17. $(40, 40)$; consistent **19.** $\left(-\frac{6}{35}, \frac{43}{35}\right)$; consistent

21. $\left(\frac{18}{5}, \frac{3}{5}\right)$; consistent **23.** $\left(\frac{19}{7}, -\frac{2}{7}\right)$; consistent

25. $\begin{cases} x + y = 13 \\ x - y = 3 \end{cases}$; $(8, 5)$ **27.** $\begin{cases} 2r + s = 8 \\ r - s = 7 \end{cases}$; $(5, -2)$

29. No solution; inconsistent

31. $\left(a, \frac{5}{6}a - \frac{1}{2}\right)$, where a is any real number; consistent

33. $\left(\frac{90}{31}, -\frac{67}{31}\right)$; consistent **35.** $\left(-\frac{1}{2}, 1\right)$ **37.** $(-2, -9)$

39. $(-0.874, 0.417), (0, 1)$ **41.** $(20, 20)$

43. Answers will vary. Sample answers:

(a) $\begin{cases} 3x + y = 18 \\ x + y = 6 \end{cases}$ (b) $\begin{cases} -2x + 8y = 38 \\ 4x + y = -8 \end{cases}$

(c) $\begin{cases} 8x - y = 1 \\ -4x + 4y = -25 \end{cases}$

45. No. The solution of the system is (79,400, 398).

47. 550 miles per hour; 50 miles per hour

49. $6\frac{2}{3}$ gallons of 20% solution

$3\frac{1}{3}$ gallons of 50% solution

51. \$8000 at 4.75%, \$17,000 at 5.5%

53. $x \approx 309{,}091$ units; $p \approx \$25.09$

55. $x = 2{,}000{,}000$ units; $p = \$100.00$

57. (a) Fast-food

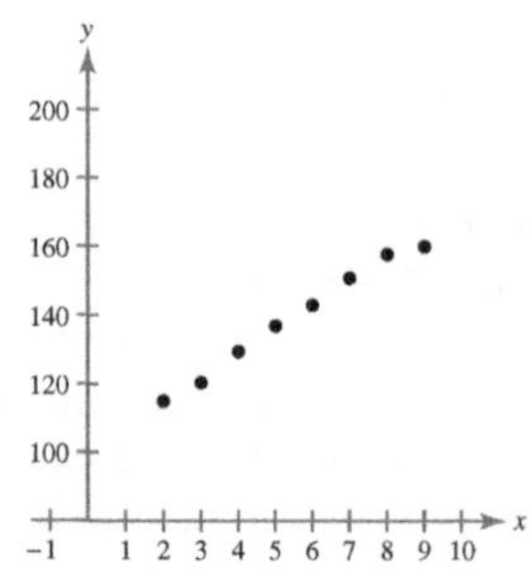

$y = 6.79x + 101.8$

Full-service

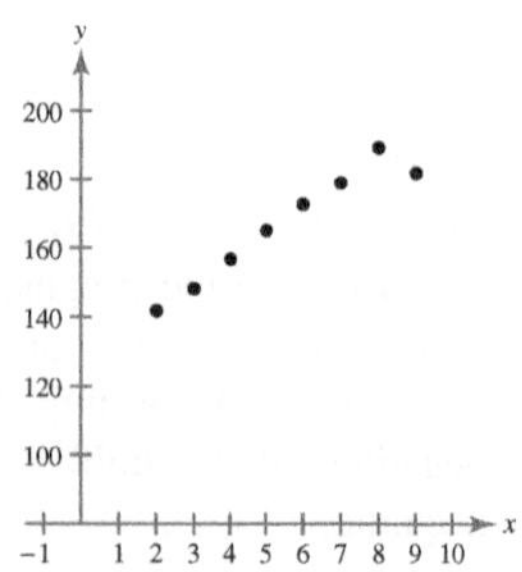

$y = 6.67x + 130.3$

(b) In the near future, fast-food sales will not equal full-service sales. According to the models, fast-food sales will equal full-service sales in the year 2237. These models, therefore, are not appropriate to use for predicting when fast-food sales will equal full-service sales.

59. 1380 units at \$810.60 **61.** $y = 0.97x + 2.1$

63. $y = 0.318x + 4.061$

65. (a) $y = 2.031t + 384.59$

(b)

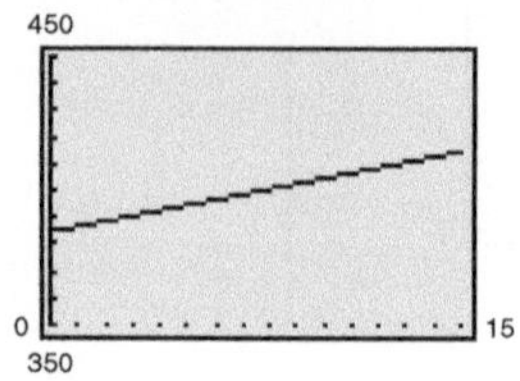

404.9

(c) You obtain the same model: $y = 2.031t + 384.59$.

67. Answers will vary.

Section 5.4 *(page 241)*

Skills Warm Up *(page 241)*

1. Line **2.** Parabola **3.** Circle **4.** Parabola

5. Line **6.** Circle **7.** $(1, 1)$ **8.** $(2, 0)$

9. $(2, 1), \left(-\frac{5}{2}, -\frac{5}{4}\right)$ **10.** $(2, 3), (3, 2)$

1. d **2.** b **3.** a **4.** c **5.** f **6.** e

7.

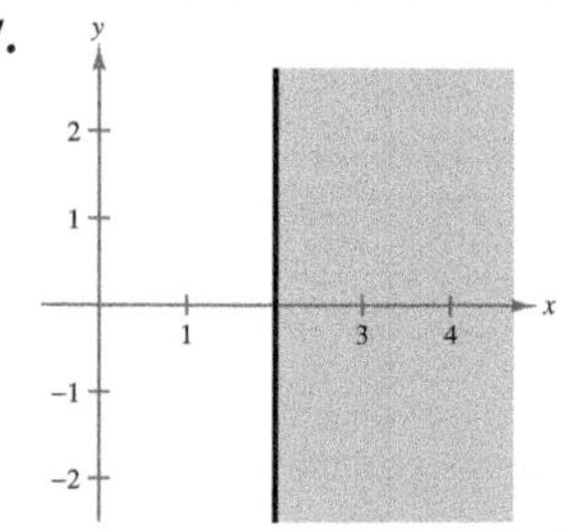

9.

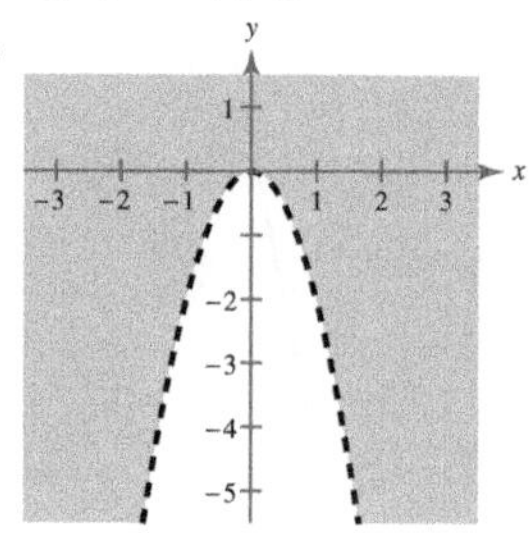

11.

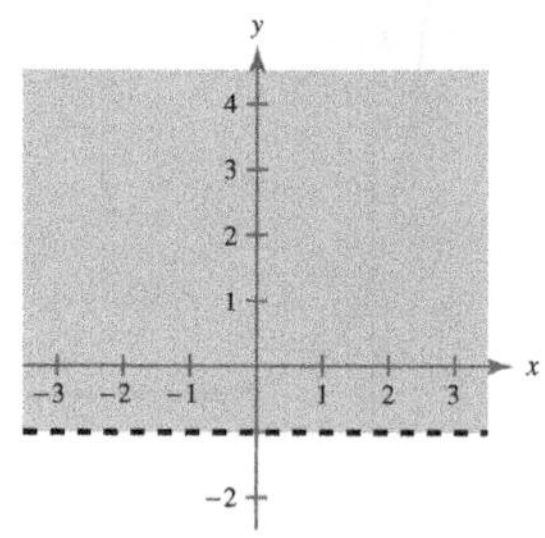

13.

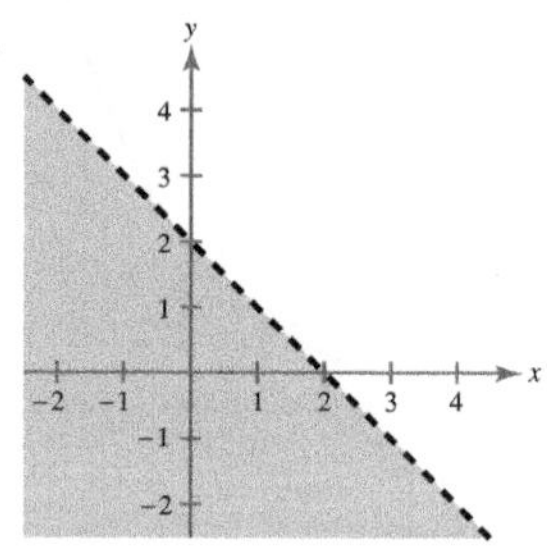

15.

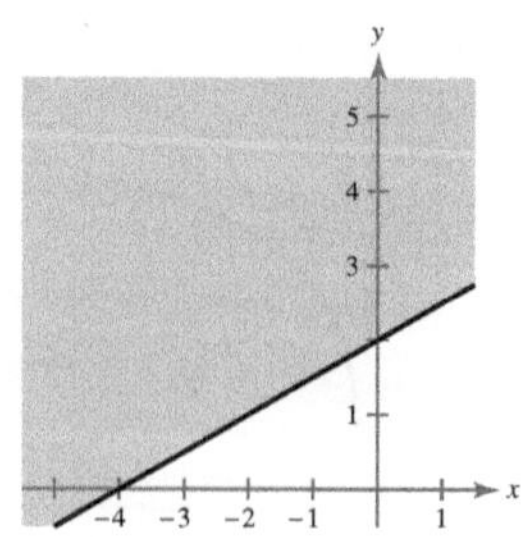

17.

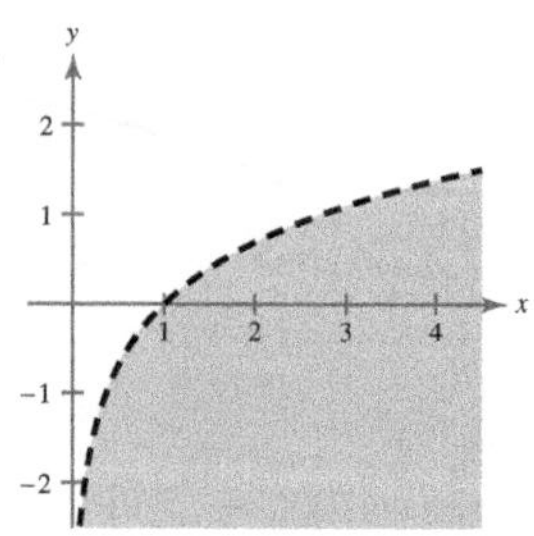

19. 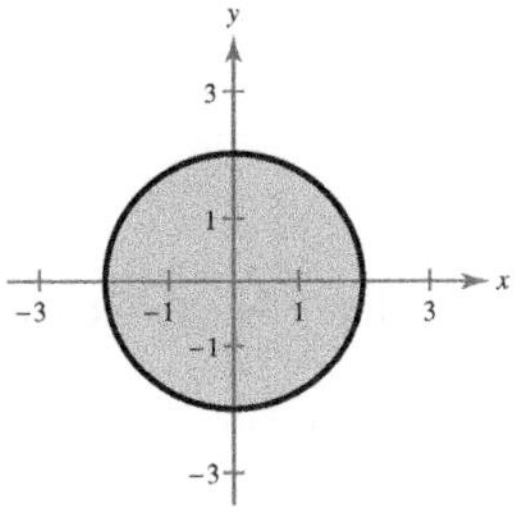

21. 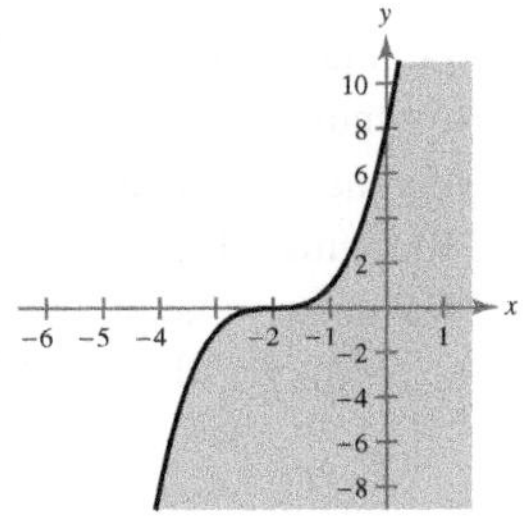

23. 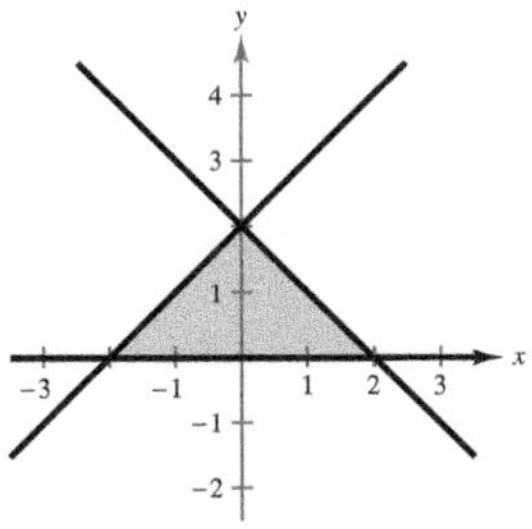

25. 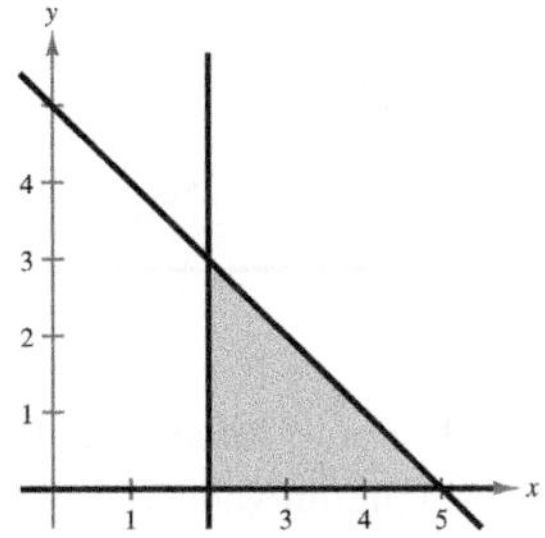

27. 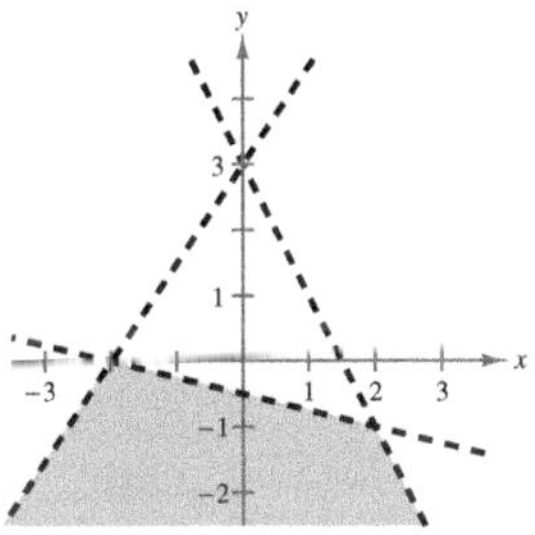

29. 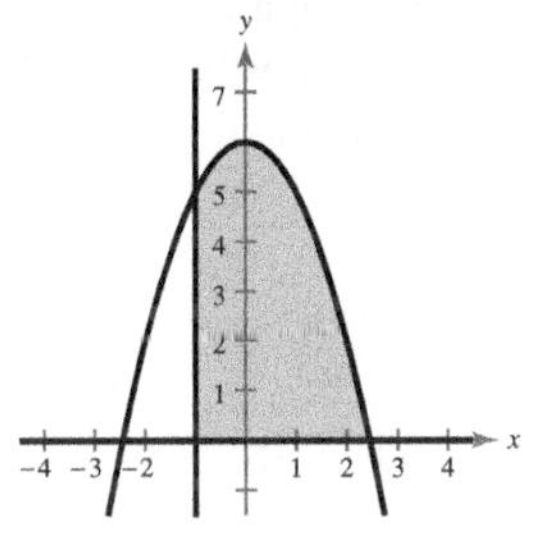

31. 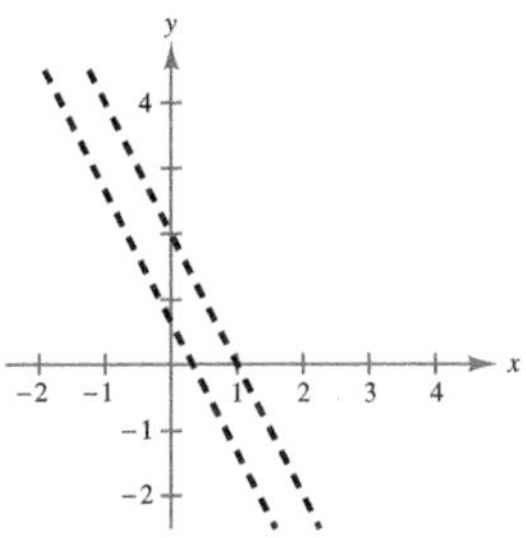

33. 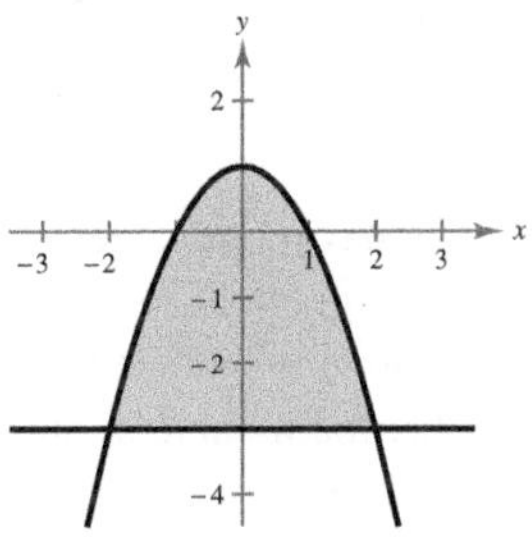

35. 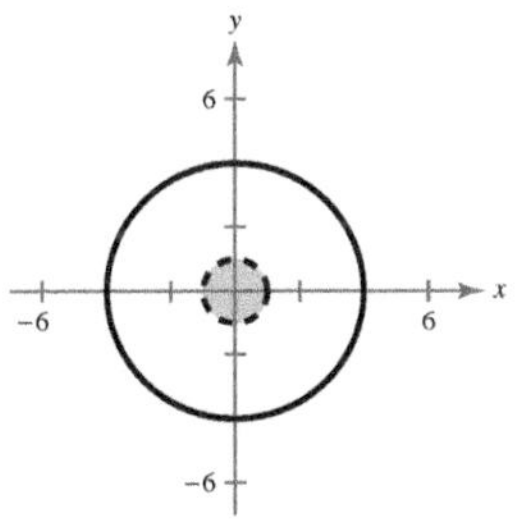

37. 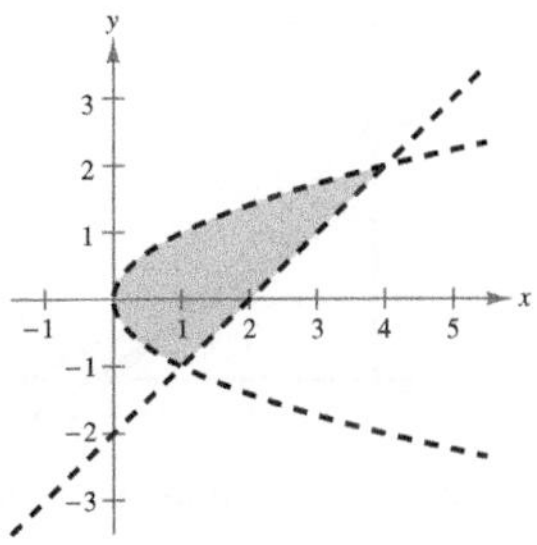

39. 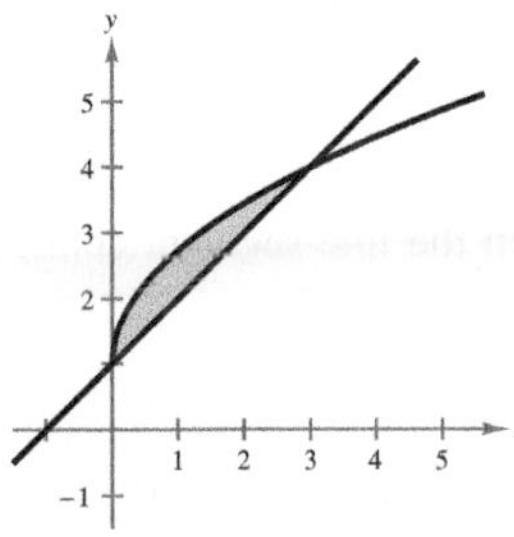

41. 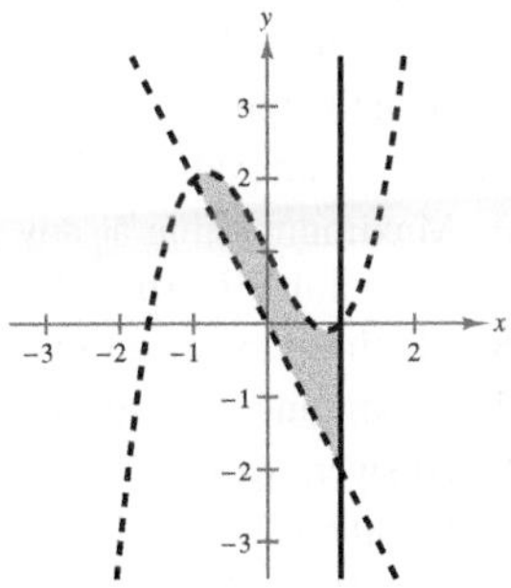

43. 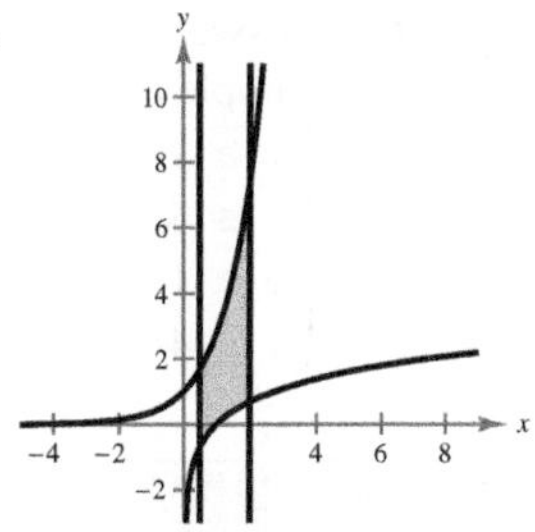

45. $y \le \frac{3}{2}x,\ y \ge \frac{3}{2}x - \frac{3}{2},\ 0 \le y \le 3$

47. $y \le \frac{4}{3}x,\ y \ge 0,\ y \le -4x + 16$

49. $x^2 + y^2 \le 16,\ x \ge 0,\ y \ge 0$

51. (a) $2x + \frac{3}{2}y \le 18,\ \frac{3}{2}x + \frac{3}{4}y \le 12,\ x \ge 0,\ y \ge 0$

(b) 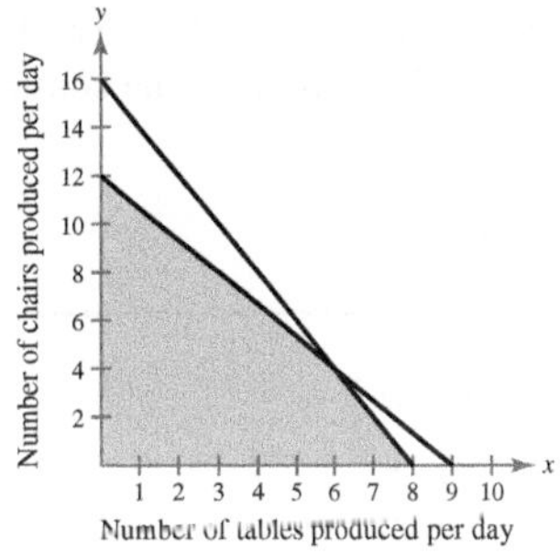

53. Consumer surplus: \$4,777,001.41
Producer surplus: \$477,545.60

55. Consumer surplus: \$40,000,000
Producer surplus: \$20,000,000

57. The consumer surplus and producer surplus are equal when the slope of the demand equation is the negative of the slope of the supply equation.

59. (a) $x + y \le 30{,}000$
$x \ge 7000$
$y \ge 7000$
$x - 2y \ge 0$

(b) 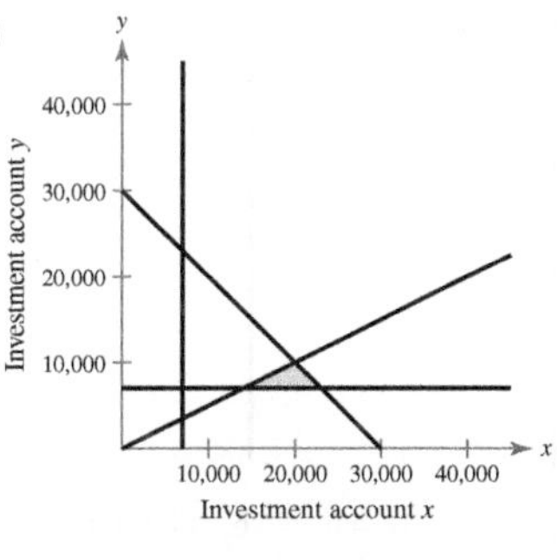

61. (a) $20x + 15y \ge 400$
$10x + 20y \ge 250$
$15x + 20y \ge 220$
$x \ge 0$
$y \ge 0$

(b) 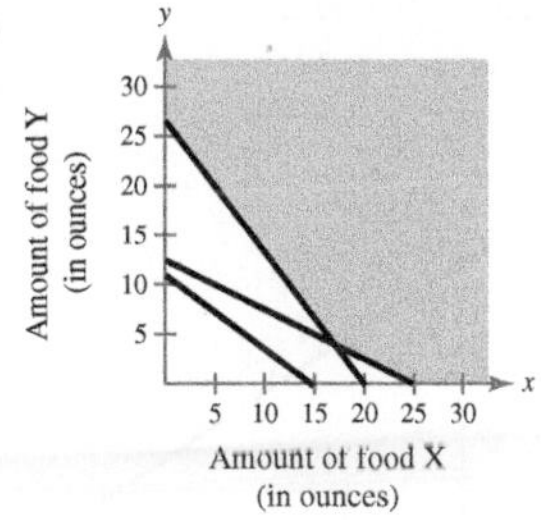

63. (a) 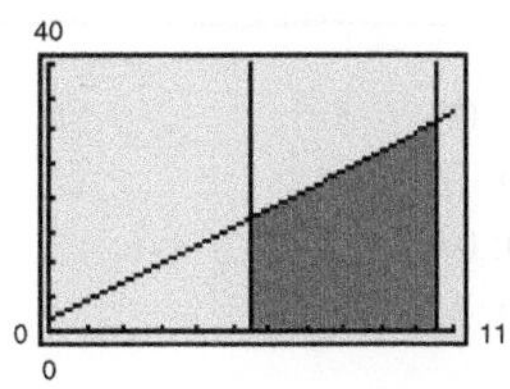

(b) 121 nesting pairs

CHAPTER 5

65. The graph is a half-line on the real number line; on the rectangular coordinate system, the graph is a half-plane.

67. Answers will vary. Sample answer:
$$\begin{cases} x \geq 1 \\ x \leq 4 \\ y \geq 1 \\ y \leq 4 \end{cases}$$

69. (a) $\begin{cases} \pi y^2 - \pi x^2 \geq 10 \\ y > x \\ x > 0 \end{cases}$ (b)

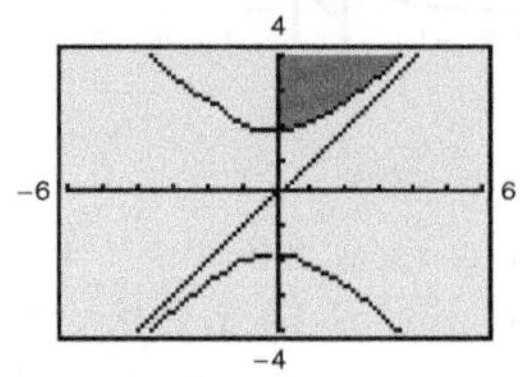

(c) The line is an asymptote to the boundary. The larger the circles, the closer the radii can be with the constraint still satisfied.

Section 5.5 *(page 250)*

Skills Warm Up *(page 250)*

1.

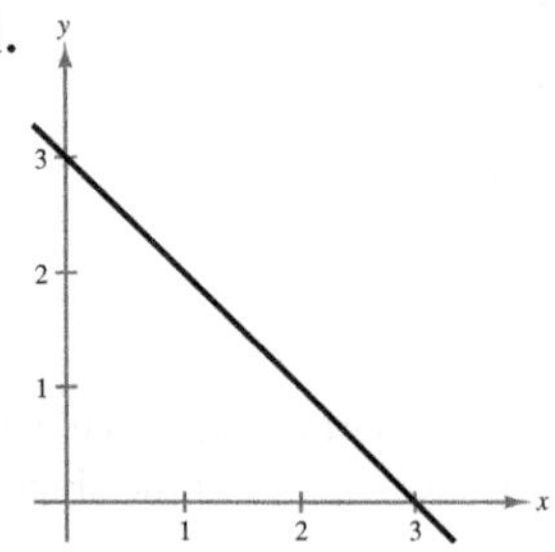

2.

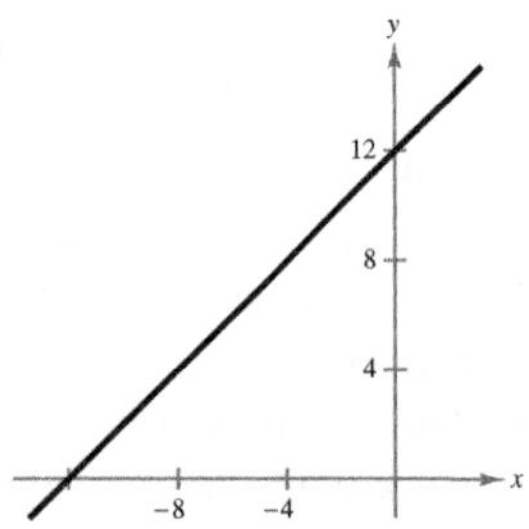

3.

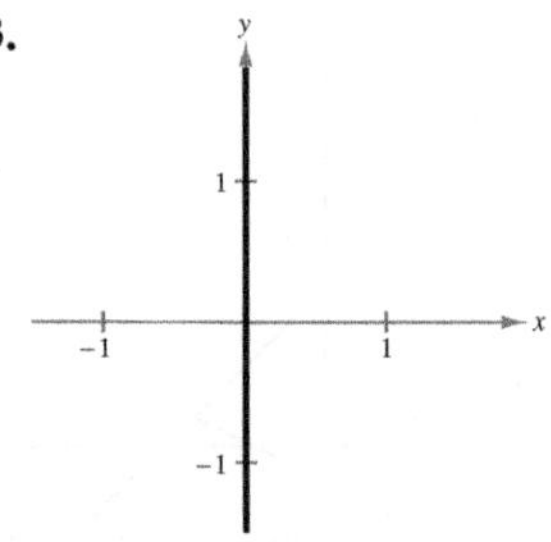

4.

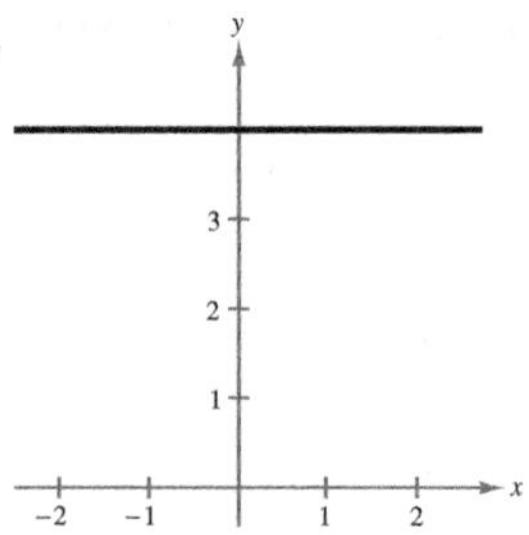

5. (0, 4) **6.** (12, 0) **7.** (3, 1) **8.** (2, 5)

9.

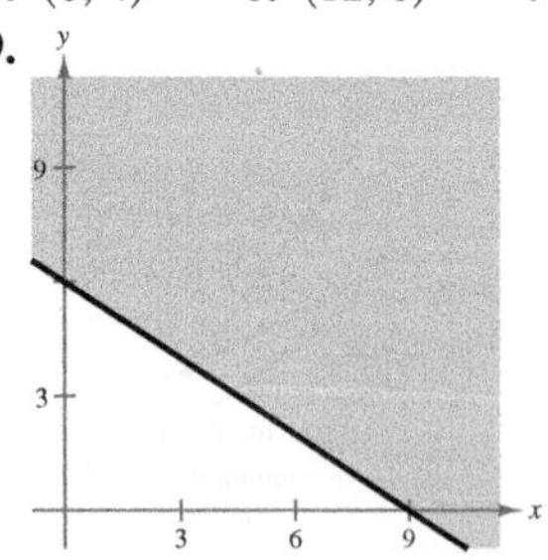

10.

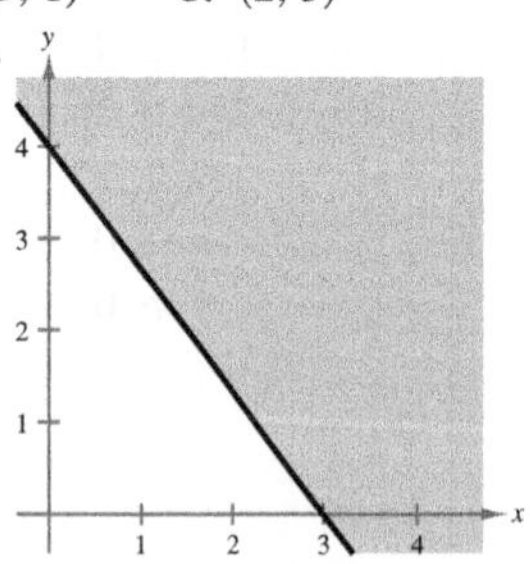

1. Minimum value at (0, 0): 0
Maximum value at (0, 5): 20

3. Minimum value at (0, 0): 0
Maximum value at (5, 0): 45

5. Minimum value at (0, 0): 0
Maximum value at (3, 4): 26

7. Minimum value at (0, 0): 0
Maximum value at (4, 0): 20

9. Minimum value at (0, 0): 0
Maximum value at any point on the line segment connecting the points (5, 0) and (0, 3): 30

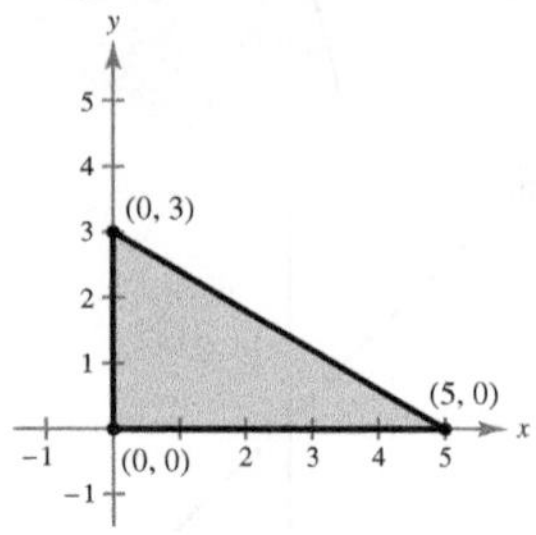

11. Minimum value at (0, 0): 0
Maximum value at (5, 0): 40
Same graph as in Answer 9

13. Minimum value at (5, 3): 35
No maximum value

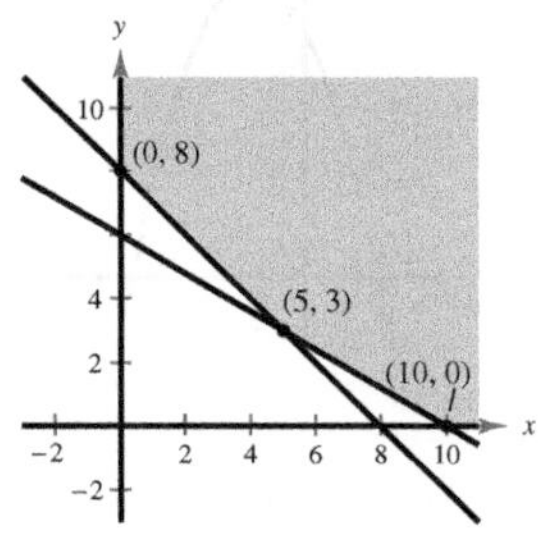

15. Minimum value at (10, 0): 20
No maximum value
Same graph as in Answer 13

17. Minimum value at (0, 0): 0
Maximum value at any point on the line segment connecting the points (0, 20) and (10, 15): 120

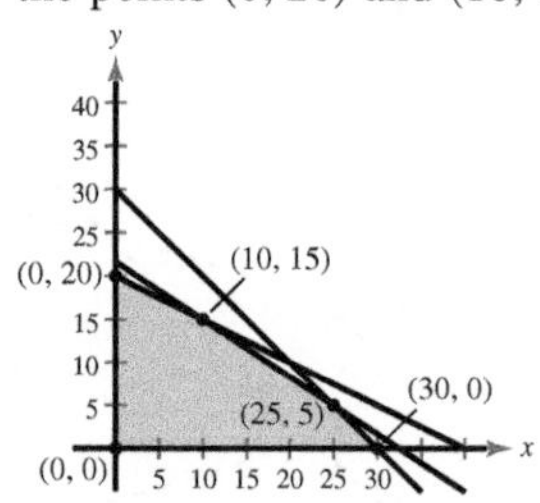

19. Minimum value at (0, 0): 0
Maximum value at any point on the line segment connecting the points (25, 5) and (30, 0): 30
Same graph as in Answer 17

21. Maximum value at (3, 6): 12

23. Maximum value at any point on the line segment connecting the points (0, 10) and (3, 6): 30

25. Maximum value at (5, 4): 58

27. Maximum value at (8, 0): 120

29. Answers will vary. Sample answer: $z = 2x + 15y$

31. Answers will vary. Sample answer: $z = -x$

33. Answers will vary.
Sample answer:
$z = x + 7y$

35. Answers will vary.
Sample answer:
$z = 6x + 5y$

37. Crop A: 60 acres
Crop B: 90 acres
\$33,150

39. Brand X: 3 bags
Brand Y: 6 bags
\$240

41. Model A: 0 bicycles
Model B: 1600 bicycles
\$120,000

43. 12 audits and 0 tax returns

45. Television: None
Newspaper: \$4,800,000
300 million people

47. Type A: \$62,500
Type B: \$187,500
\$26,875

49. Model A: 929 units
Model B: 77 units
\$99,445

51. z has a maximum value (of 5) at any point on the line segment connecting the vertices $(2, 0)$ and $\left(\frac{20}{19}, \frac{45}{19}\right)$.

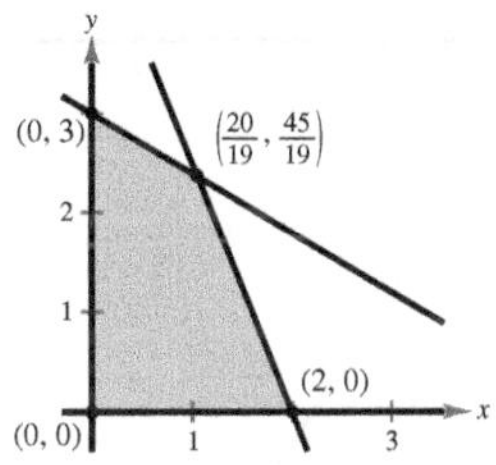

53. The constraint $x \leq 10$ is extraneous. The maximum value of z, $z = 14$, occurs at $(0, 7)$.

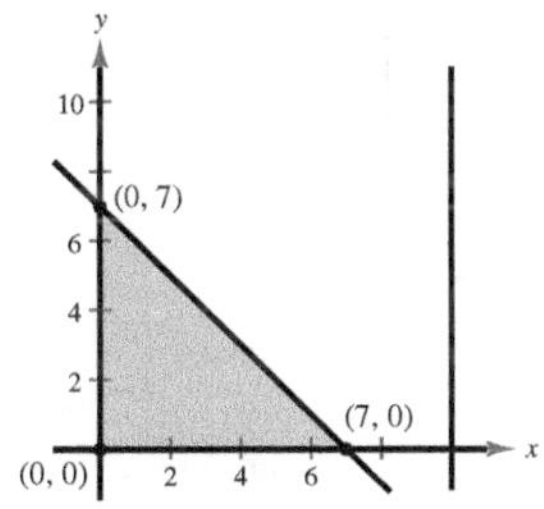

55. The constraint $2x + y \leq 4$ is extraneous. The maximum value of z, $z = 4$, occurs at $(0, 1)$.

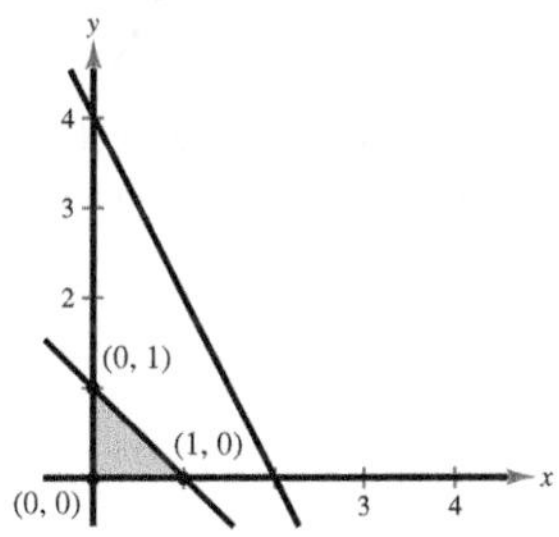

57. No. You need more information about the feasible region to determine whether the minimum also occurs at $(0, 0)$.

59. Yes; The objective function also has the maximum value at any point on the line segment connecting the two vertices, so there are an infinite number of points that produce the maximum value.

Review Exercises *(page 257)*

1. $(-2, 4)$ **3.** $(8, -10)$ **5.** $(8, 6), (0, 10)$

7. $(1, 9), (1.5, 8.75)$

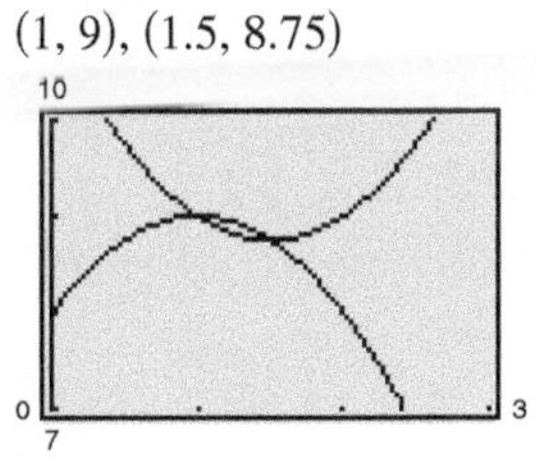

9. 800 plants **11.** During the fourth month of the new format

13. $(3, -5)$ **15.** $\left(a, \frac{4}{3}a - \frac{10}{3}\right)$, where a is any real number

17. Inconsistent **19.** $(8, 9)$

21. The graph is a point. Solution: $(-1, 1)$

23. (a) $\begin{cases} 0.1x + 0.5y = 0.25(12) \\ x + y = 12 \end{cases}$

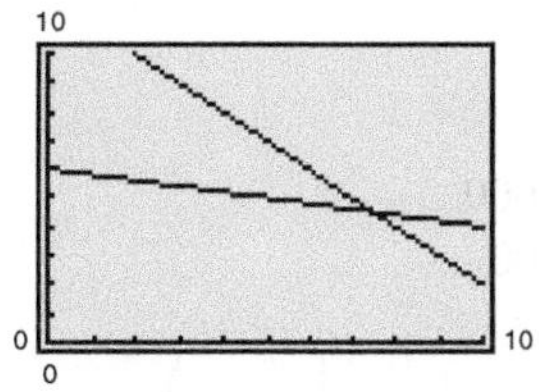

(b) 7.5 gallons of 10% solution
4.5 gallons of 50% solution

25. $y = 1.01x + 1.54$ **27.** $x = 71{,}429$ units, $p = \$22.71$

29. $(12, 5, 2)$ **31.** $(2, -1, 3)$

33. $\left(\frac{1}{5}a + \frac{8}{5}, -\frac{6}{5}a + \frac{42}{5}, a\right)$, a is any real number

35. Inconsistent **37.** $y = 2x^2 + x - 6$

39. $x^2 + y^2 - 4x + 2y - 4 = 0$

41. \$200,000 in certificates of deposit
\$100,000 in municipal bonds
\$15,000 in blue-chip stocks
\$185,000 in growth or speculative stocks

43. $y = 0.114x^2 + 0.80x + 1.6$

45.

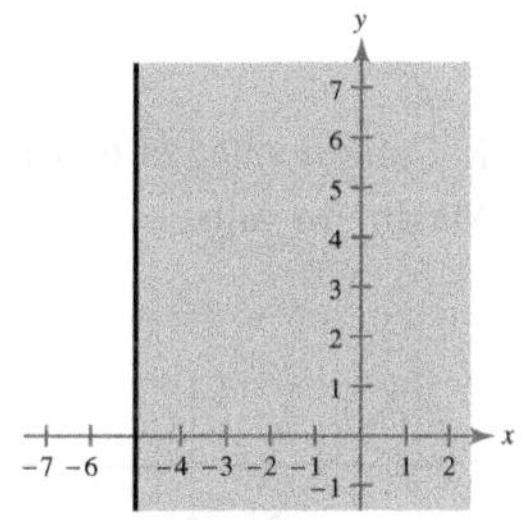

47.

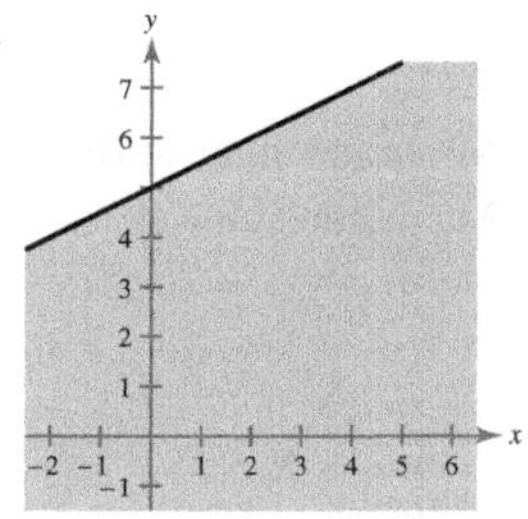

49.

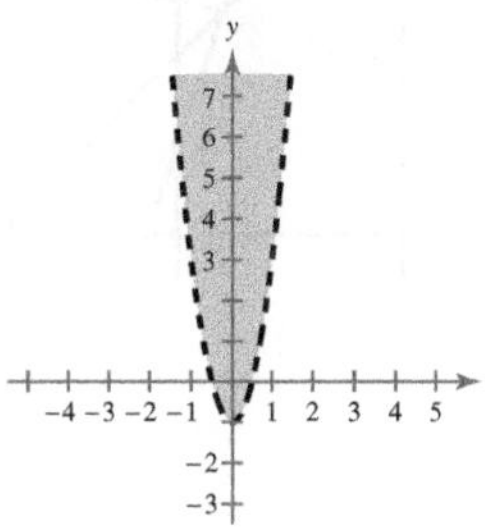

51.

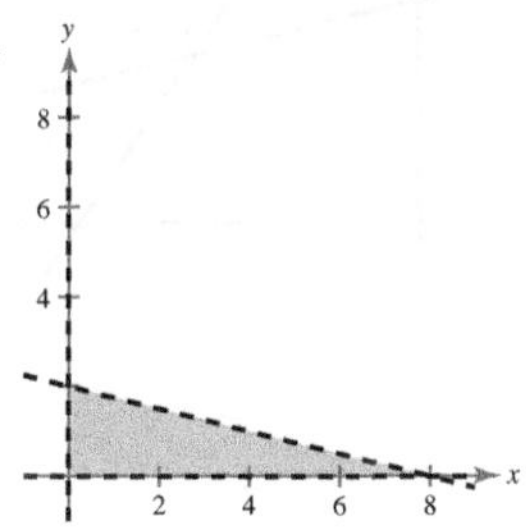

53.

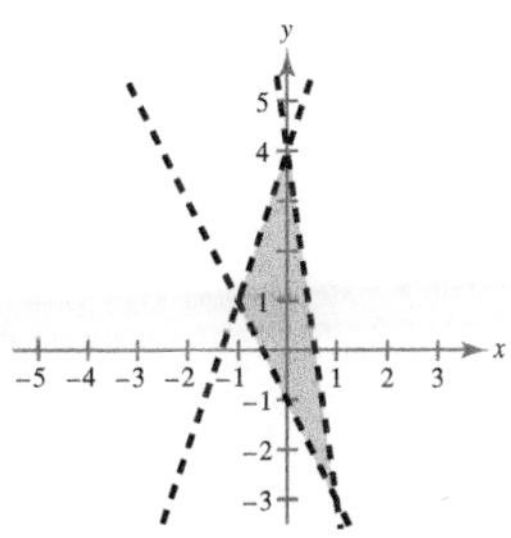

55.

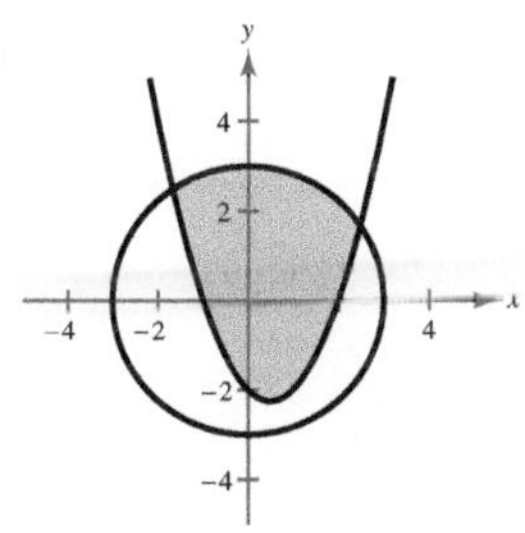

57.

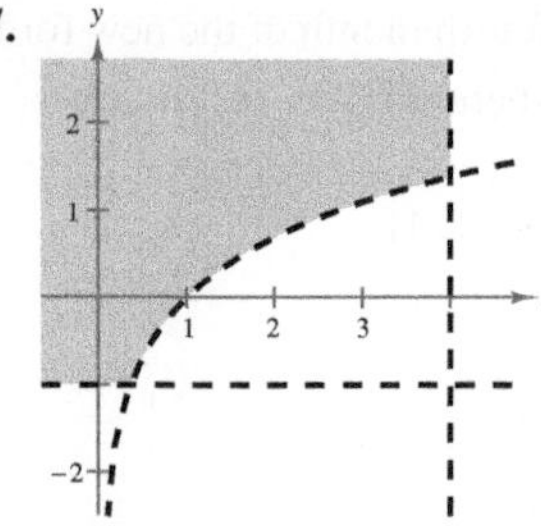

59. $\begin{cases} y \le \frac{5}{2}x - \frac{3}{2} \\ y \le \frac{1}{5}x + \frac{27}{5} \\ y \ge \frac{5}{2}x - 13 \\ y \ge \frac{1}{5}x + \frac{4}{5} \end{cases}$

61. Consumer surplus: $4,500,000
Producer surplus: $9,000,000

63. $\begin{cases} x \ge 2y \\ 100x + 150y \le 4000 \\ x \ge 4 \\ y \ge 2 \end{cases}$

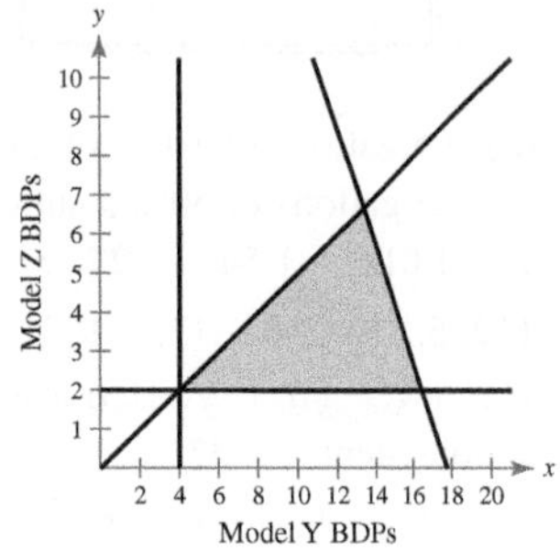

65. Minimum value at (0, 0): 0
Maximum value at (0, 8): 48

67. Minimum value at (0, 0): 0
Maximum value at any point on the line segment connecting (25, 35) and (50, 15): 550

69. Minimum value at (0, 0): 0
Maximum value at (4, 3): 48

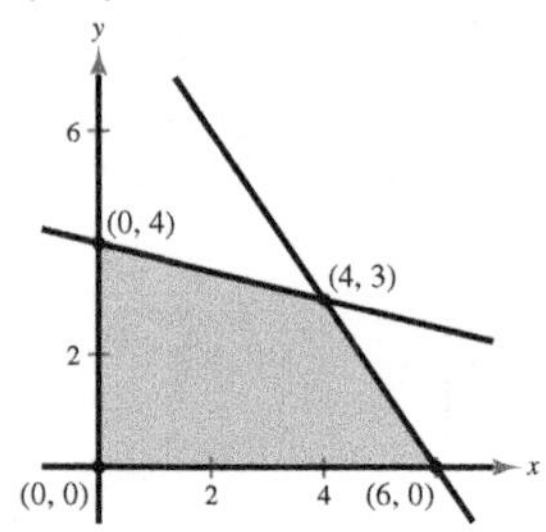

71. Minimum value at (0, 0): 0
Maximum value at (4, 5): 47

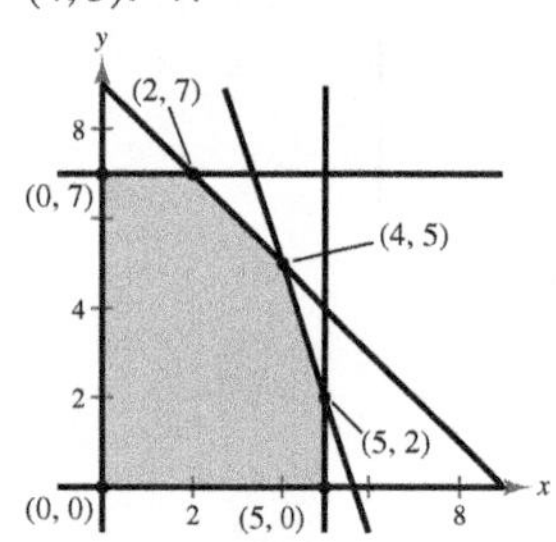

73. 369 units of model A
538 units of model B
$117,600

75. $525 model: 0 units
$675 model: 305 units
$38,125

77. 8 audits
0 tax returns
$20,000

Test Yourself *(page 262)*

1. (2, 4) **2.** (−2, 5), (3, 0) **3.** (−1, 6), (2, 0)
4. (3.36, −1.32) **5.** (2, −1, 3) **6.** (2, −3, 4)
7. $60,000 at 2.3%
$20,000 at 3.1% **8.** (50,000, 34)
9. (a) $y = 0.593x^2 + 4.95x + 46.7$ (b) ≈66.9 million adults

10.

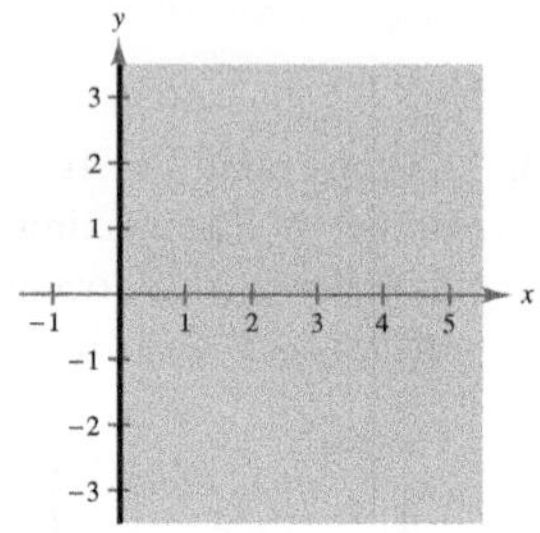

11.

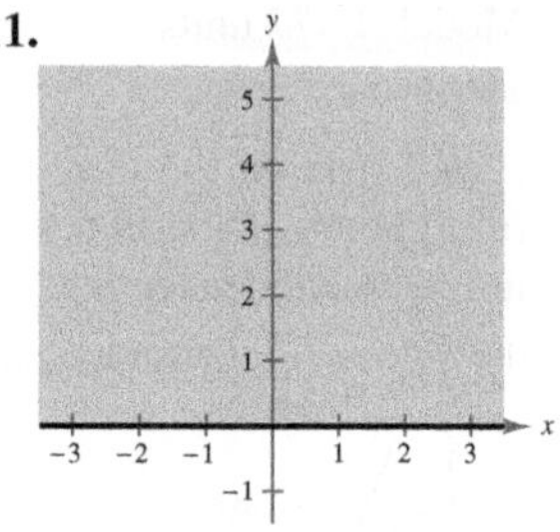

12.

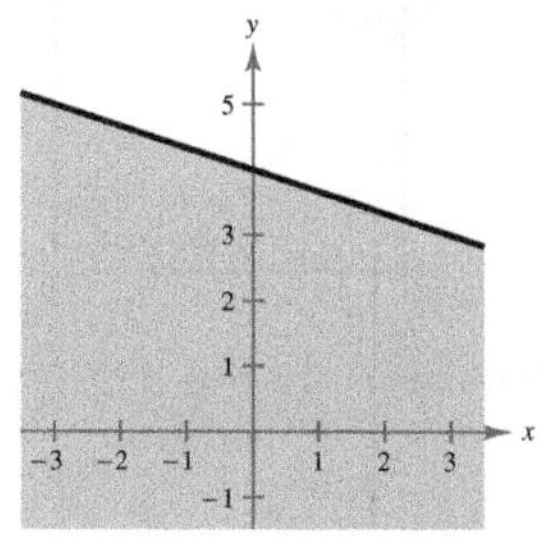

13.

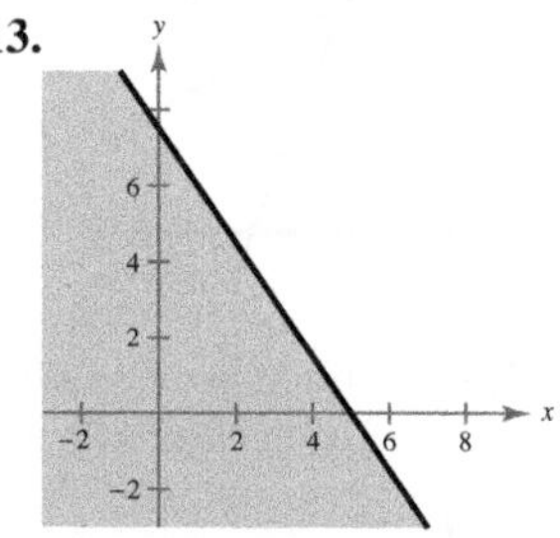

14.

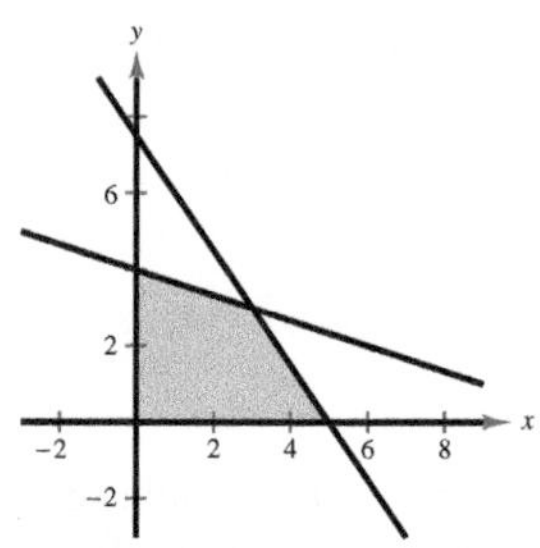

15. Minimum value at (0, 0): 0
Maximum value at (3, 3): 39

Answers to Checkpoints

Chapter 1

Checkpoints for Section 1.1

1 Identity **2** (a) 2 (b) -1 **3** (a) -11 (b) $\frac{11}{3}$
4 Infinitely many solutions **5** No solution **6** 30
7 No solution **8** 6 **9** 0.794 **10** 2003 ($t \approx 3$)

Checkpoints for Section 1.2

1 \$950 **2** $p = 6800 + 0.05(6800)$ **3** 20%
4 1.5% **5** 15 feet × 45 feet
6 1.1 hours **7** 32 feet
8 \$1200 was invested at 1% and \$1300 was invested at 2%.
9 About 1.27 feet

Checkpoints for Section 1.3

1 (a) $-3, 4$ (b) $-6, -4$ **2** 0, 2 **3** -2
4 ± 2 **5** $-3, 5$ **6** 6 feet × 14 feet
7 3 seconds **8** ≈ 8.5 feet **9** 2010 ($t \approx 10.80$)

Checkpoints for Section 1.4

1 One real solution **2** $-1 \pm \sqrt{3}$ **3** $\frac{1}{3}$
4 $-0.831, 1.511$ **5** 2:00 P.M. **6** ≈ 5.1 seconds

Checkpoints for Section 1.6

1 (a) $2 \le x < 7$; bounded (b) $-\infty < x < 6$; unbounded
2 $x < 1$
3 $x \ge -6$

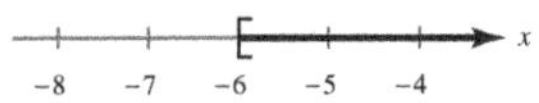

4 $-1 \le x < 2$

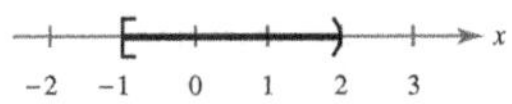

5 $-9 \le x \le 5$

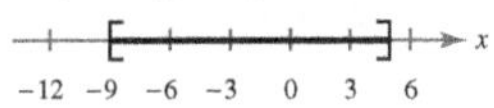

6 $x < -4$ or $x > 2$

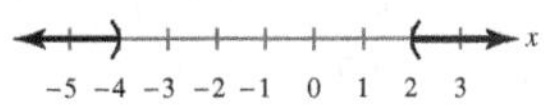

7 More than 3.5 miles **8** At most $12\frac{1}{2}$ weeks
9 Undercharged by as much as \$0.08 or overcharged by as much as \$0.08

Chapter 2

Checkpoints for Section 2.2

1 $\frac{1}{2}$
2 (a) $y = -2x + 8$ (b) $y = \frac{3}{2}x + 9$ (c) $y = 1$
3 $y = 0.2x + 1$
4 (a)

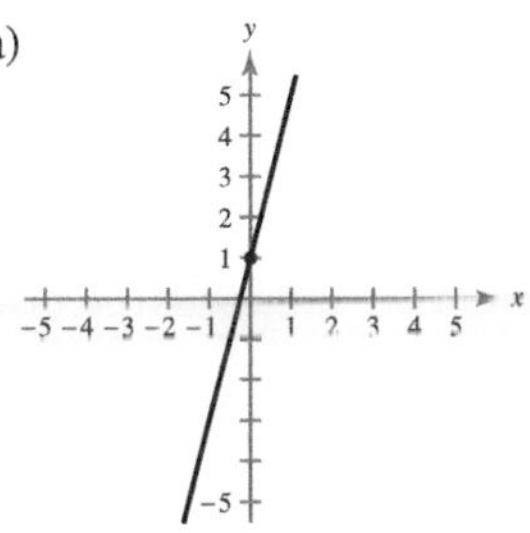

(b)

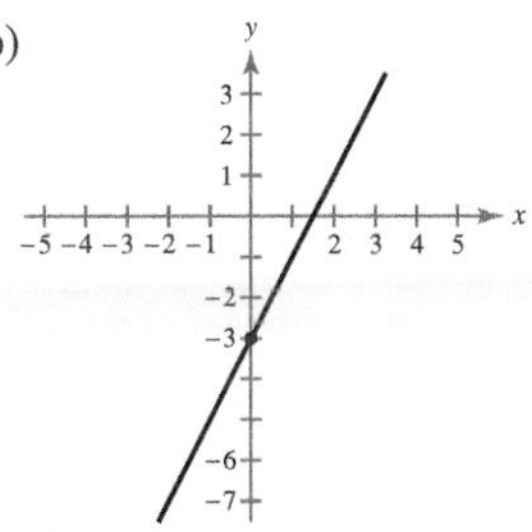

5 (a) $y = 3x - 2$ (b) $y = 3x + 6$ **6** $y = -4x + 4$

Checkpoints for Section 2.3

1 The model approximates the weight of the puppy best for $t = 2$ months and worst for $t = 10$ months.
2 $y = 0.06x$ **3** $y = 33.84615x$ **4** 6 P.M.
5 836,821 people **6** $V = -195t + 2300$
7 (a) $y = 1.97x + 121.4$ (b)

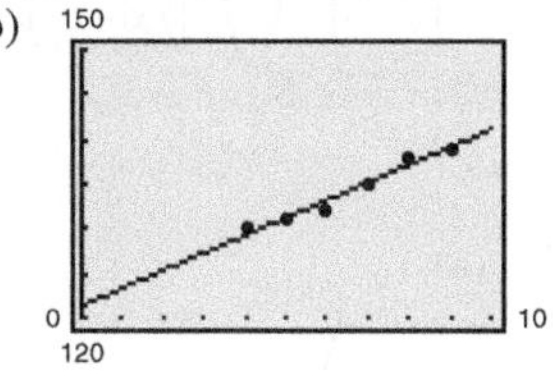

(c) $\approx$ 143,000,000 tax returns

8

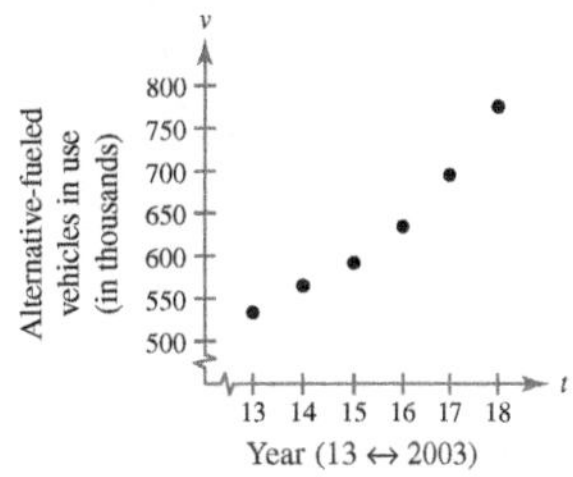

Answers will vary. Sample answer: $v = 50t - 150$

Checkpoints for Section 2.4

1 (a) No (b) Yes (c) No
2 (a) Yes (b) No **3** -3 **4** 10, 2
5 All real numbers **6** $V = 4\pi h^3$ **7** No
8

t	9	10	11	12	13
p	160.2	168.9	177.6	186.2	194.9

t	14	15	16	17	18	19
p	221.9	239.3	247.3	246.0	235.3	215.3

Checkpoints for Section 2.5

1 Domain: $(-\infty, \infty)$ **2** (a) Yes (b) Yes
Range: $[-3, \infty)$
3 Decreasing on $\left(-\infty, -\frac{3}{2}\right)$ and increasing on $\left(-\frac{3}{2}, \infty\right)$
4 (2, 2) **5** (a)

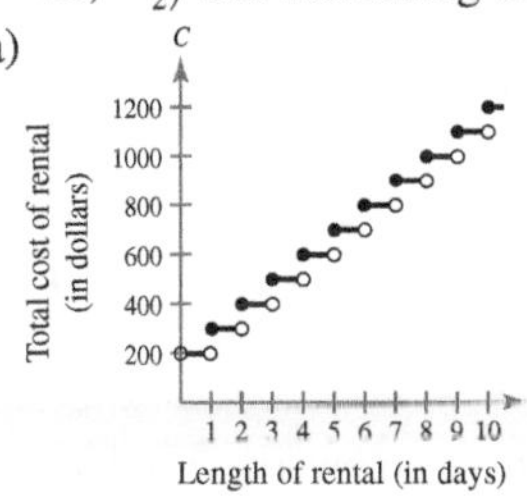

(b) 8 days or less

6 Neither

Chapter 3

Checkpoints for Section 3.1

1

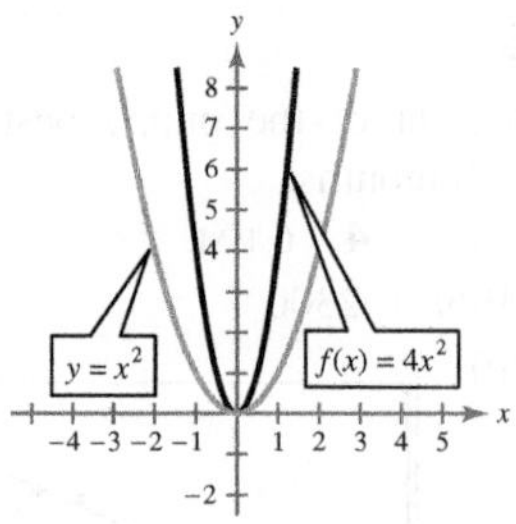

Compared with the graph of $y = x^2$, each output of $f(x) = 4x^2$ vertically stretches the graph by a factor of 4.

2 (a)

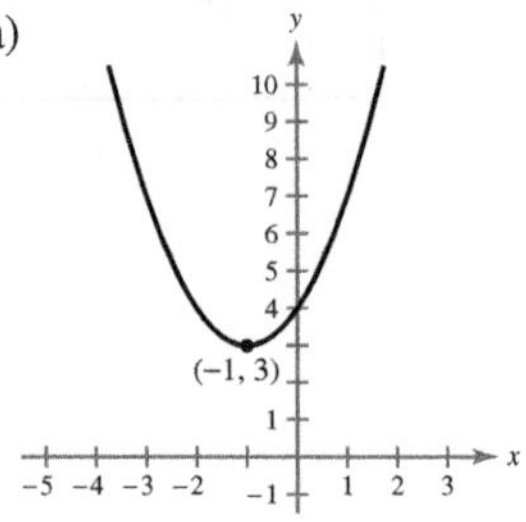

Vertex: $(-1, 3)$

(b)

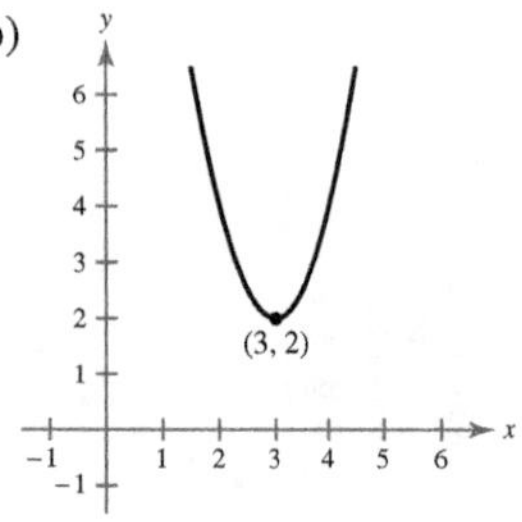

Vertex: $(3, 2)$

3

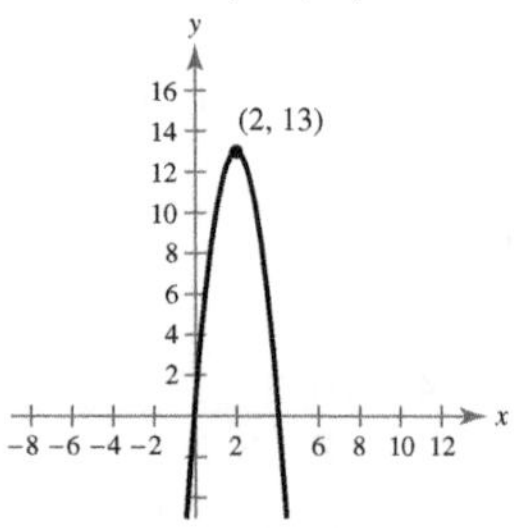

Vertex: $(2, 13)$

4 $f(x) = x^2 - 6x + 13$

5 ≈ 39.7 feet

6 30 moth sites; Answers will vary.

7 30 units

Chapter 4

Checkpoints for Section 4.1

1 $f^{-1}(x) = 6x$ **2** $f^{-1}(x) = x - 10$

3 $(\sqrt[3]{x - 6})^3 + 6 = x$; $\sqrt[3]{(x^3 + 6) - 6} = x$

4 $g(x)$ **5** $f^{-1}(x) = \dfrac{x - 5}{4}$

6

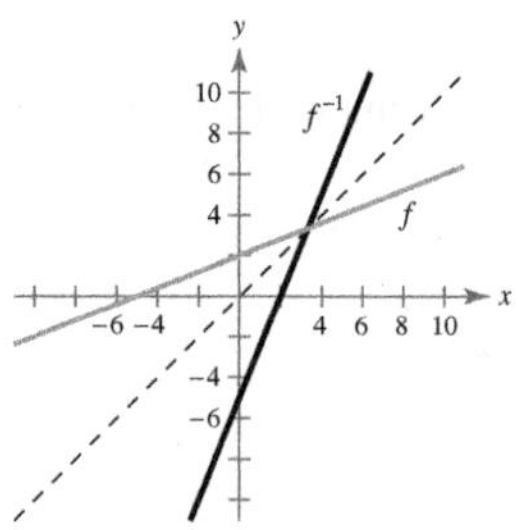

7

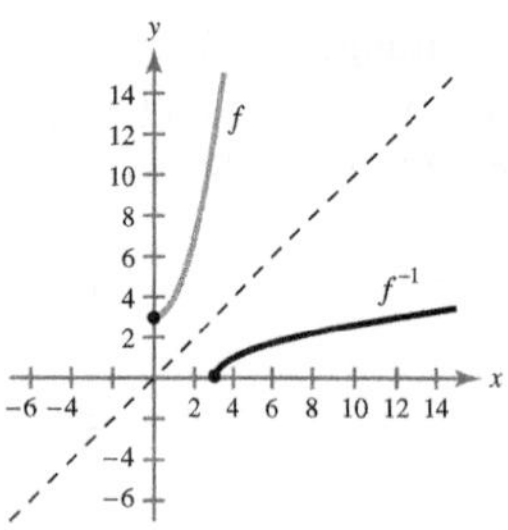

8 (a) f does not have an inverse function.
(b) f has an inverse function.

Checkpoints for Section 4.2

1 3.918

2

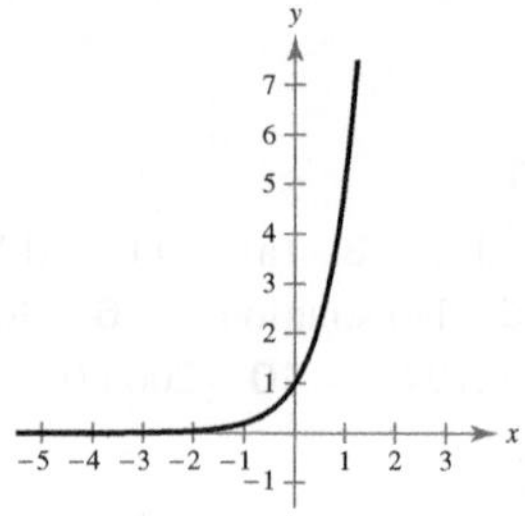

3

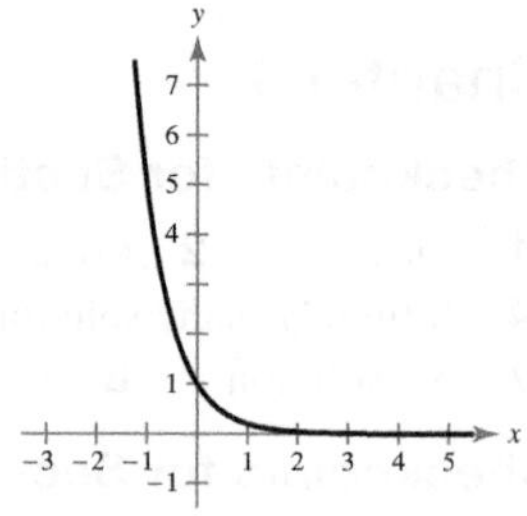

4

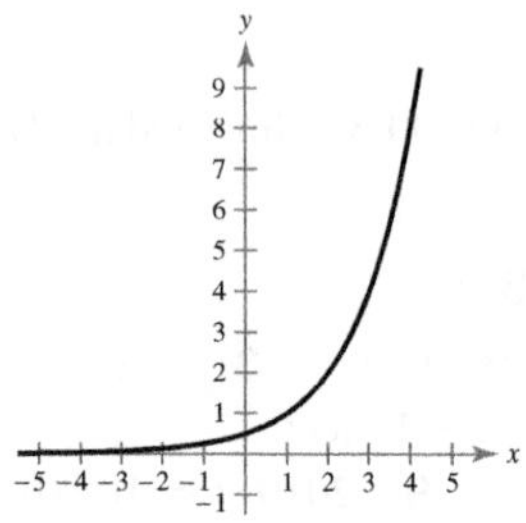

5 403.4287935; 0.0301973834

6

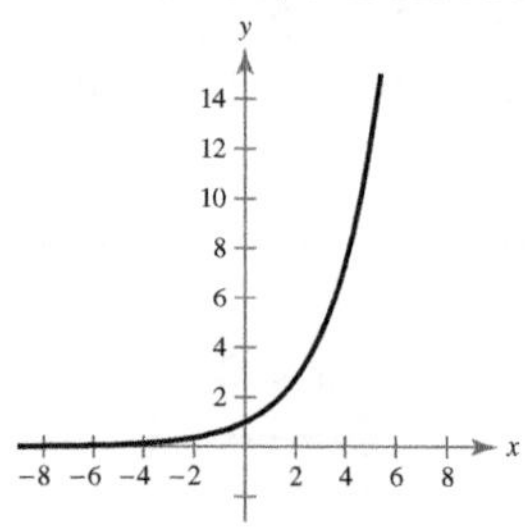

7 (a) \$7927.75 (b) \$7935.08 (c) \$7938.78

8 ≈ 9.971 pounds

Checkpoints for Section 4.3

1 (a) 2 (b) -2 **2** 2.301 **3** 0

4

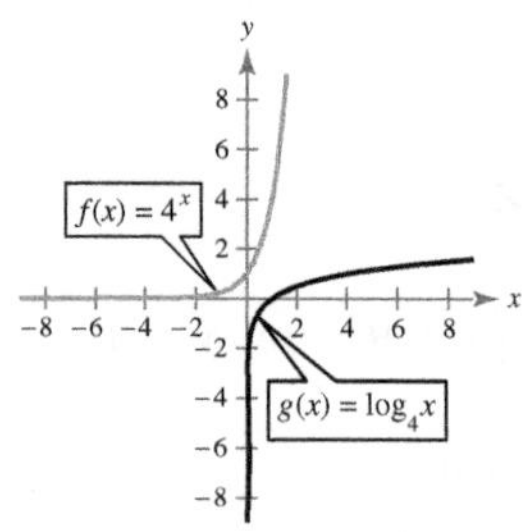

5

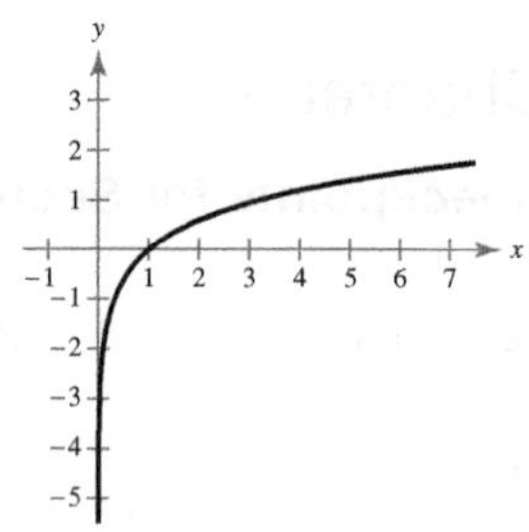

6

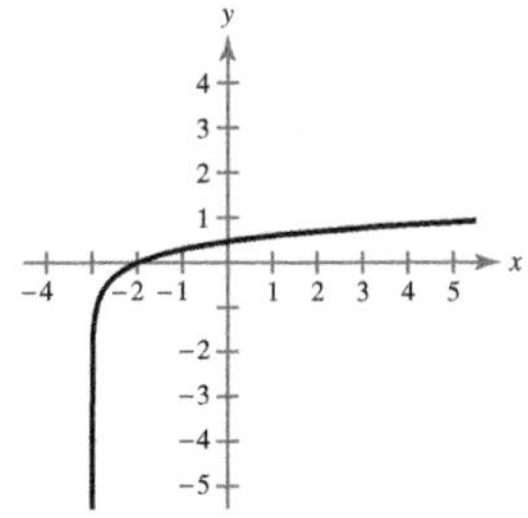

7 (a) 7 (b) 0 (c) $\frac{3}{4}$ **8** -2.303 **9** $(-5, \infty)$

10 103.0 in., 114.0 in., 128.2 in., 132.8 in., 141.7 in.

Checkpoints for Section 4.4

1 1.936 **2** 1.936 **3** $2 \log_{10} 5 - \log_{10} 3$

4 $-\ln \frac{2}{e} = -(\ln 2 - \ln e)$
$= -\ln 2 + \ln e$
$= 1 - \ln 2$
5 $\ln 2 + \ln m + 2 \ln n$

6 $\log_{10} \frac{(x+1)^2}{(x-1)^3}$ **7** $\ln y = \frac{1}{2} \ln x$ **8** 30 decibels

Checkpoints for Section 4.5

1 (a) 4 (b) 216 (c) 5

2 (a) $\log_6 84 \approx 2.473$ (b) $\log_{10} 38 \approx 1.580$

3 $\ln 15 \approx 2.708$ **4** $\frac{1}{2}(7 + \log_4 24) \approx 4.646$

5 $\ln 3 \approx 1.099$, $\ln 4 \approx 1.386$ **6** (a) 5 (b) $\frac{81}{2}$

7 3 **8** $e^4 \approx 54.598$ **9** 4 **10** ≈ 2.70 years

11 2001 ($t \approx 11.01$)

Checkpoints for Section 4.6

1 2013 ($t \approx 23.89$) **2** $y = 3e^{0.19617x}$ **3** ≈ 7671 years

4

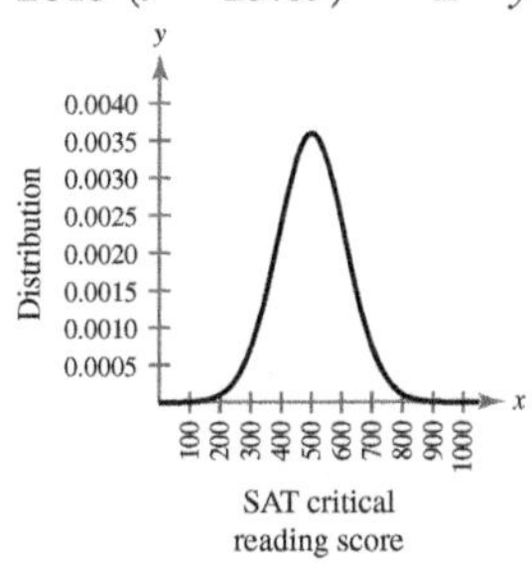

Average score: 501

5 9 days **6** $\approx 31{,}622{,}777$; 100 times greater

7 1×10^{-5} mole of hydrogen per liter

8 Logistic growth model

Chapter 5

Checkpoints for Section 5.1

1 (5, 1)

2 \$4000 is invested at 3% and \$11,000 is invested at 5%.

3 $(-4, -7), (2, 5)$ **4** No solution **5** (4, 0)

6 ≈ 4762 pairs of shoes **7** 2006 ($t \approx 6.7$)

Checkpoints for Section 5.2

1 $\left(2, -\frac{1}{3}\right)$ **2** (3, 2) **3** $(-3, -3)$

4 $(1, -4)$ **5** No solution **6** $(a, -a - 5)$

7 Speed of plane: ≈ 471.18 miles per hour
Speed of wind: ≈ 16.63 miles per hour

8 (2,000,000, 290)

Checkpoints for Section 5.4

1

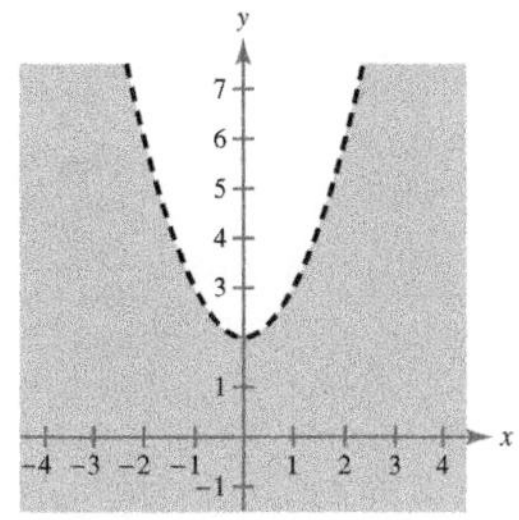

2

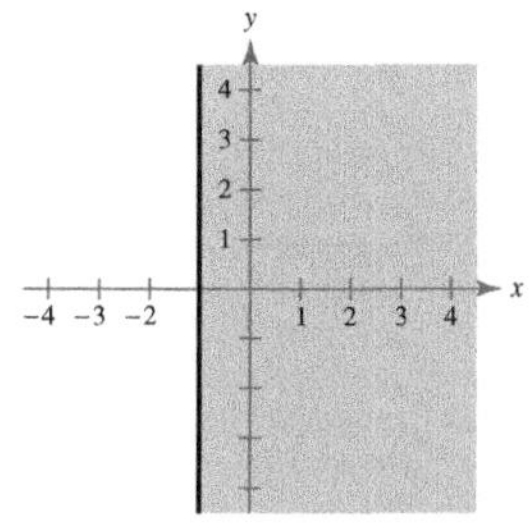

3

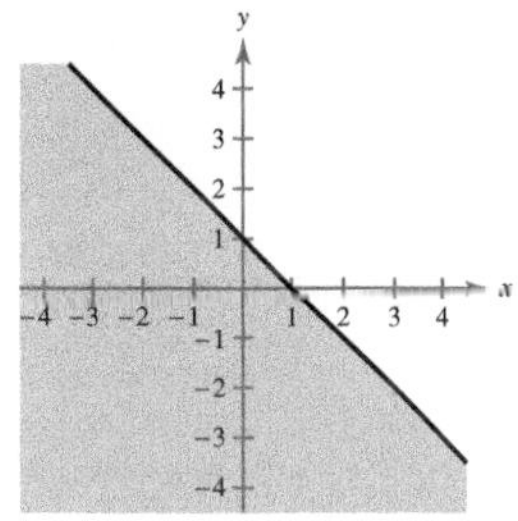

4

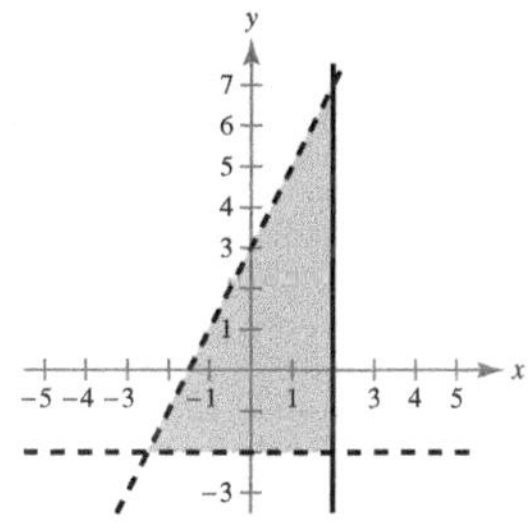

5

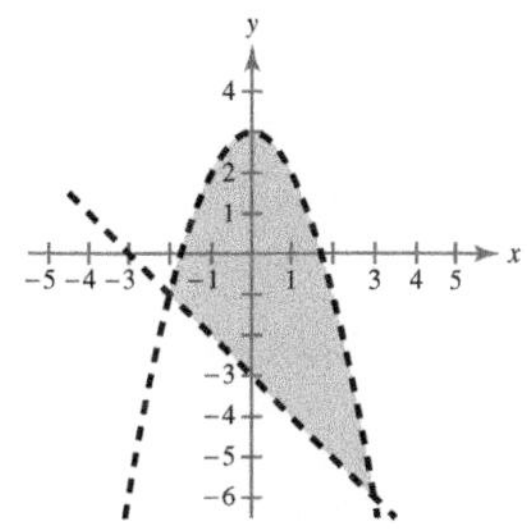

6 Consumer surplus: \$60,000,000
Producer surplus: \$60,000,000

7 No. The combination of 4 cups of dietary drink X and 1 cup of dietary drink Y does not meet all the minimum daily requirements.

Checkpoints for Section 5.5

1 Maximum value at (0, 3): 9

2 (a) Maximum value at (27, 0): 135
(b) Minimum value at (0, 0): 0

3 No maximum value

4 The maximum profit would be \$2925, and it would occur at monthly production levels of 1050 units of product I and 150 units of product II.

5 The minimum cost would be \$0.56 per day, and it would occur when 1 cup of drink X and 4 cups of drink Y were consumed each day.

Answers to Tech Tutors

Section 1.2 *(page 17)*

\$143.75

Section 1.4 *(page 38)*

1.854, −4.854

Section 2.2 *(page 69)*

$y = 2.2x + 11.3$

Section 4.6 *(page 190)*

$C = 138.78e^{0.167t}$; the models are the same.

Section 5.1 *(page 213)*

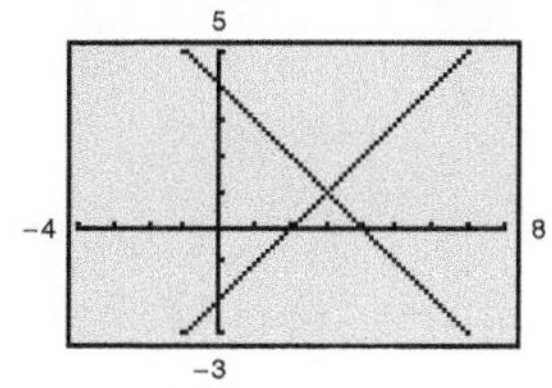

Solution: (3, 1)

Section 5.2 *(page 225)*

(16, 14)

1

Paul Spinelli/Contributor/Major League Baseball/Getty Images

Bettmann/Historical/Corbis

Chance favors the prepared mind.

—Louis Pasteur

Statistical techniques are tools of thought . . . not substitutes for thought.

—Abraham Kaplan

Louis Pasteur (1822–1895) is the founder of modern bacteriology. At age 57, Pasteur was studying cholera. He accidentally left some bacillus culture unattended in his laboratory during the summer. In the fall, he injected laboratory animals with this bacilli. To his surprise, the animals did not die—in fact, they thrived and were resistant to cholera.

When the final results were examined, it is said that Pasteur remained silent for a minute and then exclaimed, as if he had seen a vision, "Don't you see they have been vaccinated!" Pasteur's work ultimately saved many human lives.

Most of the important decisions in life involve incomplete information. Such decisions often involve so many complicated factors that a complete analysis is not practical or even possible. We are often forced into the position of making a guess based on limited information.

As the first quote reminds us, our chances of success are greatly improved if we have a "prepared mind." The statistical methods you will learn in this book will help you achieve a prepared mind for the study of many different fields. The second quote reminds us that statistics is an important tool, but it is not a replacement for an in-depth knowledge of the field to which it is being applied.

The authors of this book want you to understand and enjoy statistics. The reading material will *tell you* about the subject. The examples will *show you* how it works. To understand, however, you must *get involved*. Guided exercises, calculator and computer applications, section and chapter problems, and writing exercises are all designed to get you involved in the subject. As you grow in your understanding of statistics, we believe you will enjoy learning a subject that has a world full of interesting applications.

GETTING STARTED

PREVIEW QUESTIONS

Why is statistics important? (SECTION 1.1)

What is the nature of data? (SECTION 1.1)

How can you draw a random sample? (SECTION 1.2)

What are other sampling techniques? (SECTION 1.2)

How can you design ways to collect data? (SECTION 1.3)

National News/ZUMA Press/Newscom

FOCUS PROBLEM

Where Have All the Fireflies Gone?

A feature article in *The Wall Street Journal* discusses the disappearance of fireflies. In the article, Professor Sara Lewis of Tufts University and other scholars express concern about the decline in the worldwide population of fireflies.

There are a number of possible explanations for the decline, including habitat reduction of woodlands, wetlands, and open fields; pesticides; and pollution. Artificial nighttime lighting might interfere with the Morse-code-like mating ritual of the fireflies. Some chemical companies pay a bounty for fireflies because the insects contain two rare chemicals used in medical research and electronic detection systems used in spacecraft.

Courtesy of Corrinne and Charles Brase

Adapted from Ohio State University Firefly Files logo

What does any of this have to do with statistics?

The truth, at this time, is that no one really knows (a) how much the world firefly population has declined or (b) how to explain the decline. The population of all fireflies is simply too large to study in its entirety.

In any study of fireflies, we must rely on incomplete information from samples. Furthermore, from these samples we must draw realistic conclusions that have statistical integrity. This is the kind of work that makes use of statistical methods to determine ways to collect, analyze, and investigate data.

Suppose you are conducting a study to compare firefly populations exposed to normal daylight/darkness conditions with firefly populations exposed to continuous light (24 hours a day). You set up two firefly colonies in a laboratory environment. The two colonies are identical except that one colony is exposed to normal daylight/darkness conditions and the other is exposed to continuous light. Each colony is populated with the same number of mature fireflies. After 72 hours, you count the number of living fireflies in each colony.

After completing this chapter, you will be able to answer the following questions.

(a) Is this an experiment or an observation study? Explain.

(b) Is there a control group? Is there a treatment group?

For online student resources, visit the Brase/Brase, *Understanding Basic Statistics*, 7th edition web site at http://www.cengagebrain.com

(c) What is the variable in this study?
(d) What is the level of measurement (nominal, interval, ordinal, or ratio) of the variable? (See Problem 11 of the Chapter 1 Review Problems.)

SECTION 1.1 What Is Statistics?

FOCUS POINTS

- Identify variables in a statistical study.
- Distinguish between quantitative and qualitative variables.
- Identify populations and samples.
- Distinguish between parameters and statistics.
- Determine the level of measurement.
- Compare descriptive and inferential statistics.

INTRODUCTION

Decision making is an important aspect of our lives. We make decisions based on the information we have, our attitudes, and our values. Statistical methods help us examine information. Moreover, statistics can be used for making decisions when we are faced with uncertainties. For instance, if we wish to estimate the proportion of people who will have a severe reaction to a flu shot without giving the shot to everyone who wants it, statistics provides appropriate methods. Statistical methods enable us to look at information from a small collection of people or items and make inferences about a larger collection of people or items.

Procedures for analyzing data, together with rules of inference, are central topics in the study of statistics.

Statistics

Statistics is the study of how to collect, organize, analyze, and interpret numerical information from data.

The subject of statistics is multifaceted. The following definition of statistics is found in the *International Encyclopedia of Statistical Science*, edited by Miodrag Lovric. Professor David Hand of Imperial College London—the president of the Royal Statistical Society—presents the definition in his article "Statistics: An Overview."

Statistics is both the science of uncertainty and the technology of extracting information from data.

The statistical procedures you will learn in this book should supplement your built-in system of inference—that is, the results from statistical procedures and good sense should dovetail. Of course, statistical methods themselves have no power to work miracles. These methods can help us make some decisions, but not all conceivable decisions. Remember, even a properly applied statistical procedure is no more accurate than the data, or facts, on which it is based. Finally, statistical results should be interpreted by one who understands not only the methods, but also the subject matter to which they have been applied.

The general prerequisite for statistical decision making is the gathering of data. First, we need to identify the individuals or objects to be included in the study and the characteristics or features of the individuals that are of interest.

Individuals
Variable

> **Individuals** are the people or objects included in the study.
> A **variable** is a characteristic of the individual to be measured or observed.

For instance, if we want to do a study about the people who have climbed Mt. Everest, then the individuals in the study are all people who have actually made it to the summit. One variable might be the height of such individuals. Other variables might be age, weight, gender, nationality, income, and so on. Regardless of the variables we use, we would not include measurements or observations from people who have not climbed the mountain.

The variables in a study may be *quantitative* or *qualitative* in nature.

Quantitative variable
Qualitative variable

> A **quantitative variable** has a value or numerical measurement for which operations such as addition or averaging make sense. A **qualitative variable** describes an individual by placing the individual into a category or group, such as male or female.

For the Mt. Everest climbers, variables such as height, weight, age, or income are *quantitative* variables. *Qualitative variables* involve nonnumerical observations such as gender or nationality. Sometimes qualitative variables are referred to as *categorical variables.*

Categorical variable

Another important issue regarding data is their source. Do the data comprise information from *all* individuals of interest, or from just *some* of the individuals?

Population data
Sample data

> In **population data,** the data are from *every* individual of interest.
> In **sample data,** the data are from *only some* of the individuals of interest.

It is important to know whether the data are population data or sample data. Data from a specific population are fixed and complete. Data from a sample may vary from sample to sample and are *not* complete.

Population parameter

Sample statistic

> A **population parameter** is a numerical measure that describes an aspect of a population.
> A **sample statistic** is a numerical measure that describes an aspect of a sample.

For instance, if we have data from *all* the individuals who have climbed Mt. Everest, then we have population data. The proportion of males in the *population* of all climbers who have conquered Mt. Everest is an example of a *parameter.*

On the other hand, if our data come from just some of the climbers, we have sample data. The proportion of male climbers in the *sample* is an example of a *statistic.* Note that different samples may have different values for the proportion of male climbers. One of the important features of sample statistics is that they can vary from sample to sample, whereas population parameters are fixed for a given population.

LOOKING FORWARD

In later chapters we will use information based on a sample and sample statistics to estimate population parameters (Chapter 8) or make decisions about the value of population parameters (Chapters 9 and 10).

EXAMPLE 1

USING BASIC TERMINOLOGY

The Hawaii Department of Tropical Agriculture is conducting a study of ready-to-harvest pineapples in an experimental field.

(a) The pineapples are the *objects* (individuals) of the study. If the researchers are interested in the individual weights of pineapples in the field, then the *variable* consists of weights. At this point, it is important to specify units of measurement

Gallo Images/Landbouweekblad/Getty Images

and degrees of accuracy of measurement. The weights could be measured to the nearest ounce or gram. Weight is a *quantitative* variable because it is a numerical measure. If weights of *all* the ready-to-harvest pineapples in the field are included in the data, then we have a *population*. The average weight of all ready-to-harvest pineapples in the field is a *parameter.*

(b) Suppose the researchers also want data on taste. A panel of tasters rates the pineapples according to the categories "poor," "acceptable," and "good." Only some of the pineapples are included in the taste test. In this case, the *variable* is taste. This is a *qualitative* or *categorical* variable. Because only some of the pineapples in the field are included in the study, we have a *sample*. The proportion of pineapples in the sample with a taste rating of "good" is a *statistic.*

Throughout this text, you will encounter *guided exercises* embedded in the reading material. These exercises are included to give you an opportunity to work immediately with new ideas. The questions guide you through appropriate analysis. Cover the answers on the right side (an index card will fit this purpose). After you have thought about or written down *your own response,* check the answers. If there are several parts to an exercise, check each part before you continue. You should be able to answer most of these exercise questions, but don't skip them—they are important.

GUIDED EXERCISE 1 | USING BASIC TERMINOLOGY

Television station QUE wants to know the proportion of TV owners in Virginia who watch the station's new program at least once a week. The station asks a group of 1000 TV owners in Virginia if they watch the program at least once a week.

(a) Identify the individuals of the study and the variable.

→ The individuals are the 1000 TV owners surveyed. The variable is the response "does" or "does not" watch the new program at least once a week.

(b) Do the data comprise a sample? If so, what is the underlying population?

→ The data comprise a sample of the population of responses from all TV owners in Virginia.

(c) Is the variable qualitative or quantitative?

→ Qualitative—the categories are the two possible responses, "does" or "does not" watch the program.

(d) Identify a quantitative variable that might be of interest.

→ Age or income might be of interest.

(e) Is the proportion of viewers in the sample who watch the new program at least once a week a statistic or a parameter?

→ Statistic—the proportion is computed from sample data.

LEVELS OF MEASUREMENT: NOMINAL, ORDINAL, INTERVAL, RATIO

We have categorized data as either qualitative or quantitative. Another way to classify data is according to one of the four *levels of measurement.* These levels indicate the type of arithmetic that is appropriate for the data, such as ordering, taking differences, or taking ratios.

Levels of measurement

LEVELS OF MEASUREMENT

Nominal level

The **nominal level of measurement** applies to data that consist of names, labels, or categories. There are no implied criteria by which the data can be ordered from smallest to largest.

Ordinal level

The **ordinal level of measurement** applies to data that can be arranged in order. However, differences between data values either cannot be determined or are meaningless.

Interval level

The **interval level of measurement** applies to data that can be arranged in order. In addition, differences between data values are meaningful.

Ratio level

The **ratio level of measurement** applies to data that can be arranged in order. In addition, both differences between data values and ratios of data values are meaningful. Data at the ratio level have a true zero.

EXAMPLE 2

LEVELS OF MEASUREMENT

Identify the type of data.

(a) Taos, Acoma, Zuni, and Cochiti are the names of four Native American pueblos from the population of names of all Native American pueblos in Arizona and New Mexico.

SOLUTION: These data are at the *nominal* level. Notice that these data values are simply names. By looking at the name alone, we cannot determine if one name is "greater than or less than" another. Any ordering of the names would be numerically meaningless.

(b) In a high school graduating class of 319 students, Jim ranked 25th, June ranked 19th, Walter ranked 10th, and Julia ranked 4th, where 1 is the highest rank.

SOLUTION: These data are at the *ordinal* level. Ordering the data clearly makes sense. Walter ranked higher than June. Jim had the lowest rank, and Julia the highest. However, numerical differences in ranks do not have meaning. The difference between June's and Jim's ranks is 6, and this is the same difference that exists between Walter's and Julia's ranks. However, this difference doesn't really mean anything significant. For instance, if you looked at grade point average, Walter and Julia may have had a large gap between their grade point averages, whereas June and Jim may have had closer grade point averages. In any ranking system, it is only the relative standing that matters. Computed differences between ranks are meaningless.

(c) Body temperatures (in degrees Celsius) of trout in the Yellowstone River.

SOLUTION: These data are at the *interval* level. We can certainly order the data, and we can compute meaningful differences. However, for Celsius-scale temperatures, there is not an inherent starting point. The value 0°C may seem to be a starting point, but this value does not indicate the state of "no heat." Furthermore, it is not correct to say that 20°C is twice as hot as 10°C.

(d) Length of trout swimming in the Yellowstone River.

SOLUTION: These data are at the *ratio* level. An 18-inch trout is three times as long as a 6-inch trout. Observe that we can divide 6 into 18 to determine a meaningful *ratio* of trout lengths.

In summary, there are four levels of measurement. The nominal level is considered the lowest, and in ascending order we have the ordinal, interval, and ratio levels. In general, calculations based on a particular level of measurement may not be appropriate for a lower level.

PROCEDURE

HOW TO DETERMINE THE LEVEL OF MEASUREMENT

The levels of measurement, listed from lowest to highest, are nominal, ordinal, interval, and ratio. To determine the level of measurement of data, state the *highest level* that can be justified for the entire collection of data. Consider which calculations are suitable for the data.

Level of Measurement	Suitable Calculation
Nominal	We can put the data into categories.
Ordinal	We can order the data from smallest to largest or "worst" to "best." Each data value can be *compared* with another data value.
Interval	We can order the data and also take the differences between data values. At this level, it makes sense to compare the differences between data values. For instance, we can say that one data value is 5 more than or 12 less than another data value.
Ratio	We can order the data, take differences, and also find the ratio between data values. For instance, it makes sense to say that one data value is twice as large as another.

WHAT DOES THE LEVEL OF MEASUREMENT TELL US?

The level of measurement tells us which arithmetic processes are appropriate for the data. This is important because different statistical processes require various kinds of arithmetic. In some instances all we need to do is count the number of data that meet specified criteria. In such cases nominal (and higher) data levels are all appropriate. In other cases we need to order the data, so nominal data would not be suitable. Many other statistical processes require division, so data need to be at the ratio level. Just keep the nature of the data in mind before beginning statistical computations.

GUIDED EXERCISE 2 | LEVELS OF MEASUREMENT

The following describe different data associated with a state senator. For each data entry, indicate the corresponding *level of measurement.*

(a) The senator's name is Sam Wilson. → Nominal level

(b) The senator is 58 years old. → Ratio level. Notice that age has a meaningful zero. It makes sense to give age ratios. For instance, Sam is twice as old as someone who is 29.

Continued

GUIDED EXERCISE 2 *continued*

(c) The years in which the senator was elected to the Senate are 2000, 2006, and 2012.

→ Interval level. Dates can be ordered, and the difference between dates has meaning. For instance, 2006 is 6 years later than 2000. However, ratios do not make sense. The year 2000 is not twice as large as the year 1000. In addition, the year 0 does not mean "no time."

(d) The senator's total taxable income last year was $878,314.

→ Ratio level. It makes sense to say that the senator's income is 10 times that of someone earning $87,831.40.

(e) The senator surveyed his constituents regarding his proposed water protection bill. The choices for response were strong support, support, neutral, against, or strongly against.

→ Ordinal level. The choices can be ordered, but there is no meaningful numerical difference between two choices.

(f) The senator's marital status is "married."

→ Nominal level

(g) A leading news magazine claims the senator is ranked seventh for his voting record on bills regarding public education.

→ Ordinal level. Ranks can be ordered, but differences between ranks may vary in meaning.

CRITICAL THINKING

"Data! Data! Data!" he cried impatiently. "I can't make bricks without clay." Sherlock Holmes said these words in *The Adventure of the Copper Beeches* by Sir Arthur Conan Doyle.

Reliable statistical conclusions require reliable data. This section has provided some of the vocabulary used in discussing data. As you read a statistical study or conduct one, pay attention to the nature of the data and the ways they were collected.

When you select a variable to measure, be sure to specify the process and requirements for measurement. For example, if the variable is the weight of ready-to-harvest pineapples, specify the unit of weight, the accuracy of measurement, and maybe even the particular scale to be used. If some weights are in ounces and others in grams, the data are fairly useless.

Another concern is whether or not your measurement instrument truly measures the variable. Just asking people if they know the geographic location of the island nation of Fiji may not provide accurate results. The answers may reflect the fact that the respondents want you to think they are knowledgeable. Asking people to locate Fiji on a map may give more reliable results.

The level of measurement is also an issue. You can put numbers into a calculator or computer and do all kinds of arithmetic. However, you need to judge whether the operations are meaningful. For ordinal data such as restaurant rankings, you can't conclude that a 4-star restaurant is "twice as good" as a 2-star restaurant, even though the number 4 is twice 2.

Continued

CRITICAL THINKING *continued*

Are the data from a sample, or do they comprise the entire population? Sample data can vary from one sample to another! This means that if you are studying the same statistic from two different samples of the same size, the data values may be different. In fact, the ways in which sample statistics vary among different samples of the same size will be the focus of our study from Section 7.4 on.

INTERPRETATION **When you work with sample data, carefully consider the population from which they are drawn. Observations and analysis of the sample are applicable to only the population from which the sample is drawn.**

Descriptive statistics

Inferential statistics

Descriptive statistics involves methods of organizing, picturing, and summarizing information from samples or populations.
Inferential statistics involves methods of using information from a sample to draw conclusions regarding the population.

LOOKING FORWARD

The purpose of collecting and analyzing data is to obtain information. Statistical methods provide us with tools to obtain information from data. These methods break into two branches.

We will look at methods of descriptive statistics in Chapters 2, 3, and 4. These methods may be applied to data from samples or populations.

Sometimes we do not have access to an entire population. At other times, the difficulties or expense of working with the entire population is prohibitive. In such cases, we will use inferential statistics together with probability. These are the topics of Chapters 5 through 11.

VIEWPOINT The First Measured Century

The 20th century saw measurements of aspects of American life that had never been systematically studied before. Social conditions involving crime, sex, food, fun, religion, and work were numerically investigated. The measurements and survey responses taken over the entire century reveal unsuspected statistical trends. The First Measured Century *is a book by Caplow, Hicks, and Wattenberg. It is also a PBS documentary available on video. For more information, visit the Brase/Brase statistics site at* ***http://www.cengagebrain.com*** *and find the link to the PBS* First Measured Century *documentary.*

SECTION 1.1 PROBLEMS

1. *Statistical Literacy* In a statistical study what is the difference between an individual and a variable?

2. *Statistical Literacy* Are data at the nominal level of measurement quantitative or qualitative?

3. *Statistical Literacy* What is the difference between a parameter and a statistic?

4. *Statistical Literacy* For a set population, does a parameter ever change? If there are three different samples of the same size from a set population, is it possible to get three different values for the same statistic?

5. *Critical Thinking* Numbers are often assigned to data that are categorical in nature.
   (a) Consider these number assignments for category items describing electronic ways of expressing personal opinions:

   1 = Twitter; 2 = e-mail; 3 = text message; 4 = Facebook; 5 = blog

   Are these numerical assignments at the ordinal data level or higher? Explain.
   (b) Consider these number assignments for category items describing usefulness of customer service:

   1 = not helpful; 2 = somewhat helpful; 3 = very helpful;
   4 = extremely helpful

   Are these numerical assignments at the ordinal data level? Explain. What about at the interval level or higher? Explain.

6. *Interpretation* Lucy conducted a survey asking some of her friends to specify their favorite type of TV entertainment from the following list of choices:

   sitcom; reality; documentary; drama; cartoon; other

   Do Lucy's observations apply to *all* adults? Explain. From the description of the survey group, can we draw any conclusions regarding age of participants, gender of participants, or education level of participants?

7. *Marketing: Fast Food* A national survey asked 1261 U.S. adult fast-food customers which meal (breakfast, lunch, dinner, snack) they ordered.
   (a) Identify the variable.
   (b) Is the variable quantitative or qualitative?
   (c) What is the implied population?

8. *Advertising: Auto Mileage* What is the average miles per gallon (mpg) for all new cars? Using *Consumer Reports,* a random sample of 35 new cars gave an average of 21.1 mpg.
   (a) Identify the variable.
   (b) Is the variable quantitative or qualitative?
   (c) What is the implied population?

9. *Ecology: Wetlands* Government agencies carefully monitor water quality and its effect on wetlands (Reference: *Environmental Protection Agency Wetland Report* EPA 832-R-93-005). Of particular concern is the concentration of nitrogen in water draining from fertilized lands. Too much nitrogen can kill fish and wildlife. Twenty-eight samples of water were taken at random from a lake. The nitrogen concentration (milligrams of nitrogen per liter of water) was determined for each sample.
   (a) Identify the variable.
   (b) Is the variable quantitative or qualitative?
   (c) What is the implied population?

10. *Archaeology: Ireland* The archaeological site of Tara is more than 4000 years old. Tradition states that Tara was the seat of the high kings of Ireland. Because of its archaeological importance, Tara has received extensive study (Reference: *Tara: An Archaeological Survey* by Conor Newman, Royal Irish Academy, Dublin). Suppose an archaeologist wants to estimate the density of ferromagnetic artifacts in the Tara region. For this purpose, a random sample

of 55 plots, each of size 100 square meters, is used. The number of ferromagnetic artifacts for each plot is determined.
(a) Identify the variable.
(b) Is the variable quantitative or qualitative?
(c) What is the implied population?

11. *Student Life: Levels of Measurement* Categorize these measurements associated with student life according to level: nominal, ordinal, interval, or ratio.
(a) Length of time to complete an exam
(b) Time of first class
(c) Major field of study
(d) Course evaluation scale: poor, acceptable, good
(e) Score on last exam (based on 100 possible points)
(f) Age of student

12. *Business: Levels of Measurement* Categorize these measurements associated with a robotics company according to level: nominal, ordinal, interval, or ratio.
(a) Salesperson's performance: below average, average, above average
(b) Price of company's stock
(c) Names of new products
(d) Temperature (°F) in CEO's private office
(e) Gross income for each of the past 5 years
(f) Color of product packaging

13. *Fishing: Levels of Measurement* Categorize these measurements associated with fishing according to level: nominal, ordinal, interval, or ratio.
(a) Species of fish caught: perch, bass, pike, trout
(b) Cost of rod and reel
(c) Time of return home
(d) Guidebook rating of fishing area: poor, fair, good
(e) Number of fish caught
(f) Temperature of water

14. *Education: Teacher Evaluation* If you were going to apply *statistical methods* to analyze teacher evaluations, which question form, A or B, would be better?

Form A: In your own words, tell how this teacher compares with other teachers you have had.

Form B: Use the following scale to rank your teacher as compared with other teachers you have had.

1	2	3	4	5
worst	**below average**	**average**	**above average**	**best**

15. *Critical Thinking* You are interested in the weights of backpacks students carry to class and decide to conduct a study using the backpacks carried by 30 students.
(a) Give some instructions for weighing the backpacks. Include unit of measure, accuracy of measure, and type of scale.
(b) Do you think each student asked will allow you to weigh his or her backpack?
(c) Do you think telling students ahead of time that you are going to weigh their backpacks will make a difference in the weights?

SECTION **1.2**

Random Samples

FOCUS POINTS

- Explain the importance of random samples.
- Construct a simple random sample using random numbers.
- Simulate a random process.
- Describe stratified sampling, cluster sampling, systematic sampling, multistage sampling, and convenience sampling.

SIMPLE RANDOM SAMPLES

Eat lamb—20,000 coyotes can't be wrong!

This slogan is sometimes found on bumper stickers in the western United States. The slogan indicates the trouble that ranchers have experienced in protecting their flocks from predators. Based on their experience with this sample of the coyote population, the ranchers concluded that *all* coyotes are dangerous to their flocks and should be eliminated! The ranchers used a special poison bait to get rid of the coyotes. Not only was this poison distributed on ranch land, but with government cooperation, it also was distributed widely on public lands.

The ranchers found that the results of the widespread poisoning were not very beneficial. The sheep-eating coyotes continued to thrive while the general population of coyotes and other predators declined. What was the problem? The sheep-eating coyotes that the ranchers had observed were not a representative sample of all coyotes. Modern methods of predator control, however, target the sheep-eating coyotes. To a certain extent, the new methods have come about through a closer examination of the sampling techniques used.

In this section, we will examine several widely used sampling techniques. One of the most important sampling techniques is a *simple random sample.*

Simple random sample

A **simple random sample** of n measurements from a population is a subset of the population selected in such a manner that every sample of size n from the population has an equal chance of being selected.

In a simple random sample, not only does every sample of the specified size have an equal chance of being selected, but every individual of the population also has an equal chance of being selected. However, the fact that each individual has an equal chance of being selected does not necessarily imply a simple random sample. Remember, for a simple random sample, every sample of the given size must also have an equal chance of being selected.

IMPORTANT FEATURES OF A SIMPLE RANDOM SAMPLE

For a simple random sample

- Every sample of specified size n from the population has an equal chance of being selected.
- No researcher bias occurs in the items selected for the sample.
- A random sample may not always reflect the diversity of the population. For instance, from a population of 10 cats and 10 dogs, a random sample of size 6 could consist of all cats.

GUIDED EXERCISE 3 SIMPLE RANDOM SAMPLE

Is open space around metropolitan areas important? Players of the Colorado Lottery might think so, since some of the proceeds of the game go to fund open space and outdoor recreational space. To play the game, you pay $1 and choose any six different numbers from the group of numbers 1 through 42. If your group of six numbers matches the winning group of six numbers selected by simple random sampling, then you are a winner of a grand prize of at least $1.5 million.

(a) Is the number 25 as likely to be selected in the winning group of six numbers as the number 5?

→ Yes. Because the winning numbers constitute a simple random sample, each number from 1 through 42 has an equal chance of being selected.

(b) Could all the winning numbers be even?

→ Yes, since six even numbers is one of the possible groups of six numbers.

(c) Your friend always plays the numbers

1 2 3 4 5 6

Could she ever win?

→ Yes. In a simple random sample, the listed group of six numbers is *as likely as any* of the 5,245,786 possible groups of six numbers to be selected as the winner. (See Section 5.3 to learn how to compute the number of possible groups of six numbers that can be selected from 42 numbers.)

How do we get random samples? Suppose you need to know if the emission systems of the latest shipment of Toyotas satisfy pollution-control standards. You want to pick a random sample of 30 cars from this shipment of 500 cars and test them. One way to pick a random sample is to number the cars 1 through 500. Write these numbers on cards, mix up the cards, and then draw 30 numbers. The sample will consist of the cars with the chosen numbers. If you mix the cards sufficiently, this procedure produces a random sample.

Random-number table

An easier way to select the numbers is to use a *random-number table.* You can make one yourself by writing the digits 0 through 9 on separate cards and mixing up these cards in a hat. Then draw a card, record the digit, return the card, and mix up the cards again. Draw another card, record the digit, and so on. Table 1 in the Appendix is a ready-made random-number table (adapted from Rand Corporation, *A Million Random Digits with 100,000 Normal Deviates*). Let's see how to pick our random sample of 30 Toyotas by using this random-number table.

EXAMPLE 3

RANDOM-NUMBER TABLE

Use a random-number table to pick a random sample of 30 cars from a population of 500 cars.

SOLUTION: Again, we assign each car a different number between 1 and 500, inclusive. Then we use the random-number table to choose the sample. Table 1 in the Appendix has 50 rows and 10 blocks of five digits each; it can be thought of as a solid mass of digits that has been broken up into rows and blocks for user convenience.

You read the digits by beginning anywhere in the table. We dropped a pin on the table, and the head of the pin landed in row 15, block 5. We'll begin there and list all the digits in that row. If we need more digits, we'll move on to row 16, and so on. The digits we begin with are

99281 59640 15221 96079 09961 05371

Since the highest number assigned to a car is 500, and this number has three digits, we regroup our digits into blocks of 3:

992 815 964 015 221 960 790 996 105 371

To construct our random sample, we use the first 30 car numbers we encounter in the random-number table when we start at row 15, block 5. We skip the first three groups—992, 815, and 964—because these numbers are all too large. The next group of three digits is 015, which corresponds to 15. Car number 15 is the first car included in our sample, and the next is car number 221. We skip the next three groups and then include car numbers 105 and 371. To get the rest of the cars in the sample, we continue to the next line and use the random-number table in the same fashion. If we encounter a number we've used before, we skip it.

COMMENT When we use the term *(simple) random sample,* we have very specific criteria in mind for selecting the sample. One proper method for selecting a simple random sample is to use a computer- or calculator-based random-number generator or a table of random numbers as we have done in the example. The term *random* should not be confused with *haphazard*!

LOOKING FORWARD

The runs test for randomness discussed in *Understandable Statistics*, 11th edition, by Brase and Brase, shows how to determine if two symbols are randomly mixed in an ordered list of symbols.

PROCEDURE

HOW TO DRAW A RANDOM SAMPLE

1. Number all members of the population sequentially.
2. Use a table, calculator, or computer to select random numbers from the numbers assigned to the population members.
3. Create the sample by using population members with numbers corresponding to those randomly selected.

LOOKING FORWARD

Simple random samples are key components in methods of inferential statistics that we will study in Chapters 8–11. In fact, in order to draw conclusions about a population, the methods we will study *require* that we have simple random samples from the populations of interest.

Another important use of random-number tables is in *simulation.* We use the word *simulation* to refer to the process of providing numerical imitations of "real" phenomena. Simulation methods have been productive in studying a diverse array of subjects such as nuclear reactors, cloud formation, cardiology (and medical science in general), highway design, production control, shipbuilding, airplane design, war games, economics, and electronics. A complete list would probably include something from every aspect of modern life. In Guided Exercise 4 we'll perform a brief simulation.

Simulation

A **simulation** is a numerical facsimile or representation of a real-world phenomenon.

GUIDED EXERCISE 4 SIMULATION

Use a random-number table to simulate the outcomes of tossing a balanced (that is, fair) penny 10 times.

(a) How many outcomes are possible when you toss a coin once? → Two—heads or tails

Continued

GUIDED EXERCISE 4 *continued*

(b) There are several ways to assign numbers to the two outcomes. Because we assume a fair coin, we can assign an even digit to the outcome "heads" and an odd digit to the outcome "tails." Then, starting at block 3 of row 2 of Table 1 in the Appendix, list the first 10 single digits.

→ 7 1 5 4 9 4 4 8 4 3

(c) What are the outcomes associated with the 10 digits?

→ T T T H T H H H H T

(d) If you start in a different block and row of Table 1 in the Appendix, will you get the same sequence of outcomes?

→ It is possible, but not very likely. (In Section 5.3 you will learn how to determine that there are 1024 possible sequences of outcomes for 10 tosses of a coin.)

TECH NOTES

Sampling with replacement

Most statistical software packages, spreadsheet programs, and statistical calculators generate random numbers. In general, these devices sample with replacement. *Sampling with replacement* means that although a number is selected for the sample, it is *not removed* from the population. Therefore, the same number may be selected for the sample more than once. If you need to sample without replacement, generate more items than you need for the sample. Then sort the sample and remove duplicate values. Specific procedures for generating random samples using the TI-84Plus/TI-83Plus/TI-*n*spire (with TI-84Plus keypad) calculator, Excel 2010, Minitab, and SPSS are shown in Using Technology at the end of this chapter. More details are given in the separate *Technology Guides* for each of these technologies.

OTHER SAMPLING TECHNIQUES

Although we will assume throughout this text that (simple) random samples are used, other methods of sampling are also widely used. Appropriate statistical techniques exist for these sampling methods, but they are beyond the scope of this text.

Stratified sampling

One of these sampling methods is called *stratified sampling.* Groups or classes inside a population that share a common characteristic are called *strata* (plural of *stratum*). For example, in the population of all undergraduate college students, some strata might be freshmen, sophomores, juniors, or seniors. Other strata might be men or women, in-state students or out-of-state students, and so on. In the method of stratified sampling, the population is divided into at least two distinct strata. Then a (simple) random sample of a certain size is drawn from each stratum, and the information obtained is carefully adjusted or weighted in all resulting calculations.

The groups or strata are often sampled in proportion to their actual percentages of occurrence in the overall population. However, other (more sophisticated) ways to determine the optimal sample size in each stratum may give the best results. In general, statistical analysis and tests based on data obtained from stratified samples are somewhat different from techniques discussed in an introductory course in statistics. Such methods for stratified sampling will not be discussed in this text.

Systematic sampling

Another popular method of sampling is called *systematic sampling.* In this method, it is assumed that the elements of the population are arranged in some natural sequential order. Then we select a (random) starting point and select every *k*th element for our sample. For example, people lining up to buy rock concert tickets are "in order." To generate a systematic sample of these people (and ask questions regarding topics such as age, smoking habits, income level, etc.), we could include every fifth person in line. The "starting" person is selected at random from the first five.

The advantage of a systematic sample is that it is easy to get. However, there are dangers in using systematic sampling. When the population is repetitive or cyclic in nature, systematic sampling should not be used. For example, consider a fabric mill that produces dress material. Suppose the loom that produces the material makes a mistake every 17th yard, but we check only every 16th yard with an automated electronic scanner. In this case, a random starting point may or may not result in detection of fabric flaws before a large amount of fabric is produced.

Cluster sampling

Cluster sampling is a method used extensively by government agencies and certain private research organizations. In cluster sampling, we begin by dividing the demographic area into sections. Then we randomly select sections or clusters. Every member of the cluster is included in the sample. For example, in conducting a survey of school children in a large city, we could first randomly select five schools and then include all the children from each selected school.

Multistage sampling

Often a population is very large or geographically spread out. In such cases, samples are constructed through a *multistage sampling design* of several stages, with the final stage consisting of clusters. For instance, the government Current Population Survey interviews about 60,000 households across the United States each month by means of a multistage sample design.

For the Current Population Survey, the first stage consists of selecting samples of large geographic areas that do not cross state lines. These areas are further broken down into smaller blocks, which are stratified according to ethnic and other factors. Stratified samples of the blocks are then taken. Finally, housing units in each chosen block are broken into clusters of nearby housing units. A random sample of these clusters of housing units is selected, and each household in the final cluster is interviewed.

Convenience sampling

Convenience sampling simply uses results or data that are conveniently and readily obtained. In some cases, this may be all that is available, and in many cases, it is better than no information at all. However, convenience sampling does run the risk of being severely biased. For instance, consider a newsperson who wishes to get the "opinions of the people" about a proposed seat tax to be imposed on tickets to all sporting events. The revenues from the seat tax will then be used to support the local symphony. The newsperson stands in front of a concert hall and surveys the first five people exiting after a symphony performance who will cooperate. This method of choosing a sample will produce some opinions, and perhaps some human interest stories, but it certainly has bias. It is hoped that the city council will not use these opinions as the sole basis for a decision about the proposed tax. It is good advice to be very cautious indeed when the data come from the method of convenience sampling.

SAMPLING TECHNIQUES

Random sampling: Use a simple random sample from the entire population.

Stratified sampling: Divide the entire population into distinct subgroups called strata. The strata are based on a specific characteristic such as age, income, education level, and so on. All members of a stratum share the specific characteristic. Draw random samples from each stratum.

Systematic sampling: Number all members of the population sequentially. Then, from a starting point selected at random, include every *k*th member of the population in the sample.

Cluster sampling: Divide the entire population into pre-existing segments or clusters. The clusters are often geographic. Make a random selection of clusters. Include every member of each selected cluster in the sample.

Multistage sampling: Use a variety of sampling methods to create successively smaller groups at each stage. The final sample consists of clusters.

Convenience sampling: Create a sample by using data from population members that are readily available.

CRITICAL THINKING

Sampling frame

We call the list of individuals from which a sample is actually selected the *sampling frame*. Ideally, the sampling frame is the entire population. However, from a practical perspective, not all members of a population may be accessible. For instance, using a telephone directory as the sample frame for residential telephone contacts would not include unlisted numbers.

Undercoverage

When the sample frame does not match the population, we have what is called *undercoverage*. In demographic studies, undercoverage could result if the homeless, fugitives from the law, and so forth, are not included in the study.

A **sampling frame** is a list of individuals from which a sample is actually selected.

Undercoverage results from omitting population members from the sample frame.

Sampling error

In general, even when the sampling frame and the population match, a sample is not a perfect representation of a population. Therefore, information drawn from a sample may not exactly match corresponding information from the population. To the extent that sample information does not match the corresponding population information, we have an error, called a *sampling error.*

Nonsampling error

A **sampling error** is the difference between measurements from a sample and corresponding measurements from the respective population. It is caused by the fact that the sample does not perfectly represent the population.

A **nonsampling error** is the result of poor sample design, sloppy data collection, faulty measuring instruments, bias in questionnaires, and so on.

Sampling errors do not represent mistakes! They are simply the consequences of using samples instead of populations. However, be alert to nonsampling errors, which may sometimes occur inadvertently.

VIEWPOINT Extraterrestrial Life?

Do you believe intelligent life exists on other planets? Using methods of random sampling, a Fox News opinion poll found that about 54% of all U.S. men do believe in intelligent life on other planets, whereas only 47% of U.S. women believe there is such life. How could you conduct a random survey of students on your campus regarding belief in extraterrestrial life?

SECTION 1.2 PROBLEMS

1. *Statistical Literacy* Explain the difference between a stratified sample and a cluster sample.
2. *Statistical Literacy* Explain the difference between a simple random sample and a systematic sample.

3. *Statistical Literacy* Marcie conducted a study of the cost of breakfast cereal. She recorded the costs of several boxes of cereal. However, she neglected to take into account the number of servings in each box. Someone told her not to worry because she just had some sampling error. Comment on that advice.

4. *Statistical Literacy* A random sample of students who use the college recreation center were asked if they approved increasing student fees for all students in order to add a climbing wall to the recreation center. Describe the sample frame. Does the sample frame include all students enrolled in the college? Explain.

5. *Interpretation* In a random sample of 50 students from a large university, all the students were between 18 and 20 years old. Can we conclude that the entire population of students at the university is between 18 and 20 years old? Explain.

6. *Interpretation* A campus performance series features plays, music groups, dance troops, and stand-up comedy. The committee responsible for selecting the performance groups include three students chosen at random from a pool of volunteers. This year the 30 volunteers came from a variety of majors. However, the three students for the committee were all music majors. Does this fact indicate there was bias in the selection process and that the selection process was not random? Explain.

7 *Critical Thinking* Greg took a random sample of size 100 from the population of current season ticket holders to State College men's basketball games. Then he took a random sample of size 100 from the population of current season ticket holders to State College women's basketball games.
(a) What sampling technique (stratified, systematic, cluster, multistage, convenience, random) did Greg use to sample from the population of current season ticket holders to all State College basketball games played by either men or women?
(b) Is it appropriate to pool the samples and claim to have a random sample of size 200 from the population of current season ticket holders to all State College home basketball games played by either men or women? Explain.

8. *Critical Thinking* Consider the students in your statistics class as the population and suppose they are seated in four rows of 10 students each. To select a sample, you toss a coin. If it comes up heads, you use the 20 students sitting in the first two rows as your sample. If it comes up tails, you use the 20 students sitting in the last two rows as your sample.
(a) Does every student have an equal chance of being selected for the sample? Explain.
(b) Is it possible to include students sitting in row 3 with students sitting in row 2 in your sample? Is your sample a simple random sample? Explain.
(c) Describe a process you could use to get a simple random sample of size 20 from a class of size 40.

9. *Critical Thinking* Suppose you are assigned the number 1, and the other students in your statistics class call out consecutive numbers until each person in the class has his or her own number. Explain how you could get a random sample of four students from your statistics class.
(a) Explain why the first four students walking into the classroom would not necessarily form a random sample.
(b) Explain why four students coming in late would not necessarily form a random sample.
(c) Explain why four students sitting in the back row would not necessarily form a random sample.
(d) Explain why the four tallest students would not necessarily form a random sample.

10. *Critical Thinking* In each of the following situations, the sampling frame does not match the population, resulting in undercoverage. Give examples of population members that might have been omitted.
    (a) The population consists of all 250 students in your large statistics class. You plan to obtain a simple random sample of 30 students by using the sampling frame of students present next Monday.
    (b) The population consists of all 15-year-olds living in the attendance district of a local high school. You plan to obtain a simple random sample of 200 such residents by using the student roster of the high school as the sampling frame.

11. *Sampling: Random* Use a random-number table to generate a list of 10 random numbers between 1 and 99. Explain your work.

12. *Sampling: Random* Use a random-number table to generate a list of eight random numbers from 1 to 976. Explain your work.

13. *Sampling: Random* Use a random-number table to generate a list of six random numbers from 1 to 8615. Explain your work.

14. *Simulation: Coin Toss* Use a random-number table to simulate the outcomes of tossing a quarter 25 times. Assume that the quarter is balanced (i.e., fair).

15. *Computer Simulation: Roll of a Die* A die is a cube with dots on each face. The faces have 1, 2, 3, 4, 5, or 6 dots. The table below is a computer simulation (from the software package Minitab) of the results of rolling a fair die 20 times.

    DATA DISPLAY

ROW	C1	C2	C3	C4	C5	C6	C7	C8	C9	C10
1	5	2	2	2	5	3	2	3	1	4
2	3	2	4	5	4	5	3	5	3	4

    (a) Assume that each number in the table corresponds to the number of dots on the upward face of the die. Is it appropriate that the same number appears more than once? Why? What is the outcome of the fourth roll?
    (b) If we simulate more rolls of the die, do you expect to get the same sequence of outcomes? Why or why not?

16. *Simulation: Birthday Problem* Suppose there are 30 people at a party. Do you think any two share the same birthday? Let's use the random-number table to simulate the birthdays of the 30 people at the party. Ignoring leap year, let's assume that the year has 365 days. Number the days, with 1 representing January 1, 2 representing January 2, and so forth, with 365 representing December 31. Draw a random sample of 30 days (with replacement). These days represent the birthdays of the people at the party. Were any two of the birthdays the same? Compare your results with those obtained by other students in the class. Would you expect the results to be the same or different?

17. *Education: Test Construction* Professor Gill is designing a multiple-choice test. There are to be 10 questions. Each question is to have five choices for answers. The choices are to be designated by the letters *a, b, c, d,* and *e.* Professor Gill wishes to use a random-number table to determine which letter choice should correspond to the correct answer for a question. Using the number correspondence 1 for *a,* 2 for *b,* 3 for *c,* 4 for *d,* and 5 for *e,* use a random-number table to determine the letter choice for the correct answer for each of the 10 questions.

18. *Education: Test Construction* Professor Gill uses true–false questions. She wishes to place 20 such questions on the next test. To decide whether to place a true statement or a false statement in each of the 20 questions, she uses a random-number table. She selects 20 digits from the table. An even digit tells her to use a true statement. An odd digit tells her to use a false statement. Use a

random-number table to pick a sequence of 20 digits, and describe the corresponding sequence of 20 true–false questions. What would the test key for your sequence look like?

19. *Sampling Methods: Benefits Package* An important part of employee compensation is a benefits package, which might include health insurance, life insurance, child care, vacation days, retirement plan, parental leave, bonuses, etc. Suppose you want to conduct a survey of benefits packages available in private businesses in Hawaii. You want a sample size of 100. Some sampling techniques are described below. Categorize each technique as *simple random sample, stratified sample, systematic sample, cluster sample,* or *convenience sample.*
    (a) Assign each business in the Island Business Directory a number, and then use a random-number table to select the businesses to be included in the sample.
    (b) Use postal ZIP Codes to divide the state into regions. Pick a random sample of 10 ZIP Code areas and then include all the businesses in each selected ZIP Code area.
    (c) Send a team of five research assistants to Bishop Street in downtown Honolulu. Let each assistant select a block or building and interview an employee from each business found. Each researcher can have the rest of the day off after getting responses from 20 different businesses.
    (d) Use the Island Business Directory. Number all the businesses. Select a starting place at random, and then use every 50th business listed until you have 100 businesses.
    (e) Group the businesses according to type: medical, shipping, retail, manufacturing, financial, construction, restaurant, hotel, tourism, other. Then select a random sample of 10 businesses from each business type.

20. *Sampling Methods: Health Care* Modern Managed Hospitals (MMH) is a national for-profit chain of hospitals. Management wants to survey patients discharged this past year to obtain patient satisfaction profiles. They wish to use a sample of such patients. Several sampling techniques are described below. Categorize each technique as *simple random sample, stratified sample, systematic sample, cluster sample,* or *convenience sample.*
    (a) Obtain a list of patients discharged from all MMH facilities. Divide the patients according to length of hospital stay (2 days or less, 3–7 days, 8–14 days, more than 14 days). Draw simple random samples from each group.
    (b) Obtain lists of patients discharged from all MMH facilities. Number these patients, and then use a random-number table to obtain the sample.
    (c) Randomly select some MMH facilities from each of five geographic regions, and then include all the patients on the discharge lists of the selected hospitals.
    (d) At the beginning of the year, instruct each MMH facility to survey every 500th patient discharged.
    (e) Instruct each MMH facility to survey 10 discharged patients this week and send in the results.

Ariel Skelley/Blend Images/Getty Images

2

Fyodor Savintsev/Itar-Tass/Corbis

World History Archive/Alamy

In dwelling upon the vital importance of sound observation, it must never be lost sight of what observation is for. It is not for the sake of piling up miscellaneous information or curious facts, but for the sake of saving life and increasing health and comfort.

—Florence Nightingale,
Notes on Nursing

Florence Nightingale (1820–1910) has been described as a "passionate statistician" and a "relevant statistician." She viewed statistics as a science that allowed one to transcend his or her narrow individual experience and aspire to the broader service of humanity. She was one of the first nurses to use graphic representation of statistics, illustrating with charts and diagrams how improved sanitation decreased the rate of mortality. Her statistical reports about the appalling sanitary conditions at Scutari (the main British hospital during the Crimean War) were taken very seriously by the English Secretary at War, Sidney Herbert. When sanitary reforms recommended by Nightingale were instituted in military hospitals, the mortality rate dropped from an incredible 42.7% to only 2.2%.

ORGANIZING DATA

© iStockphoto.com/Sportstock

PREVIEW QUESTIONS

What are histograms? When are they used? (SECTION 2.1)

What are common distribution shapes? (SECTION 2.1)

How can you select graphs appropriate for given data sets? (SECTION 2.2)

How can you quickly order data and, at the same time, reveal the distribution shape? (SECTION 2.3)

FOCUS PROBLEM

Say It with Pictures

Edward R. Tufte, in his book *The Visual Display of Quantitative Information,* presents a number of guidelines for producing good graphics. According to the criteria, a graphical display should

- show the data;
- induce the viewer to think about the substance of the graphic rather than about the methodology, the design, the technology, or other production devices;
- avoid distorting what the data have to say.

Vandystadt/Science Source

As an example of a graph that violates some of the criteria, Tufte includes a graphic that appeared in a well-known newspaper. Figure 2-1(a), on the next page, shows a figure similar to the problem graphic, whereas part (b) of the figure shows a better rendition of the data display.

After completing this chapter, you will be able to answer the following questions.

(a) Look at the graph in Figure 2-1(a). Is it essentially a bar graph? Explain. What are some of the flaws of Figure 2-1(a) as a bar graph?

(b) Examine Figure 2-1(b), which shows the same information. Is it essentially a time-series graph? Explain. In what ways does the second graph seem to display the information in a clearer manner?

(See Problem 5 of the Chapter 2 Review Problems.)

For online student resources, visit the Brase/Brase, *Understanding Basic Statistics*, 7th edition web site at http://www.cengagebrain.com

FIGURE 2-1

(a) Fuel Economy Standards for Autos

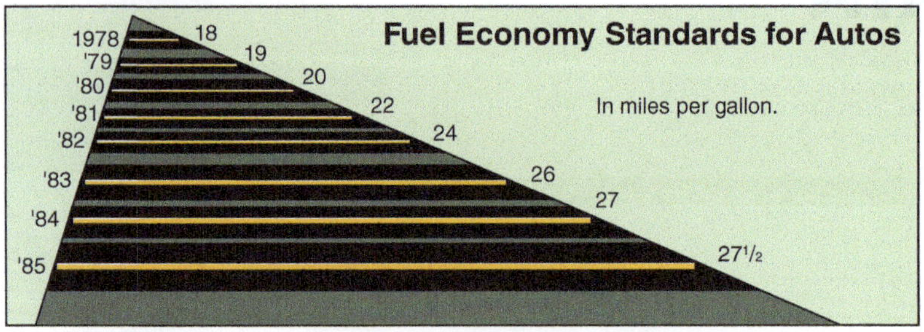

Set by Congress and supplemented by the Transportation Department.

Source: Redrawn from *The Visual Display of Quantitative Information* by Edward R. Tufte, p. 58.

Note: Fuel-economy standards for passenger cars remained at 27.5 miles per gallon through 2010. Mileage standards are set to increase to 54.5 miles per gallon by 2025.

(b) **Required Fuel Economy Standards: New Cars Built From 1978 To 1985**

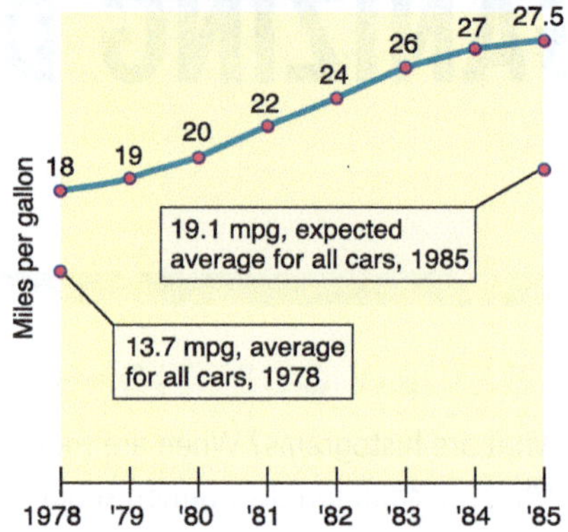

Source: Redrawn from *The Visual Display of Quantitive Information* by Edward R. Tufte, p. 58.

SECTION **2.1**

Frequency Distributions, Histograms, and Related Topics

FOCUS POINTS

- Organize raw data using a frequency table.
- Construct histograms, relative-frequency histograms, and ogives.
- Recognize basic distribution shapes: uniform, symmetric, skewed, and bimodal.
- Interpret graphs in the context of the data setting.

FREQUENCY TABLES

When we have a large set of quantitative data, it's useful to organize it into smaller intervals or *classes* and count how many data values fall into each class. A frequency table does just that.

Frequency table

A **frequency table** partitions data into classes or intervals of equal width and shows how many data values are in each class. The classes or intervals are constructed so that each data value falls into exactly one class.

Constructing a frequency table involves a number of steps. Example 1 demonstrates the steps.

EXAMPLE 1

FREQUENCY TABLE

A task force to encourage car pooling did a study of one-way commuting distances of workers in the downtown Dallas area. A random sample of 60 of these workers was taken. The commuting distances of the workers in the sample are given in Table 2-1. Make a frequency table for these data.

SOLUTION:

(a) First decide how many classes you want. Five to 15 classes are usually used. If you use fewer than five classes, you risk losing too much information. If you use more than 15 classes, the data may not be sufficiently summarized. Let the

Bob Turner/Art Directors & TRIP/Alamy

TABLE 2-1 **One-Way Commuting Distances (in Miles) for 60 Workers in Downtown Dallas**

13	47	10	3	16	20	17	40	4	2
7	25	8	21	19	15	3	17	14	6
12	45	1	8	4	16	11	18	23	12
6	2	14	13	7	15	46	12	9	18
34	13	41	28	36	17	24	27	29	9
14	26	10	24	37	31	8	16	12	16

spread of the data and the purpose of the frequency table be your guides when selecting the number of classes. In the case of the commuting data, let's use *six* classes.

Class width

(b) Next, find the *class width* for the six classes.

PROCEDURE

HOW TO FIND THE CLASS WIDTH (INTEGER DATA)

1. Compute $\dfrac{\text{Largest data value} - \text{smallest data value}}{\text{Desired number of classes}}$
2. Increase the computed value to the next highest whole number.

Note: To ensure that all the classes taken together cover the data, we need to increase the result of Step 1 to the *next whole number,* even if Step 1 produced a whole number. For instance, if the calculation in Step 1 produces the value 4, we make the class width 5.

To find the class width for the commuting data, we observe that the largest distance commuted is 47 miles and the smallest is 1 mile. Using six classes, the class width is 8, since

$$\text{Class width} = \frac{47 - 1}{6} \approx 7.7 \text{ (increase to 8)}$$

(c) Now we determine the data range for each class.

Class limits

The **lower class limit** is the lowest data value that can fit in a class. The **upper class limit** is the highest data value that can fit in a class. The **class width** is the difference between the *lower* class limit of one class and the *lower* class limit of the next class.

The smallest commuting distance in our sample is 1 mile. We use this *smallest* data value as the lower class limit of the *first* class. Because the class width is 8, we add 8 to 1 to find that the *lower* class limit for the *second* class is 9. Following this pattern, we establish *all* the *lower class limits*. Then we fill in the *upper class limits* so that the classes span the entire range of data. Table 2-2, on the next page, shows the upper and lower class limits for the commuting distance data.

(d) Now we are ready to tally the commuting distance data into the six classes and find the frequency for each class.

TABLE 2-2 **Frequency Table of One-Way Commuting Distances for 60 Downtown Dallas Workers (Data in Miles)**

Class Limits Lower–Upper	Class Boundaries Lower–Upper	Tally	Frequency	Class Midpoint
1–8	0.5–8.5	𝍸 𝍸 ||||	14	4.5
9–16	8.5–16.5	𝍸 𝍸 𝍸 𝍸 |	21	12.5
17–24	16.5–24.5	𝍸 𝍸 |	11	20.5
25–32	24.5–32.5	𝍸 |	6	28.5
33–40	32.5–40.5	||||	4	36.5
41–48	40.5–48.5	||||	4	44.5

PROCEDURE

HOW TO TALLY DATA

Tallying data is a method of counting data values that fall into a particular class or category.

To tally data into classes of a frequency table, examine each data value. Determine which class contains the data value and make a tally mark or vertical stroke (|) beside that class. For ease of counting, each fifth tally mark of a class is placed diagonally across the prior four marks (𝍸).

Class frequency

The *class frequency* for a class is the number of tally marks corresponding to that class.

Table 2-2 shows the tally and frequency of each class.

Class midpoint or class mark

(e) The center of each class is called the *midpoint* (or *class mark*). The midpoint is often used as a representative value of the entire class. The midpoint is found by adding the lower and upper class limits of one class and dividing by 2.

$$\text{Midpoint} = \frac{\text{Lower class limit} + \text{upper class limit}}{2}$$

Table 2-2 shows the class midpoints.

Class boundaries

(f) There is a space between the upper limit of one class and the lower limit of the next class. The halfway points of these intervals are called *class boundaries*. These are shown in Table 2-2.

PROCEDURE

HOW TO FIND CLASS BOUNDARIES (INTEGER DATA)

To find **upper class boundaries,** add 0.5 unit to the upper class limits.

To find **lower class boundaries,** subtract 0.5 unit from the lower class limits.

Problems 21 and 22 show the procedure for finding the class width, class limits, and class boundaries for noninteger decimal data.

Relative frequency

Basic frequency tables show how many data values fall into each class. It's also useful to know the *relative frequency* of a class. The relative frequency of a class is the proportion of all data values that fall into that class. To find the relative

TABLE 2-3 Relative Frequencies of One-Way Commuting Distances

Class	Frequency f	Relative Frequency f/n
1–8	14	$14/60 \approx 0.23$
9–16	21	$21/60 \approx 0.35$
17–24	11	$11/60 \approx 0.18$
25–32	6	$6/60 \approx 0.10$
33–40	4	$4/60 \approx 0.07$
41–48	4	$4/60 \approx 0.07$

frequency of a particular class, divide the class frequency f by the total of all frequencies n (sample size).

$$\text{Relative frequency} = \frac{f}{n} = \frac{\text{Class frequency}}{\text{Total of all frequencies}}$$

Table 2-3 shows the relative frequencies for the commuter data of Table 2-1. Because we already have the frequency table (Table 2-2), the relative-frequency table is obtained easily. The sample size is $n = 60$. Notice that the sample size is the total of all the frequencies. Therefore, the relative frequency for the first class (the class from 1 to 8) is

$$\text{Relative frequency} = \frac{f}{n} = \frac{14}{60} \approx 0.23$$

> **LOOKING FORWARD**
>
> Sorting data into classes can be somewhat tedious. It is an easier task if the data are first ordered from smallest to largest. Stem-and-leaf diagrams, presented in Section 2.3, provide a convenient way to order data by hand.

The symbol $\approx$ means "approximately equal to." We use the symbol because we rounded the relative frequency. Relative frequencies for the other classes are computed in a similar way.

The total of the relative frequencies should be 1. However, rounded results may make the total slightly higher or lower than 1.

Let's summarize the procedure for making a frequency table that includes relative frequencies.

PROCEDURE

HOW TO MAKE A FREQUENCY TABLE

1. Determine the number of classes and the corresponding class width.
2. Create the distinct classes. We use the convention that the *lower class limit* of the first class is the smallest data value. Add the class width to this number to get the *lower class limit* of the next class.
3. Fill in *upper class limits* to create distinct classes that accommodate all possible data values from the data set.
4. Tally the data into classes. Each data value should fall into exactly one class. Total the tallies to obtain each *class frequency.*
5. Compute the *midpoint* (class mark) for each class.
6. Determine the *class boundaries.*

PROCEDURE

HOW TO MAKE A RELATIVE-FREQUENCY TABLE

Relative-frequency table

First make a frequency table. Then, for each class, compute the *relative frequency, f/n,* where f is the class frequency and n is the total sample size.

HISTOGRAMS AND RELATIVE-FREQUENCY HISTOGRAMS

Histogram
Relative-frequency histogram

Histograms and *relative-frequency histograms* provide effective visual displays of data organized into frequency tables. In these graphs, we use bars to represent each class, where the width of the bar is the class width. For histograms, the height of the bar is the class frequency; for relative-frequency histograms, the height of the bar is the relative frequency of that class.

Problems 23, 24, and 25 introduce dotplots, which are diagrams that show the frequency with which *each* data value occurs.

PROCEDURE

HOW TO MAKE A HISTOGRAM OR A RELATIVE-FREQUENCY HISTOGRAM

1. Make a frequency table (including relative frequencies) with the designated number of classes.
2. Place class boundaries on the horizontal axis and frequencies or relative frequencies on the vertical axis.
3. For each class of the frequency table, draw a bar whose width extends between corresponding class boundaries. For histograms, the height of each bar is the corresponding class frequency. For relative-frequency histograms, the height of each bar is the corresponding class-relative frequency.

EXAMPLE 2

HISTOGRAM AND RELATIVE-FREQUENCY HISTOGRAM

Make a histogram and a relative-frequency histogram with six bars for the data in Table 2-1 showing one-way commuting distances.

SOLUTION: The first step is to make a frequency table and a relative-frequency table with six classes. We'll use Table 2-2 and Table 2-3. Figures 2-2 and 2-3 show the histogram and relative-frequency histogram. In both graphs, class boundaries are marked on the horizontal axis. For each class of the frequency table, make a corresponding bar with horizontal width extending from the lower boundary to the upper boundary of the respective class. For a histogram, the height of each bar is the corresponding class frequency. For a relative-frequency histogram, the height of

FIGURE 2-2

Histogram for Dallas Commuters: One-Way Commuting Distances

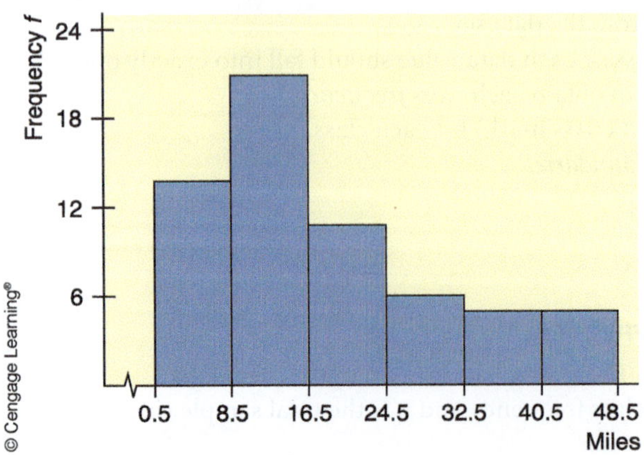

FIGURE 2-3

Relative-Frequency Histogram for Dallas Commuters: One-Way Commuting Distances

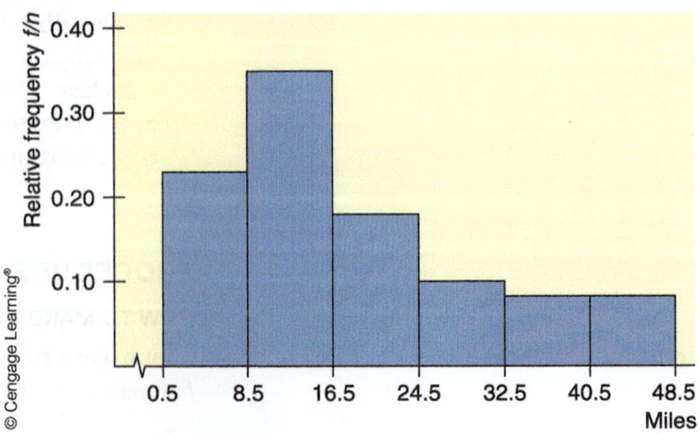

each bar is the corresponding relative frequency. Notice that the basic shapes of the graphs are the same. The only difference involves the vertical axis. The vertical axis of the histogram shows frequencies, while the vertical axis of the relative-frequency histogram shows relative frequencies.

COMMENT The use of class boundaries in histograms assures us that the bars of the histogram touch and that no data fall on the boundaries. Both of these features are important. But a histogram displaying class boundaries may look awkward. For instance, the mileage range of 8.5 to 16.5 miles shown in Figure 2-2 isn't as natural a choice as a mileage range of 8 to 16 miles. For this reason, many magazines and newspapers do not use class boundaries as labels on a histogram. Instead, some use lower class limits as labels, with the convention that *a data value falling on the class limit is included in the next higher class (class to the right of the limit)*. Another convention is to label midpoints instead of class boundaries. Determine the default convention being used before creating frequency tables and histograms on a computer.

GUIDED EXERCISE 1 | HISTOGRAM AND RELATIVE-FREQUENCY HISTOGRAM

An irate customer called Dollar Day Mail Order Company 40 times during the last two weeks to see why his order had not arrived. Each time he called, he recorded the length of time he was put "on hold" before being allowed to talk to a customer service representative. See Table 2-4.

TABLE 2-4 **Length of Time on Hold, in Minutes**

1	5	5	6	7	4	8	7	6	5
5	6	7	6	6	5	8	9	9	10
7	8	11	2	4	6	5	12	13	6
3	7	8	8	9	9	10	9	8	9

(a) What are the largest and smallest values in Table 2-4? If we want five classes in a frequency table, what should the class width be?

→ The largest value is 13; the smallest value is 1. The class width is

$$\frac{13-1}{5} = 2.4 \approx 3$$

Note: Increase the value to 3.

(b) Complete the following frequency table.

TABLE 2-5 **Time on Hold**

Class Limits Lower-Upper	Tally	Frequency	Midpoint
1–3	____	____	____
4–____	____	____	____
____–9	____	____	____
____–____	____	____	____
____–____	____	____	____

→ TABLE 2-6 **Completion of Table 2-5**

Class Limits Lower-Upper	Tally	Frequency	Midpoint
1–3	|||	3	2
4–6	卌 卌 卌	15	5
7–9	卌 卌 卌 ||	17	8
10–12	||||	4	11
13–15			1	14

(c) Recall that the class boundary is halfway between the upper limit of one class and the lower limit of the next. Use this fact to find the class boundaries in Table 2-7 and to complete the partial histogram in Figure 2-4.

Continued

GUIDED EXERCISE 1 *continued*

TABLE 2-7 **Class Boundaries**

Class Limits	Class Boundaries
1–3	0.5–3.5
4–6	3.5–6.5
7–9	6.5–___
10–12	___–___
13–15	___–___

➡ TABLE 2-8 **Completion of Table 2-7**

Class Limits	Class Boundaries
1–3	0.5–3.5
4–6	3.5–6.5
7–9	6.5–9.5
10–12	9.5–12.5
13–15	12.5–15.5

FIGURE 2-4

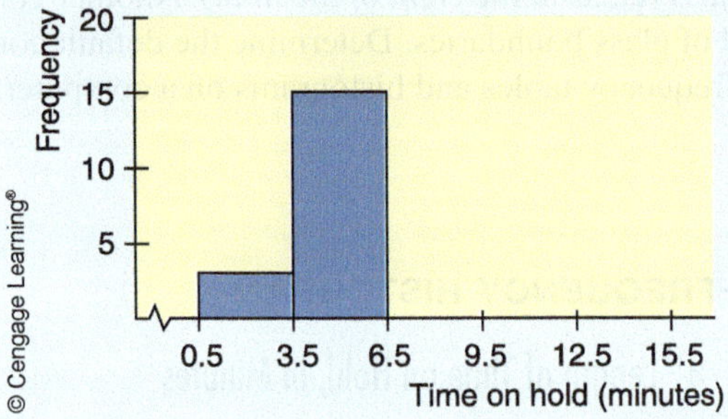

➡ FIGURE 2-5 Completion of Figure 2-4

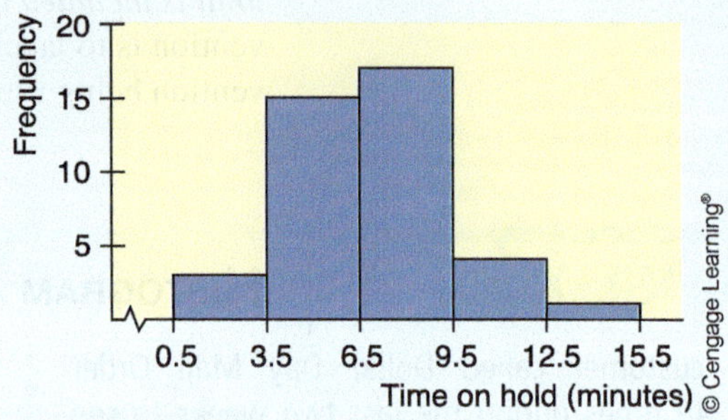

(d) Compute the relative class frequency *f/n* for each class in Table 2-9 and complete the partial relative-frequency histogram in Figure 2-6.

TABLE 2-9 **Relative Class Frequency**

Class	*f/n*
1–3	3/40 = 0.075
4–6	15/40 = 0.375
7–9	___
10–12	___
13–15	___

➡ TABLE 2-10 **Completion of Table 2-9**

Class	*f/n*
1–3	0.075
4–6	0.375
7–9	0.425
10–12	0.100
13–15	0.025

FIGURE 2-6

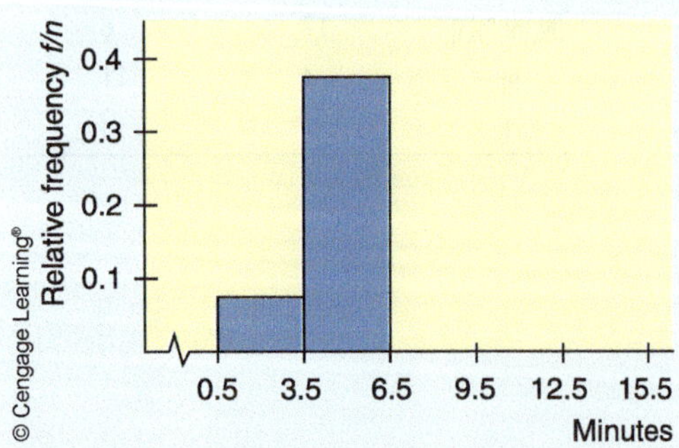

➡ FIGURE 2-7 Completion of Figure 2-6

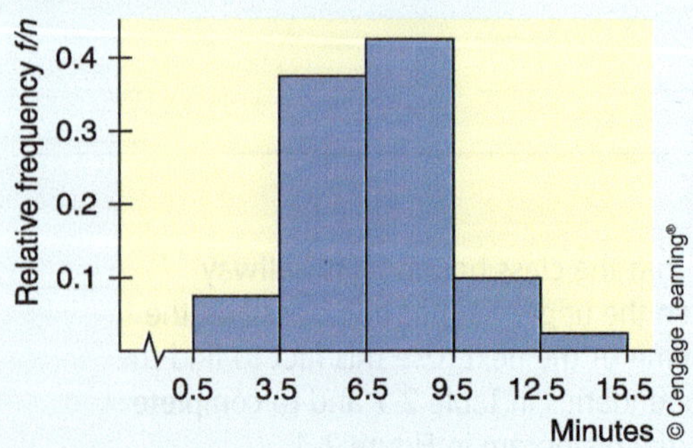

FIGURE 2-8
Types of Histograms

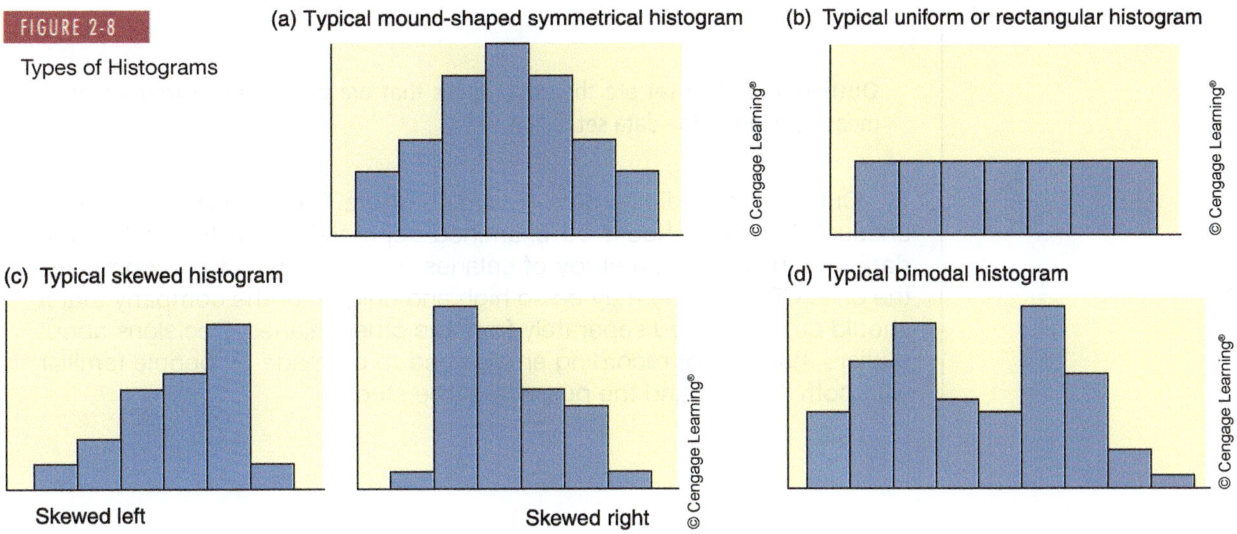

LOOKING FORWARD

We will see relative-frequency distributions again when we study probability in Chapter 5. There we will see that if a random sample is large enough, then we can estimate the probability of an event by the relative frequency of the event. The relative-frequency distribution then can be interpreted as a *probability distribution.* Such distributions will form the basis of our work in inferential statistics.

DISTRIBUTION SHAPES

Histograms are valuable and useful tools. If the raw data came from a random sample of population values, the histogram constructed from the sample values should have a distribution shape that is reasonably similar to that of the population.

Several terms are commonly used to describe histograms and their associated population distributions.

Mound-shaped symmetrical distribution

Uniform distribution

Skewed left

Skewed right

Bimodal distribution

(a) ***Mound-shaped symmetrical:*** This term refers to a mound-shaped histogram in which both sides are (more or less) the same when the graph is folded vertically down the middle. Figure 2-8(a) shows a typical mound-shaped symmetrical histogram.

(b) ***Uniform or rectangular:*** These terms refer to a histogram in which every class has equal frequency. From one point of view, a uniform distribution is symmetrical with the added property that the bars are of the same height. Figure 2-8(b) illustrates a typical histogram with a uniform shape.

(c) ***Skewed left or skewed right:*** These terms refer to a histogram in which one tail is stretched out longer than the other. The direction of skewness is on the side of the *longer* tail. So, if the longer tail is on the left, we say the histogram is skewed to the left. Figure 2-8(c) shows a typical histogram skewed to the left and another skewed to the right.

(d) ***Bimodal:*** This term refers to a histogram in which the two classes with the largest frequencies are separated by at least one class. The top two frequencies of these classes may have slightly different values. This type of situation sometimes indicates that we are sampling from two different populations. Figure 2-8(d) illustrates a typical histogram with a bimodal shape.

LOOKING FORWARD

Normal and Student's *t* distributions are two important probability distributions used extensively in inferential statistics (Chapters 7–10). Both of these distributions are mound-shaped symmetrical. Another useful probability distribution is the Chi-square distribution (Chapter 11), which is skewed right.

CRITICAL THINKING

A bimodal distribution shape might indicate that the data are from two different populations. For instance, a histogram showing the heights of a random sample of adults is likely to be bimodal because two populations, male and female, were combined.

If there are gaps in the histogram between bars at either end of the graph, the data set might include *outliers*.

Continued

CRITICAL THINKING *continued*

Outliers

Outliers in a data set are the data values that are very different from other measurements in the data set.

Outliers may indicate data-recording errors. Valid outliers may be so unusual that they should be examined separately from the rest of the data. For instance, in a study of salaries of employees at one company, the chief CEO's salary may be so high and unique for the company that it should be considered separately from the other salaries. Decisions about outliers that are not recording errors need to be made by people familiar with both the field and the purpose of the study.

WHAT DO HISTOGRAMS AND RELATIVE FREQUENCY HISTOGRAMS TELL US?

Histograms and relative frequency histograms show us how the data are distributed. By looking at such graphs, we can tell

- if the data distribution is more symmetric, skewed, or bimodal;
- if there are possible outliers;
- which data intervals contain the most data;
- how spread out the data are.

In the next chapter we will look at measures of center and spread of data. Histograms help us visualize such measures.

TECH NOTES The TI-84Plus/TI-83Plus/TI-*n*spire calculators, Excel 2010, and Minitab all create histograms. However, each technology automatically selects the number of classes to use. In Using Technology at the end of this chapter, you will see instructions for specifying the number of classes yourself and for generating histograms such as those we create "by hand."

CUMULATIVE-FREQUENCY TABLES AND OGIVES

Sometimes we want to study cumulative totals instead of frequencies. Cumulative frequencies tell us how many data values are smaller than an upper class boundary. Once we have a frequency table, it is a fairly straightforward matter to add a column of cumulative frequencies.

Cumulative frequency

The **cumulative frequency** for a class is the sum of the frequencies for *that class* and *all previous classes.*

Ogive

An *ogive* (pronounced "oh-jīve") is a graph that displays cumulative frequencies.

PROCEDURE

HOW TO MAKE AN OGIVE

1. Make a frequency table showing class boundaries and cumulative frequencies.
2. For each class, make a dot over the *upper class boundary* at the height of the cumulative class frequency. The coordinates of the dots are (upper class boundary, cumulative class frequency). Connect these dots with line segments.
3. By convention, an ogive begins on the horizontal axis at the lower class boundary of the first class.

EXAMPLE 3

CUMULATIVE-FREQUENCY TABLE AND OGIVE

Aspen, Colorado, is a world-famous ski area. If the daily high temperature is above 40°F, the surface of the snow tends to melt. It then freezes again at night. This can result in a snow crust that is icy. It also can increase avalanche danger.

Table 2-11 gives a summary of daily high temperatures (°F) in Aspen during the 151-day ski season.

Ph. Royer/Science Source

TABLE 2-11 **High Temperatures During the Aspen Ski Season (°F)**

Class Boundaries			
Lower	Upper	Frequency	Cumulative Frequency
10.5	20.5	23	23
20.5	30.5	43	66 (sum 23 + 43)
30.5	40.5	51	117 (sum 66 + 51)
40.5	50.5	27	144 (sum 117 + 27)
50.5	60.5	7	151 (sum 144 + 7)

(a) The cumulative frequency for a class is computed by adding the frequency of that class to the frequencies of previous classes. Table 2-11 shows the cumulative frequencies.

(b) To draw the corresponding ogive, we place a dot at cumulative frequency 0 on the lower class boundary of the first class. Then we place dots over the *upper class boundaries* at the height of the cumulative class frequency for the corresponding class. Finally, we connect the dots. Figure 2-9 shows the corresponding ogive.

(c) Looking at the ogive, estimate the total number of days with a high temperature lower than or equal to 40°F.

FIGURE 2-9

Ogive for Daily High Temperatures (°F) During Aspen Ski Season

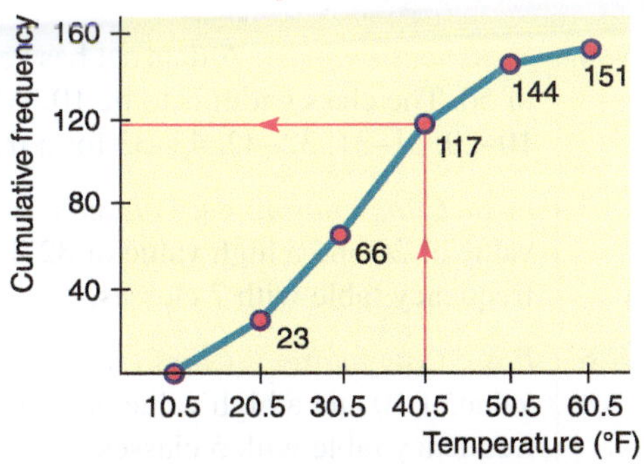

© Cengage Learning®

SOLUTION: Following the red lines on the ogive in Figure 2-9, we see that 117 days have had high temperatures of no more than 40°F.

WHAT DOES AN OGIVE TELL US?

An ogive (also known as a cumulative-frequency diagram) tells us

- how many data are less than the indicated value on the horizontal axis;
- how slowly or rapidly the data values accumulate over the range of the data.

In addition, the vertical scale can be changed to cumulative percentages by dividing the cumulative frequencies by the total number of data. Then we can tell what percentage of data are below values specified on the horizontal axis.

VIEWPOINT Mush, You Huskies!

In 1925, the village of Nome, Alaska, had a terrible diphtheria epidemic. Serum was available in Anchorage but had to be brought to Nome by dogsled over the 1161-mile Iditarod Trail. Since 1973, the Iditarod Dog Sled Race from Anchorage to Nome has been an annual sporting event, with a current purse of more than $600,000. Winning times range from just more than 9 days to more than 20 days.

To collect data on winning times, visit the Brase/Brase statistics site at ***http://www.cengagebrain.com*** *and find the link to the Iditarod. Make a frequency distribution for these times.*

SECTION 2.1 PROBLEMS

1. *Statistical Literacy* What is the difference between a class boundary and a class limit?

2. *Statistical Literacy* A data set has values ranging from a low of 10 to a high of 52. What's wrong with using the class limits 10–19, 20–29, 30–39, 40–49 for a frequency table?

3. *Statistical Literacy* A data set has values ranging from a low of 10 to a high of 50. What's wrong with using the class limits 10–20, 20–30, 30–40, 40–50 for a frequency table?

4. *Statistical Literacy* A data set has values ranging from a low of 10 to a high of 50. The class width is to be 10. What's wrong with using the class limits 10–20, 21–31, 32–42, 43–53 for a frequency table with a class width of 10?

5. *Basic Computation: Class Limits* A data set with whole numbers has a low value of 20 and a high value of 82. Find the class width and class limits for a frequency table with 7 classes.

6. *Basic Computation: Class Limits* A data set with whole numbers has a low value of 10 and a high value of 120. Find the class width and class limits for a frequency table with 5 classes.

7. *Interpretation* You are manager of a specialty coffee shop and collect data throughout a full day regarding waiting time for customers from the time they enter the shop until the time they pick up their order.
   (a) What type of distribution do you think would be most desirable for the waiting times: skewed right, skewed left, mound-shaped symmetrical? Explain.
   (b) What if the distribution for waiting times were bimodal? What might be some explanations?

8. *Critical Thinking* A web site rated 100 colleges and ranked the colleges from 1 to 100, with a rank of 1 being the best. Each college was ranked, and there were no ties. If the ranks were displayed in a histogram, what would be the shape of the histogram: skewed, uniform, mound-shaped?

9. *Critical Thinking* Look at the histogram in Figure 2-10(a), which shows mileage, in miles per gallon (mpg), for a random selection of passenger cars (Reference: *Consumer Reports*).
   (a) Is the shape of the histogram essentially bimodal?
   (b) Jose looked at the raw data and discovered that the 54 data values included both the city and the highway mileages for 27 cars. He used the city mileages for the 27 cars to make the histogram in Figure 2-10(b). Using this information and Figure 2-10, parts (a) and (b), construct a histogram for the highway mileages of the same cars. Use class boundaries 16.5, 20.5, 24.5, 28.5, 32.5, 36.5, and 40.5.

FIGURE 2-10

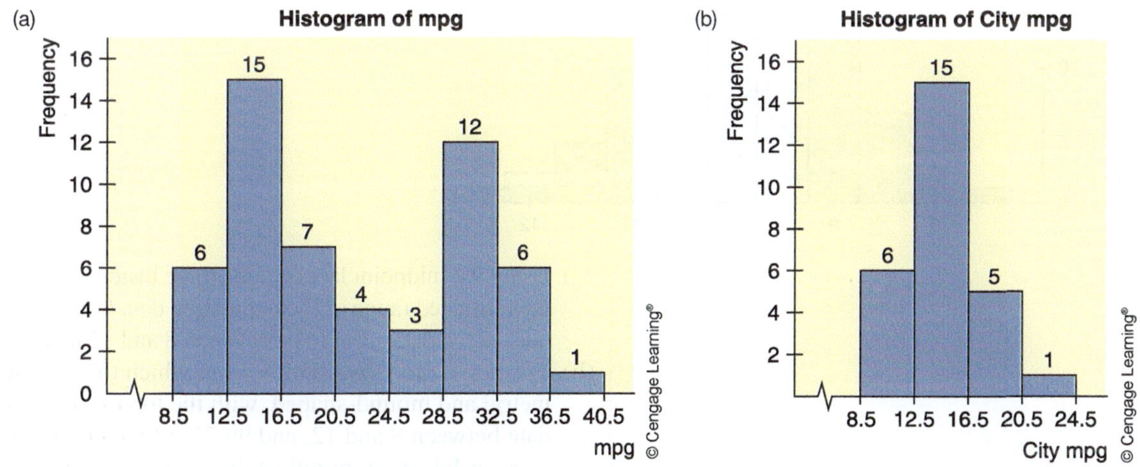

10. *Critical Thinking* The following data represent salaries, in thousands of dollars, for employees of a small company. Notice that the data have been sorted in increasing order.

54	55	55	57	57	59	60	65	65	65	66	68	68
69	69	70	70	70	75	75	75	75	77	82	82	82
88	89	89	91	91	97	98	98	98	280			

   (a) Make a histogram using the class boundaries 53.5, 99.5, 145.5, 191.5, 237.5, 283.5.
   (b) Look at the last data value. Does it appear to be an outlier? Could this be the owner's salary?
   (c) Eliminate the high salary of 280 thousand dollars. Make a new histogram using the class boundaries 53.5, 62.5, 71.5, 80.5, 89.5, 98.5. Does this histogram reflect the salary distribution of most of the employees better than the histogram in part (a)?

11. *Interpretation* Histograms of random sample data are often used as an indication of the shape of the underlying population distribution. The histograms on the next page are based on random samples of size 30, 50, and 100 from the same population.

(i) Sample of size 30

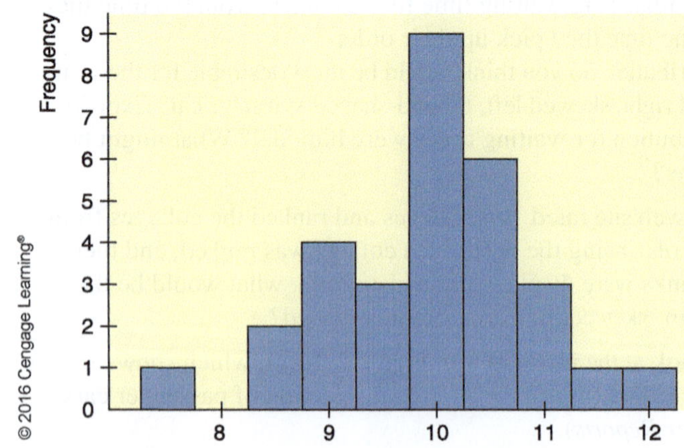

(ii) Sample of size 50

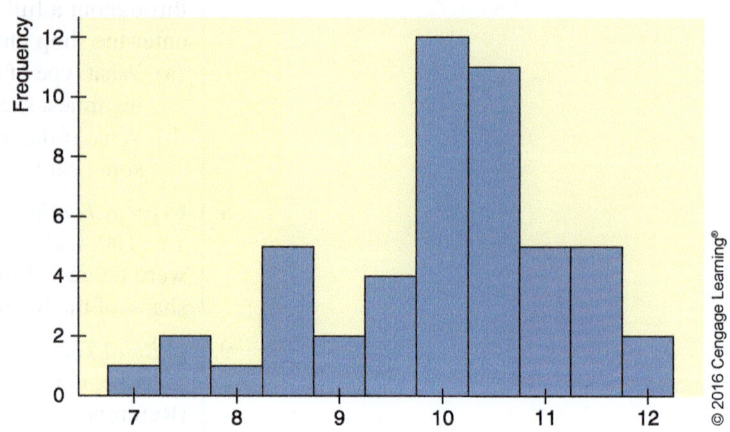

(iii) Sample of size 100

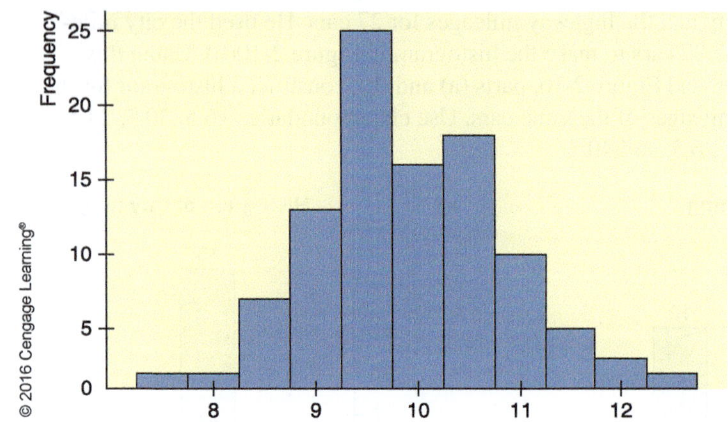

(a) Using the midpoint labels of the three histograms, what would you say about the estimated range of the population data from smallest to largest? Does the bulk of the data seem to be between 8 and 12 in all three histograms?

(b) The population distribution from which the samples were drawn is symmetric and mound-shaped, with the top of the mound at 10, 95% of the data between 8 and 12, and 99.7% of data between 7 and 13. How well does each histogram reflect these characteristics?

12. *Interpretation* The following histograms are based on different random samples of size 100 drawn from the same population.

(i)

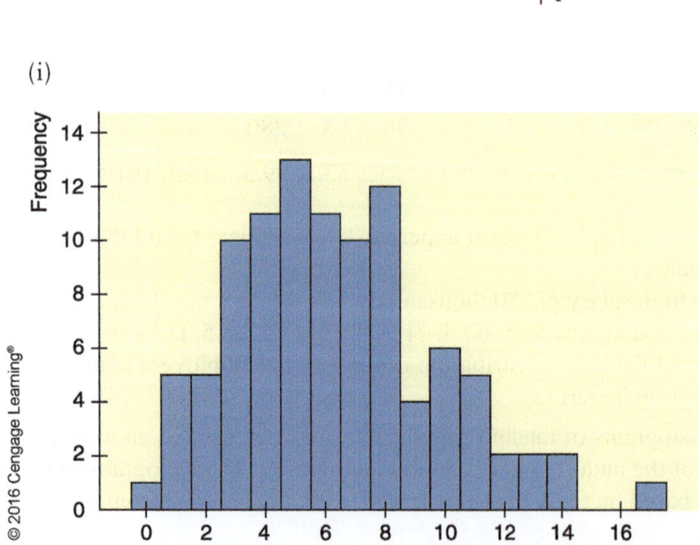

(ii)

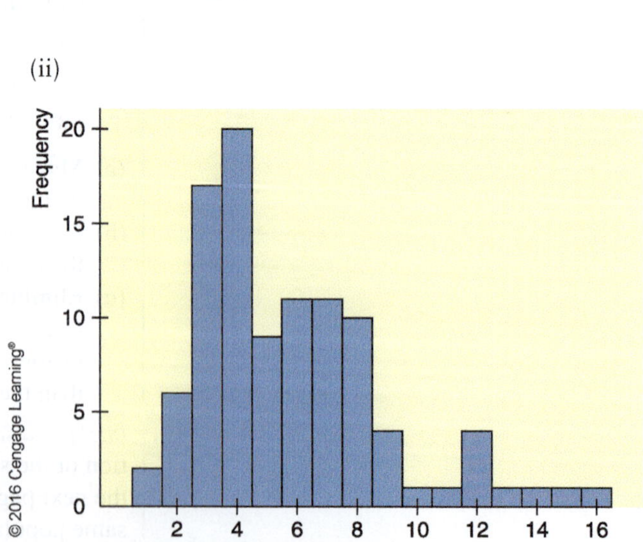

(iii)

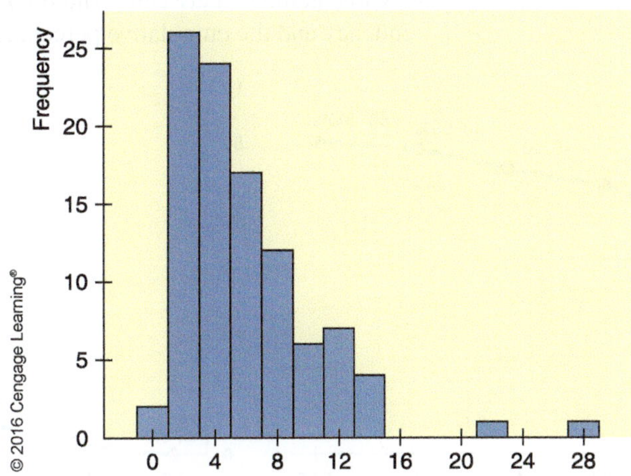

(a) Identify the midpoint of the class with the highest frequency in each of the three histograms.
(b) Using the class midpoints, what is the range of data shown in each histogram?
(c) Which of the histograms are more clearly skewed right?
(d) Based on your study of random sample in Chapter 1, is it surprising to see the variations in the samples as displayed in the histograms? The original population from which the samples were drawn is skewed right with a high frequency near 4. Do all three random samples seem to reflect these properties equally well?

13. *Interpretation* The ogives shown are based on U.S. Census data and show the average personal income per capita for each of the 50 states. The data are rounded to the nearest thousand dollars.

(i) Ogive

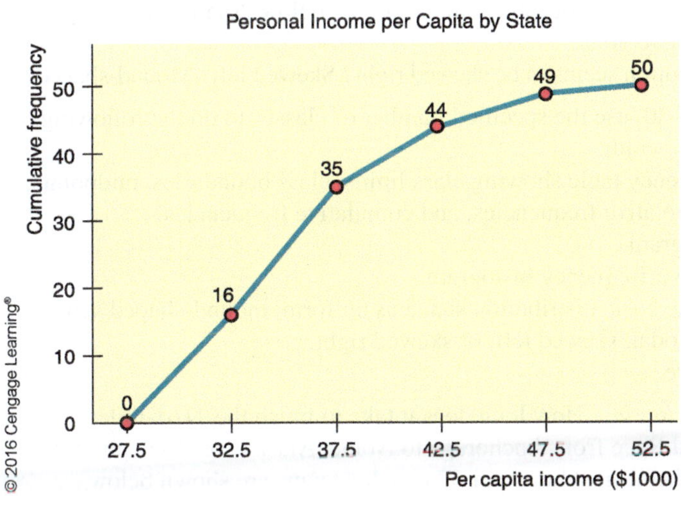

(ii) Ogive Showing Cumulative Percentage of Data

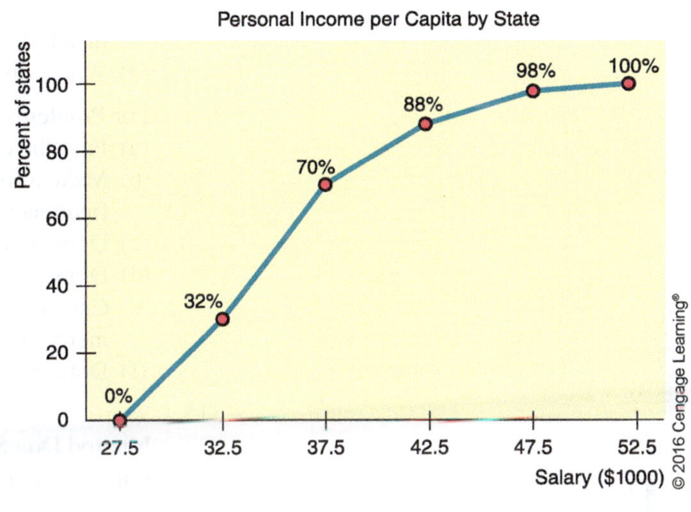

(a) How were the percentages shown in graph (ii) computed?
(b) How many states have average per capita income less than 37.5 thousand dollars?
(c) How many states have average per capita income between 42.5 and 52.5 thousand dollars?
(d) What percentage of the states have average per capita income more than 47.5 thousand dollars?

14. *Critical Thinking* The following ogives come from different distributions of 50 whole numbers between 1 and 60. Labels on each point give the cumulative frequency and the cumulative percentage of data.

(i)

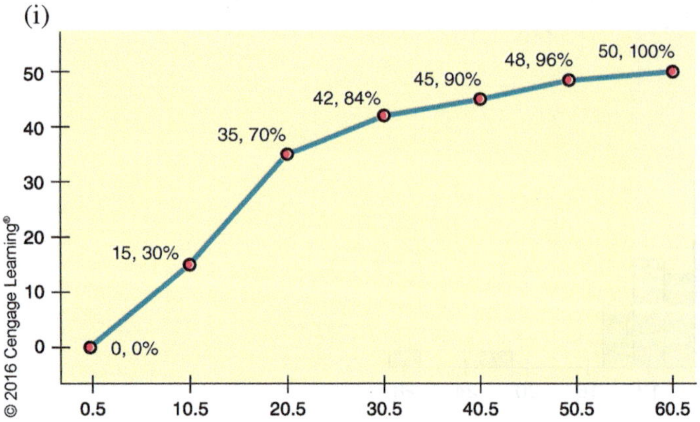

(ii)

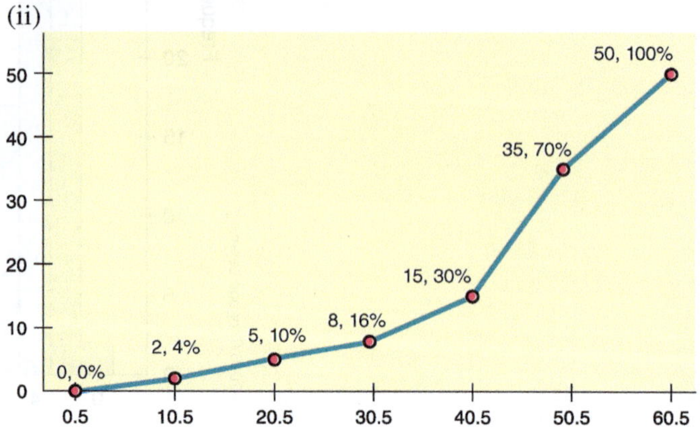

(iii)

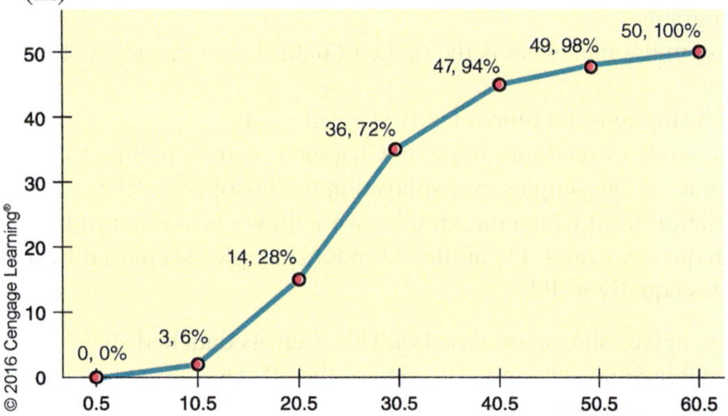

(a) In which distribution does the most data fall below 20.5?
(b) In which distribution does the most data fall below 40.5?
(c) In which distribution does the amount of data below 20.5 most closely match that above 30.5?
(d) Which distribution seems to be skewed right? Skewed left? Mound-shaped?

For Problems 15–20, use the specified number of classes to do the following.
(a) Find the class width.
(b) Make a frequency table showing class limits, class boundaries, midpoints, frequencies, relative frequencies, and cumulative frequencies.
(c) Draw a histogram.
(d) Draw a relative-frequency histogram.
(e) Categorize the basic distribution shape as uniform, mound-shaped symmetrical, bimodal, skewed left, or skewed right.
(f) Draw an ogive.

15. *Sports: Dog Sled Racing* How long does it take to finish the 1161-mile Iditarod Dog Sled Race from Anchorage to Nome, Alaska (*see* Viewpoint)? Finish times (to the nearest hour) for 57 dogsled teams are shown below.

261	271	236	244	279	296	284	299	288	288	247	256
338	360	341	333	261	266	287	296	313	311	307	307
299	303	277	283	304	305	288	290	288	289	297	299
332	330	309	328	307	328	285	291	295	298	306	315
310	318	318	320	333	321	323	324	327			

Use five classes.

16. *Medical: Glucose Testing* The following data represent glucose blood levels (mg/100 ml) after a 12-hour fast for a random sample of 70 women (Reference: *American Journal of Clinical Nutrition,* Vol. 19, pp. 345–351). *Note:* These data are also available for download at the Online Study Center.

45	66	83	71	76	64	59	59
76	82	80	81	85	77	82	90
87	72	79	69	83	71	87	69
81	76	96	83	67	94	101	94
89	94	73	99	93	85	83	80
78	80	85	83	84	74	81	70
65	89	70	80	84	77	65	46
80	70	75	45	101	71	109	73
73	80	72	81	63	74		

Use six classes.

17. *Medical: Tumor Recurrence* Certain kinds of tumors tend to recur. The following data represent the lengths of time, in months, for a tumor to recur after chemotherapy (Reference: D. P. Byar, *Journal of Urology,* Vol. 10, pp. 556–561). *Note:* These data are also available for download at the Online Study Center.

19	18	17	1	21	22	54	46	25	49
50	1	59	39	43	39	5	9	38	18
14	45	54	59	46	50	29	12	19	36
38	40	43	41	10	50	41	25	19	39
27	20								

Use five classes.

18. *Archaeology: New Mexico* The Wind Mountain excavation site in New Mexico is an important archaeological location of the ancient Native American Anasazi culture. The following data represent depths (in cm) below surface grade at which significant artifacts were discovered at this site (Reference: A. I. Woosley and A. J. McIntyre, *Mimbres Mogollon Archaeology,* University of New Mexico Press). *Note:* These data are also available for download at the Online Study Center.

85	45	75	60	90	90	115	30	55	58
78	120	80	65	65	140	65	50	30	125
75	137	80	120	15	45	70	65	50	45
95	70	70	28	40	125	105	75	80	70
90	68	73	75	55	70	95	65	200	75
15	90	46	33	100	65	60	55	85	50
10	68	99	145	45	75	45	95	85	65
65	52	82							

Use seven classes.

19. *Environment: Gasoline Consumption* The following data represent highway fuel consumption in miles per gallon (mpg) for a random sample of 55 models of passenger cars (Source: Environmental Protection Agency). *Note:* These data are also available for download at the Online Study Center.

30	27	22	25	24	25	24	15
35	35	33	52	49	10	27	18
20	23	24	25	30	24	24	24
18	20	25	27	24	32	29	27
24	27	26	25	24	28	33	30
13	13	21	28	37	35	32	33
29	31	28	28	25	29	31	

Use five classes.

20. *Advertising: Readability* "Readability Levels of Magazine Ads," by F. K. Shuptrine and D. D. McVicker, is an article in the *Journal of Advertising Research.* (For more information, visit the Brase/Brase statistics site at **http://www.cengagebrain.com** and find the link to DASL, the Carnegie Mellon University Data and Story Library. Look in Data Subjects under Consumer and then Magazine Ads Readability file.) The following is a list of the number of three-syllable (or longer) words in advertising copy of randomly selected magazine advertisements.

34	21	37	31	10	24	39	10	17	18	32
17	3	10	6	5	6	6	13	22	25	3
5	2	9	3	0	4	29	26	5	5	24
15	3	8	16	9	10	3	12	10	10	10
11	12	13	1	9	43	13	14	32	24	15

Use eight classes.

21. *Expand Your Knowledge: Decimal Data* The following data represent tonnes of wheat harvested each year (1894–1925) from Plot 19 at the Rothamsted Agricultural Experiment Stations, England.

2.71	1.62	2.60	1.64	2.20	2.02	1.67	1.99	2.34	1.26	1.31
1.80	2.82	2.15	2.07	1.62	1.47	2.19	0.59	1.48	0.77	2.04
1.32	0.89	1.35	0.95	0.94	1.39	1.19	1.18	0.46	0.70	

(a) Multiply each data value by 100 to "clear" the decimals.
(b) Use the standard procedures of this section to make a frequency table and histogram with your whole-number data. Use six classes.
(c) Divide class limits, class boundaries, and class midpoints by 100 to get back to your original data values.

22. *Decimal Data: Batting Averages* The following data represent baseball batting averages for a random sample of National League players near the end of the baseball season. The data are from the baseball statistics section of the *Denver Post.*

0.194	0.258	0.190	0.291	0.158	0.295	0.261	0.250	0.181
0.125	0.107	0.260	0.309	0.309	0.276	0.287	0.317	0.252
0.215	0.250	0.246	0.260	0.265	0.182	0.113	0.200	

(a) Multiply each data value by 1000 to "clear" the decimals.

(b) Use the standard procedures of this section to make a frequency table and histogram with your whole-number data. Use five classes.

(c) Divide class limits, class boundaries, and class midpoints by 1000 to get back to your original data.

23. *Expand Your Knowledge: Dotplot* Another display technique that is somewhat similar to a histogram is a *dotplot*. In a dotplot, the data values are displayed along the horizontal axis. A dot is then plotted over each data value in the data set.

PROCEDURE

HOW TO MAKE A DOTPLOT

Display the data along a horizontal axis. Then plot each data value with a dot or point above the corresponding value on the horizontal axis. For repeated data values, stack the dots.

Dotplot

The next display shows a *dotplot* generated by Minitab (➤**Graph** ➤**Dotplot**) for the number of licensed drivers per 1000 residents by state, including the District of Columbia (Source: U.S. Department of Transportation).

Dotplot for Licensed Drivers per 1000 Residents

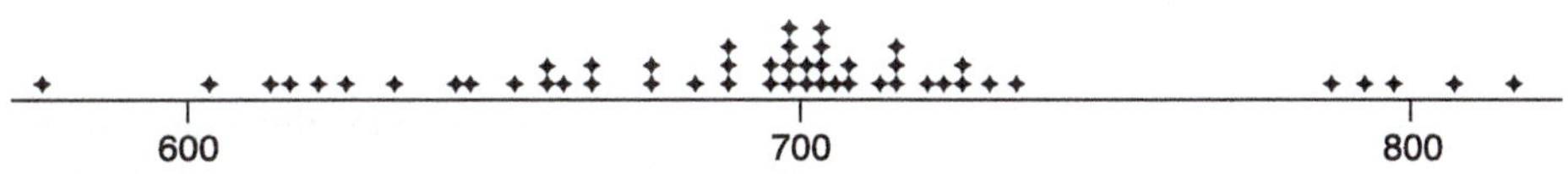

Source: U.S. Department of Transportation

(a) From the dotplot, how many states have 600 or fewer licensed drivers per 1000 residents?

(b) About what percentage of the states (50 states and District of Columbia) seem to have close to 800 licensed drivers per 1000 residents?

(c) Consider the intervals 550 to 650, 650 to 750, and 750 to 850 licensed drivers per 1000 residents. In which interval do most of the states fall?

24. *Dotplot: Dog Sled Racing* Make a dotplot for the data in Problem 15 regarding the finish time (number of hours) for the Iditarod Dog Sled Race. Compare the dotplot to the histogram of Problem 15.

25. *Dotplot: Tumor Recurrence* Make a dotplot for the data in Problem 17 regarding the recurrence of tumors after chemotherapy. Compare the dotplot to the histogram of Problem 17.

SECTION **2.2**

Bar Graphs, Circle Graphs, and Time-Series Graphs

FOCUS POINTS

- Determine types of graphs appropriate for specific data.
- Construct bar graphs, Pareto charts, circle graphs, and time-series graphs.
- Interpret information displayed in graphs.

Histograms provide a useful visual display of the distribution of data. However, the data must be quantitative. In this section, we examine other types of graphs, some of which are suitable for qualitative or category data as well.

Bar graph

Let's start with *bar graphs*. These are graphs that can be used to display quantitative or qualitative data.

FEATURES OF A BAR GRAPH

1. Bars can be vertical or horizontal.
2. Bars are of uniform width and uniformly spaced.
3. The lengths of the bars represent values of the variable being displayed, the frequency of occurrence, or the percentage of occurrence. The same measurement scale is used for the length of each bar.
4. The graph is well annotated with title, labels for each bar, and vertical scale or actual value for the length of each bar.

EXAMPLE 4

BAR GRAPH

Figure 2-11 shows two bar graphs depicting the life expectancies for men and women born in the designated year. Let's analyze the features of these graphs.

Cluster bar graph

SOLUTION: The graphs are called *cluster bar graphs* because there are two bars for each year of birth. One bar represents the life expectancy for men, and the other represents the life expectancy for women. The height of each bar represents the life expectancy (in years).

Changing scale

An important feature illustrated in Figure 2-11(b) is that of a *changing scale.* Notice that the scale between 0 and 65 is compressed. The changing scale amplifies the apparent difference between life spans for men and women, as well as the increase in life spans from those born in 1980 to the projected span of those born in 2010.

CHANGING SCALE

Whenever you use a change in scale in a graphic, warn the viewer by using a squiggle ∿ on the changed axis. Sometimes, if a single bar is unusually long, the bar length is compressed with a squiggle in the bar itself.

FIGURE 2-11

Life Expectancy

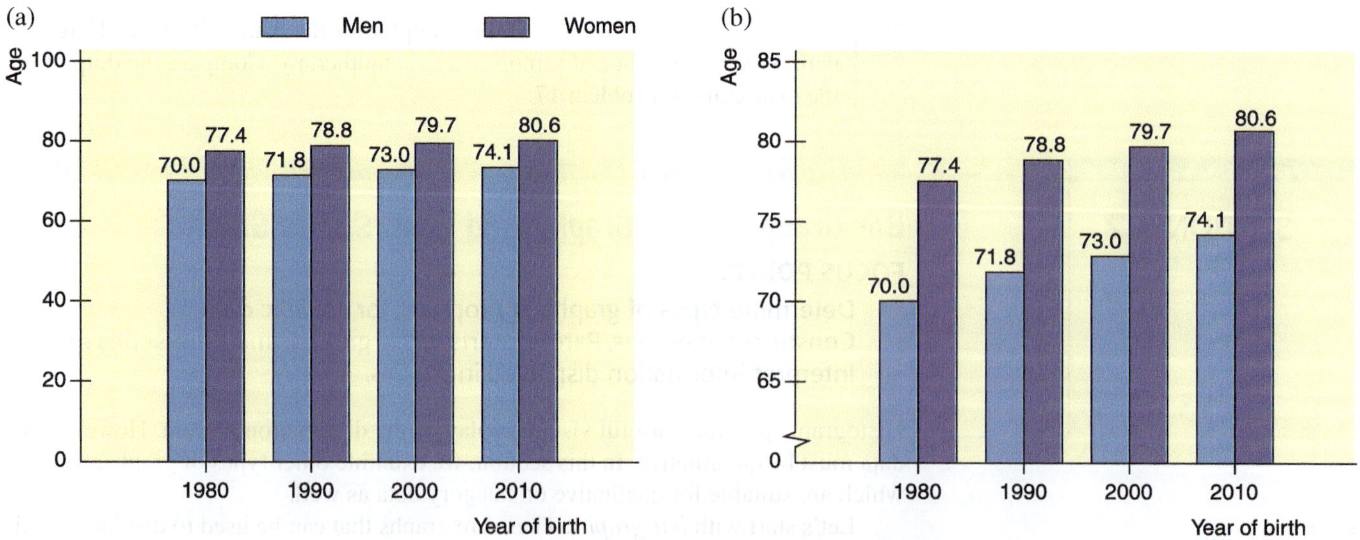

Source: U.S. Census Bureau

Source: U.S. Census Bureau

LOOKING FORWARD

Cluster bar graphs displaying percentages are particularly useful visuals for showing how different populations fit different categories. For instance, in Example 2 in Chapter 11, we see a cluster graph showing preferences of males and females for dogs, cats, other pets, or no pets. This graph displays the sample data for an inferential test called a *test of homogeneity*.

Quality control is an important aspect of today's production and service industries. Dr. W. Edwards Deming was one of the developers of total quality management (TQM). In his book *Out of Crisis,* he outlines many strategies for monitoring and improving service and production industries. In particular, Dr. Deming recommends the use of some statistical methods to organize and analyze data from industries so sources of problems can be identified and then corrected. *Pareto* (pronounced "Pah-rāy-tō") *charts* are among the many techniques used in quality-control programs.

Pareto chart

A **Pareto chart** is a bar graph in which the bar height represents frequency of an event. In addition, the bars are arranged from left to right according to decreasing height.

GUIDED EXERCISE 2 | PARETO CHARTS

This exercise is adapted from *The Deming Management Method* by Mary Walton. Suppose you want to arrive at college 15 minutes before your first class so that you can feel relaxed when you walk into class. An early arrival time also allows room for unexpected delays. However, you always find yourself arriving "just in time" or slightly late. What causes you to be late? Charlotte made a list of possible causes and then kept a checklist for 2 months (Table 2-12). On some days more than one item was checked because several events occurred that caused her to be late.

TABLE 2-12 **Causes for Lateness (September–October)**

Cause	Frequency
Snoozing after alarm goes off	15
Car trouble	5
Too long over breakfast	13
Last-minute studying	20
Finding something to wear	8
Talking too long with roommate	9
Other	3

(a) Make a Pareto chart showing the causes for lateness. Be sure to label the causes, and draw the bars using the same vertical scale.

FIGURE 2-12 Pareto Chart: Conditions That Might Cause Lateness.

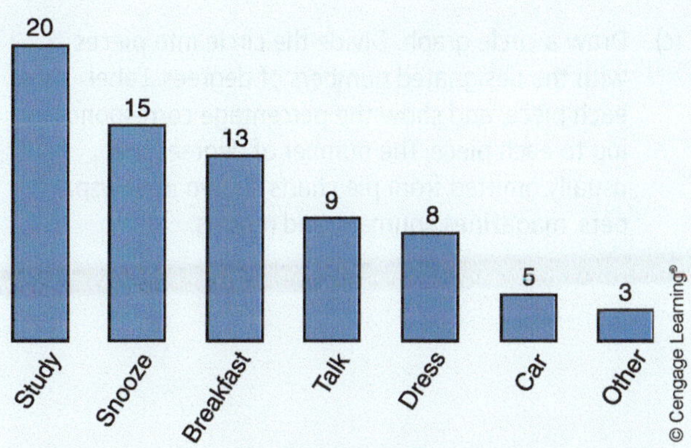

(b) Looking at the Pareto chart, what recommendations do you have for Charlotte?

According to the chart, rearranging study time, or getting up earlier to allow for studying, would cure her most frequent cause for lateness. Repairing the car might be important, but for getting to campus early, it would not be as effective as adjusting study time.

Circle graph or pie chart

Donut pie charts are a popular variation of pie charts. Problem 15 of this section discusses donut pie charts.

Another popular pictorial representation of data is the *circle graph* or *pie chart.* It is relatively safe from misinterpretation and is especially useful for showing the division of a total quantity into its component parts. The total quantity, or 100%, is represented by the entire circle. Each wedge of the circle represents a component part of the total. These proportional segments are usually labeled with corresponding percentages of the total. Guided Exercise 3 shows how to make a circle graph.

In a **circle graph** or **pie chart,** wedges of a circle visually display proportional parts of the total population that share a common characteristic.

GUIDED EXERCISE 3 | CIRCLE GRAPH

How long do we spend talking on the telephone after hours (at home after 5 P.M.)? The results from a recent survey of 500 people (as reported in *USA Today*) are shown in Table 2-13. We'll make a circle graph to display these data.

TABLE 2-13 **Time Spent on Home Telephone After 5 P.M.**

Time	Number	Fractional Part	Percentage	Number of Degrees
Less than ½ hour	296	296/500	59.2	59.2% × 360° ≈ 213°
½ hour to 1 hour	83	83/500	16.6	16.6% × 360° ≈ 60°
More than 1 hour	121	____	____	____
Total	____		____	____

(a) Fill in the missing parts in Table 2-13 for "More than 1 hour." Remember that the central angle of a circle is 360°. Round to the nearest degree.

➡ For "More than 1 hour," Fractional Part = 121/500; Percentage = 24.2%; Number of Degrees = 24.2% × 360° ≈ 87°. The symbol ≈ means "approximately equal."

(b) Fill in the totals. What is the total number of responses? Do the percentages total 100% (within rounding error)? Do the number of degrees total 360° (within rounding error)?

➡ The total number of responses is 500. The percentages total 100%. You must have such a total in order to create a circle graph. The number of degrees total 360°.

(c) Draw a circle graph. Divide the circle into pieces with the designated numbers of degrees. Label each piece, and show the percentage corresponding to each piece. The number of degrees are usually omitted from pie charts shown in newspapers, magazines, journals, and reports.

➡ **FIGURE 2-13** Hours on Home Telephone After 5 P.M.

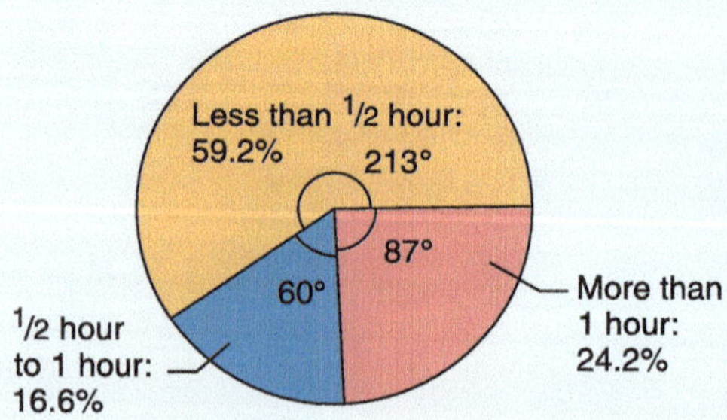

Suppose you begin an exercise program that involves walking or jogging for 30 minutes. You exercise several times a week but monitor yourself by logging the distance you cover in 30 minutes each Saturday. How do you display these data in a meaningful way? Making a bar chart showing the frequency of distances you cover might be interesting, but it does not really show how the distance you cover in 30 minutes has changed over time. A graph showing the distance covered on each date will let you track your performance over time.

Time-series graph

We will use a *time-series graph.* A time-series graph is a graph showing data measurements in chronological order. To make a time-series graph, we put time on the horizontal scale and the variable being measured on the vertical scale. In a basic time-series graph, we connect the data points by line segments.

In a **time-series graph,** data are plotted in order of occurrence at regular intervals over a period of time.

EXAMPLE 5

TIME-SERIES GRAPH

Suppose you have been in the walking/jogging exercise program for 20 weeks, and for each week you have recorded the distance you covered in 30 minutes. Your data log is shown in Table 2-14.

Myrleen Pearson/PhotoEdit

TABLE 2-14 **Distance (in Miles) Walked/Jogged in 30 Minutes**

Week	1	2	3	4	5	6	7	8	9	10
Distance	1.5	1.4	1.7	1.6	1.9	2.0	1.8	2.0	1.9	2.0
Week	11	12	13	14	15	16	17	18	19	20
Distance	2.1	2.1	2.3	2.3	2.2	2.4	2.5	2.6	2.4	2.7

(a) Make a time-series graph.

SOLUTION: The data are appropriate for a time-series graph because they represent the same measurement (distance covered in a 30-minute period) taken at different times. The measurements are also recorded at equal time intervals (every week). To make our time-series graph, we list the weeks in order on the horizontal scale. Above each week, plot the distance covered that week on the vertical scale. Then connect the dots. Figure 2-14 shows the time-series graph. Be sure the scales are labeled.

FIGURE 2-14

Time-Series Graph of Distance (in miles) Jogged in 30 Minutes

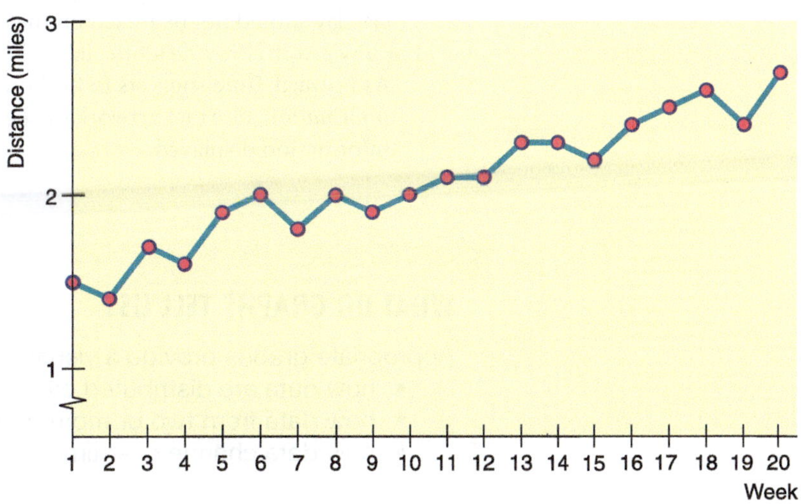

(b) From looking at Figure 2-14, can you detect any patterns?

SOLUTION: There seems to be an upward trend in distance covered. The distances covered in the last few weeks are about a mile farther than those for the first few weeks. However, we cannot conclude that this trend will continue. Perhaps you have reached your goal for this training activity and now wish to maintain a distance of about 2.5 miles in 30 minutes.

Time series

Data sets composed of similar measurements taken at regular intervals over time are called *time series*. Time series are often used in economics, finance, sociology, medicine, and any other situation in which we want to study or monitor a similar measure over a period of time. A time-series graph can reveal some of the main features of a time series.

Time-series data consist of measurements of the same variable for the same subject taken at regular intervals over a period of time.

CRITICAL THINKING

We've seen several styles of graphs. Which kinds are suitable for a specific data collection?

PROCEDURE

HOW TO DECIDE WHICH TYPE OF GRAPH TO USE

Bar graphs are useful for quantitative or qualitative data. With qualitative data, the frequency or percentage of occurrence can be displayed. With quantitative data, the measurement itself can be displayed, as was done in the bar graph showing life expectancy. Watch that the measurement scale is consistent or that a jump scale squiggle is used.

Pareto charts identify the frequency of events or categories in decreasing order of frequency of occurrence.

Circle graphs display how a *total* is dispersed into several categories. The circle graph is very appropriate for qualitative data, or any data for which percentage of occurrence makes sense. Circle graphs are most effective when the number of categories or wedges is 10 or fewer.

Time-series graphs display how data change over time. It is best if the units of time are consistent in a given graph. For instance, measurements taken every day should not be mixed on the same graph with data taken every week.

For any graph: Provide a title, label the axes, and identify units of measure. As Edward Tufte suggests in his book *The Visual Display of Quantitative Information*, don't let artwork or skewed perspective cloud the clarity of the information displayed.

WHAT DO GRAPHS TELL US?

Appropriate graphs provide a visual summary of data that tells us

- how data are distributed over several categories or data intervals;
- how data from two or more data sets compare;
- how data change over time.

TECH NOTES *Bar graphs, circle graphs, and time-series graphs*

TI-84Plus/TI-83Plus/TI-*n*spire (with TI-84Plus keypad) Only time-series. Place consecutive values 1 through the number of time segments in list L1 and corresponding data in L2. Press **Stat Plot** and highlight an *xy* line plot.

Excel 2010 First enter the data into the spreadsheet. Then click the **Insert Tab** and select the type of chart you want to create. A variety of bar graphs, pie charts, and line graphs that can be used as time-series graphs are available. Use the **Design Tab** and the **Layout Tab** on the **Chart Tools** menu to access options such as title, axis labels, etc., for your chart.

Minitab Use the menu selection **Graph.** Select the desired option and follow the instructions in the dialogue boxes.

VIEWPOINT Do Ethical Standards Vary by the Situation?

The Lutheran Brotherhood conducted a national survey and found that nearly 60% of all U.S. adults claim that ethics vary by the situation; 33% claim that there is only one ethical standard; and 7% are not sure. How could you draw a circle graph to make a visual impression of Americans' views on ethical standards?

SECTION 2.2 PROBLEMS

1. *Interpretation* Consider graph (a) of Reasons People Like Texting on Cell Phones, based on a GfK Roper survey of 1000 adults.

Reasons People Like Texting on Cell Phones

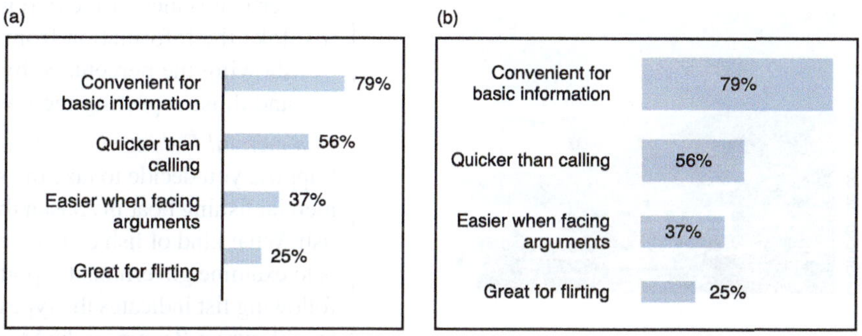

   (a) Do you think respondents could select more than one response? Explain.
   (b) Could the same information be displayed in a circle graph? Explain.
   (c) Is graph (a) a Pareto chart?

2. *Interpretation* Look at graph (b) of Reasons People Like Texting on Cell Phones. Is this a proper bar graph? Explain.

3. *Critical Thinking* A personnel office is gathering data regarding working conditions. Employees are given a list of five conditions that they might want to see improved. They are asked to select the one item that is most critical to them. Which type of graph, circle graph or Pareto chart, would be most useful for displaying the results of the survey? Why?

4. *Critical Thinking* Your friend is thinking about buying shares of stock in a company. You have been tracking the closing prices of the stock shares for the past 90 trading days. Which type of graph for the data, histogram or time-series, would be best to show your friend? Why?

5. *Education: Does College Pay Off?* It is costly in both time and money to go to college. Does it pay off? According to the Bureau of the Census, the answer is yes. The average annual income (in thousands of dollars) of a *household* headed by a person with the stated education level is as follows: 21.6 if ninth grade is the highest level achieved, 39.6 for high school graduates, 56.8 for those holding associate degrees, 75.6 for those with bachelor's degrees, 91.7 for those with master's degrees, and 120.9 for those with doctoral degrees. Make a bar graph showing household income for each education level.

6. *Interpretation* Consider the two graphs depicting the influence of advertisements on making large purchases for two different age groups, those 18–34 years old and those 45–54 years old (based on a Harris Poll of about 2500 adults aged 18 or older). *Note:* Other responses such as "not sure" and "not applicable" were also possible.

Influence of Advertising on Most Recent Large Purchase

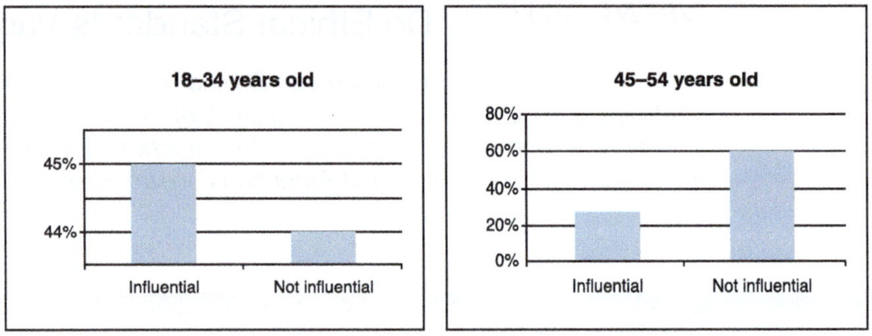

(a) Taking a quick glance at the graphs, Jenna thought that there was very little difference (maybe less than 1%) in the percentage of the two age groups who said that ads were influential. How would you change the graphs so that Jenna would not be misled so easily? *Hint:* Look at the vertical scales of the two graphs.

(b) Take the information from the two graphs and make a cluster bar graph showing the percentage by age group reporting to be influenced by ads and those reporting they were not influenced by ads.

7. *Commercial Fishing: Gulf of Alaska* It's not an easy life, but it's a good life! Suppose you decide to take the summer off and sign on as a deck hand for a commercial fishing boat in Alaska that specializes in deep-water fishing for groundfish. What kind of fish can you expect to catch? One way to answer this question is to examine government reports on groundfish caught in the Gulf of Alaska. The following list indicates the types of fish caught annually in thousands of metric tons (Source: *Report on the Status of U.S. Living Marine Resources,* National Oceanic and Atmospheric Administration): flatfish, 36.3; Pacific cod, 68.6; sablefish, 16.0; Walleye pollock, 71.2; rockfish, 18.9. Make a Pareto chart showing the annual harvest for commercial fishing in the Gulf of Alaska.

8. *Archaeology: Ireland* Commercial dredging operations in ancient rivers occasionally uncover archaeological artifacts of great importance. One such artifact is Bronze Age spearheads recovered from ancient rivers in Ireland. A recent study gave the following information regarding discoveries of ancient bronze spearheads in Irish rivers.

River	Bann	Blackwater	Erne	Shannon	Barrow
No. of spearheads	19	8	15	33	14

(Based on information from *Crossing the Rubicon, Bronze Age Studies 5,* Lorraine Bourke, Department of Archaeology, National University of Ireland, Galway.)

(a) Make a Pareto chart for these data.

(b) Make a circle graph for these data.

9. *Lifestyle: Hide the Mess!* A survey of 1000 adults (reported in *USA Today*) uncovered some interesting housekeeping secrets. When unexpected company comes, where do we hide the mess? The survey showed that 68% of the respondents toss their mess into the closet, 23% shove things under the bed, 6% put things into the bathtub, and 3% put the mess into the freezer. Make a circle graph to display this information.

10. *Education: College Professors' Time* How do college professors spend their time? *The National Education Association Almanac of Higher Education* gives the following average distribution of professional time allocation: teaching, 51%; research, 16%; professional growth, 5%; community service, 11%; service to the college, 11%; and consulting outside the college, 6%. Make a pie chart showing the allocation of professional time for college professors.

11. *FBI Report: Hawaii* In the Aloha state, you are very unlikely to be murdered! However, it is considerably more likely that your house might be burgled, your car might be stolen, or you might be punched in the nose. That said, Hawaii is still a great place to vacation or, if you are very lucky, to live. The following numbers represent the crime rates per 100,000 population in Hawaii: murder, 2.6; rape, 33.4; robbery, 93.3; house burglary, 911.6; motor vehicle theft, 550.7; assault, 125.3 (Source: *Crime in the United States,* U.S. Department of Justice, Federal Bureau of Investigation).
(a) Display this information in a Pareto chart, showing the crime rate for each category.
(b) Could the information as reported be displayed as a circle graph? Explain. *Hint:* Other forms of crime, such as arson, are not included in the information. In addition, some crimes might occur together.

12. *Driving: Bad Habits* Driving would be more pleasant if we didn't have to put up with the bad habits of other drivers. *USA Today* reported the results of a Valvoline Oil Company survey of 500 drivers, in which the drivers marked their complaints about other drivers. The top complaints turned out to be tailgating, marked by 22% of the respondents; not using turn signals, marked by 19%; being cut off, marked by 16%; other drivers driving too slowly, marked by 11%; and other drivers being inconsiderate, marked by 8%. Make a Pareto chart showing percentage of drivers listing each stated complaint. Could this information as reported be put in a circle graph? Why or why not?

13. *Ecology: Lakes* Pyramid Lake, Nevada, is described as the pride of the Paiute Indian Nation. It is a beautiful desert lake famous for very large trout. The elevation of the lake surface (feet above sea level) varies according to the annual flow of the Truckee River from Lake Tahoe. The U.S. Geological Survey provided the following data:

Year	Elevation
1986	3817
1987	3815
1988	3810
1989	3812
1990	3808
1991	3803

Year	Elevation
1992	3798
1993	3797
1994	3795
1995	3797
1996	3802
1997	3807

Year	Elevation
1998	3811
1999	3816
2000	3817

Make a time-series graph displaying the data. For more information, visit the Brase/Brase statistics site at **http://www.cengagebrain.com** and find the link to the Pyramid Lake Fisheries.

Ryan McVay/Photodisc/Jupiter Images

14. *Vital Statistics: Height* How does average height for boys change as boys get older? According to *Physician's Handbook,* the average heights at different ages are as follows:

Age (years)	0.5	1	2	3	4	5	6	7
Height (inches)	26	29	33	36	39	42	45	47

Age (years)	8	9	10	11	12	13	14
Height (inches)	50	52	54	56	58	60	62

Make a time-series graph for average height for ages 0.5 through 14 years.

15. *Expand Your Knowledge Donut Pie Charts* The book *The Wall Street Journal. Guide to Information Graphics* by Dona M. Wong gives strategies for using graphs and charts to display information effectively. One popular graph discussed is the *donut pie chart*. The donut pie chart is simply a pie chart with the center removed. A recent Harris Poll asked adults about their opinions regarding whether books should be banned from libraries because of social, language, violent, sexual, or religious content. The responses by education level to the question "Do you think that there are any books which should be banned completely?" are shown in the following donut pie charts.

(i) (ii)

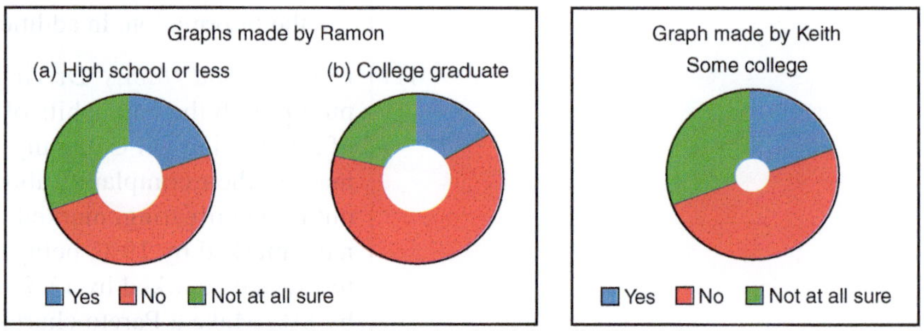

(a) What feature of Keith's graph makes it difficult to visually compare the responses of those with some college to those shown in the other graphs? How would you change Keith's graph for easier comparison?
(b) *Interpretation* Compare the two graphs made by Ramon. At which of the two education levels is the "no" response more frequent?

16 *Technology: Cars* The following cluster bar graph shows responses from different age groups to questions regarding connectivity and tracking technology found in new cars. A recent Harris Poll asked respondents how much they agreed or disagreed with statements that they

(1) worry that the technologies cause too much distraction and are dangerous;
(2) worry about letting companies know too much about location and driving habits;
(3) worry that insurance rates could increase because of knowledge of driving habits;
(4) think the technologies make driving more enjoyable;
(5) feel safer with the technologies;
(6) feel it is important to stay connected when in vehicle.

The graph shows the percentage of respondents in each age category who agree strongly or somewhat agree to each of the six statements.

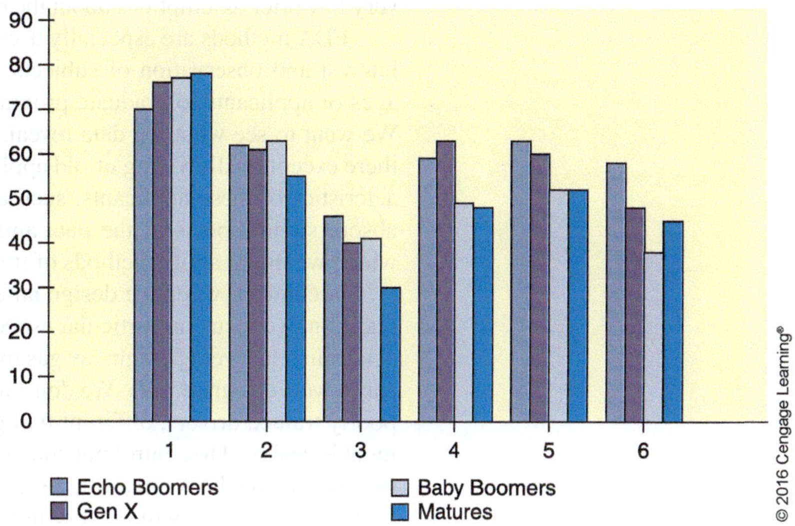

(a) *Interpretation* Which statement has the highest rate of agreement for all four age groups?

(b) *Interpretation* Which age group expresses least worry about insurance companies raising their rates because of the driving habit information collected by the technologies?

(c) *Interpretation* Which age group has the highest percentage of those who find the technologies make driving more enjoyable?

SECTION **2.3**

Stem-and-Leaf Displays

FOCUS POINTS

- Construct a stem-and-leaf display from raw data.
- Use a stem-and-leaf display to visualize data distribution.
- Compare a stem-and-leaf display to a histogram.

EXPLORATORY DATA ANALYSIS

EDA

Together with histograms and other graphics techniques, the stem-and-leaf display is one of many useful ways of studying data in a field called *exploratory data analysis* (often abbreviated as *EDA*). John W. Tukey wrote one of the definitive books on the subject, *Exploratory Data Analysis* (Addison-Wesley). Another very useful reference for EDA techniques is the book *Applications, Basics, and Computing of Exploratory Data Analysis* by Paul F. Velleman and David C. Hoaglin (Duxbury Press). Exploratory data analysis techniques are particularly useful for detecting patterns and extreme data values. They are designed to help us explore a data set, to ask questions we had not thought of before, or to pursue leads in many directions.

EDA techniques are similar to those of an explorer. An explorer has a general idea of destination but is always alert for the unexpected. An explorer needs to assess situations quickly and often simplify and clarify them. An explorer makes pictures—that is, maps showing the relationships of landscape features. The aspects of rapid implementation, visual displays such as graphs and charts, data simplification, and robustness (that is, analysis that is not influenced much by extreme data values) are key ingredients of EDA techniques. In addition, these techniques are good for exploration because they require very few prior assumptions about the data.

EDA methods are especially useful when our data have been gathered for general interest and observation of subjects. For example, we may have data regarding the ages of applicants to graduate programs. We don't have a specific question in mind. We want to see what the data reveal. Are the ages fairly uniform or spread out? Are there exceptionally young or old applicants? If there are, we might look at other characteristics of these applicants, such as field of study. EDA methods help us quickly absorb some aspects of the data and then may lead us to ask specific questions to which we might apply methods of traditional statistics.

In contrast, when we design an experiment to produce data to answer a specific question, we focus on particular aspects of the data that are useful to us. If we want to determine the average highway gas mileage of a specific sports car, we use that model car in well-designed tests. We don't need to worry about unexpected road conditions, poorly trained drivers, different fuel grades, sudden stops and starts, etc. Our experiment is designed to control outside factors. Consequently, we do not need to "explore" our data as much. We can often make valid assumptions about the data. Methods of traditional statistics will be very useful to analyze such data and answer our specific questions.

STEM-AND-LEAF DISPLAY

In this text, we will introduce two EDA techniques: stem-and-leaf displays and, in Section 3.3, box-and-whisker plots. Let's first look at a stem-and-leaf display.

Stem-and-leaf display

A **stem-and-leaf display** is a method of exploratory data analysis that is used to rank-order and arrange data into groups.

We know that frequency distributions and histograms provide a useful organization and summary of data. However, in a histogram, we lose most of the specific data values. A stem-and-leaf display is a device that organizes and groups data but allows us to recover the original data if desired. In the next example, we will make a stem-and-leaf display.

EXAMPLE 6

STEM-AND-LEAF DISPLAY

Many airline passengers seem weighted down by their carry-on luggage. Just how much weight are they carrying? The carry-on luggage weights in pounds for a random sample of 40 passengers returning from a vacation to Hawaii were recorded (see Table 2-15).

TABLE 2-15 **Weights of Carry-On Luggage in Pounds**

30	27	12	42	35	47	38	36	27	35
22	17	29	3	21	0	38	32	41	33
26	45	18	43	18	32	31	32	19	21
33	31	28	29	51	12	32	18	21	26

FIGURE 2-15

Stem-and-Leaf Displays of Airline Carry-On Luggage Weights

(a) Leaves Not Ordered

3 | 2 represents 32 lb

Stem	Leaves
0	3 0
1	2 7 8 8 9 2 8
2	7 7 2 9 1 6 1 8 9 1 6
3	0 5 8 6 5 8 2 3 2 1 2 3 1 2
4	2 7 1 5 3
5	1

(b) Final Display with Leaves Ordered

3 | 2 represents 32 lb

Stem	Leaves
0	0 3
1	2 2 7 8 8 8 9
2	1 1 1 2 6 6 7 7 8 9 9
3	0 1 1 2 2 2 2 3 3 5 5 6 8 8
4	1 2 3 5 7
5	1

Stem
Leaf

To make a stem-and-leaf display, we break the digits of each data value into *two* parts. The left group of digits is called a *stem,* and the remaining group of digits on the right is called a *leaf.* We are free to choose the number of digits to be included in the stem.

The weights in our example consist of two-digit numbers. For a two-digit number, the stem selection is obviously the left digit. In our case, the tens digits will form the stems, and the units digits will form the leaves. For example, for the weight 12, the stem is 1 and the leaf is 2. For the weight 18, the stem is again 1, but the leaf is 8. In the stem-and-leaf display, we list each possible stem once on the left and all its leaves in the same row on the right, as in Figure 2-15(a). Finally, we order the leaves as shown in Figure 2-15(b).

Figure 2-15 shows a stem-and-leaf display for the weights of carry-on luggage. From the stem-and-leaf display in Figure 2-15, we see that two bags weighed 27 lb, one weighed 3 lb, one weighed 51 lb, and so on. We see that most of the weights were in the 30-lb range, only two were less than 10 lb, and six were over 40 lb. Note that the lengths of the lines containing the leaves give the visual impression that a sideways histogram would present.

As a final step, we need to indicate the scale. This is usually done by indicating the value represented by the stem and one leaf.

There are no firm rules for selecting the group of digits for the stem. But whichever group you select, you must list all the possible stems from smallest to largest in the data collection.

LOOKING FORWARD

Ordering or sorting data is an essential first step to finding the median or center value of the data distribution. We will discuss the median in Section 3.1. The data also need to be ordered before we can construct a box-and-whisker plot (Section 3.3). After a little practice, you will find that stem-and-leaf diagrams provide a very efficient way of ordering data by hand.

PROCEDURE

HOW TO MAKE A STEM-AND-LEAF DISPLAY

1. Divide the digits of each data value into two parts. The leftmost part is called the *stem* and the rightmost part is called the *leaf.*
2. Align all the stems in a vertical column from smallest to largest. Draw a vertical line to the right of all the stems.
3. Place all the leaves with the same stem in the same row as the stem, and arrange the leaves in increasing order.
4. Use a label to indicate the magnitude of the numbers in the display. We include the decimal position in the label rather than with the stems or leaves.

GUIDED EXERCISE 4 Stem-and-Leaf Display

What does it take to win at sports? If you're talking about basketball, one sportswriter gave the answer. He listed the winning scores of the conference championship games over the last 35 years. The scores for those games follow below.

AP Images/Rogelio Solis

132	118	124	109	104	101	125	83	99
131	98	125	97	106	112	92	120	103
111	117	135	143	112	112	116	106	117
119	110	105	128	112	126	105	102	

To make a stem-and-leaf display, we'll use the first *two* digits as the stems (see Figure 2-16). Notice that the distribution of scores is fairly symmetrical.

(a) Use the first *two* digits as the stem. Then order the leaves. Provide a label that shows the meaning and units of the first stem and first leaf.

FIGURE 2-16 Winning Scores

08 | 3 represents 083 or 83 points

Stem	Leaves
08	3
09	2 7 8 9
10	1 2 3 4 5 5 6 6 9
11	0 1 2 2 2 2 6 7 7 8 9
12	0 4 5 5 6 8
13	1 2 5
14	3

(b) Looking at the distribution, would you say that it is fairly symmetrical?

Yes. Notice that stem 11 has the most data.

WHAT DO STEM-AND-LEAF DISPLAYS TELL US?

Stem-and-leaf displays give a visual display that

- shows us all the data (or truncated data) in order from smallest to largest;
- helps us spot extreme data values or clusters of data values;
- displays the shape of the data distribution.

COMMENT Stem-and-leaf displays organize the data, let the data analyst spot extreme values, and are easy to create. In fact, they can be used to organize data so that frequency tables are easier to make. However, at this time, histograms are used more often in formal data presentations, and stem-and-leaf displays are used by data analysts to gain initial insights about the data.

TECH NOTES *Stem-and-leaf display*

TI-84Plus/TI-83Plus/TI-*nspire* Does not support stem-and-leaf displays. You can sort the data by using keys **Stat ➤ Edit ➤ 2:SortA.**

Excel 2010 Enter your data and select the data you want to sort. On the **Home** ribbon, click the **Sort and Filter** button in the **Editing** group of the ribbon and select the desired sorting option.

Minitab Use the menu selections **Graph ➤ Stem-and-Leaf** and fill in the dialogue box.

Minitab Stem-and-Leaf Display (for Data in Guided Exercise 4)

```
Stem-and-Leaf of Scores      N=35
Leaf Unit=1.0

   1        8    3
   5        9    2789
  14       10    123455669
 (11)      11    01222267789
  10       12    045568
   4       13    125
   1       14    3
```


The values shown in the left column represent depth. Numbers above the value in parentheses show the cumulative number of values from the top to the stem of the middle value. Numbers below the value in parentheses show the cumulative number of values from the bottom to the stem of the middle value. The number in parentheses shows how many values are on the same line as the middle value.

CRITICAL THINKING

Stem-and-leaf displays show each of the original or truncated data values. By looking at the display "sideways," you can see the distribution shape of the data. If there are large gaps between stems containing leaves, especially at the top or bottom of the display, the data values in the first or last lines may be outliers. Outliers should be examined carefully to see if they are data errors or simply unusual data values. Someone very familiar with the field of study as well as the purpose of the study should decide how to treat outliers.

Problems 5 and 6 show how to split a stem.

Problem 10 discusses back-to-back stem-and-leaf displays.

VIEWPOINT What Does It Take to Win?

Scores for NFL Super Bowl games can be found at the NFL web site. Visit the Brase/Brase statistics site at **http://www.cengagebrain.com** *and find the link to the NFL. Once at the NFL web site, follow the links to the Super Bowl. Of special interest in football statistics is the spread, or difference, between scores of the winning and losing teams. If the spread is too large, the game can appear to be lopsided, and TV viewers become less interested in the game (and accompanying commercial ads). Make a stem-and-leaf display of the spread for the NFL Super Bowl games and analyze the results.*

SECTION 2.3 PROBLEMS

1. *Cowboys: Longevity* How long did *real* cowboys live? One answer may be found in the book *The Last Cowboys* by Connie Brooks (University of New Mexico Press). This delightful book presents a thoughtful sociological study of cowboys in west Texas and southeastern New Mexico around the year 1890. A sample of 32 cowboys gave the following years of longevity:

58	52	68	86	72	66	97	89	84	91	91
92	66	68	87	86	73	61	70	75	72	73
85	84	90	57	77	76	84	93	58	47	

(a) Make a stem-and-leaf display for these data.
(b) *Interpretation* Consider the following quote from Baron von Richthofen in his *Cattle Raising on the Plains of North America:* "Cowboys are to be found among the sons of the best families. The truth is probably that most were not a drunken, gambling lot, quick to draw and fire their pistols." Does the data distribution of longevity lend credence to this quote?

2. *Ecology: Habitat* Wetlands offer a diversity of benefits. They provide a habitat for wildlife, spawning grounds for U.S. commercial fish, and renewable timber resources. In the last 200 years, the United States has lost more than half its wetlands. *Environmental Almanac* gives the percentage of wetlands lost in each state in the last 200 years. For the lower 48 states, the percentage loss of wetlands per state is as follows:

46	37	36	42	81	20	73	59	35	50
87	52	24	27	38	56	39	74	56	31
27	91	46	9	54	52	30	33	28	35
35	23	90	72	85	42	59	50	49	
48	38	60	46	87	50	89	49	67	

Make a stem-and-leaf display of these data. Be sure to indicate the scale. How are the percentages distributed? Is the distribution skewed? Are there any gaps?

3. *Health Care: Hospitals* The American Medical Association Center for Health Policy Research, in its publication *State Health Care Data: Utilization, Spending, and Characteristics*, included data, by state, on the number of community hospitals and the average patient stay (in days). The data are shown in the table. Make a stem-and-leaf display of the data for the average length of stay in days. Comment about the general shape of the distribution.

State	No. of Hospitals	Average Length of Stay	State	No. of Hospitals	Average Length of Stay	State	No. of Hospitals	Average Length of Stay
Alabama	119	7.0	Kentucky	107	6.9	New Mexico	37	5.5
Alaska	16	5.7	Louisiana	136	6.7	Oregon	66	5.3
Arizona	61	5.5	Maine	38	7.2	Ohio	193	6.6
Arkansas	88	7.0	Maryland	51	6.8	Oklahoma	113	6.7
California	440	6.0	Massachusetts	101	7.0	Pennsylvania	236	7.5
Colorado	71	6.8	Michigan	175	7.3	Rhode Island	12	6.9
Connecticut	35	7.4	Minnesota	148	8.7	S. Carolina	68	7.1
Delaware	8	6.8	Mississippi	102	7.2	S. Dakota	52	10.3
Dist. of Columbia	11	7.5	Missouri	133	7.4	Tennessee	122	6.8
Florida	227	7.0	Montana	53	10.0	Texas	421	6.2
Georgia	162	7.2	N. Carolina	117	7.3	Utah	42	5.2
Hawaii	19	9.4	N. Dakota	47	11.1	Vermont	15	7.6
Idaho	41	7.1	Nebraska	90	9.6	Virginia	98	7.0
Indiana	113	6.6	Nevada	21	6.4	Washington	92	5.6
Illinois	209	7.3	New Hampshire	27	7.0	W. Virginia	59	7.1
Iowa	123	8.4	New York	231	9.9	Wisconsin	129	7.3
Kansas	133	7.8	New Jersey	96	7.6	Wyoming	27	8.5

4. *Health Care: Hospitals* Using the number of hospitals per state listed in the table in Problem 3, make a stem-and-leaf display for the number of community hospitals per state. Which states have an unusually high number of hospitals?

5. *Expand Your Knowledge: Split Stem* The Boston Marathon is the oldest and best-known U.S. marathon. It covers a route from Hopkinton, Massachusetts, to downtown Boston. The distance is approximately 26 miles. Visit the Brase/Brase statistics site at **http://www.cengagebrain.com** and find the link to the Boston Marathon. Search the marathon site to find a wealth of information about the history of the race. In particular, the site gives the winning times for the Boston Marathon. They are all over 2 hours. The following data are the minutes over 2 hours for the winning male runners:

1961–1980

23	23	18	19	16	17	15	22	13	10
18	15	16	13	9	20	14	10	9	12

1981–2000

9	8	9	10	14	7	11	8	9	8
11	8	9	7	9	9	10	7	9	9

(a) Make a stem-and-leaf display for the minutes over 2 hours of the winning times for the years 1961 to 1980. Use two lines per stem.

PROCEDURE

HOW TO SPLIT A STEM

When a stem has many leaves, it is useful to split the stem into two lines or more. For two lines per stem, place leaves 0 to 4 on the first line and leaves 5 to 9 on the next line.

(b) Make a stem-and-leaf display for the minutes over 2 hours of the winning times for the years 1981 to 2000. Use two lines per stem.

(c) *Interpretation* Compare the two distributions. How many times under 15 minutes are in each distribution?

6. *Split Stem: Golf* The U.S. Open Golf Tournament was played at Congressional Country Club, Bethesda, Maryland, with prizes ranging from \$465,000 for first place to \$5000. Par for the course was 70. The tournament consisted of four rounds played on different days. The scores for each round of the 32 players who placed in the money (more than \$17,000) were given on a web site. For more information, visit the Brase/Brase statistics site at **http://www.cengagebrain.com** and find the link to golf. The scores for the first round were as follows:

71	65	67	73	74	73	71	71	74	73	71
70	75	71	72	71	75	75	71	71	74	75
66	75	75	75	71	72	72	73	71	67	

The scores for the fourth round for these players were as follows:

69	69	73	74	72	72	70	71	71	70	72
73	73	72	71	71	71	69	70	71	72	73
74	72	71	68	69	70	69	71	73	74	

(a) Make a stem-and-leaf display for the first-round scores. Use two lines per stem. (See Problem 5.)
(b) Make a stem-and-leaf display for the fourth-round scores. Use two lines per stem.
(c) *Interpretation* Compare the two distributions. How do the highest scores compare? How do the lowest scores compare?

Are cigarettes bad for people? Cigarette smoking involves tar, carbon monoxide, and nicotine. The first two are definitely not good for a person's health, and the last ingredient can cause addiction. Problems 7, 8, and 9 refer to Table 2-16, which was taken from the web site maintained by the *Journal of Statistics Education.* For more information, visit the Brase/Brase statistics site at **http://www.cengagebrain.com** and find the link to the *Journal of Statistics Education.* Follow the links to the cigarette data.

TABLE 2-16 **Milligrams of Tar, Nicotine, and Carbon Monoxide (CO) per One Cigarette**

Brand	Tar	Nicotine	CO	Brand	Tar	Nicotine	CO
Alpine	14.1	0.86	13.6	MultiFilter	11.4	0.78	10.2
Benson & Hedges	16.0	1.06	16.6	Newport Lights	9.0	0.74	9.5
Bull Durham	29.8	2.03	23.5	Now	1.0	0.13	1.5
Camel Lights	8.0	0.67	10.2	Old Gold	17.0	1.26	18.5
Carlton	4.1	0.40	5.4	Pall Mall Lights	12.8	1.08	12.6
Chesterfield	15.0	1.04	15.0	Raleigh	15.8	0.96	17.5
Golden Lights	8.8	0.76	9.0	Salem Ultra	4.5	0.42	4.9
Kent	12.4	0.95	12.3	Tareyton	14.5	1.01	15.9
Kool	16.6	1.12	16.3	True	7.3	0.61	8.5
L&M	14.9	1.02	15.4	Viceroy Rich Light	8.6	0.69	10.6
Lark Lights	13.7	1.01	13.0	Virginia Slims	15.2	1.02	13.9
Marlboro	15.1	0.90	14.4	Winston Lights	12.0	0.82	14.9
Merit	7.8	0.57	10.0				

Source: Federal Trade Commission, USA (public domain).

7. *Health: Cigarette Smoke* Use the data in Table 2-16 to make a stem-and-leaf display for milligrams of tar per cigarette smoked. Are there any outliers?

8. *Health: Cigarette Smoke* Use the data in Table 2-16 to make a stem-and-leaf display for milligrams of carbon monoxide per cigarette smoked. Are there any outliers?

9. *Health: Cigarette Smoke* Use the data in Table 2-16 to make a stem-and-leaf display for milligrams of nicotine per cigarette smoked. In this case, truncate the measurements at the tenths position and use two lines per stem (see Problem 5, part a).

Back-to-back stem plot

10. *Expand Your Knowledge: Back-to-Back Stem Plot* In archaeology, the depth (below surface grade) at which artifacts are found is very important. Greater depths sometimes indicate older artifacts, perhaps from a different archaeological period. Figure 2-17 is a *back-to-back stem plot* showing the depths of artifact locations at two different archaeological sites. These sites are from similar geographic locations. Notice that the stems are in the center of the diagram. The leaves for Site I artifact depths are shown to the left of the

stem, while the leaves for Site II are to the right of the stem (see *Mimbres Mogollon Archaeology* by A. I. Woosley and A. J. McIntyre, University of New Mexico Press).

(a) What are the least and greatest depths of artifact finds at Site I? at Site II?
(b) Describe the data distribution of depths of artifact finds at Site I and at Site II.
(c) *Interpretation* At Site II, there is a gap in the depths at which artifacts were found. Does the Site II data distribution suggest that there might have been a period of no occupation?

FIGURE 2-17

Depth (in cm) of Artifact Location

5 | 2 | 0 = 25 cm at Site I and 20 cm at Site II

Site I		Site II
5	2	0 5 5
5 0	3	0 0 0 0 5 5
5 5 5 5	4	0 0 0 5
5 0	5	0 0 5 5
5 5 5 5 5 0	6	0 0 0 5 5 5
5 5 5 5 5 5 0 0	7	
5 5 0 0 0	8	
5 5 0 0	9	
5 5	10	
0	11	0 0 5 5 5 5
	12	0 0 0 0 5

3

Christopher Raphael/Warner Bros./Everett Collection

Imagno/Hulton Archive/Getty Images

While the individual man is an insolvable puzzle, in the aggregate he becomes a mathematical certainty. You can, for example, never foretell what any one man will do, but you can say with precision what an average number will be up to.

—Arthur Conan Doyle, *The Sign of Four*

Sherlock Holmes spoke the words quoted here to his colleague, Dr. Watson, as the two were unraveling a mystery. The detective was implying that if a single member is drawn at random from a population, we cannot predict *exactly* what that member will look like. However, there are some "average" features of the entire population that an individual is likely to possess. The degree of certainty with which we would expect to observe such average features in any individual depends on our knowledge of the variations among individuals in the population. Sherlock Holmes has led us to two of the most important statistical concepts: average and variation.

AVERAGES AND VARIATION

PREVIEW QUESTIONS

What are commonly used measures of central tendency? What do they tell you? (SECTION 3.1)

How do variance and standard deviation measure data spread? Why is this important? (SECTION 3.2)

How do you make a box-and-whisker plot, and what does it tell about the spread of the data? (SECTION 3.3)

Nancy Sheehan/PhotoEdit

FOCUS PROBLEM

The Educational Advantage

Is it really worth all the effort to get a college degree? From a philosophical point of view, the love of learning is sufficient reason to get a college degree. However, the U.S. Census Bureau also makes another relevant point. Annually, college graduates (bachelor's degree) earn on average \$23,788 more than high school graduates. This means college graduates earn about 78.5% more than high school graduates, and according to "Education Pays" on the next page, the gap in earnings is increasing. Furthermore, as the College Board indicates, for most Americans college remains relatively affordable.

After completing this chapter, you will be able to answer the following questions.

(a) Does a college degree *guarantee* someone an 78.5% increase in earnings over a high school degree? Remember, we are using only *averages* from census data.

(b) Using census data (not shown in "Education Pays"), it is estimated that the standard deviation of college-graduate earnings is about \$8500. Compute a 75% Chebyshev confidence interval centered on the mean (\$54,091) for bachelor's degree earnings.

(c) How much does college tuition cost? That depends, of course, on where you go to college. The 10th percentile, 25th percentile, median (50th percentile), 75th percentile, and 90th percentile for annual tuition and required fees at public 4-year (in-state) and private 4-year not-for-profit institutions are shown next to "Education pays." This information is from the National Center for Education Statistics. Using these data, what is the range of annual tuition and fees for the group between the 25th percentile and the 75th

Myrleen Ferguson Cate/PhotoEdit

For online student resources, visit the Brase/Brase, *Understanding Basic Statistics*, 7th edition web site at http://www.cengagebrain.com

percentile for each type of institution? About what percent of private not-for-profit 4-year colleges have annual tuition and fees between \$34,417 and \$40,082? What percent of public 4-year colleges have annual tuition and fees at or below \$6,780?

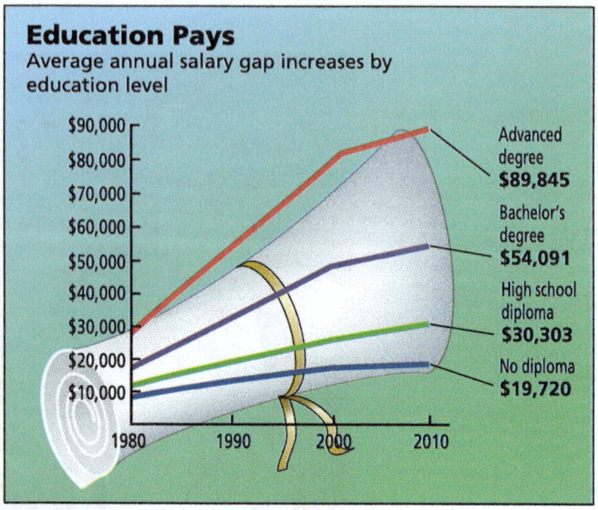

Source: Census Bureau

Annual Undergraduate Tuition and Fees (2010–2011)
Percentiles for Public 4-Year Colleges (for In-State Students)

10th	25th	50th	75th	90th
\$4,336	\$5,105	\$6,780	\$8,689	\$11,029

Percentiles for Private Not-for-Profit 4-Year Colleges

10th	25th	50th	75th	90th
\$12,220	\$19,854	\$27,100	\$34,417	\$40,082

SECTION **3.1**

Measures of Central Tendency: Mode, Median, and Mean

FOCUS POINTS

- Compute mean, median, and mode from raw data.
- Interpret what mean, median, and mode tell you.
- Explain how mean, median, and mode can be affected by extreme data values.
- What is a trimmed mean? How do you compute it?
- Compute a weighted average.

The average price of an ounce of gold is \$1350. The Zippy car averages 39 miles per gallon on the highway. A survey showed the average shoe size for women is size 9.

In each of the preceding statements, *one* number is used to describe the entire sample or population. Such a number is called an *average.* There are many ways to compute averages, but we will study only three of the major ones.

Average

The easiest average to compute is the *mode.*

Mode

The **mode** of a data set is the value that occurs most frequently. *Note:* If a data set has no single value that occurs more frequently than any other, then that data set has no mode.

EXAMPLE 1

Mode

Count the letters in each word of this sentence and give the mode. The numbers of letters in the words of the sentence are

5 3 7 2 4 4 2 4 8 3 4 3 4

Scanning the data, we see that 4 is the mode because more words have 4 letters than any other number. For larger data sets, it is useful to order—or sort—the data before scanning them for the mode.

Not every data set has a mode. For example, if Professor Fair gives equal numbers of A's, B's, C's, D's, and F's, then there is no modal grade. In addition, the mode is not very stable. Changing just one number in a data set can change the mode dramatically. However, the mode is a useful average when we want to know the most frequently occurring data value, such as the most frequently requested shoe size.

Median

Another average that is useful is the *median,* or central value, of an ordered distribution. When you are given the median, you know there are an equal number of data values in the ordered distribution that are above it and below it.

PROCEDURE

HOW TO FIND THE MEDIAN

The **median** is the central value of an ordered distribution. To find it,

1. Order the data from smallest to largest.
2. For an *odd* number of data values in the distribution,

$$\text{Median} = \text{Middle data value}$$

3. For an *even* number of data values in the distribution,

$$\text{Median} = \frac{\text{Sum of middle two values}}{2}$$

EXAMPLE 2

MEDIAN

What do barbecue-flavored potato chips cost? According to *Consumer Reports,* Vol. 66, No. 5, the prices per ounce in cents of the rated chips are

19 19 27 28 18 35

(a) To find the median, we first order the data, and then note that there are an even number of entries. So the median is constructed using the two middle values.

18 19 19 27 28 35

(19 and 27: middle values)

$$\text{Median} = \frac{19 + 27}{2} = 23 \text{ cents}$$

(b) According to *Consumer Reports,* the brand with the lowest overall taste rating costs 35 cents per ounce. Eliminate that brand, and find the median price per ounce for the remaining barbecue-flavored chips. Again order the data. Note that there are an odd number of entries, so the median is simply the middle value.

18 19 19 27 28

(19: middle values)

$$\text{Median} = \text{middle value} = 19 \text{ cents}$$

(c) One ounce of potato chips is considered a small serving. Is it reasonable to budget about \$10.45 to serve the barbecue-flavored chips to 55 people?

Yes, since the median price of the chips is 19 cents per small serving. This budget for chips assumes that there is plenty of other food!

The median uses the *position* rather than the specific value of each data entry. If the extreme values of a data set change, the median usually does not change. This is why the median is often used as the average for house prices. If one mansion costing several million dollars sells in a community of much-lower-priced homes, the median selling price for houses in the community would be affected very little, if at all.

GUIDED EXERCISE 1 MEDIAN AND MODE

Belleview College must make a report to the budget committee about the average credit hour load a full-time student carries. (A 12-credit-hour load is the minimum requirement for full-time status. For the same tuition, students may take up to 20 credit hours.) A random sample of 40 students yielded the following information (in credit hours):

17	12	14	17	13	16	18	20	13	12
12	17	16	15	14	12	12	13	17	14
15	12	15	16	12	18	20	19	12	15
18	14	16	17	15	19	12	13	12	15

(a) Organize the data from smallest to largest number of credit hours.

→ 12 12 12 12 12 12 12 12 12 12
13 13 13 13 14 14 14 14 15 (15)
(15) 15 15 15 16 16 16 16 17 17
17 17 17 18 18 18 19 19 20 20

(b) Since there are an ______ (odd, even) number of values, we add the two middle values and divide by 2 to get the median. What is the median credit hour load?

→ There are an even number of entries. The two middle values are circled in part (a).

$$\text{Median} = \frac{15 + 15}{2} = 15$$

(c) What is the mode of this distribution? Is it different from the median? *Interpretation* If the budget committee is going to fund the college according to the average student credit hour load (more money for higher loads), which of these two averages do you think the college will report?

→ The mode is 12. It is different from the median. Since the median is higher, the college will probably report it and indicate that the average being used is the median.

Note: For small ordered data sets, we can easily scan the set to find the *location* of the median. However, for large ordered data sets of size *n*, it is convenient to have a formula to find the middle of the data set.

> For an ordered data set of size n,
>
> $$\text{Position of the middle value} = \frac{n + 1}{2}$$

For instance, if $n = 99$ then the middle value is the $(99 + 1)/2$ or 50th data value in the ordered data. If $n = 100$, then $(100 + 1)/2 = 50.5$ tells us that the two middle values are in the 50th and 51st positions.

Mean

An average that uses the exact value of each entry is the *mean* (sometimes called the *arithmetic mean*). To compute the mean, we add the values of all the entries and then divide by the number of entries.

$$\text{Mean} = \frac{\text{Sum of all entries}}{\text{Number of entries}}$$

The mean is the average usually used to compute a test average.

EXAMPLE 3

MEAN

To graduate, Linda needs at least a B in biology. She did not do very well on her first three tests; however, she did well on the last four. Here are her scores:

58 67 60 84 93 98 100

Jose Carillo/PhotoEdit

Compute the mean and determine if Linda's grade will be a B (80 to 89 average) or a C (70 to 79 average).

SOLUTION:

$$\text{Mean} = \frac{\text{Sum of scores}}{\text{Number of scores}} = \frac{58 + 67 + 60 + 84 + 93 + 98 + 100}{7}$$

$$= \frac{560}{7} = 80$$

Since the average is 80, Linda will get the needed B.

COMMENT When we compute the mean, we sum the given data. There is a convenient notation to indicate the sum. Let x represent any value in the data set. Then the notation

Σx (read "the sum of all given x values")

Summation symbol, Σ

means that we are to sum all the data values. In other words, we are to sum all the entries in the distribution. The *summation symbol* Σ means *sum the following* and is capital sigma, the *S* of the Greek alphabet.

Sample mean, $\bar{x}$
Population mean, μ

The symbol for the mean of a *sample* distribution of x values is denoted by $\bar{x}$ (read "x bar"). If your data comprise the entire *population,* we use the symbol μ (lowercase Greek letter mu, pronounced "mew") to represent the mean.

PROCEDURE

HOW TO FIND THE MEAN

1. Compute Σx; that is, find the sum of all the data values.
2. Divide the sum total by the number of data values.

Sample statistic $\bar{x}$	Population parameter μ
$\bar{x} = \frac{\Sigma x}{n}$	$\mu = \frac{\Sigma x}{N}$

where n = number of data values in the sample
N = number of data values in the population

When data consist of rates of change, the *harmonic mean* is the appropriate average to use. Problem 29 discusses the harmonic mean.

Problem 30 discusses the *geometric mean*, which is a useful average for percentages, ratios, or growth rates.

LOOKING FORWARD

In our future work with inferential statistics, we will use the mean $\bar{x}$ from a random sample to estimate the population parameter μ (Chapter 8) or to make decisions regarding the value of μ (Chapter 9).

CALCULATOR NOTE It is very easy to compute the mean on *any* calculator: Simply add the data values and divide the total by the number of data. However, on calculators with a statistics mode, you place the calculator in that mode, *enter* the data, and then press the key for the mean. The key is usually designated $\bar{x}$. Because the formula for the population mean is the same as that for the sample mean, the same key gives the value for μ.

WHAT DO AVERAGES TELL US?

An average provides a one-number summary of a data set.

- The **mode** tells us the single data value that occurs most frequently in the data set. The value of the mode is completely determined by the data value that occurs most frequently. If no data value occurs more frequently than all the other data values, there is no mode. The specific values of the less frequently occurring data do not change the mode.
- The **median** tells us the middle value of a data set that has been arranged in order from smallest to largest. The median is affected by only the relative position of the data values. For instance, if a data value above the median (or above the middle two values of a data set with an even number of data) is changed to another value above the median, the median itself does not change.
- The **mean** tells us the value obtained by adding up *all* the data and dividing by the number of data. As such, the mean can change if just one data value changes. On the other hand, if data values change, but the sum of the data remains the same, the mean will not change.

We have seen three averages: the mode, the median, and the mean. For later work, the mean is the most important. A disadvantage of the mean, however, is that it can be affected by exceptional values.

Resistant measure

A *resistant measure* is one that is not influenced by extremely high or low data values. The mean is not a resistant measure of center because we can make the mean as large as we want by changing the size of only one data value. The median, on the other hand, is more resistant. However, a disadvantage of the median is that it is not sensitive to the specific size of a data value.

Trimmed mean

A measure of center that is more resistant than the mean but still sensitive to specific data values is the *trimmed mean*. A trimmed mean is the mean of the data values left after "trimming" a specified percentage of the smallest and largest data values from the data set. Usually a 5% trimmed mean is used. This implies that we trim the lowest 5% of the data as well as the highest 5% of the data. A similar procedure is used for a 10% trimmed mean.

PROCEDURE

HOW TO COMPUTE A 5% TRIMMED MEAN

1. Order the data from smallest to largest.
2. Delete the bottom 5% of the data and the top 5% of the data. *Note:* If the calculation of 5% of the number of data values does not produce a whole number, *round* to the nearest integer.
3. Compute the mean of the remaining 90% of the data.

GUIDED EXERCISE 2 | MEAN AND TRIMMED MEAN

Barron's Profiles of American Colleges, 19th Edition, lists average class size for introductory lecture courses at each of the profiled institutions. A sample of 20 colleges and universities in California showed class sizes for introductory lecture courses to be

(14) 20 20 20 20 23 25 30 30 30
35 35 35 40 40 42 50 50 80 (80)

(a) Compute the mean for the entire sample. → Add all the values and divide by 20:

$$\bar{x} = \frac{\Sigma x}{n} = \frac{719}{20} \approx 36.0$$

(b) Compute a 5% trimmed mean for the sample. → The data are already ordered. Since 5% of 20 is 1, we eliminate one data value from the bottom of the list and one from the top. These values are circled in the data set. Then we take the mean of the remaining 18 entries.

$$5\% \text{ trimmed mean} = \frac{\Sigma x}{n} = \frac{625}{18} \approx 34.7$$

(c) Find the median of the original data set. → Note that the data are already ordered.

$$\text{Median} = \frac{30 + 35}{2} = 32.5$$

(d) Find the median of the 5% trimmed data set. Does the median change when you trim the data? → The median is still 32.5. Notice that trimming the same number of entries from both ends leaves the middle position of the data set unchanged.

(e) Is the trimmed mean or the original mean closer to the median? → The trimmed mean is closer to the median.

TECH NOTES

Minitab, Excel 2010, and TI-84Plus/TI-83Plus/TI-*n*spire calculators all provide the mean and median of a data set. Minitab and Excel 2010 also provide the mode. The TI-84Plus/TI-83Plus/TI-*n*spire calculators sort data, so you can easily scan the sorted data for the mode. Minitab provides the 5% trimmed mean, as does Excel 2010.

All this technology is a wonderful aid for analyzing data. However, *a measurement is worthless if you do not know what it represents or how a change in data values might affect the measurement.* The defining formulas and procedures for computing the measures tell you a great deal about the measures. Even if you use a calculator to evaluate all the statistical measures, pay attention to the information the formulas and procedures give you about the components or features of the measurement.

CRITICAL THINKING

Data types and averages

In Chapter 1, we examined four levels of data: nominal, ordinal, interval, and ratio. The mode (if it exists) can be used with all four levels, including nominal. For instance, the modal color of all passenger cars sold last year might be blue. The median may be used with data at the ordinal level or

Continued

CRITICAL THINKING *continued*

above. If we ranked the passenger cars in order of customer satisfaction level, we could identify the median satisfaction level. For the mean, our data need to be at the interval or ratio level (although there are exceptions in which the mean of ordinal-level data is computed). We can certainly find the mean model year of used passenger cars sold or the mean price of new passenger cars.

Another issue of concern is that of taking the average of averages. For instance, if the values \$520, \$640, \$730, \$890, and \$920 represent the mean monthly rents for five different apartment complexes, we can't say that \$740 (the mean of the five numbers) is the mean monthly rent of all the apartments. We need to know the number of apartments in each complex before we can determine an average based on the number of apartments renting at each designated amount.

Distribution shapes and averages

In general, when a data distribution is mound-shaped symmetrical, the values for the mean, median, and mode are the same or almost the same. For skewed-left distributions, the mean is less than the median and the median is less than the mode. For skewed-right distributions, the mode is the smallest value, the median is the next largest, and the mean is the largest. Figure 3-1 shows the general relationships among the mean, median, and mode for different types of distributions.

WEIGHTED AVERAGE

Weighted average

Sometimes we wish to average numbers, but we want to assign more importance, or weight, to some of the numbers. For instance, suppose your professor tells you that your grade will be based on a midterm and a final exam, each of which is based on 100 possible points. However, the final exam will be worth 60% of the grade and the midterm only 40%. How could you determine an average score that would reflect these different weights? The average you need is the *weighted average*.

$$\text{Weighted average} = \frac{\Sigma xw}{\Sigma w}$$

where x is a data value and w is the weight assigned to that data value. The sum is taken over all data values.

FIGURE 3-1

Distribution Types and Averages

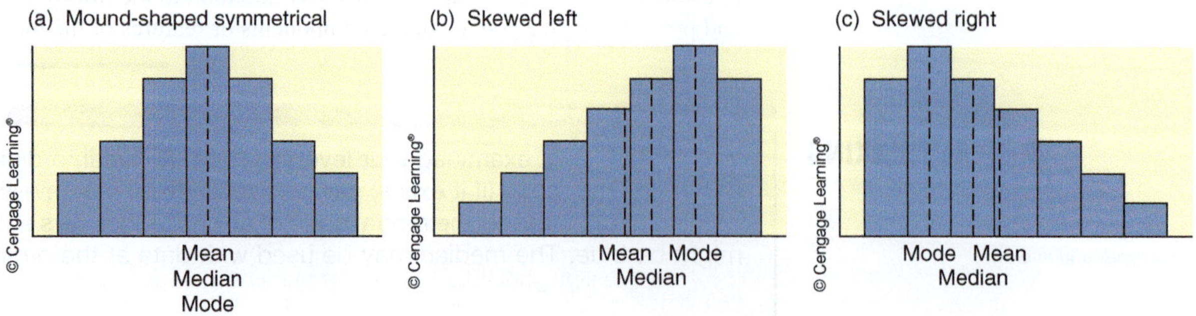

EXAMPLE 4

WEIGHTED AVERAGE

Suppose your midterm test score is 83 and your final exam score is 95. Using weights of 40% for the midterm and 60% for the final exam, compute the weighted average of your scores. If the minimum average for an A is 90, will you earn an A?

SOLUTION: By the formula, we multiply each score by its weight and add the results together. Then we divide by the sum of all the weights. Converting the percentages to decimal notation, we get

$$\text{Weighted average} = \frac{83(0.40) + 95(0.60)}{0.40 + 0.60}$$

$$= \frac{33.2 + 57}{1} = 90.2$$

Your average is high enough to earn an A.

TECH NOTES

The TI-84Plus/TI-83Plus/TI-*n*spire calculators directly support weighted averages. Both Excel 2010 and Minitab can be programmed to provide the averages.

TI-84Plus/TI-83Plus/TI-*n*spire (with TI-84 Plus keypad) Enter the data into one list, such as L1, and the corresponding weights into another list, such as L2. Then press **Stat ➤ Calc ➤ 1: 1-Var Stats**. Enter the list containing the data, followed by a comma and the list containing the weights.

VIEWPOINT What's Wrong with Pitching Today?

One way to answer this question is to look at averages. Batting averages and average hits per game are shown for selected years from 1901 to 2000 (Source: The Wall Street Journal*).*

Year	1901	1920	1930	1941	1951	1961	1968	1976	1986	2000
B.A.	0.277	0.284	0.288	0.267	0.263	0.256	0.231	0.256	0.262	0.276
Hits	19.2	19.2	20.0	18.4	17.9	17.3	15.2	17.3	17.8	19.1

A quick scan of the averages shows that batting averages and average hits per game are virtually the same as they were almost 100 years ago. It seems there is nothing wrong *with today's pitching! So what's changed? For one thing, the rules have changed! The strike zone is considerably smaller than it once was, and the pitching mound is lower. Both give the hitter an advantage over the pitcher. Even so, pitchers don't give up hits with any greater frequency than they did a century ago (look at the averages). However, modern hits go much farther, which is something a pitcher can't control.*

SECTION 3.1 PROBLEMS

1. *Statistical Literacy* Consider the mode, median, and mean. Which average represents the middle value of a data distribution? Which average represents the most frequent value of a distribution? Which average takes all the specific values into account?

2. *Statistical Literacy* What symbol is used for the arithmetic mean when it is a sample statistic? What symbol is used when the arithmetic mean is a population parameter?

3. *Statistical Literacy* Look at the formula for the mean. List the two arithmetic procedures that are used to compute the mean.

4. *Statistical Literacy* In order to find the median of a data set, what do we do first with the data?

5. *Basic Computation: Mean, Median, Mode* Find the mean, median, and mode of the data set

   8 2 7 2 6

6. *Basic Computation: Mean, Median, Mode* Find the mean, median, and mode of the data set

   10 12 20 15 20

7. *Basic Computations Mean, Median, Mode* Find the mean, median, and mode of the data set

   8 2 7 2 6 5

8. *Critical Thinking* Consider a data set with at least three data values. Suppose the highest value is increased by 10 and the lowest is decreased by 5.
   (a) Does the mean change? Explain
   (b) Does the median change? Explain
   (c) Is it possible for the mode to change? Explain.

9. *Critical Thinking* Consider a data set with at least three data values. Suppose the highest value is increased by 10 and the lowest is decreased by 10.
   (a) Does the mean change? Explain
   (b) Does the median change? Explain
   (c) Is it possible for the mode to change? Explain.

10. *Critical Thinking* If a data set has an even number of data, is it true or false that the median is never equal to a value in the data set? Explain.

11. *Critical Thinking* When a distribution is mound-shaped symmetrical, what is the general relationship among the values of the mean, median, and mode?

12. *Critical Thinking* Consider the following types of data that were obtained from a random sample of 49 credit card accounts. Identify all the averages (mean, median, or mode) that can be used to summarize the data.
   (a) Outstanding balance on each account
   (b) Name of credit card (e.g., MasterCard, Visa, American Express, etc.)
   (c) Dollar amount due on next payment

13. *Critical Thinking* Consider the numbers

   2 3 4 5 5

   (a) Compute the mode, median, and mean.
   (b) If the numbers represent codes for the colors of T-shirts ordered from a catalog, which average(s) would make sense?
   (c) If the numbers represent one-way mileages for trails to different lakes, which average(s) would make sense?
   (d) Suppose the numbers represent survey responses from 1 to 5, with 1 = disagree strongly, 2 = disagree, 3 = agree, 4 = agree strongly, and 5 = agree very strongly. Which averages make sense?

14. *Critical Thinking* Consider two data sets.

   Set A: $n = 5; \bar{x} = 10$ Set B: $n = 50; \bar{x} = 10$

   (a) Suppose the number 20 is included as an additional data value in Set A. Compute $\bar{x}$ for the new data set. *Hint:* $\Sigma x = n\bar{x}$. To compute $\bar{x}$ for the new data set, add 20 to Σx of the original data set and divide by 6.

(b) Suppose the number 20 is included as an additional data value in Set B. Compute $\bar{x}$ for the new data set.
(c) Why does the addition of the number 20 to each data set change the mean for Set A more than it does for Set B?

15. *Interpretation* A job-performance evaluation form has these categories:

1 = excellent; 2 = good; 3 = satisfactory; 4 = poor; 5 = unacceptable

Based on 15 client reviews, one employee had

median rating of 4; mode rating of 1

The employee was pleased that most clients had rated her as excellent. The supervisor said improvement was needed because at least half the clients had rated the employee at the poor or unacceptable level. Comment on the different perspectives.

16. *Critical Thinking: Data Transformation* In this problem, we explore the effect on the mean, median, and mode of adding the same number to each data value. Consider the data set 2, 2, 3, 6, 10.
(a) Compute the mode, median, and mean.
(b) Add 5 to each of the data values. Compute the mode, median, and mean.
(c) Compare the results of parts (a) and (b). In general, how do you think the mode, median, and mean are affected when the same constant is added to each data value in a set?

17. *Critical Thinking: Data Transformation* In this problem, we explore the effect on the mean, median, and mode of multiplying each data value by the same number. Consider the data set 2, 2, 3, 6, 10.
(a) Compute the mode, median, and mean.
(b) Multiply each data value by 5. Compute the mode, median, and mean.
(c) Compare the results of parts (a) and (b). In general, how do you think the mode, median, and mean are affected when each data value in a set is multiplied by the same constant?
(d) Suppose you have information about average heights of a random sample of airplane passengers. The mode is 70 inches, the median is 68 inches, and the mean is 71 inches. To convert the data into centimeters, multiply each data value by 2.54. What are the values of the mode, median, and mean in centimeters?

18. *Critical Thinking* Consider a data set of 15 distinct measurements with mean *A* and median *B*.
(a) If the highest number were increased, what would be the effect on the median and mean? Explain.
(b) If the highest number were decreased to a value still larger than *B*, what would be the effect on the median and mean?
(c) If the highest number were decreased to a value smaller than *B*, what would be the effect on the median and mean?

19. *Environmental Studies: Death Valley* How hot does it get in Death Valley? The following data are taken from a study conducted by the National Park Service, of which Death Valley is a unit. The ground temperatures (°F) were taken from May to November in the vicinity of Furnace Creek.

146	152	168	174	180	178	179
180	178	178	168	165	152	144

Compute the mean, median, and mode for these ground temperatures.

20. *Ecology: Wolf Packs* How large is a wolf pack? The following information is from a random sample of winter wolf packs in regions of Alaska, Minnesota, Michigan, Wisconsin, Canada, and Finland (Source: *The Wolf*, by L. D. Mech, University of Minnesota Press). Winter pack size:

13	10	7	5	7	7	2	4	3
2	3	15	4	4	2	8	7	8

Compute the mean, median, and mode for the size of winter wolf packs.

21. *Medical: Injuries* The Grand Canyon and the Colorado River are beautiful, rugged, and sometimes dangerous. Thomas Myers is a physician at the park clinic in Grand Canyon Village. Dr. Myers has recorded (for a 5-year period) the number of visitor injuries at different landing points for commercial boat trips down the Colorado River in both the Upper and Lower Grand Canyon (Source: *Fateful Journey* by Myers, Becker, and Stevens).

Upper Canyon: Number of Injuries per Landing Point Between North Canyon and Phantom Ranch

2 3 1 1 3 4 6 9 3 1 3

Lower Canyon: Number of Injuries per Landing Point Between Bright Angel and Lava Falls

8 1 1 0 6 7 2 14 3 0 1 13 2 1

Fuse/Getty Images

(a) Compute the mean, median, and mode for injuries per landing point in the Upper Canyon.
(b) Compute the mean, median, and mode for injuries per landing point in the Lower Canyon.
(c) Compare the results of parts (a) and (b).
(d) The Lower Canyon stretch had some extreme data values. Compute a 5% trimmed mean for this region, and compare this result to the mean for the Upper Canyon computed in part (a).

22. *Football: Age of Professional Players* How old are professional football players? The 11th Edition of *The Pro Football Encyclopedia* gave the following information. Random sample of pro football player ages in years:

24	23	25	23	30	29	28	26	33	29
24	37	25	23	22	27	28	25	31	29
25	22	31	29	22	28	27	26	23	21
25	21	25	24	22	26	25	32	26	29

(a) Compute the mean, median, and mode of the ages.
(b) *Interpretation* Compare the averages. Does one seem to represent the age of the pro football players most accurately? Explain.

© iStockphoto.com/adwalsh

23. *Leisure: Maui Vacation* How expensive is Maui? If you want a vacation rental condominium (up to four people), visit the Brase/Brase statistics site at **http://www.cengagebrain.com,** find the link to Maui, and then search for accommodations. The *Maui News* gave the following costs in dollars per day for a random sample of condominiums located throughout the island of Maui.

89	50	68	60	375	55	500	71	40	350
60	50	250	45	45	125	235	65	60	130

(a) Compute the mean, median, and mode for the data.
(b) Compute a 5% trimmed mean for the data, and compare it with the mean computed in part (a). Does the trimmed mean more accurately reflect the general level of the daily rental costs?
(c) *Interpretation* If you were a travel agent and a client asked about the daily cost of renting a condominium on Maui, what average would you use? Explain. Is there any other information about the costs that you think might be useful, such as the spread of the costs?

24. *Basic Computation: Weighted Average* Find the weighted average of a data set where

    10 has a weight of 5; 20 has a weight of 3; 30 has a weight of 2

25. *Basic Computation: Weighted Average* Find the weighted average of a data set where

    10 has a weight of 2; 20 has a weight of 3; 30 has a weight of 5

26. *Grades: Weighted Average* In your biology class, your final grade is based on several things: a lab score, scores on two major tests, and your score on the final exam. There are 100 points available for each score. However, the lab score is worth 25% of your total grade, each major test is worth 22.5%, and the final exam is worth 30%. Compute the weighted average for the following scores: 92 on the lab, 81 on the first major test, 93 on the second major test, and 85 on the final exam.

27. *Merit Pay Scale: Weighted Average* At General Hospital, nurses are given performance evaluations to determine eligibility for merit pay raises. The supervisor rates the nurses on a scale of 1 to 10 (10 being the highest rating) for several activities: promptness, record keeping, appearance, and bedside manner with patients. Then an average is determined by giving a weight of 2 for promptness, 3 for record keeping, 1 for appearance, and 4 for bedside manner with patients. What is the average rating for a nurse with ratings of 9 for promptness, 7 for record keeping, 6 for appearance, and 10 for bedside manner?

28. *EPA: Wetlands* Where does all the water go? According to the Environmental Protection Agency (EPA), in a typical wetland environment, 38% of the water is outflow, 47% is seepage, 7% evaporates, and 8% remains as water volume in the ecosystem (Reference: U.S. Environmental Protection Agency Case Studies Report 832-R-93-005). Chloride compounds as residuals from residential areas are a problem for wetlands. Suppose that in a particular wetland environment the following concentrations (mg/L) of chloride compounds were found: outflow, 64.1; seepage, 75.8; remaining due to evaporation, 23.9; in the water volume, 68.2.
    (a) Compute the weighted average of chlorine compound concentration (mg/L) for this ecological system.
    (b) Suppose the EPA has established an average chlorine compound concentration target of no more than 58 mg/L. Comment on whether this wetlands system meets the target standard for chlorine compound concentration.

29. *Expand Your Knowledge: Harmonic Mean* When data consist of rates of change, such as speeds, the *harmonic mean* is an appropriate measure of central tendency. For n data values,

Harmonic mean

$$\text{Harmonic mean} = \frac{n}{\Sigma \frac{1}{x}}, \quad \text{assuming no data value is 0}$$

Suppose you drive 60 miles per hour for 100 miles, then 75 miles per hour for 100 miles. Use the harmonic mean to find your average speed.

30. *Expand Your Knowledge: Geometric Mean* When data consist of percentages, ratios, growth rates, or other rates of change, the *geometric mean* is a useful measure of central tendency. For n data values,

Geometric mean

$$\text{Geometric mean} = \sqrt[n]{\text{product of the } n \text{ data values}}, \quad \text{assuming all data values are positive}$$

To find the *average growth factor* over 5 years of an investment in a mutual fund with growth rates of 10% the first year, 12% the second year, 14.8% the third year, 3.8% the fourth year, and 6% the fifth year, take the geometric mean of 1.10, 1.12, 1.148, 1.038, and 1.16. Find the average growth factor of this investment.

Note that for the same data, the relationships among the harmonic, geometric, and arithmetic means are harmonic mean $\leq$ geometric mean $\leq$ arithmetic mean (Source: *Oxford Dictionary of Statistics*).

SECTION 3.2 Measures of Variation

FOCUS POINTS

- Find the range, variance, and standard deviation.
- Compute the coefficient of variation from raw data. Why is the coefficient of variation important?
- Apply Chebyshev's theorem to raw data. What does a Chebyshev interval tell us?

An average is an attempt to summarize a set of data using just one number. As some of our examples have shown, an average taken by itself may not always be very meaningful. We need a statistical cross-reference that measures the spread of the data.

The *range* is one such measure of variation.

Range

The **range** is the difference between the largest and smallest values of a data distribution.

EXAMPLE 5

RANGE

A large bakery regularly orders cartons of Maine blueberries. The average weight of the cartons is supposed to be 22 ounces. Random samples of cartons from two suppliers were weighed. The weights in ounces of the cartons were

Supplier I:	17	22	22	22	27
Supplier II:	17	19	20	27	27

(a) Compute the range of carton weights from each supplier.

$$\text{Range} = \text{Largest value} - \text{Smallest value}$$
$$\text{Supplier I range} = 27 - 17 = 10 \text{ ounces}$$
$$\text{Supplier II range} = 27 - 17 = 10 \text{ ounces}$$

(b) Compute the mean weight of cartons from each supplier. In both cases the mean is 22 ounces.

Daniel Dempster Photography/Alamy

Blueberry patch

(c) Look at the two samples again. The samples have the same range and mean. How do they differ? The bakery uses one carton of blueberries in each blueberry muffin recipe. It is important that the cartons be of consistent weight so that the muffins turn out right.

Supplier I provides more cartons that have weights closer to the mean. Or, put another way, the weights of cartons from Supplier I are more clustered around the mean. The bakery might find Supplier I more satisfactory.

As we see in Example 5, although the range tells the difference between the largest and smallest values in a distribution, it does not tell us how much other values vary from one another or from the mean.

VARIANCE AND STANDARD DEVIATION

Variance
Standard deviation

We need a measure of the distribution or spread of data around an expected value (either $\bar{x}$ or μ). *Variance* and *standard deviation* provide such measures. Formulas and rationale for these measures are described in the next Procedure display. Then, examples and guided exercises show how to compute and interpret these measures.

As we will see later, the formulas for variance and standard deviation differ slightly, depending on whether we are using a sample or the entire population.

PROCEDURE

HOW TO COMPUTE THE SAMPLE VARIANCE AND SAMPLE STANDARD DEVIATION

Sum of squares

Sample variance, s^2

Quantity	Description
x	The variable x represents a **data value** or outcome.
Mean $\bar{x} = \frac{\Sigma x}{n}$	This is the **average of the data values,** or what you "expect" to happen the next time you conduct the statistical experiment. Note that n is the sample size.
$x - \bar{x}$	This is the **difference** between what happened and what you expected to happen. This represents a "deviation" away from what you "expect" and is a measure of risk.
$\Sigma(x - \bar{x})^2$	The expression $\Sigma(x - \bar{x})^2$ is called the **sum of squares.** The $(x - \bar{x})$ quantity is squared to make it nonnegative. The sum is over all the data. If you don't square $(x - \bar{x})$, then the sum $\Sigma(x - \bar{x})$ is equal to 0 because the negative values cancel the positive values. This occurs even if some $(x - \bar{x})$ values are large, indicating a large deviation or risk.
Sum of squares $\Sigma(x - \bar{x})^2$ or $\Sigma x^2 - \frac{(\Sigma x)^2}{n}$	This is an **algebraic simplification of the sum of squares** that is easier to compute. The **defining formula** for the sum of squares is the upper one. The **computation formula** for the sum of squares is the lower one. Both formulas give the same result.
Sample variance $s^2 = \frac{\Sigma(x - \bar{x})^2}{n - 1}$ or $s^2 = \frac{\Sigma x^2 - (\Sigma x)^2/n}{n - 1}$	The **sample variance is s^2.** The variance can be thought of as a kind of average of the $(x - \bar{x})^2$ values. However, for technical reasons, we divide the sum by the quantity $n - 1$ rather than n. This gives us the best mathematical estimate for the sample variance. The **defining formula** for the variance is the upper one. The **computation formula** for the variance is the lower one. Both formulas give the same result.

Continued

Sample standard deviation, s

Sample standard deviation	This is the **sample standard deviation, s.** Why do we take the square root? Well, if the original x units were, say, days or dollars, then the s^2 units would be days squared or dollars squared (wow, what's that?). We take the square root to return to the original units of the data measurements. The standard deviation can be thought of as a measure of variability or risk. Larger values of s imply greater variability in the data. The **defining formula** for the standard deviation is the upper one. The **computation formula** for the standard deviation is the lower one. Both formulas give the same result.
$s = \sqrt{\frac{\Sigma(x - \bar{x})^2}{n - 1}}$ or $s = \sqrt{\frac{\Sigma x^2 - (\Sigma x)^2/n}{n - 1}}$	

COMMENT Why is s called a *sample standard* deviation? First, it is computed from sample data. Then why do we use the word *standard* in the name? We know s is a measure of deviation or risk. You should be aware that there are other statistical measures of risk that we have not yet mentioned. However, s is the one that everyone uses, so it is called the "standard" (like standard time).

In statistics, the sample standard deviation and sample variance are used to describe the spread of data about the mean $\bar{x}$. The next example shows how to find these quantities by using the defining formulas. Guided Exercise 3 shows how to use the computation formulas.

As you will discover, for "hand" calculations, the computation formulas for s^2 and s are much easier to use. However, the defining formulas for s^2 and s emphasize the fact that the variance and standard deviation are based on the differences between each data value and the mean.

DEFINING FORMULAS (SAMPLE STATISTIC)

$$\text{Sample variance} = s^2 = \frac{\Sigma(x - \bar{x})^2}{n - 1} \tag{1}$$

$$\text{Sample standard deviation} = s = \sqrt{\frac{\Sigma(x - \bar{x})^2}{n - 1}} \tag{2}$$

where x is a member of the data set, $\bar{x}$ is the mean, and n is the number of data values. The sum is taken over all data values.

COMPUTATION FORMULAS (SAMPLE STATISTIC)

$$\text{Sample variance} = s^2 = \frac{\Sigma x^2 - (\Sigma x)^2/n}{n - 1} \tag{3}$$

$$\text{Sample standard deviation} = s = \sqrt{\frac{\Sigma x^2 - (\Sigma x)^2/n}{n - 1}} \tag{4}$$

where x is a member of the data set and n is the number of data values. The sum is taken over all data values.

EXAMPLE 6

SAMPLE STANDARD DEVIATION (DEFINING FORMULA)

Big Blossom Greenhouse was commissioned to develop an extra large rose for the Rose Bowl Parade. A random sample of blossoms from Hybrid A bushes yielded the following diameters (in inches) for mature peak blooms.

2 3 3 8 10 10

Use the defining formula to find the sample variance and standard deviation.

SOLUTION: Several steps are involved in computing the variance and standard deviation. A table will be helpful (see Table 3-1). Since $n = 6$, we take the sum of the entries in the first column of Table 3-1 and divide by 6 to find the mean $\bar{x}$.

$$\bar{x} = \frac{\Sigma x}{n} = \frac{36}{6} = 6.0 \text{ inches}$$

TABLE 3-1 **Diameters of Rose Blossoms (in inches)**

Column I x	Column II $x - \bar{x}$	Column III $(x - \bar{x})^2$
2	$2 - 6 = -4$	$(-4)^2 = 16$
3	$3 - 6 = -3$	$(-3)^2 = 9$
3	$3 - 6 = -3$	$(-3)^2 = 9$
8	$8 - 6 = 2$	$(2)^2 = 4$
10	$10 - 6 = 4$	$(4)^2 = 16$
10	$10 - 6 = 4$	$(4)^2 = 16$
$\Sigma x = 36$		$\Sigma(x - \bar{x})^2 = 70$

Tony Freeman/PhotoEdit

Using this value for $\bar{x}$, we obtain Column II. Square each value in the second column to obtain Column III, and then add the values in Column III. To get the sample variance, divide the sum of Column III by $n - 1$. Since $n = 6$, $n - 1 = 5$.

$$s^2 = \frac{\Sigma(x - \bar{x})^2}{n - 1} = \frac{70}{5} = 14$$

Now obtain the sample standard deviation by taking the square root of the variance.

$$s = \sqrt{s^2} = \sqrt{14} \approx 3.74$$

(Use a calculator to compute the square root. Because of rounding, we use the approximately equal symbol, $\approx$.)

GUIDED EXERCISE 3

SAMPLE STANDARD DEVIATION (COMPUTATION FORMULA)

Big Blossom Greenhouse gathered another random sample of mature peak blooms from Hybrid B. The six blossoms had the following widths (in inches):

5 5 5 6 7 8

(a) Again, we will construct a table so that we can find the mean, variance, and standard deviation more easily. In this case, what is the value of n? Find the sum of Column I in Table 3-2 on the next page, and compute the mean.

→ $n = 6$. The sum of Column I is $\Sigma x = 36$, so the mean is

$$\bar{x} = \frac{36}{6} = 6$$

Continued

GUIDED EXERCISE 3 *continued*

TABLE 3-2 Complete Columns I and II

I x	II x^2
5	____
5	____
5	____
6	____
7	____
8	____
$\Sigma x =$ ____	$\Sigma x^2 =$ ____

TABLE 3-3 Completion of Table 3-2

I x	II x^2
5	25
5	25
5	25
6	36
7	49
8	64
$\Sigma x = 36$	$\Sigma x^2 = 224$

(b) What is the value of n? of $n - 1$? Use the computation formula to find the sample variance s^2. *Note:* Be sure to distinguish between Σx^2 and $(\Sigma x)^2$. For Σx^2, you square the x values first and then sum them. For $(\Sigma x)^2$, you sum the x values first and then square the result.

→ $n = 6; n - 1 = 5.$

$$s^2 = \frac{\Sigma x^2 - (\Sigma x)^2/n}{n - 1}$$

$$= \frac{224 - (36)^2/6}{5} = \frac{8}{5} = 1.6$$

(c) Use a calculator to find the square root of the variance. Is this the standard deviation?

→ $s = \sqrt{s^2} = \sqrt{1.6} \approx 1.26$

Yes.

Let's summarize and compare the results of Guided Exercise 3 and Example 6. The greenhouse found the following blossom diameters for Hybrid A and Hybrid B:

Hybrid A: Mean, 6.0 inches; standard deviation, 3.74 inches

Hybrid B: Mean, 6.0 inches; standard deviation, 1.26 inches

Interpretation In both cases, the means are the same: 6 inches. But the first hybrid has a larger standard deviation. This means that the blossoms of Hybrid A are less consistent than those of Hybrid B. If you want a rosebush that occasionally has 10-inch blooms and occasionally has 2-inch blooms, use the first hybrid. But if you want a bush that consistently produces roses close to 6 inches across, use Hybrid B.

ROUNDING NOTE Rounding errors cannot be completely eliminated, even if a computer or calculator does all the computations. However, software and calculator routines are designed to minimize the errors. If the mean is rounded, the value of the standard deviation will change slightly, depending on how much the mean is rounded. If you do your calculations "by hand" or reenter intermediate values into a calculator, try to carry one or two more digits than occur in the original data. If your resulting answers vary slightly from those in this text, do not be overly concerned. The text answers are computer- or calculator-generated.

WHAT DO MEASURES OF VARIATION TELL US?

Measures of variation give information about the spread of the data.

- The **range** tells us the difference between the highest data value and the lowest. It tells us about the spread of data but does not tell us if most of the data is or is not closer to the mean.

- The **sample standard deviation** is based on the difference between *each* data value and the mean of the data set. The magnitude of each data value enters into the calculation. The formula tells us to compute the difference between each data value and the mean, square each difference, add up all the squares, divide by $n - 1$, and then take the square root of the result. The standard deviation gives an average of data spread around the mean. The larger the standard deviation, the more spread out the data are around the mean. A smaller standard deviation indicates that the data tend to be closer to the mean.
- The **variance** tells us the square of standard deviation. As such, it is also a measure of data spread around the mean.

Population mean, μ
Population variance, σ^2
Population standard deviation, σ

In most applications of statistics, we work with a random sample of data rather than the entire population of *all* possible data values. However, if we have data for the entire population, we can compute the *population mean* μ, *population variance* σ^2, and *population standard deviation* σ (lowercase Greek letter sigma) using the following formulas:

POPULATION PARAMETERS

$$\text{Population mean} = \mu = \frac{\Sigma x}{N}$$

$$\text{Population variance} = \sigma^2 = \frac{\Sigma(x - \mu)^2}{N}$$

$$\text{Population standard deviation} = \sigma = \sqrt{\frac{\Sigma(x - \mu)^2}{N}}$$

where N is the number of data values in the population and x represents the individual data values of the population.

What if you do not have access to raw data, but you do have a histogram or frequency table of the data and you want to estimate the mean and standard deviation? Problems 22–25 address grouped data, with the formulas shown before Problem 22.

Problem 26 discusses moving averages, and explores the difference in the standard deviations of raw data compared to the standard deviation of the moving average of the raw data.

We note that the formula for μ is the same as the formula for $\bar{x}$ (the sample mean) and that the formulas for σ^2 and σ are the same as those for s^2 and s (sample variance and sample standard deviation), except that the population size N is used instead of $n - 1$. Also, μ is used instead of $\bar{x}$ in the formulas for σ^2 and σ.

Population size, N

In the formulas for s and σ, we use $n - 1$ to compute s and N to compute σ. Why? The reason is that N (capital letter) represents the *population size,* whereas n (lowercase letter) represents the sample size. Since a random sample usually will not contain extreme data values (large or small), we divide by $n - 1$ in the formula for s to make s a little larger than it would have been had we divided by n. Courses in advanced theoretical statistics show that this procedure will give us the best possible estimate for the standard deviation σ. In fact, s is called the *unbiased estimate* for σ. If we have the population of all data values, then extreme data values are, of course, present, so we divide by N instead of $N - 1$.

When we use stratified sampling, how do we determine the sample size from each stratum so that we get a more accurate estimate for the total population mean μ of a specified measurement? Problems 28, 29, and 30 show how the sample standard deviation of the measurement from each stratum help determine optimal samples sizes for the estimate of the total population mean μ.

COMMENT The computation formula for the population standard deviation is

$$\sigma = \sqrt{\frac{\Sigma x^2 - (\Sigma x)^2/N}{N}}$$

We've seen that the standard deviation (sample or population) is a measure of data spread. We will use the standard deviation extensively in later chapters.

TECH NOTES

Most scientific or business calculators have a statistics mode and provide the mean and sample standard deviation directly. The TI-84Plus/TI-83Plus/TI-*n*spire calculators, Excel 2010, and Minitab provide the median and several other measures as well.

Many technologies display only the sample standard deviation s. You can quickly compute σ if you know s by using the formula

$$\sigma = s\sqrt{\frac{n-1}{n}}$$

The mean given in displays can be interpreted as the sample mean $\bar{x}$ or the population mean μ as appropriate.

The three displays, show output for the hybrid rose data of Guided Exercise 3.

TI-84Plus/TI-83Plus/TI-*n*spire (with TI-84Plus Keypad) Display Press **STAT ➤ CALC ➤ 1:1-Var Stats.** S_x is the sample standard deviation. σ_x is the population standard deviation.

```
1-Var Stats
 x̄=6
 Σx=36
 Σx²=224
 Sx=1.264911064
 σx=1.154700538
↓n=6
█
```


Excel 2010 Display Click the **Data Tab**. Then in the **Analysis Group** click **Data Analysis**. Select **Descriptive Statistics**. In the dialogue box, fill in the input range with the column and row numbers containing your data. Then check the Summary Statistics box.

Column 1	
Mean	6
Standard Error	0.516398
Median	5.5
Mode	5
Standard Deviation	1.264911
Sample Variance	1.6
Kurtosis	–0.78125
Skewness	0.889391
Range	3
Minimum	5
Maximum	8
Sum	36
Count	6

Minitab Display Menu choices: **Stat ➤ Basic Statistics ➤ Display Descriptive Statistics.** StDev is the sample standard deviation. TrMean is a 5% trimmed mean.

```
N          Mean       Median       TrMean       StDev      SEMean
6          6.000      5.500        6.000        1.265      0.516
Minimum         Maximum         Q1           Q3
5.000           8.000           5.000        7.250
```

Now let's look at two immediate applications of the standard deviation. The first is the coefficient of variation, and the second is Chebyshev's theorem.

COEFFICIENT OF VARIATION

Coefficient of variation, *CV*

A disadvantage of the standard deviation as a comparative measure of variation is that it depends on the units of measurement. This means that it is difficult to use the standard deviation to compare measurements from different populations. For this reason, statisticians have defined the *coefficient of variation,* which expresses the standard deviation as a percentage of the sample or population mean.

If $\bar{x}$ and s represent the sample mean and sample standard deviation, respectively, then the sample **coefficient of variation** *CV* is defined to be

$$CV = \frac{s}{\bar{x}} \cdot 100\%$$

If μ and σ represent the population mean and population standard deviation, respectively, then the population coefficient of variation *CV* is defined to be

$$CV = \frac{\sigma}{\mu} \cdot 100\%$$

Notice that the numerator and denominator in the definition of *CV* have the same units, so *CV* itself has no units of measurement. This gives us the advantage of being able to directly compare the variability of two different populations using the coefficient of variation.

In the next example and guided exercise, we will compute the *CV* of a population and of a sample and then compare the results.

EXAMPLE 7

COEFFICIENT OF VARIATION

The Trading Post on Grand Mesa is a small, family-run store in a remote part of Colorado. The Grand Mesa region contains many good fishing lakes, so the Trading Post sells spinners (a type of fishing lure). The store has a very limited selection of spinners. In fact, the Trading Post has only eight different types of spinners for sale. The prices (in dollars) are

2.10 1.95 2.60 2.00 1.85 2.25 2.15 2.25

Since the Trading Post has only eight different kinds of spinners for sale, we consider the eight data values to be the *population.*

Ron Watts/Flirt/Corbis

(a) Use a calculator with appropriate statistics keys to verify that for the Trading Post data, $\mu \approx \$2.14$ and $\sigma \approx \$0.22$.

SOLUTION: Since the computation formulas for $\bar{x}$ and μ are identical, most calculators provide the value of $\bar{x}$ only. Use the output of this key for μ. The computation formulas for the sample standard deviation s and the population standard deviation σ are slightly different. Be sure that you use the key for σ (sometimes designated as σ_n or σ_x).

(b) Compute the *CV* of prices for the Trading Post and comment on the meaning of the result.

SOLUTION:

$$CV = \frac{\sigma}{\mu} \times 100\% = \frac{0.22}{2.14} \times 100\% = 10.28\%$$

Interpretation The coefficient of variation can be thought of as a measure of the spread of the data relative to the average of the data. Since the Trading Post is very small, it carries a small selection of spinners that are all priced similarly. The *CV* tells us that the standard deviation of the spinner prices is only 10.28% of the mean.

GUIDED EXERCISE 4 COEFFICIENT OF VARIATION

Cabela's in Sidney, Nebraska, is a very large outfitter that carries a broad selection of fishing tackle. It markets its products nationwide through a catalog service. A random sample of 10 spinners from Cabela's extensive spring catalog gave the following prices (in dollars):

1.69 1.49 3.09 1.79 1.39 2.89 1.49 1.39 1.49 1.99

(a) Use a calculator with sample mean and sample standard deviation keys to compute $\bar{x}$ and *s*.

→ $\bar{x} = \$1.87$ and $s \approx \$0.62$.

(b) Compute the *CV* for the spinner prices at Cabela's.

→ $CV = \frac{s}{\bar{x}} \times 100\% = \frac{0.62}{1.87} \times 100\% = 33.16\%$

(c) *Interpretation* Compare the mean, standard deviation, and *CV* for the spinner prices at the Grand Mesa Trading Post (Example 7) and Cabela's. Comment on the differences.

→ The *CV* for Cabela's is more than three times the *CV* for the Trading Post. Why? First, because of the remote location, the Trading Post tends to have somewhat higher prices (larger μ). Second, the Trading Post is very small, so it has a rather limited selection of spinners with a smaller variation in price.

CHEBYSHEV'S THEOREM

From our earlier discussion about standard deviation, we recall that the spread or dispersion of a set of data about the mean will be small if the standard deviation is small, and it will be large if the standard deviation is large. If we are dealing with a symmetrical, bell-shaped distribution, then we can make very definite statements about the proportion of the data that must lie within a certain number of standard deviations on either side of the mean. This will be discussed in detail in Chapter 7 when we talk about normal distributions.

However, the concept of data spread about the mean can be expressed quite generally for *all data distributions* (skewed, symmetric, or other shapes) by using the remarkable theorem of Chebyshev.

Chebyshev's theorem

CHEBYSHEV'S THEOREM

For *any* set of data (either population or sample) and for any constant *k* greater than 1, the proportion of the data that must lie within *k* standard deviations on either side of the mean is *at least*

$$1 - \frac{1}{k^2}$$

Put another way: For sample data with mean $\bar{x}$ and standard deviation *s*, at least $1 - 1/k^2$ (fractional part) of data must fall between $\bar{x} - ks$ and $\bar{x} + ks$.

RESULTS OF CHEBYSHEV'S THEOREM

For *any* set of data:

- *at least* 75% of the data fall in the interval from $\mu - 2\sigma$ to $\mu + 2\sigma$.
- *at least* 88.9% of the data fall in the interval from $\mu - 3\sigma$ to $\mu + 3\sigma$.
- *at least* 93.8% of the data fall in the interval from $\mu - 4\sigma$ to $\mu + 4\sigma$.

The results of Chebyshev's theorem can be derived by using the theorem and a little arithmetic. For instance, if we create an interval $k = 2$ standard deviations on either side of the mean, Chebyshev's theorem tells us that

$$1 - \frac{1}{2^2} = 1 - \frac{1}{4} = \frac{3}{4} \text{ or } 75\%$$

is the minimum percentage of data in the $\mu - 2\sigma$ to $\mu + 2\sigma$ interval.

Notice that Chebyshev's theorem refers to the *minimum* percentage of data that must fall within the specified number of standard deviations of the mean. If the distribution is mound-shaped, an even *greater* percentage of data will fall into the specified intervals (see the Empirical Rule in Section 7.1).

WHAT DOES CHEBYSHEV'S THEOREM TELL US?

Chebyshev's theorem applies to *any* distribution. It tells us

- the *minimum* percentage of data that falls between the mean and any specified number of standard deviations on either side of the mean.
- a minimum of 88.9% of the data falls between the values 3 standard deviations below the mean and 3 standard deviations above the mean. This implies that a maximum of 11.1% of data fall beyond 3 standard deviations of the mean. Such values might be suspect outliers, particularly for a mound-shaped symmetric distribution.

EXAMPLE 8

CHEBYSHEV'S THEOREM

Students Who Care is a student volunteer program in which college students donate work time to various community projects such as planting trees. Professor Gill is the faculty sponsor for this student volunteer program. For several years, Dr. Gill has kept a careful record of $x =$ total number of work hours volunteered by a student in the program each semester. For a random sample of students in the program, the mean number of hours was $\bar{x} = 29.1$ hours each semester, with a standard deviation of $s = 1.7$ hours each semester. Find an interval A to B for the number of hours volunteered into which at least 75% of the students in this program would fit.

Max Wanger/Blink/Corbis

SOLUTION: According to results of Chebyshev's theorem, at least 75% of the data must fall within 2 standard deviations of the mean. Because the mean is $\bar{x} = 29.1$ and the standard deviation is $s = 1.7$, the interval is

$$\bar{x} - 2s \text{ to } \bar{x} + 2s$$
$$29.1 - 2(1.7) \text{ to } 29.1 + 2(1.7)$$
$$25.7 \text{ to } 32.5$$

At least 75% of the students would fit into the group that volunteered from 25.7 to 32.5 hours each semester.

GUIDED EXERCISE 5 Chebyshev Interval

The *East Coast Independent News* periodically runs ads in its classified section offering a month's free subscription to those who respond. In this way, management can get a sense about the number of subscribers who read the classified section each day. Over a period of 2 years, careful records have been kept. The mean number of responses per ad is $\bar{x} = 525$ with standard deviation $s = 30$.

Determine a Chebyshev interval about the mean in which at least 88.9% of the data fall.

By Chebyshev's theorem, at least 88.9% of the data fall into the interval

$\bar{x} - 3s$ to $\bar{x} + 3s$

Because $\bar{x} = 525$ and $s = 30$, the interval is

$525 - 3(30)$ to $525 + 3(30)$

or from 435 to 615 responses per ad.

LOOKING FORWARD

In Chapter 7 we will study normal distributions. For normal distributions we can make much more precise statements about the percentage of data falling within 1, 2, or 3 standard deviations of the mean. In fact, the empirical rule (see Section 7.1) tells us that for normal distributions, about 68% of the data fall within 1 standard deviation of the mean, about 95% fall within 2 standard deviations of the mean, and almost all (99.7%) fall within 3 standard deviations of the mean.

CRITICAL THINKING

Averages such as the mean are often referred to in the media. However, an average by itself does not tell much about the way data are distributed about the mean. Knowledge about the standard deviation or variance, along with the mean, gives a much better picture of the data distribution.

Chebyshev's theorem tells us that no matter what the data distribution looks like, at least 75% of the data will fall within 2 standard deviations of the mean. As we will see in Chapter 7, when the distribution is mound-shaped and symmetrical, about 95% of the data are within 2 standard deviations of the mean. Data values beyond 2 standard deviations from the mean are less common than those closer to the mean.

Outlier

In fact, one indicator that a data value might be an *outlier* is that it is more than 2.5 standard deviations from the mean (Source: *Oxford Dictionary of Statistics*, Oxford University Press).

VIEWPOINT Socially Responsible Investing

Make a difference and make money! Socially responsible mutual funds tend to screen out corporations that sell tobacco, weapons, and alcohol, as well as companies that are environmentally unfriendly. In addition, these funds screen out companies that use child labor in sweatshops. There are 68 socially responsible funds tracked by the Social Investment Forum. For more information, visit the Brase/Brase statistics site at **http://www.cengagebrain.com** *and find the link to social investing.*

How do these funds rate compared to other funds? One way to answer this question is to study the annual percent returns of the funds using both the mean *and* standard deviation. *(See Problem 20 of this section.)*

SECTION 3.2 PROBLEMS

1. *Statistical Literacy* Which average—mean, median, or mode—is associated with the standard deviation?

2. *Statistical Literacy* What is the relationship between the variance and the standard deviation for a sample data set?

3. *Statistical Literacy* When computing the standard deviation, does it matter whether the data are sample data or data comprising the entire population? Explain.

4. *Statistical Literacy* What symbol is used for the standard deviation when it is a sample statistic? What symbol is used for the standard deviation when it is a population parameter?

5. *Basic Computation: Range, Standard Deviation* Consider the data set

   2 3 4 5 6

   (a) Find the range.
   (b) Use the defining formula to compute the sample standard deviation s.
   (c) Use the defining formula to compute the population standard deviation σ.

6. *Basic Computation: Range, Standard Deviation* Consider the data set

   1 2 3 4 5

   (a) Find the range.
   (b) Use the defining formula to compute the sample standard deviation s.
   (c) Use the defining formula to compute the population standard deviation σ.

7. *Critical Thinking* For a given data set in which not all data values are equal, which value is smaller, s or σ? Explain.

8. *Critical Thinking* Consider two data sets with equal sample standard deviations. The first data set has 20 data values that are not all equal, and the second has 50 data values that are not all equal. For which data set is the difference between s and σ greater? Explain. *Hint:* Consider the relationship $\sigma = s\sqrt{(n-1)/n}$.

9. *Critical Thinking* Each of the following data sets has a mean of $\bar{x} = 10$.

   (i) 8 9 10 11 12 (ii) 7 9 10 11 13 (iii) 7 8 10 12 13

   (a) Without doing any computations, order the data sets according to increasing value of standard deviations.
   (b) Why do you expect the difference in standard deviations between data sets (i) and (ii) to be greater than the difference in standard deviations between data sets (ii) and (iii)? *Hint:* Consider how much the data in the respective sets differ from the mean.

10. *Critical Thinking: Data Transformation* In this problem, we explore the effect on the standard deviation of adding the same constant to each data value in a data set. Consider the data set 5, 9, 10, 11, 15.

   (a) Use the defining formula, the computation formula, or a calculator to compute s.
   (b) Add 5 to each data value to get the new data set 10, 14, 15, 16, 20. Compute s.

(c) Compare the results of parts (a) and (b). In general, how do you think the standard deviation of a data set changes if the same constant is added to each data value?

11. *Critical Thinking: Data Transformation* In this problem, we explore the effect on the standard deviation of multiplying each data value in a data set by the same constant. Consider the data set 5, 9, 10, 11, 15.
(a) Use the defining formula, the computation formula, or a calculator to compute s.
(b) Multiply each data value by 5 to obtain the new data set 25, 45, 50, 55, 75. Compute s.
(c) Compare the results of parts (a) and (b). In general, how does the standard deviation change if each data value is multiplied by a constant c?
(d) You recorded the weekly distances you bicycled in miles and computed the standard deviation to be $s = 3.1$ miles. Your friend wants to know the standard deviation in kilometers. Do you need to redo all the calculations? Given 1 mile $\approx$ 1.6 kilometers, what is the standard deviation in kilometers?

12. *Critical Thinking: Outliers* One indicator of an outlier is that an observation is more than 2.5 standard deviations from the mean. Consider the data value 80.
(a) If a data set has mean 70 and standard deviation 5, is 80 a suspect outlier?
(b) If a data set has mean 70 and standard deviation 3, is 80 a suspect outlier?

13. *Basic Computation: Variance, Standard Deviation* Given the sample data

x:	23	17	15	30	25

(a) Find the range.
(b) Verify that $\Sigma x = 110$ and $\Sigma x^2 = 2568$.
(c) Use the results of part (b) and appropriate computation formulas to compute the sample variance s^2 and sample standard deviation s.
(d) Use the defining formulas to compute the sample variance s^2 and sample standard deviation s.
(e) Suppose the given data comprise the entire population of all x values. Compute the population variance σ^2 and population standard deviation σ.

14. *Basic Computation: Coefficient of Variation, Chebyshev Interval* Consider sample data with $\bar{x} = 15$ and $s = 3$.
(a) Compute the coefficient of variation.
(b) Compute a 75% Chebyshev interval around the sample mean.

15. *Basic Computation: Coefficient of Variation, Chebyshev Interval* Consider population data with $\mu = 20$ and $\sigma = 2$.
(a) Compute the coefficient of variation.
(b) Compute an 88.9% Chebyshev interval around the population mean.

16. *Investing: Stocks and Bonds* Do bonds reduce the overall risk of an investment portfolio? Let x be a random variable representing annual percent return for Vanguard Total Stock Index (all stocks). Let y be a random variable representing annual return for Vanguard Balanced Index (60% stock and 40% bond). For the past several years, we have the following data (Reference: Morningstar Research Group, Chicago).

x:	11	0	36	21	31	23	24	−11	−11	−21
y:	10	−2	29	14	22	18	14	−2	−3	−10

(a) Compute Σx, Σx^2, Σy, and Σy^2.
(b) Use the results of part (a) to compute the sample mean, variance, and standard deviation for x and for y.

NASA Images

(c) Compute a 75% Chebyshev interval around the mean for x values and also for y values. Use the intervals to compare the two funds.

(d) ***Interpretation:*** Compute the coefficient of variation for each fund. Use the coefficients of variation to compare the two funds. If s represents risks and $\bar{x}$ represents expected return, then $s/\bar{x}$ can be thought of as a measure of risk per unit of expected return. In this case, why is a smaller *CV* better? Explain.

17. *Space Shuttle: Epoxy* Kevlar epoxy is a material used on the NASA space shuttles. Strands of this epoxy were tested at the 90% breaking strength. The following data represent time to failure (in hours) for a random sample of 50 epoxy strands (Reference: R. E. Barlow, University of California, Berkeley). Let x be a random variable representing time to failure (in hours) at 90% breaking strength. *Note:* These data are also available for download at the Online Study Center.

0.54	1.80	1.52	2.05	1.03	1.18	0.80	1.33	1.29	1.11
3.34	1.54	0.08	0.12	0.60	0.72	0.92	1.05	1.43	3.03
1.81	2.17	0.63	0.56	0.03	0.09	0.18	0.34	1.51	1.45
1.52	0.19	1.55	0.02	0.07	0.65	0.40	0.24	1.51	1.45
1.60	1.80	4.69	0.08	7.89	1.58	1.64	0.03	0.23	0.72

(a) Find the range.

(b) Use a calculator to verify that $\Sigma x = 62.11$ and $\Sigma x^2 \approx 164.23$.

(c) Use the results of part (b) to compute the sample mean, variance, and standard deviation for the time to failure.

(d) ***Interpretation*** Use the results of part (c) to compute the coefficient of variation. What does this number say about time to failure? Why does a small *CV* indicate more consistent data, whereas a larger *CV* indicates less consistent data? Explain.

18. *Archaeology: Ireland* The Hill of Tara in Ireland is a place of great archaeological importance. This region has been occupied by people for more than 4000 years. Geomagnetic surveys detect subsurface anomalies in the earth's magnetic field. These surveys have led to many significant archaeological discoveries. After collecting data, the next step is to begin a statistical study. The following data measure magnetic susceptibility (centimeter-gram-second $\times 10^{-6}$) on two of the main grids of the Hill of Tara (Reference: *Tara: An Archaeological Survey* by Conor Newman, Royal Irish Academy, Dublin).

Grid E: x variable

13.20	5.60	19.80	15.05	21.40	17.25	27.45
16.95	23.90	32.40	40.75	5.10	17.75	28.35

Grid H: y variable

11.85	15.25	21.30	17.30	27.50	10.35	14.90
48.70	25.40	25.95	57.60	34.35	38.80	41.00
31.25						

(a) Compute Σx, Σx^2, Σy, and Σy^2.

(b) Use the results of part (a) to compute the sample mean, variance, and standard deviation for x and for y.

(c) Compute a 75% Chebyshev interval around the mean for x values and also for y values. Use the intervals to compare the magnetic susceptibility on the two grids. Higher numbers indicate higher magnetic susceptibility. However, extreme values, high or low, could mean an anomaly and possible archaeological treasure.

(d) ***Interpretation*** Compute the sample coefficient of variation for each grid. Use the *CV*s to compare the two grids. If s represents variability in the signal (magnetic susceptibility) and $\bar{x}$ represents the expected level of the signal, then $s/\bar{x}$ can be thought of as a measure of the variability per unit of expected signal. Remember, a considerable variability in the signal (above or below average) might indicate buried artifacts. Why, in this case, would a large *CV* be better, or at least more exciting? Explain.

19. *Wildlife: Mallard Ducks and Canada Geese* For mallard ducks and Canada geese, what percentage of nests are successful (at least one offspring survives)? Studies in Montana, Illinois, Wyoming, Utah, and California gave the following percentages of successful nests (Reference: *The Wildlife Society Press,* Washington, D.C.).

x: Percentage success for mallard duck nests

56 85 52 13 39

y: Percentage success for Canada goose nests

24 53 60 69 18

(a) Use a calculator to verify that $\Sigma x = 245$; $\Sigma x^2 = 14{,}755$; $\Sigma y = 224$; and $\Sigma y^2 = 12{,}070$.
(b) Use the results of part (a) to compute the sample mean, variance, and standard deviation for x, the percent of successful mallard nests.
(c) Use the results of part (a) to compute the sample mean, variance, and standard deviation for y, the percent of successful Canada goose nests.
(d) ***Interpretation*** Use the results of parts (b) and (c) to compute the coefficient of variation for successful mallard nests and Canada goose nests. Write a brief explanation of the meaning of these numbers. What do these results say about the nesting success rates for mallards compared to those of Canada geese? Would you say one group of data is more or less consistent than the other? Explain.

20. *Investing: Socially Responsible Mutual Funds* Pax World Balanced is a highly respected, socially responsible mutual fund of stocks and bonds (see Viewpoint). Vanguard Balanced Index is another highly regarded fund that represents the entire U.S. stock and bond market (an index fund). The mean and standard deviation of annualized percent returns are shown below. The annualized mean and standard deviation are based on the years 1993 through 2002 (Source: Morningstar).

Pax World Balanced: $\bar{x} = 9.58\%$; $s = 14.05\%$

Vanguard Balanced Index: $\bar{x} = 9.02\%$; $s = 12.50\%$

(a) *Interpretation* Compute the coefficient of variation for each fund. If $\bar{x}$ represents return and s represents risk, then explain why the coefficient of variation can be taken to represent risk per unit of return. From this point of view, which fund appears to be better? Explain.
(b) *Interpretation* Compute a 75% Chebyshev interval around the mean for each fund. Use the intervals to compare the two funds. As usual, past performance does not guarantee future performance.

21. *Medical: Physician Visits* In some reports, the mean and coefficient of variation are given. For instance, in *Statistical Abstract of the United States,* 116th Edition, one report gives the average number of physician visits by males per year. The average reported is 2.2, and the reported coefficient of variation is 1.5%. Use this information to determine the standard deviation of the annual number of visits to physicians made by males.

Grouped data

Approximating $\bar{x}$ and s from grouped data

EXPAND YOUR KNOWLEDGE: GROUPED DATA

When data are grouped, such as in a frequency table or histogram, we can estimate the mean and standard deviation by using the following formulas. Notice that all data values in a given class are treated as though each of them equals the midpoint x of the class.

Sample Mean for a Frequency Distribution

$$\bar{x} = \frac{\Sigma xf}{n} \qquad (5)$$

Sample Standard Deviation for a Frequency Distribution

$$s = \sqrt{\frac{\Sigma(x - \bar{x})^2 f}{n - 1}} \qquad (6)$$

Computation Formula for the Sample Standard Deviation

$$s = \sqrt{\frac{\Sigma x^2 f - (\Sigma xf)^2/n}{n - 1}} \qquad (7)$$

where

x is the midpoint of a class,

f is the number of entries in that class,

n is the total number of entries in the distribution, and $n = \Sigma f$.

The summation Σ is over all classes in the distribution.

Use formulas (5) and (6) or (5) and (7) to solve Problems 22–25. To use formulas (5) and (6) to evaluate the sample mean and standard deviation, use the following column heads:

Midpoint x	Frequency f	xf	$(x - \bar{x})$	$(x - \bar{x})^2$	$(x - \bar{x})^2 f$

For formulas (5) and (7), use these column heads:

Midpoint x	Frequency f	xf	x^2	$x^2 f$

Note: On the TI-84Plus/TI-83Plus/TI-*n*spire (with TI-84Plus keypad) calculators, enter the midpoints in column $\mathbf{L_1}$ and the frequencies in column $\mathbf{L_2}$. Then use **1-VarStats** $\mathbf{L_1, L_2}$.

22. *Grouped Data: Anthropology* What was the age distribution of prehistoric Native Americans? Extensive anthropologic studies in the southwestern United States gave the following information about a prehistoric extended family group of 80 members on what is now the Navajo Reservation in northwestern New Mexico (Source: Based on information taken from *Prehistory in the Navajo Reservation District*, by F. W. Eddy, Museum of New Mexico Press).

Age range (years)	1–10*	11–20	21–30	31 and over
Number of individuals	34	18	17	11

*Includes infants.

For this community, estimate the mean age expressed in years, the sample variance, and the sample standard deviation. For the class 31 and over, use 35.5 as the class midpoint.

23. *Grouped Data: Shoplifting* What is the age distribution of adult shoplifters (21 years of age or older) in supermarkets? The following is based on information taken from the National Retail Federation. A random sample of 895 incidents of shoplifting gave the following age distribution:

Age range (years)	21–30	31–40	41 and over
Number of shoplifters	260	348	287

Estimate the mean age, sample variance, and sample standard deviation for the shoplifters. For the class 41 and over, use 45.5 as the class midpoint.

24. *Grouped Data: Hours of Sleep per Day* Alexander Borbely is a professor at the University of Zurich Medical School, where he is director of the sleep laboratory. The histogram in Figure 3-2 is based on information from his book *Secrets of Sleep*. The histogram displays hours of sleep per day for a random sample of 200 subjects. Estimate the mean hours of sleep, standard deviation of hours of sleep, and coefficient of variation.

FIGURE 3-2

Hours of Sleep Each Day (24-hour period)

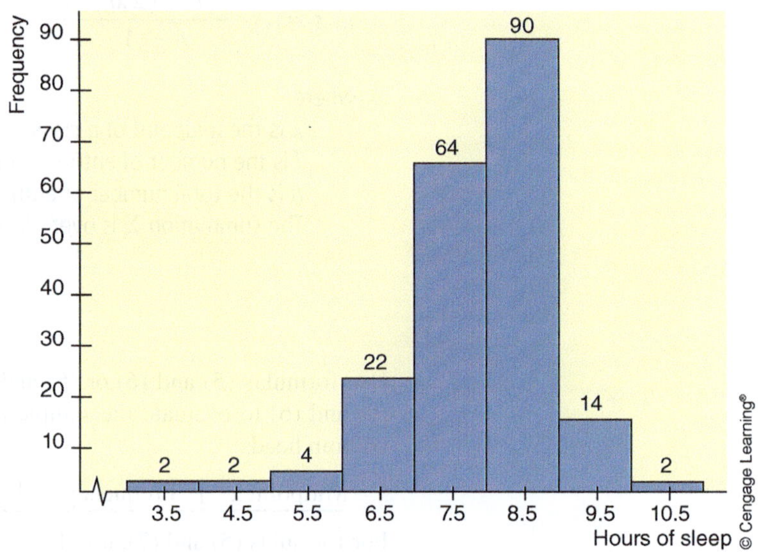

25. *Grouped Data: Business Administration* What are the big corporations doing with their wealth? One way to answer this question is to examine profits as percentage of assets. A random sample of 50 Fortune 500 companies gave the following information (Source: Based on information from *Fortune 500*, Vol. 135, No. 8).

Profit as percentage of assets	8.6–12.5	12.6–16.5	16.6–20.5	20.6–24.5	24.6–28.5
Number of companies	15	20	5	7	3

Estimate the sample mean, sample variance, and sample standard deviation for profit as percentage of assets.

Moving average

26. *Expand Your Knowledge: Moving Averages* You do not need a lot of money to invest in a mutual fund. However, if you decide to put some money into an investment, you are usually advised to leave it in for (at least) several years. Why? Because good years tend to cancel out bad years, giving you a better overall return with less risk. To see what we mean, let's use a 3-year *moving average* on the Calvert Social Balanced Fund (a socially responsible fund).

Year	1990	1991	1992	1993	1994	1995	1996	1997	1998	1999	2000
% Return	1.78	17.79	7.46	5.95	−4.74	25.85	9.03	18.92	17.49	6.80	−2.38

Source: Morningstar

(a) Use a calculator with mean and standard deviation keys to verify that the mean annual return for all 11 years is approximately 9.45%, with standard deviation 9.57%.

(b) To compute a 3-year moving average for 1992, we take the data values for 1992 and the prior two years and average them. To compute a 3-year moving average for 1993, we take the data values for 1993 and the prior two years and average them. Verify that the following 3-year moving averages are correct.

Year	1992	1993	1994	1995	1996	1997	1998	1999	2000
3-year moving average	9.01	10.40	2.89	9.02	10.05	17.93	15.15	14.40	7.30

(c) Use a calculator with mean and standard deviation keys to verify that for the 3-year moving average, the mean is 10.68% with sample standard deviation 4.53%.

(d) ***Interpretation*** Compare the results of parts (a) and (c). Suppose we take the point of view that risk is measured by standard deviation. Is the risk (standard deviation) of the 3-year moving average considerably smaller? This is an example of a general phenomenon that will be studied in more detail in Chapter 7.

27. *Brain Teaser: Sum of Squares* If you like mathematical puzzles or love algebra, try this! Otherwise, just trust that the computational formula for the sum of squares is correct. We have a sample of x values. The sample size is n. Fill in the details for the following steps.

$$\begin{aligned}\Sigma(x-\bar{x})^2 &= \Sigma x^2 - 2\bar{x}\Sigma x + n\bar{x}^2 \\ &= \Sigma x^2 - 2n\bar{x}^2 + n\bar{x}^2 \\ &= \Sigma x^2 - \frac{(\Sigma x)^2}{n}\end{aligned}$$

28. *Expand Your Knowledge: Stratified Sampling and Students in Sixth Grade* This is a technique to break down the variation of a random variable into useful components (called stratum) in order to decrease experimental variation and increase accuracy of results. It has been found that a more accurate estimate of population mean μ can often be obtained by taking measurements from naturally occurring subpopulations and combining the results using weighted averages. For example, suppose an accurate estimate of the mean weight of sixth grade students is desired for a large school system. Suppose (for cost reasons) we can only take a random sample of $m = 100$ students. Instead of taking a simple random sample of 100 students from the entire population of all sixth grade students we use stratified sampling as follows. The school system under study consists three large schools. School A has $N_1 = 310$ sixth grade students, School B has $N_2 = 420$ sixth grade students, and School C has $N_3 = 516$ sixth grade students. This is a total population of 1246 sixth grade students in our study and we have strata consisting of the 3 schools. A preliminary study in each school with relatively small sample size has given estimates for the sample standard deviation s of sixth grade student weights in each school. These are shown in the following table

School A	School B	School C
$N_1 = 310$	$N_2 = 420$	$N_3 = 516$
$s_1 = 3$ lb	$s_2 = 12$ lb	$s_3 = 6$ lb

How many students should we randomly choose from each school for a best estimate μ for the population mean weight? A lot of mathematics goes into the answer. Forunately, Bill Williams of Bell Laboratories wrote a book called *A Sampler on Sampling* (John Wiley and Sons, publisher), which provides an answer. Let n_1 be the number of students randomly chosen from School A, n_2 be the number chosen from School B, and n_3 be the number chosen from School C. This means our total sample size will be $m = n_1 + n_2 + n_3$. What is the formula for n_i? A popular and widely used technique is the following.

$$n_i = \left[\frac{N_i s_i}{N_1 s_1 + N_2 s_2 + N_3 s_3}\right] m$$

The n_i are usually not whole numbers, so we need to round to the nearest whole number. This formula allocates more students to schools that have a larger population of sixth graders and/or have larger sample standard deviations. Remember, this is a popular and widely used technique for stratified sampling. It is not an absolute rule. There are other methods of stratified sampling also in use. In general practice, according to Bill Williams, the use of naturally occurring strata seems to reduce overall variability in measurements by about 20% compared to simple random samples taken from the entire (unstratified) population.

Now suppose you have taken a random sample size n_i from each appropriate school and you got a sample mean weight $\bar{x}_i$ from each school. How do you get the best estimate for population mean weight μ of the all 1246 students? The answer is, we use a weighted average.

$$\mu \approx \frac{n_1}{m}\bar{x}_1 + \frac{n_2}{m}\bar{x}_2 + \frac{n_3}{m}\bar{x}_3$$

COMMENT: This is an example with three strata. Applications with any number of strata can be solved in a similar way with obvious extensions of formulas.

(a) Compute the size of the random samples n_1, n_2, n_3 to be taken from each school. Round each sample size to the nearest whole number and make sure they add up to $m = 100$.

(b) Suppose you took the appropriate random sample from each school and you got the following average student weights: $\bar{x}_1 = 82$ lb, $\bar{x}_2 = 115$ lb, $\bar{x}_3 = 90$ lb. Compute your best estimate for the population mean weight μ.

29. *Expand Your Knowledge: Stratified Sampling and Politics*
Three local districts are "swing" districts for an upcoming election on a contentious political issue. A survey will be conducted in which voters will be asked to rate their opinion regarding this issue on a scale of 0 (strongly oppose) to 10 (strongly support). A small preliminary random sample from each district was used to estimate the sample standard deviation s of responses for the district. The following table shows the number of voters N in each district and the sample standard deviation s of strength of support in each district.

District 1	District 2	District 3
$N_1 = 1525$	$N_2 = 917$	$N_3 = 2890$
$s_1 = 2.2$	$s_2 = 1.4$	$s_3 = 3.3$

We have a total population of 5332 voters and 3 strata (districts.)

The group doing the survey has enough funding to obtain a random sample of $m = 250$ total responses from all the districts.

(a) Compute the size of the random samples n_1, n_2, n_3 to taken from each district. Round each sample size to the nearest whole number and make sure they add up to $m = 250$. *Hint*: see problem 28.

(b) Suppose you took the appropriate random samples from each district and you got the following average political support measures (scale 0 to 10): $\overline{x_1} = 6.2, \overline{x_2} = 3.1\ \overline{x_3} = 8.5$. Compute your best estimate for the population mean μ (scale 0 to 10) of voter support for the issue from the total population of these three districts. *Hint*: see problem 28.

30. *Expand Your Knowledge: Stratified Sampling, No Preliminary Study, Police Identification* Police are tested for their ability to correctly recognize and identify a suspect based on a witness or victim's verbal description of the suspect. Scores on the identification test range from 0 to 100 (perfect score). Three cities in Massachusetts are under study. The mean score for all police in the three cities is requested. However, funding will only permit $m = 150$ police to be tested. There is no preliminary study to estimate sample standard deviation of scores in each city. City A has $N_1 = 183$ police, City B has $N_2 = 371$ police, and City C has $N_3 = 255$ police.

If we have no preliminary study for sample standard deviations, we can use a method called *proportional sampling* (also called *representative sampling*). The sample size n_1 for each stratum (city police department) is given by

$$n_i = \left[\frac{N_i}{N_1 + N_2 + N_3}\right]m$$

In most cases n_i is not a whole number so we round to the nearest whole number.

Remember our total sample size is $m = n_1 + n_2 + n_3$.

(a) Use the method of proportional sampling to compute n_1, n_2, and n_3, the size of the random sample to be taken from each of the three cities. Round each n_i to the nearest whole number and make sure $m = n_1 + n_2 + n_3$.

(b) Suppose you actually conducted the specified number of tests in each city and obtained the following result: $\overline{x_1} = 96$ is the mean test score from city A, $\overline{x_2} = 85$ is the mean test score from City B, and $\overline{x_3} = 88$ is the mean test score from City C. Use the weighted average

$$\mu \approx \frac{n_1}{m}\overline{x_1} + \frac{n_2}{m}\overline{x_2} + \frac{n_3}{m}\overline{x_3}$$

to get your best estimate for the population mean test score of all police in all three cities.

COMMENT: Estimates for μ are usually better if we have a preliminary study with reasonably accurate estimates for the sample standard deviation s_i of each stratum (*see* Problem 28 and 29). However, in the case where there is no preliminary study we can use proportional sampling and still obtain good results.

SECTION **3.3**

Percentiles and Box-and-Whisker Plots

FOCUS POINTS

- Interpret the meaning of percentile scores.
- Compute the median, quartiles, and five-number summary from raw data.
- Make a box-and-whisker plot. Interpret the results.
- Describe how a box-and-whisker plot indicates spread of data about the median.

We've seen measures of central tendency and spread for a set of data. The arithmetic mean $\overline{x}$ and the standard deviation s will be very useful in later work. However, because they each utilize every data value, they can be heavily influenced by one or two extreme data values. In cases where our data distributions are heavily skewed or even bimodal, we often get a better summary of the distribution by utilizing relative position of data rather than exact values.

Recall that the median is an average computed by using relative position of the data. If we are told that 81 is the median score on a biology test, we know that after the data have been ordered, 50% of the data fall at or below the median value of 81.

Percentile

The median is an example of a *percentile;* in fact, it is the 50th percentile. The general definition of the *P*th percentile follows.

For whole numbers P (where $1 \leq P \leq 99$), the Pth **percentile** of a distribution is a value such that P% of the data fall at or below it and $(100 - P)$% of the data fall at or above it.

In Figure 3-3, we see the 60th percentile marked on a histogram. We see that 60% of the data lie below the mark and 40% lie above it.

FIGURE 3-3

A Histogram with the 60th Percentile Shown

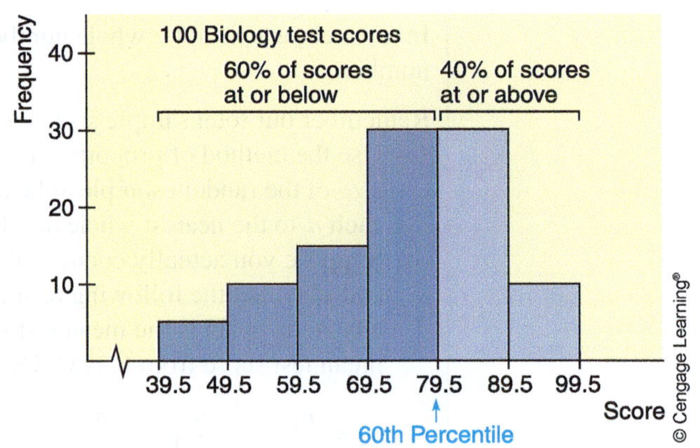

GUIDED EXERCISE 6 Percentiles

You took the English achievement test to obtain college credit in freshman English by examination.

(a) If your score is at the 89th percentile, what percentage of scores are at or below yours?

→ The percentile means that 89% of the scores are at or below yours.

(b) If the scores ranged from 1 to 100 and your raw score is 95, does this necessarily mean that your score is at the 95th percentile?

→ No, the percentile gives an indication of relative position of the scores. The determination of your percentile has to do with the number of scores at or below yours. If everyone did very well and only 80% of the scores fell at or below yours, you would be at the 80th percentile even though you got 95 out of 100 points on the exam.

There are 99 percentiles, and in an ideal situation, the 99 percentiles divide the data set into 100 equal parts. (See Figure 3-4.) However, if the number of data elements is not exactly divisible by 100, the percentiles will not divide the data into equal parts.

FIGURE 3-4

Percentiles

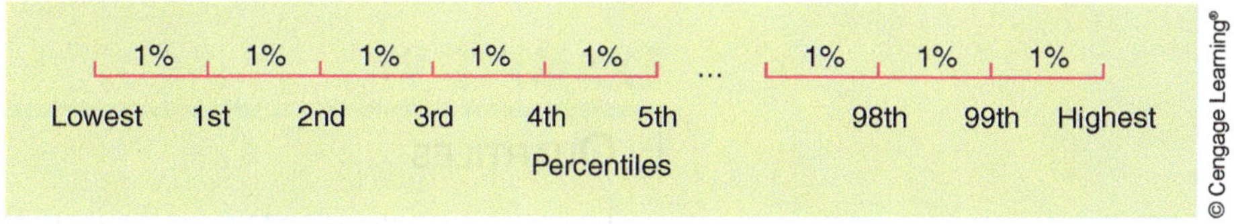

There are several widely used conventions for finding percentiles. They lead to slightly different values for different situations, but these values are close together. For all conventions, the data are first *ranked* or ordered from smallest to largest. A natural way to find the *P*th percentile is to then find a value such that *P*% of the data fall at or below it. This will not always be possible, so we take the nearest value satisfying the criterion. It is at this point that there is a variety of processes to determine the exact value of the percentile.

Quartiles

We will not be very concerned about exact procedures for evaluating percentiles in general. However, *quartiles* are special percentiles used so frequently that we want to adopt a specific procedure for their computation.

Quartiles are those percentiles that divide the data into fourths. The *first quartile* Q_1 is the 25th percentile, the *second quartile* Q_2 is the median, and the *third quartile* Q_3 is the 75th percentile. (See Figure 3-5.)

FIGURE 3-5

Quartiles

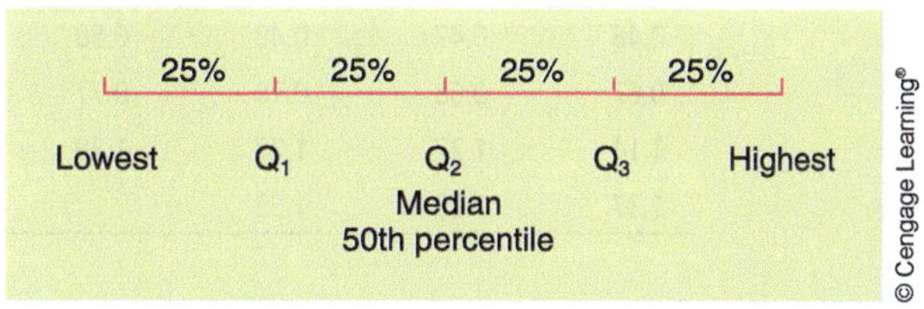

Again, several conventions are used for computing quartiles, but the following convention utilizes the median and is widely adopted.

PROCEDURE

HOW TO COMPUTE QUARTILES

1. Order the data from smallest to largest.
2. Find the median. This is the second quartile.
3. The first quartile Q_1 is then the median of the lower half of the data; that is, it is the median of the data falling *below* the Q_2 position (and not including Q_2).
4. The third quartile Q_3 is the median of the upper half of the data; that is, it is the median of the data falling *above* the Q_2 position (and not including Q_2).

In short, all we do to find the quartiles is find three medians.

Interquartile range, *IQR*

The median, or second quartile, is a popular measure of the center utilizing relative position. A useful measure of data spread utilizing relative position is the *interquartile range (IQR)*. It is simply the difference between the third and first quartiles.

$$\textbf{Interquartile range} = Q_3 - Q_1$$

The interquartile range tells us the spread of the middle half of the data. Now let's look at an example to see how to compute all of these quantities.

EXAMPLE 9

QUARTILES

In a hurry? On the run? Hungry as well? How about an ice cream bar as a snack? Ice cream bars are popular among all age groups. *Consumer Reports* did a study of ice cream bars. Twenty-seven bars with taste ratings of at least "fair" were listed, and cost per bar was included in the report. Just how much does an ice cream bar cost? The inflation adjusted data, expressed in dollars, appear in Table 3-4. As you can see, the cost varies quite a bit, partly because the bars are not of uniform size.

TABLE 3-4 **Cost of Ice Cream Bars (in dollars)**

1.29	1.37	1.30	0.80	0.67	1.33	1.37	1.37
1.27	0.93	0.63	0.80	0.27	1.38	0.77	1.14
1.53	0.55	0.80	0.70	0.36	0.65	0.47	0.68
0.50	0.48	0.46					

TABLE 3-5 **Ordered Cost of Ice Cream Bars (in dollars)**

0.46	0.47	0.48	0.50	0.55	0.63	0.63	0.65
0.67	0.68	0.70	0.77	0.80	0.80	0.80	0.93
1.14	1.27	1.27	1.29	1.30	1.33	1.37	1.37
1.37	1.38	1.53					

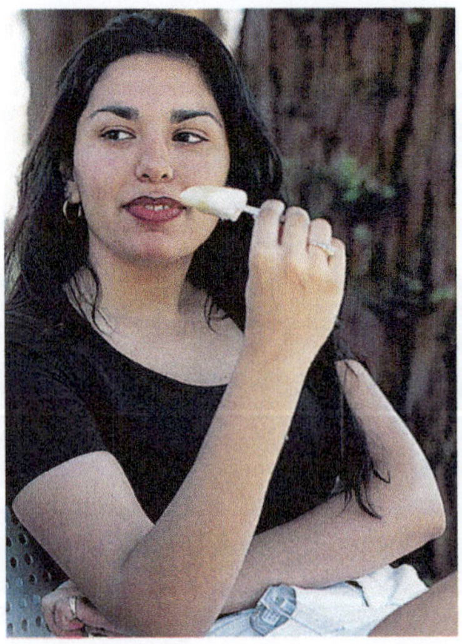

Gary Conner/PhotoEdit

(a) Find the quartiles.

SOLUTION: We first order the data from smallest to largest. Table 3-5 shows the data in order. Next, we find the median. Since the number of data values is 27, there are an odd number of data, and the median is simply the center or 14th value. The value is shown boxed in Table 3-5.

$$\text{Median} = Q_2 = 0.80$$

There are 13 values below the median position, and Q_1 is the median of these values. It is the middle or seventh value and is shaded in Table 3-5.

$$\text{First quartile} = Q_1 = 0.63$$

There are also 13 values above the median position. The median of these is the seventh value from the right end. This value is also shaded in Table 3-5.

$$\text{Third quartile} = Q_3 = 1.30$$

(b) Find the interquartile range.

SOLUTION:

$$\begin{aligned} IQR &= Q_3 - Q_1 \\ &= 1.30 - 0.63 \\ &= 0.67 \end{aligned}$$

This means that the middle half of the data has a cost spread of 67¢.

GUIDED EXERCISE 7 QUARTILES

Many people consider the number of calories in an ice cream bar as important as, if not more important than, the cost. The *Consumer Reports* article also included the calorie count of the rated ice cream bars (Table 3-6). There were 22 vanilla-flavored bars rated. Again, the bars varied in size, and some of the smaller bars had fewer calories. The calorie counts for the vanilla bars follow.

TABLE 3-6 **Calories in Vanilla-Flavored Ice Cream Bars**

342	377	319	353	295
234	294	286	377	182
310	439	111	201	182
197	209	147	190	151
131	151			

(a) Our first step is to order the data. Do so.

→ TABLE 3-7 **Ordered Data**

111	131	147	151	151	182
182	190	197	201	209	234
286	294	295	310	319	342
353	377	377	439		

(b) There are 22 data values. Find the median.

→ Average the 11th and 12th data values boxed together in Table 3-7.

$$\text{Median} = \frac{209 + 234}{2}$$
$$= 221.5$$

(c) How many values are below the median position? Find Q_1.

→ Since the median lies halfway between the 11th and 12th values, there are 11 values below the median position. Q_1 is the median of these values.

$Q_1 = 182$

(d) There are the same number of data above as below the median. Use this fact to find Q_3.

→ Q_3 is the median of the upper half of the data. There are 11 values in the upper portion.

$Q_3 = 319$

(e) Find the interquartile range and comment on its meaning.

→ $IQR = Q_3 - Q_1$
$= 319 - 182$
$= 137$

The middle portion of the data has a spread of 137 calories.

BOX-AND-WHISKER PLOTS

Five-number summary

The quartiles together with the low and high data values give us a very useful *five-number summary* of the data and their spread.

FIVE-NUMBER SUMMARY

Lowest value, Q_1, median, Q_3, highest value

Box-and-whisker plot

We will use these five numbers to create a graphic sketch of the data called a *box-and-whisker plot*. Box-and-whisker plots provide another useful technique from exploratory data analysis (EDA) for describing data.

PROCEDURE

HOW TO MAKE A BOX-AND-WHISKER PLOT

1. Draw a vertical scale to include the lowest and highest data values.
2. To the right of the scale, draw a box from Q_1 to Q_3.
3. Include a solid line through the box at the median level.
4. Draw vertical lines, called *whiskers*, from Q_1 to the lowest value and from Q_3 to the highest value.

Whiskers

FIGURE 3-6

Box-and-Whisker Plot

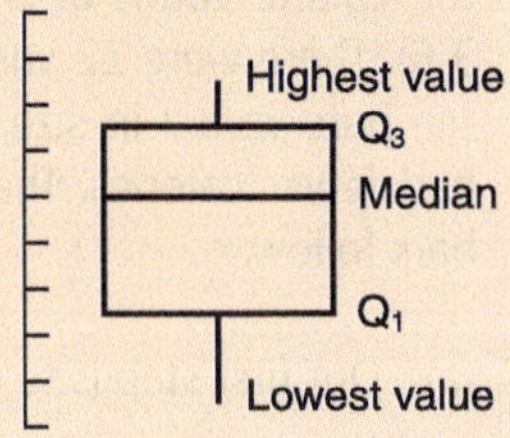

The next example demonstrates the process of making a box-and-whisker plot.

EXAMPLE 10

Box-and-Whisker Plot

Using the data from Guided Exercise 7, make a box-and-whisker plot showing the calories in vanilla-flavored ice cream bars. Use the plot to make observations about the distribution of calories.

(a) In Guided Exercise 7, we ordered the data (see Table 3-7) and found the values of the median, Q_1, and Q_3. From this previous work we have the following five-number summary:

$$\text{low value} = 111;\ Q_1 = 182;\ \text{median} = 221.5;\ Q_3 = 319;\ \text{high value} = 439$$

(b) We select an appropriate vertical scale and make the plot (Figure 3-7).

(c) ***Interpretation*** A quick glance at the box-and-whisker plot reveals the following:

(i) The box tells us where the middle half of the data lies, so we see that half of the ice cream bars have between 182 and 319 calories, with an interquartile range of 137 calories.

(ii) The median is slightly closer to the lower part of the box. This means that the lower calorie counts are more concentrated. The calorie counts above the median are more spread out, indicating that the distribution is slightly skewed toward the higher values.

(iii) The upper whisker is longer than the lower, which again emphasizes skewness toward the higher values.

FIGURE 3-7

Box-and-Whisker Plot for Calories in Vanilla-Flavored Ice Cream Bars

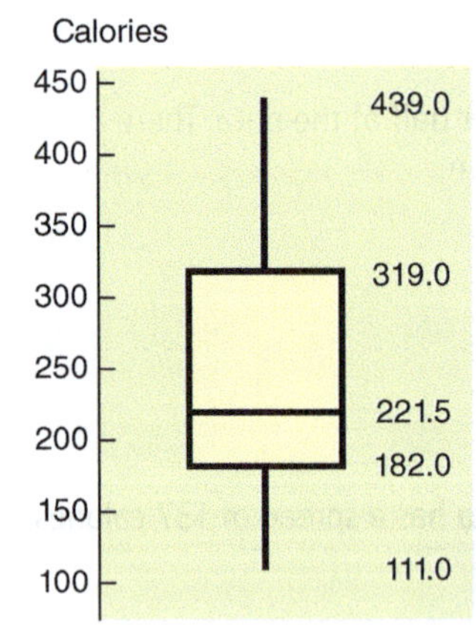

COMMENT In exploratory data analysis, *hinges* rather than quartiles are used to create the box. Hinges are computed in a manner similar to the method used to compute quartiles. However, in the case of an odd number of data values, include the median itself in both the lower and upper halves of the data (see *Applications, Basics, and Computing of Exploratory Data Analysis* by Paul Velleman and David Hoaglin, Duxbury Press). This has the effect of shrinking the box and moving the ends of the box slightly toward the median. For an even number of data, the quartiles as we computed them equal the hinges.

GUIDED EXERCISE 8 | Box-and-whisker plot

The Renata College Development Office sent salary surveys to alumni who graduated 2 and 5 years ago. The voluntary responses received are summarized in the box-and-whisker plots shown in Figure 3-8

(a) From Figure 3-8, estimate the median and extreme values of salaries of alumni graduating 2 years ago. In what range are the middle half of the salaries?

→ The median seems to be about \$44,000. The extremes are about \$33,000 and \$54,000. The middle half of the salaries fall between \$40,000 and \$47,000.

FIGURE 3-8

Box-and-Whisker Plots for Alumni Salaries (in thousands of dollars)

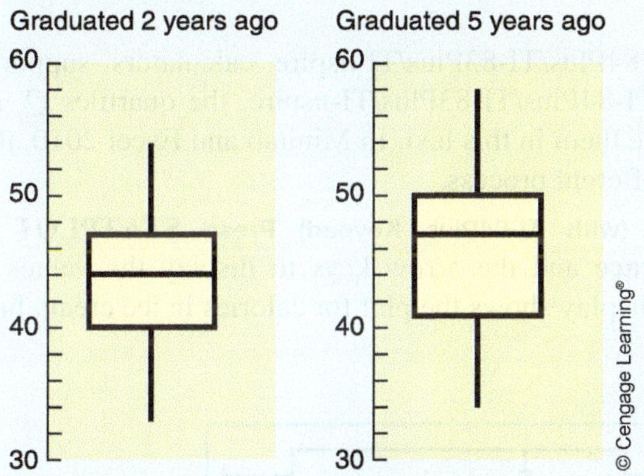

(b) From Figure 3-8, estimate the median and the extreme values of salaries of alumni graduating 5 years ago. What is the location of the middle half of the salaries?

→ The median seems to be \$47,000. The extremes are \$34,000 and \$58,000. The middle half of the data is enclosed by the box with low side at \$41,000 and high side at \$50,000.

(c) *Interpretation* Compare the two box-and-whisker plots and make comments about the salaries of alumni graduating 2 and 5 years ago.

→ The salaries of the alumni graduating 5 years ago have a larger range. They begin slightly higher than and extend to levels about \$4000 above the salaries of those graduating 2 years ago. The middle half of the data is also more spread out, with higher boundaries and a higher median.

CRITICAL THINKING

Problem 12 discusses how to identify possible outliers on a box-and-whisker plot.

Box-and-whisker plots provide a graphic display of the spread of data about the median. The box shows the location of the middle half of the data. One-quarter of the data is located along each whisker.

To the extent that the median is centered in the box and the whiskers are about the same length, the data distribution is symmetric around the median. If the median line is near one end of the box, the data are skewed toward the other end of the box.

We have developed the skeletal box-and-whisker display. Other variations include *fences,* which are marks placed on either side of the box to represent various portions of data. Values that lie beyond the fences are *outliers.* Problem 12 of this section discusses some criteria for locating fences and identifying outliers.

WHAT DOES A BOX-AND-WHISKER PLOT TELL US?

A box-and-whisker plot is a visual display of data spread around the *median*. It tells us

- the high value, low value, first quartile, median, and fourth quartile;
- how the data are spread around the median;
- the location of the middle half of the data;
- if there are outliers (*see* Problem 12 of this section).

TECH NOTES *Box-and-Whisker Plot*

Both Minitab and the TI-84Plus/TI-83Plus/TI-*n*spire calculators support box-and-whisker plots. On the TI-84Plus/TI-83Plus/TI-*n*spire, the quartiles Q_1 and Q_3 are calculated as we calculate them in this text. In Minitab and Excel 2010, they are calculated using a slightly different process.

TI-84Plus/TI-83Plus/TI-*n*spire (with TI-84Plus Keypad) Press **STATPLOT ➤On.** Highlight box plot. Use **Trace** and the arrow keys to display the values of the five-number summary. The display shows the plot for calories in ice cream bars.

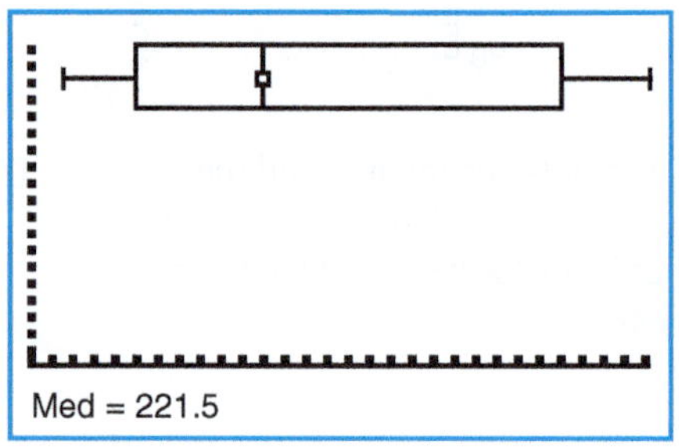

Excel 2010 Does not produce box-and-whisker plot. However, each value of the five-number summary can be found. On the **Home** ribbon, click the **Insert Function** f_x. In the dialogue box, select **Statistical** as the category and scroll to **Quartile**. In the dialogue box, enter the data location and then enter the number of the value you want. For instance, enter 1 in the quartile box for the first quartile.

Minitab Press **Graph ➤ Boxplot.** In the dialogue box, set Display to IQRange Box.

VIEWPOINT Is Shorter Higher?

Can you estimate a person's height from the pitch of his or her voice? Is a soprano shorter than an alto? Is a bass taller than a tenor? A statistical study of singers in the New York Choral Society provided information. For more information, visit the Brase/Brase statistics site at **http://www.cengagebrain.com** *and find the link to DASL, the Carnegie Mellon University Data and Story Library. From Data Subjects, select Music and then Singers. Methods of this chapter can be used with new methods we will learn in Chapters 4 and 8–11 to examine such questions from a statistical point of view.*

SECTION 3.3 PROBLEMS

1. *Statistical Literacy* Angela took a general aptitude test and scored in the 82nd percentile for aptitude in accounting. What percentage of the scores were at or below her score? What percentage were above?

2. *Statistical Literacy* One standard for admission to Redfield College is that the student rank in the upper quartile of his or her graduating high school class. What is the minimal percentile rank of a successful applicant?

3. *Critical Thinking* The town of Butler, Nebraska, decided to give a teacher-competency exam and defined the passing scores to be those in the 70th percentile or higher. The raw test scores ranged from 0 to 100. Was a raw score of 82 necessarily a passing score? Explain.

4. *Critical Thinking* Clayton and Timothy took different sections of Introduction to Economics. Each section had a different final exam. Timothy scored 83 out of 100 and had a percentile rank in his class of 72. Clayton scored 85 out of 100 but his percentile rank in his class was 70. Who performed better with respect to the rest of the students in the class, Clayton or Timothy? Explain your answer.

5. *Basic Computation: Five-Number Summary, Interquartile Range* Consider the following ordered data:

   2 5 5 6 7 7 8 9 10

   (a) Find the low, Q_1, median, Q_3, high.
   (b) Find the interquartile range.
   (c) Make a box-and-whisker plot.

6. *Basic Computation: Five-Number Summary, Interquartile Range* Consider the following ordered data:

   2 5 5 6 7 8 8 9 10 12

   (a) Find the low, Q_1, median, Q_3 high.
   (b) Find the interquartile range.
   (c) Make a box-and-whisker plot.

7. *Health Care: Nurses* At Center Hospital there is some concern about the high turnover of nurses. A survey was done to determine how long (in months) nurses had been in their current positions. The responses (in months) of 20 nurses were

23	2	5	14	25	36	27	42	12	8
7	23	29	26	28	11	20	31	8	36

   Make a box-and-whisker plot of the data. Find the interquartile range.

8. *Health Care: Staff* Another survey was done at Center Hospital to determine how long (in months) clerical staff had been in their current positions. The responses (in months) of 20 clerical staff members were

25	22	7	24	26	31	18	14	17	20
31	42	6	25	22	3	29	32	15	72

   (a) Make a box-and-whisker plot. Find the interquartile range.
   (b) Compare this plot with the one in Problem 7. Discuss the locations of the medians, the locations of the middle halves of the data banks, and the distances from Q_1 and Q_3 to the extreme values.

9. *Sociology: College Graduates* What percentage of the general U.S. population have bachelor's degrees? The *Statistical Abstract of the United States,* 120th Edition, gives the percentage of bachelor's degrees by state. For convenience, the data are sorted in increasing order.

17	18	18	18	19	20	20	20	21	21
21	21	22	22	22	22	22	22	23	23
24	24	24	24	24	24	24	24	25	26
26	26	26	26	26	27	27	27	27	27
28	28	29	31	31	32	32	34	35	38

(a) Make a box-and-whisker plot and find the interquartile range.
(b) Illinois has a bachelor's degree percentage rate of about 26%. Into what quartile does this rate fall?

10. *Sociology: High School Dropouts* What percentage of the general U.S. population are high school dropouts? The *Statistical Abstract of the United States,* 120th Edition, gives the percentage of high school dropouts by state. For convenience, the data are sorted in increasing order.

5	6	7	7	7	7	8	8	8	8
8	9	9	9	9	9	9	9	10	10
10	10	10	10	10	10	11	11	11	11
11	11	11	11	12	12	12	12	13	13
13	13	13	13	14	14	14	14	14	15

(a) Make a box-and-whisker plot and find the interquartile range.
(b) Wyoming has a dropout rate of about 7%. Into what quartile does this rate fall?

11. *Auto Insurance: Interpret Graphs* *Consumer Reports* rated automobile insurance companies and listed annual premiums for top-rated companies in several states. Figure 3-9 shows box-and-whisker plots for annual premiums for urban customers (married couple with one 17-year-old son) in three states. The box-and-whisker plots in Figure 3-9 were all drawn using the same scale on a TI-84Plus/TI-83Plus/TI-*n*spire calculator.

FIGURE 3-9

Insurance Premium (annual, urban)

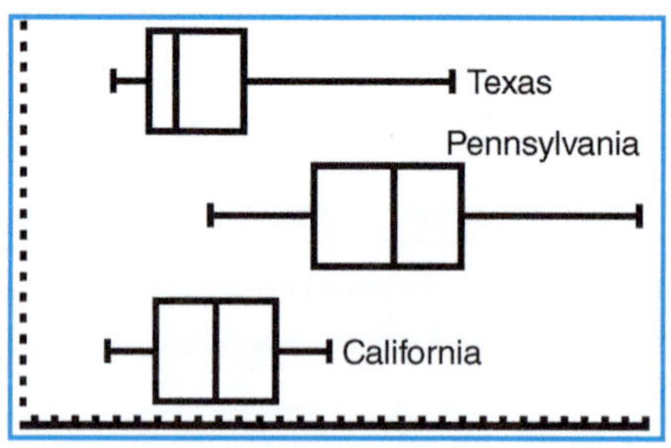

(a) Which state has the lowest premium? the highest?
(b) Which state has the highest median premium?
(c) Which state has the smallest range of premiums? the smallest interquartile range?
(d) Figure 3-10 gives the five-number summaries generated on the TI-84Plus/TI-83Plus/TI-*n*spire calculators for the box-and-whisker plots of Figure 3-9. Match the five-number summaries to the appropriate box-and-whisker plots.

FIGURE 3-10

Five-Number Summaries for Insurance Premiums

(a)

```
1-Var Stats
↑n=10
 minX=2382
 Q1=2758
 Med=2991
 Q3=3652
 maxX=5715
```


(b)

```
1-Var Stats
↑n=10
 minX=3314
 Q1=4326
 Med=5116.5
 Q3=5801
 maxX=7527
```


(c)

```
1-Var Stats
↑n=10
 minX=2323
 Q1=2801
 Med=3377.5
 Q3=3966
 maxX=4482
```


12. *Expand Your Knowledge: Outliers* Some data sets include values so high or so low that they seem to stand apart from the rest of the data. These data are called *outliers.* Outliers may represent data collection errors, data entry errors, or simply valid but unusual data values. It is important to identify outliers in the data set and examine the outliers carefully to determine if they are in error. One way to detect outliers is to use a box-and-whisker plot. Data values that fall beyond the limits,

Lower limit: $Q_1 - 1.5 \times (IQR)$

Upper limit: $Q_3 + 1.5 \times (IQR)$

where *IQR* is the interquartile range, are suspected outliers. In the computer software package Minitab, values beyond these limits are plotted with asterisks (*).

Students from a statistics class were asked to record their heights in inches. The heights (as recorded) were

65	72	68	64	60	55	73	71	52	63	61	74
69	67	74	50	4	75	67	62	66	80	64	65

(a) Make a box-and-whisker plot of the data.
(b) Find the value of the interquartile range (*IQR*).
(c) Multiply the *IQR* by 1.5 and find the lower and upper limits.
(d) Are there any data values below the lower limit? above the upper limit? List any suspected outliers. What might be some explanations for the outliers?

4

4.1 Scatter Diagrams and Linear Correlation

Note: This chapter may be delayed until after Chapter 9 Hypothesis Testing, and then covered right before Section 11.4, Inferences Relating to Linear Regression. Chapters 5–10 and Sections 11.1–11.3 do not depend on this material.

Portrait of Rene Descartes c. 1649 (after) Frans Hals/Louvre/ Giraudon/The Bridgeman Art Library

When it is not in our power to determine what is true, we ought to follow what is most probable.
—René Descartes

It is important to realize that statistics and probability do not deal in the realm of certainty. If there is any realm of human knowledge where genuine certainty exists, you may be sure that our statistical methods are not needed there. In most human endeavors, and in almost all of the natural world around us, the element of chance happenings cannot be avoided. When we cannot expect something with true certainty, we must rely on probability to be our guide. In this chapter, we will study regression, correlation, and forecasting. One of the tools we use is a scatter plot. René Decartes (1596–1650) was the first mathematician to systematically use rectangular coordinate plots. For this reason, such a coordinate axis is called a Cartesian axis.

CORRELATION AND REGRESSION

Drive Images/Alamy Limited

PREVIEW QUESTIONS

How can you use a scatter diagram to visually estimate the degree of linear correlation of two random variables? (SECTION 4.1)

How do you compute the correlation coefficient and what does it tell you about the strength of the linear relationship between two random variables? (SECTION 4.1)

What is the least-squares criterion? How do you find the equation of the least-squares line? (SECTION 4.2)

What is the coefficient of determination and what does it tell you about explained variation of y in a random sample of data pairs (x, y)? (SECTION 4.2)

© Iofoto/Shutterstock.com

FOCUS PROBLEM

Changing Populations and Crime Rate

Is the crime rate higher in neighborhoods where people might not know each other very well? Is there a relationship between crime rate and population change? If so, can we make predictions based on such a relationship? Is the relationship statistically significant? Is it possible to predict crime rates from population changes?

Denver is a city that has had a lot of growth and consequently a lot of population change in recent years. Sociologists studying population changes and crime rates could find a wealth of information in Denver statistics. Let x be a random variable representing percentage change in neighborhood population in the past few years, and let y be a random variable representing crime rate (crimes per 1000 population). A random sample of six Denver neighborhoods gave the following information (Source: *Neighborhood Facts*, The Piton Foundation). To find out more about the Piton Foundation, visit the Brase/Brase statistics site at **http://www.cengagebrain.com** and find the link to the Piton Foundation.

For online student resources, visit the Brase/Brase, *Understanding Basic Statistics*, 7th edition web site at http://www.cengagebrain.com

x	29	2	11	17	7	6
y	173	35	132	127	69	53

Using information presented in this chapter, you will be able to analyze the relationship between the variables x and y using the following tools.

- Scatter diagram
- Sample correlation coefficient and coefficient of determination
- Least-squares line equation
- Predictions for y using the least-squares line

(See Problem 10 in the Chapter Review Problems.)

SECTION 4.1 Scatter Diagrams and Linear Correlation

FOCUS POINTS

- Make a scatter diagram.
- Visually estimate the location of the "best-fitting" line for a scatter diagram.
- Use sample data to compute the sample correlation coefficient r.
- Investigate the meaning of the correlation coefficient r.

Paired data values

Studies of correlation and regression of two variables usually begin with a graph of *paired data values* (x, y). We call such a graph a *scatter diagram*.

Scatter diagram
Explanatory variable
Response variable

A **scatter diagram** is a graph in which data pairs (x, y) are plotted as individual points on a grid with horizontal axis x and vertical axis y. We call x the **explanatory variable** and y the **response variable**.

By looking at a scatter diagram of data pairs, you can observe whether there seems to be a linear relationship between the x and y values.

EXAMPLE 1

SCATTER DIAGRAM

Phosphorous is a chemical used in many household and industrial cleaning compounds. Unfortunately, phosphorous tends to find its way into surface water, where it can kill fish, plants, and other wetland creatures. Phosphorous-reduction programs are required by law and are monitored by the Environmental Protection Agency (EPA) (Reference: *EPA Case Study 832-R-93-005*).

A random sample of eight sites in a California wetlands study gave the following information about phosphorous reduction in drainage water. In this study, x is a random variable that represents phosphorous concentration (in 100 mg/L) at the inlet of a passive biotreatment facility, and y is a random variable that represents total phosphorous concentration (in 100 mg/L) at the outlet of the passive biotreatment facility.

x	5.2	7.3	6.7	5.9	6.1	8.3	5.5	7.0
y	3.3	5.9	4.8	4.5	4.0	7.1	3.6	6.1

FIGURE 4-1

Phosphorous Reduction (100 mg/L)

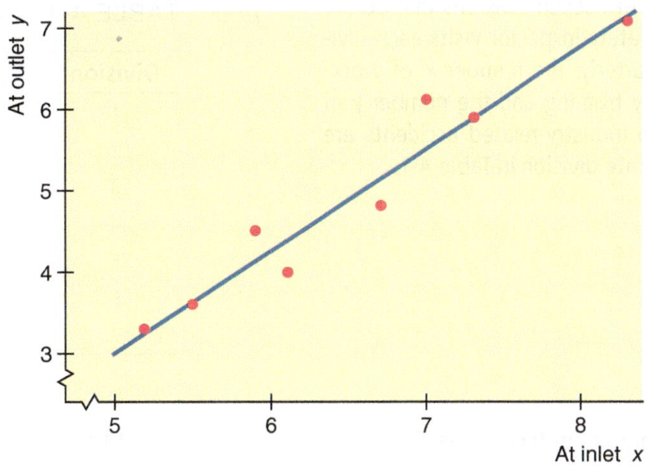

(a) Make a scatter diagram for these data.

SOLUTION: Figure 4-1 shows points corresponding to the given data pairs. These plotted points constitute the scatter diagram. To make the diagram, first scan the data and decide on an appropriate scale for each axis. Figure 4-1 shows the scatter diagram (points) along with a line segment showing the basic trend. Notice a "jump scale" on both axes.

(b) ***Interpretation*** Comment on the relationship between x and y shown in Figure 4-1.

SOLUTION: By inspecting the figure, we see that smaller values of x are associated with smaller values of y and larger values of x tend to be associated with larger values of y. Roughly speaking, the general trend seems to be reasonably well represented by an upward-sloping line segment, as shown in the diagram.

Of course, it is possible to draw many curves close to the points in Figure 4-1, but a straight line is the simplest and most widely used in elementary studies of paired data. We can draw many lines in Figure 4-1, but in some sense, the "best" line should be the one that comes closest to each of the points of the scatter diagram. To single out one line as the "best-fitting line," we must find a mathematical criterion for this line and a formula representing the line. This will be done in Section 4.2 using the *method of least squares*.

Introduction to linear correlation

Another problem precedes that of finding the "best-fitting line." That is the problem of determining how well the points of the scatter diagram are suited for fitting *any* line. Certainly, if the points are a very poor fit to any line, there is little use in trying to find the "best" line.

No linear correlation

If the points of a scatter diagram are located so that *no* line is realistically a "good" fit, we then say that the points possess *no linear correlation*. We see some examples of scatter diagrams for which there is no linear correlation in Figure 4-2.

FIGURE 4-2

Scatter Diagrams with No Linear Correlation

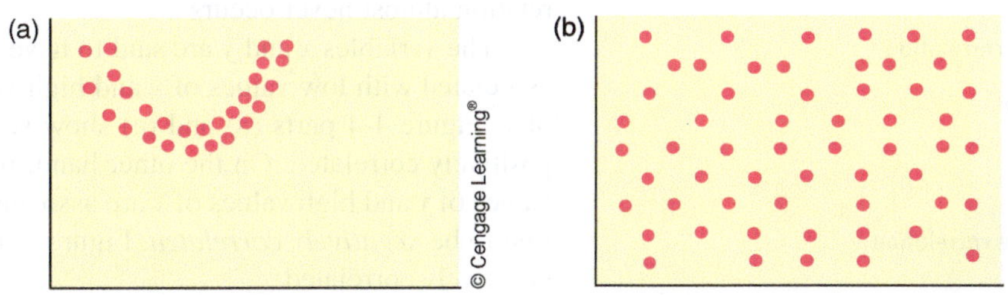

GUIDED EXERCISE 1 Scatter diagram

A large industrial plant has seven divisions that do the same type of work. A safety inspector visits each division of 20 workers quarterly. The number x of work-hours devoted to safety training and the number y of work-hours lost due to industry-related accidents are recorded for each separate division in Table 4-1.

TABLE 4-1 **Safety Report**

Division	x	y
1	10.0	80
2	19.5	65
3	30.0	68
4	45.0	55
5	50.0	35
6	65.0	10
7	80.0	12

(a) Make a scatter diagram for these pairs. Place the x values on the horizontal axis and the y values on the vertical axis.

FIGURE 4-3 Scatter Diagram for Safety Report

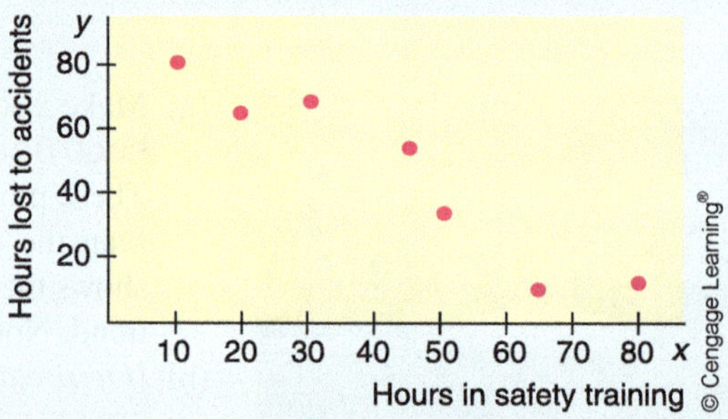

(b) As the number of hours spent on safety training increases, what happens to the number of hours lost due to industry-related accidents?

In general, as the number of hours in safety training goes up, the number of hours lost due to accidents goes down.

(c) ***Interpretation*** Does a line fit the data reasonably well?

A line fits reasonably well.

(d) Draw a line that you think "fits best."

Use a downward-sloping line that lies close to the points. Later, you will find the equation of the line that is a "best fit."

If the points seem close to a straight line, we say the linear correlation is moderate to high, depending on how close the points lie to the line. If all the points do, in fact, lie on a line, then we have *perfect linear correlation*. In Figure 4-4, we see some diagrams with perfect linear correlation. In statistical applications, perfect linear correlation almost never occurs.

Perfect linear correlation

Positive correlation

The variables x and y are said to have *positive correlation* if low values of x are associated with low values of y and high values of x are associated with high values of y. Figure 4-4 parts (a) and (c) show scatter diagrams in which the variables are positively correlated. On the other hand, if low values of x are associated with high values of y and high values of x are associated with low values of y, the variables are said to be *negatively correlated*. Figure 4-4 parts (b) and (d) show variables that are negatively correlated.

Negative correlation

FIGURE 4-4

Scatter Diagrams with Moderate and Perfect Linear Correlation

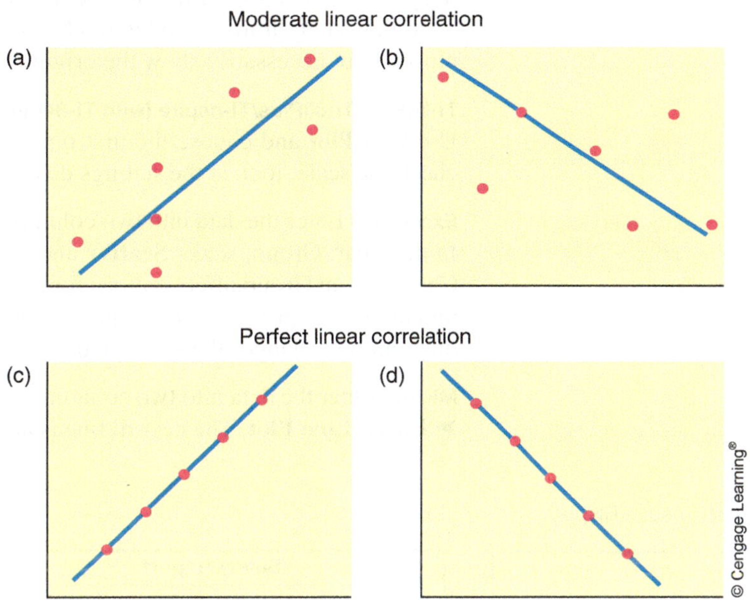

Scatter diagrams for the same data look different from one another when they are graphed using different scales. Problem 19 at the end of this section explores how changing scales affect the look of a scatter diagram.

GUIDED EXERCISE 2 | SCATTER DIAGRAM AND LINEAR CORRELATION

Examine the scatter diagrams in Figure 4-5 and then answer the following questions.

FIGURE 4-5 Scatter Diagrams

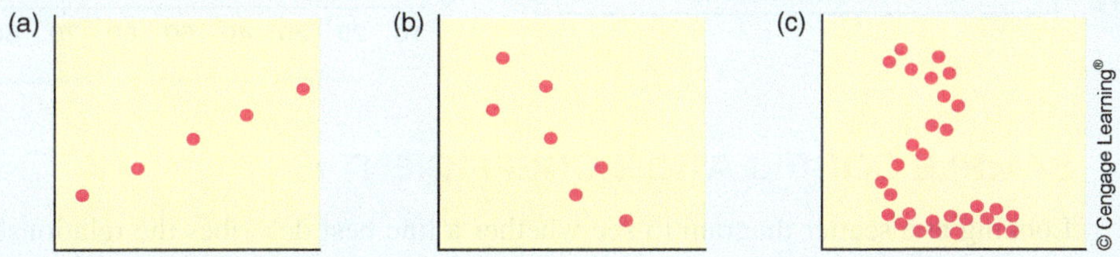

(a) Which diagram has no linear correlation? → Figure 4-5(c) has no linear correlation. No straight-line fit should be attempted.

(b) Which has perfect linear correlation? → Figure 4-5(a) has perfect linear correlation and can be fitted exactly by a straight line.

(c) Which can be reasonably fitted by a straight line? → Figure 4-5(b) can be reasonably fitted by a straight line.

WHAT DOES A SCATTER DIAGRAM TELL US?

A scatter diagram shows the relationship between two paired variables x and y. For each pair (x,y) of the data set, we plot the point on a grid with horizontal axis x and vertical axis y.

- To the extent the data points on the graph fall closer to a straight line, we can say that the relationship between x and y is more linear.
- To the extent the data pairs are scattered over the graph, we can say no linear relationship between x and y is apparent.

TECH NOTES

The TI-84Plus/TI-83Plus/TI-*n*spire calculators, Excel 2010, and Minitab all produce scatter plots. For each technology, enter the *x* values in one column and the corresponding *y* values in another column. The displays show the data from Guided Exercise 1 regarding safety training and hours lost because of accidents. Notice that the scatter plots do not necessarily show the origin.

TI-84Plus/TI-83Plus/TI-*n*spire (with TI-84Plus keypad) Enter the data into two columns. Use **Stat Plot** and choose the first type. Use option **9: ZoomStat** under **Zoom**. To check the scale, look at the settings displayed under **Window**.

Excel 2010 Enter the data into two columns. On the home screen, click the **Insert** tab. In the Chart Group, select **Scatter** and choose the first type. In the next ribbon, the Chart Layout Group offers options for including titles and axes labels. Right clicking on data points provides other options such as data labels. Changing the size of the diagram box changes the scale on the axes.

Minitab Enter the data into two columns. Use the menu selections **Stat ➤ Regression ➤ Fitted Line Plot**. The best-fit line is automatically plotted on the scatter diagram.

TI-84Plus/TI-83Plus/TI-*n*spire Display

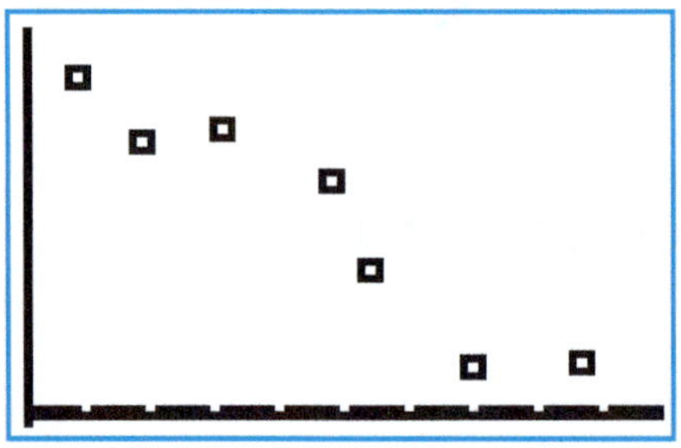

Excel Display

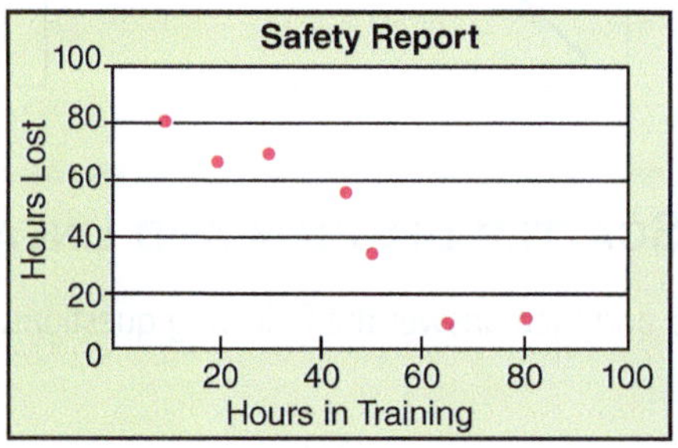

Minitab Display

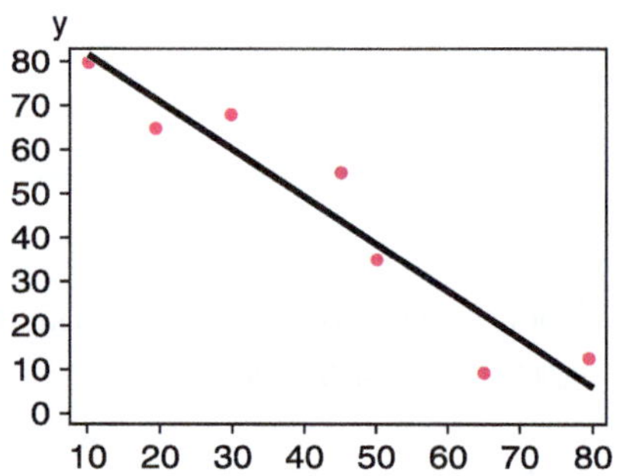

SAMPLE CORRELATION COEFFICIENT *r*

Looking at a scatter diagram to see whether a line best describes the relationship between the values of data pairs is useful. In fact, whenever you are looking for a relationship between two variables, making a scatter diagram is a good first step.

Sample correlation coefficient *r*

There is a mathematical measurement that describes the strength of the linear association between two variables. This measure is the *sample correlation coefficient r*. The full name for *r* is the *Pearson product-moment correlation coefficient*, named in honor of the English statistician Karl Pearson (1857–1936), who is credited with formulating *r*.

The **sample correlation coefficient *r*** is a numerical measurement that assesses the strength of a *linear* relationship between two variables *x* and *y*.

1. *r* is a unitless measurement between −1 and 1. In symbols, $-1 \le r \le 1$. If $r = 1$, there is perfect positive linear correlation. If $r = -1$, there is perfect negative linear correlation. If $r = 0$, there is no linear correlation. The closer *r* is to 1 or −1, the better a line describes the relationship between the two variables *x* and *y*.
2. Positive values of *r* imply that as *x* increases, *y* tends to increase. Negative values of *r* imply that as *x* increases, *y* tends to decrease.
3. The value of *r* is the same regardless of which variable is the explanatory variable and which is the response variable. In other words, the value of *r* is the same for the pairs (*x*, *y*) and the corresponding pairs (*y*, *x*).
4. The value of *r* does not change when either variable is converted to different units.

We'll develop the defining formula for r and then give a more convenient computation formula.

DEVELOPMENT OF FORMULA FOR *r*

If there is a *positive* linear relation between variables x and y, then high values of x are paired with high values of y, and low values of x are paired with low values of y. [See Figure 4-6(a).] In the case of *negative* linear correlation, high values of x are paired with low values of y, and low values of x are paired with high values of y. This relation is pictured in Figure 4-6(b). If there is *little or no linear correlation* between x and y, however, then we will find both high and low x values sometimes paired with high y values and sometimes paired with low y values. This relation is shown in Figure 4-6(c).

These observations lead us to the development of the formula for the sample correlation coefficient r. Taking *high* to mean "above the mean," we can express the relationships pictured in Figure 4-6 by considering the products

$$(x - \bar{x})(y - \bar{y})$$

If both x and y are high, both factors will be positive, and the product will be positive as well. The sign of this product will depend on the relative values of x and y compared with their respective means.

$$(x - \bar{x})(y - \bar{y}) \begin{cases} \text{is positive if } x \text{ and } y \text{ are both "high"} \\ \text{is positive if } x \text{ and } y \text{ are both "low"} \\ \\ \text{is negative if } x \text{ is "low," but } y \text{ is "high"} \\ \text{is negative if } x \text{ is "high," but } y \text{ is "low"} \end{cases}$$

In the case of positive linear correlation, most of the products $(x - \bar{x})(y - \bar{y})$ will be positive, and so will the sum over all the data pairs

$$\Sigma(x - \bar{x})(y - \bar{y})$$

For negative linear correlation, the products will tend to be negative, so the sum also will be negative. On the other hand, in the case of little, if any, linear correlation, the sum will tend to be zero.

One trouble with the preceding sum is that it increases or decreases, depending on the units of x and y. Because we want r to be unitless, we standardize both x and y of a data pair by dividing each factor $(x - \bar{x})$ by the sample standard deviation s_x and each factor $(y - \bar{y})$ by s_y. Finally, we take an average of all the products. For

FIGURE 4-6

Patterns for Linear Correlation

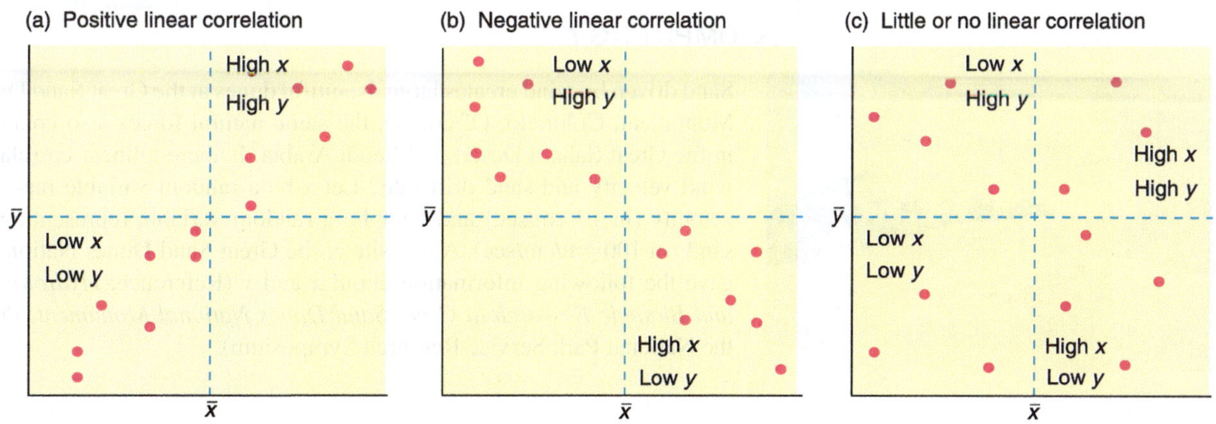

technical reasons, we take the average by dividing by $n - 1$ instead of by n. This process leads us to the desired measurement, r.

$$r = \frac{1}{n-1} \Sigma \frac{(x - \bar{x})}{s_x} \cdot \frac{(y - \bar{y})}{s_y} \tag{1}$$

COMPUTATION FORMULA FOR *r*

The defining formula for r shows how the mean and standard deviation of each variable in the data pair enter into the formulation of r. However, the defining formula is technically difficult to work with because of all the subtractions and products. A computation formula for r uses the raw data values of x and y directly.

PROCEDURE

HOW TO COMPUTE THE SAMPLE CORRELATION COEFFICIENT *r*

Requirements

Obtain a random sample of n data pairs (x, y). *Note*: For inferences (Section 11.4) regarding the correlation coefficient, the data pairs should have a *bivariate normal distribution*. This means that for a fixed value of x, the y values should have a normal distribution (or at least a mound-shaped and symmetric distribution), and for a fixed y, the x values should have their own (approximately) normal distribution.

Procedure

1. Using the data pairs, compute Σx, Σy, Σx^2, Σy^2, and Σxy.
2. With n = sample size, Σx, Σy, Σx^2, Σy^2, and Σxy, you are ready to compute the sample correlation coefficient r using the computation formula

$$r = \frac{n\Sigma xy - (\Sigma x)(\Sigma y)}{\sqrt{n\Sigma x^2 - (\Sigma x)^2}\sqrt{n\Sigma y^2 - (\Sigma y)^2}} \tag{2}$$

Be careful! The notation Σx^2 means first square x and then calculate the sum, whereas $(\Sigma x)^2$ means first sum the x values and then square the result.

When x and y values of the data pairs are exchanged, the sample correlation coefficient r remains the same. Problem 20 of this section explores this result.

Interpretation It can be shown mathematically that r is always a number between $+1$ and -1 $(-1 \le r \le +1)$. Table 4-2 gives a quick summary of some basic facts about r.

For most applications, you will use a calculator or computer software to compute r directly. However, to build some familiarity with the structure of the sample correlation coefficient, it is useful to do some calculations for yourself. Example 2 and Guided Exercise 3 show how to use the computation formula to compute r.

EXAMPLE 2

COMPUTING *r*

Sand driven by wind creates large, beautiful dunes at the Great Sand Dunes National Monument, Colorado. Of course, the same natural forces also create large dunes in the Great Sahara Desert and Saudi Arabia. Is there a linear correlation between wind velocity and sand drift rate? Let x be a random variable representing wind velocity (in 10 cm/sec) and let y be a random variable representing drift rate of sand (in 100 gm/cm/sec). A test site at the Great Sand Dunes National Monument gave the following information about x and y (Reference: *Hydrologic, Geologic, and Biologic Research at Great Sand Dunes National Monument*, Proceedings of the National Park Service Research Symposium).

TABLE 4-2 **Some Facts about the Correlation Coefficient**

If r Is	Then	The Scatter Diagram Might Look Something Like	
0	There is no linear relation among the points of the scatter diagram.	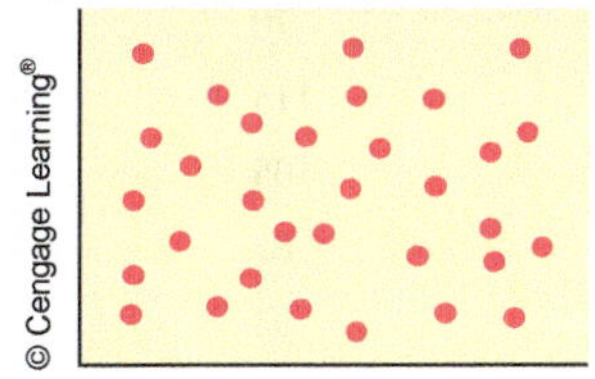	
1 or -1	There is a perfect linear relation between x and y values; all points lie on the least-squares line.	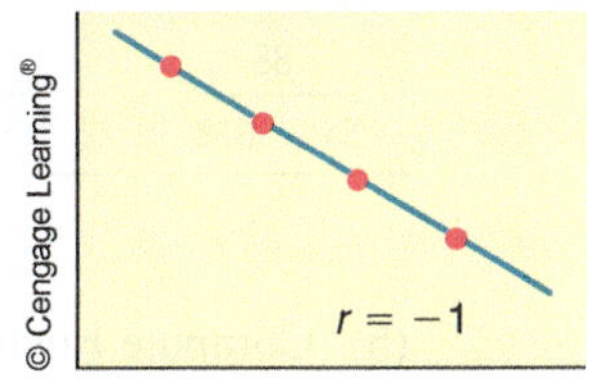	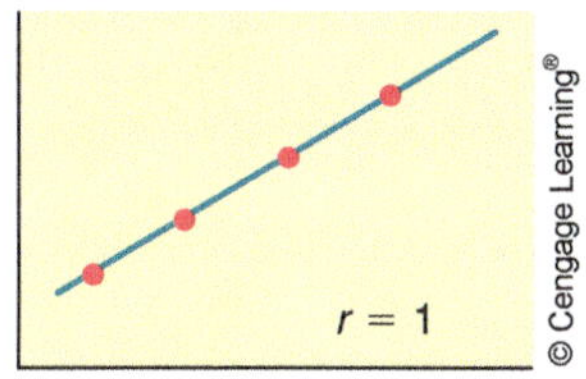
Between 0 and 1 ($0 < r < 1$)	The x and y values have a *positive correlation*. By this, we mean that *large* x values are associated with *large* y values, and *small* x values are associated with *small* y values.	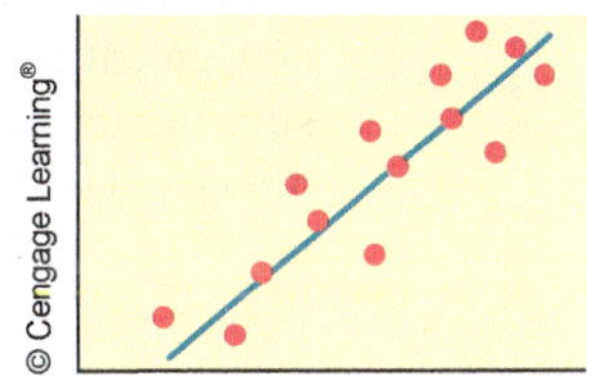	As we go from left to right, the least-squares line goes *up*.
Between -1 and 0 ($-1 < r < 0$)	The x and y values have a *negative correlation*. By this, we mean that *large* x values are associated with *small* y values, and *small* x values are associated with *large* y values.	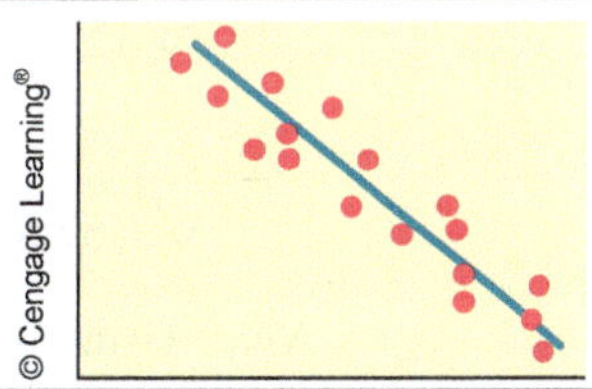	As we go from left to right, the least-squares line goes *down*.

x	70	115	105	82	93	125	88
y	3	45	21	7	16	62	12

(a) Construct a scatter diagram. Do you expect r to be positive?

SOLUTION: Figure 4-7 displays the scatter diagram. From the scatter diagram, it appears that as x values increase, y values also tend to increase. Therefore, r should be positive.

FIGURE 4-7

Wind Velocity (10 cm/sec) and Drift Rate of Sand (100 gm/cm/sec)

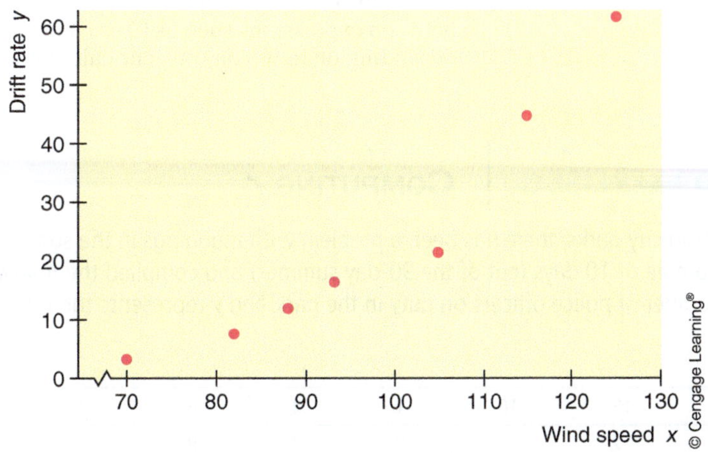

TABLE 4-3 **Computation Table**

x	y	x^2	y^2	xy
70	3	4900	9	210
115	45	13,225	2025	5175
105	21	11,025	441	2205
82	7	6724	49	574
93	16	8649	256	1488
125	62	15,625	3844	7750
88	12	7744	144	1056
$\Sigma x = 678$	$\Sigma y = 166$	$\Sigma x^2 = 67{,}892$	$\Sigma y^2 = 6768$	$\Sigma xy = 18{,}458$

(b) Compute r using the computation formula (formula 2).

SOLUTION: To find r, we need to compute Σx, Σx^2, Σy, Σy^2, and Σxy. It is convenient to organize the data in a table of five columns (Table 4-3) and then sum the entries in each column. Of course, many calculators give these sums directly. Using the computation formula for r, the sums from Table 4-3, and $n = 7$, we have

$$r = \frac{n\Sigma xy - (\Sigma x)(\Sigma y)}{\sqrt{n\Sigma x^2 - (\Sigma x)^2}\sqrt{n\Sigma y^2 - (\Sigma y)^2}} \quad (2)$$

$$= \frac{7(18{,}458) - (678)(166)}{\sqrt{7(67{,}892) - (678)^2}\sqrt{7(6768) - (166)^2}} \approx \frac{16{,}658}{(124.74)(140.78)} \approx 0.949$$

Note: Using a calculator to compute r directly gives 0.949 to three places after the decimal.

(c) ***Interpretation*** What does the value of r tell you?

SOLUTION: Since r is very close to 1, we have an indication of a strong positive linear correlation between wind velocity and drift rate of sand. In other words, we expect that higher wind speeds tend to mean greater drift rates. Because r is so close to 1, the association between the variables appears to be linear.

Because it is quite a task to compute r for even seven data pairs, the use of columns as in Example 2 is extremely helpful. Your value for r should always be between -1 and 1, inclusive. Use a scatter diagram to get a rough idea of the value of r. If your computed value of r is outside the allowable range, or if it disagrees quite a bit with the scatter diagram, recheck your calculations. Be sure you distinguish between expressions such as (Σx^2) and $(\Sigma x)^2$. Negligible rounding errors may occur, depending on how you (or your calculator) round.

GUIDED EXERCISE 3 COMPUTING r

In one of the Boston city parks, there has been a problem with muggings in the summer months. A police cadet took a random sample of 10 days (out of the 90-day summer) and compiled the following data. For each day, x represents the number of police officers on duty in the park and y represents the number of reported muggings on that day.

x	10	15	16	1	4	6	18	12	14	7
y	5	2	1	9	7	8	1	5	3	6

Continued

GUIDED EXERCISE 3 *continued*

(a) Construct a scatter diagram of x and y values.

→ Figure 4-8 shows the scatter diagram.

FIGURE 4-8 Scatter Diagram for Number of Police Officers versus Number of Muggings

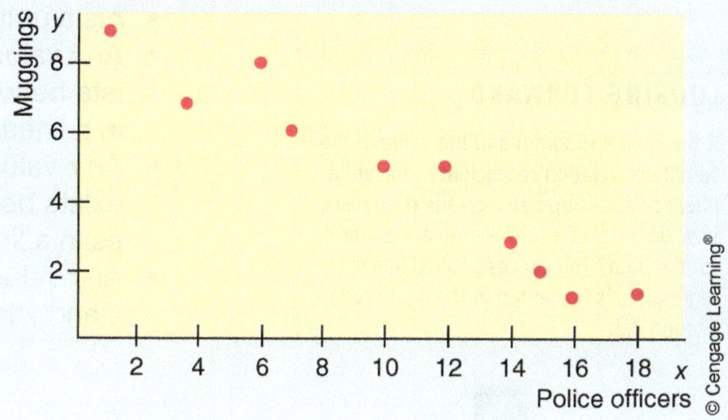

(b) From the scatter diagram, do you think the computed value of r will be positive, negative, or zero? Explain.

→ r will be negative. The general trend is that large x values are associated with small y values, and vice versa. From left to right, the least-squares line goes down.

(c) Complete Table 4-4.

TABLE 4-4

x	y	x^2	y^2	xy
10	5	100	25	50
15	2	225	4	30
16	1	256	1	16
1	9	1	81	9
4	7	16	49	28
6	8	___	___	___
18	1	___	___	___
12	5	___	___	___
14	3	___	___	___
7	6	49	36	42
$\Sigma x = 103$	$\Sigma y = 47$	$\Sigma x^2 =$ ___	$\Sigma y^2 =$ ___	$\Sigma xy =$ ___
$(\Sigma x)^2 =$ ___	$(\Sigma y)^2 =$ ___			

→ TABLE 4-5 **Completion of Table 4-4**

x	y	x^2	y^2	xy
6	8	36	64	48
18	1	324	1	18
12	5	144	25	60
14	3	196	9	42
		$\Sigma x^2 = 1347$	$\Sigma y^2 = 295$	$\Sigma xy = 343$
$(\Sigma x)^2 = 10{,}609$		$(\Sigma y)^2 = 2209$		

(d) Compute r. Alternatively, find the value of r directly by using a calculator or computer software.

→

$$r = \frac{n\Sigma xy - (\Sigma x)(\Sigma y)}{\sqrt{n\Sigma x^2 - (\Sigma x)^2}\sqrt{n\Sigma y^2 - (\Sigma y)^2}}$$

$$= \frac{10(343) - (103)(47)}{\sqrt{10(1347) - (103)^2}\sqrt{10(295) - (47)^2}}$$

$$\approx \frac{-1411}{(53.49)(27.22)} \approx -0.969$$

(e) ***Interpretation*** What does the value of r tell you about the relationship between the number of police officers and the number of muggings in the park?

→ There is a strong negative linear relationship between the number of police officers and the number of muggings. It seems that the more officers there are in the park, the fewer the number of muggings.

WHAT DOES THE CORRELATION COEFFICIENT r TELL US?

The correlation coefficient r is a sample statistic from a data set of ordered pairs (x, y). It is a measurement indicating the strength of a *linear* relationship between x and y.

- r is a unitless measurement ranging from -1 to 1.
- An r value close to 1 indicates that a positive linear relationship exists between x and y. This means that as x increases, y increases in a linear fashion.
- An r value close to -1 indicates that a negative linear relationship exists between x and y. This means that as x increases, y decreases in a linear fashion.
- An r value close to 0 indicates that the relationship (if any) between x and y is *not* linear.

LOOKING FORWARD

If the scatter diagram and the value of the sample correlation coefficient r indicate a linear relationship between the data pairs, how do we find a suitable linear equation for the data? This process, called linear regression, is presented in the next section, Section 4.2.

TECH NOTES

Most calculators that support two-variable statistics provide the value of the sample correlation coefficient r directly. Statistical software provides r, r^2, or both.

TI-84Plus/TI-83Plus/TI-*n*spire (with TI-84Plus keypad) First use **CATALOG,** find **DiagnosticOn,** and press **Enter** twice. Then, when you use **STAT, CALC,** option **8:LinReg(a + bx),** the value of r will be given (data from Example 2). In the next section, we will discuss the line $y = a + bx$ and the meaning of r^2.

Excel 2010 Excel gives the value of the sample correlation coefficient r in several outputs. One way to find the value of r is to click the **Insert** function (fx). Then in the dialogue box, select **Statistical** for the category and **Correl** for the function.

Minitab Use the menu selection **Stat ➤ Basic Statistics ➤ Correlation**.

```
LinReg
 y=a+bx
 a=-79.97763496
 b=1.070565553
 r²=.8997719968
 r=.9485631222
```


LOOKING FORWARD

When the data are ranks (without ties) instead of measurements, the Pearson product-moment correlation coefficient can be reduced to a simpler equation called the Spearman rank correlation coefficient. This coefficient is used with nonparametric methods and is discussed in *Understandable Statistics*, 11th edition, by Brase and Brase.

CRITICAL THINKING

Sample correlation compared to population correlation

CAUTIONS ABOUT CORRELATION

The correlation coefficient can be thought of as a measure of how well a linear model fits the data points on a scatter diagram. The closer r is to $+1$ or -1, the better a line "fits" the data. Values of r close to 0 indicate a poor fit to any line.

Usually a scatter diagram does not contain *all* possible data points that could be gathered. Most scatter diagrams represent only a *random sample* of data pairs taken from a very large population of all possible pairs. Because r is computed on the basis of a random sample of (x, y) pairs, we expect the values of r to vary from one sample to the next (much as the sample mean $\bar{x}$ varies from sample to sample). This brings up the question of the *significance* of r. Or, put another way, what are the chances that our random sample of data pairs indicates a high correlation when, in fact, the population's x and y values are not so strongly correlated? Right now, let's just say that the significance of r is a separate issue that will be treated in Section 11.4, where we test the *population correlation coefficient* ρ (Greek letter *rho*, pronounced "row").

Continued

CRITICAL THINKING *continued*

Problem 21 demonstrates an informal process for determining whether or not r is significant. Problem 22 explores the effect of sample size on the significance of r.

r = **sample** correlation coefficient computed from a random sample of (x, y) data pairs.

ρ = **population** correlation coefficient computed from all population data pairs (x, y).

There is a less formal way to address the significance of r using a table of "critical values" or "cut-off values" based on the r distribution and the number of data pairs. Problem 21 at the end of this section discusses this method.

The value of the sample correlation coefficient r and the strength of the linear relationship between variables is computed based on the sample data. The situation may change for measurements larger than or smaller than the data values included in the sample. For instance, for infants, there may be a high positive correlation between age in months and weight. However, that correlation might not apply for people ages 20 to 30 years.

Causation

The correlation coefficient is a mathematical tool for measuring the strength of a linear relationship between two variables. As such, it makes no implication about cause or effect. The fact that two variables tend to increase or decrease together does not mean that a change in one is *causing* a change in the other. A strong correlation between x and y is sometimes due to other (either known or unknown) variables. Such variables are called *lurking variables*.

Lurking variables

In ordered pairs (x, y), x is called the **explanatory** variable and y is called the **response** variable. When r indicates a linear correlation between x and y, changes in values of y tend to respond to changes in values of x according to a linear model. A **lurking variable** is a variable that is neither an explanatory nor a response variable. Yet a lurking variable may be responsible for changes in both x and y.

EXAMPLE 3

Causation and Lurking Variables

Over a period of years, the population of a certain town increased. It was observed that during this period the correlation between x, the number of people attending church, and y, the number of people in the city jail, was $r = 0.90$. Does going to church *cause* people to go to jail? Is there a *lurking variable* that might cause both variables x and y to increase?

SOLUTION: We hope church attendance does not cause people to go to jail! During this period, there was an increase in population. Therefore, it is not too surprising that both the number of people attending church and the number of people in jail increased. The high correlation between x and y is likely due to the lurking variable of population increase.

Correlation between averages

The correlation between two variables consisting of averages is usually higher than the correlation between two variables representing corresponding raw data. One reason is that the use of averages reduces the variation that exists between individual

Problem 23 at the end of this section explores the correlation of averages.

measurements (see Section 7.5 and the central limit theorem). A high correlation based on two variables consisting of averages does not necessarily imply a high correlation between two variables consisting of individual measurements. See Problem 23 at the end of this section.

VIEWPOINT Low on Credit, High on Cost!!!

How do you measure automobile insurance risk? One way is to use a little statistics and customer credit ratings. Insurers say statistics show that drivers who have a history of bad credit are more likely to be in serious car accidents. According to a high-level executive at Allstate Insurance Company, financial instability is an extremely powerful predictor of future insurance losses. In short, there seems to be a strong correlation between bad credit ratings and auto insurance claims. Consequently, insurance companies want to charge higher premiums to customers with bad credit ratings. Consumer advocates object strongly because they say bad credit does not cause automobile accidents, and more than 20 states prohibit or restrict the use of credit ratings to determine auto insurance premiums. Insurance companies respond by saying that your best defense is to pay your bills on time!

SECTION 4.1 PROBLEMS

Note: Answers may vary due to rounding.

1. *Statistical Literacy* When drawing a scatter diagram, along which axis is the explanatory variable placed? Along which axis is the response variable placed?

2. *Statistical Literacy* Suppose two variables are positively correlated. Does the response variable increase or decrease as the explanatory variable increases?

3. *Statistical Literacy* Suppose two variables are negatively correlated. Does the response variable increase or decrease as the explanatory variable increases?

4. *Statistical Literacy* Describe the relationship between two variables when the correlation coefficient r is
   (a) near -1.
   (b) near 0.
   (c) near 1.

5. *Critical Thinking: Linear Correlation* Look at the following diagrams. Does each diagram show high linear correlation, moderate or low linear correlation, or no linear correlation?

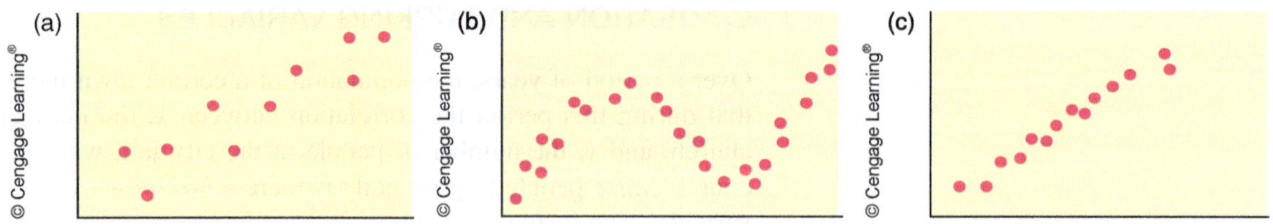

6. *Critical Thinking: Linear Correlation* Look at the following diagrams. Does each diagram show high linear correlation, moderate or low linear correlation, or no linear correlation?

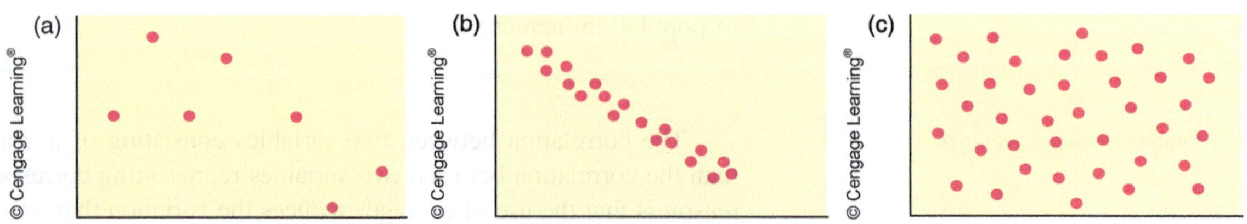

7. *Critical Thinking: Lurking Variables* Over the past few years, there has been a strong positive correlation between the annual consumption of diet soda drinks and the number of traffic accidents.
   (a) Do you think increasing consumption of diet soda drinks causes traffic accidents? Explain.
   (b) What lurking variables might be causing the increase in one or both of the variables? Explain.

8. *Critical Thinking: Lurking Variables* Over the past decade, there has been a strong positive correlation between teacher salaries and prescription drug costs.
   (a) Do you think paying teachers more causes prescription drugs to cost more? Explain.
   (b) What lurking variables might be causing the increase in one or both of the variables? Explain.

9. *Critical Thinking: Lurking Variables* Over the past 50 years, there has been a strong negative correlation between average annual income and the record time to run 1 mile. In other words, average annual incomes have been rising while the record time to run 1 mile has been decreasing.
   (a) Do you think increasing incomes cause decreasing times to run the mile? Explain.
   (b) What lurking variables might be causing the change in one or both of the variables? Explain.

10. *Critical Thinking: Lurking Variables* Over the past 30 years in the United States, there has been a strong negative correlation between the number of infant deaths at birth and the number of people over age 65.
   (a) Is the fact that people are living longer causing a decrease in infant mortalities at birth?
   (b) What lurking variables might be causing the increase in one or both of the variables? Explain.

11. *Interpretation* Trevor conducted a study and found that the correlation between the price of a gallon of gasoline and gasoline consumption has a linear correlation coefficient of -0.7. What does this result say about the relationship between price of gasoline and consumption? The study included gasoline prices ranging from \$2.70 to \$5.30 per gallon. Is it reliable to apply the results of this study to prices of gasoline higher than \$5.30 per gallon? Explain.

12. *Interpretation* Do people who spend more time on social networking sites spend more time using Twitter? Megan conducted a study and found that the correlation between the times spent on the two activities was 0.8. What does this result say about the relationship between times spent on the two activities? If someone spends more time than average on a social networking site, can you automatically conclude that he or she spends more time than average using Twitter? Explain.

13. *Veterinary Science: Shetland Ponies* How much should a healthy Shetland pony weigh? Let x be the age of the pony (in months), and let y be the average weight of the pony (in kilograms). The following information is based on data taken from *The Merck Veterinary Manual* (a reference used in most veterinary colleges).

x	3	6	12	18	24
y	60	95	140	170	185

   (a) Make a scatter diagram and draw the line you think best fits the data.
   (b) Would you say the correlation is low, moderate, or strong? positive or negative?

(c) Use a calculator to verify that $\Sigma x = 63$, $\Sigma x^2 = 1089$, $\Sigma y = 650$, $\Sigma y^2 = 95{,}350$, and $\Sigma xy = 9930$. Compute r. As x increases from 3 to 24 months, does the value of r imply that y should tend to increase or decrease? Explain.

14. *Health Insurance: Administrative Cost* The following data are based on information from *Domestic Affairs*. Let x be the average number of employees in a group health insurance plan, and let y be the average administrative cost as a percentage of claims.

x	3	7	15	35	75
y	40	35	30	25	18

(a) Make a scatter diagram and draw the line you think best fits the data.
(b) Would you say the correlation is low, moderate, or strong? positive or negative?
(c) Use a calculator to verify that $\Sigma x = 135$, $\Sigma x^2 = 7133$, $\Sigma y = 148$, $\Sigma y^2 = 4674$, and $\Sigma xy = 3040$. Compute r. As x increases from 3 to 75, does the value of r imply that y should tend to increase or decrease? Explain.

15. *Meteorology: Cyclones* Can a low barometer reading be used to predict maximum wind speed of an approaching tropical cyclone? Data for this problem are based on information taken from *Weatherwise* (Vol. 46, No. 1), a publication of the American Meteorological Society. For a random sample of tropical cyclones, let x be the lowest pressure (in millibars) as a cyclone approaches, and let y be the maximum wind speed (in miles per hour) of the cyclone.

x	1004	975	992	935	985	932
y	40	100	65	145	80	150

(a) Make a scatter diagram and draw the line you think best fits the data.
(b) Would you say the correlation is low, moderate, or strong? positive or negative?
(c) Use a calculator to verify that $\Sigma x = 5823$, $\Sigma x^2 = 5{,}655{,}779$, $\Sigma y = 580$, $\Sigma y^2 = 65{,}750$, and $\Sigma xy = 556{,}315$. Compute r. As x increases, does the value of r imply that y should tend to increase or decrease? Explain.

16. *Geology: Earthquakes* Is the magnitude of an earthquake related to the depth below the surface at which the quake occurs? Let x be the magnitude of an earthquake (on the Richter scale), and let y be the depth (in kilometers) of the quake below the surface at the epicenter. The following is based on information taken from the National Earthquake Information Service of the U.S. Geological Survey. Additional data may be found by visiting the Brase/Brase statistics site at **http://www.cengagebrain.com** and finding the link to Earthquakes.

x	2.9	4.2	3.3	4.5	2.6	3.2	3.4
y	5.0	10.0	11.2	10.0	7.9	3.9	5.5

(a) Make a scatter diagram and draw the line you think best fits the data.
(b) Would you say the correlation is low, moderate, or strong? positive or negative?
(c) Use a calculator to verify that $\Sigma x = 24.1$, $\Sigma x^2 = 85.75$, $\Sigma y = 53.5$, $\Sigma y^2 = 458.31$, and $\Sigma xy = 190.18$. Compute r. As x increases, does the value of r imply that y should tend to increase or decrease? Explain.

17. *Baseball: Batting Averages and Home Runs* In baseball, is there a linear correlation between batting average and home run percentage? Let x represent the batting average of a professional baseball player, and let y represent the player's home run percentage (number of home runs per 100 times at bat). A random

sample of $n = 7$ professional baseball players gave the following information (Reference: *The Baseball Encyclopedia*, Macmillan Publishing Company).

x	0.243	0.259	0.286	0.263	0.268	0.339	0.299
y	1.4	3.6	5.5	3.8	3.5	7.3	5.0

(a) Make a scatter diagram and draw the line you think best fits the data.
(b) Would you say the correlation is low, moderate, or high? positive or negative?
(c) Use a calculator to verify that $\Sigma x = 1.957$, $\Sigma x^2 \approx 0.553$, $\Sigma y = 30.1$, $\Sigma y^2 = 150.15$, and $\Sigma xy \approx 8.753$. Compute r. As x increases, does the value of r imply that y should tend to increase or decrease? Explain.

18. *University Crime: FBI Report* Do larger universities tend to have more property crime? University crime statistics are affected by a variety of factors. The surrounding community, accessibility given to outside visitors, and many other factors influence crime rates. Let x be a variable that represents student enrollment (in thousands) on a university campus, and let y be a variable that represents the number of burglaries in a year on the university campus. A random sample of $n = 8$ universities in California gave the following information about enrollments and annual burglary incidents (Reference: *Crime in the United States*, Federal Bureau of Investigation).

x	12.5	30.0	24.5	14.3	7.5	27.7	16.2	20.1
y	26	73	39	23	15	30	15	25

(a) Make a scatter diagram and draw the line you think best fits the data.
(b) Would you say the correlation is low, moderate, or high? positive or negative?
(c) Using a calculator, verify that $\Sigma x = 152.8$, $\Sigma x^2 = 3350.98$, $\Sigma y = 246$, $\Sigma y^2 = 10{,}030$, and $\Sigma xy = 5488.4$. Compute r. As x increases, does the value of r imply that y should tend to increase or decrease? Explain.

19. *Expand Your Knowledge: Effect of Scale on Scatter Diagram* The initial visual impact of a scatter diagram depends on the scales used on the x and y axes. Consider the following data:

x	1	2	3	4	5	6
y	1	4	6	3	6	7

(a) Make a scatter diagram using the same scale on both the x and y axes (i.e., make sure the unit lengths on the two axes are equal).
(b) Make a scatter diagram using a scale on the y axis that is twice as long as that on the x axis.
(c) Make a scatter diagram using a scale on the y axis that is half as long as that on the x axis.
(d) On each of the three graphs, draw the straight line that you think best fits the data points. How do the slopes (or directions) of the three lines appear to change? *Note:* The actual slopes will be the same; they just appear different because of the choice of scale factors.

20. *Expand Your Knowledge: Effect on r of Exchanging x and y Values* Examine the computation formula for r, the sample correlation coefficient [formulas (1) and (2) of this section].
(a) In the formula for r, if we exchange the symbols x and y, do we get a different result or do we get the same (equivalent) result? Explain.
(b) If we have a set of x and y data values and we exchange corresponding x and y values to get a new data set, should the sample correlation coefficient be the same for both sets of data? Explain.

(c) Compute the sample correlation coefficient r for each of the following data sets and show that r is the same for both.

x	1	3	4
y	2	1	6

x	2	1	6
y	1	3	4

21. *Expand Your Knowledge: Using a Table to Test* ρ The correlation coefficient r is a *sample* statistic. What does it tell us about the value of the population correlation coefficient ρ (Greek letter *rho*)? We will build the formal structure of hypothesis tests of ρ in Section 11.4. However, there is a quick way to determine if the sample evidence based on r is strong enough to conclude that there is some population correlation between the variables. In other words, we can use the value of r to determine if $\rho \neq 0$. We do this by comparing the value $|r|$ to an entry in Table 4-6. The value of α in the table gives us the probability of concluding that $\rho \neq 0$ when, in fact, $\rho = 0$ and there is no population correlation. We have two choices for α: $\alpha = 0.05$ or $\alpha = 0.01$.

PROCEDURE

HOW TO USE TABLE 4-6 TO TEST ρ

1. First compute r from a random sample of n data pairs (x, y).
2. Find the table entry in the row headed by n and the column headed by your choice of α. Your choice of α is the risk you are willing to take of mistakenly concluding that $\rho \neq 0$ when, in fact, $\rho = 0$.
3. Compare $|r|$ to the table entry.
   (a) If $|r| \geq$ table entry, then there is sufficient evidence to conclude that $\rho \neq 0$, and we say that r is **significant**. In other words, we conclude that there is some population correlation between the two variables x and y.
   (b) If $|r| <$ table entry, then the evidence is insufficient to conclude that $\rho \neq 0$, and we say that r is **not significant**. We do not have enough evidence to conclude that there is any correlation between the two variables x and y.

TABLE 4-6 **Critical Values for Correlation Coefficient r**

n	$\alpha = 0.05$	$\alpha = 0.01$	n	$\alpha = 0.05$	$\alpha = 0.01$	n	$\alpha = 0.05$	$\alpha = 0.01$
3	1.00	1.00	13	0.53	0.68	23	0.41	0.53
4	0.95	0.99	14	0.53	0.66	24	0.40	0.52
5	0.88	0.96	15	0.51	0.64	25	0.40	0.51
6	0.81	0.92	16	0.50	0.61	26	0.39	0.50
7	0.75	0.87	17	0.48	0.61	27	0.38	0.49
8	0.71	0.83	18	0.47	0.59	28	0.37	0.48
9	0.67	0.80	19	0.46	0.58	29	0.37	0.47
10	0.63	0.76	20	0.44	0.56	30	0.36	0.46
11	0.60	0.73	21	0.43	0.55			
12	0.58	0.71	22	0.42	0.54			

(a) Look at Problem 13 regarding the variables x = age of a Shetland pony and y = weight of that pony. Is the value of $|r|$ large enough to conclude that weight and age of Shetland ponies are correlated? Use $\alpha = 0.05$.

(b) Look at Problem 15 regarding the variables x = lowest barometric pressure as a cyclone approaches and y = maximum wind speed of the cyclone. Is the value of $|r|$ large enough to conclude that lowest barometric pressure and wind speed of a cyclone are correlated? Use $\alpha = 0.01$.

22. *Expand Your Knowledge: Sample Size and Significance of Correlation* In this problem, we use Table 4-6 to explore the significance of r based on different sample sizes. See Problem 21.

(a) Is a sample correlation coefficient $r = 0.820$ significant at the $\alpha = 0.01$ level based on a sample size of $n = 7$ data pairs? What about $n = 9$ data pairs?

(b) Is a sample correlation coefficient $r = 0.40$ significant at the $\alpha = 0.05$ level based on a sample size of $n = 20$ data pairs? What about $n = 27$ data pairs?

(c) Is it true that in order to be significant, an r value must be larger than 0.90? larger than 0.70? larger than 0.50? What does sample size have to do with the significance of r? Explain.

23. *Expand Your Knowledge: Correlation of Averages* Fuming because you are stuck in traffic? Roadway congestion is a costly item, in both time wasted and fuel wasted. Let x represent the *average* annual hours per person spent in traffic delays and let y represent the *average* annual gallons of fuel wasted per person in traffic delays. A random sample of eight cities showed the following data (Reference: *Statistical Abstract of the United States*, 122nd Edition).

x (hr)	28	5	20	35	20	23	18	5
y (gal)	48	3	34	55	34	38	28	9

(a) Draw a scatter diagram for the data. Verify that $\Sigma x = 154$, $\Sigma x^2 = 3712$, $\Sigma y = 249$, $\Sigma y^2 = 9959$, and $\Sigma xy = 6067$. Compute r.

The data in part (a) represent *average* annual hours lost per person and *average* annual gallons of fuel wasted per person in traffic delays. Suppose that instead of using average data for different cities, you selected one person at random from each city and measured the annual number of hours lost x for that person and the annual gallons of fuel wasted y for the same person.

x (hr)	20	4	18	42	15	25	2	35
y (gal)	60	8	12	50	21	30	4	70

(b) Compute $\bar{x}$ and $\bar{y}$ for both sets of data pairs and compare the averages. Compute the sample standard deviations s_x and s_y for both sets of data pairs and compare the standard deviations. In which set are the standard deviations for x and y larger? Look at the defining formula for r, Equation 1. Why do smaller standard deviations s_x and s_y tend to increase the value of r?

(c) Make a scatter diagram for the second set of data pairs. Verify that $\Sigma x = 161$, $\Sigma x^2 = 4583$, $\Sigma y = 255$, $\Sigma y^2 = 12{,}565$, and $\Sigma xy = 7071$. Compute r.

(d) Compare r from part (a) with r from part (c). Do the data for averages have a higher correlation coefficient than the data for individual measurements? List some reasons why you think hours lost per individual and fuel wasted per individual might vary more than the same quantities averaged over all the people in a city.

5

Nathan Griffith/Comet/Corbis

5.1 What Is Probability?

North Wind/North Wind Picture Archive

We see that the theory of probabilities is at bottom only common sense reduced to calculation; it makes us appreciate with exactitude what reasonable minds feel by a sort of instinct, often without being able to account for it.

—Pierre-Simon Laplace

This quote explains how the great mathematician Pierre-Simon Laplace (1749–1827) described the theory of mathematical probability. The discovery of the mathematical theory of probability was shared by two Frenchmen: Blaise Pascal and Pierre Fermat. These 17th-century scholars were attracted to the subject by the inquiries of the Chevalier de Méré, a gentleman gambler.

Although the first applications of probability were to games of chance and gambling, today the subject seems to pervade almost every aspect of modern life. Everything from the orbits of spacecraft to the social behavior of woodchucks is described in terms of probabilities.

ELEMENTARY PROBABILITY THEORY

PREVIEW QUESTIONS

Why would anyone study probability? *Hint:* Most big issues in life involve uncertainty. (SECTION 5.1)

What are the basic definitions and rules of probability? (SECTION 5.2)

What are counting techniques, trees, permutations, and combinations? (SECTION 5.3)

Tony Evans/Timelapse Library Ltd./Hulton Archive/Getty Images

FOCUS PROBLEM

How Often Do Lie Detectors Lie?

James Burke is an educator who is known for his interesting science-related radio and television shows aired by the British Broadcasting Corporation. His book *Chances: Risk and Odds in Everyday Life* (Virginia Books, London) contains a great wealth of fascinating information about probabilities. The following quote is from Professor Burke's book:

> *If I take a polygraph test and lie, what is the risk I will be detected?* According to some studies, there's about a 72 percent chance you will be caught by the machine.
>
> *What is the risk that if I take a polygraph test it will incorrectly say that I lied?* At least 1 in 15 will be thus falsely accused.

Both of these statements contain conditional probabilities, which we will study in Section 5.2. Information from that section will enable us to answer the following:

Suppose a person answers 10% of a long battery of questions with lies. Assume that the remaining 90% of the questions are answered truthfully.

1. Estimate the percentage of answers the polygraph will *wrongly* indicate as lies.
2. Estimate the percentage of answers the polygraph will *correctly* indicate as lies.

(See Problem 27 in Section 5.2.)

John Neubauer/PhotoEdit

For online student resources, visit the Brase/Brase, *Understanding Basic Statistics, 7th edition* web site at http://www.cengagebrain.com

SECTION **5.1**

What Is Probability?

FOCUS POINTS

- Assign probabilities to events.
- Explain how the law of large numbers relates to relative frequencies.
- Apply basic rules of probability in everyday life.
- Explain the relationship between statistics and probability.

We encounter statements given in terms of probability all the time. An excited sports announcer claims that Sheila has a 90% chance of breaking the world record in the upcoming 100-yard dash. Henry figures that if he guesses on a true–false question, the probability of getting it right is 1/2. The Right to Health lobby claims that the probability is 0.40 of getting an erroneous report from a medical laboratory at a low-cost health center. It is consequently lobbying for a federal agency to license and monitor all medical laboratories.

When we use probability in a statement, we're using a *number between 0 and 1* to indicate the likelihood of an event.

Probability is a numerical measure between 0 and 1 that describes the likelihood that an event will occur. Probabilities closer to 1 indicate that the event is more likely to occur. Probabilities closer to 0 indicate that the event is less likely to occur.

Probability of an event A, $P(A)$

$P(A)$, read "P of A," denotes the **probability of event A.**
If $P(A) = 1$, event A is certain to occur.
If $P(A) = 0$, event A is certain not to occur.

It is important to know what probability statements mean and how to determine probabilities of events, because probability is the language of inferential statistics.

PROBABILITY ASSIGNMENTS

Intuition

1. A probability assignment based on **intuition** incorporates past experience, judgment, or opinion to estimate the likelihood of an event.

Relative frequency

2. A probability assignment based on **relative frequency** uses the formula

$$\text{Probability of event} = \text{relative frequency} = \frac{f}{n} \qquad (1)$$

where f is the frequency of the event occurrence in a sample of n observations.

Equally likely outcomes

3. A probability assignment based on **equally likely outcomes** uses the formula

$$\text{Probability of event} = \frac{\text{Number of outcomes favorable to event}}{\text{Total number of outcomes}} \qquad (2)$$

EXAMPLE 1

PROBABILITY ASSIGNMENT

Consider each of the following events, and determine how the probability is assigned.

(a) A sports announcer claims that Sheila has a 90% chance of breaking the world record in the 100-yard dash.

SOLUTION: It is likely the sports announcer used intuition based on Sheila's past performance.

(b) Henry figures that if he guesses on a true–false question, the probability of getting it right is 0.50.

SOLUTION: In this case there are two possible outcomes: Henry's answer is either correct or incorrect. Since Henry is guessing, we assume the outcomes are equally likely. There are $n = 2$ equally likely outcomes, and only one is correct. By formula (2),

$$P(\text{correct answer}) = \frac{\text{Number of favorable outcomes}}{\text{Total number of outcomes}} = \frac{1}{2} = 0.50$$

(c) The Right to Health lobby claims that the probability of getting an erroneous medical laboratory report is 0.40, based on a random sample of 200 laboratory reports, of which 80 were erroneous.

SOLUTION: Formula (1) for relative frequency gives the probability, with sample size $n = 200$ and number of errors $f = 80$.

$$P(\text{error}) = \text{relative frequency} = \frac{f}{n} = \frac{80}{200} = 0.40$$

We've seen three ways to assign probabilities: intuition, relative frequency, and—when outcomes are equally likely—a formula. Which do we use? Most of the time it depends on the information that is at hand or that can be feasibly obtained. Our choice of methods also depends on the particular problem. In Guided Exercise 1, you will see three different situations, and you will decide the best way to assign the probabilities. *Remember, probabilities are numbers between 0 and 1, so don't assign probabilities outside this range.*

GUIDED EXERCISE 1 DETERMINE A PROBABILITY

Assign a probability to the indicated event on the basis of the information provided. Indicate the technique you used: intuition, relative frequency, or the formula for equally likely outcomes.

(a) A random sample of 500 students at Hudson College were surveyed and it was determined that 375 wear glasses or contact lenses. Estimate the probability that a Hudson College student selected at random wears corrective lenses.

→ In this case we are given a sample size of 500, and we are told that 375 of these students wear corrective lenses. It is appropriate to use a relative frequency for the desired probability:

$$P(\text{student needs corrective lenses}) = \frac{f}{n} = \frac{375}{500} = 0.75$$

Continued

GUIDED EXERCISE 1 *continued*

(b) The Friends of the Library hosts a fundraising barbecue. George is on the cleanup committee. There are four members on this committee, and they draw lots to see who will clean the grills. Assuming that each member is equally likely to be drawn, what is the probability that George will be assigned the grill-cleaning job?

➡ There are four people on the committee, and each is equally likely to be drawn. It is appropriate to use the formula for equally likely events. George can be drawn in only one way, so there is only one outcome favorable to the event.

$$P(\text{George}) = \frac{\text{No. of favorable outcomes}}{\text{Total no. of outcomes}} = \frac{1}{4} = 0.25$$

(c) Joanna photographs whales for Sea Life Adventure Films. On her next expedition, she is to film blue whales feeding. Based on her knowledge of the habits of blue whales, she is almost certain she will be successful. What specific number do you suppose she estimates for the probability of success?

➡ Since Joanna is almost certain of success, she should make the probability close to 1. We could say $P(\text{success})$ is above 0.90 but less than 1. This probability assignment is based on intuition.

The technique of using the relative frequency of an event as the probability of that event is a common way of assigning probabilities and will be used a great deal in later chapters. The underlying assumption we make is that if events occurred a certain percentage of times in the past, they will occur about the same percentage of times in the future. In fact, this assumption can be strengthened to a very general statement called the *law of large numbers*.

Law of large numbers

LAW OF LARGE NUMBERS

In the long run, as the sample size increases and increases, the relative frequencies of outcomes get closer and closer to the theoretical (or actual) probability value.

The law of large numbers is the reason businesses such as health insurance, automobile insurance, and gambling casinos can exist and make a profit.

No matter how we compute probabilities, it is useful to know what outcomes are possible in a given setting. For instance, if you are going to decide the probability that Hardscrabble will win the Kentucky Derby, you need to know which other horses will be running.

To determine the possible outcomes for a given setting, we need to define a *statistical experiment*.

Statistical experiment

Event

Simple event

Sample space

A **statistical experiment** or **statistical observation** can be thought of as any random activity that results in a definite outcome.

An **event** is a collection of one or more outcomes of a statistical experiment or observation.

A **simple event** is one particular outcome of a statistical experiment.

The set of all simple events constitutes the **sample space** of an experiment.

EXAMPLE 2

Using a Sample Space

Human eye color is controlled by a single pair of genes (one from the father and one from the mother) called a *genotype*. Brown eye color, B, is dominant over blue eye color, ℓ. Therefore, in the genotype Bℓ, consisting of one brown gene B and one blue gene ℓ, the brown gene dominates. A person with a Bℓ genotype has brown eyes.

If both parents have brown eyes and have genotype Bℓ, what is the probability that their child will have blue eyes? What is the probability the child will have brown eyes?

Bonnie Kamin/PhotoEdit

SOLUTION: To answer these questions, we need to look at the sample space of all possible eye-color genotypes for the child. They are given in Table 5-1.

TABLE 5-1 Eye Color Genotypes for Child

	Mother	
Father	**B**	**ℓ**
B	BB	Bℓ
ℓ	ℓB	$\ell\ell$

According to genetics theory, the four possible genotypes for the child are equally likely. Therefore, we can use Formula (2) to compute probabilities. Blue eyes can occur only with the $\ell\ell$ genotype, so there is only one outcome favorable to blue eyes. By formula (2),

$$P(\text{blue eyes}) = \frac{\text{Number of favorable outcomes}}{\text{Total number of outcomes}} = \frac{1}{4}$$

Brown eyes occur with the three remaining genotypes: BB, Bℓ, and ℓB. By formula (2),

$$P(\text{brown eyes}) = \frac{\text{Number of favorable outcomes}}{\text{Total number of outcomes}} = \frac{3}{4}$$

GUIDED EXERCISE 2

Using a Sample Space

Professor Gutierrez is making up a final exam for a course in literature of the southwest. He wants the last three questions to be of the true–false type. To guarantee that the answers do not follow his favorite pattern, he lists all possible true–false combinations for three questions on slips of paper and then picks one at random from a hat.

(a) Finish listing the outcomes in the given sample space.

TTT	FTT	TFT	____
TTF	FTF	TFF	____

➡ The missing outcomes are FFT and FFF.

(b) What is the probability that all three items will be false? Use the formula

$$P(\text{all F}) = \frac{\text{No. of favorable outcomes}}{\text{Total no. of outcomes}}$$

➡ There is only one outcome, FFF, favorable to all false, so

$$P(\text{all F}) = \frac{1}{8}$$

Continued

GUIDED EXERCISE 2 *continued*

(c) What is the probability that exactly two items will be true?

→ There are three outcomes that have exactly two true items: TTF, TFT, and FTT. Thus,

$$P(\text{two T}) = \frac{\text{No. of favorable outcomes}}{\text{Total no. of outcomes}} = \frac{3}{8}$$

There is another important point about probability assignments of simple events.

The **sum** of the probabilities of all simple events in a sample space must equal 1.

We can use this fact to determine the probability that an event will not occur. For instance, if you think the probability is 0.65 that you will win a tennis match, you assume the probability is 0.35 that your opponent will win.

The *complement* of an event A is the event that A *does not occur*. We use the notation A^c to designate the complement of event A. Figure 5-1 shows the event A and its complement A^c.

Notice that the two distinct events A and A^c make up the entire sample space. Therefore, the sum of their probabilities is 1.

Complement of an event A^c

The **complement of event** A is the event that A *does not occur*. A^c designates the complement of event A. Furthermore,

1. $P(A) + P(A^c) = 1$
2. $P(\text{event } A \text{ does } \textit{not} \text{ occur}) = P(A^c) = 1 - P(A)$ (3)

EXAMPLE 3

COMPLEMENT OF AN EVENT

The probability that a college student who has not received a flu shot will get the flu is 0.45. What is the probability that a college student will *not* get the flu if the student has not had the flu shot?

SOLUTION: In this case, we have

$P(\text{will get flu}) = 0.45$

$P(\text{will } \textit{not} \text{ get flu}) = 1 - P(\text{will get flu}) = 1 - 0.45 = 0.55$

FIGURE 5-1

The Event A and Its Complement A^c

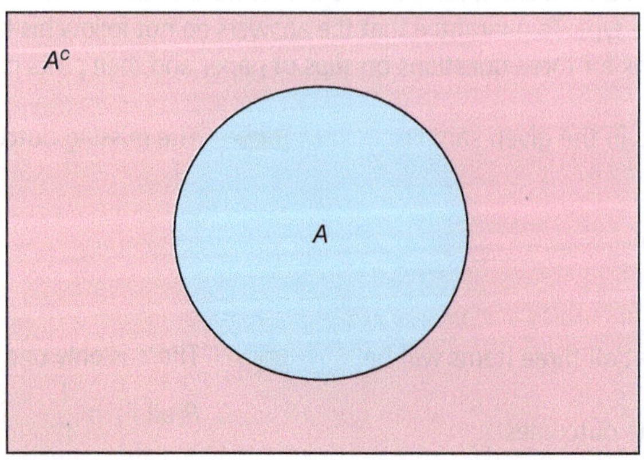

GUIDED EXERCISE 3 Complement of an Event

A veterinarian tells you that if you breed two cream-colored guinea pigs, the probability that an offspring will be pure white is 0.25. What is the probability that an offspring will not be pure white?

(a) $P(\text{pure white}) + P(\textit{not}\text{ pure white}) =$ ______ → 1

(b) $P(\textit{not}\text{ pure white}) =$ ______ → $1 - 0.25$, or 0.75

LOOKING FORWARD

The complement rule has special application in Chapter 6. There we study probabilities associated with binomial experiments. These are experiments that have only two outcomes, "success" and "failure." One feature of these experiments is that the outcomes can be categorized according to the number of "successes" out of n independent trials.

Simple Events of Sample Space of a Binomial Experiment
0 successes, 1 success, 2 successes, 3 successes, . . . , n successes

A convenient way to compute the probability of "at least one" success is to observe that the complement of "at least one success" is the event "0 successes." Then, by the complement rule

$$P(\text{at least one success}) = 1 - P(0\text{ successes})$$

Problems 10, 11, and 12 of this section show a similar use of the complement rule.

Sometimes, instead of being told the probability of an event, we are given the odds for the event. Problems 21 and 22 of this section discuss odds in favor and odds against an event.

SUMMARY: SOME IMPORTANT FACTS ABOUT PROBABILITY

1. A **statistical experiment** or **statistical observation** is any random activity that results in a definite outcome. A **simple event** consists of one and only one outcome of the experiment. The **sample space** is the set of all simple events. An **event *A*** is any subset of the sample space.
2. The probability of an event *A* is denoted by $\boldsymbol{P(A)}$.
3. The probability of an event is a number between 0 and 1. The closer to 1 the probability is, the more likely it is the event will occur. The closer to 0 the probability is, the less likely it is the event will occur.
4. The sum of the probabilities of all simple events in a sample space is 1.
5. Probabilities can be assigned by using intuition, relative frequencies, or the formula for equally likely outcomes. Additional ways to assign probabilities will be introduced in later chapters.
6. The **complement** of an event *A* is denoted by $\boldsymbol{A^c}$. So, A^c is the event that *A* does not occur.
7. $P(A) + P(A^c) = 1$

INTERPRETING PROBABILITIES

As stressed in item 3 in the summary about probabilities, the closer the probability is to 1, the more likely the event is to occur. The closer the probability is to 0, the less likely the event is to occur. However, just because the probability of an event is very high, it is not a certainty that the event will occur. Likewise, even though the probability of an event is very low, the event might still occur.

Events with low probability but big consequences are of special concern. Such events are sometimes referred to as "black swan events," as discussed in the very popular book *The Black Swan* by Nassim Nicholas Taleb. In his varied careers Taleb has been a stock trader, author, and professor at the University of Massachusetts. His special interests are mathematical problems of luck, uncertainty, probability, and general knowledge. In his book *The Black Swan*, Taleb discusses "the impact of the highly improbable." If an event has (a) little impact, bearing, or meaning on a person's life and (b) very low probability, then (c) it seems reasonable to ignore the event.

However, some of the really big mistakes in a person's life can result from misjudging either (a) the size of an event's impact or (b) the likelihood the event will occur. An event of great importance cannot be ignored even if it has a low probability of occurrence.

So what should a person do?

Taleb describes himself as a hyperskeptic when others are gullible, and gullible when others are skeptical. In particular, he recommends being skeptical about (mentally) confirming an event when the errors of confirmation are costly. Lots and lots of data should not be thought of as confirming an event when a simple instance can disconfirm an important and costly event. Taleb recommends that one be skeptical in the face of wild randomness (and big consequences) and gullible when the randomness (and consequences) are mild. One way to estimate the "randomness" of a situation is to estimate its probability. Estimating the "consequences" of an event in your own life must be a personal decision.

WHAT DOES THE PROBABILITY OF AN EVENT TELL US?

- The probability of an event *A* tells us the likelihood that event *A* will occur. If the probability is 1, the event *A* is certain to occur. If the probability is 0, the event *A* will not occur. Probabilities closer to 1 indicate the event *A* is more likely to occur, but it might not. Probabilities closer to 0 indicate event *A* is less likely to occur, but it might.
- The probability of event *A* applies only in the context of conditions surrounding the sample space containing event *A*. For example, consider the event *A* that a freshman student graduates with a bachelor's degree in 4 years or less. Do events in the sample space apply to entering freshmen in all colleges and universities in the United States or just a particular college? Are majors specified? Are minimal SAT or ACT scores specified? All these conditions of events in the sample space can affect the probability of the event. It would be inappropriate to apply the probability of event *A* to students outside those described in the sample space.
- If we know the probability of event *A*, then we can easily compute the probability of event *not A* in the context of the same sample space. $P(\text{not } A) = 1 - P(A)$.

PROBABILITY RELATED TO STATISTICS

We conclude this section with a few comments on the nature of statistics versus probability. Although statistics and probability are closely related fields of mathematics, they are nevertheless separate fields. It can be said that probability is the medium through which statistical work is done. In fact, if it were not for probability theory, inferential statistics would not be possible.

Put very briefly, probability is the field of study that makes statements about what will occur when samples are drawn from a *known population*. Statistics is the field of study that describes how samples are to be obtained and how inferences are to be made about *unknown populations*.

A simple but effective illustration of the difference between these two subjects can be made by considering how we treat the following examples.

EXAMPLE OF A PROBABILITY APPLICATION

Condition: We *know* the exact makeup of the *entire* population.

Example: Given 3 green marbles, 5 red marbles, and 4 white marbles in a bag, draw 6 marbles at random from the bag. What is the probability that none of the marbles is red?

EXAMPLE OF A STATISTICAL APPLICATION

Condition: We have only *samples* from an otherwise *unknown* population.

Example: Draw a random sample of 6 marbles from the (unknown) population of all marbles in a bag and observe the colors. Based on the sample results, make a conjecture about the colors and numbers of marbles in the entire population of all marbles in the bag.

In another sense, probability and statistics are like flip sides of the same coin. On the probability side, you know the overall description of the population. The central problem is to compute the likelihood that a specific outcome will happen. On the statistics side, you know only the results of a sample drawn from the population. The central problem is to describe the sample (descriptive statistic) and to draw conclusions about the population based on the sample results (inferential statistics).

In statistical work, the inferences we draw about an unknown population are not claimed to be absolutely correct. Since the population remains unknown (in a theoretical sense), we must accept a "best guess" for our conclusions and act using the most probable answer rather than absolute certainty.

Probability is the topic of this chapter. However, we will not study probability just for its own sake. Probability is a wonderful field of mathematics, but we will study mainly the ideas from probability that are needed for a proper understanding of statistics.

VIEWPOINT | What Makes a Good Teacher?

A survey of 735 students at nine colleges in the United States was taken to determine instructor behaviors that help students succeed. Data from this survey can be found by visiting the Brase/Brase statistics site at **http://www.cengagebrain.com** *and finding the link to DASL, the Carnegie Mellon University Data and Story Library. Once at the DASL site, select Data Subjects, then Psychology, and then Instructor Behavior. You can estimate the probability of how a student would respond (very positive, neutral, very negative) to different instructor behaviors. For example, more than 90% of the students responded "very positive" to the instructor's use of real-world examples in the classroom.*

SECTION 5.1 PROBLEMS

1. *Statistical Literacy* List three methods of assigning probabilities.

2. *Statistical Literacy* Suppose the newspaper states that the probability of rain today is 30%. What is the complement of the event "rain today"? What is the probability of the complement?

3. *Statistical Literacy* What is the probability of
   (a) an event A that is certain to occur?
   (b) an event B that is impossible?

4. *Statistical Literacy* What is the law of large numbers? If you were using the relative frequency of an event to estimate the probability of the event, would it be better to use 100 trials or 500 trials? Explain.

5. *Interpretation* A Harris Poll indicated that of those adults who drive and have a cell phone, the probability that a driver between the ages of 18 and 24 sends or reads text messages is 0.51. Can this probability be applied to *all* drivers with cell phones? Explain.

6. *Interpretation* According to a recent Harris Poll of adults with pets, the probability that the pet owner cooks especially for the pet either frequently or occasionally is 0.24.
   (a) From this information, can we conclude that the probability a male owner cooks for the pet is the same as for a female owner? Explain.
   (b) According to the poll, the probability a male owner cooks for his pet is 0.27, whereas the probability a female owner does so is 0.22. Let's explore how such probabilities might occur. Suppose the pool of pet owners surveyed consisted of 200 pet owners, 100 of whom are male and 100 of whom are female. Of the pet owners, a total of 49 cook for their pets. Of the 49 who cook for their pets, 27 are male and 22 are female. Use relative frequencies to determine the probability a pet owner cooks for a pet, the probability a male owner cooks for his pet, and the probability a female owner cooks for her pet.

7. *Basic Computation: Probability as Relative Frequency* A recent Harris Poll survey of 1010 U.S. adults selected at random showed that 627 consider the occupation of firefighter to have very great prestige. Estimate the probability (to the nearest hundredth) that a U.S. adult selected at random thinks the occupation of firefighter has very great prestige.

8. *Basic Computation: Probability of Equally Likely Events* What is the probability that a day of the week selected at random will be a Wednesday?

9. *Interpretation* An investment opportunity boasts that the chance of doubling your money in 3 years is 95%. However, when you research the details of the investment, you estimate that there is a 3% chance that you could lose the entire investment. Based on this information, are you certain to make money on this investment? Are there risks in this investment opportunity?

10. *Interpretation* A sample space consists of 4 simple events: A, B, C, D. Which events comprise the complement of A? Can the sample space be viewed as having two events, A and A^c? Explain.

11. *Critical Thinking* Consider a family with 3 children. Assume the probability that one child is a boy is 0.5 and the probability that one child is a girl is also 0.5, and that the events "boy" and "girl" are independent.
    (a) List the equally likely events for the gender of the 3 children, from oldest to youngest.
    (b) What is the probability that all 3 children are male? Notice that the complement of the event "all three children are male" is "at least one of the children is female." Use this information to compute the probability that at least one child is female.

12. *Critical Thinking* Consider the experiment of tossing a fair coin 3 times. For each coin, the possible outcomes are heads or tails.
    (a) List the equally likely events of the sample space for the three tosses.
    (b) What is the probability that all three coins come up heads? Notice that the complement of the event "3 heads" is "at least one tail." Use this information to compute the probability that there will be at least one tail.

13. *Critical Thinking* On a single toss of a fair coin, the probability of heads is 0.5 and the probability of tails is 0.5. If you toss a coin twice and get heads on the first toss, are you guaranteed to get tails on the second toss? Explain.

14. *Critical Thinking*
    (a) Explain why −0.41 cannot be the probability of some event.
    (b) Explain why 1.21 cannot be the probability of some event.
    (c) Explain why 120% cannot be the probability of some event.
    (d) Can the number 0.56 be the probability of an event? Explain.

15. *Probability Estimate: Wiggle Your Ears* Can you wiggle your ears? Use the students in your statistics class (or a group of friends) to estimate the percentage of people who can wiggle their ears. How can your result be thought of as an estimate for the probability that a person chosen at random can wiggle his or her ears? *Comment*: National statistics indicate that about 13% of Americans can wiggle their ears (Source: Bernice Kanner, *Are You Normal?*, St. Martin's Press, New York).

16. *Probability Estimate: Raise One Eyebrow* Can you raise one eyebrow at a time? Use the students in your statistics class (or a group of friends) to estimate the percentage of people who can raise one eyebrow at a time. How can your result be thought of as an estimate for the probability that a person chosen at random can raise one eyebrow at a time? *Comment:* National statistics indicate that about 30% of Americans can raise one eyebrow at a time (see source in Problem 15).

17. *Myers–Briggs: Personality Types* Isabel Briggs Myers was a pioneer in the study of personality types. The personality types are broadly defined according to four main preferences. Do married couples choose similar or different personality types in their mates? The following data give an indication (Source: I. B. Myers and M. H. McCaulley, *A Guide to the Development and Use of the Myers–Briggs Type Indicators*).

Similarities and Differences in a Random Sample of 375 Married Couples

Number of Similar Preferences	Number of Married Couples
All four	34
Three	131
Two	124
One	71
None	15

Suppose that a married couple is selected at random.
    (a) Use the data to estimate the probability that they will have 0, 1, 2, 3, or 4 personality preferences in common.
    (b) Do the probabilities add up to 1? Why should they? What is the sample space in this problem?

18. *General: Roll a Die*
    (a) If you roll a single die and count the number of dots on top, what is the sample space of all possible outcomes? Are the outcomes equally likely?
    (b) Assign probabilities to the outcomes of the sample space of part (a). Do the probabilities add up to 1? Should they add up to 1? Explain.
    (c) What is the probability of getting a number less than 5 on a single throw?
    (d) What is the probability of getting 5 or 6 on a single throw?

19. *Psychology: Creativity* When do creative people get their *best* ideas? *USA Today* did a survey of 966 inventors (who hold U.S. patents) and obtained the following information:

Time of Day When Best Ideas Occur

Time	Number of Inventors
6 A.M.–12 noon	290
12 noon–6 P.M.	135
6 P.M.–12 midnight	319
12 midnight–6 A.M.	222

(a) Assuming that the time interval includes the left limit and all the times up to but not including the right limit, estimate the probability that an inventor has a best idea during each time interval: from 6 A.M. to 12 noon, from 12 noon to 6 P.M., from 6 P.M. to 12 midnight, from 12 midnight to 6 A.M.
(b) Do the probabilities of part (a) add up to 1? Why should they? What is the sample space in this problem?

20. *Agriculture: Cotton* A botanist has developed a new hybrid cotton plant that can withstand insects better than other cotton plants. However, there is some concern about the germination of seeds from the new plant. To estimate the probability that a seed from the new plant will germinate, a random sample of 3000 seeds was planted in warm, moist soil. Of these seeds, 2430 germinated.
(a) Use relative frequencies to estimate the probability that a seed will germinate. What is your estimate?
(b) Use relative frequencies to estimate the probability that a seed will *not* germinate. What is your estimate?
(c) Either a seed germinates or it does not. What is the sample space in this problem? Do the probabilities assigned to the sample space add up to 1? Should they add up to 1? Explain.
(d) Are the outcomes in the sample space of part (c) equally likely?

21. *Expand Your Knowledge: Odds in Favor* Sometimes probability statements are expressed in terms of odds.

The *odds in favor* of an event A are the ratio $\frac{P(A)}{P(not\ A)} = \frac{P(A)}{P(A^c)}$.

For instance, if $P(A) = 0.60$, then $P(A^c) = 0.40$ and the odds in favor of A are

$$\frac{0.60}{0.40} = \frac{6}{4} = \frac{3}{2}, \text{written as 3 to 2 or 3:2}$$

(a) Show that if we are given the *odds in favor* of event A as $n{:}m$, the probability of event A is given by $P(A) = \frac{n}{n+m}$. *Hint:* Solve the equation $\frac{n}{m} = \frac{P(A)}{1 - P(A)}$ for $P(A)$.
(b) A telemarketing supervisor tells a new worker that the odds of making a sale on a single call are 2 to 15. What is the probability of a successful call?
(c) A sports announcer says that the odds a basketball player will make a free throw shot are 3 to 5. What is the probability the player will make the shot?

22. *Expand Your Knowledge: Odds Against* Betting odds are usually stated against the event happening (against winning).

The *odds against* event W are the ratio $\frac{P(not\ W)}{P(W)} = \frac{P(W^c)}{P(W)}$.

Cynthia HUNTER/AFP PHOTO/Getty Images

In horse racing, the betting odds are based on the probability that the horse does *not* win.

(a) Show that if we are given the *odds against* an event W as $a{:}b$, the probability of *not* W is $P(W^c) = \frac{a}{a+b}$. *Hint:* Solve the equation $\frac{a}{b} = \frac{P(W^c)}{1 - P(W^c)}$ for $P(W^c)$.

(b) In a recent Kentucky Derby, the betting odds for the favorite horse, Point Given, were 9 to 5. Use these odds to compute the probability that Point Given would lose the race. What is the probability that Point Given would win the race?

(c) In the same race, the betting odds for the horse Monarchos were 6 to 1. Use these odds to estimate the probability that Monarchos would lose the race. What is the probability that Monarchos would win the race?

(d) Invisible Ink was a long shot, with betting odds of 30 to 1. Use these odds to estimate the probability that Invisible Ink would lose the race. What is the probability the horse would win the race? For further information on the Kentucky Derby, visit the Brase/Brase statistics site at **http://www.cengagebrain.com** and find the link to the Kentucky Derby.

23. *Business: Customers* John runs a computer software store. Yesterday he counted 127 people who walked by his store, 58 of whom came into the store. Of the 58, only 25 bought something in the store.

(a) Estimate the probability that a person who walks by the store will enter the store.

(b) Estimate the probability that a person who walks into the store will buy something.

(c) Estimate the probability that a person who walks by the store will come in *and* buy something.

(d) Estimate the probability that a person who comes into the store will buy nothing.

7

One cannot escape the feeling that these mathematical formulas have an independent existence and an intelligence of their own, that they are wiser than we are, wiser even than their discoverers, that we get more out of them than was originally put into them.

—Heinrich Hertz

How can it be that mathematics, a product of human thought independent of experience, is so admirably adapted to the objects of reality?

—Albert Einstein

Heinrich Hertz (1857–1894) was a pioneer in the study of radio waves. His work and the later work of Maxwell and Marconi led the way to modern radio, television, and radar. Albert Einstein is world renowned for his great discoveries in relativity and quantum mechanics. Everyone who has worked in both mathematics and real-world applications cannot help but marvel at how the "pure thought" of the mathematical sciences can predict and explain events in other realms. In this chapter, we will study the most important type of probability distribution in all of mathematical statistics: the normal distribution. Why is the normal distribution so important? Two of the reasons are that it applies to a wide variety of situations and that other distributions tend to become normal under certain conditions.

NORMAL CURVES AND SAMPLING DISTRIBUTIONS

PREVIEW QUESTIONS

What are some characteristics of a normal distribution? What does the empirical rule tell you about data spread around the mean? (SECTION 7.1)

Can you compare apples and oranges, or maybe elephants and butterflies? In most cases, the answer is no—unless you first standardize your measurements. What are a standard normal distribution and a standard *z* score? (SECTION 7.2)

How do you convert any normal distribution to a standard normal distribution? How do you find probabilities of "standardized events"? (SECTION 7.3)

As humans, our experiences are finite and limited. Consequently, most of the important decisions in our lives are based on sample (incomplete) information. What is a probability sampling distribution? How will sampling distributions help us make good decisions based on incomplete information? (SECTION 7.4)

There is an old saying: All roads lead to Rome. In statistics, we could recast this saying: All probability distributions average out to be normal distributions (as the sample size increases). How can we take advantage of this in our study of sampling distributions? (SECTION 7.5)

The binomial and normal distributions are two of the most important probability distributions in statistics. Under certain limiting conditions, the binomial can be thought to evolve (or envelope) into the normal distribution. How can you apply this concept in the real world? (SECTION 7.6)

Many issues in life come down to success or failure. In most cases, we will not be successful all the time, so proportions of successes are very important. What is the probability sampling distribution for proportions? (SECTION 7.6)

Dana White/PhotoEdit

FOCUS PROBLEM

Impulse Buying

©Antonina Sotnykova/Shutterstock.com

The Food Marketing Institute, Progressive Grocer, New Products News, and Point of Purchaser Advertising Institute are organizations that analyze supermarket sales. One of the interesting discoveries was that the average amount of impulse buying in a grocery store is very time-dependent. As reported in *The Denver Post,* "When you dilly dally in a store for 10 unplanned minutes, you can kiss nearly $20 good-bye." For this reason, it is in the best interest of the supermarket to keep you in the store longer. In the *Post* article, it was pointed out that long checkout lines (near end-aisle displays), "samplefest" events of free tasting, video kiosks, magazine and book sections, and so on, help keep customers in the store longer. On average, a single customer who strays from his or her

For online student resources, visit the Brase/Brase, *Understanding Basic Statistics*, 7th edition web site at http://www.cengagebrain.com

grocery list can plan on impulse spending of \$20 for every 10 minutes spent wandering about in the supermarket.

Let x represent the dollar amount spent on supermarket impulse buying in a 10-minute (unplanned) shopping interval. Based on the *Post* article, the mean of the x distribution is about \$20 and the (estimated) standard deviation is about \$7.

(a) Consider a random sample of $n = 100$ customers, each of whom has 10 minutes of unplanned shopping time in a supermarket. From the central limit theorem, what can you say about the probability distribution of $\bar{x}$, the *average* amount spent by these customers due to impulse buying? Is the $\bar{x}$ distribution approximately normal? What are the mean and standard deviation of the $\bar{x}$ distribution? Is it necessary to make any assumption about the x distribution? Explain.
(b) What is the probability that $\bar{x}$ is between \$18 and \$22?
(c) Let us assume that x has a distribution that is approximately normal. What is the probability that x is between \$18 and \$22?
(d) In part (b), we used $\bar{x}$, the *average* amount spent, computed for 100 customers. In part (c), we used x, the amount spent by only *one* individual customer. The answers to parts (b) and (c) are very different. Why would this happen? In this example, $\bar{x}$ is a much more predictable or reliable statistic than x. Consider that almost all marketing strategies and sales pitches are designed for the *average* customer and *not* the *individual* customer. How does the central limit theorem tell us that the average customer is much more predictable than the individual customer? (See Problem 18 of Section 7.5.)

SECTION **7.1**

Graphs of Normal Probability Distributions

FOCUS POINTS

- **Graph a normal curve and summarize its important properties.**
- **Apply the empirical rule to solve real-world problems.**

Normal distribution

One of the most important examples of a continuous probability distribution is the *normal distribution.* This distribution was studied by the French mathematician Abraham de Moivre (1667–1754) and later by the German mathematician Carl Friedrich Gauss (1777–1855), whose work is so important that the normal distribution is sometimes called *Gaussian.* The work of these mathematicians provided a foundation on which much of the theory of statistical inference is based.

Applications of a normal probability distribution are so numerous that some mathematicians refer to it as "a veritable Boy Scout knife of statistics." However, before we can apply it, we must examine some of the properties of a normal distribution.

A rather complicated formula, presented later in this section, defines a normal distribution in terms of μ and σ, the mean and standard deviation of the population distribution. It is only through this formula that we can verify if a distribution is normal. However, we can look at the graph of a normal distribution and get a good pictorial idea of some of the essential features of any normal distribution.

Normal curve

The graph of a normal distribution is called a *normal curve.* It possesses a shape very much like the cross section of a pile of dry sand. Because of its shape, blacksmiths would sometimes use a pile of dry sand in the construction of a mold for a bell. Thus the normal curve is also called a *bell-shaped curve* (see Figure 7-1).

We see that a general normal curve is smooth and symmetrical about the vertical line extending upward from the mean μ. Notice that the highest point of the curve occurs over μ. If the distribution were graphed on a piece of sheet metal, cut out, and placed on a knife edge, the balance point would be at μ. We also see that the curve tends to level out and approach the horizontal (x axis) like a glider

FIGURE 7-1

A Normal Curve

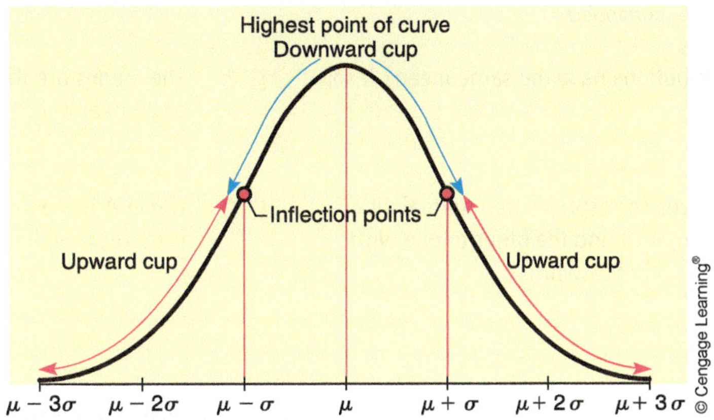

Another continuous probability distribution is discussed in the section problems. Problems 12 and 13 discuss the *uniform probability distribution*.

making a landing. However, in mathematical theory, such a glider would never quite finish its landing because a normal curve never touches the horizontal axis.

The parameter σ controls the spread of the curve. The curve is quite close to the horizontal axis at $\mu + 3\sigma$ and $\mu - 3\sigma$. Thus, if the standard deviation σ is large, the curve will be more spread out; if it is small, the curve will be more peaked. Figure 7-1 shows the normal curve cupped downward for an interval on either side of the mean μ. Then it begins to cup upward as we go to the lower part of the bell. The exact places where the *transition* between the upward and downward cupping occur are above the points $\mu + \sigma$ and $\mu - \sigma$. In the terminology of calculus, transition points such as these are called *inflection points.*

Downward cup

Upward cup

IMPORTANT PROPERTIES OF A NORMAL CURVE

Symmetry of normal curves

1. The curve is bell-shaped, with the highest point over the mean μ.
2. The curve is symmetrical about a vertical line through μ.
3. The curve approaches the horizontal axis but never touches or crosses it.
4. The inflection (transition) points between cupping upward and downward occur above $\mu + \sigma$ and $\mu - \sigma$.
5. The area under the entire curve is 1.

The parameters that control the shape of a normal curve are the mean μ and the standard deviation σ. When both μ and σ are specified, a specific normal curve is determined. In brief, μ locates the balance point and σ determines the extent of the spread.

GUIDED EXERCISE 1 — IDENTIFY μ AND σ ON A NORMAL CURVE

Look at the normal curves in Figure 7-2.

FIGURE 7-2

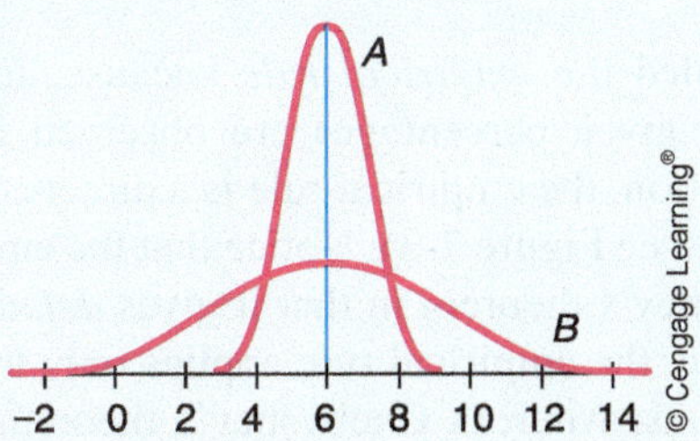

Continued

GUIDED EXERCISE 1 *continued*

(a) Do these distributions have the same mean? If so, what is it?	The means are the same, since both graphs have the high point over 6. $\mu = 6$.
(b) One of the curves corresponds to a normal distribution with $\sigma = 3$ and the other to one with $\sigma = 1$. Which curve has which σ?	Curve *A* has $\sigma = 1$ and curve *B* has $\sigma = 3$. (Since curve *B* is more spread out, it has the larger σ value.)

COMMENT The normal distribution curve is always above the horizontal axis. The area beneath the curve and above the axis is exactly 1. As such, the normal distribution curve is an example of a *density curve.* The formula used to generate the shape of the normal distribution curve is called the *normal density function.* If *x* is a normal random variable with mean μ and standard deviation σ, the formula for the normal density function is

$$f(x) = \frac{e^{(-1/2)((x-\mu)/\sigma)^2}}{\sigma\sqrt{2\pi}}$$

In this text, we will not use this formula explicitly. However, we will use tables of areas based on the normal density function.

The total area under any normal curve studied in this book will *always* be 1. The graph of the normal distribution is important because the portion of the *area* under the curve above a given interval represents the *probability* that a measurement will lie in that interval.

Empirical rule

In Section 3.2, we studied Chebyshev's theorem. This theorem gives us information about the *smallest* proportion of data that lies within 2, 3, or *k* standard deviations of the mean. This result applies to *any* distribution. However, for normal distributions, we can get a much more precise result, which is given by the *empirical rule.*

EMPIRICAL RULE

For a distribution that is symmetrical and bell-shaped (in particular, for a normal distribution):

Approximately 68% of the data values will lie within 1 standard deviation on each side of the mean.

Approximately 95% of the data values will lie within 2 standard deviations on each side of the mean.

Approximately 99.7% (or almost all) of the data values will lie within 3 standard deviations on each side of the mean.

The preceding statement is called the *empirical rule* because, for symmetrical, bell-shaped distributions, the given percentages are observed in practice. Furthermore, for the normal distribution, the empirical rule is a direct consequence of the very nature of the distribution (see Figure 7-3). Notice that the empirical rule is a stronger statement than Chebyshev's theorem in that it gives *definite percentages,* not just lower limits. Of course, the empirical rule applies only to normal or symmetrical, bell-shaped distributions, whereas Chebyshev's theorem applies to all distributions.

FIGURE 7-3

Area Under a Normal Curve

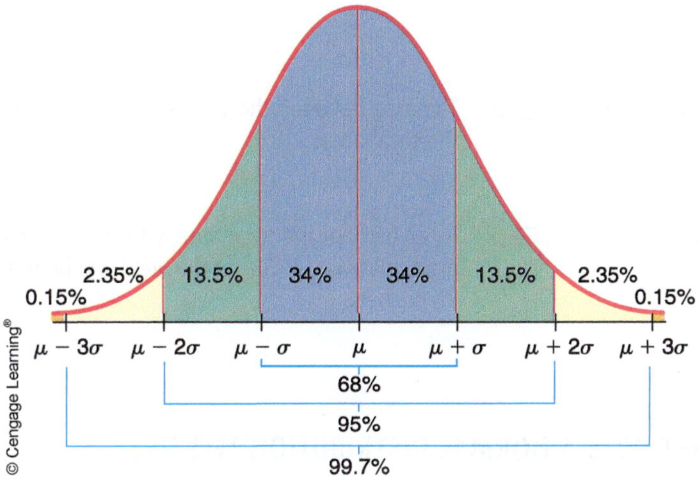

FIGURE 7-4

Distribution of Playing Times

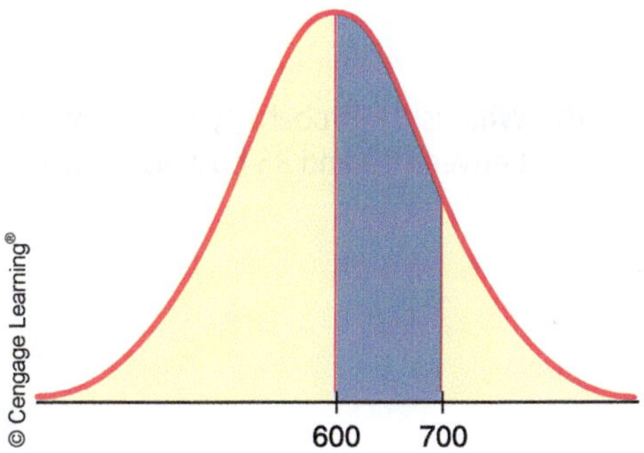

EXAMPLE 1

EMPIRICAL RULE

The playing life of a Sunshine radio is normally distributed with mean $\mu = 600$ hours and standard deviation $\sigma = 100$ hours. What is the probability that a radio selected at random will last from 600 to 700 hours?

SOLUTION: The probability that the playing life will be between 600 and 700 hours is equal to the percentage of the total area under the curve that is shaded in Figure 7-4. Since $\mu = 600$ and $\mu + \sigma = 600 + 100 = 700$, we see that the shaded area is simply the area between μ and $\mu + \sigma$. The area from μ to $\mu + \sigma$ is 34% of the total area. This tells us that the probability a Sunshine radio will last between 600 and 700 playing hours is about 0.34.

GUIDED EXERCISE 2

EMPIRICAL RULE

The yearly wheat yield per acre on a particular farm is normally distributed with mean $\mu = 35$ bushels and standard deviation $\sigma = 8$ bushels.

(a) Shade the area under the curve in Figure 7-5 that represents the probability that an acre will yield between 19 and 35 bushels.

→ See Figure 7-6.

FIGURE 7-5

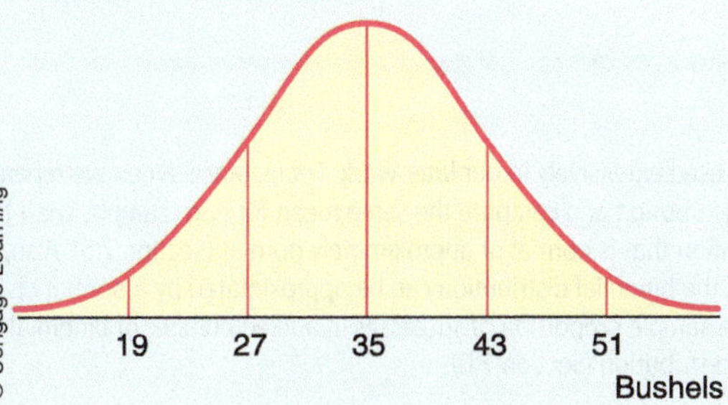

FIGURE 7-6 Completion of Figure 7-5

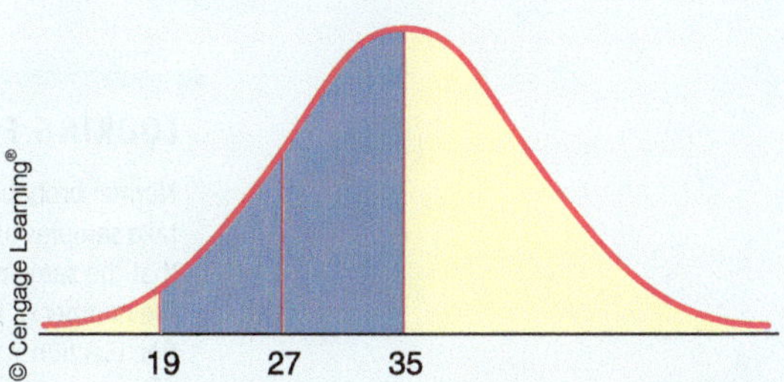

Continued

GUIDED EXERCISE 2 *continued*

(b) Is the area the same as the area between $\mu - 2\sigma$ and μ?	Yes, since $\mu = 35$ and $\mu - 2\sigma = 35 - 2(8) = 19$.
(c) Use Figure 7-3 to find the percentage of area over the interval between 19 and 35.	The area between the values $\mu - 2\sigma$ and μ is 47.5% of the total area.
(d) What is the probability that the yield will be between 19 and 35 bushels per acre?	It is 47.5% of the total area, which is 1. Therefore, the probability is 0.475 that the yield will be between 19 and 35 bushels.

WHAT DOES A NORMAL DISTRIBUTION TELL US?

If a continuous random variable has a normal distribution, then

- the area under the entire distribution is 1.
- the area over a specific interval of values from *a* to *b* is the probability that a randomly selected value falls between *a* and *b*.
- the distribution is symmetrical and mound-shaped and is centered over μ.
- most of the data (99.7%) range from $\mu - 3\sigma$ to $\mu + 3\sigma$.

TECH NOTES

We can graph normal distributions using the TI-84Plus/TI-83Plus/TI-*n*spire calculators, Excel 2010, and Minitab. In each technology, set the range of *x* values between $\mu - 3.5\sigma$ and $\mu + 3.5\sigma$. Then use the built-in normal density functions to generate the corresponding *y* values.

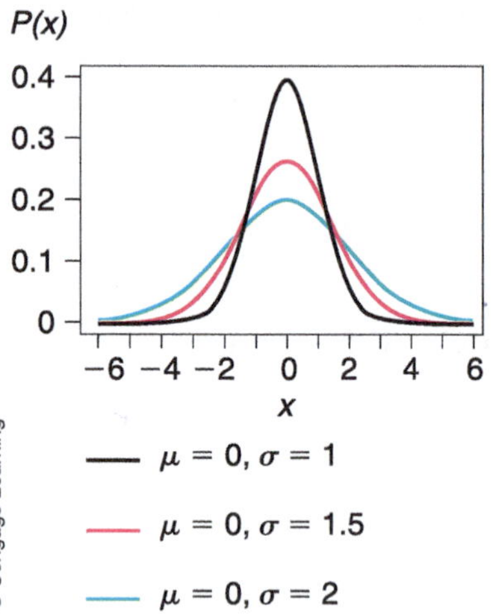

TI-84Plus/TI-83Plus/TI-*n*spire (with TI-84Plus keypad) Press the Y = key. Then, under **DISTR,** select **1:normalpdf** (x,μ,σ) and fill in desired μ and σ values. Press the **WINDOW** key. Set **Xmin** to $\mu - 3\sigma$ and **Xmax** to $\mu + 3\sigma$. Finally, press the **ZOOM** key and select option **0:ZoomFit.**

Excel 2010 In one column, enter *x* values from $\mu - 3.5\sigma$ to $\mu + 3.5\sigma$ in increments of 0.2σ. In the next column, generate *y* values by using the ribbon choices **Insert function** (f_x)

In the dialogue box, select **Statistical** for the Category and then for the Function, select **NORM.DIST** (x,μ,σ,false). Next click the **Insert** tab and in the **Charts** group, click **Scatter**. Select the scatter diagram with smooth lines (third grouping).

Minitab In one column, enter *x* values from -3.5σ to 3.5σ in increments of 0.2σ. In the next column, enter *y* values by using the menu choices **Calc ➤ Probability Distribution ➤ Normal.** Fill in the dialogue box. Next, use menu choices **Graph ➤ Plot.** Fill in the dialogue box. Under Display, select connect.

LOOKING FORWARD

Normal probability distributions will be used extensively in our later work. For instance, when we repeatedly take samples of the same size from a distribution and compute the same mean for each sample, we'll find that the sample means follow a distribution that is normal or approximately normal (Section 7.5). Also, when the number of trials is sufficiently large, the binomial distribution can be approximated by a normal distribution (Section 7.6). The distribution of the sample proportion of successes in a fixed number of binomial trials also can be approximated by a normal distribution (Section 7.6).

COMMENT Uniform Distribution

Normal distributions are the central topic of this chapter. Normal distributions are very important in general probability and statistics. However, there are other, more specialized distributions that also have many applications. One such distribution is the *uniform distribution*. To learn more about this distribution, please see Problems 12 and 13 at the end of this section.

VIEWPOINT In Control? Out of Control?

If you care about quality, you also must care about control! Dr. Walter Shewhart invented control charts when he was working for Bell Laboratories. The great contribution of control charts is that they separate variation into two sources: (1) random or chance causes (in control) and (2) special or assignable causes (out of control). A process is said to be in statistical control *when it is no longer afflicted with special or assignable causes. The performance of a process that is in statistical control is predictable. Predictability and quality control tend to be closely associated.*

(Source: Adapted from the classic text *Statistical Methods from the Viewpoint of Quality Control*, by W. A. Shewhart, with foreword by W. E. Deming, Dover Publications.)

SECTION 7.1 PROBLEMS

1. *Statistical Literacy* Which, if any, of the curves in Figure 7-7 look(s) like a normal curve? If a curve is not a normal curve, tell why.

2. *Statistical Literacy* Look at the normal curve in Figure 7-8, and find μ, $\mu + \sigma$, and σ.

FIGURE 7-7

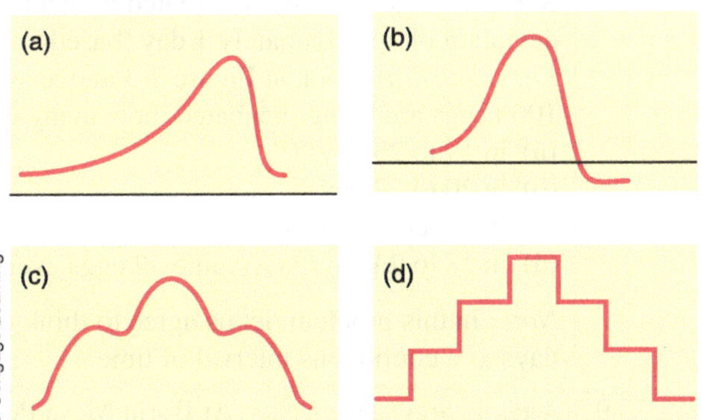

FIGURE 7-8

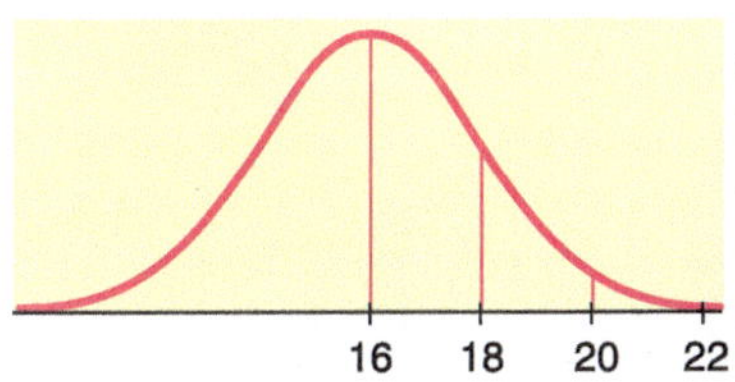

3. *Critical Thinking* Look at the two normal curves in Figures 7-9 and 7-10. Which has the larger standard deviation? What is the mean of the curve in Figure 7-9? What is the mean of the curve in Figure 7-10?

FIGURE 7-9

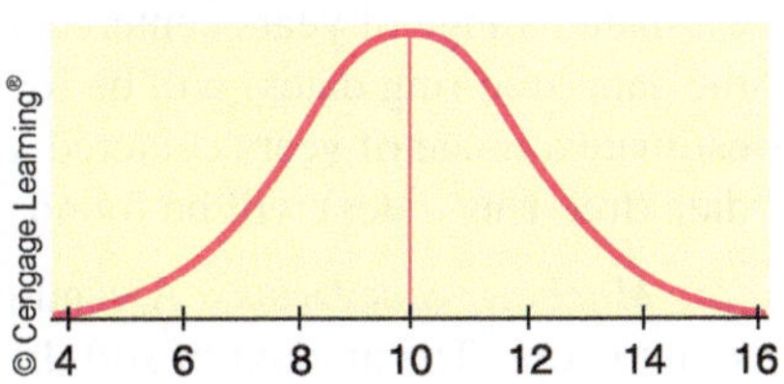

FIGURE 7-10

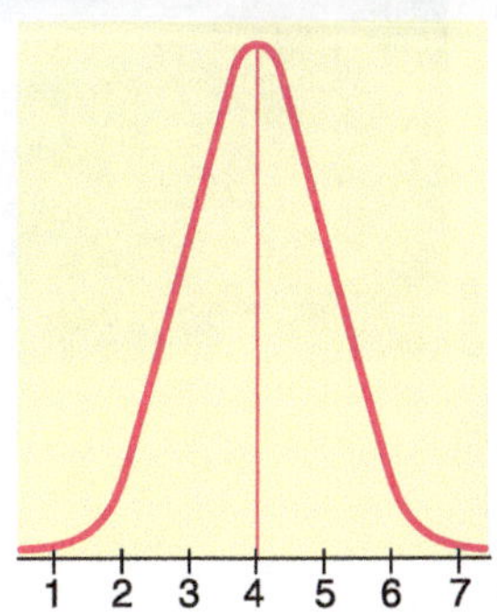

4. *Critical Thinking* Sketch a normal curve
   (a) with mean 15 and standard deviation 2.
   (b) with mean 15 and standard deviation 3.
   (c) with mean 12 and standard deviation 2.
   (d) with mean 12 and standard deviation 3.
   (e) Consider two normal curves. If the first one has a larger mean than the second one, must it have a larger standard deviation as well? Explain your answer.

5. *Basic Computation: Empirical Rule* What percentage of the area under the normal curve lies
   (a) to the left of μ?
   (b) between $\mu - \sigma$ and $\mu + \sigma$?
   (c) between $\mu - 3\sigma$ and $\mu + 3\sigma$?

6. *Basic Computation: Empirical Rule* What percentage of the area under the normal curve lies
   (a) to the right of μ?
   (b) between $\mu - 2\sigma$ and $\mu + 2\sigma$?
   (c) to the right of $\mu + 3\sigma$?

7. *Distribution: Heights of Coeds* Assuming that the heights of college women are normally distributed with mean 65 inches and standard deviation 2.5 inches (based on information from *Statistical Abstract of the United States,* 112th Edition), answer the following questions. *Hint:* Use Problems 5 and 6 and Figure 7-3.
   (a) What percentage of women are taller than 65 inches?
   (b) What percentage of women are shorter than 65 inches?
   (c) What percentage of women are between 62.5 inches and 67.5 inches?
   (d) What percentage of women are between 60 inches and 70 inches?

8. *Distribution: Rhode Island Red Chicks* The incubation time for Rhode Island Red chicks is normally distributed with a mean of 21 days and standard deviation of approximately 1 day (based on information from *World Book Encyclopedia*). Look at Figure 7-3 and answer the following questions. If 1000 eggs are being incubated, how many chicks do we expect will hatch
   (a) in 19 to 23 days?
   (b) in 20 to 22 days?
   (c) in 21 days or fewer?
   (d) in 18 to 24 days? (Assume all eggs eventually hatch.)

   Note: In this problem, let us agree to think of a single day or a succession of days as a continuous interval of time.

Raul Touzon/National Geographic Creative

9. *Archaeology: Tree Rings* At Burnt Mesa Pueblo, archaeological studies have used the method of tree-ring dating in an effort to determine when prehistoric people lived in the pueblo. Wood from several excavations gave a mean of (year) 1243 with a standard deviation of 36 years (*Bandelier Archaeological Excavation Project: Summer 1989 Excavations at Burnt Mesa Pueblo,* edited by Kohler, Washington State University Department of Anthropology). The distribution of dates was more or less mound-shaped and symmetrical about the mean. Use the empirical rule to
   (a) estimate a range of years centered about the mean in which about 68% of the data (tree-ring dates) will be found.
   (b) estimate a range of years centered about the mean in which about 95% of the data (tree-ring dates) will be found.
   (c) estimate a range of years centered about the mean in which almost all the data (tree-ring dates) will be found.

10. *Vending Machine: Soft Drinks* A vending machine automatically pours soft drinks into cups. The amount of soft drink dispensed into a cup is normally

distributed with a mean of 7.6 ounces and standard deviation of 0.4 ounce. Examine Figure 7-3 and answer the following questions.
(a) Estimate the probability that the machine will overflow an 8-ounce cup.
(b) Estimate the probability that the machine will not overflow an 8-ounce cup.
(c) The machine has just been loaded with 850 cups. How many of these do you expect will overflow when served?

11. *Pain Management: Laser Therapy* "Effect of Helium-Neon Laser Auriculotherapy on Experimental Pain Threshold" is the title of an article in the journal *Physical Therapy* (Vol. 70, No. 1, pp. 24–30). In this article, laser therapy was discussed as a useful alternative to drugs in pain management of chronically ill patients. To measure pain threshold, a machine was used that delivered low-voltage direct current to different parts of the body (wrist, neck, and back). The machine measured current in milliamperes (mA). The pretreatment experimental group in the study had an average threshold of pain (pain was first detectable) at $\mu = 3.15$ mA with standard deviation $\sigma = 1.45$ mA. Assume that the distribution of threshold pain, measured in milliamperes, is symmetrical and more or less mound-shaped. Use the empirical rule to
(a) estimate a range of milliamperes centered about the mean in which about 68% of the experimental group had a threshold of pain.
(b) estimate a range of milliamperes centered about the mean in which about 95% of the experimental group had a threshold of pain.

12. *Expand Your Knowledge: Continuous Uniform Probability Distribution* Let α and β be any two constants such that $\alpha < \beta$. Suppose we choose a point x at random in the interval from α to β. In this context the phrase *at random* is taken to mean that the point x is as likely to be chosen from one particular part of the interval as any other part. Consider the rectangle.

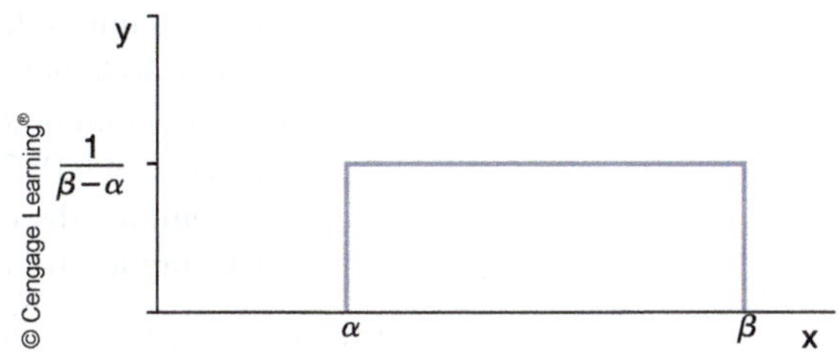

The base of the rectangle has length $\beta - \alpha$ and the height of the rectangle is $1/(\beta - \alpha)$, so the area of the rectangle is 1. As such, this rectangle's top can be thought of as part of a probability density curve. Since we specify that x must lie between α and β, the probability of a point occurring outside the interval $[\alpha, \beta]$ is, by definition, 0. From a geometric point of view, x chosen at random from α to β means we are equally likely to land anywhere in the interval from α to β. For this reason, the top of the (rectangle's) density curve is flat or uniform.

Now suppose that a and b are numbers such that $\alpha \le a < b \le \beta$. What is the probability that a number x chosen at random from α to β will fall in the interval $[a, b]$? Consider the graph

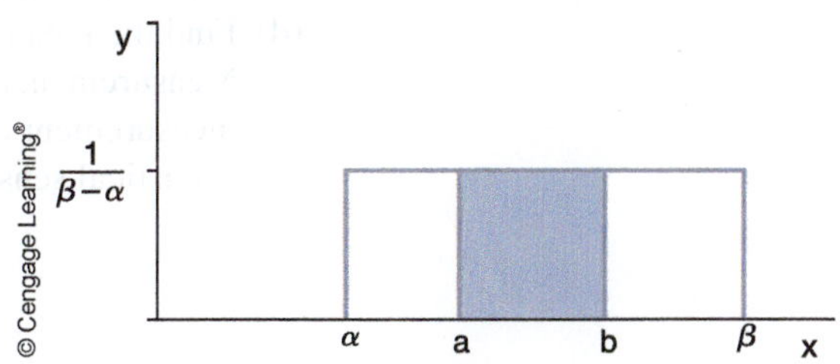

Because x is chosen at random from $[\alpha, \beta]$, the area of the rectangle that lies above $[a, b]$ is the probability that x lies in $[a, b]$. This area is

$$P(a < x < b) = \frac{b - a}{\beta - \alpha}$$

Continuous uniform distribution
Rectangular distribution

In this way we can assign a probability to any interval inside $[\alpha, \beta]$. This probability distribution is called the *continuous uniform distribution* (also called the rectangular distribution). Using some extra mathematics, it can be shown that if x is a random variable with this distribution, then the mean and standard deviation of x are

$$\mu = \frac{\alpha + \beta}{2} \quad \text{and} \quad \sigma = \frac{\beta - \alpha}{\sqrt{12}}$$

Sedimentation experiments are very important in the study of biology, medicine, hydrodynamics, petroleum engineering, civil engineering, and so on. The size (diameter) of approximately spherical particles is important since larger particles hinder and sometimes block the movement of smaller particles. Usually the size of sediment particles follows a uniform distribution (Reference: Y. Zimmels, "Theory of Kindred Sedimentation of Polydisperse Mixtures," *AIChE Journal*, Vol. 29, No. 4, pp. 669–676).

Suppose a veterinary science experiment injects very small, spherical pellets of low-level radiation directly into an animal's bloodstream. The purpose is to attempt to cure a form of recurring cancer. The pellets eventually dissolve and pass through the animal's system. Diameters of the pellets are uniformly distributed from 0.015 mm to 0.065 mm. If a pellet enters an artery, what is the probability that it will be the following sizes?

(a) 0.050 mm or larger. *Hint*: All particles are between 0.015 mm and 0.065 mm, so larger than 0.050 means $0.050 \le x \le 0.065$.
(b) 0.040 mm or smaller
(c) between 0.035 mm and 0.055 mm
(d) Compute the mean size of the particles.
(e) Compute the standard deviation of particle size.

13. *Uniform Distribution: Measurement Errors* Measurement errors from instruments are often modeled using the uniform distribution (see Problem 12). To determine the range of a large public address system, acoustical engineers use a method of triangulation to measure the shock waves sent out by the speakers. The time at which the waves arrive at the sensors must be measured accurately. In this context, a negative error means the signal arrived too early. A positive error means the signal arrived too late. Measurement errors in reading these times have a uniform distribution from -0.05 to $+0.05$ microseconds. (Reference: J. Perruzzi and E. Hilliard, "Modeling Time Delay Measurement Errors," *Journal of the Acoustical Society of America*, Vol. 75, No. 1, pp. 197–201.) What is the probability that such measurements will be in error by

(a) less than $+0.03$ microsecond (i.e., $-0.05 \le x < 0.03$)?
(b) more than -0.02 microsecond?
(c) between -0.04 and $+0.01$ microsecond?
(d) Find the mean and standard deviation of measurement errors. Measurements from an instrument are called *unbiased* if the mean of the measurement errors is zero. Would you say the measurements for these acoustical sensors are unbiased? Explain.

SECTION **7.2**

Standard Units and Areas Under the Standard Normal Distribution

FOCUS POINTS

- Given μ and σ, convert raw data to z scores.
- Given μ and σ, convert z scores to raw data.
- Graph the standard normal distribution, and find areas under the standard normal curve.

z SCORES AND RAW SCORES

Normal distributions vary from one another in two ways: The mean μ may be located anywhere on the x axis, and the bell shape may be more or less spread according to the size of the standard deviation σ. The differences among the normal distributions cause difficulties when we try to compute the area under the curve in a specified interval of x values and, hence, the probability that a measurement will fall into that interval.

It would be a futile task to try to set up a table of areas under the normal curve for each different μ and σ combination. We need a way to standardize the distributions so that we can use *one* table of areas for *all* normal distributions. We achieve this standardization by considering how many standard deviations a measurement lies from the mean. In this way, we can compare a value in one normal distribution with a value in another, different normal distribution. The next situation shows how this is done.

Suppose Tina and Jack are in two different sections of the same course. Each section is quite large, and the scores on the midterm exams of each section follow a normal distribution. In Tina's section, the average (mean) was 64 and her score was 74. In Jack's section, the mean was 72 and his score was 82. Both Tina and Jack were pleased that their scores were each 10 points above the average of each respective section. However, the fact that each was 10 points above average does not really tell us how each did *with respect to the other students in the section.* In Figure 7-11, we see the normal distribution of grades for each section.

Tina's 74 was higher than most of the other scores in her section, while Jack's 82 is only an upper-middle score in his section. Tina's score is far better with respect to her class than Jack's score is with respect to his class.

FIGURE 7-11

Distributions of Midterm Scores

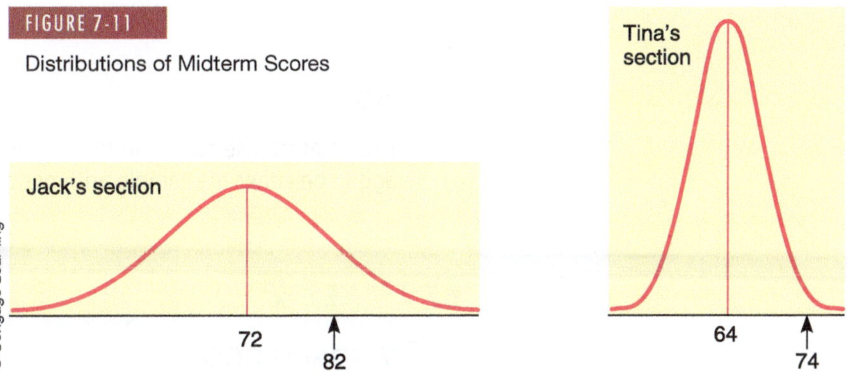

The preceding situation demonstrates that it is not sufficient to know the difference between a measurement (x value) and the mean of a distribution. We need also to consider the spread of the curve, or the standard deviation. What we really want to know is the number of standard deviations between a measurement and the mean. This "distance" takes both μ and σ into account.

We can use a simple formula to compute the number z of standard deviations between a measurement x and the mean μ of a normal distribution with standard deviation σ:

$$\begin{pmatrix}\text{Number of standard deviations}\\ \text{between the measurement and}\\ \text{the mean}\end{pmatrix} = \left(\frac{\text{Difference between the measurement and the mean}}{\text{Standard deviation}}\right)$$

z value, z score, or standard score

The ***z* value** or ***z* score** (also known as standard score) gives the number of standard deviations between the original measurement x and the mean μ of the x distribution.

$$z = \frac{x - \mu}{\sigma}$$

TABLE 7-1 ***x* Values and Corresponding *z* Values**

x Value in Original Distribution	Corresponding *z* Value or Standard Unit
$x = \mu$	$z = 0$
$x > \mu$	$z > 0$
$x < \mu$	$z < 0$

The mean is a special value of a distribution. Let's see what happens when we convert $x = \mu$ to a z value:

$$z = \frac{x - \mu}{\sigma} = \frac{\mu - \mu}{\sigma} = 0$$

The mean of the original distribution is always zero, in *standard units*. This makes sense because the mean is zero standard variations from itself.

Standard units

An x value in the original distribution that is *above* the mean μ has a corresponding z value that is *positive*. Again, this makes sense because a measurement above the mean would be a positive number of standard deviations from the mean. Likewise, an x value *below* the mean has a *negative* z value. (See Table 7-1.)

WHAT DOES A STANDARD SCORE TELL US?

A *standard score* or *z score* of a measurement tells us the number of standard deviations the measurement is from the mean.

- A standard score close to zero tells us the measurement is near the mean of the distribution.
- A positive standard score tells us the measurement is above the mean.
- A negative standard score tells us the measurement is below the mean.

NOTE

Unless otherwise stated, in the remainder of the book we will take the word *average* to be either the sample arithmetic mean $\bar{x}$ or the population mean μ.

EXAMPLE 2

STANDARD SCORE

A pizza parlor franchise specifies that the average (mean) amount of cheese on a large pizza should be 8 ounces and the standard deviation only 0.5 ounce. An inspector picks out a large pizza at random in one of the pizza parlors and finds that it is made with 6.9 ounces of cheese. Assume that the amount of cheese on a pizza follows a normal distribution. If the amount of cheese is below the mean by more than 3 standard deviations, the parlor will be in danger of losing its franchise.

How many standard deviations from the mean is 6.9? Is the pizza parlor in danger of losing its franchise?

SOLUTION: Since we want to know the number of standard deviations from the mean, we want to convert 6.9 to standard z units.

$$z = \frac{x - \mu}{\sigma} = \frac{6.9 - 8}{0.5} = -2.20$$

Interpretation The amount of cheese on the selected pizza is only 2.20 standard deviations below the mean. The fact that z is negative indicates that the amount of cheese is 2.20 standard deviations *below* the mean. The parlor will not lose its franchise based on this sample.

Raw score, x

We have seen how to convert from x measurements to standard units z. We can easily reverse the process to find the original *raw score* x if we know the mean and standard deviation of the original x distribution. Simply solve the z score formula for x.

Given an x distribution with mean μ and standard deviation σ, the **raw score x** corresponding to a z score is

$$x = z\sigma + \mu$$

GUIDED EXERCISE 3 | STANDARD SCORE AND RAW SCORE

Rod figures that it takes an average (mean) of 17 minutes with a standard deviation of 3 minutes to drive from home, park the car, and walk to an early morning class.

(a) One day it took Rod 21 minutes to get to class. How many standard deviations from the average is that? Is the z value positive or negative? Explain why it should be either positive or negative.

→ The number of standard deviations from the mean is given by the z value:

$$z = \frac{x - \mu}{\sigma} = \frac{21 - 17}{3} \approx 1.33$$

The z value is positive. We should expect a positive z value, since 21 minutes is *more* than the mean of 17.

(b) What commuting time corresponds to a standard score of $z = -2.5$? ***Interpretation*** Could Rod count on making it to class in this amount of time or less?

→ $x = z\sigma + \mu$
$= (-2.5)(3) + 17$
$= 9.5$ minutes

No, commute times at or less than 2.5 standard deviations below the mean are rare.

LOOKING FORWARD

The basic structure of the formula for the standard score of a distribution is very general. When we verbalize the formula, we see it is

$$z = \frac{\text{measurement} - \text{mean of the distribution}}{\text{standard deviation of the distribution}}$$

We will see this general formula used again and again. In particular, when we look at sampling distributions for the mean (Section 7.4) and when we use the normal approximation of the binomial distribution (Section 7.6), we'll see this formula. We'll also see it when we discuss the sampling distribution for proportions (Section 7.6). Further uses occur in computations for confidence intervals (Chapter 8) and hypothesis testing (Chapters 9 and 10).

STANDARD NORMAL DISTRIBUTION

If the original distribution of x *values is normal,* then the corresponding z *values have a normal distribution as well.* The z distribution has a mean of 0 and a standard deviation of 1. The normal curve with these properties has a special name.

Standard normal distribution

> The **standard normal distribution** is a normal distribution with mean $\mu = 0$ and standard deviation $\sigma = 1$ (Figure 7-12).

Any normal distribution of x values can be converted to the standard normal distribution by converting all x values to their corresponding z values. The resulting standard distribution will always have mean $\mu = 0$ and standard deviation $\sigma = 1$.

FIGURE 7-12

The Standard Normal Distribution ($\mu = 0, \sigma = 1$)

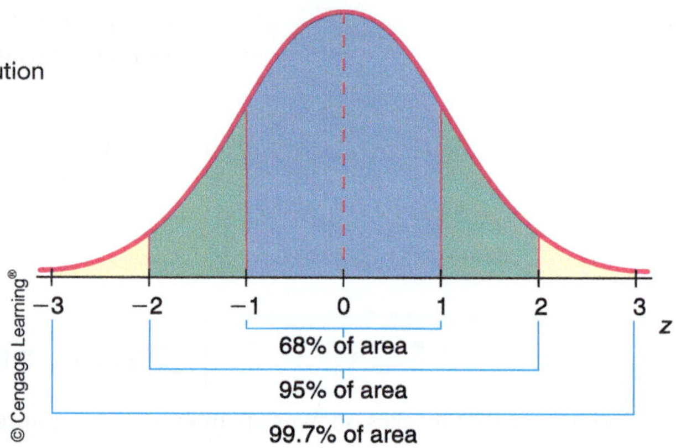

WHAT DOES THE STANDARD NORMAL DISTRIBUTION TELL US?

When we have the standard normal distribution, we know

- the mean is 0.
- the standard deviation is 1.
- any normal distribution can be converted to a standard normal distribution by converting all the measurements to standard z scores.

AREAS UNDER THE STANDARD NORMAL CURVE

Area under the standard normal curve

We have seen how to convert *any* normal distribution to the *standard* normal distribution. We can change any x value to a z value and back again. But what is the advantage of all this work? The advantage is that there are extensive tables that show the *area under the standard normal curve* for almost any interval along the z axis. The areas are important because each area is equal to the *probability* that the measurement of an item selected at random falls in this interval. Thus, the *standard* normal distribution can be a tremendously helpful tool.

USING A STANDARD NORMAL DISTRIBUTION TABLE

Using a table to find areas and probabilities associated with the standard normal distribution is a fairly straightforward activity. However, it is important to first observe the range of z values for which areas are given. This range is usually depicted in a picture that accompanies the table.

Left-tail style table

In this text, *we will use the left-tail style table*. This style table gives cumulative areas to the left of a specified z. Determining other areas under the curve utilizes the fact that the area under the entire curve is 1. Taking advantage of the symmetry of the normal distribution is also useful. The procedures you learn for using the left-tail style normal distribution table apply directly to *cumulative* normal distribution areas found on calculators and in computer software packages such as Excel 2010 and Minitab.

EXAMPLE 3

STANDARD NORMAL DISTRIBUTION TABLE

Use Table 3 of the Appendix to find the described areas under the standard normal curve.

(a) Find the area under the standard normal curve to the left of $z = -1.00$.

SOLUTION: First, shade the area to be found on the standard normal distribution curve, as shown in Figure 7-13. Notice that the z value we are using is negative. This means that we will look at the portion of Table 3 of the Appendix for which the z values are negative. In the upper-left corner of the table we see the letter z. The column under z gives us the units value and tenths value for z. The other column headings indicate the hundredths value of z. Table entries give areas under the standard normal curve to the left of the listed z values. To find the area to the left of $z = -1.00$, we use the row headed by -1.0 and then move to the column headed by the hundredths position, .00. This entry is shaded in Table 7-2. We see that the area is 0.1587.

FIGURE 7-13

Area to the Left of $z = -1.00$

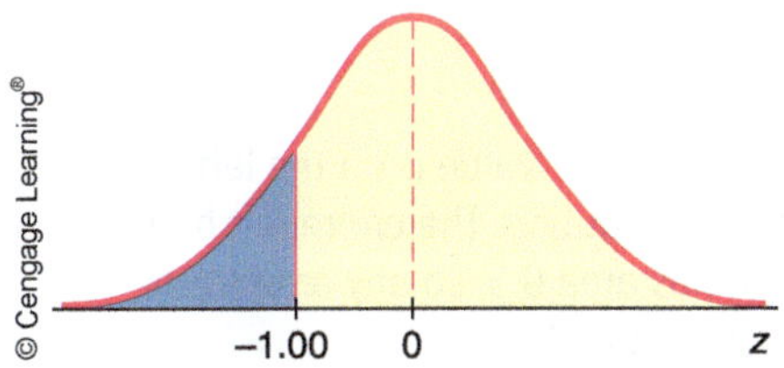

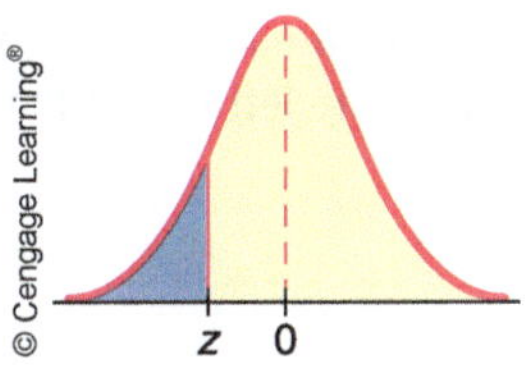

TABLE 7-2 **Excerpt from Table 3 of the Appendix Showing Negative z Values**

z	.00	.01	...	.07	.08	.09
−3.4	.0003	.0003	...	.0003	.0003	.0002
⋮						
−1.1	.1357	.1335	...	.1210	.1190	.1170
−1.0	.1587	.1562	...	.1423	.1401	.1379
−0.9	.1841	.1814	...	.1660	.1635	.1611
⋮						
−0.0	.5000	.4960	...	.4721	.4681	.4641

FIGURE 7-14

Area to the Left of $z = 1.18$

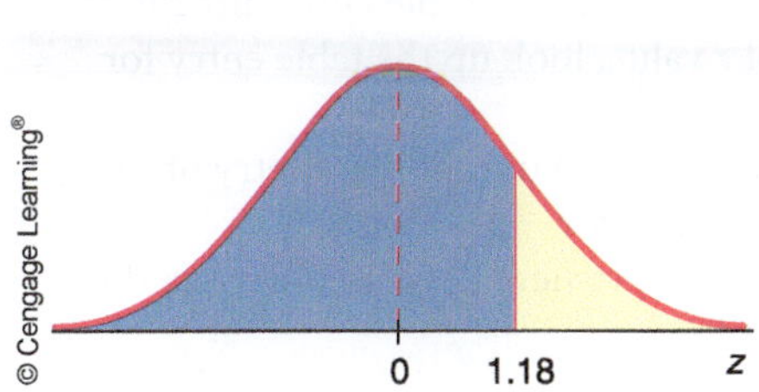

(b) Find the area to the left of $z = 1.18$, as illustrated in Figure 7-14.

SOLUTION: In this case, we are looking for an area to the left of a positive z value, so we look in the portion of Table 3 that shows positive z values. Again, we first sketch the area to be found on a standard normal curve, as shown in Figure 7-14. Look in the row headed by 1.1 and move to the column headed by .08. The desired area is shaded (see Table 7-3 on the next page). We see that the area to the left of 1.18 is 0.8810.

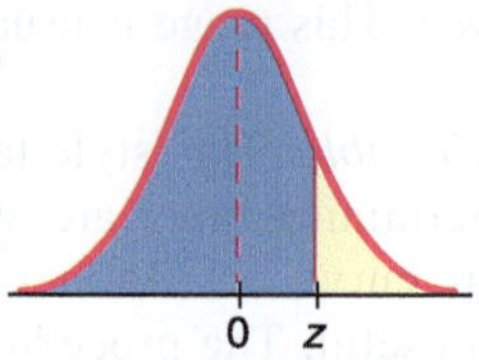

TABLE 7-3 **Excerpt from Table 3 of the Appendix Showing Positive *z* Values**

z	.00	.01	.02	...	.08	.09
0.0	.5000	.5040	.5080	...	.5319	.5359
⋮						
0.9	.8159	.8186	.8212	...	.8365	.8389
1.0	.8413	.8438	.8461	...	.8599	.8621
1.1	.8643	.8665	.8686	...	.8810	.8830
⋮						
3.4	.9997	.9997	.9997	...	.9997	.9998

GUIDED EXERCISE 4 USING THE STANDARD NORMAL DISTRIBUTION TABLE

Table 3, Areas of a Standard Normal Distribution, is located in the Appendix as well as in the endpapers of the text. Spend a little time studying the table, and then answer these questions.

(a) As *z* values increase, do the areas to the left of *z* increase?

→ Yes. As *z* values increase, we move to the right on the normal curve, and the areas increase.

(b) If a *z* value is negative, is the area to the left of *z* less than 0.5000?

→ Yes. Remember that a negative *z* value is on the left side of the standard normal distribution. The entire left half of the normal distribution has area 0.5, so any area to the left of $z = 0$ will be less than 0.5.

(c) If a *z* value is positive, is the area to the left of *z* greater than 0.5000?

→ Yes. Positive *z* values are on the right side of the standard normal distribution, and any area to the left of a positive *z* value includes the entire left half of the normal distribution.

Using Table 3 to find other areas

Table 3 gives areas under the standard normal distribution that are to the *left of a z value.* How do we find other areas under the standard normal curve?

PROCEDURE

HOW TO USE A LEFT-TAIL STYLE STANDARD NORMAL DISTRIBUTION TABLE

1. For areas to the left of a specified *z* value, use the table entry directly.
2. For areas to the right of a specified *z* value, look up the table entry for *z* and subtract the area from 1.
   Note: Another way to find the same area is to use the symmetry of the normal curve and look up the table entry for $-z$.
3. For areas between two *z* values, z_1 and z_2 (where $z_2 > z_1$), *subtract* the table area for z_1 from the table area for z_2.

Figure 7-15 illustrates the procedure for using Table 3, Areas of a Standard Normal Distribution, to find any specified area under the standard normal distribution. Again, it is useful to sketch the area in question before you use Table 3.

COMMENT Notice that the z values shown in Table 3 of the Appendix are formatted to the hundredths position. It is convenient to *round or format z values to the hundredths position* before using the table. The areas are all given to four places after the decimal, so give your answers to four places after the decimal.

COMMENT The smallest z value shown in Table 3 is -3.49, while the largest value is 3.49. These values are, respectively, far to the left and far to the right on the standard normal distribution, with very little area beyond either value. We will follow the common convention of treating any area to the left of a z value smaller than -3.49 as 0.000. Similarly, we will consider any area to the right of a z value greater than 3.49 as 0.000. We understand that there is some area in these extreme tails. However, these areas are each less than 0.0002. Now let's get real about this! Some very specialized applications, beyond the scope of this book, do need to measure areas and corresponding probabilities in these extreme tails. But in most practical applications, *we follow the convention of treating the areas in the extreme tails as zero.*

FIGURE 7-15

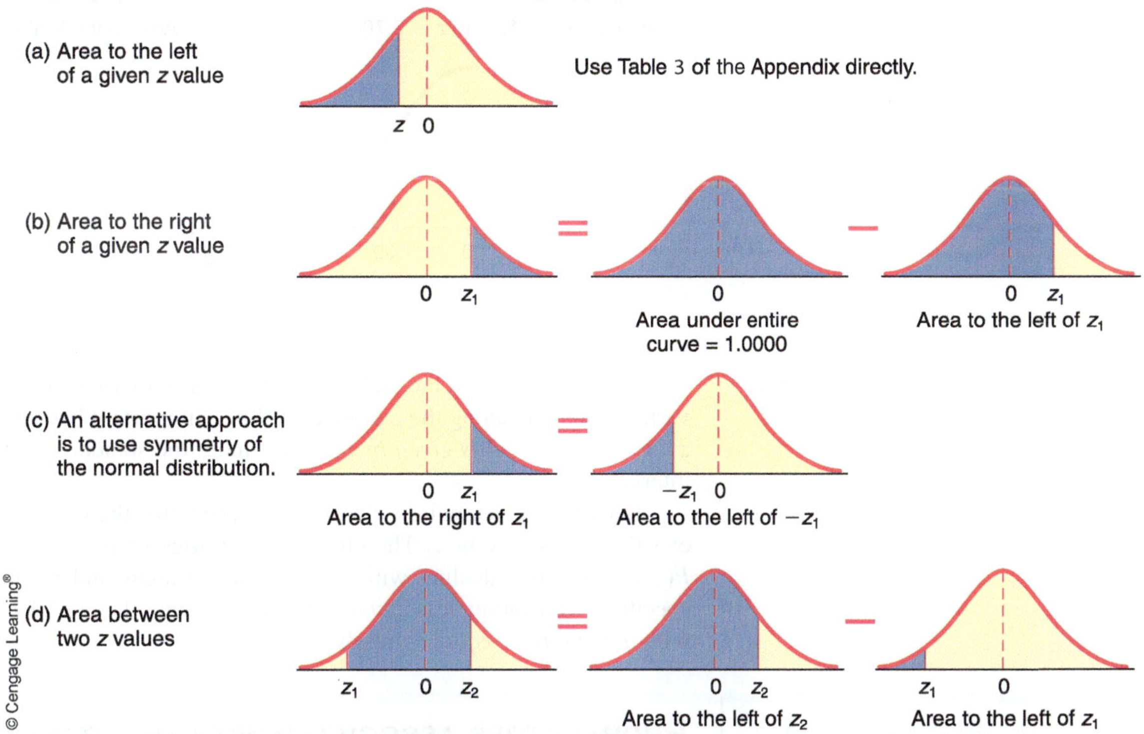

CONVENTION FOR USING TABLE 3 OF THE APPENDIX

1. Treat any area to the left of a z value smaller than -3.49 as 0.000.
2. Treat any area to the left of a z value greater than 3.49 as 1.000.

EXAMPLE 4

USING TABLE TO FIND AREAS

Use Table 3 of the Appendix to find the specified areas.

(a) Find the area between $z = 1.00$ and $z = 2.70$.

SOLUTION: First, sketch a diagram showing the area (see Figure 7-16). Because we are finding the area between two z values, we subtract corresponding table entries.

$$\begin{aligned}(\text{Area between } 1.00 \text{ and } 2.70) &= (\text{Area left of } 2.70) - (\text{Area left of } 1.00)\\ &= 0.9965 - 08413\\ &= 0.1552\end{aligned}$$

(b) Find the area to the right of $z = 0.94$.

SOLUTION: First, sketch the area to be found (see Figure 7-17).

$$\begin{aligned}(\text{Area to right of } 0.94) &= (\text{Area under entire curve}) - (\text{Area to left of } 0.94)\\ &= 1.000 - 0.8264\\ &= 0.1736\end{aligned}$$

Alternatively,

$$\begin{aligned}(\text{Area to right of } 0.94) &= (\text{Area to left of } -0.94)\\ &= 0.1736\end{aligned}$$

FIGURE 7-16

Area from $z = 1.00$ to $z = 2.70$

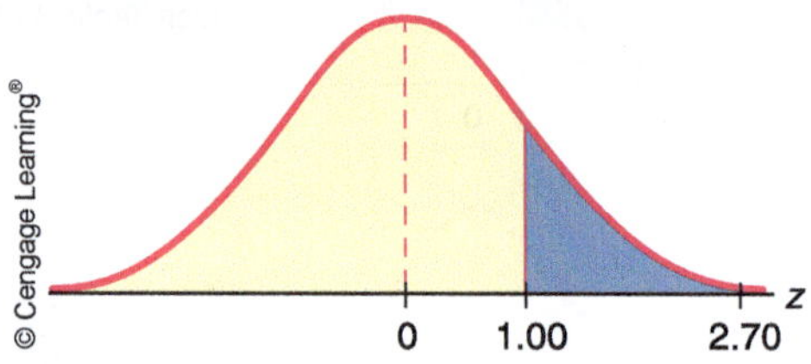

FIGURE 7-17

Area to the Right of $z = 0.94$

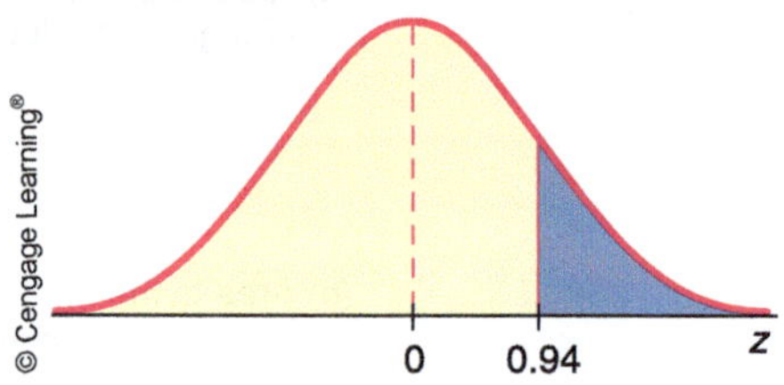

Probabilities associated with the standard normal distribution

We have practiced the skill of finding areas under the standard normal curve for various intervals along the z axis. This skill is important because *the probability* that z lies in an interval *is given by the area* under the standard normal curve above that interval.

Because the normal distribution is continuous, there is no area under the curve exactly over a specific z. Therefore, probabilities such as $P(z \geq z_1)$ are the same as $P(z > z_1)$. When dealing with probabilities or areas under a normal curve that are specified with inequalities, *strict inequality* symbols can be used *interchangeably* with *inequality or equal* symbols.

GUIDED EXERCISE 5 PROBABILITIES ASSOCIATED WITH THE STANDARD NORMAL DISTRIBUTION

Let z be a random variable with a standard normal distribution.

(a) $P(z \geq 1.15)$ refers to the probability that z values lie to the right of 1.15. Shade the corresponding area under the standard normal curve (Figure 7-18) and find $P(z \geq 1.15)$.

FIGURE 7-18 Area to Be Found

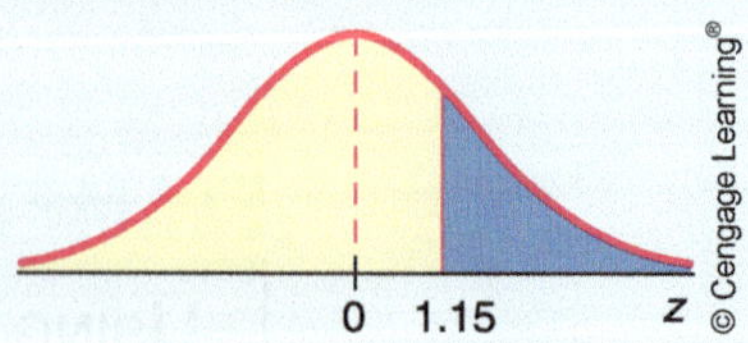

$P(z \geq 1.15) = 1.000 - P(z \leq 1.15) = 1.000 - 0.8749 = 0.1251$

Alternatively,

$P(z \geq 1.15) = P(z \leq -1.15) = 0.1251$

Continued

GUIDED EXERCISE 5 *continued*

(b) Find $P(-1.78 \leq z \leq 0.35)$. First, sketch the area under the standard normal curve corresponding to the area (Figure 7-19).

FIGURE 7-19 Area to Be Found

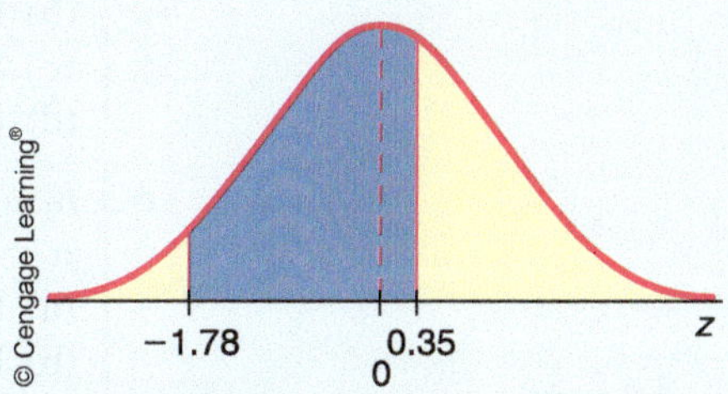

$$P(-1.78 \leq z \leq 0.35) = P(z \leq 0.35) - P(z \leq -1.78)$$
$$= 0.6368 - 0.0375 = 0.5993$$

TECH NOTES The TI-84Plus/TI-83Plus/TI-*n*spire calculators, Excel 2010, and Minitab all provide cumulative areas under any normal distribution, including the standard normal. The Tech Note of Section 7.3 shows examples.

VIEWPOINT Mighty Oaks from Little Acorns Grow!

Just how big is that acorn? What if we compare it with other acorns? Is that oak tree taller than an average oak tree? How does it compare with other oak trees? What do you mean, this oak tree has a larger geographic range? Compared with what? Answers to questions such as these can be given only if we resort to standardized statistical units. Can you compare a single oak tree with an entire forest of oak trees? The answer is yes, if you use standardized z scores. *For more information about sizes of acorns, oak trees, and geographic locations, visit the Brase/Brase statistics site at* **http://www.cengagebrain.com** *and find the link to DASL, the Carnegie Mellon University Data and Story Library. From the DASL site, find Biology under Data Subjects, and select Acorns.*

SECTION 7.2 PROBLEMS

In these problems, assume that all distributions are *normal*. In all problems in Chapter 7, *average* is always taken to be the arithmetic mean $\bar{x}$ or μ.

1. *Statistical Literacy* What does a standard score measure?

2. *Statistical Literacy* Does a raw score less than the mean correspond to a positive or negative standard score? What about a raw score greater than the mean?

3. *Statistical Literacy* What is the value of the standard score for the mean of a distribution?

4. *Statistical Literacy* What are the values of the mean and standard deviation of a standard normal distribution?

5. *Basic Computation: z Score and Raw Score* A normal distribution has $\mu = 30$ and $\sigma = 5$.
(a) Find the z score corresponding to $x = 25$.
(b) Find the z score corresponding to $x = 42$.
(c) Find the raw score corresponding to $z = -2$.
(d) Find the raw score corresponding to $z = 1.3$.

6. *Basic Computation: z Score and Raw Score* A normal distribution has $\mu = 10$ and $\sigma = 2$.
(a) Find the z score corresponding to $x = 12$.
(b) Find the z score corresponding to $x = 4$.
(c) Find the raw score corresponding to $z = 1.5$.
(d) Find the raw score corresponding to $z = -1.2$.

7. *Critical Thinking* Consider the following scores:
(i) Score of 40 from a distribution with mean 50 and standard deviation 10
(ii) Score of 45 from a distribution with mean 50 and standard deviation 5

How do the two scores compare relative to their respective distributions?

8. *Critical Thinking* Raul received a score of 80 on a history test for which the class mean was 70 with standard deviation 10. He received a score of 75 on a biology test for which the class mean was 70 with standard deviation 2.5. On which test did he do better relative to the rest of the class?

9. *z Scores: First Aid Course* The college physical education department offered an advanced first aid course last semester. The scores on the comprehensive final exam were normally distributed, and the z scores for some of the students are shown below:

Robert, 1.10	Juan, 1.70	Susan, −2.00
Joel, 0.00	Jan, −0.80	Linda, 1.60.

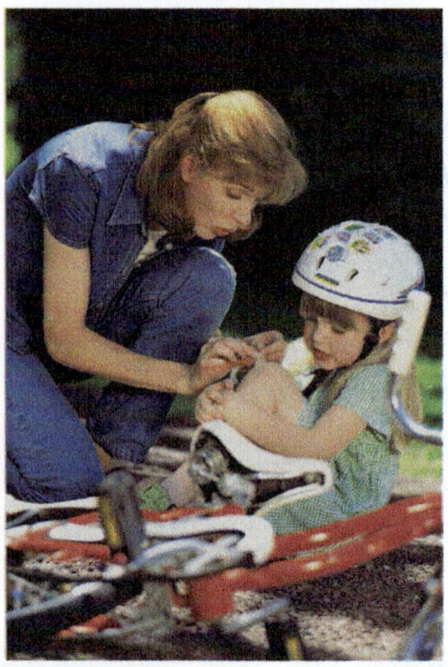
Comstock/Stockbyte/Jupiter Images

(a) Which of these students scored above the mean?
(b) Which of these students scored on the mean?
(c) Which of these students scored below the mean?
(d) If the mean score was $\mu = 150$ with standard deviation $\sigma = 20$, what was the final exam score for each student?

10. *z Scores: Fawns* Fawns between 1 and 5 months old in Mesa Verde National Park have a body weight that is approximately normally distributed with mean $\mu = 27.2$ kilograms and standard deviation $\sigma = 4.3$ kilograms (based on information from *The Mule Deer of Mesa Verde National Park,* by G. W. Mierau and J. L. Schmidt, Mesa Verde Museum Association). Let x be the weight of a fawn in kilograms. Convert each of the following x intervals to z intervals.
(a) $x < 30$ (b) $19 < x$ (c) $32 < x < 35$

Convert each of the following z intervals to x intervals.
(d) $-2.17 < z$ (e) $z < 1.28$ (f) $-1.99 < z < 1.44$
(g) *Interpretation* If a fawn weighs 14 kilograms, would you say it is an unusually small animal? Explain using z values and Figure 7-12.
(h) *Interpretation* If a fawn is unusually large, would you say that the z value for the weight of the fawn will be close to 0, −2, or 3? Explain.

11. *z Scores: Red Blood Cell Count* Let x = red blood cell (RBC) count in millions per cubic millimeter of whole blood. For healthy females, x has an approximately normal distribution with mean $\mu = 4.8$ and standard deviation $\sigma = 0.3$ (based on information from *Diagnostic Tests with Nursing Implications,* edited by S. Loeb, Springhouse Press). Convert each of the following x intervals to z intervals.
(a) $4.5 < x$ (b) $x < 4.2$ (c) $4.0 < x < 5.5$

Convert each of the following z intervals to x intervals.
(d) $z < -1.44$ (e) $1.28 < z$ (f) $-2.25 < z < -1.00$
(g) *Interpretation* If a female had an RBC count of 5.9 or higher, would that be considered unusually high? Explain using z values and Figure 6-15.

12. *Normal Curve: Tree Rings* Tree-ring dates were used extensively in archaeological studies at Burnt Mesa Pueblo (*Bandelier Archaeological Excavation Project: Summer 1989 Excavations at Burnt Mesa Pueblo,* edited by Kohler, Washington State University Department of Anthropology). At one site on the mesa, tree-ring dates (for many samples) gave a mean date of μ_1 = year 1272 with standard deviation $\sigma_1 = 35$ years. At a second, removed site, the tree-ring dates gave a mean of μ_2 = year 1122 with standard deviation $\sigma_2 = 40$ years. Assume that both sites had dates that were approximately normally distributed. In the first area, an object was found and dated as x_1 = year 1250. In the second area, another object was found and dated as x_2 = year 1234.
(a) Convert both x_1 and x_2 to z values, and locate both of these values under the standard normal curve of Figure 7-12.
(b) *Interpretation* Which of these two items is the more unusual as an archaeological find in its location?

Basic Computation: Finding Areas Under the Standard Normal Curve In Problems 13–30, sketch the areas under the standard normal curve over the indicated intervals and find the specified areas.

13. To the right of $z = 0$
14. To the left of $z = 0$
15. To the left of $z = -1.32$
16. To the left of $z = -0.47$
17. To the left of $z = 0.45$
18. To the left of $z = 0.72$
19. To the right of $z = 1.52$
20. To the right of $z = 0.15$
21. To the right of $z = -1.22$
22. To the right of $z = -2.17$
23. Between $z = 0$ and $z = 3.18$
24. Between $z = 0$ and $z = -1.93$
25. Between $z = -2.18$ and $z = 1.34$
26. Between $z = -1.40$ and $z = 2.03$
27. Between $z = 0.32$ and $z = 1.92$
28. Between $z = 1.42$ and $z = 2.17$
29. Between $z = -2.42$ and $z = -1.77$
30. Between $z = -1.98$ and $z = -0.03$

Basic Computation: Finding Probabilities In Problems 31–50, let z be a random variable with a standard normal distribution. Find the indicated probability, and shade the corresponding area under the standard normal curve.

31. $P(z \le 0)$
32. $P(z \ge 0)$
33. $P(z \le -0.13)$
34. $P(z \le -2.15)$
35. $P(z \le 1.20)$
36. $P(z \le 3.20)$
37. $P(z \ge 1.35)$
38. $P(z \ge 2.17)$
39. $P(z \ge -1.20)$
40. $P(z \ge -1.50)$
41. $P(-1.20 \le z \le 2.64)$
42. $P(-2.20 \le z \le 1.40)$
43. $P(-2.18 \le z \le -0.42)$
44. $P(-1.78 \le z \le -1.23)$
45. $P(0 \le z \le 1.62)$
46. $P(0 \le z \le 0.54)$
47. $P(-0.82 \le z \le 0)$
48. $P(-2.37 \le z \le 0)$
49. $P(-0.45 \le z \le 2.73)$
50. $P(-0.73 \le z \le 3.12)$

SECTION 7.3 Areas Under Any Normal Curve

FOCUS POINTS

- Compute the probability of "standardized events."
- Find a z score from a given normal probability (inverse normal).
- Use the inverse normal to solve guarantee problems.

NORMAL DISTRIBUTION AREAS

In many applied situations, the original normal curve is not the standard normal curve. Generally, there will not be a table of areas available for the original normal curve. This does not mean that we cannot find the probability that a measurement x will fall into an interval from a to b. What we must do is *convert* the original measurements x, a, and b to z values.

To compute a z score, we need to know the standard deviation of the distribution. But what if we don't know σ? If we have a range of data values, we can estimate σ. Problems 31 to 34 show how this is done.

PROCEDURE

HOW TO WORK WITH NORMAL DISTRIBUTIONS

To find areas and probabilities for a random variable x that follows a normal distribution with mean μ and standard deviation σ, convert x values to z values using the formula

$$z = \frac{x - \mu}{\sigma}$$

Then use Table 3 of the Appendix to find corresponding areas and probabilities.

EXAMPLE 5

NORMAL DISTRIBUTION PROBABILITY

Let x have a normal distribution with $\mu = 10$ and $\sigma = 2$. Find the probability that an x value selected at random from this distribution is between 11 and 14. In symbols, find $P(11 \leq x \leq 14)$.

SOLUTION: Since probabilities correspond to areas under the distribution curve, we want to find the area under the x curve above the interval from $x = 11$ to $x = 14$. To do so, we will convert the x values to standard z values and then use Table 3 of the Appendix to find the corresponding area under the standard curve.

We use the formula

$$z = \frac{x - \mu}{\sigma}$$

to convert the given x interval to a z interval.

$$z_1 = \frac{11 - 10}{2} = 0.50 \qquad \text{(Use } x = 11, \mu = 10, \sigma = 2.)$$

$$z_2 = \frac{14 - 10}{2} = 2.00 \qquad \text{(Use } x = 14, \mu = 10, \sigma = 2.)$$

The corresponding areas under the x and z curves are shown in Figure 7-20. From Figure 7-20 we see that

$$\begin{aligned} P(11 \leq x \leq 14) &= P(0.50 \leq z \leq 2.00) \\ &= P(z \leq 2.00) - P(z \leq 0.50) \end{aligned}$$

$= 0.9772 - 0.6915$ (From Table 3 of the Appendix)
$= 0.2857$

Interpretation The probability is 0.2857 that an x value selected at random from a normal distribution with mean 10 and standard deviation 2 lies between 11 and 14.

FIGURE 7-20

Corresponding Areas Under the x Curve and z Curve

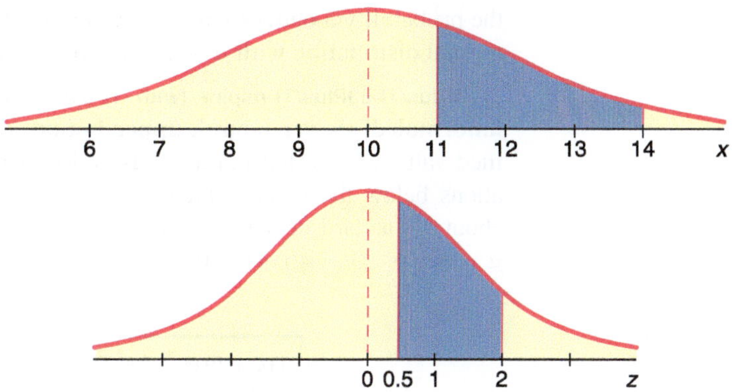

GUIDED EXERCISE 6 | NORMAL DISTRIBUTION PROBABILITY

The life span of a rechargeable battery is the time before the battery must be replaced because it no longer holds a charge. One tablet computer model has a battery with a life span that is normally distributed with a mean of 2.3 years and a standard deviation of 0.4 year. What is the probability that the battery will have to be replaced during the guarantee period of 2 years?

(a) Let x represent the battery life span. The statement that the battery needs to be replaced during the 2-year guarantee period means the life span is less than 2 years, or $x \le 2$. Convert this statement to a statement about z.

→ $z = \dfrac{x - \mu}{\sigma} = \dfrac{2 - 2.3}{0.4} = -0.75$

So, $x \le 2$ means $z \le -0.75$.

(b) Indicate the area to be found in Figure 7-21. Does this area correspond to the probability that $z \le -0.75$?

→ See Figure 7-22.
Yes, the shaded area does correspond to the probability that $z \le -0.75$.

FIGURE 7-21

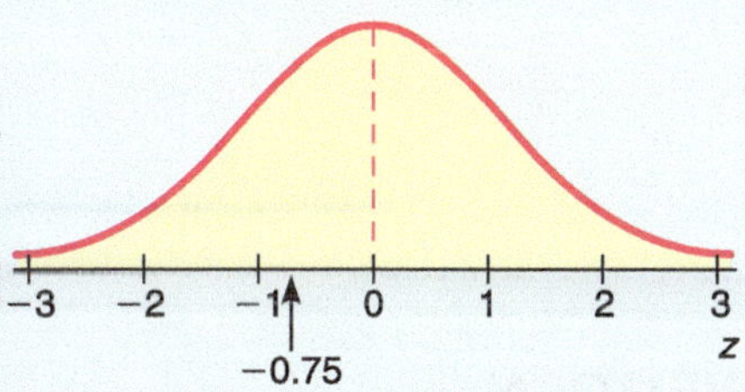

FIGURE 7-22 $z \le -0.75$

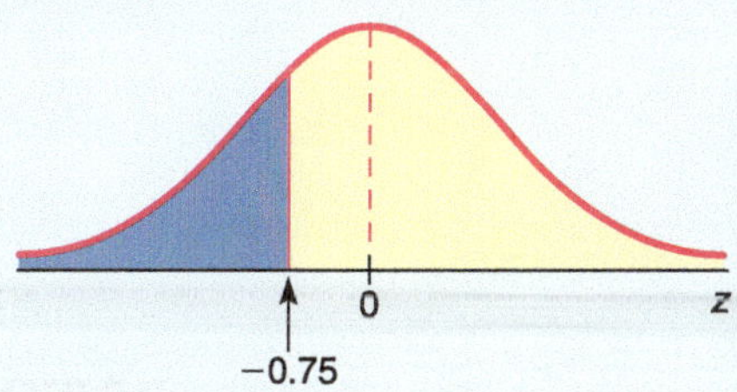

(c) Use Table 3 of the Appendix to find $P(z \le -0.75)$.

→ 0.2266

(d) ***Interpretation*** What is the probability that the battery will fail before the end of the guarantee period? [*Hint:* $P(x \le 2) = P(z \le -0.75)$.]

→ The probability is
$P(x \le 2) = P(z \le -0.75)$
$= 0.2266$
This means that the company will repair or replace about 23% of the batteries.

TECH NOTES The TI-84Plus/TI-83Plus/TI-*n*spire calculators, Excel 2010, and Minitab all provide areas under any normal distribution. Excel 2010 and Minitab give the left-tail area to the left of a specified x value. The TI-84Plus/TI-83Plus/TI-*n*spire has you specify an interval from a lower bound to an upper bound and provides the area under the normal curve for that interval. For example, to solve Guided Exercise 6 regarding the probability a battery will fail during the guarantee period, we find $P(x \leq 2)$ for a normal distribution with $\mu = 2.3$ and $\sigma = 0.4$.

TI-84Plus/TI-83Plus/TI-*n*spire (with TI-84Plus keypad) Press the **DISTR** key, select **2:normalcdf (lower bound, upper bound, μ, σ)** and press Enter. Type in the specified values. For a left-tail area, use a lower bound setting at about 4 standard deviations below the mean. Likewise, for a right-tail area, use an upper bound setting about 4 standard deviations above the mean. For our example, use a lower bound of $\mu - 4\sigma = 2.3 - 4(0.4) = 0.7$.

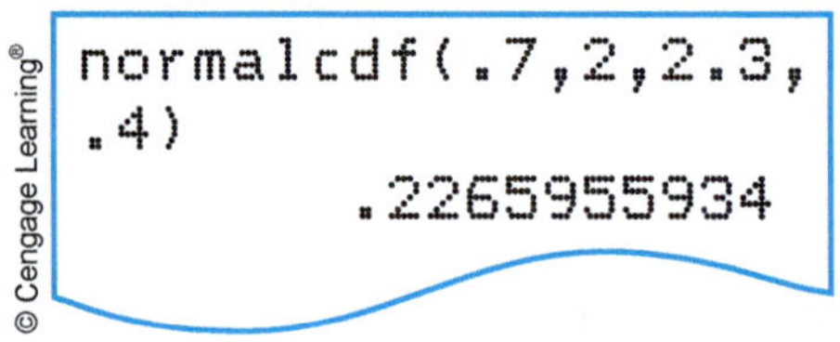

Excel 2010 Select **Insert Function** (fx) In the dialogue box, select **Statistical** for the Category and then for the Function, select **NORM.DIST**. Fill in the dialogue box using True for cumulative.

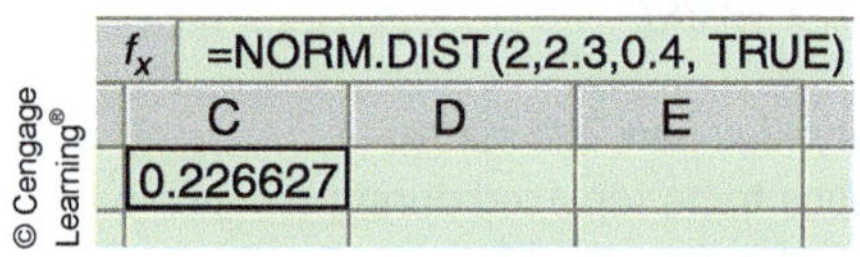

Minitab Use the menu selection **Calc ➤ Probability Distribution ➤ Normal.** Fill in the dialogue box, marking cumulative.

```
Cumulative Distribution Function
Normal with mean = 2.3 and
standard deviation = 0.4
       x    P(X< = x)
     2.0    0.2266
```

INVERSE NORMAL DISTRIBUTION

Finding z or x, given a probability

Inverse normal probability distribution

Sometimes we need to find z or x values that correspond to a given area under the normal curve. This situation arises when we want to specify a guarantee period such that a given percentage of the total products produced by a company last at least as long as the duration of the guarantee period. In such cases, we use the standard normal distribution table "in reverse." When we look up an area and find the corresponding z value, we are using the *inverse normal probability distribution.*

EXAMPLE 6

FIND x, GIVEN PROBABILITY

Magic Video Games, Inc., sells an expensive video games package. Because the package is so expensive, the company wants to advertise an impressive guarantee for the life expectancy of its computer control system. The guarantee policy will refund the full purchase price if the computer fails during the guarantee period. The research department has done tests that show that the mean life for the computer is 30 months, with standard deviation of 4 months. The computer life is normally distributed. How long can the guarantee period be if management does not want to refund the purchase price on more than 7% of the Magic Video packages?

SOLUTION: Let us look at the distribution of lifetimes for the computer control system, and shade the portion of the distribution in which the computer lasts fewer months than the guarantee period. (See Figure 7-23.)

If a computer system lasts fewer months than the guarantee period, a full-price refund will have to be made. The lifetimes requiring a refund are in the shaded region in Figure 7-23. This region represents 7% of the total area under the curve.

We can use Table 3 of the Appendix to find the z value such that 7% of the total area under the *standard* normal curve lies to the left of the z value. Then we convert the z value to its corresponding x value to find the guarantee period.

We want to find the z value with 7% of the area under the standard normal curve to the left of z. Since we are given the area in a left tail, we can use Table 3 of the Appendix directly to find z. The area value is 0.0700. However, this area is not in our table, so we use the closest area, which is 0.0694, and the corresponding z value of $z = -1.48$ (see Table 7-4).

Pierre Arsenault/Alamy

FIGURE 7-23

7% of the Computers Have a Lifetime Less Than the Guarantee Period

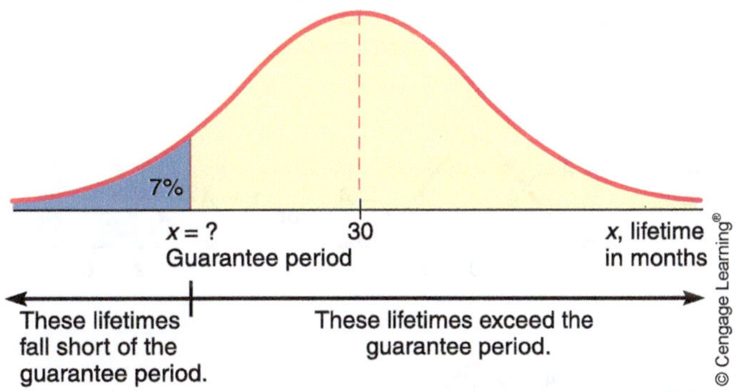

TABLE 7-4 Excerpt from Table 3 of the Appendix

z	.00	...	.07		.08	.09
⋮						
−1.4	.0808		.0708		.0694	.0681
				↑ 0.0700		

To translate this value back to an x value (in months), we use the formula

$$
\begin{aligned}
x &= z\sigma + \mu \\
&= -1.48(4) + 30 \quad \text{(Use } \sigma = 4 \text{ months and } \mu = 30 \text{ months.)} \\
&= 24.08 \text{ months}
\end{aligned}
$$

Interpretation The company can guarantee the Magic Video Games package for $x = 24$ months. For this guarantee period, they expect to refund the purchase price of no more than 7% of the video games packages.

Example 6 had us find a z value corresponding to a given area to the left of z. What if the specified area is to the right of z or between $-z$ and z? Figure 7-24 shows us how to proceed.

COMMENT When we use Table 3 of the Appendix to find a z value corresponding to a given area, we usually use the nearest area value rather than interpolating between values. However, when the area value given is exactly halfway between two area values of the table, we use the z value halfway between the z values of the corresponding table areas. Example 7 demonstrates this procedure. However, this interpolation convention is not always used, especially if the area is changing slowly, as it does in the tail ends of the distribution. *When the z value corresponding to an area is smaller than -2, the standard convention is to use the z value corresponding to the smaller area. Likewise, when the z value is larger than 2, the standard convention is to use the z value corresponding to the larger area.* We will see an example of this special case in Example 1 of Chapter 8.

FIGURE 7-24

Inverse Normal: Use Table 3 of the Appendix to Find z Corresponding to a Given Area A ($0 < A < 1$)

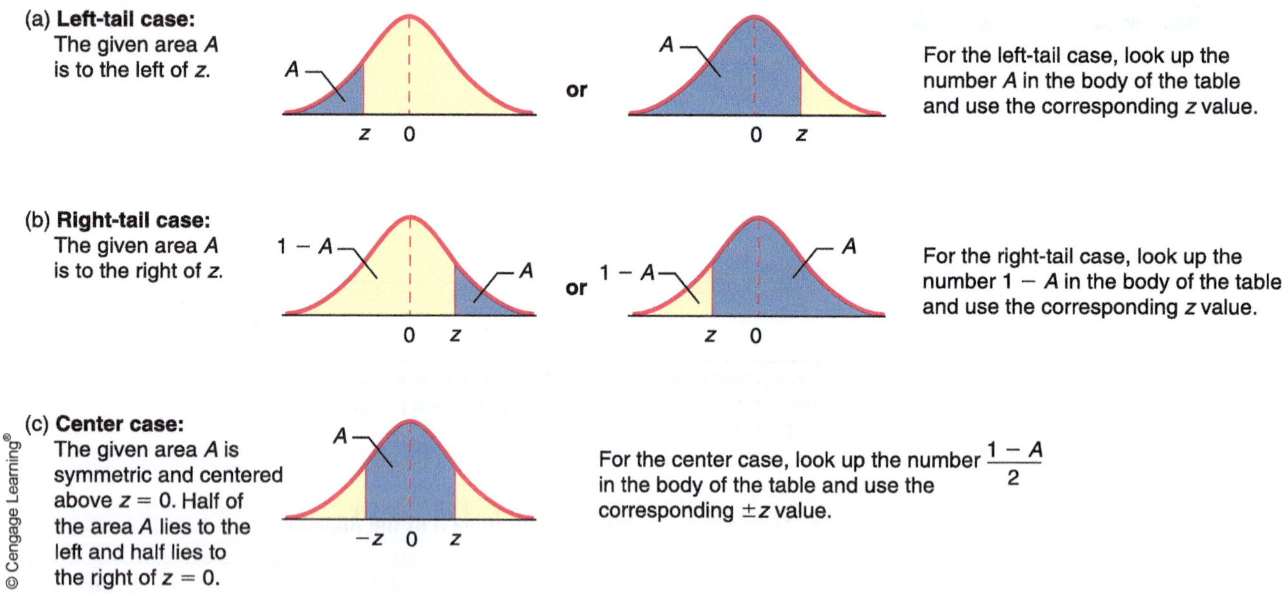

EXAMPLE 7

FIND Z

Find the z value such that 90% of the area under the standard normal curve lies between $-z$ and z.

SOLUTION: Sketch a picture showing the described area (see Figure 7-25).

FIGURE 7-25

Area between $-z$ and z Is 90%

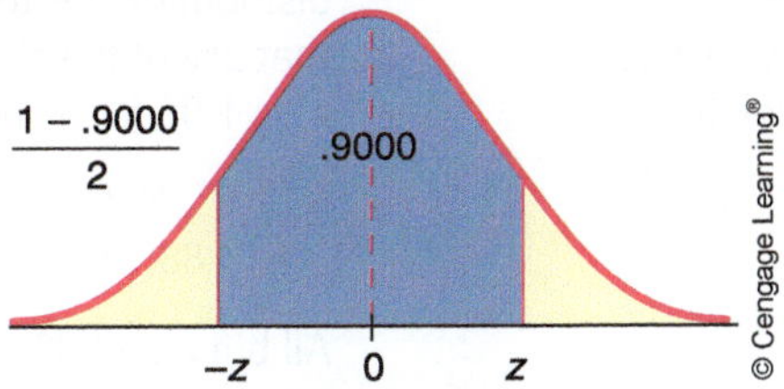

We find the corresponding area in the left tail.

$$(\text{Area left of } -z) = \frac{1 - 0.9000}{2}$$
$$= 0.0500$$

Looking in Table 7-5, we see that 0.0500 lies exactly between areas 0.0495 and 0.0505. The halfway value between $z = -1.65$ and $z = -1.64$ is $z = -1.645$. Therefore, we conclude that 90% of the area under the standard normal curve lies between the z values -1.645 and 1.645.

TABLE 7-5 **Excerpt from Table 3 of the Appendix**

z	...	.04		.05
:				
−1.6		.0505		.0495
			↑ 0.0500	

GUIDED EXERCISE 7 | FIND z

Find the z value such that 3% of the area under the standard normal curve lies to the right of z.

(a) Draw a sketch of the standard normal distribution showing the described area (Figure 7-26).

FIGURE 7-26 3% of Total Area Lies to the Right of z

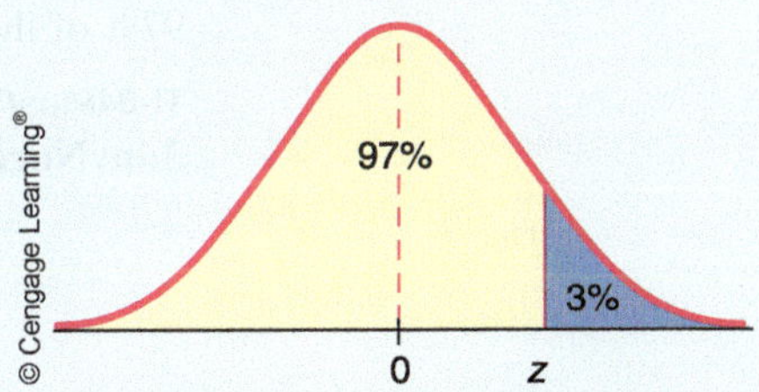

(b) Find the area to the left of z.

Area to the left of $z = 1 - 0.0300 = 0.9700$.

(c) Look up the area in Table 7-6 and find the corresponding z.

The closest area is 0.9699. This area is to the left of $z = 1.88$.

TABLE 7-6 **Excerpt from Table 3 of the Appendix**

z	.00	.01	.02	.03	.04	.05	.06	.07	.08	.09
1.8	.9641	.9649	.9656	.9664	.9671	.9678	.9686	.9693	.9699	.9706
1.9	.9713	.9719	.9726	.9732	.9738	.9744	.9750	.9756	.9761	.9767

Continued

GUIDED EXERCISE 7 *continued*

(d) Suppose the time to complete a test is normally distributed with $\mu = 40$ minutes and $\sigma = 5$ minutes. After how many minutes can we expect all but about 3% of the tests to be completed?

→ We are looking for an x value such that 3% of the normal distribution lies to the right of x. In part (c), we found that 3% of the standard normal curve lies to the right of $z = 1.88$. We convert $z = 1.88$ to an x value.

$$x = z\sigma + \mu$$
$$= 1.88(5) + 40 = 49.4 \text{ minutes}$$

All but about 3% of the tests will be complete after 50 minutes.

(e) Use Table 7-7 to find a z value such that 3% of the area under the standard normal curve lies to the left of z.

→ The closest area is 0.0301. This is the area to the left of $z = -1.88$.

TABLE 7-7 **Excerpt from Table 3 of the Appendix**

z	.00	.01	.02	.03	.04	.05	.06	.07	.08	.09
−1.9	.0287	.0281	.0274	.0268	.0262	.0256	.0250	.0244	.0239	.0233
−1.8	.0359	.0351	.0344	.0336	.0329	.0322	.0314	.0307	.0301	.0294

(f) Compare the z value of part (c) with the z value of part (e). Is there any relationship between the z values?

→ One z value is the negative of the other. This result is expected because of the symmetry of the normal distribution.

TECH NOTES When we are given a z value and we find an area to the left of z, we are using a normal distribution function. When we are given an area to the left of z and we find the corresponding z, we are using an inverse normal distribution function. The TI-84Plus/TI-83Plus/TI-*n*spire calculators, Excel 2010, and Minitab all have inverse normal distribution functions for any normal distribution. For instance, to find an x value from a normal distribution with mean 40 and standard deviation 5 such that 97% of the area lies to the left of x, use the described instructions.

TI-84Plus/TI-83Plus/TI-*n*spire (with TI-84Plus keypad) Press the **DISTR** key and select **3:invNorm(area,μ,σ)**.

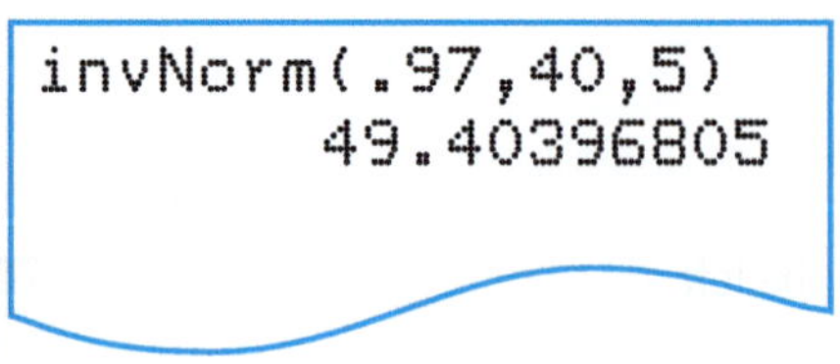

Excel 2010 Select **Insert Function** (f_x) In the dialogue box, select **Statistical** for the Category and then for the Function, select **NORM.INV**. Fill in the dialogue box.

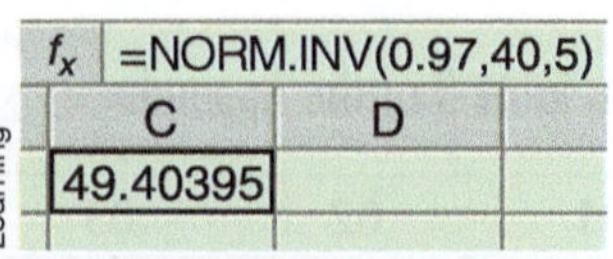

LOOKING FORWARD

In our work with confidence intervals (Chapter 8), we will use inverse probability distributions for the normal distribution and for the Student's *t* distribution (a similar distribution introduced in Chapter 8). Just as in Example 7, we'll use inverse probability distributions to identify values such that 90%, 95%, or 99% of the area under the distribution graph centered over the mean falls between the values.

Minitab Use the menu selection **Calc ➤ Probability Distribution ➤ Normal.** Fill in the dialogue box, marking Inverse Cumulative.

```
Inverse Cumulative Distribution Function
Normal with mean = 40.000 and
  standard deviation = 5.00000
          P(X< = x)        x
              0.9700   49.4040
```

CRITICAL THINKING

CHECKING FOR NORMALITY

How can we tell if data follow a normal distribution? There are several checks we can make. The following procedure lists some guidelines.

PROCEDURE

HOW TO DETERMINE WHETHER DATA HAVE A NORMAL DISTRIBUTION

The following guidelines represent some useful devices for determining whether or not data follow a normal distribution.

1. **Histogram:** Make a histogram. For a normal distribution, the histogram should be roughly bell-shaped.
2. **Outliers:** For a normal distribution, there should not be more than one outlier. One way to check for outliers is to use a box-and-whisker plot. Recall that outliers are those data values that are

   above Q_3 by an amount greater than 1.5 × interquartile range
   below Q_1 by an amount greater than 1.5 × interquartile range

3. **Skewness:** Normal distributions are symmetric. One measure of skewness for sample data is given by Pearson's index:

   $$\text{Pearson's index} = \frac{3(\bar{x} - \text{median})}{s}$$

   An index value greater than 1 or less than −1 indicates skewness. Skewed distributions are not normal.

4. **Normal quantile plot (or normal probability plot):** This plot is provided through statistical software on a computer or graphing calculator. The Using Technology feature, Application 1, at the end of this chapter gives a brief description of how such plots are constructed. The section also gives commands for producing such plots on the TI-84Plus/TI-83Plus/TI-*n*spire calculators, Minitab, or SPSS.

   Examine a normal quantile plot of the data.

   If the points lie close to a straight line, the data come from a distribution that is approximately normal.

   If the points do not lie close to a straight line or if they show a pattern that is not a straight line, the data are likely to come from a distribution that is not normal.

EXAMPLE 8

ASSESSING NORMALITY

Consider the following data, which are rounded to the nearest integer.

19 19 19 16 21 14 23 17 19 20 18 24 20 13 16
17 19 18 19 17 21 24 18 23 19 21 22 20 20 20
24 17 20 22 19 22 21 18 20 22 16 15 21 23 21
18 18 20 15 25

(a) Look at the histogram and box-and-whisker plot generated by Minitab in Figure 7-27 and comment about normality of the data from these indicators.

FIGURE 7-27

Histogram and Box-and-Whisker Plot

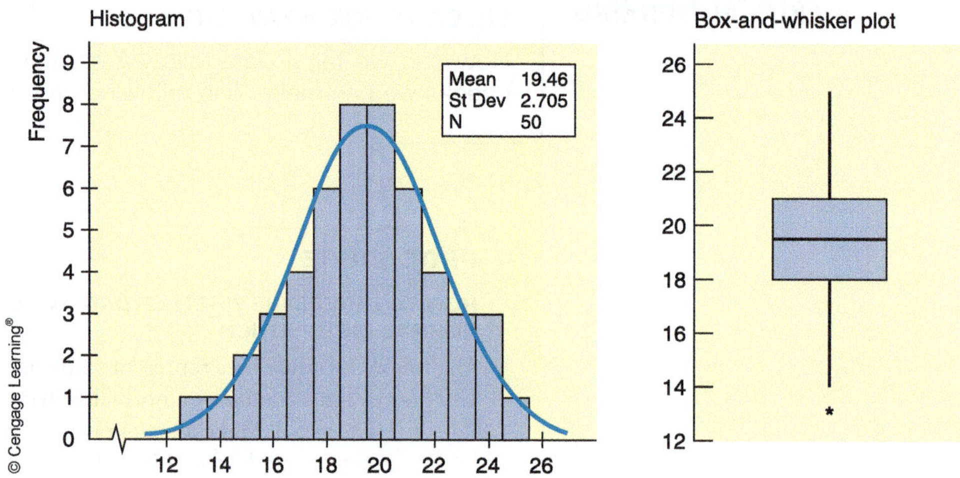

SOLUTION: Note that the histogram is approximately normal. The box-and-whisker plot shows just one outlier. Both of these graphs indicate normality.

(b) Use Pearson's index to check for skewness.

SOLUTION: Summary statistics from Minitab:

```
Variable   N  N*    Mean  Se Mean  StDev  Minimum      Q1  Median      Q3
C2        50   0  19.460    0.382  2.705   13.000  18.000  19.500  21.000
Variable   Maximum
C2          25.000
```

We see that $\bar{x} = 19.46$, median $= 19.5$, and $s = 2.705$.

$$\text{Pearson's index} = \frac{3(19.46 - 19.5)}{2.705} \approx -0.04$$

Since the index is between -1 and 1, we detect no skewness. The data appear to be symmetric.

(c) Look at the normal quantile plot in Figure 7-28 and comment on normality.

FIGURE 7-28

Normal Quantile Plot

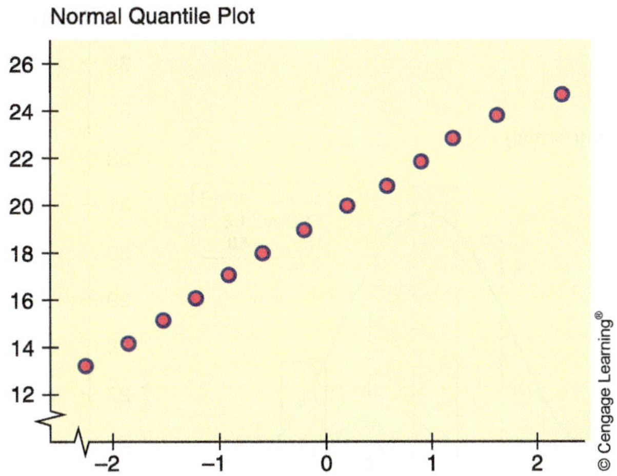

SOLUTION: The data fall close to a straight line, so the data appear to come from a normal distribution.

(d) ***Interpretation*** Interpret the results.

SOLUTION: The histogram is roughly bell-shaped, there is only one outlier, Pearson's index does not indicate skewness, and the points on the normal quantile plot lie fairly close to a straight line. It appears that the data are likely from a distribution that is approximately normal.

VIEWPOINT Want to Be an Archaeologist?

Each year about 4500 students work with professional archaeologists in scientific research at the Crow Canyon Archaeological Center, Cortez, Colorado. In fact, Crow Canyon was included in The Princeton Review Guide to America's Top 100 Internships. *The nonprofit, multidisciplinary program at Crow Canyon enables students and laypeople with little or no archaeology background to get started in archaeological research. The only requirement is that you be interested in Native American culture and history. By the way, a knowledge of introductory statistics could come in handy for this internship. For more information about the program, visit the Brase/Brase statistics site at* **http://www.cengagebrain.com** *and find the link to Crow Canyon.*

SECTION 7.3 PROBLEMS

1. *Statistical Literacy* Consider a normal distribution with mean 30 and standard deviation 2. What is the probability a value selected at random from this distribution is greater than 30?

2. *Statistical Literacy* Suppose 5% of the area under the standard normal curve lies to the right of z. Is z positive or negative?

3. *Statistical Literacy* Suppose 5% of the area under the standard normal curve lies to the left of z. Is z positive or negative?

4. *Critical Thinking: Normality* Consider the following data. The summary statistics, histogram, and normal quantile plot were generated by Minitab.

27	27	27	28	28	28	28	28	28	29	29	29	29	29	29
29	29	29	29	30	30	30	30	30	30	30	30	30	30	30
30	31	31	31	31	31	31	31	31	32	32	32	32	33	33
33	33	33	34	34										

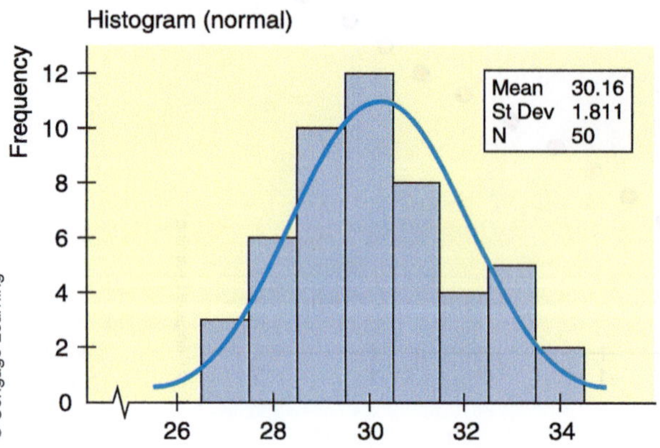

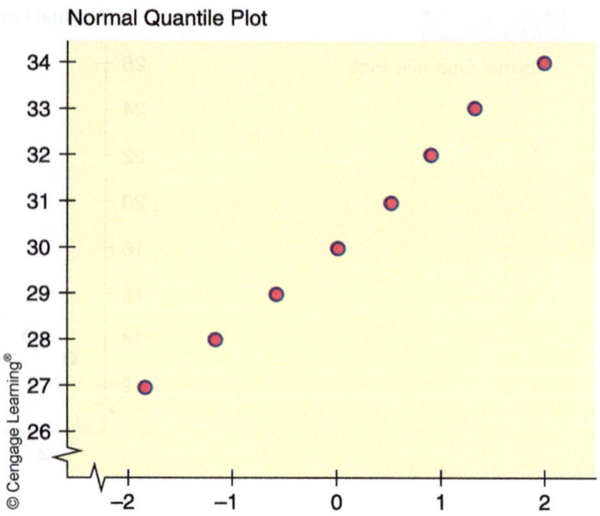


```
Variable   N  N*    Mean  Se Mean  StDev  Minimum      Q1  Median      Q3
Data      50   0  30.160    0.256  1.811   27.000  29.000  30.000  31.000
Variable    Maximum
Data         24.000
```

(a) Does the histogram indicate normality for the data distribution? Explain.
(b) Does the normal quantile plot indicate normality for the data distribution? Explain.
(c) Compute the interquartile range and check for outliers.
(d) Compute Pearson's index. Does the index value indicate skewness?
(e) Using parts (a) through (d), would you say the data are from a normal distribution?

Basic Computation: Find Probabilities In Problems 5–14, assume that x has a normal distribution with the specified mean and standard deviation. Find the indicated probabilities.

5. $P(3 \le x \le 6);\ \mu = 4;\ \sigma = 2$
6. $P(10 \le x \le 26);\ \mu = 15;\ \sigma = 4$
7. $P(50 \le x \le 70);\ \mu = 40;\ \sigma = 15$
8. $P(7 \le x \le 9);\ \mu = 5;\ \sigma = 1.2$
9. $P(8 \le x \le 12);\ \mu = 15;\ \sigma = 3.2$
10. $P(40 \le x \le 47);\ \mu = 50;\ \sigma = 15$
11. $P(x \ge 30);\ \mu = 20;\ \sigma = 3.4$
12. $P(x \ge 120);\ \mu = 100;\ \sigma = 15$
13. $P(x \ge 90);\ \mu = 100;\ \sigma = 15$
14. $P(x \ge 2);\ \mu = 3;\ \sigma = 0.25$

Basic Computation: Find z Values In Problems 15–24, find the z value described and sketch the area described.

15. Find z such that 6% of the standard normal curve lies to the left of z.
16. Find z such that 5.2% of the standard normal curve lies to the left of z.
17. Find z such that 55% of the standard normal curve lies to the left of z.
18. Find z such that 97.5% of the standard normal curve lies to the left of z.
19. Find z such that 8% of the standard normal curve lies to the right of z.
20. Find z such that 5% of the standard normal curve lies to the right of z.
21. Find z such that 82% of the standard normal curve lies to the right of z.

22. Find z such that 95% of the standard normal curve lies to the right of z.

23. Find the z value such that 98% of the standard normal curve lies between $-z$ and z.

24. Find the z value such that 95% of the standard normal curve lies between $-z$ and z.

25. *Medical: Blood Glucose* A person's blood glucose level and diabetes are closely related. Let x be a random variable measured in milligrams of glucose per deciliter (1/10 of a liter) of blood. After a 12-hour fast, the random variable x will have a distribution that is approximately normal with mean $\mu = 85$ and standard deviation $\sigma = 25$ (Source: *Diagnostic Tests with Nursing Implications,* edited by S. Loeb, Springhouse Press). *Note:* After 50 years of age, both the mean and standard deviation tend to increase. What is the probability that, for an adult (under 50 years old) after a 12-hour fast,
(a) x is more than 60?
(b) x is less than 110?
(c) x is between 60 and 110?
(d) x is greater than 140 (borderline diabetes starts at 140)?

26. *Medical: Blood Protoplasm* Porphyrin is a pigment in blood protoplasm and other body fluids that is significant in body energy and storage. Let x be a random variable that represents the number of milligrams of porphyrin per deciliter of blood. In healthy adults, x is approximately normally distributed with mean $\mu = 38$ and standard deviation $\sigma = 12$ (see reference in Problem 25). What is the probability that
(a) x is less than 60?
(b) x is greater than 16?
(c) x is between 16 and 60?
(d) x is more than 60? (This may indicate an infection, anemia, or another type of illness.)

27. *Archaeology: Hopi Village* Thickness measurements of ancient prehistoric Native American pot shards discovered in a Hopi village are approximately normally distributed, with a mean of 5.1 millimeters (mm) and a standard deviation of 0.9 mm (Source: *Homol'ovi II: Archaeology of an Ancestral Hopi Village, Arizona,* edited by E. C. Adams and K. A. Hays, University of Arizona Press). For a randomly found shard, what is the probability that the thickness is
(a) less than 3.0 mm?
(b) more than 7.0 mm?
(c) between 3.0 mm and 7.0 mm?

28. *Law Enforcement: Police Response Time* Police response time to an emergency call is the difference between the time the call is first received by the dispatcher and the time a patrol car radios that it has arrived at the scene (based on information from *The Denver Post*). Over a long period of time, it has been determined that the police response time has a normal distribution with a mean of 8.4 minutes and a standard deviation of 1.7 minutes. For a randomly received emergency call, what is the probability that the response time will be
(a) between 5 and 10 minutes?
(b) less than 5 minutes?
(c) more than 10 minutes?

29. *Guarantee: Batteries* Quick Start Company makes 12-volt car batteries. After many years of product testing, the company knows that the average life of a Quick Start battery is normally distributed, with a mean of 45 months and a standard deviation of 8 months.

(a) If Quick Start guarantees a full refund on any battery that fails within the 36-month period after purchase, what percentage of its batteries will the company expect to replace?

(b) *Inverse Normal Distribution* If Quick Start does not want to make refunds for more than 10% of its batteries under the full-refund guarantee policy, for how long should the company guarantee the batteries (to the nearest month)?

30. *Guarantee: Watches* Accrotime is a manufacturer of quartz crystal watches. Accrotime researchers have shown that the watches have an average life of 28 months before certain electronic components deteriorate, causing the watch to become unreliable. The standard deviation of watch lifetimes is 5 months, and the distribution of lifetimes is normal.

(a) If Accrotime guarantees a full refund on any defective watch for 2 years after purchase, what percentage of total production should the company expect to replace?

(b) *Inverse Normal Distribution* If Accrotime does not want to make refunds on more than 12% of the watches it makes, how long should the guarantee period be (to the nearest month)?

31. *Expand Your Knowledge: Estimating the Standard Deviation* *Consumer Reports* gave information about the ages at which various household products are replaced. For example, color TVs are replaced at an average age of $\mu = 8$ years after purchase, and the (95% of data) range was from 5 to 11 years. Thus, the range was $11 - 5 = 6$ years. Let x be the age (in years) at which a color TV is replaced. Assume that x has a distribution that is approximately normal.

(a) The empirical rule (Section 7.1) indicates that for a symmetrical and bell-shaped distribution, approximately 95% of the data lies within two standard deviations of the mean. Therefore, a 95% range of data values extending from $\mu - 2\sigma$ to $\mu + 2\sigma$ is often used for "commonly occurring" data values. Note that the interval from $\mu - 2\sigma$ to $\mu + 2\sigma$ is 4σ in length. This leads to a "rule of thumb" for estimating the standard deviation from a 95% range of data values.

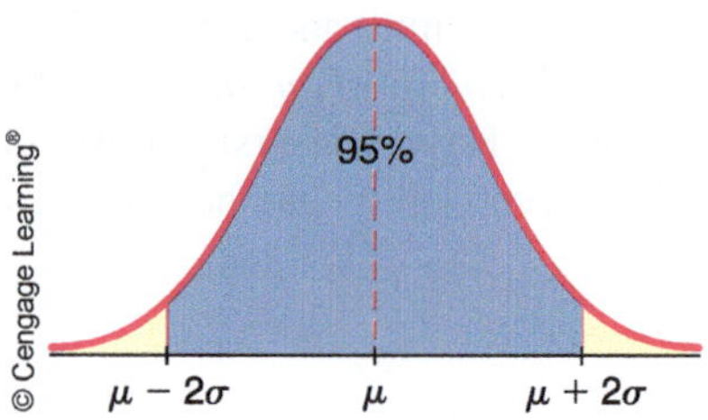

ESTIMATING THE STANDARD DEVIATION

For a symmetric, bell-shaped distribution,

$$\textbf{standard deviation} \approx \frac{\text{range}}{4} \approx \frac{\text{high value} - \text{low value}}{4}$$

where it is estimated that about 95% of the commonly occurring data values fall into this range.

Use this "rule of thumb" to approximate the standard deviation of x values, where x is the age (in years) at which a color TV is replaced.

(b) What is the probability that someone will keep a color TV more than 5 years before replacement?

(c) What is the probability that someone will keep a color TV fewer than 10 years before replacement?

(d) *Inverse Normal Distribution* Assume that the average life of a color TV is 8 years with a standard deviation of 1.5 years before it breaks. Suppose that a company guarantees color TVs and will replace a TV that breaks while under guarantee with a new one. However, the company does not want to replace more than 10% of the TVs under guarantee. For how long should the guarantee be made (rounded to the nearest tenth of a year)?

32. *Estimating the Standard Deviation: Refrigerator Replacement* *Consumer Reports* indicated that the average life of a refrigerator before replacement is $\mu = 14$ years with a (95% of data) range from 9 to 19 years. Let x = age at which a refrigerator is replaced. Assume that x has a distribution that is approximately normal.
(a) Find a good approximation for the standard deviation of x values. *Hint:* See Problem 31.
(b) What is the probability that someone will keep a refrigerator fewer than 11 years before replacement?
(c) What is the probability that someone will keep a refrigerator more than 18 years before replacement?
(d) *Inverse Normal Distribution* Assume that the average life of a refrigerator is 14 years, with the standard deviation given in part (a) before it breaks. Suppose that a company guarantees refrigerators and will replace a refrigerator that breaks while under guarantee with a new one. However, the company does not want to replace more than 5% of the refrigerators under guarantee. For how long should the guarantee be made (rounded to the nearest tenth of a year)?

33. *Estimating the Standard Deviation: Veterinary Science* The resting heart rate for an adult horse should average about $\mu = 46$ beats per minute with a (95% of data) range from 22 to 70 beats per minute, based on information from the *Merck Veterinary Manual* (a classic reference used in most veterinary colleges). Let x be a random variable that represents the resting heart rate for an adult horse. Assume that x has a distribution that is approximately normal.
(a) Estimate the standard deviation of the x distribution. *Hint:* See Problem 31.
(b) What is the probability that the heart rate is fewer than 25 beats per minute?
(c) What is the probability that the heart rate is greater than 60 beats per minute?
(d) What is the probability that the heart rate is between 25 and 60 beats per minute?
(e) *Inverse Normal Distribution* A horse whose resting heart rate is in the upper 10% of the probability distribution of heart rates may have a secondary infection or illness that needs to be treated. What is the heart rate corresponding to the upper 10% cutoff point of the probability distribution?

34. *Estimating the Standard Deviation: Veterinary Science* How much should a healthy kitten weigh? A healthy 10-week-old (domestic) kitten should weigh an average of $\mu = 24.5$ ounces with a (95% of data) range from 14 to 35 ounces. (See reference in Problem 33.) Let x be a random variable that represents the weight (in ounces) of a healthy 10-week-old kitten. Assume that x has a distribution that is approximately normal.
(a) Estimate the standard deviation of the x distribution. *Hint:* See Problem 31.
(b) What is the probability that a healthy 10-week-old kitten will weigh less than 14 ounces?
(c) What is the probability that a healthy 10-week-old kitten will weigh more than 33 ounces?
(d) What is the probability that a healthy 10-week-old kitten will weigh between 14 and 33 ounces?
(e) *Inverse Normal Distribution* A kitten whose weight is in the bottom 10% of the probability distribution of weights is called *undernourished.* What is the cutoff point for the weight of an undernourished kitten?

35. *Insurance: Satellites* A relay microchip in a telecommunications satellite has a life expectancy that follows a normal distribution with a mean of 90 months and a standard deviation of 3.7 months. When this computer-relay microchip malfunctions, the entire satellite is useless. A large

Max Dannenbaum/Stone/Getty Images

London insurance company is going to insure the satellite for $50 million. Assume that the only part of the satellite in question is the microchip. All other components will work indefinitely.

(a) *Inverse Normal Distribution* For how many months should the satellite be insured to be 99% confident that it will last beyond the insurance date?

(b) If the satellite is insured for 84 months, what is the probability that it will malfunction before the insurance coverage ends?

(c) If the satellite is insured for 84 months, what is the expected loss to the insurance company?

(d) If the insurance company charges $3 million for 84 months of insurance, how much profit does the company expect to make?

Answers and Key Steps

TO ODD-NUMBERED PROBLEMS

CHAPTER 1

Section 1.1

1. An individual is a member of the population of interest. A variable is an aspect of an individual subject or object being measured.
3. A parameter is a numerical measurement describing data from a population. A statistic is a numerical measurement describing data from a sample.
5. (a) Nominal level. There is no apparent order relationship among responses.
(b) Ordinal level. There is an increasing relationship from worst to best level of service. The interval between service levels is not meaningful, nor are ratios.
7. (a) Response regarding meal ordered at fast-food restaurants. (b) Qualitative. (c) Responses for *all* adult fast-food customers in the U.S.
9. (a) Nitrogen concentration (mg nitrogen/L water). (b) Quantitative. (c) Nitrogen concentration (mg nitrogen/L water) in the entire lake.
11. (a) Ratio. (b) Interval. (c) Nominal. (d) Ordinal. (e) Ratio. (f) Ratio.
13. (a) Nominal. (b) Ratio. (c) Interval. (d) Ordinal. (e) Ratio. (f) Interval.
15. Answers vary.
(a) For example: Use pounds. Round weights to the nearest pound. Since backpacks might weigh as much as 30 pounds, you might use a high-quality bathroom scale. (b) Some students may not allow you to weigh their backpacks for privacy reasons, etc. (c) Possibly. Some students may want to impress you with the heaviness of their backpacks, or they may be embarrassed about the "junk" they have stowed inside and thus may clean out their backpacks.

Section 1.2

1. In a stratified sample, random samples from each stratum are included. In a cluster sample, the clusters to be included are selected at random and then all members of each selected cluster are included.
3. The advice is wrong. A sampling error accounts only for the difference in results based on the use of a sample rather than of the entire population.
5. No, even though the sample is random, some students younger than 18 or older than 20 may not have been included in the sample.
7. (a) Stratified. (b) No, because each pooled sample would have 100 season ticket holders for men's basketball games and 100 for women's basketball games. Samples with, say, 125 ticket holders for men's basketball games and 75 for women's games are not possible.
9. Use a random-number table to select four distinct numbers corresponding to people in your class. (a) Reasons may vary. For instance, the first four students may make a special effort to get to class on time. (b) Reasons may vary. For instance, four students who come in late might all be nursing students enrolled in an anatomy and physiology class that meets the hour before in a faraway building. They may be more motivated than other students to complete a degree requirement. (c) Reasons may vary. For instance, four students sitting in the back row might be less inclined to participate in class discussions. (d) Reasons may vary. For instance, the tallest students might all be male.
11. Answers vary. Use groups of two digits.
13. Select a starting place in the table and group the digits in groups of four. Scan the table by rows and include the first six groups with numbers between 0001 and 8615.
15. (a) Yes, when a die is rolled several times, the same number may appear more than once. Outcome on the fourth roll is 2. (b) No, for a fair die, the outcomes are random.
17. Since there are five possible outcomes for each question, read single digits from a random-number table. Select a starting place and proceed until you have 10 digits from 1 to 5. Repetition is required. The correct answer for each question will be the letter choice corresponding to the digit chosen for that question.
19. (a) Simple random sample. (b) Cluster sample. (c) Convenience sample. (d) Systematic sample. (e) Stratified sample.

Chapter 1 Review

1. Because of the requirement that each number appear only once in any row, column, or box, it would be very inefficient to use a random-number table to select the numbers. It's better to simply look at existing numbers, list possibilities that meet the requirement, and eliminate numbers that don't work.
3. (a) Stratified. (b) Students on your campus with work-study jobs. (c) Hours scheduled; quantitative; ratio. (d) Rating of applicability of work experience to future employment; qualitative; ordinal. (e) Statistic. (f) 60%; The people choosing not to respond may have some characteristics, such as not working many hours, that would bias the study. (g) No. The sample frame is restricted to one campus.

5. Assign digits so that 3 out of the 10 digits 0 through 9 correspond to the answer "Yes" and 7 of the digits correspond to the answer "No." One assignment is digits 0, 1, and 2 correspond to "Yes," while digits 3, 4, 5, 6, 7, 8, and 9 correspond to "No." Starting with line 1, block 1 of Table 1, this assignment of digits gives the sequence No, Yes, No, No, Yes, No, No.
7. (a) Observational study. (b) Experiment.
9. Possible directions on survey questions: Give height in inches, give age as of last birthday, give GPA to one decimal place, and so forth. Think about the types of responses you wish to have on each question.
11. (a) Experiment, since a treatment is imposed on one colony. (b) The control group receives normal daylight/darkness conditions. The treatment group has light 24 hours per day. (c) The number of fireflies living at the end of 72 hours. (d) Ratio.

CHAPTER 2

Section 2.2

1. (a) Yes, the percentages total more than 100%.
   (b) No, in a circle graph the percentages must total 100% (within rounding error).
   (c) Yes, the graph is organized in order from most frequently selected reason to least.
3. Pareto chart, because it shows the items in order of importance to the greatest number of employees.
5. Highest Level of Education and Average Annual Household Income (in thousands of dollars).

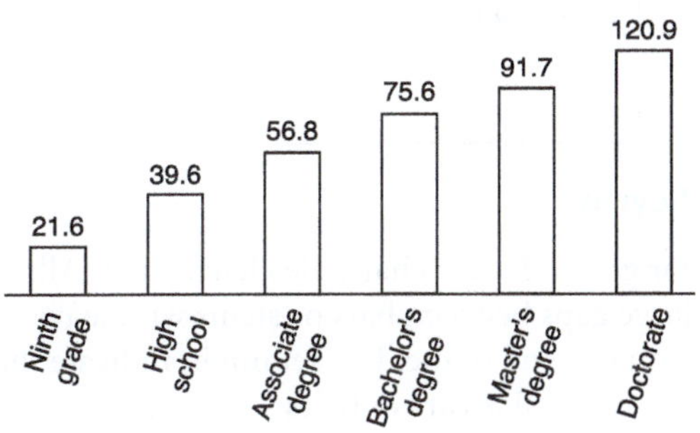

7. Annual Harvest (1000 Metric Tons)—Pareto Chart

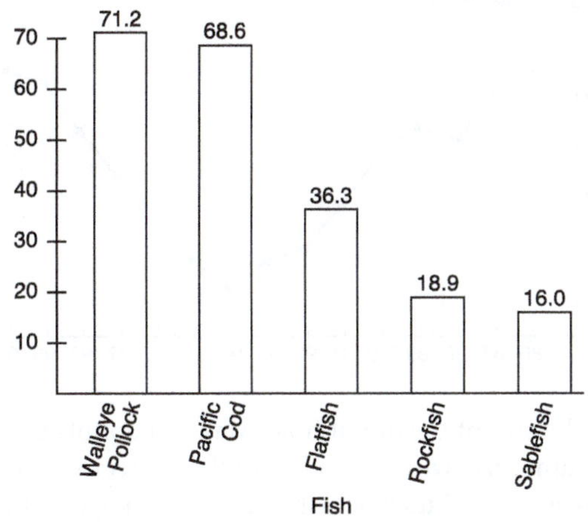

9. Where We Hide the Mess

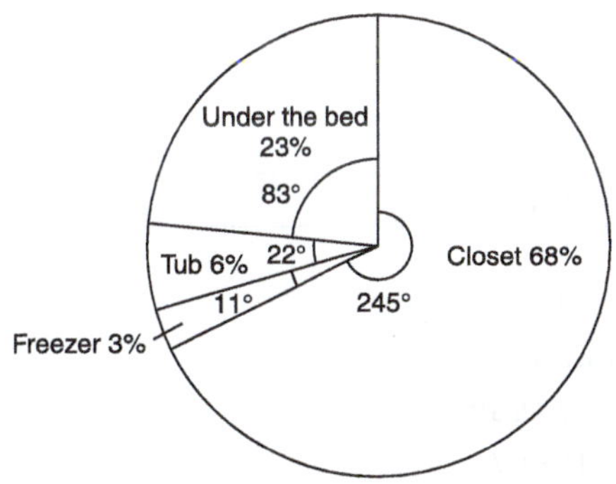

11. (a) Hawaii Crime Rate per 100,000 Population

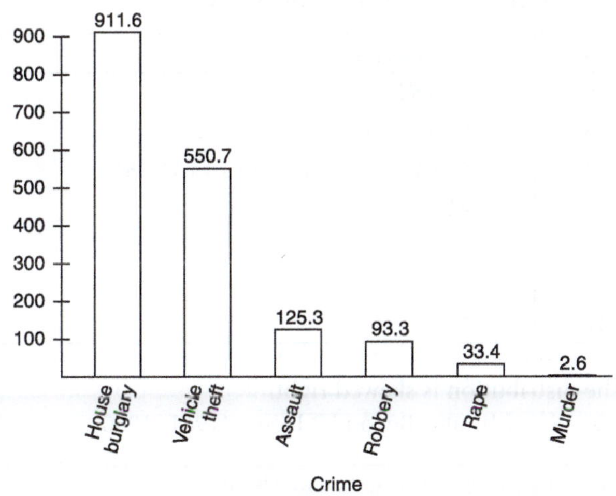

(b) A circle graph is not appropriate because the data do not reflect all types of crime. Also, the same person may have been the victim of more than one crime.

13. Elevation of Pyramid Lake Surface—Time Plot

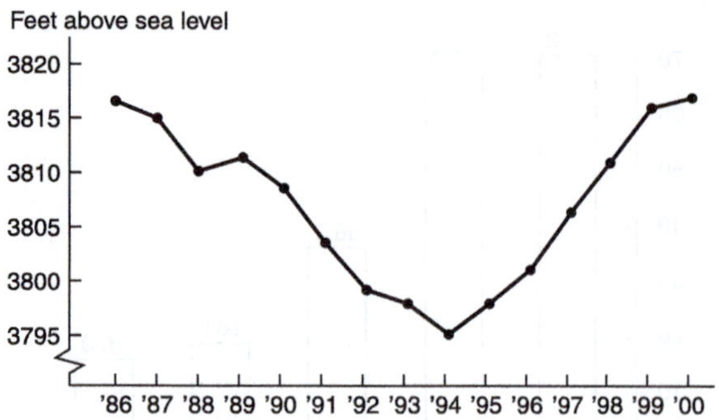

15. (a) The size of the donut hole. Make all donuts exactly the same size, with the radius of the respective holes the same as well. Data labels showing percentages for each response would also be useful. (b) College graduates have a higher frequency of "no" response than those having only high school or less.

Section 2.3

1. (a) Longevity of Cowboys

4	7 = 47 years
4	7
5	2 7 8 8
6	1 6 6 8 8
7	0 2 2 3 3 5 6 7
8	4 4 4 5 6 6 7 9
9	0 1 1 2 3 7

(b) Yes, certainly these cowboys lived long lives.

3. Average Length of Hospital Stay

5	2 = 5.2 years
5	2 3 5 5 6 7
6	0 2 4 6 6 7 7 8 8 8 8 9 9
7	0 0 0 0 0 0 1 1 1 2 2 2 3 3 3 3 4 4 5 5 6 6 8
8	4 5 7
9	4 6 9
10	0 3
11	1

The distribution is skewed right.

5. (a) Minutes Beyond 2 Hours (1961–1980)

0	9 = 9 minutes past 2 hours
0	9 9
1	0 0 2 3 3 4
1	5 5 6 6 7 8 8 9
2	0 2 3 3

(b) Minutes Beyond 2 Hours (1981–2000)

0	7 = 7 minutes past 2 hours
0	7 7 7 8 8 8 8 9 9 9 9 9 9 9
1	0 0 1 1 4

(c) In more recent years, the winning times have been closer to 2 hours, with all the times between 7 and 14 minutes over 2 hours. In the earlier period, more than half the times were more than 2 hours and 14 minutes.

7. Milligrams of Tar per Cigarette

1	0 = 1.0 mg tar		
1	0	11	4
2		12	0 4 8
3		13	7
4	1 5	14	1 5 9
5		15	0 1 2 8
6		16	0 6
7	3 8	17	0
8	0 6 8		
9	0		
10		29	8

The value 29.8 may be an outlier.

9. Milligrams of Nicotine per Cigarette

0	1 = 0.1 milligram
0	1 4 4
0	5 6 6 6 7 7 7 8 8 9 9 9
1	0 0 0 0 0 0 0 1 2
1	
2	0

Chapter 2 Review

1. (a) Bar graph, Pareto chart, pie chart. (b) All.
3. Any large gaps between bars or stems with leaves at the beginning or end of the data set might indicate that the extreme data values are outliers.
5. (a) Yes, with lines used instead of bars. However, because of the perspective nature of the drawing, the lengths of the bars do not represent the mileages. Thus, the scale for each bar changes. (b) Yes. The scale does not change, and the viewer is not distracted by the graphic of the highway.
7. Problems with Tax Returns

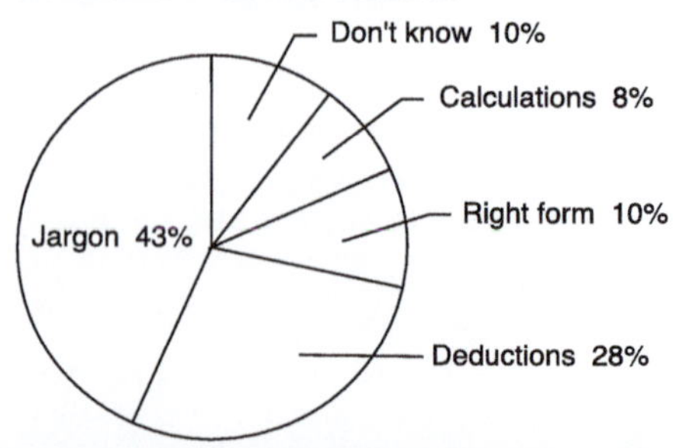

9. (a) Class width = 11.

Class Limits	Class Boundaries	Midpoint	Frequency	Relative Frequency	Cumulative Frequency
69–79	68.5–79.5	74	2	0.03	2
80–90	79.5–90.5	85	3	0.05	5
91–101	90.5–101.5	96	8	0.13	13
102–112	101.5–112.5	107	19	0.32	32
113–123	112.5–123.5	118	22	0.37	54
124–134	123.5–134.5	129	3	0.05	57
135–145	134.5–145.5	140	3	0.05	60

(b, c) Trunk Circumference (mm)—Histogram, Relative-Frequency Histogram

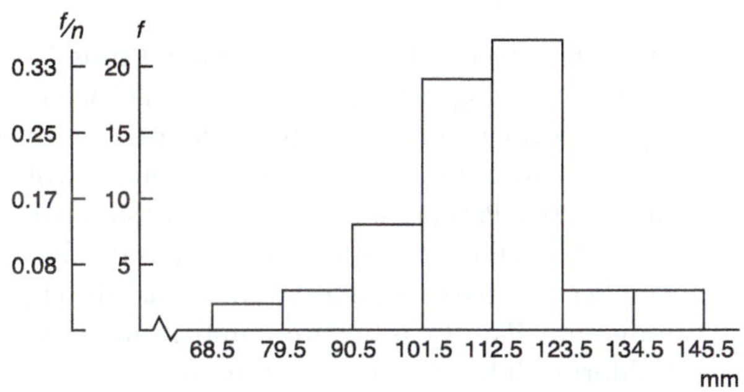

(d) Skewed slightly left.

(e) Trunk Circumference (mm)—Ogive

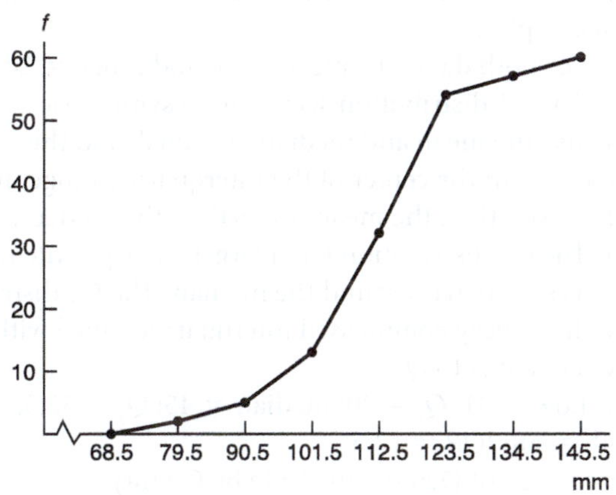

11. (a) 1240s had 40 data values. (b) 75. (c) From 1203 to 1212. Little if any repairs or new construction.

CHAPTER 3

Section 3.1

1. Median; mode; mean.
3. First add up all the data values, then divide by the number of data.
5. $\bar{x} = 5$; median = 6; mode = 2.
7. $\bar{x} = 5$; median = 5.5; mode = 2.
9. (a) No, the sum of the data does not change. (b) No, changing extreme data values does not change the median (c) Yes, depending on which data value occurs most frequently after the data are changed.
11. Mean, median, and mode are approximately equal.
13. (a) Mode = 5; median = 4; mean = 3.8. (b) Mode. (c) Mean, median, and mode. (d) Mode, median.
15. The supervisor has a legitimate concern because at least half the clients rated the employee below satisfactory. From the information given, it seems that this employee is very inconsistent in her performance.
17. (a) Mode = 2; median = 3; mean = 4.6. (b) Mode = 10; median = 15; mean = 23. (c) Corresponding values are 5 times the original averages. In general, multiplying each data value by a constant *c* results in the mode, median, and mean changing by a factor of *c*. (d) Mode = 177.8 cm; median = 172.72 cm; mean = 180.34 cm.
19. $\bar{x} \approx 167.3$°F; median = 171°F; mode = 178°F.
21. (a) $\bar{x} \approx 3.27$; median = 3; mode = 3. (b) $\bar{x} \approx 4.21$; median = 2; mode = 1. (c) Lower Canyon mean is greater; median and mode are less. (d) Trimmed mean = 3.75 and is closer to Upper Canyon mean.
23. (a) $\bar{x}$ = \$136.15; median = \$66.50; mode = \$60. (b) 5% trimmed mean ≈ \$121.28; yes, but still higher than the median. (c) Median. The low and high prices would be useful.
25. 23.
27. $\Sigma wx = 85$; $\Sigma w = 10$; weighted average = 8.5.
29. Approx. 66.67 mph.

Section 3.2

1. Mean.
3. Yes. For the sample standard deviation *s*, the sum $\Sigma(x - \bar{x})^2$ is divided by $n - 1$, where *n* is the sample size. For the population standard deviation σ, the sum $\Sigma(x - \mu)^2$ is divided by *N*, where *N* is the population size.
5. (a) Range is 4. (b) $s \approx 1.58$. (c) $\sigma \approx 1.41$.
7. For a data set in which not all data values are equal, σ is less than *s*. The reason is that to compute σ, we divide the sum of the squares by *n*, and to compute *s* we divide by the smaller number $n - 1$.
9. (a) (i), (ii), (iii). (b) The data change between data sets (i) and (ii) increased the sum of squared differences $\Sigma(x - \bar{x})^2$ by 10, whereas the data change between data sets (ii) and (iii) increased the sum of squared differences $\Sigma(x - \bar{x})^2$ by only 6.
11. (a) $s \approx 3.6$. (b) $s \approx 18.0$. (c) When each data value is multiplied by 5, the standard deviation is five times greater than that of the original data set. In general, multiplying each data value by the same constant *c* results in the standard deviation being $|c|$ times as large. (d) No. Multiply 3.1 miles by 1.6 kilometers/mile to obtain $s \approx 4.96$ kilometers.

13. (a) 15. (b) Use a calculator. (c) 37; 6.08. (d) 37; 6.08. (e) $\sigma^2 \approx 29.59$; $\sigma \approx 5.44$.
15. (a) $CV = 10\%$. (b) 14 to 26.
17. (a) 7.87. (b) Use a calculator. (c) $\bar{x} \approx 1.24$; $s^2 \approx 1.78$; $s \approx 1.33$. (d) $CV \approx 107\%$.
The standard deviation of the time to failure is just slightly larger than the average time.
19. (a) Use a calculator. (b) $\bar{x} = 49$; $s^2 \approx 687.49$; $s \approx 26.22$. (c) $\bar{y} = 44.8$; $s^2 \approx 508.50$; $s \approx 22.55$.
(d) Mallard nests, $CV \approx 53.5\%$; Canada goose nests, $CV \approx 50.3\%$. The CV gives the ratio of the standard deviation to the mean; the CV for mallard nests is slightly higher.
21. Since $CV = s/\bar{x}$, then $s = CV(\bar{x})$; $s = 0.033$.
23. Midpoints: 25.5, 35.5, 45.5; $\bar{x} \approx 35.80$; $s^2 \approx 61.1$; $s \approx 7.82$.
25. Midpoints: 10.55, 14.55, 18.55, 22.55, 26.55; $\bar{x} \approx 15.6$; $s^2 \approx 23.4$; $s \approx 4.8$.
29. (a) $n_1 \approx 59$; $n_2 \approx 23$; $n_3 \approx 168$.
(b) $\mu \approx 7.46$.

Section 3.3

1. 82% or more of the scores were at or below Angela's score; 18% or fewer of the scores were above Angela's score.
3. No, the score 82 might have a percentile rank less than 70.
5. (a) Low = 2; $Q_1 = 5$; median = 7; $Q_3 = 8.5$; high = 10. (b) $IQR = 3.5$. (c) Box-and-Whisker Plot

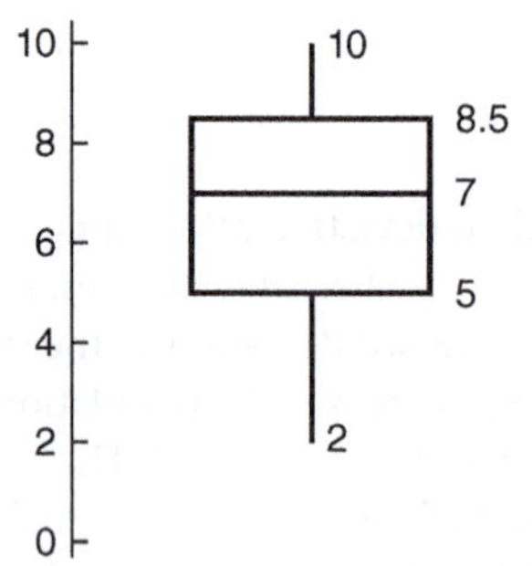

7. Low = 2; $Q_1 = 9.5$; median = 23; $Q_3 = 28.5$; high = 42; $IQR = 19$.
Nurses' Length of Employment (months)

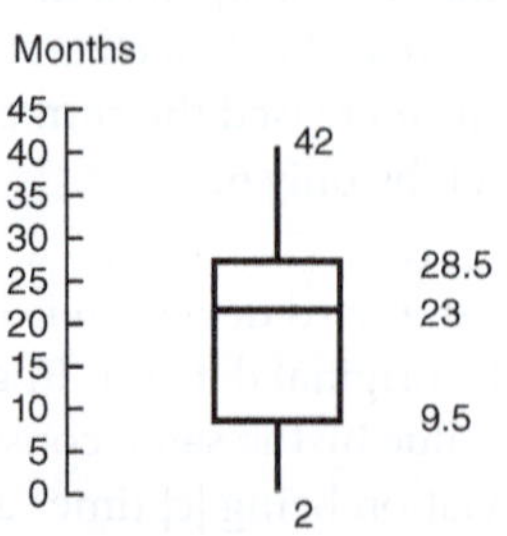

9. (a) Low = 17; $Q_1 = 22$; median = 24; $Q_3 = 27$; high = 38; $IQR = 5$. (b) Third quartile, since it is between the median and Q_3.
Bachelor's Degree Percentage by State

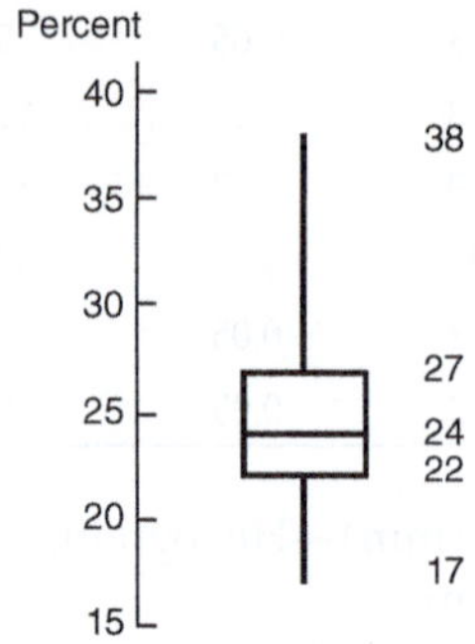

11. (a) California has the lowest premium. Pennsylvania has the highest. (b) Pennsylvania has the highest median premium. (c) California has the smallest range. Texas has the smallest interquartile range. (d) Part (a) is the five-number summary for Texas. It has the smallest IQR. Part (b) is the five-number summary for Pennsylvania. It has the largest minimum. Part (c) is the five-number summary for California. It has the lowest minimum.

Chapter 3 Review

1. (a) Variance and standard deviation. (b) Box-and-whisker plot.
3. (a) For both data sets, mean = 20 and range = 24.
(b) The C1 distribution seems more symmetric because the mean and median are equal, and the median is in the center of the interquartile range. In the C2 distribution, the mean is less than the median.
(c) The C1 distribution has a larger interquartile range that is symmetric around the median. The C2 distribution has a very compressed interquartile range with the median equal to Q_3.
5. (a) Low = 31; $Q_1 = 40$; median = 45; $Q_3 = 52.5$; high = 68; $IQR = 12.5$.
Percentage of Democratic Vote by County

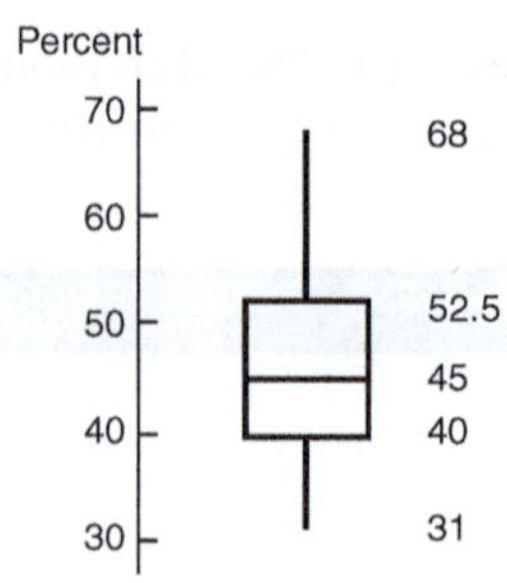

(b) Class width = 8.

Class	Midpoint	f
31–38	34.5	11
39–46	42.5	24
47–54	50.5	15
55–62	58.5	7
63–70	66.5	3

$\bar{x} \approx 46.1$; $s \approx 8.64$; 28.82 to 63.38.

(c) $\bar{x} \approx 46.15$; $s \approx 8.63$.

7. Mean weight = 156.25 pounds.
9. No.
   (b) \$37,091 to \$71,091.
   (c) Public 4-year \$5,105 to \$8,689; private 4-year \$19,854 to \$34,417; 15%; 50%.
11. $\Sigma w = 16$, $\Sigma wx = 121$, average = 7.56.

CHAPTER 4

Section 4.1

1. Explanatory variable is placed along horizontal axis, usually x axis. Response variable is placed along vertical axis, usually y axis.
3. Decreases.
5. (a) Moderate. (b) None. (c) High.
7. (a) No. (b) Increasing population might be a lurking variable causing both variables to increase.
9. (a) No. (b) One lurking variable responsible for average annual income increases is inflation. Better training might be a lurking variable responsible for shorter times to run the mile.
11. The correlation coefficient is moderate and negative. It suggests that as gasoline prices increase, consumption decreases, and the relationship is moderately linear. It is risky to apply these results to gasoline prices much higher than \$5.30 per gallon. It could be that many of the discretionary and technical means of reducing consumption have already been applied, so consumers cannot reduce their consumption much more.
13. (a) Ages and Average Weights of Shetland Ponies

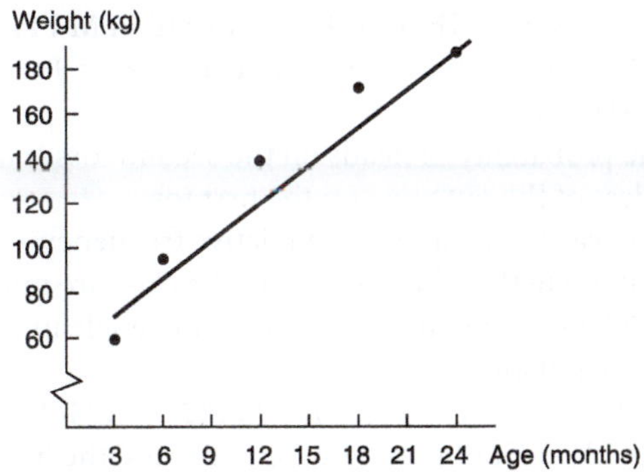

Line slopes upward.
(b) Strong; positive. (c) $r \approx 0.972$; increase.

15. (a) Lowest Barometric Pressure and Maximum Wind Speed for Tropical Cyclones

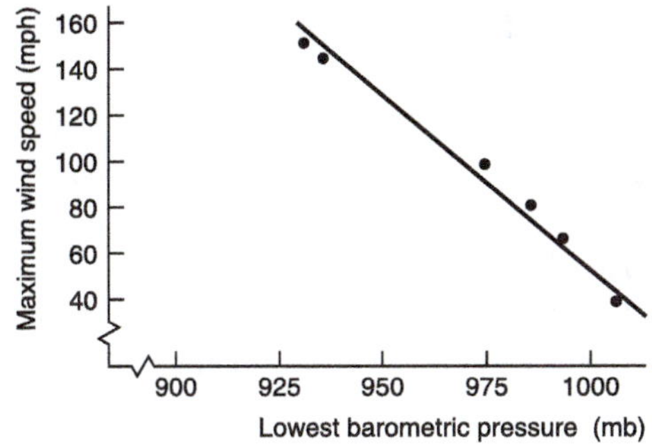

Line slopes downward.
(b) Strong; negative. (c) $r \approx -0.990$; decrease.

17. (a) Batting Average and Home Run Percentage

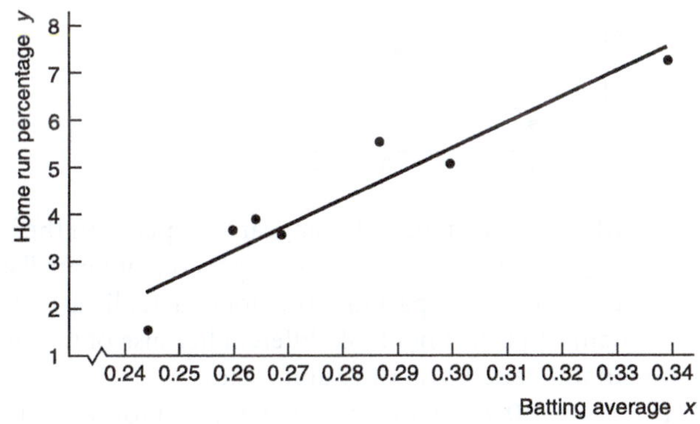

Line slopes upward.
(b) High; positive. (c) $r \approx 0.948$; increase.

19. (a) Unit Length on y Same as That on x

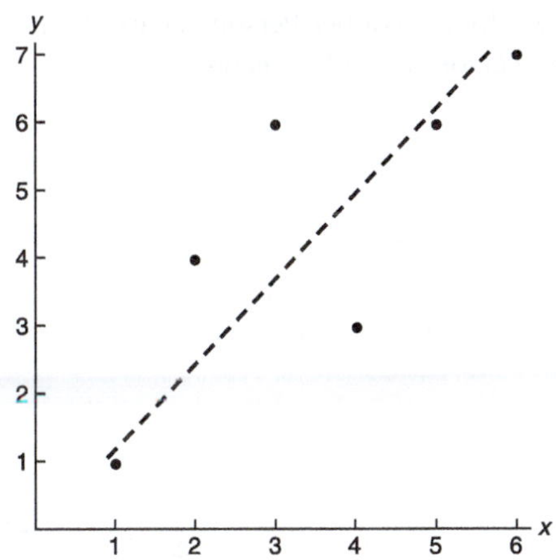

(b) Unit Length on y Twice That on x

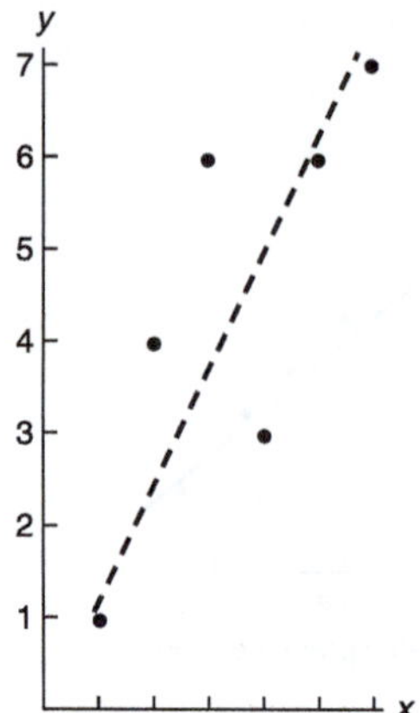

(c) Unit Length on y Half That on x

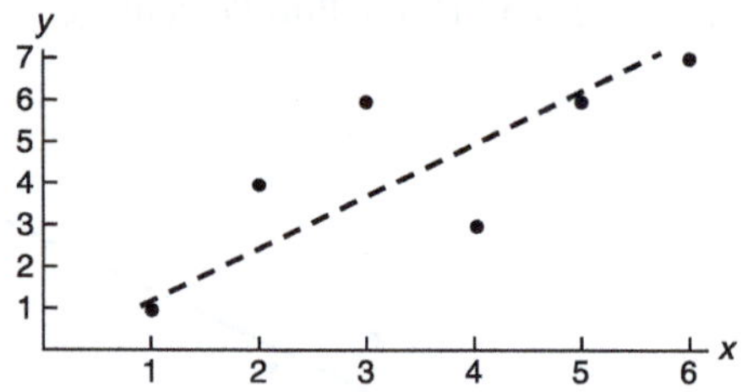

(d) The line in part (b) appears steeper than the line in part (a), whereas the line in part (c) appears flatter than the line in part (a). The slopes actually are all the same, but the lines look different because of the change in unit lengths on the y and x axes.

21. (a) $r \approx 0.972$ with $n = 5$ is significant for $\alpha = 0.05$. For this α, we conclude that age and weight of Shetland ponies are correlated. (b) $r \approx -0.990$ with $n = 6$ is significant for $\alpha = 0.01$. For this α, we conclude that lowest barometric pressure reading and maximum wind speed for cyclones are correlated.

23. (a) Average Hours Lost per Person versus Average Fuel Wasted per Person in Traffic Delays

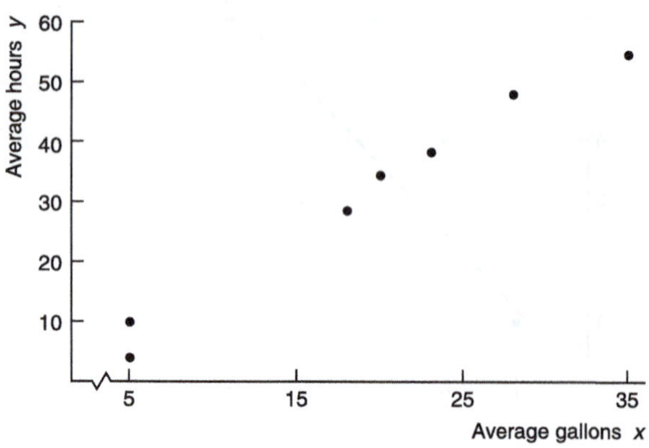

$r \approx 0.991$.

(b) For variables based on averages, $\bar{x} = 19.25$ hr; $s_x \approx 10.33$ hr; $\bar{y} = 31.13$ gal; $s_y \approx 17.76$ gal. For variables based on single individuals, $\bar{x} = 20.13$ hr; $s_x \approx 13.84$ hr; $\bar{y} = 31.87$ gal; $s_y \approx 25.18$ gal. Dividing by larger numbers results in a smaller value.

(c) Hours Lost per Person versus Fuel Wasted per Person in Traffic Delays

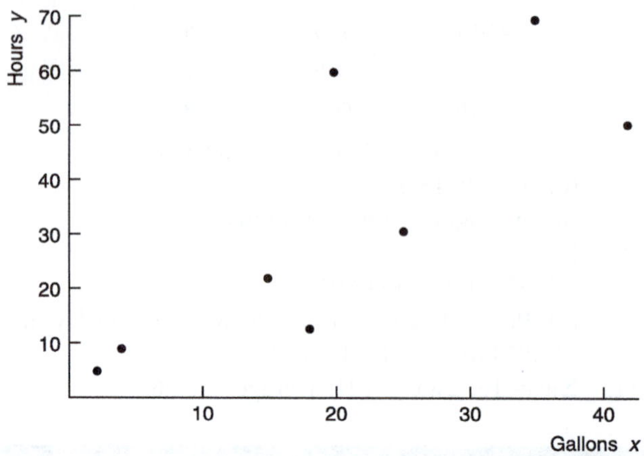

$r \approx 0.794$.

(d) Yes; by the central limit theorem, the $\bar{x}$ distribution has a smaller standard deviation than the corresponding x distribution.

CHAPTER 5

Section 5.1

1. Equally likely outcomes, relative frequency, intuition.
3. (a) 1. (b) 0.
5. No, the probability was stated for drivers in the age range from 18 to 24. We have no information for other age groups. Other age groups may not behave the same way as the 18- to 24-year-olds.
7. $627/1010 \approx 0.62$.
9. Although the probability is high that you will make money, it is not completely certain that you will. In fact, there is a small chance that you could lose your entire investment. If you can afford to lose all of the investment, it might be worthwhile to invest because there is a high chance of doubling your money.
11. (a) MMM MMF MFM MFF FMM FMF FFM FFF. (b) $P(\text{MMM}) = 1/8$. $P(\text{at least one female}) = 1 - P(\text{MMM}) = 7/8$.
13. No. The probability of heads on the second toss is 0.50 regardless of the outcome on the first toss.
15. Answers vary. Probability as a relative frequency. One concern is whether the students in the class are more or less adept at wiggling their ears than people in the general population.
17. (a) $P(0) = 15/375$; $P(1) = 71/375$; $P(2) = 124/375$; $P(3) = 131/375$; $P(4) = 34/375$. (b) Yes, the listed numbers of similar preferences form the sample space.

19. (a) P(best idea 6 A.M.–12 noon) $= 290/966 \approx 0.30$; P(best idea 12 noon–6 P.M.) $= 135/966 \approx 0.14$; P(best idea 6 P.M.–12 midnight) $= 319/966 \approx 0.33$; P(best idea 12 midnight–6 A.M.) $= 222/966 \approx 0.23$. (b) The probabilities add up to 1. They should add up to 1 (within rounding errors), provided the intervals do not overlap and each inventor chose only one interval. The sample space is the set of four time intervals.
21. (b) $P(\text{success}) = 2/17 \approx 0.118$. (c) P(make shot) $= 3/8$ or 0.375.
23. (a) P(enter if walks by) $= 58/127 \approx 0.46$. (b) P(buy if entered) $= 25/58 \approx 0.43$. (c) P(walk in and buy) $= 25/127 \approx 0.20$. (d) $P(\text{not buy}) = 1 - P(\text{buy}) \approx 1 - 0.43 = 0.57$.

CHAPTER 7

Section 7.1

1. (a) No, it's skewed. (b) No, it crosses the horizontal axis. (c) No, it has three peaks. (d) No, the curve is not smooth.
3. Figure 7-9 has the larger standard deviation. The mean of Figure 7-9 is $\mu = 10$. The mean of Figure 7-10 is $\mu = 4$.
5. (a) 50%. (b) 68%. (c) 99.7%.
7. (a) 50%. (b) 50%. (c) 68%. (d) 95%.
9. (a) From 1207 to 1279. (b) From 1171 to 1315. (c) From 1135 to 1351.
11. (a) From 1.70 mA to 4.60 mA. (b) From 0.25 mA to 6.05 mA.
13. (a) 0.8000. (b) 0.7000. (c) 0.5000. (d) $\mu = 0$; $\sigma \approx 0.0289$. Since $\mu = 0$, the measurements are unbiased.

Section 7.2

1. The number of standard deviations from the mean.
3. 0.
5. (a) -1. (b) 2.4. (c) 20. (d) 36.5.
7. They are the same, since both are 1 standard deviation below the mean.
9. (a) Robert, Juan, and Linda each scored above the mean. (b) Joel scored on the mean. (c) Susan and Jan scored below the mean. (d) Robert, 172; Juan, 184; Susan, 110; Joel, 150; Jan, 134; Linda, 182.
11. (a) $-1.00 < z$. (b) $z < -2.00$. (c) $-2.67 < z < 2.33$. (d) $x < 4.4$. (e) $5.2 < x$. (f) $4.1 < x < 4.5$. (g) A red blood cell count of 5.9 or higher corresponds to a standard z score of 3.67. Practically no data values occur this far above the mean. Such a count would be considered unusually high for a healthy female.

13. 0.5000. 15. 0.0934. 17. 0.6736. 19. 0.0643.
21. 0.8888. 23. 0.4993. 25. 0.8953. 27. 0.3471.
29. 0.0306. 31. 0.5000. 33. 0.4483. 35. 0.8849.
37. 0.0885. 39. 0.8849. 41. 0.8808. 43. 0.3226.
45. 0.4474. 47. 0.2939. 49. 0.6704.

Section 7.3

1. 0.50.
3. Negative.
5. $P(3 \le x \le 6) = P(-0.50 \le z \le 1.00) = 0.5328$.
7. $P(50 \le x \le 70) = P(0.67 \le z \le 2.00) = 0.2286$.
9. $P(8 \le x \le 12) = P(-2.19 \le z \le -0.94) = 0.1593$.
11. $P(x \ge 30) = P(z \ge 2.94) = 0.0016$.
13. $P(x \ge 90) = P(z \ge -0.67) = 0.7486$.
15. -1.555. 17. 0.13. 19. 1.41. 21. -0.92.
23. ± 2.33.
25. (a) $P(x > 60) = P(z > -1) = 0.8413$. (b) $P(x < 110) = P(z < 1) = 0.8413$. (c) $P(60 \le x \le 110) = P(-1.00 \le z \le 1.00) = 0.8413 - 0.1587 = 0.6826$. (d) $P(x > 140) = P(z > 2.20) = 0.0139$.
27. (a) $P(x < 3.0 \text{ mm}) = P(z < -2.33) = 0.0099$. (b) $P(x > 7.0 \text{ mm}) = P(z > 2.11) = 0.0174$. (c) $P(3.0 \text{ mm} < x < 7.0 \text{ mm}) = P(-2.33 < z < 2.11) = 0.9727$.
29. (a) $P(x < 36 \text{ months}) = P(z < -1.13) = 0.1292$. The company will replace 13% of its batteries. (b) $P(z < z_0) = 10\%$ for $z_0 = -1.28$; $x = -1.28(8) + 45 = 34.76$. Guarantee the batteries for 35 months.
31. (a) According to the empirical rule, about 95% of the data lies between $\mu - 2\sigma$ and $\mu + 2\sigma$. Since this interval is 4σ wide, we have $4\sigma \approx 6$ years, so $\sigma \approx 1.5$ years. (b) $P(x > 5) = P(z > -2.00) = 0.9772$. (c) $P(x < 10) = P(z < 1.33) = 0.9082$. (d) $P(z < z_0) = 0.10$ for $z_0 = -1.28$; $x = -1.28(1.5) + 8 = 6.08$ years. Guarantee the TVs for about 6.1 years.
33. (a) $\sigma \approx 12$ beats/minute. (b) $P(x < 25) = P(z < -1.75) = 0.0401$. (c) $P(x > 60) = P(z > 1.17) = 0.1210$. (d) $P(25 \le x \le 60) = P(-1.75 \le z \le 1.17) = 0.8389$. (e) $P(z \le z_0) = 0.90$ for $z_0 = 1.28$; $x = 1.28(12) + 46 = 61.36$ beats/minute. A heart rate of 61 beats/minute corresponds to the 90% cutoff point of the distribution.
35. (a) $P(z \ge z_0) = 0.99$ for $z_0 = -2.33$; $x = -2.33(3.7) + 90 \approx 81.38$ months. Guarantee the microchips for 81 months. (b) $P(x \le 84) = P(z \le -1.62) = 0.0526$. (c) Expected loss $= (50{,}000{,}000)(0.0526) = \$2{,}630{,}000$. (d) Profit $= \$370{,}000$.

Chapter 7 Review

1. Normal probability distributions are distributions of continuous random variables. They are symmetric about the mean and bell-shaped. Most of the data fall within 3 standard deviations of the mean. The mean and median are the same.
3. Normal; $\mu_{\bar{x}} = 10$; $\sigma_{\bar{x}} = 1$.
5. The points lie close to a straight line.
7. $\sigma_{\bar{x}} = \sigma/\sqrt{n}$.
9. (a) A normal distribution. (b) The mean μ of the x distribution. (c) $\sigma/\sqrt{n}$, where σ is the standard deviation of the x distribution. (d) They will both be approximately normal with the same mean, but the standard deviations will be $\sigma/\sqrt{50}$ and $\sigma/\sqrt{100}$, respectively.
11. (a) 0.9821. (b) 0.3156. (c) 0.2977.
13. 1.645.
15. (a) 0.8665. (b) 0.7330.
17. (a) 0.0166. (b) 0.975.
19. (a) 0.9772. (b) 17.3 hours.
21. $P(98 \le \bar{x} \le 102) = P(-1.33 \le z \le 1.33) = 0.8164$.
23. (a) Yes, np and nq both exceed 5.
    (b) $\mu_{\hat{p}} = 0.4$; $\sigma_{\hat{p}} \approx 0.1$

ANSWERS TO SELECTED
Even-Numbered Problems

CHAPTER 2

Even-numbered answers not included here appear in the margins of the chapters, next to the problems.

Section 2.2

6. (b)

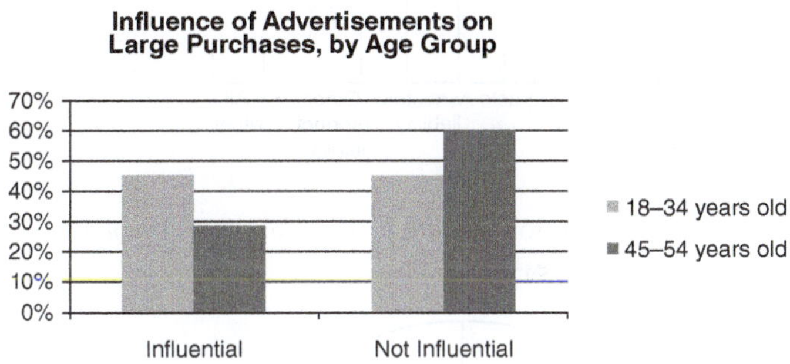

8. (a) Number of Spearheads—Pareto Chart

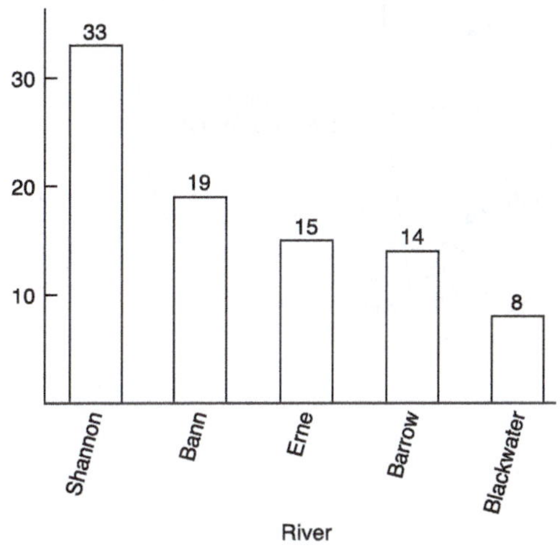

(b) Number of Spearheads—Circle Graph

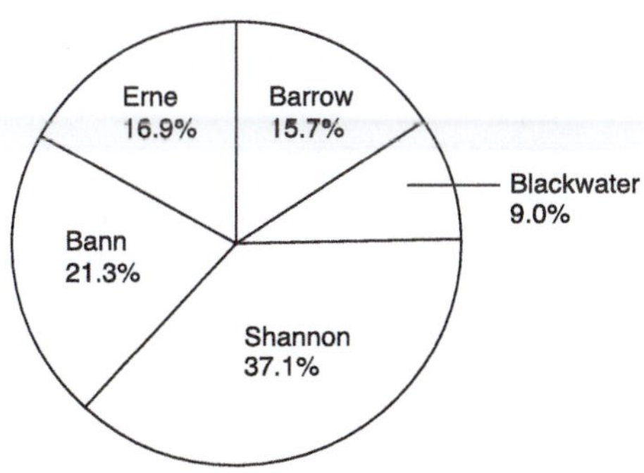

10. How College Professors Spend Their Time—Circle Graph

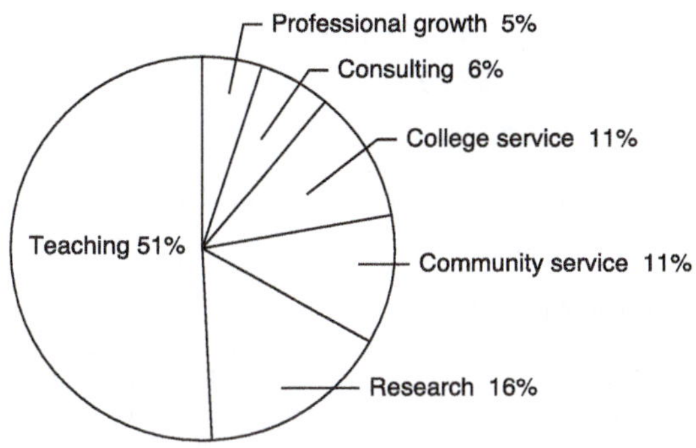

12. Driving Problems—Pareto Chart

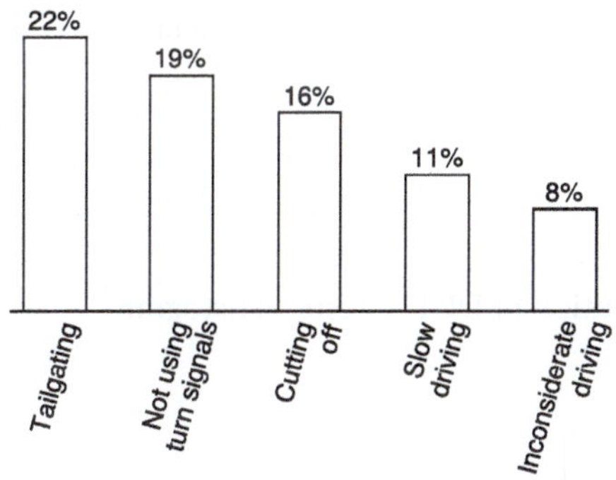

No. The total is not 100%, and it is not clear if respondents could mark more than one problem.

14. Changes in Boys' Height with Age—Time-Series Graph

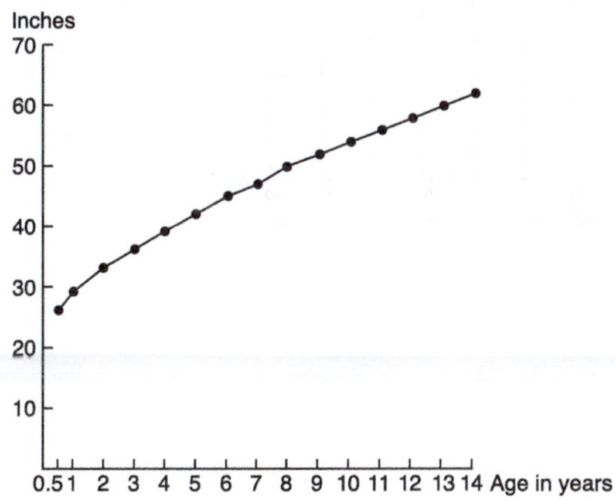

Chapter 2 Review

8. (a)

Age of DUI Arrests

1	6 = 16 years
1	6 8
2	0 1 1 2 2 2 3 4 4 5 6 6 6 7 7 7 9
3	0 0 1 1 2 3 4 4 5 5 6 7 8 9
4	0 0 1 3 5 6 7 7 9 9
5	1 3 5 6 8
6	3 4

(b) Class width = 7.

Age Distribution of DUI Arrests

Class Limits	Class Boundaries	Midpoint	Frequency	Relative Frequency	Cumulative Frequency
16–22	15.5–22.5	19	8	0.16	8
23–29	22.5–29.5	26	11	0.22	19
30–36	29.5–36.5	33	11	0.22	30
37–43	36.5–43.5	40	7	0.14	37
44–50	43.5–50.5	47	6	0.12	43
51–57	50.5–57.5	54	4	0.08	47
58–64	57.5–64.5	61	3	0.06	50

(c) Age Distribution of DUI Arrests—Histogram

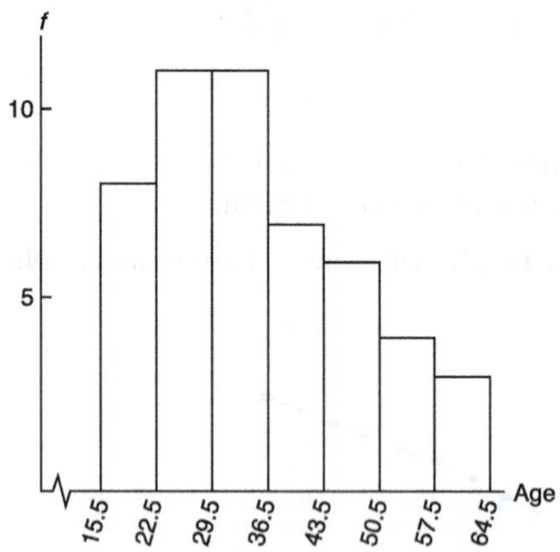

10. (a) Distribution of Civil Justice Caseloads Involving Businesses—Pareto Chart

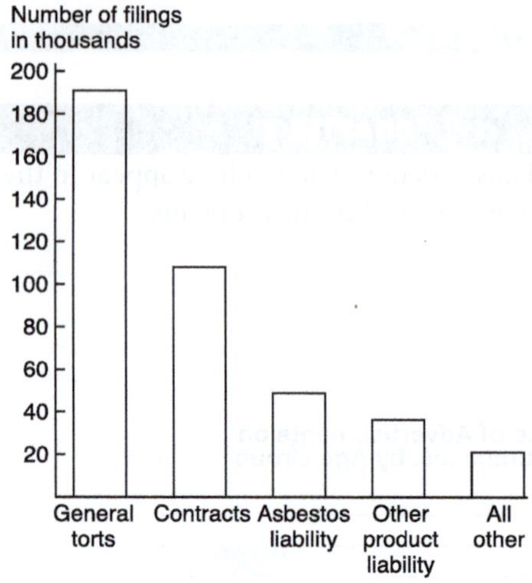

(b) Distribution of Civil Justice Caseloads Involving Businesses—Pie Chart

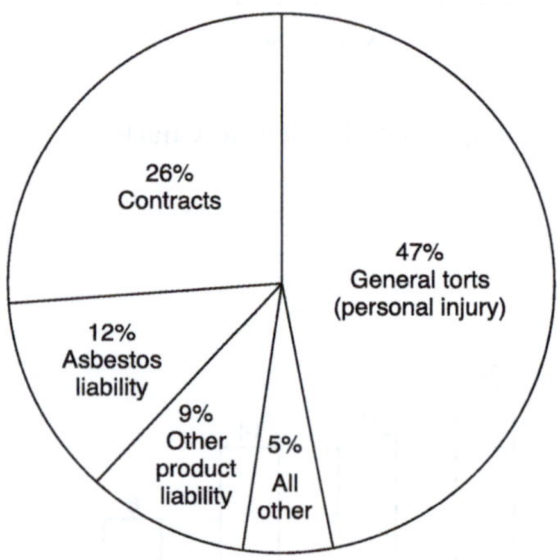

CHAPTER 3

Section 3.3

6. (c) Box-and-Whisker Plot

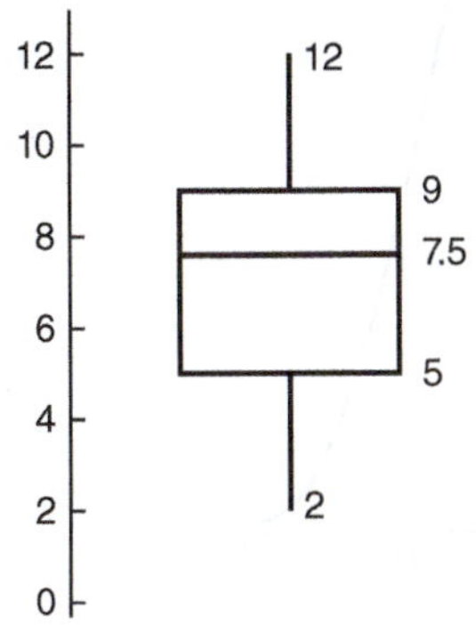

8. (a) Low = 3; Q_1 = 16; median = 23; Q_3 = 30; high = 72; *IQR* = 14.
Clerical Staff Length of Employment (months)

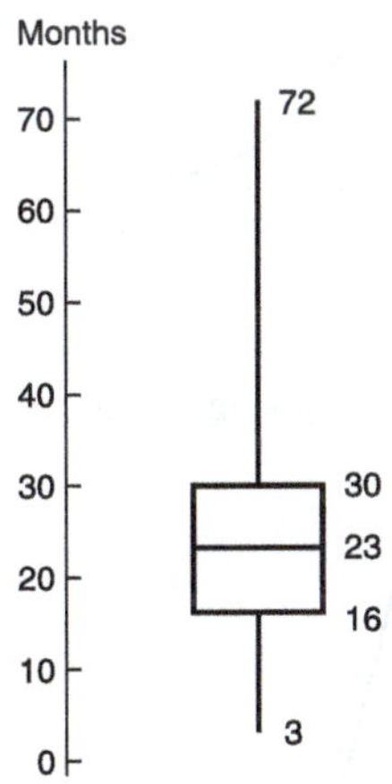

10. (a) Low = 5; Q_1 = 9; median = 10; Q_3 = 12; high = 15; *IQR* = 3.
(b) First quartile, since it is below Q_1.
High School Dropout Percentage by State

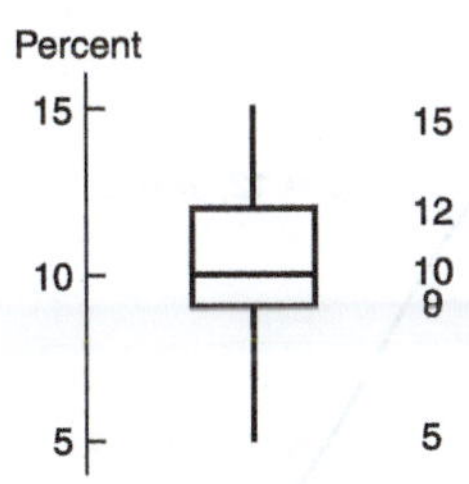

12. (a) Low value = 4; Q_1 = 61.5; median = 65.5; Q_3 = 71.5; high value = 80.
(b) *IQR* = 10.
(c) Lower limit = 46.5; upper limit = 86.5.
(d) Yes, the value 4 is below the lower limit and is probably an error. Our guess is that one of the students is 4 feet tall and listed height in feet instead of inches. There are no values above the upper limit.
Students' Heights (inches)

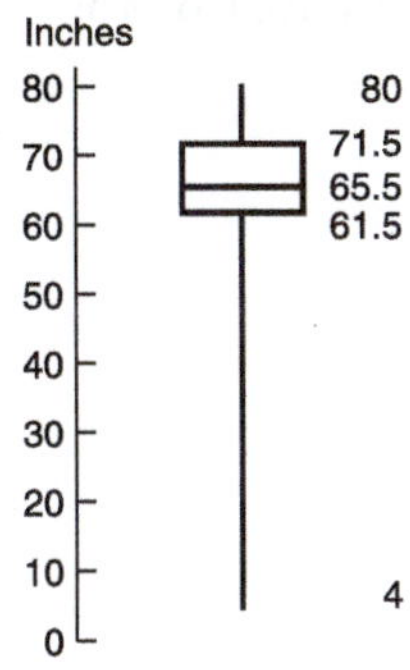

Chapter 3 Review

8. (a) Low = 7.8; Q_1 = 14.2; (kilograms) median = 20.25; Q_3 = 23.8; high = 29.5.
(b) *IQR* = 9.6 kilograms.
(d) Yes, the lower half shows slightly more spread.
Maize Harvest

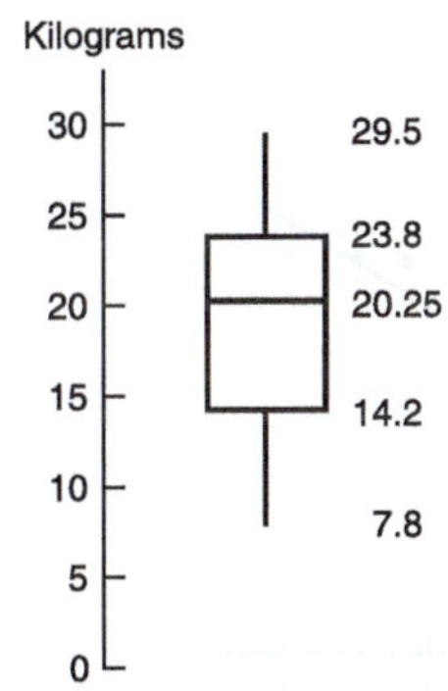

10. (a) Low = 6; Q_1 = 10; median = 11; Q_3 = 13; high = 16; *IQR* = 3.
Soil Water Content

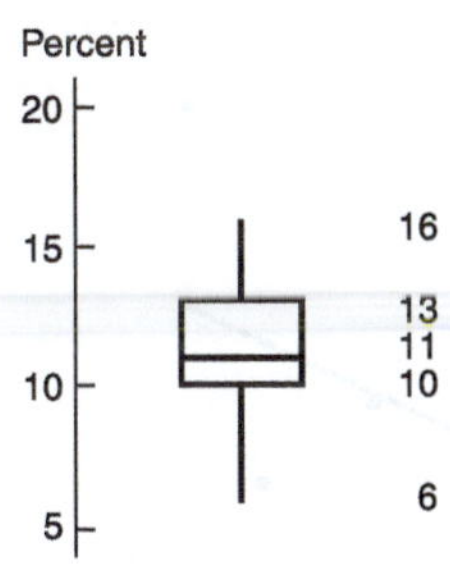

CHAPTER 4

Section 4.1

14. (a) Group Health Insurance Plans: Average Number of Employees versus Administrative Costs as a Percentage of Claims

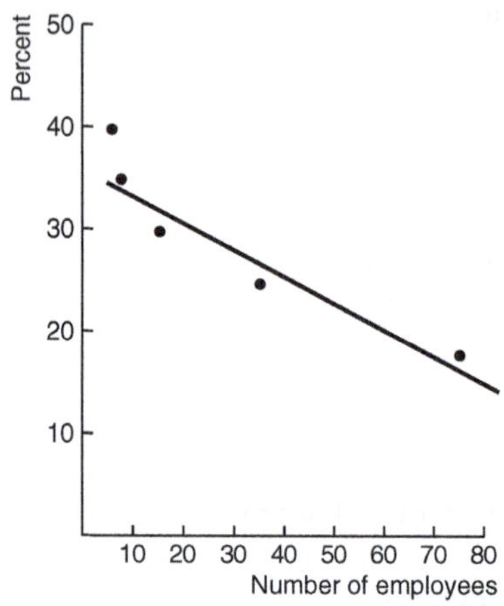

16. (a) Magnitude (Richter Scale) and Depth (km) of Earthquakes

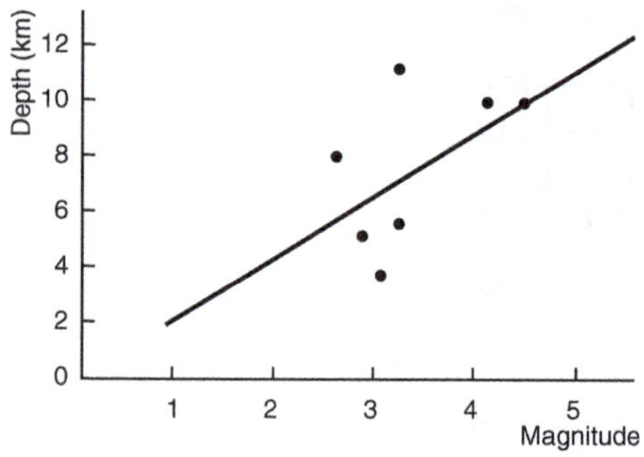

18. (a) Student Enrollment (in thousands) versus Number of Burglaries

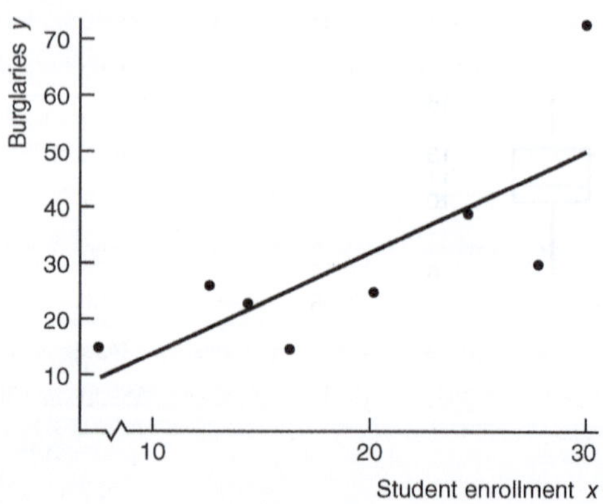

CHAPTER 7

Section 7.1

4. (a) Normal Curve

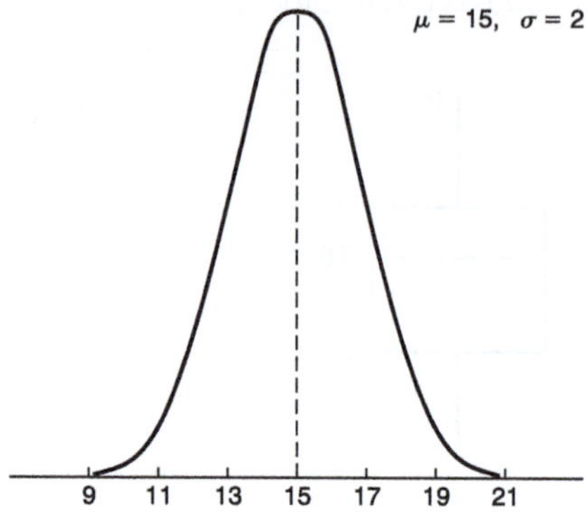

(b) Normal Curve

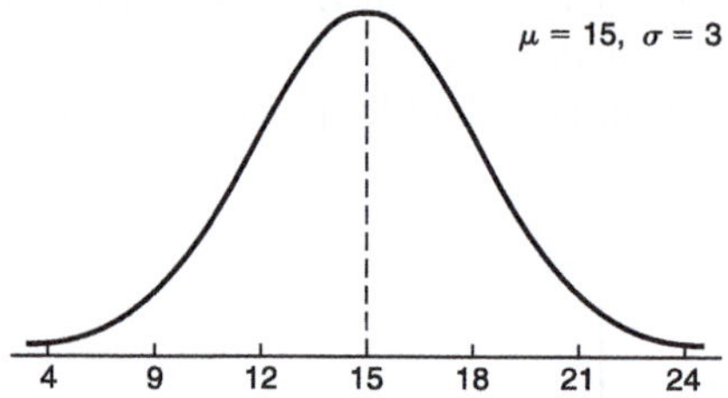

(c) Normal Curve

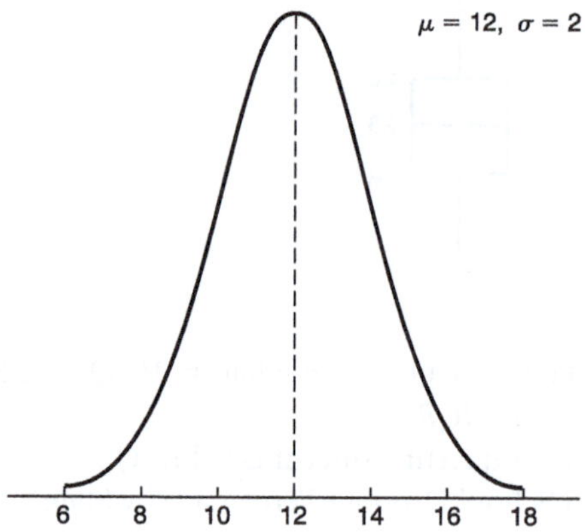

(d) Normal Curve

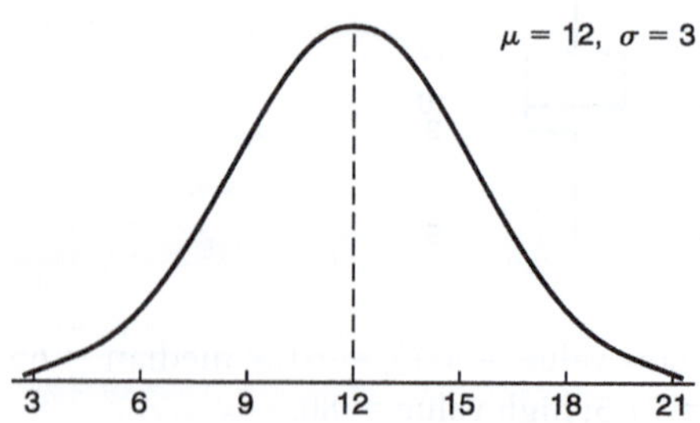

Appendix A: Random Number Table

1. Random Numbers
2. Binomial Probability Distribution $C_{n,r}p^r q^{n-r}$
3. Areas of a Standard Normal Distribution
4. Critical Values for Student's t Distribution
5. The χ^2 Distribution

TABLE 1 Random Numbers

92630	78240	19267	95457	53497	23894	37708	79862	76471	66418
79445	78735	71549	44843	26104	67318	00701	34986	66751	99723
59654	71966	27386	50004	05358	94031	29281	18544	52429	06080
31524	49587	76612	39789	13537	48086	59483	60680	84675	53014
06348	76938	90379	51392	55887	71015	09209	79157	24440	30244
28703	51709	94456	48396	73780	06436	86641	69239	57662	80181
68108	89266	94730	95761	75023	48464	65544	96583	18911	16391
99938	90704	93621	66330	33393	95261	95349	51769	91616	33238
91543	73196	34449	63513	83834	99411	58826	40456	69268	48562
42103	02781	73920	56297	72678	12249	25270	36678	21313	75767
17138	27584	25296	28387	51350	61664	37893	05363	44143	42677
28297	14280	54524	21618	95320	38174	60579	08089	94999	78460
09331	56712	51333	06289	75345	08811	82711	57392	25252	30333
31295	04204	93712	51287	05754	79396	87399	51773	33075	97061
36146	15560	27592	42089	99281	59640	15221	96079	09961	05371
29553	18432	13630	05529	02791	81017	49027	79031	50912	09399
23501	22642	63081	08191	89420	67800	55137	54707	32945	64522
57888	85846	67967	07835	11314	01545	48535	17142	08552	67457
55336	71264	88472	04334	63919	36394	11196	92470	70543	29776
10087	10072	55980	64688	68239	20461	89381	93809	00796	95945
34101	81277	66090	88872	37818	72142	67140	50785	21380	16703
53362	44940	60430	22834	14130	96593	23298	56203	92671	15925
82975	66158	84731	19436	55790	69229	28661	13675	99318	76873
54827	84673	22898	08094	14326	87038	42892	21127	30712	48489
25464	59098	27436	89421	80754	89924	19097	67737	80368	08795
67609	60214	41475	84950	40133	02546	09570	45682	50165	15609
44921	70924	61295	51137	47596	86735	35561	76649	18217	63446
33170	30972	98130	95828	49786	13301	36081	80761	33985	68621
84687	85445	06208	17654	51333	02878	35010	67578	61574	20749
71886	56450	36567	09395	96951	35507	17555	35212	69106	01679
00475	02224	74722	14721	40215	21351	08596	45625	83981	63748
25993	38881	68361	59560	41274	69742	40703	37993	03435	18873
92882	53178	99195	93803	56985	53089	15305	50522	55900	43026

Continued

TABLE 1 *continued*

25138	26810	07093	15677	60688	04410	24505	37890	67186	62829
84631	71882	12991	83028	82484	90339	91950	74579	03539	90122
34003	92326	12793	61453	48121	74271	28363	66561	75220	35908
53775	45749	05734	86169	42762	70175	97310	73894	88606	19994
59316	97885	72807	54966	60859	11932	35265	71601	55577	67715
20479	66557	50705	26999	09854	52591	14063	30214	19890	19292
86180	84931	25455	26044	02227	52015	21820	50599	51671	65411
21451	68001	72710	40261	61281	13172	63819	48970	51732	54113
98062	68375	80089	24135	72355	95428	11808	29740	81644	86610
01788	64429	14430	94575	75153	94576	61393	96192	03227	32258
62465	04841	43272	68702	01274	05437	22953	18946	99053	41690
94324	31089	84159	92933	99989	89500	91586	02802	69471	68274
05797	43984	21575	09908	70221	19791	51578	36432	33494	79888
10395	14289	52185	09721	25789	38562	54794	04897	59012	89251
35177	56986	25549	59730	64718	52630	31100	62384	49483	11409
25633	89619	75882	98256	02126	72099	57183	55887	09320	73463
16464	48280	94254	45777	45150	68865	11382	11782	22695	41988

Source: MILLION RANDOM DIGITS WITH 100,000 NORMAL DEVIATES. By Staff. Copyright 1983 by RAND CORPORATION. Reproduced with permission of RAND CORPORATION in the formats Textbook and Other Book via Copyright Clearance Center.

Appendix B: Areas of a Standard Normal Distribution

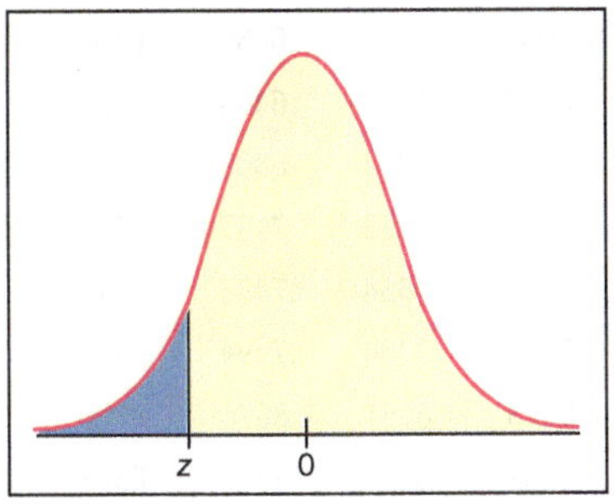

The table entry for z is the area to the left of z.

TABLE 3 Areas of a Standard Normal Distribution

(a) Table of Areas to the Left of z

z	.00	.01	.02	.03	.04	.05	.06	.07	.08	.09
−3.4	.0003	.0003	.0003	.0003	.0003	.0003	.0003	.0003	.0003	.0002
−3.3	.0005	.0005	.0005	.0004	.0004	.0004	.0004	.0004	.0004	.0003
−3.2	.0007	.0007	.0006	.0006	.0006	.0006	.0006	.0005	.0005	.0005
−3.1	.0010	.0009	.0009	.0009	.0008	.0008	.0008	.0008	.0007	.0007
−3.0	.0013	.0013	.0013	.0012	.0012	.0011	.0011	.0011	.0010	.0010
−2.9	.0019	.0018	.0018	.0017	.0016	.0016	.0015	.0015	.0014	.0014
−2.8	.0026	.0025	.0024	.0023	.0023	.0022	.0021	.0021	.0020	.0019
−2.7	.0035	.0034	.0033	.0032	.0031	.0030	.0029	.0028	.0027	.0026
−2.6	.0047	.0045	.0044	.0043	.0041	.0040	.0039	.0038	.0037	.0036
−2.5	.0062	.0060	.0059	.0057	.0055	.0054	.0052	.0051	.0049	.0048
−2.4	.0082	.0080	.0078	.0075	.0073	.0071	.0069	.0068	.0066	.0064
−2.3	.0107	.0104	.0102	.0099	.0096	.0094	.0091	.0089	.0087	.0084
−2.2	.0139	.0136	.0132	.0129	.0125	.0122	.0119	.0116	.0113	.0110
−2.1	.0179	.0174	.0170	.0166	.0162	.0158	.0154	.0150	.0146	.0143
−2.0	.0228	.0222	.0217	.0212	.0207	.0202	.0197	.0192	.0188	.0183
−1.9	.0287	.0281	.0274	.0268	.0262	.0256	.0250	.0244	.0239	.0233
−1.8	.0359	.0351	.0344	.0336	.0329	.0322	.0314	.0307	.0301	.0294
−1.7	.0446	.0436	.0427	.0418	.0409	.0401	.0392	.0384	.0375	.0367
−1.6	.0548	.0537	.0526	.0516	.0505	.0495	.0485	.0475	.0465	.0455
−1.5	.0668	.0655	.0643	.0630	.0618	.0606	.0594	.0582	.0571	.0559
−1.4	.0808	.0793	.0778	.0764	.0749	.0735	.0721	.0708	.0694	.0681
−1.3	.0968	.0951	.0934	.0918	.0901	.0885	.0869	.0853	.0838	.0823
−1.2	.1151	.1131	.1112	.1093	.1075	.1056	.1038	.1020	.1003	.0985
−1.1	.1357	.1335	.1314	.1292	.1271	.1251	.1230	.1210	.1190	.1170
−1.0	.1587	.1562	.1539	.1515	.1492	.1469	.1446	.1423	.1401	.1379
−0.9	.1841	.1814	.1788	.1762	.1736	.1711	.1685	.1660	.1635	.1611
−0.8	.2119	.2090	.2061	.2033	.2005	.1977	.1949	.1922	.1894	.1867
−0.7	.2420	.2389	.2358	.2327	.2296	.2266	.2236	.2206	.2177	.2148
−0.6	.2743	.2709	.2676	.2643	.2611	.2578	.2546	.2514	.2483	.2451
−0.5	.3085	.3050	.3015	.2981	.2946	.2912	.2877	.2843	.2810	.2776
−0.4	.3446	.3409	.3372	.3336	.3300	.3264	.3228	.3192	.3156	.3121
−0.3	.3821	.3783	.3745	.3707	.3669	.3632	.3594	.3557	.3520	.3483
−0.2	.4207	.4168	.4129	.4090	.4052	.4013	.3974	.3936	.3897	.3859
−0.1	.4602	.4562	.4522	.4483	.4443	.4404	.4364	.4325	.4286	.4247
−0.0	.5000	.4960	.4920	.4880	.4840	.4801	.4761	.4721	.4681	.4641

For values of z less than −3.49, use 0.000 to approximate the area.

Continued

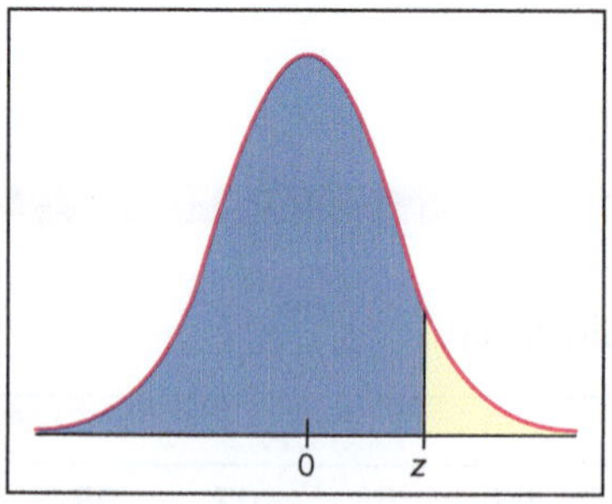

The table entry for z is the area to the left of z.

TABLE 3a *continued*

z	.00	.01	.02	.03	.04	.05	.06	.07	.08	.09
0.0	.5000	.5040	.5080	.5120	.5160	.5199	.5239	.5279	.5319	.5359
0.1	.5398	.5438	.5478	.5517	.5557	.5596	.5636	.5675	.5714	.5753
0.2	.5793	.5832	.5871	.5910	.5948	.5987	.6026	.6064	.6103	.6141
0.3	.6179	.6217	.6255	.6293	.6331	.6368	.6406	.6443	.6480	.6517
0.4	.6554	.6591	.6628	.6664	.6700	.6736	.6772	.6808	.6844	.6879
0.5	.6915	.6950	.6985	.7019	.7054	.7088	.7123	.7157	.7190	.7224
0.6	.7257	.7291	.7324	.7357	.7389	.7422	.7454	.7486	.7517	.7549
0.7	.7580	.7611	.7642	.7673	.7704	.7734	.7764	.7794	.7823	.7852
0.8	.7881	.7910	.7939	.7967	.7995	.8023	.8051	.8078	.8106	.8133
0.9	.8159	.8186	.8212	.8238	.8264	.8289	.8315	.8340	.8365	.8389
1.0	.8413	.8438	.8461	.8485	.8508	.8531	.8554	.8577	.8599	.8621
1.1	.8643	.8665	.8686	.8708	.8729	.8749	.8770	.8790	.8810	.8830
1.2	.8849	.8869	.8888	.8907	.8925	.8944	.8962	.8980	.8997	.9015
1.3	.9032	.9049	.9066	.9082	.9099	.9115	.9131	.9147	.9162	.9177
1.4	.9192	.9207	.9222	.9236	.9251	.9265	.9279	.9292	.9306	.9319
1.5	.9332	.9345	.9357	.9370	.9382	.9394	.9406	.9418	.9429	.9441
1.6	.9452	.9463	.9474	.9484	.9495	.9505	.9515	.9525	.9535	.9545
1.7	.9554	.9564	.9573	.9582	.9591	.9599	.9608	.9616	.9625	.9633
1.8	.9641	.9649	.9656	.9664	.9671	.9678	.9686	.9693	.9699	.9706
1.9	.9713	.9719	.9726	.9732	.9738	.9744	.9750	.9756	.9761	.9767
2.0	.9772	.9778	.9783	.9788	.9793	.9798	.9803	.9808	.9812	.9817
2.1	.9821	.9826	.9830	.9834	.9838	.9842	.9846	.9850	.9854	.9857
2.2	.9861	.9864	.9868	.9871	.9875	.9878	.9881	.9884	.9887	.9890
2.3	.9893	.9896	.9898	.9901	.9904	.9906	.9909	.9911	.9913	.9916
2.4	.9918	.9920	.9922	.9925	.9927	.9929	.9931	.9932	.9934	.9936
2.5	.9938	.9940	.9941	.9943	.9945	.9946	.9948	.9949	.9951	.9952
2.6	.9953	.9955	.9956	.9957	.9959	.9960	.9961	.9962	.9963	.9964
2.7	.9965	.9966	.9967	.9968	.9969	.9970	.9971	.9972	.9973	.9974
2.8	.9974	.9975	.9976	.9977	.9977	.9978	.9979	.9979	.9980	.9981
2.9	.9981	.9982	.9982	.9983	.9984	.9984	.9985	.9985	.9986	.9986
3.0	.9987	.9987	.9987	.9988	.9988	.9989	.9989	.9989	.9990	.9990
3.1	.9990	.9991	.9991	.9991	.9992	.9992	.9992	.9992	.9993	.9993
3.2	.9993	.9993	.9994	.9994	.9994	.9994	.9994	.9995	.9995	.9995
3.3	.9995	.9995	.9995	.9996	.9996	.9996	.9996	.9996	.9996	.9997
3.4	.9997	.9997	.9997	.9997	.9997	.9997	.9997	.9997	.9997	.9998

For z values greater than 3.49, use 1.000 to approximate the area.

TABLE 3 *continued*

(b) Confidence Interval Critical Values z_c

Level of Confidence c	Critical Value z_c
0.70, or 70%	1.04
0.75, or 75%	1.15
0.80, or 80%	1.28
0.85, or 85%	1.44
0.90, or 90%	1.645
0.95, or 95%	1.96
0.98, or 98%	2.33
0.99, or 99%	2.58

TABLE 3 *continued*

(c) Hypothesis Testing, Critical Values z_0

Level of Significance	$\alpha = 0.05$	$\alpha = 0.01$
Critical value z_0 for a left-tailed test	−1.645	−2.33
Critical value z_0 for a right-tailed test	1.645	2.33
Critical values $\pm z_0$ for a two-tailed test	±1.96	±2.58

FREQUENTLY USED FORMULAS

n = sample size N = population size f = frequency

Chapter 2

Class width $= \dfrac{\text{high} - \text{low}}{\text{number of classes}}$ (increase to next integer)

Class midpoint $= \dfrac{\text{upper limit} + \text{lower limit}}{2}$

Lower boundary = lower boundary of previous class + class width

Chapter 3

Sample mean $\bar{x} = \dfrac{\Sigma x}{n}$

Population mean $\mu = \dfrac{\Sigma x}{N}$

Weighted average $= \dfrac{\Sigma xw}{\Sigma w}$

Range = largest data value − smallest data value

Sample standard deviation $s = \sqrt{\dfrac{\Sigma(x - \bar{x})^2}{n - 1}}$

Computation formula $s = \sqrt{\dfrac{\Sigma x^2 - (\Sigma x)^2/n}{n - 1}}$

Population standard deviation $\sigma = \sqrt{\dfrac{\Sigma(x - \mu)^2}{N}}$

Sample variance s^2

Population variance σ^2

Sample coefficient of variation $CV = \dfrac{s}{\bar{x}} \cdot 100\%$

Sample mean for grouped data $\bar{x} = \dfrac{\Sigma xf}{n}$

Sample standard deviation for grouped data

$$s = \sqrt{\frac{\Sigma(x - \bar{x})^2 f}{n - 1}} = \sqrt{\frac{\Sigma x^2 f - (\Sigma xf)^2/n}{n - 1}}$$

Chapter 4

Regression and Correlation

Pearson product moment correlation coefficient

$$r = \frac{n\Sigma xy - (\Sigma x)(\Sigma y)}{\sqrt{n\Sigma x^2 - (\Sigma x)^2}\sqrt{n\Sigma y^2 - (\Sigma y)^2}}$$

Least-squares line $\hat{y} = a + bx$

where $b = \dfrac{n\Sigma xy - (\Sigma x)(\Sigma y)}{n\Sigma x^2 - (\Sigma x)^2}$

$a = \bar{y} - b\bar{x}$

Coefficient of determination $= r^2$

Chapter 5

Probability of the complement of event A
$P(A^c) = 1 - P(A)$

Multiplication rule for independent events
$P(A \text{ and } B) = P(A) \cdot P(B)$

General multiplication rules
$P(A \text{ and } B) = P(A) \cdot P(B|A)$
$P(A \text{ and } B) = P(B) \cdot P(A|B)$

Addition rule for mutually exclusive events
$P(A \text{ or } B) = P(A) + P(B)$

General addition rule
$P(A \text{ or } B) = P(A) + P(B) - P(A \text{ and } B)$

Permutation rule $P_{n,r} = \dfrac{n!}{(n - r)!}$

Combination rule $C_{n,r} = \dfrac{n!}{r!(n - r)!}$

Chapter 6

Mean of a discrete probability distribution $\mu = \Sigma xP(x)$

Standard deviation of a discrete probability distribution

$\sigma = \sqrt{\Sigma(x - \mu)^2 P(x)}$

For Binomial Distributions

r = number of successes; p = probability of success; $q = 1 - p$

Binomial probability distribution $P(r) = C_{n,r}p^r q^{n-r}$

Mean $\mu = np$

Standard deviation $\sigma = \sqrt{npq}$

Chapter 7

Raw score $x = z\sigma + \mu$ Standard score $z = \dfrac{x - \mu}{\sigma}$

Mean of $\bar{x}$ distribution $\mu_{\bar{x}} = \mu$

Standard deviation of $\bar{x}$ distribution $\sigma_{\bar{x}} = \dfrac{\sigma}{\sqrt{n}}$

Standard score for $\bar{x}$; $z = \dfrac{\bar{x} - \mu}{\sigma/\sqrt{n}}$

Mean of $\hat{p}$ distribution $\mu_{\hat{p}} = p$

Standard deviation for $\hat{p}$; $\sigma_{\hat{p}} = \sqrt{\dfrac{pq}{n}}$; $q = 1 - p$

Chapter 8

Confidence Interval

for μ

$\bar{x} - E < \mu < \bar{x} + E$

where $E = z_c \dfrac{\sigma}{\sqrt{n}}$ when σ is known

$$E = t_c \frac{s}{\sqrt{n}} \text{ when } \sigma \text{ is unknown}$$
with $d.f. = n - 1$

for $p(np > 5$ and $n(1 - p) > 5)$

$$\hat{p} - E < p < \hat{p} + E$$

$$\text{where } E = z_c \sqrt{\frac{\hat{p}(1-\hat{p})}{n}}$$

$$\hat{p} = \frac{r}{n}$$

Sample Size for Estimating

$$\text{means } n = \left(\frac{z_c \sigma}{E}\right)^2$$

proportions

$$n = p(1-p)\left(\frac{z_c}{E}\right)^2 \text{ with preliminary estimate for } p$$

$$n = \frac{1}{4}\left(\frac{z_c}{E}\right)^2 \text{ without preliminary estimate for } p$$

Chapter 9

Sample Test Statistics for Tests of Hypotheses

for μ (σ known) $\quad z = \dfrac{\bar{x} - \mu}{\sigma/\sqrt{n}}$

for μ (σ unknown) $\quad t = \dfrac{\bar{x} - \mu}{s/\sqrt{n}}$; $d.f. = n - 1$

for p ($np > 5$ and $nq > 5$) $\quad z = \dfrac{\hat{p} - p}{\sqrt{pq/n}}$

where $q = 1 - p$; $\hat{p} = r/n$

Chapter 10

Sample Test Statistics for Tests of Hypotheses

for paired differences d $\quad t = \dfrac{\bar{d} - \mu_d}{s_d/\sqrt{n}}$; $d.f. = n - 1$

for difference of means, σ_1 and σ_2 known

$$z = \frac{\bar{x}_1 - \bar{x}_2}{\sqrt{\frac{\sigma_1^2}{n_1} + \frac{\sigma_2^2}{n_2}}}$$

for difference of means, σ_1 or σ_2 unknown

$$t = \frac{\bar{x}_1 - \bar{x}_2}{\sqrt{\frac{s_1^2}{n_1} + \frac{s_2^2}{n_2}}}$$

$d.f.$ = smaller of $n_1 - 1$ and $n_2 - 1$

(*Note:* Software uses Satterthwaite's approximation for degrees of freedom $d.f.$)

for difference of proportions

$$z = \frac{\hat{p}_1 - \hat{p}_2}{\sqrt{\frac{\bar{p}\bar{q}}{n_1} + \frac{\bar{p}\bar{q}}{n_2}}}$$

where $\bar{p} = \dfrac{r_1 + r_2}{n_1 + n_2}$ and $\bar{q} = 1 - \bar{p}$

$\hat{p}_1 = r_1/n_1$; $\hat{p}_2 = r_2/n_2$

Confidence Interval

for $\mu_1 - \mu_2$ (independent samples)

$$(\bar{x}_1 - \bar{x}_2) - E < \mu_1 - \mu_2 < (\bar{x}_1 - \bar{x}_2) + E$$

$$\text{where } E = z_c \sqrt{\frac{\sigma_1^2}{n_1} + \frac{\sigma_2^2}{n_2}} \text{ when } \sigma_1 \text{ and } \sigma_2 \text{ are known}$$

$$E = t_c \sqrt{\frac{s_1^2}{n_1} + \frac{s_2^2}{n_2}} \text{ when } \sigma_1 \text{ or } \sigma_2 \text{ is unknown}$$

with $d.f.$ = smaller of $n_1 - 1$ and $n_2 - 1$

(*Note:* Software uses Satterthwaite's approximation for degrees of freedom $d.f.$)

for difference of proportions $p_1 - p_2$

$$(\hat{p}_1 - \hat{p}_2) - E < p_1 - p_2 < (\hat{p}_1 - \hat{p}_2) + E$$

$$\text{where } E = z_c \sqrt{\frac{\hat{p}_1\hat{q}_1}{n_1} + \frac{\hat{p}_2\hat{q}_2}{n_2}}$$

$\hat{p}_1 = r_1/n_1$; $\hat{p}_2 = r_2/n_2$

$\hat{q}_1 = 1 - \hat{p}_1$; $\hat{q}_2 = 1 - \hat{p}_2$

Chapter 11

$$\chi^2 = \sum \frac{(O - E)^2}{E} \text{ where O = observed frequency}$$

For tests of independence and tests of homogeneity

$$E = \frac{(\text{row total})(\text{column total})}{\text{sample size}}$$

For goodness of fit test E = (given percent)(sample size)

Tests of independence $d.f. = (R - 1)(C - 1)$

Test of homogeneity $d.f. = (R - 1)(C - 1)$

Goodness of fit $d.f.$ = (number of categories) − 1

Sample test statistic for σ^2

$$\chi^2 = \frac{(n-1)s^2}{\sigma^2} \text{ with } d.f. = n - 1$$

Linear Regression

Sample test statistic for r

$$t = \frac{r\sqrt{n-2}}{\sqrt{1 - r^2}} \text{ with } d.f. = n - 2$$

Standard error of estimate $S_e = \sqrt{\dfrac{\Sigma y^2 - a\Sigma y - b\Sigma xy}{n - 2}}$

Confidence interval for y: $\hat{y} - E < y < \hat{y} + E$

$$\text{where } E = t_c S_e \sqrt{1 + \frac{1}{n} + \frac{n(x - \bar{x})^2}{n\Sigma x^2 - (\Sigma x)^2}}$$

with $d.f. = n - 2$

Sample test statistic for slope b

$$t = \frac{b}{S_e}\sqrt{\Sigma x^2 - \frac{1}{n}(\Sigma x)^2} \text{ with } d.f. = n - 2$$

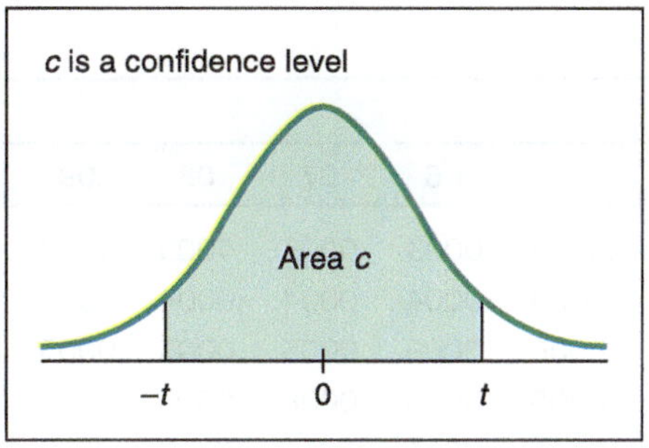

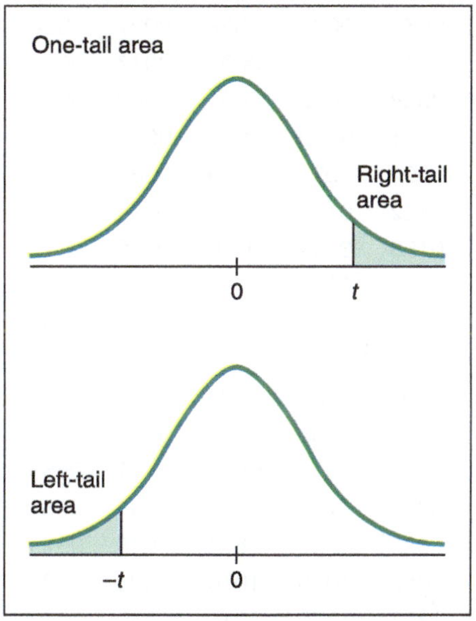

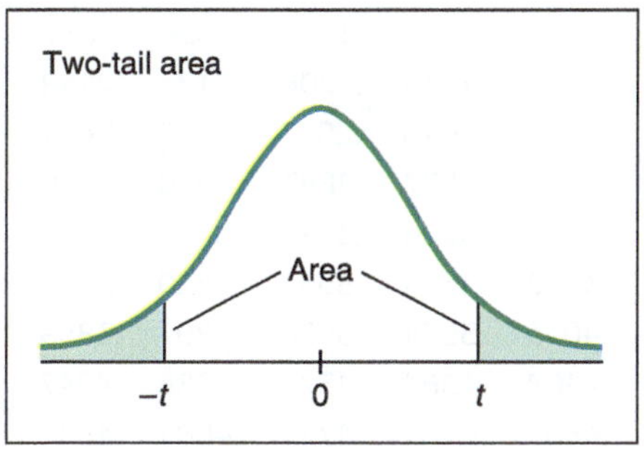

Critical Values for Student's *t* Distribution

one-tail area	0.250	0.125	0.100	0.075	0.050	0.025	0.010	0.005	0.0005
two-tail area	0.500	0.250	0.200	0.150	0.100	0.050	0.020	0.010	0.0010
d.f.* \ *c	0.500	0.750	0.800	0.850	0.900	0.950	0.980	0.990	0.999
1	1.000	2.414	3.078	4.165	6.314	12.706	31.821	63.657	636.619
2	0.816	1.604	1.886	2.282	2.920	4.303	6.965	9.925	31.599
3	0.765	1.423	1.638	1.924	2.353	3.182	4.541	5.841	12.924
4	0.741	1.344	1.533	1.778	2.132	2.776	3.747	4.604	8.610
5	0.727	1.301	1.476	1.699	2.015	2.571	3.365	4.032	6.869
6	0.718	1.273	1.440	1.650	1.943	2.447	3.143	3.707	5.959
7	0.711	1.254	1.415	1.617	1.895	2.365	2.998	3.499	5.408
8	0.706	1.240	1.397	1.592	1.860	2.306	2.896	3.355	5.041
9	0.703	1.230	1.383	1.574	1.833	2.262	2.821	3.250	4.781
10	0.700	1.221	1.372	1.559	1.812	2.228	2.764	3.169	4.587
11	0.697	1.214	1.363	1.548	1.796	2.201	2.718	3.106	4.437
12	0.695	1.209	1.356	1.538	1.782	2.179	2.681	3.055	4.318
13	0.694	1.204	1.350	1.530	1.771	2.160	2.650	3.012	4.221
14	0.692	1.200	1.345	1.523	1.761	2.145	2.624	2.977	4.140
15	0.691	1.197	1.341	1.517	1.753	2.131	2.602	2.947	4.073
16	0.690	1.194	1.337	1.512	1.746	2.120	2.583	2.921	4.015
17	0.689	1.191	1.333	1.508	1.740	2.110	2.567	2.898	3.965
18	0.688	1.189	1.330	1.504	1.734	2.101	2.552	2.878	3.922
19	0.688	1.187	1.328	1.500	1.729	2.093	2.539	2.861	3.883
20	0.687	1.185	1.325	1.497	1.725	2.086	2.528	2.845	3.850
21	0.686	1.183	1.323	1.494	1.721	2.080	2.518	2.831	3.819
22	0.686	1.182	1.321	1.492	1.717	2.074	2.508	2.819	3.792
23	0.685	1.180	1.319	1.489	1.714	2.069	2.500	2.807	3.768
24	0.685	1.179	1.318	1.487	1.711	2.064	2.492	2.797	3.745
25	0.684	1.198	1.316	1.485	1.708	2.060	2.485	2.787	3.725
26	0.684	1.177	1.315	1.483	1.706	2.056	2.479	2.779	3.707
27	0.684	1.176	1.314	1.482	1.703	2.052	2.473	2.771	3.690
28	0.683	1.175	1.313	1.480	1.701	2.048	2.467	2.763	3.674
29	0.683	1.174	1.311	1.479	1.699	2.045	2.462	2.756	3.659
30	0.683	1.173	1.310	1.477	1.697	2.042	2.457	2.750	3.646
35	0.682	1.170	1.306	1.472	1.690	2.030	2.438	2.724	3.591
40	0.681	1.167	1.303	1.468	1.684	2.021	2.423	2.704	3.551
45	0.680	1.165	1.301	1.465	1.679	2.014	2.412	2.690	3.520
50	0.679	1.164	1.299	1.462	1.676	2.009	2.403	2.678	3.496
60	0.679	1.162	1.296	1.458	1.671	2.000	2.390	2.660	3.460
70	0.678	1.160	1.294	1.456	1.667	1.994	2.381	2.648	3.435
80	0.678	1.159	1.292	1.453	1.664	1.990	2.374	2.639	3.416
100	0.677	1.157	1.290	1.451	1.660	1.984	2.364	2.626	3.390
500	0.675	1.152	1.283	1.442	1.648	1.965	2.334	2.586	3.310
1000	0.675	1.151	1.282	1.441	1.646	1.962	2.330	2.581	3.300
∞	**0.674**	**1.150**	**1.282**	**1.440**	**1.645**	**1.960**	**2.326**	**2.576**	**3.291**

For degrees of freedom (*d.f.*) not in the table, use the closest *d.f.* that is *smaller.*

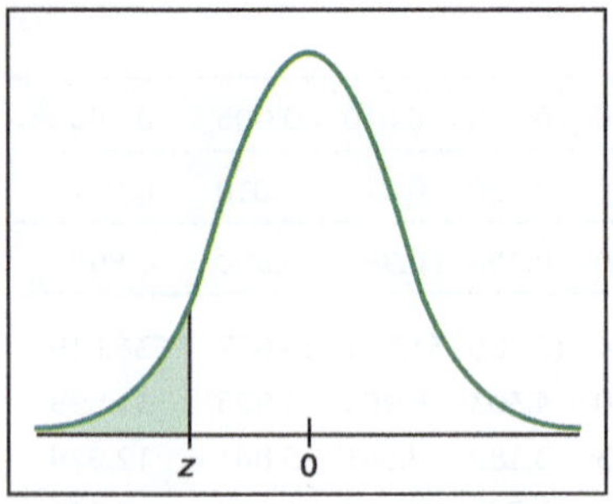

Table entry for z is the area to the left of z.

Areas of a Standard Normal Distribution

(a) Table of Areas to the Left of z

z	.00	.01	.02	.03	.04	.05	.06	.07	.08	.09
−3.4	.0003	.0003	.0003	.0003	.0003	.0003	.0003	.0003	.0003	.0002
−3.3	.0005	.0005	.0005	.0004	.0004	.0004	.0004	.0004	.0004	.0003
−3.2	.0007	.0007	.0006	.0006	.0006	.0006	.0006	.0005	.0005	.0005
−3.1	.0010	.0009	.0009	.0009	.0008	.0008	.0008	.0008	.0007	.0007
−3.0	.0013	.0013	.0013	.0012	.0012	.0011	.0011	.0011	.0010	.0010
−2.9	.0019	.0018	.0018	.0017	.0016	.0016	.0015	.0015	.0014	.0014
−2.8	.0026	.0025	.0024	.0023	.0023	.0022	.0021	.0021	.0020	.0019
−2.7	.0035	.0034	.0033	.0032	.0031	.0030	.0029	.0028	.0027	.0026
−2.6	.0047	.0045	.0044	.0043	.0041	.0040	.0039	.0038	.0037	.0036
−2.5	.0062	.0060	.0059	.0057	.0055	.0054	.0052	.0051	.0049	.0048
−2.4	.0082	.0080	.0078	.0075	.0073	.0071	.0069	.0068	.0066	.0064
−2.3	.0107	.0104	.0102	.0099	.0096	.0094	.0091	.0089	.0087	.0084
−2.2	.0139	.0136	.0132	.0129	.0125	.0122	.0119	.0116	.0113	.0110
−2.1	.0179	.0174	.0170	.0166	.0162	.0158	.0154	.0150	.0146	.0143
−2.0	.0228	.0222	.0217	.0212	.0207	.0202	.0197	.0192	.0188	.0183
−1.9	.0287	.0281	.0274	.0268	.0262	.0256	.0250	.0244	.0239	.0233
−1.8	.0359	.0351	.0344	.0336	.0329	.0322	.0314	.0307	.0301	.0294
−1.7	.0446	.0436	.0427	.0418	.0409	.0401	.0392	.0384	.0375	.0367
−1.6	.0548	.0537	.0526	.0516	.0505	.0495	.0485	.0475	.0465	.0455
−1.5	.0668	.0655	.0643	.0630	.0618	.0606	.0594	.0582	.0571	.0559
−1.4	.0808	.0793	.0778	.0764	.0749	.0735	.0721	.0708	.0694	.0681
−1.3	.0968	.0951	.0934	.0918	.0901	.0885	.0869	.0853	.0838	.0823
−1.2	.1151	.1131	.1112	.1093	.1075	.1056	.1038	.1020	.1003	.0985
−1.1	.1357	.1335	.1314	.1292	.1271	.1251	.1230	.1210	.1190	.1170
−1.0	.1587	.1562	.1539	.1515	.1492	.1469	.1446	.1423	.1401	.1379
−0.9	.1841	.1814	.1788	.1762	.1736	.1711	.1685	.1660	.1635	.1611
−0.8	.2119	.2090	.2061	.2033	.2005	.1977	.1949	.1922	.1894	.1867
−0.7	.2420	.2389	.2358	.2327	.2296	.2266	.2236	.2206	.2177	.2148
−0.6	.2743	.2709	.2676	.2643	.2611	.2578	.2546	.2514	.2483	.2451
−0.5	.3085	.3050	.3015	.2981	.2946	.2912	.2877	.2843	.2810	.2776
−0.4	.3446	.3409	.3372	.3336	.3300	.3264	.3228	.3192	.3156	.3121
−0.3	.3821	.3783	.3745	.3707	.3669	.3632	.3594	.3557	.3520	.3483
−0.2	.4207	.4168	.4129	.4090	.4052	.4013	.3974	.3936	.3897	.3859
−0.1	.4602	.4562	.4522	.4483	.4443	.4404	.4364	.4325	.4286	.4247
−0.0	.5000	.4960	.4920	.4880	.4840	.4801	.4761	.4721	.4681	.4641

For values of z less than −3.49, use 0.000 to approximate the area.

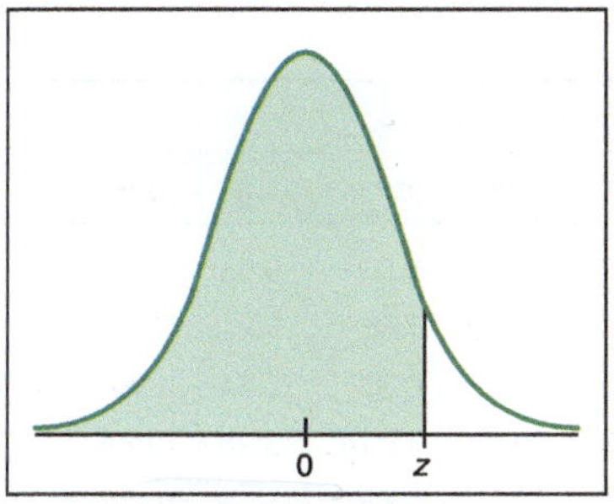

Table entry for z is the area to the left of z.

Areas of a Standard Normal Distribution *continued*

z	.00	.01	.02	.03	.04	.05	.06	.07	.08	.09
0.0	.5000	.5040	.5080	.5120	.5160	.5199	.5239	.5279	.5319	.5359
0.1	.5398	.5438	.5478	.5517	.5557	.5596	.5636	.5675	.5714	.5753
0.2	.5793	.5832	.5871	.5910	.5948	.5987	.6026	.6064	.6103	.6141
0.3	.6179	.6217	.6255	.6293	.6331	.6368	.6406	.6443	.6480	.6517
0.4	.6554	.6591	.6628	.6664	.6700	.6736	.6772	.6808	.6844	.6879
0.5	.6915	.6950	.6985	.7019	.7054	.7088	.7123	.7157	.7190	.7224
0.6	.7257	.7291	.7324	.7357	.7389	.7422	.7454	.7486	.7517	.7549
0.7	.7580	.7611	.7642	.7673	.7704	.7734	.7764	.7794	.7823	.7852
0.8	.7881	.7910	.7939	.7967	.7995	.8023	.8051	.8078	.8106	.8133
0.9	.8159	.8186	.8212	.8238	.8264	.8289	.8315	.8340	.8365	.8389
1.0	.8413	.8438	.8461	.8485	.8508	.8531	.8554	.8577	.8599	.8621
1.1	.8643	.8665	.8686	.8708	.8729	.8749	.8770	.8790	.8810	.8830
1.2	.8849	.8869	.8888	.8907	.8925	.8944	.8962	.8980	.8997	.9015
1.3	.9032	.9049	.9066	.9082	.9099	.9115	.9131	.9147	.9162	.9177
1.4	.9192	.9207	.9222	.9236	.9251	.9265	.9279	.9292	.9306	.9319
1.5	.9332	.9345	.9357	.9370	.9382	.9394	.9406	.9418	.9429	.9441
1.6	.9452	.9463	.9474	.9484	.9495	.9505	.9515	.9525	.9535	.9545
1.7	.9554	.9564	.9573	.9582	.9591	.9599	.9608	.9616	.9625	.9633
1.8	.9641	.9649	.9656	.9664	.9671	.9678	.9686	.9693	.9699	.9706
1.9	.9713	.9719	.9726	.9732	.9738	.9744	.9750	.9756	.9761	.9767
2.0	.9772	.9778	.9783	.9788	.9793	.9798	.9803	.9808	.9812	.9817
2.1	.9821	.9826	.9830	.9834	.9838	.9842	.9846	.9850	.9854	.9857
2.2	.9861	.9864	.9868	.9871	.9875	.9878	.9881	.9884	.9887	.9890
2.3	.9893	.9896	.9898	.9901	.9904	.9906	.9909	.9911	.9913	.9916
2.4	.9918	.9920	.9922	.9925	.9927	.9929	.9931	.9932	.9934	.9936
2.5	.9938	.9940	.9941	.9943	.9945	.9946	.9948	.9949	.9951	.9952
2.6	.9953	.9955	.9956	.9957	.9959	.9960	.9961	.9962	.9963	.9964
2.7	.9965	.9966	.9967	.9968	.9969	.9970	.9971	.9972	.9973	.9974
2.8	.9974	.9975	.9976	.9977	.9977	.9978	.9979	.9979	.9980	.9981
2.9	.9981	.9982	.9982	.9983	.9984	.9984	.9985	.9985	.9986	.9986
3.0	.9987	.9987	.9987	.9988	.9988	.9989	.9989	.9989	.9990	.9990
3.1	.9990	.9991	.9991	.9991	.9992	.9992	.9992	.9992	.9993	.9993
3.2	.9993	.9993	.9994	.9994	.9994	.9994	.9994	.9995	.9995	.9995
3.3	.9995	.9995	.9995	.9996	.9996	.9996	.9996	.9996	.9996	.9997
3.4	.9997	.9997	.9997	.9997	.9997	.9997	.9997	.9997	.9997	.9998

For z values greater than 3.49, use 1.000 to approximate the area.

Areas of a Standard Normal Distribution *continued*

(b) Confidence Interval Critical Values z_c	
Level of Confidence c	Critical Value z_c
0.70, or 70%	1.04
0.75, or 75%	1.15
0.80, or 80%	1.28
0.85, or 85%	1.44
0.90, or 90%	1.645
0.95, or 95%	1.96
0.98, or 98%	2.33
0.99, or 99%	2.58

Areas of a Standard Normal Distribution *continued*

(c) Hypothesis Testing, Critical Values z_0		
Level of Significance	$\alpha = 0.05$	$\alpha = 0.01$
Critical value z_0 for a left-tailed test	−1.645	−2.33
Critical value z_0 for a right-tailed test	1.645	2.33
Critical values $\pm z_0$ for a two-tailed test	±1.96	±2.58

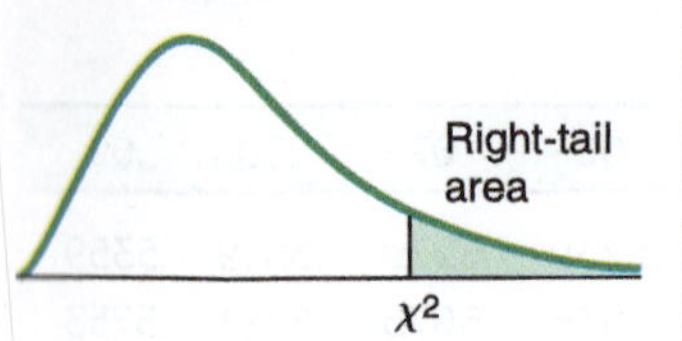

For *d.f.* ≥ 3

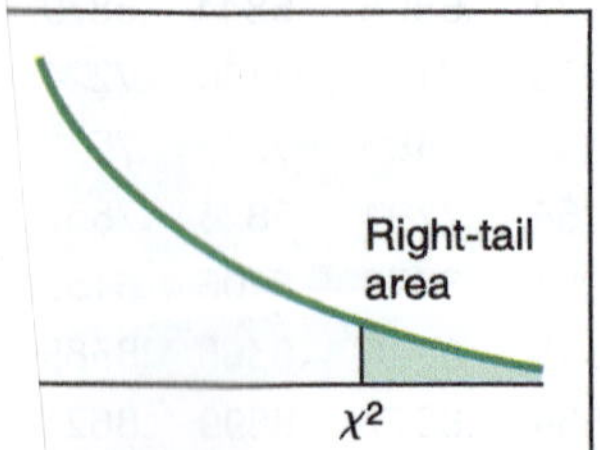

d.f. = 1 or 2

The χ^2 Distribution

	Right-Tail Area									
d.f.	.995	.990	.975	.950	.900	.100	.050	.025	.010	.005
1	0.0^4393	0.0^3157	0.0^3982	0.0^2393	0.0158	2.71	3.84	5.02	6.63	7.88
2	0.0100	0.0201	0.0506	0.103	0.211	4.61	5.99	7.38	9.21	10.60
3	0.072	0.115	0.216	0.352	0.584	6.25	7.81	9.35	11.34	12.84
4	0.207	0.297	0.484	0.711	1.064	7.78	9.49	11.14	13.28	14.86
5	0.412	0.554	0.831	1.145	1.61	9.24	11.07	12.83	15.09	16.75
6	0.676	0.872	1.24	1.64	2.20	10.64	12.59	14.45	16.81	18.55
7	0.989	1.24	1.69	2.17	2.83	12.02	14.07	16.01	18.48	20.28
8	1.34	1.65	2.18	2.73	3.49	13.36	15.51	17.53	20.09	21.96
9	1.73	2.09	2.70	3.33	4.17	14.68	16.92	19.02	21.67	23.59
10	2.16	2.56	3.25	3.94	4.87	15.99	18.31	20.48	23.21	25.19
11	2.60	3.05	3.82	4.57	5.58	17.28	19.68	21.92	24.72	26.76
12	3.07	3.57	4.40	5.23	6.30	18.55	21.03	23.34	26.22	28.30
13	3.57	4.11	5.01	5.89	7.04	19.81	22.36	24.74	27.69	29.82
14	4.07	4.66	5.63	6.57	7.79	21.06	23.68	26.12	29.14	31.32
15	4.60	5.23	6.26	7.26	8.55	22.31	25.00	27.49	30.58	32.80
16	5.14	5.81	6.91	7.96	9.31	23.54	26.30	28.85	32.00	34.27
17	5.70	6.41	7.56	8.67	10.09	24.77	27.59	30.19	33.41	35.72
18	6.26	7.01	8.23	9.39	10.86	25.99	28.87	31.53	34.81	37.16
19	6.84	7.63	8.91	10.12	11.65	27.20	30.14	32.85	36.19	38.58
20	7.43	8.26	8.59	10.85	12.44	28.41	31.41	34.17	37.57	40.00
21	8.03	8.90	10.28	11.59	13.24	29.62	32.67	35.48	38.93	41.40
22	8.64	9.54	10.98	12.34	14.04	30.81	33.92	36.78	40.29	42.80
23	9.26	10.20	11.69	13.09	14.85	32.01	35.17	38.08	41.64	44.18
24	9.89	10.86	12.40	13.85	15.66	33.20	36.42	39.36	42.98	45.56
25	10.52	11.52	13.12	14.61	16.47	34.38	37.65	40.65	44.31	46.93
26	11.16	12.20	13.84	15.38	17.29	35.56	38.89	41.92	45.64	48.29
27	11.81	12.88	14.57	16.15	18.11	36.74	40.11	43.19	46.96	49.64
28	12.46	13.56	15.31	16.93	18.94	37.92	41.34	44.46	48.28	50.99
29	13.21	14.26	16.05	17.71	19.77	39.09	42.56	45.72	49.59	52.34
30	13.79	14.95	16.79	18.49	20.60	40.26	43.77	46.98	50.89	53.67
40	20.71	22.16	24.43	26.51	29.05	51.80	55.76	59.34	63.69	66.77
50	27.99	29.71	32.36	34.76	37.69	63.17	67.50	71.42	76.15	79.49
60	35.53	37.48	40.48	43.19	46.46	74.40	79.08	83.30	88.38	91.95
70	43.28	45.44	48.76	51.74	55.33	85.53	90.53	95.02	100.4	104.2
80	51.17	53.54	57.15	60.39	64.28	96.58	101.9	106.6	112.3	116.3
90	59.20	61.75	65.65	69.13	73.29	107.6	113.1	118.1	124.1	128.3
100	67.33	70.06	74.22	77.93	82.36	118.5	124.3	129.6	135.8	140.2

Source: From H. L. Herter, *Biometrika*, June 1964. Printed by permission of the Biometrika Trustees.

0.0^4393 means that there are 4 0's between the decimal and 3; $0.0^4393 = 0.0000393$.